FOUNDATIONS OF BIOLOGY

Foundations

FOUNDATIONS OF BIOLOGY PROGRAM

William D. McElroy and Carl P. Swanson, *Editors*

Roy A. Gallant, *Editorial Adviser*

Prentice-Hall of Canada Ltd.
Scarborough Ontario

# of Biology

PART 1
Animal and Plant Diversity
Neal D. Buffaloe
State College of Arkansas

PART 2
Modern Cell Biology
William D. McElroy and Carl P. Swanson
The Johns Hopkins University

PART 3
The Green Plant
Arthur W. Galston
Yale University

PART 4
Human Physiology
Robert I. Macey
University of California, Berkeley

Photo by Y. Haneda

ABOUT THE COVER

Luminous toad stools of the genus *Mycena* growing on a dead log. The top photograph was taken by ordinary sun illumination while the lower one was photographed in the dark in order to demonstrate the bioluminescent nature of the plant.

FOUNDATIONS OF BIOLOGY PROGRAM
William D. McElroy and Carl P. Swanson, Editors

FOUNDATIONS OF BIOLOGY
William D. McElroy, Carl P. Swanson, Neal D. Buffaloe, Arthur W. Galston, Robert I. Macey

© 1968 by PRENTICE-HALL, INC.
Englewood Cliffs, New Jersey
All rights reserved. No part of this book
may be reproduced in any form or by any means
without permission in writing from the publisher.

Current printing (last digit):
2 3 4 5 6 7 8 9 0   JD   80 79 78 77 76 75
Library of Congress Catalog Card Number: 68-16119
Printed in Canada

Designer   Merrill Haber

Illustrations   Prepared by Joseph M. Sedacca, Robert Bryant, Juan Barberis, scientific illustration specialists who are also employed as graphic designers at The American Museum of Natural History, New York, N.Y.

Picture Research   Gabriele Wunderlich

ISBN 0-13-329565-6

ABOUT THIS PROGRAM

A few years ago the series editors of this biology program were involved in the preparation of a paperback series entitled *Foundations of Modern Biology*. The success of that series led us to explore the possibilities of producing a similar series organized on the basis of a slightly different approach and organization. Extensive inquiry among biology teachers indicated that such a program would fill a real need. With that encouragement, we have planned the present FOUNDATIONS OF BIOLOGY PROGRAM.

Realizing that the subject matter and philosophy of biology are an extremely important part of the liberal education of every citizen, we felt that a biology program should be varied, yet pertinent. It should also do the following: 1. convey something of the meaning, scope, and excitement of biological science as a significant perspective from which to view the world; 2. provide an acquaintance with the world of living things, and of the relationships of one organism to others; 3. provide a knowledge of the structure and function of organisms and of populations; and 4. provide a knowledge of man: his history as an organism, his relation to other organisms, his rise to a position of dominance in the biological world, and the ways in which he functions as an animal and as a human being. In general, these are our goals in the four parts comprising this program. While each part stands on its own as a coherent unit, each is related to the other three.

Animal and Plant Diversity starts the program. We believe that the most effective approach to the teaching of biology is one that introduces the student to the biological world with which he is most familiar and which he can comprehend with a minimum of formalized knowledge. From his intuitive starting point, the student's familiarity with the world around him can be broadened and deepened. *Animal and Plant Diversity* accomplishes this purpose.

Modern Cell Biology, Part 2, stresses the approaches and emphases of biology today: the structure, function, and biochemistry of cells, and the ideas of inheritance, development, and evolution, which are enriched by a knowledge of cellular biology.

The Green Plant, Part 3, plays a particularly important role in the program since plants dominate our lives in many ways: as ecological elements of our environment, as sources of food, fiber, and lumber, and as the principal converters of solar energy, on which the great majority of living things depend. No significant introduction to biology can neglect them.

Human Physiology, Part 4, focuses on man—his metabolic machinery and the regulatory systems that make him operate. Each of us, as student and enlightened and responsible citizen, needs to know himself, for it is out of the long and involved ancestry of man, and out of his structure and behavior, that his humanity and cultural development have emerged.

All of us—authors and editors alike—are grateful for the excellent advice and constructive criticism so generously offered by the many teachers who helped in the preparation of this program. Their familiarity with the varied needs of students has been extremely valuable to us. Those who have been particularly helpful, and who deserve our particular thanks, are Edwin M. Fields, Carol L. Crow, Vincent J. Silluzio, Elizabeth A. Simendinger, Irwin Spear, and R. W. Van Norman.

—The Editors

CONTENTS

PART ONE
ANIMAL AND PLANT DIVERSITY

    About Part One   1

1 The Diversity of Living Things   2

2 How Animals and Plants are Classified   6

    Old Systems of Classification   7
    Modern Taxonomy and the Species Problem   8
    A Difference Between Plants and Animals   14

3 Adaptation: How Organisms Respond to their Environment   18

    The Biological Meaning of Adaptation   18
    The Origin of Adaptations: Evolution   20
    The Biological Basis of Evolution   26
    Some Special Examples of Adaptation   32

4 The Animal Kingdom   36

    Protozoans (Phylum PROTOZOA)   38
    Sponges (Phylum PARAZOA)   40
    Cnidarians (Phylum CNIDARIA)   41
    Flatworms (Phylum PLATYHELMINTHES)   42
    Nematodes (Phylum ASCHELMINTHES)   44

Mollusks (Phylum MOLLUSCA) 45
Annelids (Phylum ANNELIDA) 46
Insects (Class INSECTA, Phylum ARTHROPODA) 48
Crustaceans (Class CRUSTACAE, Phylum ARTHROPODA) 49
Arachnids (Class ARACHNIDA, Phylum ARTHROPODA) 51
Centipedes (Class CHILOPODA, Phylum ARTHROPODA) 52
Millipedes (Class DIPLOPODA, Phylum ARTHROPODA) 52
Echinoderms (Phylum ECHINODERMATA) 53
Protochordates (Classes UROCHORDATA, CEPHALOCHORDATA; Phylum CHORDATA) 55
Fishes (Classes AGNATHA, CHONDRICHTHYES, OSTEICHTHYES; Phylum CHORDATA) 57
Amphibians (Class AMPHIBIA, Phylum CHORDATA) 58
Reptiles (Class REPTILIA, Phylum CHORDATA) 60
Birds (Class AVES, Phylum CHORDATA) 62
Mammals (Class MAMMALIA, Phylum CHORDATA) 64
The Evolution of Animals 66

5 The Plant Kingdom 72

How Plants are Classified 72
Algae 74
Blue-green Algae (Phylum CYANOPHYTA) 75
The Green Algae (Phylum CHLOROPHYTA) 76
Brown Algae (Phylum PHAEOPHYTA) 77
Red Algae (Phylum PHAEOPHYTA) 77
Fungi 82
Bacteria (Phylum SCHIZOMYCOPHYTA) 82
Slime Molds (Phylum MYXOMYCOPHYTA) 82
True Fungi (Phylum EUMYCOPHYTA) 84
Lichens 86
Liverworts (Phylum BRYOPHYTA) 87
Mosses (Phylum BRYOPHYTA) 89
Ferns (Class FILICINEAE, Phylum TRACHEOPHYTA) 91
Seed Plants (Classes GYMNOSPERMAE and ANGIOSPERMAE, Phylum TRACHEOPHYTA) 94
Plants and Animals Through the Ages 102
The Evolution of Plants 103

6 Diversity, Similarity, and the Human Viewpoint 106

Order in Diversity 107
Man's Place in Nature 109

PART TWO
MODERN CELL BIOLOGY **113**

About Part Two **115**

7 Biology in a Modern World **118**

The Growth of Science   119
The Ways of Science   121
The Science of Biology   123
Scientific Method   128
Why Study Biology?   131

8 The Cell **134**

How the Cell Concept was Formed   136
Cells Under the Microscope   138
Structure of Cells under the Light Microscope   141
Possible Exceptions to the Cell Theory   144
Structure of Cells under the Electron Microscope   145
The Bacterial Cell   150

9 General Cell Features **152**

Cell Size   153
Cell Shape   157
Cell Number   161
Cell Death   163

10 Cell Division **168**

Roottip Cells in Division   170
Cell Division in Animal Cells   173
Time Sequence of Cell Division   175

11 The Nucleus—Control Center of the Cell **178**

The Controlling Element in the Nucleus   180
The Chemistry of Chromosomes   182

12 Atoms and Molecules **186**

The Structure of Atoms   187
How Atoms Combine   190

13 The Chemistry of Biological Compounds **194**

What is Water   194
Carbon Compounds   196
Important Chemical Groups   198

Chemical Reactions 199
Oxidation-Reduction 200
Acids and Bases 200
Concentration of H^+ 201

14 Major Compounds of Cells — 204

The Role of Carbohydrates 204
The Lipids 207
Proteins 208
The Structure of Proteins 209
Growth Requirements of Organisms 211
Autotrophic Organisms 213
Heterotrophic Organisms 214
Aerobic and Anaerobic Organisms—Oxygen as a Nutrient 215
Inorganic Salts 216

15 Metabolic Properties of Cells — 220

The Physical and Chemical Nature of Protoplasm 220
Protoplasm and Colloidal Systems 221
The Cell Membrane 223
Permeability 223
Osmosis 223
Active Transport 226
Pinocytosis and Phagocytosis 229
Enzymes 230
Effect of Temperature on an Enzyme Reaction 231
How an Enzyme Works 233

16 Metabolism and Energy — 236

Oxidation and Energy 237
Alcoholic Fermentation 240
Phosphorylation of Glucose 242
Muscle Metabolism—Glycolysis 246
Oxidation of Reduced DPN by Other Substances 249
Carbon Dioxide Fixation 250
Fatty Acid Oxidation 253
Amino Acid Metabolism 254

17 Light and Life — 258

Photosynthesis and the Atmosphere 260
The Role of Light 260
The Rate of Photosynthesis 263
The Production of Chlorophyll 263
Photosynthesis as an Energy Source 264
Vision 265
Bioluminescence 267
Effect of Light on Biological Processes 268

18 DNA—The Molecule of Life — 272

The Structure of DNA   274
A Molecular Model of DNA   276
Proof of DNA Replication   277
DNA-RNA Protein Chain of Relationships   281
Genes and Enzymes   286

19 Control of Cellular Metabolism — 290

Effect of Nutrients on Enzyme Synthesis   291
Enzyme Repression and Feedback Inhibition   292
Protein Structure and Enzyme Activity   294
Enzyme Complexes   295
Biological Membranes and Control Systems   297

20 Inheritance of a Trait — 302

The Life Cycle of *Neurospora*   303
The Inheritance Test   304
Random Assortment of Genes   307

21 Meiosis and its Relation to Sexual Reproduction — 312

The Stages of Meiosis   313
Reproduction in Animals   318
The Egg   318
The Sperm   320
Fertilization   321
Reproduction in Plants   321
Fertilization   323

22 Inheritance in a Diploid Organism — 326

The 3:1 ratio   326
9:3:3:1   330
Chance and Probability in Inheritance   331

23 Linkage, Crossing-Over, and Gene Maps — 336

Linkage and Crossing-Over in *Drosophila*   337
Proofs of Crossing-Over   340
Chromosome Maps   342

24 Sex as an Inherited Trait — 348

Sex Differentiation   351
Sex Linkage   352
How Chromosomes Determine Sex   358

xii   Contents

25   Heredity and Environment   362
         Twin Studies   368

26   Development—An Inherited Pattern   374
         Growth   377
         Differentiation   380
         Integration   386

27   The Evolution of Inherited Patterns   390
         Diversity   392
         Continuity   394
         The Theory of Evolution   394
         The Course of Evolution   399

28   Causes and Results of Evolution   406
         Darwin's Theory of Natural Selection   409
         Source of Variation   412
         The Fate of Variation   417
         Results of Variability as Adaptations   419
         The Origin of Life   424
         The Planet Earth   424
         In the Beginning   425

29   The Origins of Man   432
         The Beginnings of Man   433
         Races of Man   435
         Climate and Race   438
         Genetics of Man   439
         The Evolution of Modern Man   441

PART THREE
THE GREEN PLANT   447
         About Part Three   449

30   The Green Plant's Role in Nature   450
         The Sun—a Thermonuclear Energy Source   451
         Radiant Energy   452
         Human Population and Food Supply   454

31   The Green Plant Cell   458
         The Nucleus—Life Center of the Cell   460
         Mitochondria—the Powerplants of Cells   464

The Cell Wall   468
The Vacuole and Membranes   470
Types of Cells (Meristematic, Parenchymatous, Supporting and Conducting, Protective, Reproductive)   473

32 Plant Nutrition   478

The Biochemistry of Photosynthesis   479
The Raw Materials of Photosynthesis   481
Carbon Dioxide and the World Climate   482
The Light Reaction   485
From CO_2 to Sugar   487
Production of Oxygen   488
Mineral Nutrition   488
Uptake of Mineral Elements from the Soil   491
Nitrogen Fixation   492
How a Plant Uses Elements   494
Water and Transpiration   496
Transport of Nutrients   498

33 How Plants Grow   504

The Kinetics of Growth   506
Meristems and Tissue Organization   507
Growth Hormones   511
The Gibberellins   518
Cytokinins and Other Growth Regulators   520
Growth Inhibitors   521

34 Development of the Plant Body   526

Organ and Tissue Culture in the Study of Morphogenesis   529
Culture of Excised Roots   529
Culture of Excised Stems   531
Culture of Leaves   531
Culture of Callus Tissue   532
The Culture of Single Cells   535
Differentiation of Reproductive Organs   536
Photoperiod and Plant Growth   537
The Importance of Temperature   541
Endogenous Rhythms   543

35 Plants and Man   548

Plant Communities   549
Plants Useful to Man   553

PART FOUR
HUMAN PHYSIOLOGY 561
    About Part Four 563

36 Survival and the Internal Environment 564
    The Internal Environment 565
    Steady State 566
    Homeostasis and the Steady State 566
    Storage and Feedback as Regulators 567
    Biological Feedback Systems 570

37 Transport 574
    Pressure Forces and Transport 575
    Concentration Differences and Transport 577
    Osmosis 578
    Electrical Forces and Transport 579
    Cell Membranes 582

38 Motion 586
    Muscle Contraction 587
    Structure of the Contractile Machinery 589

39 Information Transfer 594
    The Role of Hormones 594
    The Nervous System 594
    Nature of a Nerve Message 596
    Polarization of Nerve Membranes 599
    Sensory Receptors 602
    Motor Fibers 603
    Passage of Action Potentials Between Cells 604
    Inhibition 606
    Reflex Actions 607
    Autonomic Nervous System 608

40 Circulation 612
    Heart Muscle 613
    The Pathway of Blood Flow 615
    One-Way Flow 616
    Cardiac Output 617
    Blood Pressure and Blood Flow 617
    Regulation of Local Blood Flow by Metabolic Products 619
    Regulation of Blood Pressure 620

41 Blood — 624

- Preventing Excess Loss of Blood   625
- Capillary Fluid Exchange   627
- The Lymphatic System   629
- Antibodies   630
- Blood Cells   632

42 Oxygen — 634

- How We Breathe   634
- Oxygen Transport   637
- The Role of Oxygen   638
- Transport of Carbon Dioxide   639
- Control of Oxygen Delivery   641
- Regulation of Hemoglobin   642
- Control of Breathing   643

43 Salt and Water—the Kidney — 648

- Structure of the Kidney   649
- The Formation of Urine   649
- Regulating Osmotic Pressure of the Body Fluids   652
- Structure of the Nephron   654
- Mechanism of Action of ADH   654
- Regulation of Body Fluid Volume   656
- Other Functions of the Kidney   657

44 Food — 660

- Food and Digestion   661
- Classes of Food   661
- Control of Digestive Secretions   665
- The Liver and Pancreas   667
- Absorption and Motility   668

45 Metabolism and Hormones — 672

- Function of Insulin   674
- Control of Insulin Secretion   674
- Metabolism   675
- Hormones and Energy Release   677
- Hormones and Growth   679
- Hormone Control   680
- Hormonal Mechanisms   681
- Heat and Body Temperature   681
- Heat Production   682
- Heat Loss   683
- Temperature Regulation   684
- Fever and the Body's Thermostat   684
- Response to Intense Heat   685

46 Interpreting the Environment: The Brain — 688

Development and Structure of the Central Nervous System   689
Mapping the Cortex   691
Association Areas   692
The Cerebellum   694
Sleep and Attention: the Reticular Formation   694
Memory   695
Conditioned Reflex   697
Hearing and the Ear   699
Seeing and the Eye   702

47 Reproduction — 708

The Biological Significance of Sexual Reproduction   709
The Production of Sperm   709
Cycle of Ova Production   711
Cycle of Changes in the Uterus   712
How the Ovarian and Uterine Cycles Work Together   713
Gonadotrophic Hormones   714
When the Uterine Cycles Stop   715
Implantation and the Placenta   717
Placenta Hormones   718
Birth and the Challenge of the New Environment   718

Epilogue — 723

Glossary — 725

Index — 731

PART ONE
Animal and Plant Diversity

ABOUT THE PHOTOGRAPH

To perceive nature with the photographer's eye is to stretch the sense of vision to its limits. Dennis Brokaw, who took this spectacular photograph of a southern California grass spider described his picture in these words: "No amount of skill and expertise in the handling of flashlamps can produce the photographs made by existing light, for it is in that light only that we live and perceive."

Photograph: Dennis Brokaw, From National Audubon Society

ABOUT PART ONE

In "Nature's Endless Variety" we dealt generally with the diversity existing in nature. From galaxies to DNA, from viruses to man, diversity is the result of the processes of evolution operating at many levels. Our focus on diversity will continue in Part 1 of the book, but here we will direct it toward the larger biological aspects of nature. In doing so, we want to accomplish several aims.

First, we wish to renew our acquaintance with what is familiar to us among plants and animals; at the same time we wish to broaden our knowledge of less familiar forms. Second, we want to explore the relationships existing among organisms and to gain an insight into the fact that a historical continuity exists among all organisms. There are gaps in our knowledge, and the systems of classification do not reveal all of the relationships linking one group of organisms with another, but we recognize that at different times during the past certain plants and animals have flourished and then died while others have given rise to the groups living today.

Third, we wish to show that all plants and animals do not occupy the same environment. Each species, if it is to survive as a species, must adapt to its environment or become extinct. Much of the diversity we observe makes good sense only when we consider it in terms of the life habits of the species.

Our object in Part 1, then, is to enlarge your acquaintance with the world of living things and to help you view nature with a more perceptive eye.

1 THE DIVERSITY OF LIVING THINGS

Esther Bubley

Man's environment consists of a great many kinds of things. There are inanimate objects, both natural and man made, and the various manifestations of energy that surround us, such as light, heat, and radio waves. Most of us are in close contact with other people and, indirectly, we meet them through their ideas when we read or watch television. Furthermore, other living forms figure prominently in our environment (Fig. 1.1).

A great many fields of learning help us to know and understand our environment, and, of course, such knowledge and understanding are very important to us. After all, we are going to encounter objects, forms of energy, people, and various plants and animals as long as we live. Regardless of a person's present and future activities in his society, a knowledge and an understanding of his surroundings will do much to equip him for living successfully in that society.

Fig. 1.1 A contrast in human environments. What advantages can we assume the country environment to have over the city-street environment? What are some of the biological problems faced by the astronaut in the closed ecological environment?

N.A.S.A.

In this book, we are going to examine some of the living things which are part of our environment. In considering these, you will be amazed by the seemingly limitless **diversity,** or variety, of animals and plants. If you were not already familiar with many different kinds of plants and animals, you would marvel at the wonderful variety which nature has produced. Suppose that a person who is almost totally unfamiliar with wildlife decides to spend one hour in a wooded area, one hour in a grassy field, and an hour beside a pond. It is a spring day and the plants and animals in these areas are typical, both in number and variety. If he looks closely, this person will see hundreds of different kinds of living things, or **organisms.** If he makes several sweeps with an insect net in the grassy field, he will find an even greater number; and by drawing a finely-woven net through the pond water, hundreds more organisms would be revealed. Many of them would be microscopic in size, as would many others found on debris from the field and wood.

Over the hundreds of years biologists have been collecting and studying plants and animals, they have catalogued well over 1,000,000 different kinds of animals, and about 350,000 different kinds of plants. Yet there are many more plant and animal types that remain to be described. An average of about 10,000 new animals and 5,000 new plants are described each year by biologists.

With such an array, it is small wonder that we find great diversity among animals and among plants. What is the largest living thing you can think of? (The largest living thing is not an animal, but a plant.) The Blue Whale, which is bigger than any other modern animal (about 100 feet long) is 10,000,000 times as long as the smallest living animal, which is one life stage of a certain tiny parasite called a microsporidian. As Fig. 1.2 shows, there is an even greater contrast in size between the largest and the smallest plants. Diversity occurs not only in size, but in shape, structure, function, life span, and reproduction. On the microscopic level, differences in tissues and cells can mark one kind of organism from another. At the molecular level, small differences in structure produce major differences at higher levels.

The general field of biology includes many subfields, each giving us a special way of looking at living things. The subfield of **taxonomy** is concerned with the classification of organisms according to their differences and similarities. **Ecology,** a much younger science, is the study of an organism's relationships to its environment. To the taxonomist, diversity and similarity among animals, or among plants, enable him to place certain organisms together in one group or to assign them to different

Fig. 1.2 Comparative sizes and shapes of selected plants and animals. The sizes given represent either typical or maximum measurements through the longest axis of the plant or animal body. As an exercise, compute the height of a sequoia tree if it were magnified as highly as the bacterium shown here.

groups. For example, all animals that have hair, secrete milk, and have certain other things in common are placed in a group called **Mammalia**.

The ecologist is interested in those features and habits of an animal or plant that enable it to survive in a given environment. For example, over the past 25 years more than half of the sheep in an area of Australia called Pilbara have died out. Meanwhile, kangaroos in the area have flourished. An ecologist who studied this area discovered that over the years the sheep gradually depleted the choice food plants. This permitted spiky grass called *spinifex*, which is a poor food for sheep, to spread. With only spinifex as their major diet, the adult female sheep could not produce enough milk to keep their lambs alive. The kangaroos, however, thrived on spinifex and increased in number. By studying the plants, animals, and other elements making up a given environment, and by understanding how they act on each other, the ecologist adds a new dimension to the study of natural history. In the chapters that follow, we shall deal with the diversity of living things from the taxonomist's view and from the view of the ecologist.

FOR THOUGHT AND DISCUSSION

1. Make a list of 15 different animals, writing down their names as quickly as you can think of them. Time yourself, and record the time that it took you. Now repeat the exercise for 15 different plants. Which list took you longer to compile? If there was a marked difference in the time required, how do you account for it?
2. Three biologists are asked to make a survey of a certain small desert. Of the three, one is a plant taxonomist, another is an animal taxonomist, and the third is an ecologist. Their only instructions are to "study the area." In general terms, how do you think each person should proceed?
3. If your library has the first and third books listed below, read the first chapter of either (or both if time permits). Make a list of new concepts that you gain about science and biology.

SELECTED READINGS

Buffaloe and Throneberry *Principles of Biology.* (2nd ed.) Englewood Cliffs, N.J.: Prentice-Hall, Inc., 1967.

Froman, Robert. *Wanted: Amateur Scientists.* New York: McKay Publishing Co., 1963.

    This is an excellent, short book in which the author discusses opportunities for the amateur scientist and his important role in modern research. It is of particular value to the beginning student of the natural sciences.

Weisz, P. B. *The Science of Biology* (2nd. ed.). New York: McGraw-Hill Book Co., Inc., 1963.

2 HOW ANIMALS AND PLANTS ARE CLASSIFIED

In Chapter 1, our discussion of plant and animal diversity was quite general. We spoke of different "kinds" of organisms, but we did not give "kind" a specific meaning. This term is too vague to be of much use in biology, although it does tell us that the organisms to which we refer are alike in some way. Instead of "kind," biologists say **species** (Fig. 2.1). How much alike do organisms have to be in order to qualify as a species? In what ways do domesticated cats differ from wildcats? How do wild dogs differ from wolves? Before such questions as these can be considered meaningfully, it is necessary to find out how, over the years, scientists have come to classify living things.

Fig. 2.1 Giraffes and zebras shown grazing in their native Rhodesia. Although these two animals have many similar characteristics, they are obviously different enough to be regarded as separate kinds, or species. Without knowing any more about giraffes and zebras than you can learn by studying this photograph, can you list three differences between them?

Courtesy of the American Museum of Natural History

Old Systems of Classification

From ancient writings we learn that men have long recognized a fundamental difference between plants and animals and placed organisms in one of two great *kingdoms*. The Greek philosopher Aristotle (384–322 B.C.) seems to have been the first person to attempt to devise a detailed classification of organisms. He worked chiefly with animals and classified several hundred varieties. Although it is more accurate to describe Aristotle's attempts at classification as a listing rather than a system, it was a beginning, nevertheless. In addition, he inspired one of his students, Theophrastus, to make a similar study of plants.

From the time of Aristotle until the 1700's, there was remarkably little improvement over the "lists" drawn up by the Greek scholar. Even the better attempts to classify living things were not very successful, partly because communication between naturalists was so poor. It was not until the middle of the 18th Century that a truly excellent system was devised, and it was due to the genius and persistence of its inventor that it became widely accepted. This man was Karl von Linné (1707–78), a Swedish botanist who is better known by the name Carolus Linnaeus.

Building on the work of others to a certain extent, Linnaeus devised such a workable system that taxonomists still use it today. There are two main principles of this system: 1. the use of Latin, or Latinized, words to name groups of organisms; and 2. the use of categories that rank organisms from broad groupings to narrow groupings. Linnaeus used Latin because it was the language of scholarship in his day, and since it still serves us quite well, the practice continues. The categories that Linnaeus defined were **kingdom, class, order, genus** (plural **genera**), and **species** (plural **species**). Organisms can be ranked in these categories in such a way that a kingdom consists of many classes, a class includes several orders, an order includes several genera, a genus includes several species, and a species consists of relatively closely related organisms.

Such a system of categories was not new; others had used these and other categories in various schemes. The real genius of Linnaeus lay in his ability to rank organisms and groups of organisms in a **natural** order. Up to this time, systems had been largely **artificial.** For example, most taxonomists before Linnaeus had classified whales, dolphins, and other aquatic mammals with the fishes, simply because of their superficial resemblance to each other. Linnaeus, however, recognized that these aquatic mammals were much more similar to the land mammals than to the fishes, and he classified them accordingly. We now interpret similarities and differences among organisms as indicating close or distant evolutionary kinship, which Linnaeus did not, but his ranking of plants and animals according to really significant traits was a step in the right direction.

Another extremely important principle that Linnaeus introduced into taxonomy was the use of a single word for the specific name. Hence, the generic and specific names of an organism became a useful and convenient **binomial** (meaning *two-name*), much like your first and last names. All human beings are classified *Homo sapiens*. *Homo* is the generic name; *sapiens* is the specific name. This highly efficient system made it possible for the first time to classify new organisms and rename the old ones in a consistent way. Linnaeus is remembered best today for his emphasis on the generic-specific names. His system is called **binomial nomenclature** ("naming with two names").

Here is an example of how this system works. Linnaeus placed all oak trees in the genus *Quercus* (which is the Latin word for oak). There are several species of oak trees, a few of which are *Quercus alba*, white oak, *Quercus rubrum*, red oak, *Quercus nigra*, water oak, and *Quercus phellos*, willow oak. These names, as well as hundreds of others given by Linnaeus, are still used today.

Modern Taxonomy and the Species Problem

Because Linnaeus was unable to predict the vast number of new plants and animals to be discovered and classified, taxonomists have had to create additional categories. Today, the major categories are kingdom, **phylum,** class, order, **family,** genus, and species. Two major groups—phylum and family—have been added to the five of Linnaeus. Furthermore, it sometimes becomes necessary to add still other categories, such as subphylum, subclass, group, tribe, and so on.

Here is an example of how the Linnaean system is used today. Starting from the most general category (kingdom) and working our way to the most specific category (species), let us classify the domestic dog. We begin by grouping the dog in the animal kingdom, or as it is expressed in Latinized form, the kingdom **Animalia.** The animal kingdom includes several phyla (plural of phylum). One of them is the phylum **Chordata,** made up of all animals that have a **notochord** at some stage of their development. Similar to a backbone, the notochord is a supporting rod-like structure that runs the length of the animal just beneath the **dorsal** (upper) body surface. There are other characteristics of this phylum, and since the dog fits them all, it is included. See Fig. 2.2.

At this point in our classification, we must include a subphylum

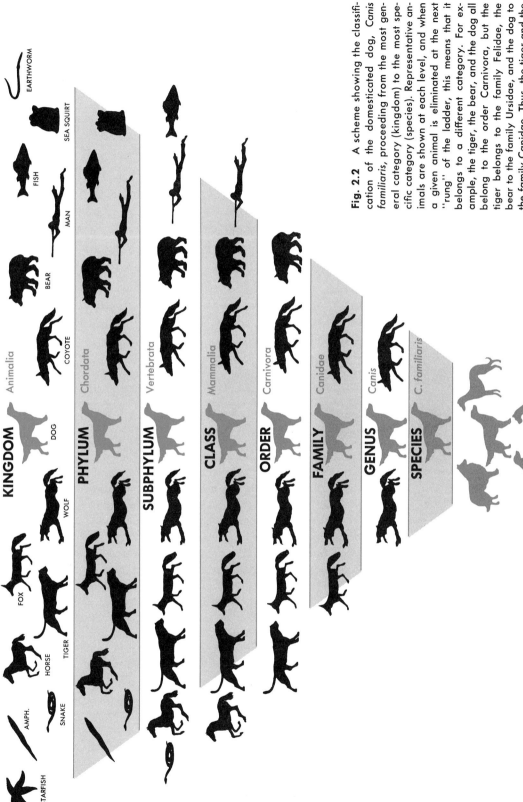

Fig. 2.2 A scheme showing the classification of the domesticated dog, *Canis familiaris*, proceeding from the most general category (kingdom) to the most specific category (species). Representative animals are shown at each level, and when a given animal is eliminated at the next "rung" of the ladder, this means that it belongs to a different category. For example, the tiger, the bear, and the dog all belong to the order Carnivora, but the tiger belongs to the family Felidae, the bear to the family Ursidae, and the dog to the family Canidae. Thus, the tiger and the bear are eliminated when we reach the rung that represents the family Canidae.

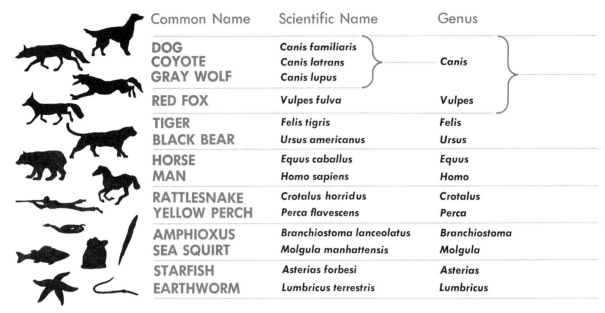

Fig. 2.3 A scheme to show the inclusive nature of taxonomic categories, using the animals shown in Figure 2.2. These animals are merely examples, of course, and do not represent a broad sampling of the animal kingdom.

to single out those chordates whose notochord is replaced by a vertebral column, or backbone, during embryonic development. Since the dog is a **vertebrate** (an animal with a vertebral column), it is included in the subphylum **Vertebrata.** There are many classes of vertebrates. One class includes all those vertebrates which produce milk to suckle their young. This is the class **Mammalia,** and our dog qualifies for this group.

It is easy to see what is happening as we proceed in this manner. Each time we move down one rung on the classification ladder we exclude a vast number of animals. At the mammal stage, however, there are still many different kinds of animals that meet the requirements met by our dog. How many other such animals can you think of? What we must do now is continue down the ladder until we reach the rung that excludes all animals except the dog.

One of the many orders of mammals is the order **Carnivora.** It includes the natural meat eaters, and the dog is one of these, along with the various cats, bears, weasels, and so on. As we move the dog down into the family grouping **Canidae,** we exclude all animals except the dog, foxes, wolves, and coyotes. The genus **Canis** includes just the dogs, wolves, and coyotes. Finally we move the dog down to the species level, **familiaris,** where it,

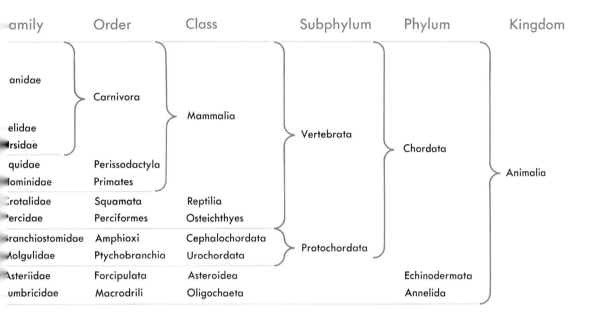

and it alone, is included. See Fig. 2.3.

It is easy to see, by this scheme, that a given category can be defined as a group of lesser categories, at least until we reach the species. A class is a group of similar orders; an order is a group of similar families, and so on. But we still have the same old problem—what do we mean by "similar?" Although we went a certain way in defining the word by showing similarities between the common dog and other animals, taxonomists need more technical definitions. Often it is necessary for them to make a final decision that is based on the judgment of one or more taxonomists. Consequently, not all botanists and zoologists are in complete agreement about the various groups that should be set up in an ideal classification of the plant and animal kingdoms—and they probably never will be.

The ideal is a completely natural system, arranged in the manner in which organisms evolved. However, the problem here is that the evidence is too fragmentary to allow for certainty in most cases. This means, in the final analysis, that taxonomic systems are natural only to a limited extent.

As we have pointed out, Linnaeus strove for a "natural" system, even though he did not view plant and animal diversity in the same light that we do today. A natural classification to a modern taxonomist is one that follows evolutionary lines. To Linnaeus it meant kinship in a figurative sense, in much the same way that we would say two automobiles are "kin" because they represent two lines produced by the same manufacturer. In spite of this difference in viewpoint, his classification of plants and animals was surprisingly "natural" by modern standards.

We have already mentioned the taxonomic problem of defining degrees of similarity. For example, just how "similar" do genera have to be in order to comprise a family? Dogs, foxes, wolves, and coyotes are similar enough to be placed in the same family, but when we move them down to the genus level, only dogs, wolves, and coyotes are similar enough to be included under *Canis*. The foxes are left behind. We have the same problem at the species level, but it is intensified because on the species level we are describing *real organisms,* not abstract categories. On the species level, the old question "How similar?" becomes even more difficult to answer than when it is asked on the class or order levels.

The fact is that no satisfactory and acceptable definition of the species as a taxonomic grouping has yet been formulated. However, there are some general and reliable rules that we can follow. Close similarity in appearance is one that is usually dependable, although one must know what similarities to look for. So far in our discussion we have mentioned only those similarities that can be determined by the naked eye. This was a necessary and good first step, but as our knowledge of living things has grown over the years as a result of improved techniques of studying organisms, we can now look for similarities and differences on the microscopic level as well.

At the cellular level, members of the same species have **chromosomes** that are usually the same in number and form. Chromosomes are structures located in the central part of a cell. They are carriers of **genes,** or determiners of hereditary traits such as height, eye color, and so on. We shall have more to say about chromosomes later. At the molecular level, the internal chemical activities of members of the same species is generally very similar. Consequently, a species is, at the very least, a group of organisms that are similar in appearance, have the same number and form of chromosomes, and have similar body chemistry.

However, this is hardly enough. Some organisms are indistinguishable from each other on this basis. They look alike, they have the same number and form of chromosomes, and their body chemistry is the same; yet, they cannot be made to interbreed. For example, some species of katydids are so much alike that they can be readily distinguished only by listening to their song patterns. In spite of their close similarities according to the criteria listed above, they do not interbreed. Sexual compatability, or interbreeding, is one of the best ways of judging the "similarity" of organisms, although it is not absolutely reliable. Some species that are not similar, according to the rules just mentioned, interbreed successfully. The horse and the donkey are one example. We can refine our rule about sexual compatability somewhat by saying that a species may be considered a group of organisms which interbreed freely in nature and which produce offspring that are capable of further successful interbreeding with each other.

The major trouble with determining the exact limits of a species is the fact that its members are not static, but changing. For example, a fairly common phenomenon in nature is the geographical separation of two populations of the same species. (See Darwin's finches on page 24.) Each undergoes basic changes on the cell level, and after a few decades, centuries, or millenia, there are two groups that may look alike but whose reproductive cells are quite different as regards their internal content. Let us assume that the geographical barrier which separated them initially is removed, say, a broad river that changed its course. They are brought back together, but interbreeding does not produce offspring. By our definition, we now have two species. Had we brought them together earlier they might have interbred but produced infertile offspring. Or they might have produced fewer offspring than before. At such times taxonomists apply the term **subspecies** to one or the other of the groups. The problem is an enormous one. The point is, however, that evolution (change) makes it impossible to define a species in hard-and-fast terms. See Fig. 2.4.

The general principles laid down by Linnaeus have served us very well over the years. The fact that more than a million species of plants and animals have been named and classified according to his system in the two centuries following Linnaeus is testimony to the value of his work.

Throughout this and other books you will find that generic and specific names are always italicized, but names in the other categories are not. This is one of the rules of the Linnaean system. Another rule is that a specific name is never used without a generic name. For instance, we never say *familiaris* to designate the common dog. We say *Canis familiaris,* or write it *C. familiaris* after having introduced the generic name once. The generic name may, however, be used alone to designate the entire genus. Notice also that the generic name always begins with a capital letter, but the specific name starts with a small letter.

Eric Hosking from National Audubon Society

Alvin Staffan from National Audubon Society

Fig. 2.4 Top: Herring gull, *Larus argentatus.* Lower photograph, American Herring gull, *Larus argentatus smithsonianus,* is a subspecies of the common Herring gull. Notice that the subspecies is given an additional name. How would you define a subspecies?

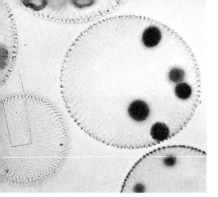

Ward's Natural Science Establishment, Inc.

Courtesy Car. Biol. Supply House

Fig. 2.5 Two common pond organisms, *Volvox* (top), and *Euglena*, which have characteristics of both plants and animals.

Fig. 2.6 Bushy stunt virus particles, photographed by means of an electron microscope, shown here magnified more than 50,000 times.

Photo courtesy Dr. R. C. Williams

A Difference Between Plants and Animals

If someone asked you to make a list of differences between plants and animals, what differences would you list? Write down three or four now. This may seem to be an easy and even useless task, since the common plants and animals are not difficult to place in their respective kingdoms. However, there are certain forms of plants and animals that are not so easily classified (for example the two organisms shown in Fig. 2.5), and the problem is not as simple as it might seem.

From an evolutionary standpoint, the first organisms were neither plants nor animals, at least if our assumptions about the origin of life are correct. Most biologists and biochemists are convinced that life began when certain complex molecules (such as nucleic acids) acquired the ability to duplicate themselves. In time, certain other molecules (such as protein) apparently became associated with them, and a given aggregate of molecules reproduced as a unit. Since they had the ability for **self-duplication,** these units can be said to exhibit life. The basic characteristic of life is that any living unit must be able to make more units like itself from chemical substances in its environment.

Some biologists believe that the first organisms resembled viruses. These extremely small units of matter are relatively simple compared with cells. A cell contains a great variety of chemical substances, whereas a virus particle consists only of a central core of nucleic acid **(deoxyribonucleic acid [DNA]** or, more rarely, a close chemical relative, **ribonucleic acid [RNA])** surrounded by a coat of protein. We shall have much more to say about DNA in Chapter 3. Viruses (Fig. 2.6) do not carry on the complex chemical activities of cells, and they cannot even reproduce, unless they manage to get inside a cell. Whether they are living or nonliving depends on one's definition of "life." Their importance to our present discussion is that life probably had its beginnings from bundles of complex molecules with a structure similar to that of viruses.

Unhappily, the fossil record does not help us in our inquiry into the earliest organisms, simply because they could not be preserved intact. Present evidence indicates that the earliest organisms existed more than two billion years ago, and that it took them hundreds of millions of years to change into the complex forms of the first plants and animals. Apparently, some of the primitive forms never evolved a great deal further. It seems that they were so well adapted to changing environments that they have persisted to the present time. Since these organisms arose and apparently remained unchanged over the millions of

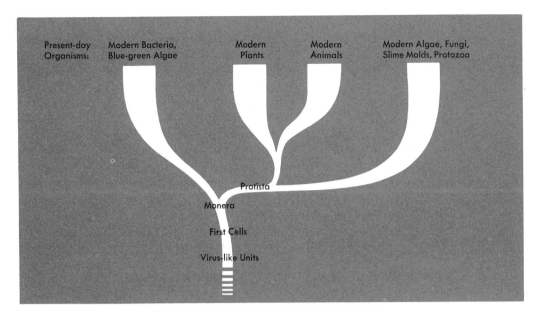

Fig. 2.7 A schematic representation of one theory of how organisms evolved. This concept suggests that four kingdoms of modern organisms should be recognized.

years during which other organisms were evolving into new forms, it is these unchanging organisms that are so difficult to pigeonhole in what we call the "plant kingdom" or "animal kingdom."

In view of the existence of these organisms, some biologists recognize two kingdoms in addition to the plant and animal kingdoms, the **Monera** and the **Protista**. The kingdom Monera includes all organisms whose cells do not have definite or clearly formed central parts called the **nucleus** (plural **nuclei**). The two major groups of organisms so constructed are the **blue-green algae** and the **bacteria** (see Chapter 5). The kingdom Protista includes all one-celled animals (the **protozoa**), one-celled **fungi** and **algae** (other than blue-green algae), and a group of unusual organisms called **slime molds.** These groups are also described later. This classification of organisms is based on the possibility that evolution proceeded along the lines indicated in Fig. 2.7.

While this system of classification has a great deal of merit, and it does make us face up to the evolutionary problems of classification, we shall be content in this book to include all organisms in the traditional plant and animal kingdoms.

Setting this restriction for ourselves, how can we proceed with a list of differences between plants and animals? What about "mobility?" Can we say that animals move about, but plants

J. Carel

Fig. 2.8 Two insectivorous plants, sundew (above) and Venus fly trap (below). These autotrophic plants are unusual because they "eat" insects.

Paul Popper Ltd.

do not? There are some animals (such as the sponges) that do not move about, and there are some plants (such as certain algae) that do; so mobility is not a valid difference. We might look for features at the cellular level that would help us. Most higher plant cells are enclosed by a rigid cell wall. The cells of most animals, on the other hand, are bounded by a soft membrane. However, there are so many exceptions that this distinction is no more valid than mobility. Most biologists feel that the nutritional habits of an organism come closest to making a valid distinction between plants and animals.

First of all, animals *eat;* that is, they take into their bodies foodstuffs that have to be digested before they can be used by the organism. Any organism which gains nutrition chiefly or entirely through eating is called a **phagotroph** (from the Greek words, *phagos,* "to eat," and *trophikos,* "nursing"). Let us turn the definition around and enlarge it slightly—any organism which is entirely or chiefly phagotrophic is an animal.

Obviously, according to this definition based on nutrition, any organism not entirely or chiefly phagotrophic is a plant. But this leaves us with two distinct modes of nutrition in the plant kingdom: 1. If a plant has **chlorophyll** (the green substance in plants), it can make its own food by using the energy of sunlight. By "food," we mean a potential energy-yielding material such as carbohydrate, fat, and protein. Such "independent" plants are called **autotrophs** (from the Greek words, *autos,* "self," and *trophikos*). 2. In contrast, plants that do not have chlorophyll must obtain their food from some outside source, as animals are obliged to do. However, unlike animals, such plants are incapable of eating as we defined the word. Instead, they must receive foodstuffs into their bodies molecule by molecule, that is, in solution. Such plants are called **heterotrophs** (from the Greek words, *hetero,* "other," and *trophikos*). Actually, there are some exceptions to these nutritional rules. For example, the plants shown in Fig. 2.8 are both autotrophic and phagotrophic. Nevertheless, a distinction is made based on nutritional habits of an organism. It does not solve all problems of distinguishing plants from animals, but it seems to work more consistently than any other single criterion.

SUMMARY

Biologists classify organisms according to a hierarchy of categories that begins with the most general group (kingdom) and goes to the most specific group (species). This system has resulted largely from the work of Linnaeus, an 18th Century botanist.

More recently, the concept of evolution has stimulated taxonomists to strive toward natural systems of classification that rely on **genetic kinship**—that is, certain characteristics passed on from parent to offspring.

Some taxonomists recognize the kingdoms Monera and Protista, in addition to the plant and animal kingdoms. Under such a scheme, the Monera and the Protista are made to include those organisms having characteristics of both plants and animals.

However, most biologists distinguish plants from animals according to the organism's mode of nutrition. They are usually able to place any organism in either the plant or the animal kingdom on this basis.

FOR THOUGHT AND DISCUSSION

1 Distinguish between an *individual* and a *species*.
2 Write an improvement of this poor definition: a species is a group of similar organisms.
3 What is a "natural" system of classification as contrasted with an "artificial" one? Why is a truly natural system difficult to achieve?
4 A tiny organism named *Chrysamoeba* has chlorophyll and is thus autotrophic, but it also eats particles of food, and is thus phagotrophic. Would you consider it a plant or an animal? Does this indicate a need for recognizing more than two kingdoms? Incidentally, what taxonomic category does the name *Chrysamoeba* represent?

SELECTED READINGS

Simpson, G. G., and W. S. Beck. *Life: An Introduction to Biology* (2nd ed.). New York: Harcourt, Brace and World, 1965.
  In this college textbook of general biology, taxonomy is placed in its proper relation to evolution.

Singer, C. *A History of Biology* (rev. ed.). New York: Henry Schuman, 1950.
  An account is given of the historical development of the Linnaean system and of its modifications since the 18th Century.

3 ADAPTATION: HOW ORGANISMS RESPOND TO THEIR ENVIRONMENT

Wisconsin Conservation Department; Madison, Wisconsin

Wisconsin Conservation Department; Madison, Wisconsin

Fig. 3.1 This bird, the ptarmigan, is pure white in winter and shows a pattern in summer. It lives in regions where snow covers the ground for much of the winter. How does this factor relate to its adaptation?

Each kind of plant and animal has a type of environment to which it is usually restricted (Fig. 3.1). Organisms are not scattered over the face of the earth in random fashion. There are good reasons why we do not find oak trees growing in the desert, or find polar bears roaming the wooded areas of North Carolina. All organisms are **adapted** to living in certain kinds of environments. Each plant and animal has structures related to its particular way of life, which is another way of saying that structure is a reflection of function.

Figure 3.2 shows a small desert animal, the kangaroo rat, which does not have to drink water at all. How it manages to survive under these conditions is a long story, but its body is structurally adapted to this kind of existence. From an evolutionary viewpoint, we do not say that the kangaroo rat has these structures *because* it lives in the desert; rather, we say, it is able to live in the desert because it has these structures. The same could be said for the structure-environment relationships of oak trees, polar bears, or any other organism.

In the preceding chapter, we considered some of the rules that biologists have established for dealing with the diversity which they find in nature. In contrasting artificial and natural systems of classification, we were concerned with how organisms differ from each other and in what ways they are similar. To put it another way, we were dealing with the "what" of nature. In this chapter let us consider the underlying causes of diversity, that is, the "how" of nature.

The Biological Meaning of Adaptation

Adaptation, in its simplest definition, is the ability of a species to survive in its particular environment. To say that organisms are adapted to their environment is to state the first axiom of ecology. If a species were *not* adapted to its environment, in order to survive it would have to move away or change. Otherwise it would become extinct.

Fossils, which are the remains or the evidences of dead organisms, show us that profound changes have occurred both in organisms and environments over many millions of years. Marine fossils found high in the Himalayas and other ranges show that before these mountains were thrust up they were once the floors of ancient seas. Fossils also tell us that a great number of plants and animals have become extinct over the millions of years during which there has been life on the Earth. Extinction seems to be the rule, not the exception. The number of living species in the animal kingdom is only one-tenth that of the fossil species. Time, then, plays an important part in adaptation. Equally important is the idea that adaptation is exhibited by species, not by individuals, for when we consider time, adaptation is more a process than a state of being. We sometimes say that an individual may "adapt" to different environments—for instance, if you move from the relatively cold climate of Montana to Florida—but this is not what the biologist means by adaptation. This is nothing more than adjustment through response to an environmental change.

Photograph courtesy: United States Department of the Interior, Fish and Wildlife Service

Fig. 3.2 A kangaroo rat shown in its native habitat. This animal is mentioned in the text because of its remarkable adaptation to desert life, but no explanation is given of those features which enable it to live without drinking water. What are these features?

Fig. 3.3 A group of prehistoric animals, including several dinosaurs, as they must have appeared some 175 million years ago. Why do you suppose these giant reptiles are no longer with us?

Peabody Museum of Natural History

Adaptation *as a process* means the ability of a group of organisms to develop, over a long period of time, certain structural and functional features which enable the group to survive and reproduce in a particular environment. The term adaptation implies change—some structural or functional feature of a group of organisms changes in such a way that it has adaptive value. Not only do we use the term *adaptation* to describe a process of change, but we frequently apply it to the result of that process. For example, the gills of fishes are an adaptation that enables these animals to receive oxygen from the water in which they live. The nectar produced by the flowers of many plants is an adaptation that ensures pollination by insects.

We say that organisms are adapted when they consistently produce enough offspring to ensure survival within their environment. If a type of organism cannot do this, as the dinosaurs (Fig. 3.3) could not, it becomes extinct. What chain of environmental events led to the extinction of dinosaurs is unknown. Yet we do know that these reptiles were successful for 100 million years.

The Origin of Adaptations: Evolution

It is one thing to observe diversity of structure and function among organisms, but it is quite another to explain it. Understanding the relationship between adaptation and environment is the problem. It is quite obvious that a fish could not live in water without gills or some equally efficient adaptation for taking oxygen from the water, and a desert plant could not survive in its environment unless it had water-conserving structures. If we say that fishes have gills *because* they live in water, or that cactus plants have fleshy, water-conserving stems *because* they live in the desert, this implies that the environment wields a modifying influence over organisms. Except in very special instances, this is not so. We must look elsewhere for answers to the question of origins of adaptation.

At least two centuries ago, naturalists began to be concerned with this problem. They began to wonder if the bewildering variety of plants and animals known to them had always existed. Had this great variety arisen full blown in nature? If not, where was the starting point? Most people who lived before the mid-1800's, including biologists, accepted a literal idea of special creation, and there they let the problem rest. But as biologists gathered more and more facts, their observations became increasingly more difficult to explain by the concept of special creation.

Among the most important new facts being brought to light

were those in the field of geology. In the early 1800's, the English geologist Charles Lyell published a paper that excited many biologists the world over. Lyell claimed that mountains, rivers, coasts, and all other land features are today changing and have always been in a constant state of change. As old mountains are worn away by erosion, new ones are thrust up out of the earth. Lyell cited evidence that such changes had been going on for thousands upon thousands of years. He was also able to show that the Earth was far older than was generally believed.

For some time geologists had had means of making reasonably accurate estimates of how long it took for certain rock and soil formations to occur. Fossil remains began to prove beyond doubt that organisms had lived on the Earth much earlier than the believers in special creation had thought. Furthermore, the fossil records revealed a wonderfully consistent trend—in general, the older the fossils, the simpler they tended to be in structure. The complex forms tended to be more recent in origin.

This discovery pointed to a process of gradual development of complexity. For example, when geologists examined the sides of a gorge, such as the Grand Canyon, they found fossils of complex plants and animals only near the top. As they examined fossils at greater depths in the gorge, the fossils were of simpler types. Proceeding toward the bottom, where the older rocks were, was like taking a reverse trip in structural complexity through the plant and animal kingdoms. See Fig. 3.4.

Before such discoveries had been made, a few biologists had conceived of a process of **evolution** to explain diversity. In essence, evolution means that nature is not static, but changing; it also means that organisms living today are the descendants of ancestors which were, at some stage of time, more simple in organization. One of the first biologists who advanced a theory of evolution was a Frenchman, Jean Baptiste Lamarck (1744–1829). In 1809, he published an account of his theory. In it, he said that evolution proceeds in relation to the use or lack of use of body parts. According to Lamarck, the continual use of a body limb, for example, strengthens and perhaps enlarges the limb, and this modification is passed on through inheritance to the next generation. Note the generations of rabbits in Fig. 3.5.

Lamarck's theory is known today as the theory of **inheritance of acquired characteristics,** and it is unacceptable to modern biologists. In fact, it never gained wide acceptance when Lamarck lived, partly because he presented his ideas at a time when very few other biologists were thinking about evolution. Furthermore, Lamarck could not support his theory by observa-

U. S. Geological Survey

Fig. 3.4 Grand Canyon, Arizona, showing fossil-bearing strata of rock. Formations such as this frequently show a continuous array of fossil organisms and their general tendency toward decreasing complexity as one proceeds from top to bottom.

1. At some point in the past, rabbits possessed rather short ears. Since their survival depended heavily upon their ability to hear an approaching predator, they stretched their ears continuously in order to hear with maximum efficiency.

2. The continual ear-stretching made an impression upon the reproductive cells, with the result that rabbits came gradually to have longer ears. These rabbits, in turn, stretched their ears, and passed the increase along to their offspring.

3. Eventually, a point was reached where ear length was sufficient to enable rabbits to survive without further stretching. At this point, ear length became stabilized.

Fig. 3.5 This is how Lamarck explained the evolution of ear length in rabbits. As you analyze the fundamentals of this explanation in the light of what you learn in this chapter, try to answer this question: Why is Lamarck's explanation not acceptable to modern biologists? Note the hypothetical steps listed above.

tion and experiment. It was, therefore, widely questioned. Although his explanation of evolution was inadequate, it did focus attention on the problem of explaining diversity in nature.

The year Lamarck published his theory a person was born who did more to popularize and clarify the concept of evolution than anyone who has lived since. He was Charles Robert Darwin (Fig. 3.6) of England (1809–82). In 1831, at age 22, Darwin set sail as ship's naturalist on H.M.S. *Beagle*. This five-year surveying voyage was to influence Darwin's thinking about evolution profoundly. It was during this time that he became aware of the exciting new discoveries being made in geology. This, in turn, influenced his thinking about the diversity and distribution of the plants and animals he observed during his voyage. He was especially impressed by what he saw on the Galapagos Islands (note Fig. 3.7) several hundred miles west of Ecuador in the Pacific Ocean. Here, as he traveled from one island to another, he noticed minute differences in related species of animals.

The finches (Fig. 3.8) fascinated him. He made a detailed study of the various finches, comparing and contrasting them. Generally, each island had its own finch species, slightly but definitely different from those of the other islands and the mainland of South America. How, Darwin wondered, did such diversity come about? The problem of the finches and other species of birds caused him to begin searching for an evolutionary explanation of diversity.

At the end of the voyage Darwin returned to England with a vast knowledge of plants and animals. As the writings of Lyell and other geologists had fascinated him, so did a book written by another English scholar. It was *An Essay On the Principle of Population,* by Thomas R. Malthus, an economist. Malthus said while human population increases geometrically, food supplies increase only arithmetically (Fig. 3.9). As a consequence, Malthus believed that there would always be a struggle for food, that in the world there would always be some people who would go hungry. Darwin was greatly influenced by this idea, and from it he formulated his theory of selection in nature.

Because he was careful and meticulous in his work, and because he was uncertain of the reception society would give to his ideas, Darwin spent more than 20 years developing his concept of evolution. It finally appeared in 1859 under the title *On The Origin of Species By Means of Natural Selection;* it was a book that eventually became established as one of the greatest influences of all time upon human thought.

Essentially, *The Origin of Species,* as the title is generally abbreviated today, set forth the idea that all present-day species of plants and animals came into being by a process of evolution. Darwin presented extensive evidence to show that evolution had occurred, and then advanced a theory explaining how it occurred. He called his theory **natural selection,** and he based it on four major points:

1. Variation exists within a species (have you ever noticed this in a litter of new-born puppies?).

Courtesy of The American Museum of Natural History

Fig. 3.6 Charles Darwin.

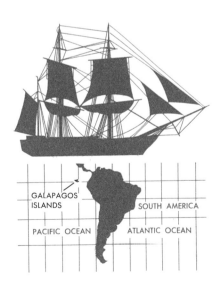

Fig. 3.7 A sketch of H.M.S. *Beagle,* on which Darwin served for five years as naturalist, and of the South American continent. Note the location of the Galapagos Islands, where Darwin made some of his most significant observations.

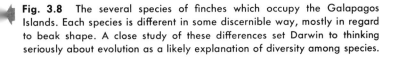

Fig. 3.8 The several species of finches which occupy the Galapagos Islands. Each species is different in some discernible way, mostly in regard to beak shape. A close study of these differences set Darwin to thinking seriously about evolution as a likely explanation of diversity among species.

2. Overproduction of offspring is characteristic of organisms (a female *Ascaris,* a parasitic worm, is capable of producing as many as 200,000 eggs a day).

3. The less fit of the variants do not survive; nature "selects" the more fit, which reproduce (a strong puppy in a litter has a better chance of reaching maturity than a weak one).

4. Favorable survival traits (for example, general constitutional strength) are inherited by offspring from their parents.

Of these four points, the first two can be seen in nature, and Darwin provided many examples. The third and fourth points were logical conclusions based on his first and second points.

Every copy of Darwin's book was sold the day the book was published. Many scientists cheered Darwin's ideas and the publication of them, but there were some who condemned the book. Those who condemned it did so on the grounds that it attacked the widely-held belief that each species had originated in a series of creative acts on the part of a divine being. In contrast, evolution implied that life and its manifestations were subject to natural law. The argument was intensified, of course, by the realization that man, himself, had to be included in such a theory of evolution. Twelve years after his first book was published, Darwin published another one, *The Descent of Man,* which left no doubts about his thinking on man's place in nature.

Since Darwin's time, the concept of evolution in general, and his theory of natural selection in particular, have received a great deal of study. Virtually all present-day biologists accept the general principle of evolution, and there is no doubt that natural selection operates all around us in the plant and animal worlds alike. Nevertheless, it was not until more was known about genes and heredity that the basic causes of evolution came to be understood in detail.

An Austrian monk, Gregor Mendel (1822–84), was the scientist whose experimental work led to an understanding of how certain traits are passed from parent to the offspring. It was not until 1900, after Mendel's death, that his work received wide recognition. Since then, it has become increasingly apparent that **genetics** (the study of how organisms inherit traits) holds the solution to the major problem of evolution that Darwin left unsolved. Darwin himself realized that his theory

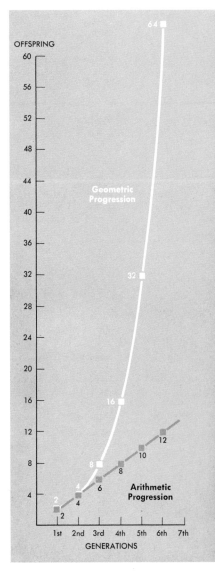

Fig. 3.9 By means of a simple graph, we can illustrate the difference in geometric and arithmetic progression. If Malthus's basic assumption is correct, it is not hard to see why life is a "struggle for existence." Darwin adapted this principle to natural selection.

would have been stronger had he been able to explain the ultimate cause of variation and how traits are passed on from parent to offspring, but there was no help for it. We have no evidence that Darwin ever heard of Mendel, and both were dead before Mendelian genetics was known and understood.

The Biological Basis of Evolution

In recent years biologists have been able to explore on the cellular and molecular levels the means by which traits are passed on from parent to offspring.

Matter (whether living or not) is composed of extremely small particles called *atoms*. Most atoms do not exist singly within matter, but are joined to one another in the form of molecules. A **molecule,** therefore, is a group of atoms held together in certain ways. A substance such as common table sugar, for example, is composed of molecules, each of which is identical to every other molecule of the sugar.

Living systems, such as any complex plant or animal, are complex mixtures of thousands of different kinds of molecules. And on a higher level, the molecules are arranged in separate small units called **cells** (Fig. 3.11). Your body is made of several billions of these microscopic units. Typically, a cell contains a kind of governing body called the nucleus, which is suspended in the rest of the material that makes up the cell. When we study the cells of different organisms closely, we find a great many cellular and molecular similarities. This means that even the cells of oak trees and birds have *some* things in common. Most of their cells are similar in size; each cell has a nucleus, and each nucleus exercises the same kind of control over the rest of its cell. On the molecular level, many of the same chemical compounds found in the cells of birds are also present in the cells of oak trees. We can carry this thought even further and say that many of the same compounds are found in the cells of *all* living things.

Cells of all living things are similar to each other in one particularly important way. The nucleus of every cell of all organisms contains **deoxyribonucleic acid** (shown in Fig. 3.12) **(DNA).** All DNA molecules are large and complex compared with most other molecules. The reason DNA molecules are important to us in this book is that they are important parts of structures known as **chromosomes.** For many years, biologists have known that chromosomes are the carriers of hereditary determiners called **genes.** However, we have learned only within recent years that a gene is a certain portion of a DNA molecule.

Actually, no one has ever seen either a DNA molecule or a

The Granger Collection

Fig. 3.10 Gregor Mendel crossed different varieties of peas and discovered certain fundamental principles of heredity. Had Darwin known these principles, his theory of evolution would have been strengthened.

Fig. 3.11 Section of a cell, highly magnified. The large central portion is the nucleus; the dark patches are composed principally of DNA.

Courtesy Dr. G. Whaley

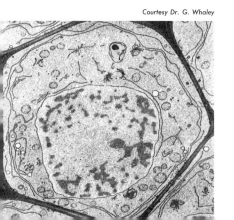

gene clearly enough to tell much about their detailed structure. However, a great many indirect observations have enabled scientists to construct a model, or picture, of a gene and DNA. A DNA molecule seems to be a kind of double helix (spiral), with chains of atoms forming connections at regular intervals. Imagine a long ladder made entirely of flexible rubber and suppose that someone twisted the ladder until it looked like the structure shown in the DNA illustration. At the present time, this is our concept of the DNA molecule. It so happens that molecules of DNA are most unusual molecules. On occasion they can duplicate themselves precisely by building other DNA molecules from materials within the cell. Because they can do this, we find identical DNA molecules in both halves (daughter cells) of a dividing cell. The DNA molecule also seems to be "marked off" in different lengths, and each length is capable of "telling" the cell to make something. Each functional length of a DNA molecule is what we have been calling a *gene*. Particular genes you have inherited from your parents are responsible for your having blue eyes or brown eyes, black hair or blonde hair, and other features. (See Fig. 3.13)

Possibly you are wondering what all of this has to do with evolution, which we set out to discuss. Again, it is giving us a detailed answer to the problem Darwin was unable to solve. Genes affect their cells by giving them certain chemical messages. If a cell happens to be dividing and forming the embryo of a new organism, it is very important that certain chemicals form in the cell in certain ways. Otherwise, a developing human being, for example, might turn out to be a grasshopper, a cantaloupe, or a formless mass of cells. It takes thousands of chemical messages, each produced by the gene segments of a DNA molecule at just the right time in embryonic development, for an organism to be formed successfully. Organisms differ, therefore, to the extent that their DNA differs. For example, all human beings have very *similar* DNA; but except for identical twins, identical triplets, and so on, no two people have *identical* DNA in their cells. This is why so few human beings look exactly alike and why, in general, variation is possible within a species of plants or animals.

Biologists can now say with certainty that *external* variation, such as hair color, height, and so on is the result of *internal* variation dictated by DNA molecules. Variation, therefore, begins on the molecular level. Perhaps you are wondering if it is ever possible for a DNA molecule to make a "mistake" by producing a version of itself which is not an absolutely faithful copy. The answer is yes, and is perhaps the most exciting part of our discussion. When a DNA molecule makes a "mistake" in

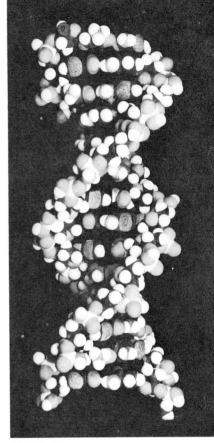

Abbott Laboratories

Fig. 3.12 A model showing the present concept of a DNA molecule (portion only). The fitted balls represent atoms. Note the double helix, as described in the text. DNA is the genetically active material of the cell.

Fig. 3.13 The common pigeon (middle figure) may give rise to variant forms. Darwin studied the artificial selection practiced by pigeon breeders of his day. Although genes and inheritance were not understood at that time, he arrived at an understanding of natural selection through his observations of pigeons and other domesticated animals.

duplicating a part of itself involving genes, we call the molecular change a **mutation**. But this seems to happen very seldom in DNA molecules—about once in 100,000 duplications for any given region, on the average.

Suppose there were a population of one species of animal in which every single member had ears six inches long. Suppose also that a mutation occurred in one of the animals and affected the gene governing ear length. Suppose further that the animal carrying the mutation produced offspring that had ears only one inch long. The offspring, in turn, would pass the "short-ear" gene on to certain of its offspring, and so on. Through mutation this population of animals would now have two varieties—one with long ears and one with short ears. What could happen to the short-ear variety through natural selection (perhaps you can guess) is extremely interesting, but this is getting ahead of our story.

As we said, once a mutation has occurred, the changed gene is transmitted as a portion of the genetic code from parent to offspring. Interestingly enough, most mutations do not benefit the organisms they affect; as a matter of fact, they frequently

cause great disturbances in the well-balanced systems of DNA. Only rarely does a favorable mutation occur. But over the millions of years that this molecular trial-and-error process has been going on, along with great environmental changes, it has been sufficiently successful to give us the diversity we have on earth today.

It is time now for us to return to the role environment plays in evolution. One thing is certain—environment does not bring about evolutionary changes, at least not as Lamarck had thought. Numerous experiments have shown beyond doubt that the cells involved in reproduction are not influenced by other cells of the organism that may be changed by the environment. But perhaps you have read that x-rays and certain other kinds of radiation cause mutations by altering genes. This is certainly so, and if a group of organisms lived in a region constantly exposed to radiation sufficiently intense to bring about a large number of mutations, then the evolutionary future of the group would be affected. But on the earth, under normal conditions, this does not happen, so our statement that environment does not bring about evolutionary changes is generally true.

The environment is a *limiting* factor in evolution. It can prevent the success of certain genetic combinations while encouraging the success of others. And this brings us to the heart of natural selection. To illustrate, let us consider a particularly well-studied case of natural selection conducted over a period of years in England.

As long ago as 1850, insect collectors noticed that most moths found in nonindustrialized areas of England were light in color; fewer than one per cent were dark. Several decades later, after these same areas were industrialized, up to 90 per cent of the moths were dark (Fig. 3.14). Now, soot was deposited on virtually everything and killed almost all the lichens that commonly grow on tree trunks. The tree trunks themselves became covered with soot and turned darker in color.

Moths often rest on the trunks of trees when they are not in flight. As long as the tree trunks were light, and supported the growth of lichens (which themselves tend to be light in color), the lighter colored moths blended with their background and thus escaped detection by predatory birds. However, when one aspect of the environment changed (the color of the tree trunks), the lighter colored moths were no longer favored. Gradually, the dark form became predominant. It has been shown that only one or two genes control the production of dark pigmentation in moths, which simplifies our analysis of the problem.

To test this apparent case of natural selection, the English biologist H. B. D. Kettlewell conducted a series of studies on light and dark forms of the peppered moth, *Biston betularia*.

Fig. 3.14 Dark and light forms of the peppered moth resting on a soot-covered tree trunk. Which of these moths has the better chance for survival in this environment? What does this tell us about the process of evolution?

Kettlewell released both light and dark moths in an industrialized area (Birmingham) and in a nonindustrialized area (Dorchetshire). He was able to show conclusively that far more light colored moths than dark ones are captured by birds on blackened tree trunks, while the reverse is true on tree trunks that are covered by lichens and are, therefore, lighter in color.

Because this is a particularly clear-cut case of natural selection, let us examine exactly what is involved. Since very few dark moths existed in the population before soot darkened the trees and other resting places of moths, the dark moths must have arisen as a result of random gene mutation. Had industrialization not taken place, the dark moths might have gone on maintaining the very small number in the particular species studied by Kettlewell, or they might have been eliminated entirely. However, as it turned out, a changed environment permitted an increasingly greater proportion of dark moths to survive, while it reduced the chances for success of the light forms. This is what is meant by natural selection. It does not produce or initiate genetic changes; this is the role of mutation. Natural selection merely tends to eliminate those individuals who are least fit, for whatever reason. The environment, then, is an important factor of natural selection in that it sets the standards for survival. (See rabbits in Fig. 3.15.)

Modern biology, with its emphasis on the cellular and molecular aspects of life, has done much to strengthen the principle that evolution is responsible for the great variety of organisms we see at the present time. Nature is not static, but ever-changing, and it is this change that we call evolution. Mutation and new gene combinations that come about through sexual reproduction allow varied DNA codes to express themselves in the make-up of an organism. The environment of that organism then plays an important role in determining whether it shall survive and reproduce, give rise to a superior and more successful race, or become extinct. Of this pattern, we are no longer in any doubt.

The differences that we see in organisms today are a reflection of some two billion years of change, both in the organisms themselves and in their environments. Whatever the origin of life may have been, it has expressed itself in the great diversity of our present world. All along the line, various groups of organisms proved successful for a time, and then disappeared. Extinction was the result whenever species were unable to live and reproduce successfully in their particular environments. So it has gone through the millions of years. Mutations arise constantly, new DNA combinations are brought into being through sexual reproduction, and organisms are continually

1. At some point in the past, ear length varied greatly among rabbits. Because rabbits are highly dependent upon hearing as a means of detecting the approach of a predator, those with short ears were at a distinct disadvantage.

2. In time, short-eared rabbits became fewer because more of them were caught and eaten. As a result, long-eared rabbits produced more and more of the offspring, passing the genes for ear length on to their progeny.

3. Since long ears constitute a successful adaptation in rabbits, those with short ears failed to survive the competition.

Fig. 3.15 This is how Darwin explained the evolution of ear length in rabbits. Compare this diagram with Fig. 3.5. Why is Darwin's explanation acceptable to modern biologists? (See steps listed above.)

put to the test of survival. Over time, natural selection determines the fate of individuals and species, and thus diversification continues. Some organisms and groups of organisms are relatively new on Earth. Others, such as the common cockroach, which lived at least 200 million years ago, have survived with very few changes.

This, then, is the biological explanation of diversity—on the molecular level of DNA, and on the broadest level of environment. Later, we shall have more to say about diversity and major pathways of evolution as they have occurred in time.

Some Special Examples of Adaptation

It is tempting to cite many examples of the curious or the unusual in nature, simply because truth is so often stranger than fiction. In this brief account, however, we must be content with a very few examples. Because of their significance in showing how adaptations arise in nature, let us now turn to three fascinating aspects of adaptation: **mimicry, camouflage,** and **convergence.**

A **mimic** is a person who imitates another person in dress, habits, mannerisms, and so on. In biology, mimicry means any close resemblance of the members of one species to another, especially when one species seems to derive some protection by appearing to be the other species. The species with some particular adaptation giving it special advantages is called the **model.** The species that derives an advantage by resembling the model is called the **mimic.**

A classic example of mimicry is found in the case of two butterflies, commonly called the Monarch and the Viceroy (Fig. 3.16). The Monarch seems to be distasteful to birds and other predators, who learn to avoid it in their search for food. The Viceroy lacks these distasteful qualities, but resembles the Monarch so closely that it is left alone by predators. The Monarch is thus the model, and the Viceroy is the mimic. Another example of mimicry is the striking resemblance of certain moths and flies, which do not sting, to wasps or bees. Mimicry is quite common among animals, and frequently the model may be a plant.

It is tempting to think that mimics deliberately copy the models in order to become better adapted to their environment. But this would be attributing to insects and other animals a degree of intelligence which they do not have, to say nothing of plants that demonstrate mimicry. We must search for the answer in some aspect of natural selection.

Mimicry seems to begin on the gene level through mutation, which is a matter of chance. Quite randomly, then, a gene mutation or a combination of mutations happens to give an individual, or a small part of a population, some resemblance to another species. Also quite by chance, in time, or immediately, that resemblance may provide protection or be of advantage in some other way to the mimics.

An advantage over other members of the population, however slight, will tip the balance within the population in favor of the mimics. Since the mimics become better adapted to their environment (which includes the model) than the less fortunate individuals who lack the protective mimicry, more mimics survive and reproduce. Eventually, unless something happens

Courtesy of The American Museum of Natural History

Fig. 3.16 Mimicry in butterflies. The Monarch, shown at the top, is the model. The Viceroy is the mimic.

to stop the trend, only the mimics in the population will survive.

Actually, this is a simplified explanation; nevertheless, it makes the point that mimicry is one important means whereby some organisms adapt to their environment. Like all adaptive mechanisms, it is evolutionary in origin and serves as an excellent example of natural selection.

To most of us, camouflage is much more obvious in nature than mimicry is. All of us have experienced the difficulty of seeing certain animals against a background of plants to which they bear a striking resemblance, or known the difficulty of seeing a fish even in clear water. Figure 3.17 shows two examples of camouflage in nature. Obviously, animals which are difficult to see in their surroundings are protected from their enemies, and the adaptive value of camouflage is unquestioned.

From the standpoint of natural selection, camouflage is a form of mimicry, and is explained in the same way. The better an organism can remain hidden from its enemies or from those organisms upon which it preys, the more successful it will be in survival and reproduction. In the case of organisms which are not very well camouflaged, other adaptive mechanisms, such as speed or the production of offensive odors, compensate for this lack.

The third example of adaptive mechanisms is an evolutionary phenomenon called **convergence.** We can define it as the development of similarities between different species which occupy similar environments. It is most striking when it occurs between species not closely related.

The shark and the porpoise, for example, belong to animal groups that are only distantly related. Their evolutionary pathways separated millions of years ago. Most animals of the group to which the porpoise belongs, the mammals, are not aquatic. All the evidence indicates that the aquatic mammals (porpoises, whales, and so on) arose from land-dwelling ancestors. Sharks, in contrast, are members of a rather primitive group of fishes. In spite of such different origins, these two animals bear a very close resemblance to each other. (Fig. 3.18)

Since the porpoise and its relatives evolved more recently than the shark group, it is tempting to call this a case of mimicry. However, mimicry involves a striking similarity between two species only. The shark and porpoise are not unique in their features. All aquatic vertebrates resemble each other in form to a great extent, and we have chosen these two merely as examples of fundamentally different groups. The point is that the porpoise, like the shark, is streamlined. This "design" feature permits the animal to move swiftly through the water—

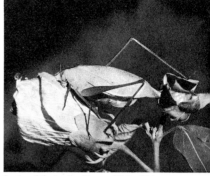

Courtesy of The American Museum of Natural History

Courtesy of The American Museum of Natural History

Fig. 3.17 Camouflage as illustrated by insects. The green color and shape of a katydid (top) make it almost indistinguishable from a leaf. The brown color and stick-like shape of the giant stick insect give it an appearance almost identical to that of brown twigs and branches.

Fig. 3.18 Convergence as it is illustrated by three distantly-related animals: a shark (top), an ichthyosaur (now extinct), and a porpoise. To what extent has environment played a role in the evolution of these three animals?

both as a means of escape from predators, and a means of catching other animals as a food source. In convergence, then, external resemblance in form between two or more different species is simply a reflection of environmental similarity, in this case, water. Perhaps our statements made earlier about environment and evolution are now clearer—a desert cactus does not have fleshy, water-conserving parts because it lives in the desert. It is able to survive in the desert because it has fleshy, water-conserving parts.

Convergence can be explained by natural selection just as mimicry can be explained by natural selection. Adaptation, no matter what form it takes, is a very complex aspect of evolution and many of the problems connected with it are not yet clearly understood. In some respects, we have not improved a great deal upon the insights of Darwin. One cannot read his *Origin of Species* without marveling at his powers of observation and his understanding of what he observed. However, by combining an ecological view with a genetical view—which Darwin was unable to do as well as we can today—biologists have come to understand the adaptation of plants and animals to an extent hardly possible only a few decades ago.

SUMMARY

Adaptation is the development of structural and functional features which occur in a sequence of organisms over a period of time, and which enable them to survive and reproduce within the limits of a particular environment. It is successful evolution.

In their adaptation, organisms have evolved a great variety of structural and functional characteristics. Many of these are developed in such a way as to be expressed in mimicry, camouflage, and convergence. Each of these phenomena can be explained by natural selection, but many aspects of adaptation are not yet clearly understood.

The diversity of living forms which we see on the Earth today is a result of some two billion years of evolution. This is a process whereby change occurs in the world of life, and as a result of this process, there appear new species which differ from their ancestral forms.

The initiating force of evolution is mutation, change that occurs at the molecular level in chromosomal DNA. Natural selection is the process whereby DNA "codes," or gene patterns, are proven successful. Diversity, as we see it reflected in the many kinds of plants and animals, is a result of diversity on the molecular level, since every organism undergoes development according to its particular DNA complement.

FOR THOUGHT AND DISCUSSION

1 It is possible to change the environment of paramecia, which are small water animals, by the gradual addition of salt to the water in which they live. After a time, they can live at salt concentrations which would kill other paramecia. In fact, they die if they are put back into the pond water. Is this adaptation? Explain.

2 Ask five individuals who have not taken a course in biology to define evolution. Write down their answers and compare them with the definition given in this chapter. Do these answers help explain why there is a great deal of misunderstanding about evolution?

3 One of the points made by Darwin in setting forth the principle of natural selection is that the fittest survive. Why would "fitness" have to be defined in relation to environment?

4 Define natural selection, basing your definition on the case of the light and dark moths described in this chapter.

5 What is DNA and why was it discussed in this chapter?

SELECTED READINGS

Dobzhansky, T. "The Genetic Basis of Evolution." *Scientific American*, Volume 182, January, 1950, page 32.
   This article explains in some detail how mutation and the recombination of genes in sexual reproduction play an important role in evolution.

Odum, E. P. *Ecology*. New York: Holt, Rinehart, and Winston, Inc. 1963.
   This small book, which is one of a series in general biology, sets forth the basic principles of ecology.

Simpson, G. G. *The Meaning of Evolution*. New Haven: Yale University Press, 1949.
   This understandable book, written by one of the outstanding students of evolution in our time, will enlarge your grasp of the subject.

Singer, C. *A History of Biology* (see reference at the end of Chapter 2).
   Those portions of this book that deal with evolution are appropriate to this chapter.

Wallace, B., and A. M. Srb. *Adaptation*, (2nd ed.). Englewood Cliffs, N.J.: Prentice-Hall, Inc., 1964.
   This is a relatively brief discussion of adaptation and evolution.

4 THE ANIMAL KINGDOM

Fig. 4.1 Carolus Linnaeus. (1707–78) Swedish founder of system of binomial nomenclature.

Courtesy of The American Museum of Natural History

Most of us know much more about animals than we know about plants. Animals move about rather freely and, for other reasons, attract our attention and interest more readily than plants do. Furthermore, we are animals, so there is a natural temptation to compare ourselves in structure and function with other members of the animal kingdom. At some point or other, all of us have commented on similarities and differences between our own body and that of some other animal. In the pages that follow, we shall attempt to present the basic concepts of animal diversity. Armed with these concepts, you will then be able to pursue further studies on your own.

In Chapter 2, you learned that the earliest attempts at animal classification were made by Aristotle, and that he described only a few hundred species. You saw Linnaeus (Fig. 4.1) invent his system of binomial nomenclature and use it to describe certain similarities among organisms. In addition to the five taxonomic categories that Linnaeus used, "phylum" and "family" were added to accommodate the thousands of new species that were being described. In general, zoologists have had more success in drawing up a list of accepted animal phyla than have botanists for their plant phyla, or *divisions* as they are sometimes called in botany. Evolutionary relationships are somewhat clearer in the animal kingdom, and structural features are such that sharper distinctions can be made between animal groups than between plant groups. Nevertheless, zoologists disagree over the number of phyla that should be recognized, and no one listing is universally accepted. However, all the major schemes of animal classification are so nearly alike that the differences are minor ones.

Because it is representative, the system devised by Lord Rothschild is presented in Table 4-1. You can see that 22 phyla are listed, and that they include approximately 1,000,000 species. This is a conservative estimate; most zoologists place

TABLE 4-1. A Classification of the Animal Kingdom

Phylum		Approximate Number of Known Species
Phylum 1.	Protozoa	30,000
Phylum 2.	Mesozoa	50
Phylum 3.	Parazoa	4,200
Phylum 4.	Cnidaria	9,600
Phylum 5.	Ctenophora	80
Phylum 6.	Platyhelminthes	15,000
Phylum 7.	Nemertina	550
Phylum 8.	Aschelminthes	12,000
Phylum 9.	Acanthocephala	300
Phylum 10.	Entoprocta	60
Phylum 11.	Polyzoa	4,000
Phylum 12.	Phoronida	15
Phylum 13.	Brachiopoda	260
Phylum 14.	Mollusca	100,000
Phylum 15.	Sipunculoidea	275
Phylum 16.	Echiuroidea	80
Phylum 17.	Annelida	7,000
Phylum 18.	Arthropoda	765,000
Phylum 19.	Chaetognatha	50
Phylum 20.	Pogonophora	43
Phylum 21.	Echinodermata	5,700
Phylum 22.	Chordata	45,000
	Approximate total:	1,000,000

Lord Rothschild, A Classification of Living Animals. New York: John Wiley and Sons, 1961. Adapted by permission.

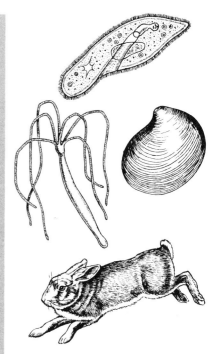

the figure closer to 1,250,000. Furthermore, on the average, some 10,000 new species of animals are described each year, which means that any estimate cannot be very accurate.

As indicated in Table 4-1, many phyla include a relatively small number of species. In fact, exactly one-half of those listed contain fewer than 1,000 species each. Three of them (phyla Mollusca, Arthropoda, and Chordata) account for nine-tenths of the entire animal kingdom! To look at the table still more closely, the phylum Arthropoda, which includes the insects, spiders, crabs, and various similar forms, is far larger than the other 21 phyla combined. As you will find in the next chapter, something similar occurs in the plant kingdom: The flowering plants (class Angiospermae) outnumber all other plants combined by about five to two (250,000 species out of a total of 350,000).

The scope of this book is such that we cannot discuss every phylum of animals. Some phyla include animals that most of us never see, and never shall see. Table 4-2 (see page 70) singles out groups that are important for our present purposes. These

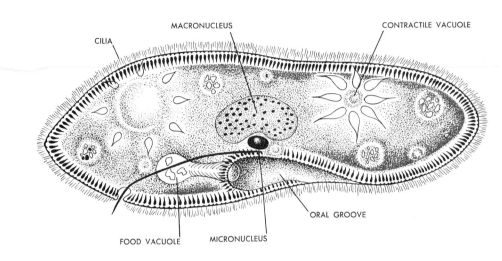

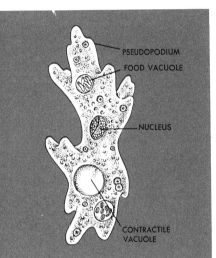

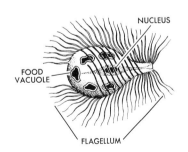

Fig. 4.2 Three representative protozoa: *Paramecium*, a ciliate (above), *Amoeba* (left), and a typical flagellate from the intestinal tract of a termite.

19 groups represent most of the known animal species, as you can see by the summary figure.

A SURVEY OF MAJOR ANIMAL TYPES

The Protozoans (Phylum PROTOZOA)

According to the **cell theory,** the fundamental unit of all living material is the cell. Most animals are composed of many cells and, therefore, are grouped in a subkingdom called the **metazoa.** Although some taxonomists recognize this group, the term is an informal one and is without taxonomic status. Metazoa simply distinguishes multicellular animals from those that are unicellular, or one-celled. This latter group has long been recognized as a single phylum, the Protozoa (Fig. 4.2).

Since the largest protozoans are barely visible to the unaided eye, we can learn little about them without a microscope. The entire group, consisting of more than 30,000 species, can be considered microscopic. As a matter of fact, most of them cannot be seen at all without the aid of a microscope, and some are so small that they are barely visible with the ordinary student microscope of 430 power. They are widespread in nature; both fresh and salt water abound with various species, and some forms live upon or within the soil. Still others are parasites of certain higher animals and are the cause of disease. For example, amoebic dysentery and malaria, both of which are widespread among humans, are caused by protozoa.

Order can be brought to the bewildering variety of protozoans by grouping the animals into classes according to their means of locomotion. Members of one class, which includes *Amoeba,* move whenever the cell contents flow in a given direction in response to a food source. An amoeba simply extends a "false foot," called a **pseudopodium,** and flows into it. Members of another class have many hair-like structures called **cilia** (singular, **cilium**) at the cell surface. Through the coordinated beating of their cilia, these animals can move rapidly through water. One of the most common ciliates, *Paramecium,* lives in most fresh-water ponds. A third class consists of protozoans that have one or more structures called **flagella** (singular, **flagellum**), which are much like cilia except that they are considerably longer.

Although many flagellates live in water, some of them live in the digestive tract of termites. These two species maintain an unusual relationship. Termites live on a diet of wood and are not able to digest cellulose, the tough cell wall material of most plants, but the protozoans can digest cellulose and do so within the intestinal tract of termites, thus benefiting the insect. Meanwhile the protozoans are provided with a suitable environment. Such a relationship is called **mutualism,** because it is a cooperative enterprise in which both species derive mutual benefit. A protozoan of the type described is shown in Fig. 4.2. In addition to the three classes of protozoans we have mentioned (*Amoeba* and its relatives, the ciliates, and the flagellates), there is a fourth class—*Sporozoa*—made up entirely of parasitic forms which do not have locomotor structures. *Plasmodium,* which causes malaria, is one such protozoan.

Although many protozoan species reproduce sexually, cell division is the usual method of reproduction. It has remained the chief reproductive method of protozoans throughout their millions of years of existence.

While protozoans are of little direct value to man, they play an extremely vital role in nature. Even though they represent a relatively small proportion of the animal kingdom in number of species, there are more *individual* protozoa than all other animals combined. In the water, they help to decompose organic matter by devouring bits of the remains of dead plants and animals. They also occupy an important place in nature's complex food web by serving as food for larger aquatic animals. Parasitic forms are of direct concern to man whenever they cause human diseases. The forms that produce disease in livestock, fish, or other animals in which man has a direct interest, are also important. Within the last few decades, protozoans have been used widely in basic biological research. Since they are at once an individual cell and an organism, it is sometimes possible to discover in them fundamental principles that can be applied to higher organisms, including man.

The Sponges (Phylum PARAZOA)

If you could see sponges in their natural habitats, (Fig. 4.3) you might find it hard to believe that they are animals. Indeed, only a few decades ago they were considered plants because they did not move about. However, they have been included in the animal kingdom ever since biologists discovered that their cells are very similar to those of certain protozoa. Apparently, the various sponges evolved from protozoans—the cells of sponges being quite different, both in structure and in function, from any cells found in the plant kingdom. Actually, a given sponge is not so much an individual organism as a *colony* of organisms, that is, a group of protozoan-like cells living together in a common skeletal structure which they secrete.

Although a few species of fresh-water sponges are known, the vast majority live in the shallower waters of the ocean, especially in the warmer latitudes. A few are found in fairly deep ocean waters. Among sponges there is considerable variation in appearance and composition of the skeleton. In the skeleton of all forms are many rod-like structures composed largely of either calcium or silicon. In the group which is of commercial importance, a protein material called **spongin** makes up a large portion of the skeleton. Long after the cells of a spongin sponge have died, this material retains its texture, and allows for the absorption of water into the various canals that penetrate the skeleton.

Sponges reproduce both sexually and asexually. In organisms that reproduce sexually, there is a union of two cells (called **gametes**) resulting in a new single cell, the **zygote.** In sexual reproduction (Fig. 4.4) one sponge is able to produce both egg cells and sperm cells, but not at the same time. This assures cross-fertilization between two different individuals. Fertilization takes place inside the body of the sponge that produces the eggs, when the sperm from another sponge enter the canal system of the egg-producing sponge. The resulting zygotes swim out of the sponge and eventually attach to some object, or to the ocean floor, where they mature.

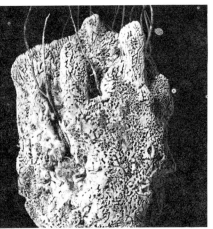

Courtesy of The American Museum of Natural History

Fig. 4.3 Some representative sponges. At first glance a sponge seems more like a plant than an animal. Why are sponges considered animals?

Asexually, some sponges produce buds, or little sponges, near their bases, and these eventually mature. However, they usually stay attached where they were formed. Large colonies of sponges may build up at a given point in this way. Sometimes, however, the mature buds may detach themselves and live independently. Figure 4.5 shows budding in sponges.

In other sponges, clusters of cells sometimes become separated from the parent sponge and are carried away by water currents. Eventually they attach themselves to some object on the ocean floor and grow to maturity. One particularly interesting thing about sponges—and about all metazoa to a certain extent—is their capacity for **regeneration.** This is the development of a new organism from a part of the body of the old organism, or it is the regrowth of a lost or injured part. A sponge may be cut into hundreds of small pieces, and if each piece is placed back in the water, it will grow into a complete sponge.

The Cnidarians (Phylum CNIDARIA) also called (Phylum COELENTERATA)

The name of this animal group, which includes jellyfish, comes from the Greek word meaning "nettle." This is because all members have cells with small barbs, or stinging threads. Each barb, or thread, carries an irritating chemical which can be damaging to other animals. The larger jellyfish, some of which are six feet or more in diameter, can inflict painful and dangerous stings upon swimmers by releasing thousands of their barbs at one time. Even small jellyfish (a few inches in diameter) should be avoided. Figure 4.6 shows one type.

The cnidarians represent a very important advancement over the protozoans and sponges. For the first time we find the **tissue** level of organization, in which there is specialization of cells and cell groups.

Although fresh-water cnidarians are common, especially a small animal known as the **hydra** (Fig. 4.7), the most familiar

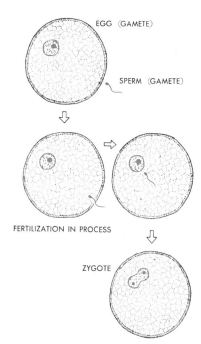

Fig. 4.4 Sexual reproduction of multicellular animals involves the union of male and female gametes to form a zygote.

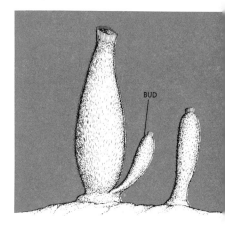

Fig. 4.5 Asexual reproduction in *Grantia*, a small sponge, by a process called budding. In most cases, the bud remains attached to the original sponge.

Fig. 4.6 A jellyfish, *Gonionemus*, with tentacles extended downward.

Courtesy Carolina Biological Supply Company

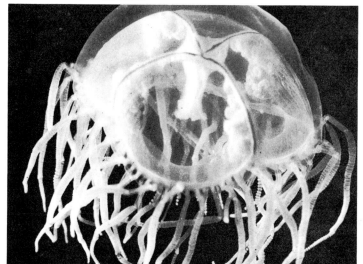

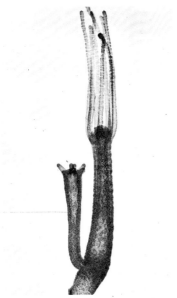

Courtesy of The American Museum of Natural History

Fig. 4.7 The hydra shown above has developed a bud which will break off and become independent. The diagram below shows a longitudinal section of a hydra.

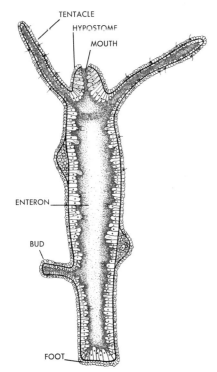

form to us is the jellyfish. Another well-known cnidarian is **coral.** What we call "coral" is really the secretion of calcium compounds by a group of animals living in a colony. After the animals die, their colony of skeletons remains, and other coral animals may build upon them. In this way tremendous deposits form over a period of years.

The fresh-water hydras, which typify the cnidarians in structure and function, abound in fresh-water ponds. Barely visible to the naked eye, they live attached to the leaves and stems of water plants. Two layers of cells enclose a digestive cavity which has a single opening to the outside of the body. Some of the cells of the outer layer have the ability to contract. In this respect they resemble the highly specialized muscle cells of more advanced animals. By virtue of this contraction, the animal is able to change the shape of its body to capture food and to move about. The hydra has a network of specialized nerve cells that enable it to coordinate its movements and respond to stimuli.

Each individual can reproduce asexually or sexually. In asexual reproduction the animal produces buds, each of which eventually breaks off and becomes an independent animal. In sexual reproduction an individual usually functions as a male at one period and as a female at another. The single egg produced in an **ovary** (female sex organ) is fertilized by one of the several sperm that may swim to it after having been produced in a **testis** (male sex organ) of another hydra. The resulting zygote eventually develops into a hydra.

To biologists, cnidarians are important because they are the least complex animals which exhibit the specialization of cells into tissues, and because in general they have a high degree of regenerative power.

The Flatworms (Phylum PLATYHELMINTHES)

At one time in the history of zoology, nearly every invertebrate animal was called a "worm." Today, we generally restrict the term to those invertebrates whose bodies are elongated in the adult stage, and which move in such a way that the front, or **anterior,** end of the body leads the way, and the rear, or **posterior,** part follows along.

The flatworms are those worms whose bodies are much wider than they are thick. One class of the three generally recognized by zoologists is composed of free-living forms. A common example of this class (Fig. 4.8) is a group of worms known as **planarians,** which are easily collected from ponds and streams. The other two classes are parasitic, and are known as **flukes** and **tapeworms.** A tapeworm is shown in Fig. 4.9.

General Biological Supply House

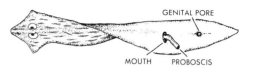

Fig. 4.8 Planarian worms. In the drawing (above) the worm is twisted to show upper and lower surfaces. Note the position of the mouth with the proboscis extended during feeding.

In the free-living flatworms there are several well-developed body systems, including muscular and nervous systems. The digestive tract is essentially a sac with a single opening, although it is much more specialized and efficient than that of the hydra and other cnidarians. The most noticeable advance over the animals mentioned so far is the difference in over-all body plan. Flatworms, like most of the other animals we will consider, are **bilaterally symmetrical** (Fig. 4.10), as opposed to the **radial symmetry** of cnidarians and sponges. A radially symmetrical animal is built on a circular plan, somewhat like a wagon wheel. A bilaterally symmetrical animal is built on a longitudinal plan. Zoologists usually define a bilaterally symmetrical body as one that can be divided by one plane so that approximate mirror images are produced. Is the human body bilateral in its symmetry?

A few species of the free-living flatworms reproduce asexually by splitting longitudinally, but sexual reproduction is the general rule both in the free-living and in the parasitic forms. As a matter of fact, asexual reproduction is so rare in animals more complex than flatworms that we shall make no further

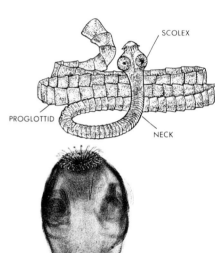

General Biological Supply House

Fig. 4.9 Diagram and photograph of a tapeworm, showing the "head" (scolex) and segments (proglottids).

Fig. 4.10 A contrast in body plans. Many of the terms used to describe bilaterally symmetrical animals are meaningless when applied to radial symmetry.

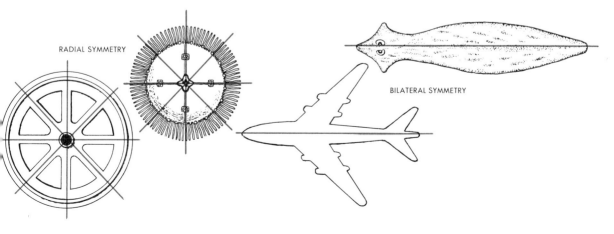

reference to it. Most flatworms have both ovaries and testes, that is, there are no separate male and female individuals among most flatworms. Planarian worms mate by exchanging sperm cells. After the eggs of each worm have been fertilized, they pass to the outside of the body. Each egg then develops into a small planarian which grows to maturity.

The flukes and tapeworms, all of which are **parasites**, are much more important to man than are the free-living flatworms. Several different species of flukes and tapeworms live within the bodies of domestic animals. They frequently do great damage, sometimes killing their hosts. Man can be a host to certain species of both forms. In certain areas of the world, especially where lax sanitation measures prevail, these two types of parasites are a major health problem.

The Nematodes (Phylum ASCHELMINTHES)

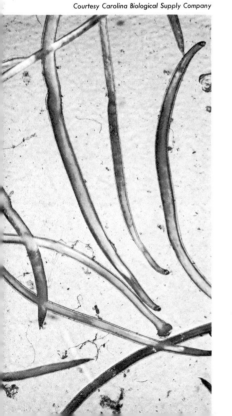

Fig. 4.11 A group of hookworms, which are representative nematode parasites. See if you can find out why hookworm disease is not nearly as common in the United States today as it was a few decades ago.

Courtesy Carolina Biological Supply Company

All the animals we shall discuss from now on have a digestive tract that begins with one opening (the **mouth**) and ends with another (the **anus**). Among the least complex of these animals are the Aschelminthes, a phylum containing a variety of forms. The most numerous of them are the nematodes, or roundworms (Fig. 4.11). Unlike the flatworms, nematodes have cylindrical bodies that are usually pointed at each end. Some are free-living and are found abundantly in fresh water and in soil. Others are parasites that use a variety of other animals, including man, as host.

Usually, a roundworm is either distinctly male or distinctly female. This is a new condition of sexuality in our survey of the animal kingdom. In each sex, the reproductive organs are in the cavity between the digestive tract and the outer body wall. During mating, sperm are transferred from the male to the female reproductive system, and eggs are fertilized inside the body of the female. Some female nematodes lay several thousands of eggs in one day. Under suitable conditions each egg develops into a small nematode worm which eventually grows to adult size.

Both plants and animals can be parasitized by nematodes. Many crop plants cannot be grown in some soils unless the nematodes are kept under control. Nematodes usually attack the roots, killing the plant or making it completely unproductive. Many roundworm parasites attack livestock animals and often produce fatal diseases. Quite a number also are known to infest the human body. How many of the following roundworms have you heard about: hookworms, trichina worms, the Guinea worm, and *Ascaris?*

The Mollusks (Phylum MOLLUSCA)

As you saw in Tables 4-1 and 4-2, this is a very large group of animals. It includes a wide variety of forms, among them clams, oysters, snails, and octopuses (Fig. 4.12). Primarily, mollusks are marine animals, although there are many fresh-water and terrestrial snails, and there are fresh-water clams. A structure called the **mantle** is what sets mollusks apart from other animals. The mantle functions as a protective covering for soft and delicate tissues. In some mollusks—snails and clams, for example—the mantle secretes a hard, calcareous shell. If a mollusk has a single shell, like a conch or a snail, it is called a **univalve.** If it has a double shell, like oysters and clams, it is called a **bivalve.**

The various body systems of mollusks show a decided increase in complexity and organization over the animals we have already studied. The nervous and digestive systems, for instance, are more complex and efficient, and for the first time we find well organized circulatory and excretory systems. The octopuses and their close relatives, the squids, are particularly advanced.

Mollusks are essentially bilateral in their symmetry. The sexes are usually separate in a given species, although there are some exceptions. Most marine bivalves are males when they are very young, but they become females when they reach maturity. Sperm are usually released in the open water, and fertilization of eggs occurs within the female body whenever sperm happen to be taken into the body with a water current. In contrast, land forms must mate since there is no medium of water to carry the sperm from male to female.

The clam, which is representative of this phylum, has a number of specializations not present in the previous phyla we have reviewed. Much of the time, the clam is half or wholly buried in mud or sand. It moves by means of a muscular appendage called a **foot,** commonly miscalled the "neck." (In the more complex mollusks—the octopus and squid, for example—the foot has changed through evolution into a group of tentacles surrounding the head.) Two strong muscles attach the soft body to the valves, and by means of these muscles the clam is able to snap the valves closed as protection against its enemies. Two tubes, called **siphons,** extend to the edge of the valves. The clam draws water in through one of the siphons and expels it through the other. Food particles enter the animal this way, and waste material is expelled. Oxygen is also carried to the gills in the mantle chamber by the inflowing current of water. A good view of a clam is provided in Fig. 4.13.

Fig. 4.12 Representative mollusks: a common snail, an octopus, and a scallop.

R. H. Noailles

Douglas P. Wilson

National Park Service, Department of Interior

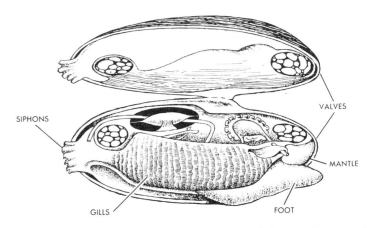

Fig. 4.13 This drawing of a clam shows some of the complex internal anatomy typical of mollusks. The four circular structures shown at the ends of the two shell-halves are points where strong muscles attach.

Sperm entering the female clam on the inflowing current of water are carried to a special pouch in the gills. Here the eggs are fertilized and develop into **larvae** (worm-like forms) called **glochidia.** Eventually the glochidia leave the female on the outward flowing current and complete their life cycle.

Other specialized organs of the clam include a stomach and intestine, blood vessels, a heart, kidneys, a urinary bladder, and a nervous system. Mollusks, as represented by the clam, are strikingly complex animals compared with those of the phyla we have seen so far.

The Annelids, or Segmented Worms (Phylum ANNELIDA)

The bodies of annelids are divided into segments. This feature is so pronounced in them that it has become the chief means of distinguishing them from other groups of animals, even though certain other animals have segmented bodies—tapeworms, for example. There are several

Fig. 4.14 Two common annelids. The earthworm (top) is shown at the surface of loose soil. The leech (bottom) is being eaten by a large diving beetle.

Fig. 4.15 This diagram shows part of the complex internal anatomy of an earthworm. Note especially the complexity of blood vessels.

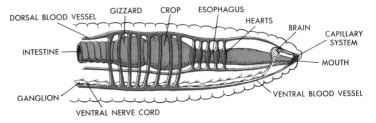

thousand different types of annelids, including marine, freshwater, terrestrial, and parasitic forms. The most familiar member of this group is the common earthworm. Another common annelid is the leech. Both of these are shown in Fig. 4.14.

The earthworm is fairly representative of this group. Externally, the segmented nature of the body is obvious, the segments often numbering well over a hundred. Internally, many of the organs and other structures of the body are repeated segment by segment, so that the animal is a series of fairly uniform "slices." In general, the body plan (Fig. 4.15) shows an advance in specialization and efficiency over the groups we have studied so far. The nervous, muscular, digestive, excretory, and circulatory systems are particularly advanced. The circulatory system, for example, consists of a series of closed tubes, or blood vessels. Blood is pumped through the vessel to other organs of the body by five pairs of **aortic arches,** or hearts. As in higher animals, food in diffusible form enters the blood through the walls of an intestine. Oxygen enters the bloodstream through the "skin," or **cuticle,** and is carried by the blood to other parts of the body. In a similar way, carbon dioxide is given off through the cuticle.

Many of the features of the earthworm are adaptations to its particular way of life and are not typical of all annelids. For example, marine annelids have eyes and other sense organs that earthworms do not have. Furthermore, earthworms do not have legs and certain other appendages that characterize marine annelids. Although the sexes are separate in marine annelids, this is not the case in earthworms. Every earthworm has both testes and ovaries. A pair of worms mate and exchange sperm. The sperm are then stored by each worm in a special receptacle for later use in fertilizing its eggs. The eggs are released from the body in groups and are deposited in the soil within cocoons. Each cocoon encloses several eggs. After a period of development, small worms emerge from these cocoons.

The annelids are not of great economic importance to man, with the exception of the earthworm. As it burrows through the soil, an earthworm forms canals through which air may pass. This permits chemical activities to occur in the ground at a much more rapid rate than would be possible otherwise. Furthermore, earthworms eat soil that is too hard to be pushed aside. As the soil passes through the worm's digestive system, a number of materials are digested or broken down into simpler forms. Through its several activities in the soil, the earthworm hastens the decomposition of dead and decaying matter, which increases soil fertility. The exact value per worm to the farmer would be difficult to calculate, but the presence of earthworms in the soil is highly beneficial.

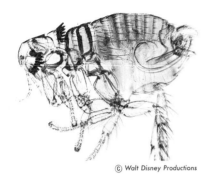

© Walt Disney Productions

P. A. Knipping

USDA

Fig. 4.16 Some representative insects. In general, insect species are rather easily distinguished because of their body structure. Top: rabbit flea. Middle: queen bee and her "court." Bottom: mayfly.

The Insects (Class INSECTA, Phylum ARTHROPODA)

The phylum Arthropoda (Fig. 4.16) is the largest in the animal kingdom. It is so large, in fact, that we will have to break it down into five smaller groups, or classes. All animals belonging to this phylum have jointed and paired appendages. They also have three jointed body segments called the **head, thorax,** and **abdomen,** although in some the head and thorax are fused. The internal organs of all arthropods (and most mollusks) are bathed in blood. This **open** circulatory system is quite different from the **closed** (blood vessel) circulatory system of the earthworm.

There are other important distinguishing features. A hard outside skeleton, called an **exoskeleton,** encloses the arthropod body. Sensory structures (particularly eyes) show a decided advance over the animals we have studied so far. In addition to other body systems being quite advanced, in all but a few arthropods the sexes are separate.

The insects, one class of this phylum, are by far the most numerous group among the Arthropoda, as shown in Table 4-2. The estimate of 700,000 known species is conservative; many zoologists use the round figure of one million. At any rate, there are more known species of insects than of all other animals combined. Insects exist in such variety that a general description of them is difficult, although certain features set them apart from other members of their phylum. They are distinguished from other arthropods by having three pairs of walking appendages in the adult stage. Most of them also have wings in the adult stage (usually two pairs). This is not true of any other invertebrate animals. The ability of insects to fly helps explain their success as an animal group. Although most insects are found in terrestrial-aerial habitats, a great many species live in water (the immature stages of many land-dwellers are also aquatic), in the soil, and as parasites. Virtually none are marine.

Some insects, such as the grasshopper and its near relatives, develop from eggs and pass only through a **nymph** stage before becoming adults. In the nymph stage they are near carbon copies of an adult, except that the nymphs are smaller. Most other species of insects go through a more complicated life cycle. Since butterflies are typical of this latter group, we shall describe their life cycle (shown in Fig. 4.17) briefly.

After the female butterfly mates with a male, she lays fertilized eggs on leaves or other vegetation. The eggs hatch into caterpillars (the larval stage), which bear no resemblance to the adult butterfly. In moths and butterflies, the larva is called a **caterpillar.** After a period of intense feeding and rapid growth, the larva spins a cocoon around itself and becomes dormant for several days, weeks, or months, depending on the species. At this stage, it is called a **pupa.** Although a pupa looks inactive, many internal changes are taking place. Finally, a fully mature butterfly emerges from the cocoon, dries its wings, and mates, thus starting the cycle anew.

The economic importance of insects ranks exceedingly high. The honey bee and the silkworm moth are probably the most beneficial of all insects, at least in a direct sense. Indirectly, many insects serve to pollinate various flowers, including those of some crop plants. Their value in this role is inestimable.

Several insects are valuable to man because they prey upon other insects. From the standpoint of damage, a great number of different forms prey upon crop plants. A swarm of locusts can completely destroy a farmer's crop in a matter of minutes. Almost every species of plant has insect enemies. It is a rare orchard or farm crop that does not have to be sprayed in order to produce a respectable yield. Various flies and lice inflict great damage on livestock animals and frequently transmit infectious diseases. Insects also are responsible for carrying many diseases to man. For example, the protozoa which cause malaria are transmitted to man by certain mosquitoes.

From an ecological viewpoint, insects are the dominant animals on Earth today. They are the "dinosaurs" of the 20th Century. Among all organisms, they are probably the only group that poses a threat to man's continued existence as a species, and we can do amazingly little to reduce the threat.

The Crustaceans (Class CRUSTACEA, Phylum ARTHROPODA)

Unlike the insects, crustaceans (Fig. 4.18) have five or more pairs of walking appendages, and two pairs of sensory appendages. The large pair are called **antennae,** and the small pair, **antennules.** Insects, you will recall, have only a single pair. The exoskeleton of crustaceans is generally heavier than that of insects. Most species are marine, although several inhabit fresh water or land. Typical marine forms are the lobsters, shrimp, and crabs. The crayfish (Fig. 4.19) is perhaps the best-known fresh-water type, and the bluish-gray sowbug (Fig. 4.18C) is a common terrestrial form.

As we have done with the other classes of this phylum, let us select one typical animal and take a detailed look at one aspect of it in order to see still another way in which this phylum exhibits complexities over the simpler animals of earlier phyla. The digestive system of the crayfish, which is a common laboratory animal, will serve as an example. Crayfish eat animals and plants, and are, therefore, said to be **omnivorous.** Several specialized appendages associated with the mouth are

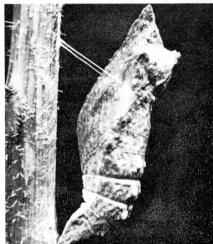

Lynwood M. Chace

Fig. 4.17 Three stages in the metamorphosis of a butterfly. Top: larvae, or "caterpillars." Middle: pupa. Bottom: adult emerging from pupa case.

Fig. 4.18 Some representative crustaceans. Left: a shrimp in natural pose. Above: a land crab walking on sand. Right: a sowbug, or "roly-poly."

used to manipulate, tear, and chew bits of food. From the mouth the food is forced into the first part of a double stomach. Teeth lining this part of the stomach, called the **gastric mill,** grind the food into smaller pieces. Stiff hairs in the second part of the stomach strain out bits of food that are too large to pass. The finely divided pieces are permitted to enter the intestine where they are broken down to a diffusible form by strong digestive juices supplied by the liver. The food, now digested, passes through the intestinal walls into the blood surrounding the intestine.

Fig. 4.19 Internal anatomy of a crayfish. Note the ventral position of the nerve cord, which is typical of invertebrate animals.

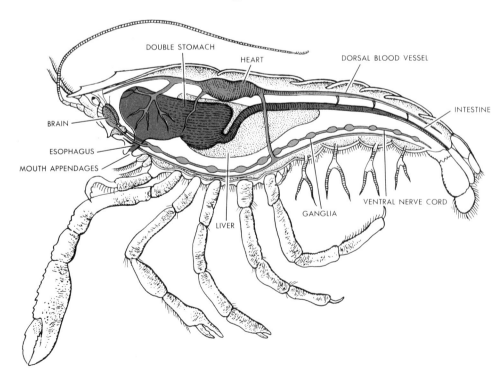

In nature, crustaceans are important in the over-all food web. Many small forms of both fresh and salt water crustaceans consume tiny plant organisms called **algae.** In turn, the crustaceans are themselves eaten by larger animals. The sole food of the blue whale, the largest living animal, is a small shrimp-like crustacean. It has been estimated that in its lifetime one blue whale eats 10,000 tons of these small animals. Other whales, other marine animals, and men also use crustaceans as a food source. In this way, crustaceans serve as an important link in the general food web of nature.

The Arachnids (Class ARACHNIDA, Phylum ARTHROPODA)

USDA Photo

Spiders and the other members of this group are distinguished from other arthropods by their four pairs of walking legs and absence of antennae. In addition, a pair of appendages just in front of the legs serve as poison claws in most arachnids. The mouth parts are not adapted for biting; hence, the bite of a spider is really a pinch of the poison claws. Other familiar forms include scorpions, harvestmen ("daddy longlegs"), ticks and mites, and the king, or horseshoe, crab. See Fig. 4.20.

The spiders are the most numerous of the arachnids. Many people fear spiders, but most spiders are entirely harmless. There are only two dangerous species in the United States. One is the black widow, whose bite is occasionally fatal, especially to small children. The other is the brown recluse, whose bite is not fatal, but which causes the flesh in the bite area to decay. The large tarantula of the southern United States is greatly feared, but its bite usually causes little more damage than pain. Although all spiders are equipped with poison claws, the amount of poison injected is usually so small and of such low toxicity to humans that it causes no more than a small amount of pain and redness.

© Walt Disney Productions

Fig. 4.20 Black widow spider, showing relative size of male (smaller) and female (larger). Below: a diving water spider beginning to submerge.

Scorpions (Fig. 4.21) are another matter. Outwardly they resemble crayfish. They are unusual in having a sting at the end of the abdomen. Although most of the 40 or so species found in the United States can deliver a painful sting to man, only two are dangerous. All of them are capable of killing animals much larger than themselves. Most scorpions of the United States are restricted to the warmer areas of the country, particularly to the Southwest.

Fig. 4.21 Scorpion with abdomen poised in the sting position.

© Walt Disney Productions

Ticks and their close relatives, the mites, are very numerous and widespread (Fig. 4.22). All are parasites on bodies of higher animals, with the exception of a few free-living mites. It is a rare bird or mammal, that is not afflicted at one time or another with some species of tick or mite.

Mites in tremendous numbers may infest a chicken. They

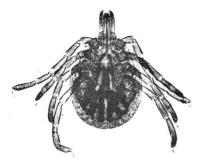

Courtesy Carolina Biological Supply Company

Fig. 4.22 A representative tick. Why are ticks considered more closely related to spiders than to insects?

Courtesy of The American Museum of Natural History

Fig. 4.23 Centipedes. How many legs do the "hundred-leggers" shown here have?

Fig. 4.24 A millipede. Can you figure out the sequence of leg motion which this animal employs as it moves along?

Courtesy of The American Museum of Natural History

live by sucking blood out of the fowl, which lowers egg production drastically and may eventually cause the chicken's death. Ticks and mites also carry microorganisms that are responsible for diseases in man, in domestic animals, and in plants. Among the more important of these diseases are Rocky Mountain spotted fever and tularemia, both of which are contracted by man and a variety of animals, and Texas fever, which is a disease of cattle. Most people have experienced an attack of chiggers, or "red-bugs," which are also a form of mite. They are extremely small, and their bite produces a red, itching welt on the skin.

The Centipedes (Class CHILOPODA, Phylum ARTHROPODA)

If the centipedes and their close relatives, the millipedes, were not clearly arthropods, we would be tempted to call them annelid worms. The centipede's elongated body has many segments, each one of which (except the first and last) has a single pair of legs. So many walking appendages has given rise to the common name "hundred-legger." The exact number of legs depends, of course, on the number of segments. On the first segment behind the head is a pair of poison claws which resemble those of spiders. A single pair of antennae projects forward from the head. See Fig. 4.23.

Centipedes invade nearly any dark place. Small forms are often seen under logs, piles of straw, or old boards. Many species invade houses. In the tropics, it is not unusual to find centipedes a foot long. These larger forms can give a painful bite. In fact, it is wise to avoid even small centipedes. One of the first things campers in the tropics learn is not to walk barefoot around temporary quarters and not to put on shoes in the morning without first making sure that a centipede has not crept in during the night. Except for being a minor menace, centipedes are of little importance to man. They do, however, eat certain harmful insects, which makes "hundred-leggers" of some value.

The Millipedes (Class DIPLOPODA, Phylum ARTHROPODA)

Millipedes are very much like centipedes in appearance and structure. Their internal anatomy is strikingly similar. However, there are enough differences to make most zoologists prefer to place the two in separate classes. One of the most obvious differences is that millipedes have *two* pairs of legs on each segment, whereas centipedes have only one pair. Furthermore, millipede legs are much shorter than those of centipedes. This seems to account for the relative slowness of movement of millipedes, while centipedes, in contrast, move rapidly. Although millipedes are sometimes called "thousand-leggers," they seldom have more than 100 segments, not all of which have appendages; see Fig. 4.24 for example.

In contrast to the centipedes, which are **carnivorous** (that is, they eat other animals), millipedes are **herbivorous** (plant eaters). The difference in rate of travel between the two types of animals is a reflection of their different feeding habits. From an evolutionary standpoint, herbivorous animals are not under pressure to move about fast—at least not in their search for food. Natural selection, then, would not weed them out for their slowness. Some slow moving animals, however, may be under pressures to escape from faster moving enemies that would eat them, but the millipede has "solved" this problem in another way. When it is disturbed, it curls up and plays dead. Furthermore, most species of millipedes have glands that produce offensive fluids, so there are few animals that care to eat them.

Millipedes are of no particular importance to man since they feed upon dead vegetation. In this respect, of course, they help to decompose plant materials in nature. Unlike the centipedes, they have no poison claws, and they are entirely harmless.

The Echinoderms (Phylum ECHINODERMATA)

You will recall that the sponges and cnidarians, which are relatively simple animals, are radially symmetrical in their body structure. You also found that later groups, relatively more complex, are bilaterally symmetrical. For years, biologists believed that radial symmetry was a reliable indicator of relative simplicity. For this reason, echinoderms were once placed far down in the scale of animal complexity. However, when embryology and biochemistry came into their own, biologists learned that: 1. echinoderms are very similar to chordates (the highest phylum consisting of the most complex animals) in the way they develop, and 2. their body fluids bear a closer chemical resemblance to those of chordates than do the body fluids of any other invertebrates.

To look again at the matter of symmetry, the larvae of echinoderms *are* bilaterally symmetrical, but on their way to becoming adults they switch over to a radial plan. Even so, modern zoologists are convinced that echinoderms are more closely related to the chordates than to the less complex invertebrates. This is another way of saying that their evolution is more closely linked to the highest animal phylum than to lower ones.

Fig. 4.25 Representative echinoderms. Top: sea urchin. Middle: starfish. Bottom: sea cucumber. Note the rough exterior surfaces and the radial symmetry.

P. A. Knipping

United States Department of the Interior, Fish and Wildlife Service

Courtesy Carolina Biological Supply Company

Exactly what are echinoderms? The name itself means "spiny skin," and this is a fairly good description of one obvious characteristic of the group. They are very rough to the touch, because their outer body wall is thoroughly covered with small, calcareous spines. Body systems of echinoderms are not as well developed as those of mollusks, annelids, and arthropods, and sense organs in particular are poorly advanced. All forms of echinoderms are marine, and they are found both in shallow and in fairly deep waters of the ocean. Some common echinoderms, sea cucumbers, sea urchins, and starfish are shown in Fig. 4.25.

In their reproduction, the sexes are separate in all but a few species. However, no mating process occurs between male and female individuals. Eggs and sperm are released into the water and fertilization takes place somewhat by chance outside the body. A zygote develops into a larva which is bilaterally symmetrical. By a very complex process of growth, the larva develops into a radially symmetrical adult.

Echinoderms are of almost no positive economic importance, although sea cucumbers are eaten in some parts of the world. Any oyster fisherman can tell you about the negative economic importance of starfish. These echinoderms prey upon mollusks to such an extent that oyster fishermen are obliged to wage war upon them to prevent the complete destruction of their oyster beds. A large starfish can easily consume five good-sized oysters in a day. It eats an oyster by folding itself around both sides of the shell, to which it attaches by its hundreds of little hydraulic suction cups called **tube feet.** Eventually, through fatigue, the oyster relaxes its muscles. Its shell is then easily opened (Fig. 4.26). The adaptable starfish then everts its own stomach and pushes it between the two halves of the shell, and proceeds to digest the fleshy body of the oyster.

Fig. 4.26 A starfish attacking an oyster. The starfish continues to pull the valves apart until the oyster's muscles relax with fatigue.

United States Department of the Interior, Fish and Wildlife Service

When oyster fishermen first became aware that they were competing with starfish, the fishermen tried to deal with the problem by catching large numbers of starfish, cutting them up, and throwing the pieces back into the water. What they did not realize is that starfish and other echinoderms have great powers of regeneration. A starfish hacked into three pieces in the right way grows into three new starfish! The hapless fishermen were simply aiding the enemy. At the present time, chemical methods are used to protect oyster beds from starfish. When fishermen catch starfish now, they either throw the animals on the beach to die in the sun, or sell them to laboratories.

Even though echinoderms are of little commercial value, and are a downright menace to the oyster industry, there is one bright spot in their existence. Since the eggs of echinoderms develop in open water, where they can be observed, they are important to embryologists. This is especially true since echinoderm development parallels so closely that of the chordates, including (to a certain extent) man himself. Some of the basic principles of embryology were discovered many years ago through study of starfish and sea urchin eggs. Furthermore, any animal that is capable of a high degree of regeneration is useful to biologists. There is still much for us to learn about the regeneration of tissues and the healing of wounds.

The Protochordates
(Classes UROCHORDATA, CEPHALOCHORDATA; Phylum CHORDATA)

At some point in evolutionary history, we are not sure where or when, a small group of animals began to undergo drastic changes. Eventually the changes gave rise to a new group of animals, the chordates, (Fig. 4.27) that were unlike anything else in the animal kingdom. There is no way of knowing with certainty what kinds of animals gave rise to the chordates. They may have been annelid-like, as some zoologists think. It seems evident that both chordates and echinoderms evolved from a common worm-like ancestor of some sort. According to fossil evidence, the change took place about 500 million years ago. Since that time chordates have gradually become the numerous and highly diversified group of animals, including man, that we know today.

The chordates kept some of the more successful features that we see in some of the other phyla—for example, bilateral symmetry, and some degree of segmentation. But there are three major structural features that chordates do not share with any other phylum of animals.

1. A structure called a **notochord** appears in the embryonic stage as a stiff rod of cells (Fig. 4.28) just beneath the dorsal surface and running parallel to the long axis of the body. In the simplest chordates the notochord remains throughout their lives. In the higher chordates the notochord disappears and is replaced in the adult stage by a series of vertebrae. It is for this structure that the phylum is named.

2. **Gill clefts** form near the anterior end of the animal during its early stages (Fig. 4.29). These clefts become slits in those chordates, such as fishes, that breathe by means of gills. In higher chordates, such as man, the gill clefts are modified into other structures.

3. All chordates have a dorsal tubular nervous system that remains with them throughout their lives.

In those animals we have studied so far, in which the

Fig. 4.27 Two common forms of protochordates, each representative of a different class. Below: a large group of sea squirts, shown somewhat smaller than natural size. Bottom: amphioxus, about natural size.

R. H. Noailles

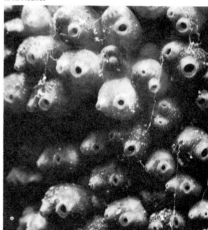

Lamont Geological Observatory

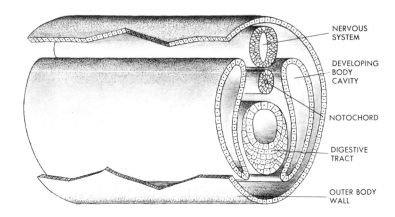

Fig. 4.28 A diagrammatic representation of development in a chordate showing relative positions of the notochord and nervous system to other structures.

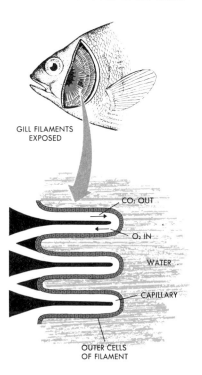

Fig. 4.29 The gaseous exchange system in fishes. The gill filaments expose a large area of blood to the water.

nervous system is at all advanced, the nerve cord lies near the ventral body wall, and is a solid mass of tissue. Thus, we find two features (notochord and gill clefts) that may or may not persist into adult stages of chordates, and one (the dorsal, tubular nervous system) that is present throughout life in all forms.

In addition to these three characteristics that set chordates apart from all other phyla of animals, there are certain others. Muscular, nervous, circulatory, excretory, and digestive systems reach a high point of specialization in the chordates. In most cases, there is a beautiful sequence of increasingly complex structures as one proceeds from the oldest to the most recent chordate animals. Sexes are almost always separate. In aquatic forms, egg fertilization usually (but not always) takes place in the open water, whereas a mating process is necessary in land-dwellers in order that sperm may be introduced into the female body by the male.

Many zoologists recognize two subphyla within the phylum Chordata: Protochordata and Vertebrata. Protochordates are those chordates whose notochords appear in development and are never replaced by a vertebral column. The notochord serves as a sort of internal skeleton in these animals. In the vertebrates, a more efficient **endoskeleton,** composed of bone, cartilage, or a combination of the two, succeeds the embryonic notochord.

The protochordates are not very numerous, and most people never see them at all. The most common forms are the sea squirts, and a small marine animal known as amphioxus. Sea squirts (tunicates) are nonmotile, jelly-like forms that grow in masses on pier supports or on the bottoms of ships. Many of them resemble boiled onions in appearance. Amphioxus is a small lance-shaped animal about an inch or more long that spends most of its life with its anterior end protruding from the floor of the ocean. It is usually found in shallow areas.

None of the protochordates are very important to man, although amphioxus is eaten by people who live along the coast of China. To biologists, however, the protochordates are a very important group. They are an important link between the invertebrate phyla that we have studied and the vertebrate members of the phylum Chordata.

Courtesy Carolina Biological Supply Company

Fig. 4.30 Lamprey attached to a solid surface by its suctorial mouth.

The Fishes (Classes AGNATHA, CHONDRICHTHYES, OSTEICHTHYES; Phylum CHORDATA)

All aquatic vertebrates that breathe by means of gills only are called fishes. There are more species of fishes than of all other vertebrates combined (Table 4-2). The fishes are very old. For well over 400 million years they have been a successful group of animals. Their oxygen needs are met by removing oxygen directly from the water around them by the use of gills. Their body systems reflect the aquatic mode of life and a great many unusual adaptations to existence in water are seen in various forms. For taxonomic purposes, three classes are generally recognized.

The class Agnatha includes those fishes whose mouths are **suctorial** and lack jaws. The most common agnathan is the lamprey, misnamed the lamprey "eel." (This is a misnomer because eels are bony fishes.) The lamprey, like all agnathans, does not have scales. Although lampreys are sometimes used for food, commercial fishermen regard them as a menace because they attack and kill more edible fishes. See Fig. 4.30.

The class Chondrichthyes is made up of fishes which have jaws and scales, and whose skeletons are composed entirely of cartilage. Although agnathans also have cartilaginous skeletons, they do not have jaws and scales. Skates, rays, and sharks are typical of this class (Figs. 4.31 and 4.32)—entirely marine and widespread in the ocean. Sharks tend to be vicious animals. At many beaches they are a threat to human life, especially if a swimmer is losing blood. Some rays have a poisonous barb on their tail. With it they can deliver a painful and dangerous sting to their victims, including humans. Other rays can deliver strong electric shocks. The meat of certain sharks and rays is edible, and shark liver is processed for its oil. In general, the cartilaginous fishes are more harmful than valuable, so far as man is concerned, chiefly because they destroy large numbers

R. H. Noailles

Fig. 4.31 A typical ray. Note that its body is flattened dorso-ventrally.

Fig. 4.32 An adult sand tiger shark in characteristic swimming pose.

New York Zoological Society Photo

Fig. 4.34 This bony fish, the largemouth black bass, is a favorite with sportsmen in a large part of the United States.

Fig. 4.33 A representative bony fish, the carp. Note mouth, adapted to bottom feeding.

Fig. 4.35 Modern lung fishes can obtain oxygen directly from the air (or from water). They employ this ability to survive out of water when the tropical streams and lakes in which they live temporarily dry up.

of valuable food fishes, crabs, and lobsters.

The class Osteichthyes (Figs. 4.33 and 4.34) includes those fishes whose skeletons are composed largely of bone, which is a characteristic of the remaining vertebrates. Members of this class usually have scales, but of a type different from those of the cartilaginous fishes. There are also other important differences.

At least 20,000 species of the 23,000 listed for all fishes (see Table 4-2) belong to the Osteichthyes. Few groups of animals show a greater variety of size, shape, color, and unusual adaptive mechanisms than do the bony fishes. In size they range from certain tropical fishes a fraction of an inch long to the Russian sturgeon which may be 20 feet or more long. They have successfully invaded virtually every known aquatic habitat. Very few are dangerous to man, although the marine barracuda, the electric fishes, and the vicious piranha of certain South American rivers are exceptions. Many marine and fresh-water forms are edible, and the economy of some countries is based largely on their fishing industries. In most years, some 10 million tons of fish are taken from the waters of the world.

The Amphibians (Class AMPHIBIA, Phylum CHORDATA)

Of the seven vertebrate classes recognized by most zoologists, three are aquatic and three are terrestrial. An intermediate group, the amphibians, have features of both aquatic and terrestrial vertebrates.

Fossil evidence indicates that amphibians arose some 300 million years ago from certain bony fishes in which lungs had evolved. These fishes also had fins which resembled limbs; apparently the animals were adapted to living in conditions of alternate flooding and drying. We know of them today from their fossil remains, and from a very few living species called the **lobe-finned** fishes.

R. H. Noailles

E. P. Haddon; U.S. Fish and Wildlife Service

The most common amphibians are the frogs, toads, and salamanders (Fig. 4.36). Typically, these animals spend their early life stages in water and breathe by means of gills, as the fishes do. Later on they develop lungs and become terrestrial. However, certain salamanders remain aquatic throughout their lives.

The most notable advance over the fishes, at least externally, is the development of legs in the place of fins. Unlike most fishes, amphibians do not have scales. Internally, there are modifications in almost all body systems (Fig. 4.37). In the circulatory system, for example, the two-chambered heart of the fishes is replaced by a three-chambered heart in amphibians. This is a reflection of the greater efficiency required in serving the lungs. Reproduction in this group resembles that of fishes more closely than that of land vertebrates. Amphibians usually lay eggs in open water where they are fertilized and develop into immature forms called **tadpoles**.

Amphibians are of very little value or importance to man. Larger forms are sometimes eaten; legs of the bullfrog, for ex-

W. P. Taylor, U.S. Fish and Wildlife Service

Fig. 4.36 Representative amphibians. Top left: spotted salamander. Top: an albino frog. Above: a toad.

Fig. 4.37 Internal anatomy of a frog. Essentially, this general pattern of structure is characteristic of all vertebrates, although organs may differ among species.

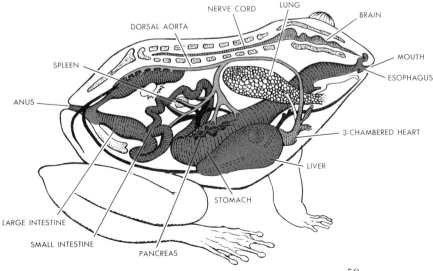

59

Fig. 4.38 Some typical reptiles. Below: a turtle sunning itself on an alligator. Middle: a lizard. Bottom: a female king snake coiled about her eggs.

Leonard Lee Rue III

E. P. Haddon, Fish and Wildlife Service

Leonard Lee Rue III

ample, are considered a delicacy. Toads have long been valued by gardeners because they eat insects, and in some countries, they have been sold for this purpose. With few exceptions, amphibians are harmless to man. They are of special interest to the biologist because they are intermediate between the fishes and terrestrial vertebrates. For this and for other reasons, frogs and salamanders are widely used in biological research.

The Reptiles (Class REPTILIA, Phylum CHORDATA)

Among the terrestrial vertebrates we find many adaptations that are related to living on land. For example, the circulatory system is powered by a four-chambered heart. In most reptiles, however, the two upper chambers (called the **atria**) are not completely separated. Another striking difference between terrestrial forms and the fishes and amphibians is their reproductive adaptations. Fertilization of eggs in the land vertebrates takes place inside the body by direct transfer of sperm from male to female. As a rule, egg fertilization is *external* in fishes and amphibians; the female releases her eggs from the body, after which the male deposits sperm upon or near them.

Since the eggs of fishes and amphibians are laid in water, the embryos of these animals have an unlimited source of vital materials such as oxygen and water molecules. They can also rid themselves of waste materials without difficulty. As a result, development is relatively uncomplicated from that standpoint. Through evolution the land vertebrates have developed a means of producing a micro-environment of water not unlike the larger water environment of fishes and amphibians. The land vertebrates have developed a membrane, the **amnion,** which is a sac enclosing the embryo within water. Embryos of reptiles, birds, and mammals all develop within an amnion. An example of this is shown in Fig. 4.39.

The slow-moving habits of most reptiles are well described by the Latin word meaning "to creep," from which the word "reptile" comes. Externally, reptiles are distinguished from amphibians by having scales and digital claws. Internally, there are many differences, as the diagram shows. The reptiles with which most of us are best acquainted are the snakes, lizards, turtles, alligators, and crocodiles (Fig. 4.38).

Reptiles are of relatively little economic importance to man. On the whole, they are simply a group of interesting animals whose place in nature today is not nearly so important as it was 150 million years ago. At that time, they were the dominant vertebrates on earth.

Some species of snakes are extremely poisonous, especially

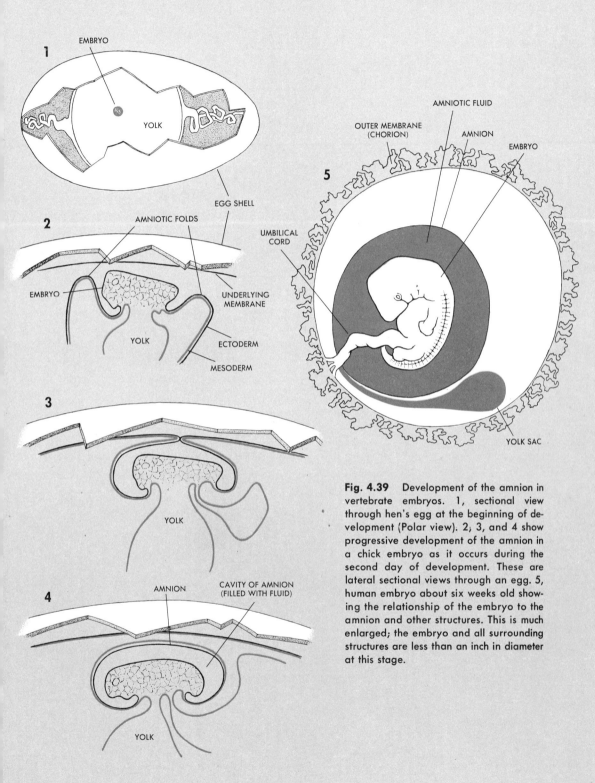

Fig. 4.39 Development of the amnion in vertebrate embryos. 1, sectional view through hen's egg at the beginning of development (Polar view). 2, 3, and 4 show progressive development of the amnion in a chick embryo as it occurs during the second day of development. These are lateral sectional views through an egg. 5, human embryo about six weeks old showing the relationship of the embryo to the amnion and other structures. This is much enlarged; the embryo and all surrounding structures are less than an inch in diameter at this stage.

certain marine and tropical forms; and in some areas alligators or crocodiles reach a sufficient size and exist in such numbers that they are quite dangerous to man. Only four types of poisonous snakes are found in the United States. These are the copperhead, the cotton mouth, the coral snake, and several species of rattlesnakes. In addition, a poisonous lizard, the Gila (pronounced hē'-la) monster, lives in the deserts of the Southwest. However, these reptiles are not nearly so dangerous as certain poisonous tropical snakes, such as the cobra, the fer-de-lance, and the bushmaster. Furthermore, certain sea-snakes are extremely poisonous.

The Birds (Class AVES, Phylum CHORDATA)

Birds (Fig. 4.40) are feathered bipeds. Only birds have feathers, and they are somewhat unusual in having only one pair of walking appendages. Birds, too, are the only vertebrates having the power of flight as a group (a very few birds, such as the ostrich, cannot fly) and many of their structural features are related to flight. Their bones, for example, are made lighter by having large central cavities, and their breast muscles are highly developed for moving their wings.

Unlike fishes, amphibians, and reptiles, birds keep a fixed body temperature. Because of this ability, birds are said to be "warm-blooded." All other vertebrates (except the mammals) are described as "cold-blooded." Actually, these are misleading terms. A lizard in the desert, for example, whose body tem-

Fig. 4.40 Representative birds. Left: a vulture. This bird is highly valuable as a scavenger. Below, left: a leghorn hen, widely raised as an egg-producer. Right: young green herons.

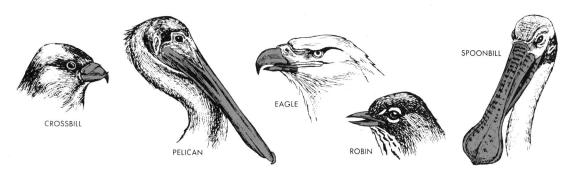

Fig. 4.41 A comparison of beak structure among various birds. Can you infer the feeding habit in each case?

perature changes with that of the surrounding air, might have a much higher temperature than a mammal living in the same environment.

Because they can fly and maintain a constant body temperature, birds have become a very successful group within the past 100 million years. We have abundant fossil evidence of flying reptiles during the time of the dinosaurs. Birds evidently evolved from such reptilian stock. Feathers and scales, by the way, are produced in much the same fashion in bird and reptile embryos, respectively. Birds still have scales on their legs as one mark of their ancestry.

Except for those modifications related to flight, the internal anatomy of birds is almost identical to that of reptiles. There are few animal groups in which more variety exists. In size, birds range from small forms such as hummingbirds to large ones such as eagles. There are also striking differences in color, wing and tail forms, beak shape, and foot structure. For example, even very closely related birds often have beaks that vary in shape (see Fig. 4.41). This variation results in completely different feeding habits, which in turn may reduce competition for food between these birds to such an extent that they can live in harmony in the same locality (see Darwin's finches, page 24).

The birds are of considerable importance to man. Certain species have been prized as food since prehistoric times. The domestic chicken of our day (developed from the wild fowl of Malay) is an important food source of meat and eggs alike. Various wild birds such as ducks and geese are hunted widely. Birds consume large numbers of harmful insects, thus benefiting the farmer and the gardener. From the recreational and aesthetic viewpoint, birds are very important. Whether or not we are interested enough in birds to join a bird-watching group, most of us enjoy hearing song birds and seeing their bright colors.

Armour and Company

© MCMLX Walt Disney Productions
New York Zoological Society Photo

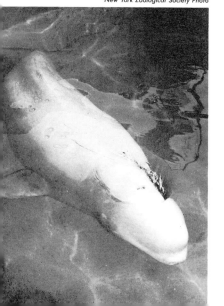

The Mammals (Class MAMMALIA, Phylum CHORDATA)

At a time when reptiles were still the dominant vertebrates, some 150 million years ago, a group of rather small animals developed. They were destined to supersede the giant lizards and develop into the most complex animals of all. These were the mammals (Fig. 4.42). At first, they were no larger than mice or rats, but they were able to maintain a constant body temperature. This plus certain other features which they developed in time (such as a more complex nervous system) enabled them to survive environmental changes which most of the reptiles of that time could not tolerate.

The mammals are those vertebrates which have hair as an external body covering, and which are nourished in their very young stages by milk secreted by the mother. During the millions of years since the mammals first arose, great diversity has taken place. While there are still some very small mammals, such as shrews, the largest animals on Earth also belong to this group. The modern blue whale, for example, is the largest animal ever known to have existed, exceeding even the size of the largest dinosaurs.

Fig. 4.42 As these photographs show, there is great variety among mammals. Top, left: Hereford cattle are a popular breed of animals domesticated by man. Opposite: a pair of jaguars (male black, female spotted). Bottom: a white whale in its water environment. Below: camels in a camel market at Kassala, in the Sudan.

United Nations

Compared with other vertebrates, the mammals show an advance in complexity and specialization in nearly every system of the body. The superior brain and nervous system of mammals make these animals more "intelligent," as a group, than any other animals. Another superior feature is their mode of reproduction. Almost all the animals we have studied in this chapter are **oviparous;** that is, they produce relatively large eggs which develop outside the body of the mother. With mammals, all except one primitive group are **viviparous;** that is, the females produce eggs which undergo both fertilization and development inside the body of the mother. Incidentally, certain animals produce large oviparous-type eggs which develop and hatch within the mother's body, and the young are thus born alive, as in the viviparous state. The pit vipers, which include the rattlesnake, the copperhead, and the cotton mouth, reproduce in this fashion. Such animals are said to be **ovoviparous.**

As in other successful animal groups we have studied, there is wide variety among the mammals. We have already seen the extremes in size. In adaptation, the variety is even greater. One group, the bats, can fly. Whales and porpoises, although they live in the ocean, are mammals. Most mammals, however, are entirely terrestrial. Altogether, zoologists recognize some 20 different orders, each of which is distinctive in some way. Among the more interesting forms are the **marsupials,** or pouched mammals (for example, the opossum and the kangaroo), the **primates** (monkeys and their kin), the elephants, the carnivores, and the various **ungulates** (hoofed mammals).

Man is more dependent directly upon other mammals than upon any other animal group. We do not know when men began domesticating such animals as the dog, the horse, and the cow, but it was long before history began to be recorded. It is a curious fact that no additional important animals have been domesticated by man within the past 5,000 years or so. Although such work animals as the horse are not as important to our present mechanized society as they were only a few decades ago, we still use them to a great extent. The mammals upon which we depend for food are of greatest importance. Milk, cheese, beef, pork, and mutton are a few of the more important mammalian products. In addition, mammals furnish us with such materials as leather and wool; even the manure of domesticated mammals is put to good use as fertilizer. To cite an entirely different sort of value, man's favorite pets are mammals. There is good evidence that the dog was the first domesticated animal (whether for companionship or for assistance in hunting, we do not know), and it has remained a popular pet.

Before we end our brief survey of mammals, let us consider

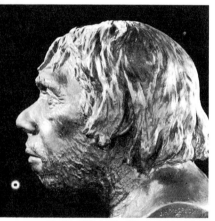

Courtesy of The American Museum of Natural History

Fig. 4.43 These reconstructions of three humans, or near-humans, who lived in the past are based on anatomical studies of fossilized skeletons. Top: Cro-Magnon man, who lived about 50,000 years ago. Middle: Java man, who existed perhaps 500,000 years ago. Above: Neanderthal man, who is thought to have occupied southern Europe more than 100,000 years ago.

the order Primates, of which man is a member. This order also includes the monkeys and apes as well as a variety of forms that are less familiar to most people (tree shrews, lemurs, galagos, tarsiers, marmosets, and so on).

Fossil and other evidence indicates that the primates arose at least 70 million years ago from mammalian ancestors that were **insectivores** (insect eaters), much like modern shrews and moles. At first, the primates were relatively small animals (many still are) and the larger forms did not evolve until several million years later. The best evidence indicates that man himself arose from a primate stock which at least 25 million years ago separated from the evolutionary lines that produced other living primates. This stock gave rise to a number of pre-men, or near-men, sometimes called "ape-men," whose fossil remains (Fig. 4.43) have been studied. The most primitive of these primates were still living at least a million years ago. By 500,000 years ago, several more advanced forms had evolved. All species of this evolutionary line are now extinct except modern man, *Homo sapiens*, who arose at least 100,000 years ago.

THE EVOLUTION OF ANIMALS

Fossil remains do not always give us a clear picture of what happened in the past, although they do tell us that evolution occurred. In many cases, we must turn to the fields of biochemistry, embryology, and anatomy (Fig. 4-44) to supplement the fossil evidence.

How can we make a general statement about the way evolution has taken place? We might arrange the phyla we have discussed in order from the Protozoa to the Chordata, and say that protozoans gave rise to sponges, sponges gave rise to cnidarians, etc. But this (see Fig. 4.45) would be wrong—just the same as saying that present-day forms have arisen from other present-day forms. This pattern, we are sure, is not reasonable in the light of our evidence. Rather, we must visualize evolution as a branching tree (Fig. 4.45), many of whose lower limbs have become lost through extinction and replacement.

In other words, the common ancestors of many present-day groups probably bore very little resemblance to any modern animals. (See Fig. 4.46 for a simplified and very generalized illustration of this evolutionary pattern.) It should be pointed out that this "tree" has been shorn of its smaller branches, many of which would illustrate groups that have become extinct. Furthermore, within a given group (such as the mammals), we can construct a tree showing the evolution of orders and families.

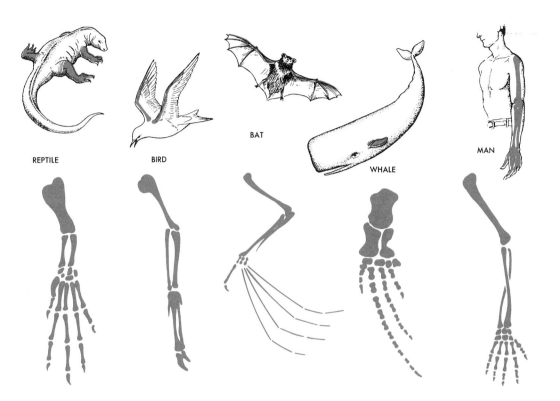

Fig. 4.44 In spite of great diversity in both structure and function, the forelimbs of all vertebrates show striking similarities in bone structure. For instance, an upper bone (humerus) is present in each type of limb shown. Such bones as the humerus of a reptile and that of a bat are said to be *homologous*. The study of homology has contributed much to our knowledge of vertebrate evolution.

Fig. 4.45 We cannot arrange present-day organisms in an order of increasing complexity and say that this is how evolution occurred. The course of evolution is more like a branching tree than a straight line.

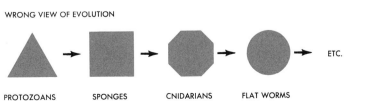

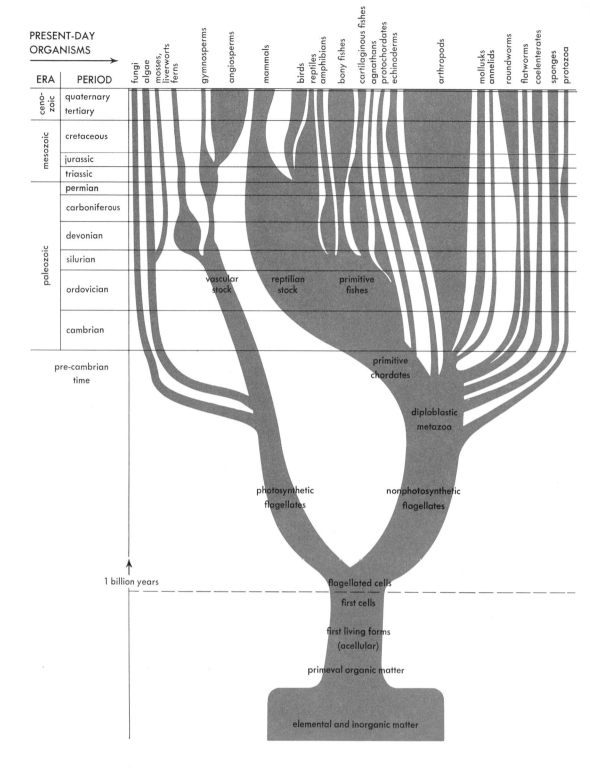

Fig. 4.46 The course of evolution, showing the periods and eras during which many modern organisms arose and showing something of the relative numbers of extant species. Because pre-Cambrian fossils are rare, that portion of this diagram is built largely upon inference.

Of the origins of animals, we know very little. In rock formations where we might expect to find the answers to our questions, the stresses of time have been too great for many fossil forms to be preserved. The first time period from which we find recognizable fossils in abundance is relatively late. It is later than the age in which all the major phyla of the present day evolved. The time chart shows this to have occurred by the beginning of the Cambrian Period, about 570 million years ago. However, it appears probable that the ancestors of animals were flagellates, and that a nonphotosynthetic line gave rise to protozoans and to the ancestors of metazoans. There is some evidence that the metazoan forms which gave rise to most phyla were very similar to the modern free-living flatworms. However, after we put aside these and certain other rather speculative ideas, evolution within the various phyla becomes more clearly illustrated.

As to the kinds of animals that were present on Earth at any given time since the beginning of the Cambrian Period, we have a reasonably clear picture. Up to the Silurian Period, which began about 435 million years ago, the most abundant animals were invertebrates, particularly certain arthropods, mollusks, and echinoderms. Fishes became numerous during the Silurian Period and began to dominate the seas. Animals apparently invaded the land during the Devonian Period, which began some 395 million years ago. These animals included the amphibians, which arose shortly before or during this time, and certain arthropods. The reptiles arose during the Mississippian Period, which began about 345 million years ago. They did not come into their own as the dominant animal group until the Mesozoic Era, which began some 225 million years ago. Small mammals appeared during this time, but they did not replace the reptiles until nearly the end of this era. In the Cenozoic Era, however, which began about 65 million years ago and continues to the present, mammalian evolution has proceeded very rapidly.

SUMMARY

We have considered as major forms 19 animal groups representing 10 phyla (see Table 4-2). The first of these, the protozoans, are unicellular. All other animals are multicellular. The sponges are little more than colonies of protozoan-like cells, and the cnidarians are the simplest animals with tissue specialization. Beginning with the flatworms, the other animals we have discussed are bilaterally symmetrical, with the exception of adult echinoderms. All animals above the flatworm level of

TABLE 4-2. A Listing of the Most Common and Numerous Animal Groups

Common Name	Phylum	Taxonomic Category	Approximate Number of Known Species
Protozoans	Protozoa	Phylum Protozoa	30,000
Sponges	Parazoa	Phylum Parazoa	4,200
Cnidarians or Coelenterates	Cnidaria	Phylum Cnidaria	9,600
Flatworms	Platyhelminthes	Phylum Platyhelminthes	15,000
Nematodes or Roundworms	Aschelminthes	Class Nematoda	10,000
Mollusks	Mollusca	Phylum Mollusca	100,000
Annelids or Segmented Worms	Annelida	Phylum Annelida	7,000
Insects	Arthropoda	Class Insecta	700,000
Crustaceans	Arthropoda	Class Crustacea	25,000
Arachnids	Arthropoda	Class Arachnida	30,000
Centipedes	Arthropoda	Class Chilopoda	3,000
Millipedes	Arthropoda	Class Diplopoda	6,000
Echinoderms	Echinodermata	Phylum Echinodermata	5,700
Protochordates	Chordata	Classes Urochordata, Cephalochordata	1,700
Fishes	Chordata	Classes Agnatha, Chondrichthyes, Osteichthyes	23,000
Amphibians	Chordata	Class Amphibia	2,000
Reptiles	Chordata	Class Reptilia	5,000
Birds	Chordata	Class Aves	8,590
Mammals	Chordata	Class Mammalia	4,500
		Approximate total:	990,000

Adapted from Lord Rothschild, A Classification of Living Animals. By permission, John Wiley and Sons, Inc.

complexity have a body through which a digestive tube runs from mouth to anus. Our consideration and description of the nematodes, mollusks, annelids, five groups of arthropods, echinoderms, and protochordates completes the survey of invertebrate animals. There are five groups of vertebrates—fishes (including three taxonomic classes), amphibians, reptiles, birds, and mammals.

Animal life became distinct from plant life approximately a billion years ago. Metazoan animals arose and evolved into the major present-day phyla some time before the Cambrian Period, which began about 570 million years ago. Since that time, great diversity has occurred, with different groups succeeding each other as the dominant animals. The land was invaded by animal life some 395 million years ago, and within 150 million years, reptiles had evolved and had become the dominant terrestrial vertebrates. They were replaced in this role by the mammals after approximately 100 million years.

FOR THOUGHT AND DISCUSSION

1 What three phyla include nine-tenths of all animals that have been classified? Which of these three phyla is the largest? What *group* of this phylum includes more animals than any other?
2 Why are the echinoderms placed where they are in the evolutionary scale, since they resemble animals that are ranked much lower?
3 Only two groups of animals, as groups, are characterized by flight. Which are they? Can you think of any animals outside these groups that can fly?
4 What is the distinction between protochordates and vertebrates?
5 Which of the animal groups that we have discussed do you consider to be of major economic importance to man? Make a list and give reasons.
6 The reptiles, birds, and mammals are sometimes called the **Amniota.** What do you suppose gave rise to this name, and of what significance is it?
7 What is regeneration? It is exhibited to a greater degree in less complex animals and to a lesser degree in more complex animals. Can you propose a reasonable explanation for this?
8 Will taxonomists ever reach a stopping point in their classification of animals? Justify your answer.

SELECTED READINGS

Blair, W. F., A. P. Blair, P. Brodkorb, F. R. Cagle, and G. A. Moore. *Vertebrates of the United States.* New York: McGraw-Hill Book Co., Inc., 1957.
   This book is essentially a taxonomic work, and is somewhat specialized. It gives a description of each of the thousands of vertebrates found in the United States.

Buchsbaum, R. *Animals Without Backbones* (2nd ed.). Chicago: The University of Chicago Press, 1948.
   A classic of descriptive invertebrate zoology, this book is beautifully written and illustrated.

Hanson, E. D. *Animal Diversity* (2nd ed.). Englewood Cliffs, N.J.: Prentice-Hall, Inc., 1964.
   The viewpoint of this book is evolutionary, rather than descriptive. It is a stimulating discussion of this aspect of zoology.

Storer, T. I. and R. L. Usinger. *General Zoology* (4th ed.). New York: McGraw-Hill Book Co., Inc., 1965.
   A thorough presentation of the animal kingdom.

5 THE PLANT KINGDOM

Fig. 5.1 The majestic redwood trees of the west coast may attain an age of thousands of years.

David Swanlund, Save-the-Redwoods League

Grasses, trees, mosses, hedges, and various small flowering plants—how many other kinds of plants can you add to this list? Probably not very many, for these five make up the bulk of terrestrial vegetation. Nevertheless, there are many other kinds of plants.

Virtually all of the plants we see around us are green plants. As you probably know, **chlorophyll** is the substance that makes green plants green. In the presence of sunlight and chlorophyll, water and carbon dioxide within a plant unite and form carbohydrates, releasing oxygen during the process. This is called **photosynthesis.** The oxygen we breathe is made by the action of green plants.

You might think that the green plants mentioned at the beginning of the chapter supply all, or nearly all, of the free oxygen we breathe, but they do not. The greater part, probably more than 80 per cent, of photosynthesis is carried on, not by terrestrial plants, like the giant redwoods of Fig. 5.1, but by small and relatively simple aquatic plants called algae, whose very existence is unknown to many people. The fact that these plants are so important in the over-all economy of nature is an indication of our need to know more about them. This is only one example of the many plant groups that are important to us, and of which most people know very little.

How Plants Are Classified

Although it seems incredible to us now, there was a time when students of **botany** (the study of plants) recognized only a few hundred plant species. Aristotle's student, Theophrastus, for example, classified about 500 plants. Linnaeus, some 2,000 years later, listed fewer than 10,000 species. As you saw in the first chapter, today we recognize about 350,000 species. However, we should remember that travel was difficult

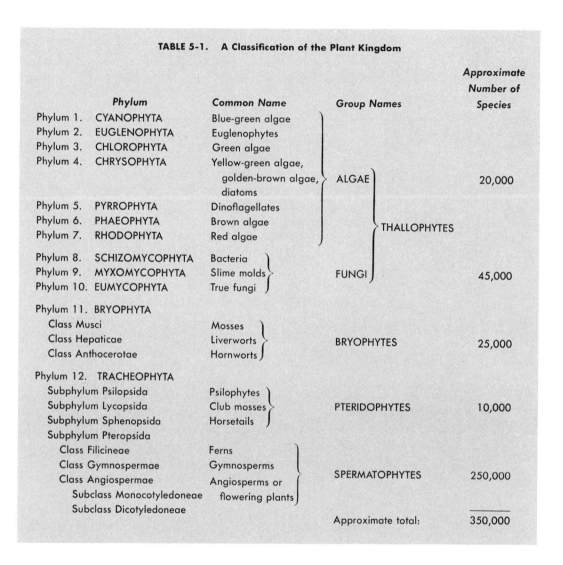

TABLE 5-1. A Classification of the Plant Kingdom

Phylum	Common Name	Group Names	Approximate Number of Species
Phylum 1. CYANOPHYTA	Blue-green algae		
Phylum 2. EUGLENOPHYTA	Euglenophytes		
Phylum 3. CHLOROPHYTA	Green algae		
Phylum 4. CHRYSOPHYTA	Yellow-green algae, golden-brown algae, diatoms	ALGAE	20,000
Phylum 5. PYRROPHYTA	Dinoflagellates		
Phylum 6. PHAEOPHYTA	Brown algae		
Phylum 7. RHODOPHYTA	Red algae	THALLOPHYTES	
Phylum 8. SCHIZOMYCOPHYTA	Bacteria		
Phylum 9. MYXOMYCOPHYTA	Slime molds	FUNGI	45,000
Phylum 10. EUMYCOPHYTA	True fungi		
Phylum 11. BRYOPHYTA			
Class Musci	Mosses		
Class Hepaticae	Liverworts	BRYOPHYTES	25,000
Class Anthocerotae	Hornworts		
Phylum 12. TRACHEOPHYTA			
Subphylum Psilopsida	Psilophytes		
Subphylum Lycopsida	Club mosses	PTERIDOPHYTES	10,000
Subphylum Sphenopsida	Horsetails		
Subphylum Pteropsida			
Class Filicineae	Ferns		
Class Gymnospermae	Gymnosperms	SPERMATOPHYTES	250,000
Class Angiospermae	Angiosperms or flowering plants		
Subclass Monocotyledoneae			
Subclass Dicotyledoneae			
		Approximate total:	350,000

in the 1700's, to say nothing of travel conditions in ancient Greece. This handicap severely limited the scope of plant investigation.

Linnaeus recognized 24 classes of plants, 23 of which included the flowering plants. Within these classes, he included orders on the basis of differences in structures related to sexual reproduction. For example, he lumped all of the plants that do not bear flowers into one class. Within a very short time his system could not accommodate the many newly-discovered species of both flowering and nonflowering plants.

Various systems succeeded each other until the late 1800's. By this time well over 100,000 species of plants were known.

Meanwhile, botanists had created a category one step above the class, which they called the **division.** Also, by this time, zoologists were using the category "phylum" at the same level; thus, the major category of the plant kingdom has traditionally been the division, while that of the animal kingdom has been the phylum. All other taxonomic categories are the same for the two kingdoms.

One system that employed four divisions of the plant kingdom became widely adopted. The divisions were as follows: 1. Thallophyta (algae and fungi), 2. Bryophyta (mosses and liverworts), 3. Pteridophyta (ferns and fern-like plants), and 4. Spermatophyta (seed plants). This system was used widely by botanists from the late 1800's until recent decades. But as the number of known plant species increased, and as more emphasis was placed on evolutionary principles, the four-divisional system became inadequate. It failed to reflect *natural relationships* among plants.

Several attempts have been made within recent years to draw up a more meaningful classification system, especially with respect to the larger categories. Unfortunately, no single system has become standard, and it will probably be some years before botanists reach a general agreement. This unhappy situation is caused partly by the difficulty of interpreting evolutionary relationships in the plant kingdom; also, evidence for these relationships is incomplete.

The most widely accepted scheme of classification is that shown in Table 5-1. According to this listing, 12 phyla are used. The use of the term "phylum" instead of "division" is interesting in view of what was said earlier. It shows a tendency among botanists to equate the major groups of the plant kingdom with those of the animal kingdom. Although a listing of plant phyla will probably mean little to you at this point, it will serve as a reference as you proceed through the chapter.

A SURVEY OF MAJOR PLANT TYPES

The Algae

Have you ever noticed that ponds often have a greenish appearance? Or, perhaps you have seen the bluish-green scum left on a patch of ground that was covered by a rain puddle. If you were to examine either the pond water or the scum with a microscope, you would see many extremely small plants. Some of them would be only single cells. Although not all algae are microscopic, most of them are much simpler in structure than the more familiar plants. It is because of their structural simplicity that we begin our survey of the plant kingdom with the algae (Fig. 5.2), which are thought to have inhabited the Earth one-and-a-half billion years ago.

As you will notice in Table 5-1 the group is quite varied, consisting of seven phyla. In addition to the microscopic, one-celled individuals, there are the marine kelps, of the brown algae, which can grow to 100 feet in length, and are very complex in structure. In general, however, the algae are rather small plants. Some phyla consist entirely of microscopic forms.

Algae have the green pigment, chlorophyll. They are widely distributed in nature, especially in marine or fresh-water habitats. However, many species are terrestrial. They live in hot springs, where the water temperature may be 85°C, in the melted water of snowbanks, in deserts, and in the polar regions. Some grow on rocks, wood, or in association with other types of plants, or with animals. You can often see excellent cultures of algae growing on the shells of turtles. There seems to be hardly any terrestrial environment to which the algae have not successfully adapted.

Fig. 5.2 The pond scum which the girl is examining consists of thousands of filaments of green algae. When viewed under a microscope, individual plants making up this scum are strikingly beautiful.

Although it is difficult to make a clear-cut distinction between the algae and other groups of plants, the major difference is one of complexity in form and structure, especially in the organs of sexual reproduction. Very few algae have parts that resemble roots, stem, leaves, or other organs of higher green plants. Let us now examine some of the more important types of algae.

The Blue-green Algae (Phylum CYANOPHYTA)

Members of this phylum have certain red and blue pigments which, in addition to chlorophyll, give them the color their name implies. Some occur as free-living individual cells while others form thread-like structures called **filaments** (Fig. 5.4). Unlike the vast majority of plant and animal cells, which have well-formed nuclei, the blue-green algae have cells in which the DNA is not contained within an inner confining membrane, or definite nucleus.

Also unlike most other organisms (Fig. 5.3), they do not reproduce sexually. New algae are produced by cell divisions of old individuals. A cell simply "pinches" itself into two hemispheres until it divides; each new cell shares parts of the loose nuclear material and the cell fluid, called **cytoplasm.** This method of cell division is much less precise than that shown by most plant cells (Fig. 5.3). For these and certain other reasons, the blue-green algae are among the least complex of all living organisms.

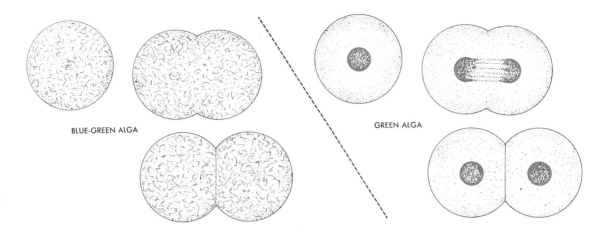

Fig. 5.3 Cells of blue-green algae (left, above) do not undergo precise nuclear division. Virtually all other cells show a process of nuclear division (called **mitosis**), exemplified at right by cell division in a green alga. (See left-hand portion of Fig. 5.10 for a detailed diagram of mitotic division.)

The Green Algae (Phylum CHLOROPHYTA)

These plants are the most varied, numerous, and widespread of all the algae. They are generally grass-green because chlorophyll is their major pigment. Like the blue-green algae, they live as unicellular forms and in filamentous colonies which frequently grow in such abundance in ponds or ditches that they appear as a frothy mass or scum.

The cells of green algae are much more highly organized than those of blue-green algae. They have definite nuclei and other specialized parts characteristic of higher plants. For instance, the pigment chlorophyll is contained in definite bodies called **chloroplasts** in the green algae and in higher green plants. In contrast, the pigments found in the cells of blue-green algae are diffused throughout the cell. The chloroplasts of green algae contain bodies called **pyrenoids**, which are storage centers for starch made during photosynthesis. Also characteristic of higher plants, the cells of green algae have walls composed chiefly of the carbohydrate cellulose.

Sexual reproduction, during which two gametes unite and form a zygote, as we saw in the animal kingdom chapter, is also characteristic of higher plants and is widespread among species of the green algae. Yet many species reproduce asexually by fission. We shall have more to say about methods of reproduction later.

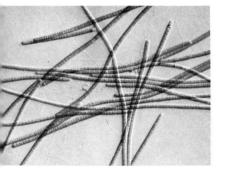

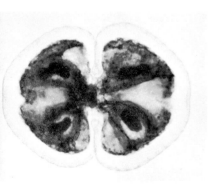

Fig. 5.4 Top photo: filaments of *Oscillatoria*, a blue-green alga. Above: *Cosmarium*, a single-celled green alga.

The Brown Algae (Phylum PHAEOPHYTA)

The dominant pigment in these algae is brown, hence their common name. Although chlorophyll is present, it is heavily masked in most species of this phylum. The cellular characteristics of brown algae are much like those of the green algae, and sexual reproduction is carried on by virtually all species. Most brown algae are relatively large and complex (Fig. 5.5).

There are no unicellular, or even very small forms. A single giant kelp ("seaweed") of the Pacific may have a total mass equal to that of a fairly large tree. The internal structure of these and other large brown algae approaches the complexity of that of many higher plants. A very few fresh water species are known. Most members of the group are marine.

The Red Algae (Phylum RHODOPHYTA)

A reddish pigment that masks the chlorophyll gives members of this phylum their common name. They are graceful forms (Fig. 5.6) that live anchored to the ocean bottom, to rocks, or to floating debris. Although most species are marine, many fresh water forms are known. Several of the red algae are rather complex, especially their sexual reproductive systems. No known member of this phylum reaches the size of the larger brown algae.

What roles do algae play in nature? And of what values are they to man? Indirectly, algae benefit man by playing a vital

Courtesy The Bronx Botanical Garden

Fig. 5.5 *Macrocystis pyrifera*, one of the larger and more complex forms of brown algae known as "kelps." See Fig. 5.11 for a smaller type of brown alga.

Fig. 5.6 *Polysiphonia*, a marine red alga.

Courtesy Carolina Biological Supply Company

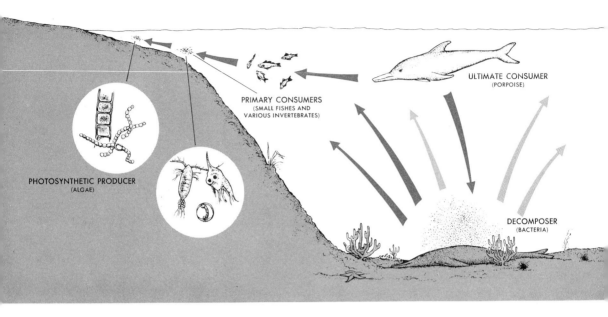

Fig. 5.7 This simplified representation of a food web in nature shows the interdependence of living forms. What is the role of sunlight in this complex web, and why is it called a "web?"

role as members of the endless **food web** (Fig. 5.7). On the first level are the photosynthetic **producers**—called "producers" because they build themselves up entirely by using the water, carbon dioxide, and minerals of their immediate environment. On the land, they are chiefly the rooted higher plants; in water, the vast majority are algae.

At the next level of aquatic food are the **primary consumers,** such as the small fishes and various invertebrate animals that consume the algae. In turn, the primary consumers are eaten by still larger animals, and so on, until some **ultimate consumer,** a porpoise, say, ends the process. Even here, however, the ultimate consumer produces waste products and eventually dies. The **decomposers,** such as bacteria, now play their role in the food web.

Consequently, the original energy that was captured by algae or other producers is made available to different organisms. About three-fourths of the Earth's surface is covered by water, and about four-fifths of all photosynthesis is carried on by algae. This means that they are the major producers in the over-all food web.

While man enjoys the indirect effects of algal photosynthesis, he also uses certain algae directly. Some species are edible. At the present time much research is being conducted to explore the possibilities of making the oceans and bodies of fresh water produce algae that can be used directly as food in large quantities. Even now, several types of algae make up a large portion of the diet of certain oriental peoples. Quite a number of valuable chemical substances come from algae. **Agar-agar** (from certain red algae) is useful in laboratories for the preparation of media used in growing microorganisms. **Algin** (from certain brown algae) is used in some food recipes. In addition to these benefits, the microscopic algae in particular have become extremely useful in genetical, nutritional, and biochemical research.

Before leaving the algae, we should describe their methods of reproduction. It is important to do so not only for the sake of understanding the group itself, but for understanding certain basic principles that will apply to the plant groups we will study next.

As we saw earlier, individuals reproduce by sexual or asexual means. Typically, sexual reproduction occurs when a relatively small sperm, or male gamete, unites with a relatively large egg, or female gamete. When this happens, the DNA of each gamete nucleus is brought into close association with that of the other. However, there are exceptions to this general rule of sexual reproduction, and our first example of sexual reproduction in algae is one of these exceptions.

Let us consider the life cycle of *Chlamydomonas eugametos*, a unicellular green alga (Fig. 5.8). Cells of this species are **motile;**

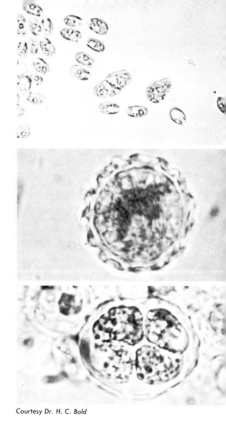

Courtesy Dr. H. C. Bold

Fig. 5.8 Top: cells of *Chlamydomonas eugametos*, some of which are paired in sexual union. Middle: a zygote of C. eugametos, much enlarged. Above: meiotic division in the zygote, showing four daughter cells.

Fig. 5.9 Life cycle of *Chlamydomonas eugametos*, sexual phase. Asexual reproduction also occurs whenever individual cells divide.

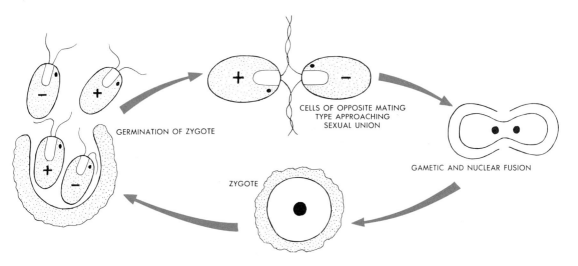

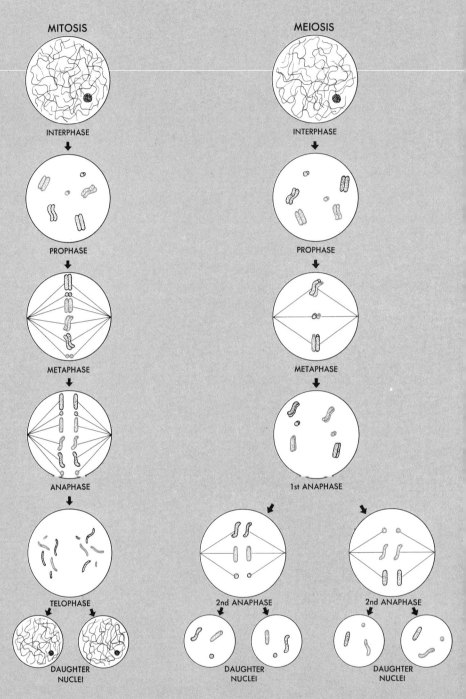

Fig. 5.10 A contrast in mitotic and meiotic nuclear division. In mitosis, each chromosome divides, thus giving each daughter cell exactly the same number of chromosomes as the mother cell. In contrast, meiotic division eventually results in daughter cells that have a chromosome representative of each *pair* present in the mother cell; that is, the number is reduced by one-half. In some way meiosis is always associated with sexual reproduction, and is characteristic of virtually all plants and animals.

that is, they have whip-like **flagella** which propel the cell through water. Although all individuals appear to be similar, by virtue of their behavior they can be assigned to mating groups arbitrarily designated "plus" and "minus." In other words, if there are "male" and "female" groups, we do not know which is which! Under normal conditions, a given cell divides and forms two daughter cells of the same mating type. This particular type of asexual reproduction, called **mitosis,** is shown in its several stages by Fig. 5.10. Each daughter cell is a new individual which may give rise to two new individuals, again by mitosis.

Sometimes, however, two cells of opposite mating type are attracted to each other. Their flagella become entangled, and, after a time, the fluid contents of the two cells mix, and their nuclei unite. The resulting single cell is called a zygote. After a while, the zygote undergoes two successive divisions, called **meiosis,** which produces four cells. Two of these are "plus" and two are "minus" (Figs. 5.9, 5.10).

Thus, *Chlamydomonas eugametos* exhibits both asexual and sexual reproduction. However, sexual reproduction here is somewhat different from that in higher plants and animals. In higher plants and animals, the gametes are individual cells themselves; in *Chlamydomonas* the entire plant, itself only a single cell, serves as a gamete. Furthermore, higher forms usually have gametes of greatly differing size, the egg being many hundreds of times larger than the sperm. In *Chlamydomonas* both gametes are similar in size.

Although the life cycle of *Chlamydomonas* is typical of a great many algae, there are other patterns of reproduction. One of them, shown by the life cycle of *Fucus vesiculosus,* is a more common method of sexual reproduction.

Courtesy General Biological Supply House

Fig. 5.11 *Fucus vesiculosus,* one of the rockweeds (brown algae), shown here slightly smaller than life size.

Fig. 5.12 Life cycle of *Fucus vesiculosus.* 1. mature female plant. If one of the bulb-shaped tips is cut in section, it appears as shown at 2. Groups of eight cells (shown at 3, with only six of the cells visible) are formed and, eventually, produce eight eggs (4). Fertilization occurs in the open sea water when a sperm from a male plant unites with an egg (5). The zygote (6) develops into a mature plant.

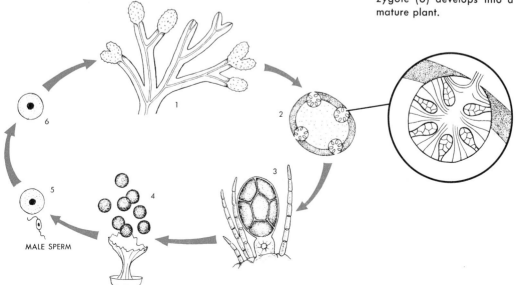

This plant (Fig. 5.11), which is found on rocks in tidal areas of the New England coast, reaches sexual maturity when large swellings appear at the tips of the plant. A plant is usually either a male or a female. A female plant produces small clusters of eggs within each swelling; a male plant produces clusters of sperm. Eventually, the sperm and eggs are released. The eggs float passively in the water. The sperm cells, which are motile, are attracted to the eggs. Eventually, one sperm penetrates an egg and fertilization takes place, thus producing a zygote which then develops into a *Fucus* plant. This is shown in diagram form in Fig. 5.12.

The Fungi

Although these plants are widespread in nature, most people are unfamiliar with them as a group. Mushrooms (Fig. 5.13) and certain molds and mildews are quite common members of the group, but there are many other kinds of fungi. Few people are aware of the very important role these plants play. Unlike virtually all other plants, fungi do not have chlorophyll. This sets the fungi apart in appearance from green plants; it also dictates a mode of nutrition which makes the fungi unique as a group. Almost all forms are heterotrophic, which means that their role in nature is one of decomposition.

For our present purposes, we shall include among the fungi two groups of plants that many botanists classify separately—the bacteria and the slime molds.

Fig. 5.13 Mushrooms are among the many different kinds of fungi. Other fungi are usually less familiar and less obvious in nature; almost all forms are important as decomposers.

The Bacteria (Phylum SCHIZOMYCOPHYTA)

Most bacteria are unicellular plants, and all of them are extremely small. If 50,000 typical bacteria were laid side by side, they would form a line only about an inch long.

Fig. 5.14 Some representative bacteria, magnified here more than 1,000 times. Left: typical bacilli. Middle: a large spirillum. Right: the cocci which cause pneumonia.

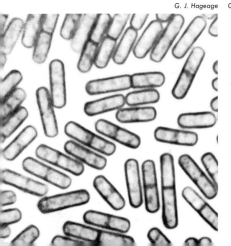

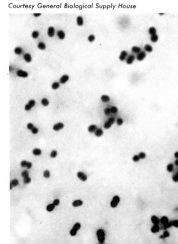

Like the cells of blue-green algae, bacteria have rather poorly organized nuclear material. This is one characteristic that distinguishes them from other fungi. Most bacteria exist as small, unicellular spheres (**cocci**), rods (**bacilli**), or spirals (**spirilla**) as shown in Fig. 5.14.

A few bacteria are **autotrophic** in their nutrition. These autotrophic forms have either certain photosynthetic pigments related to chlorophyll, or they have special mechanisms enabling them to derive energy through chemical reactions with their immediate environment. A few species are parasitic, deriving their food materials from living tissues of higher plants, or animals. If these bacteria cause disease in their host, they are said to be **pathogenic.** Typhoid fever, pneumonia, and tuberculosis are diseases produced by pathogenic bacteria.

The relatively few pathogenic bacteria have given other bacteria a bad name. Most people are surprised to learn that these organisms, as a group, do far more good in nature than harm. They exist in the untold billions in soil, water, and virtually every other medium that exists on the surface of our planet. They maintain their populations by decomposing the remains of other organisms and their products. Sometimes they attack other living organisms. Some bacterial products of metabolism are important in medicine, industry, and agriculture. Certain antibiotics and organic acids, for example, come from cultures of bacteria grown in laboratories. Such bacterial products as ammonia and nitrates, produced in the soil, are of great agricultural importance.

Courtesy A. C. Lonert

Fig. 5.15 A slime mold growing in a laboratory dish, shown here about one-half natural size.

The Slime Molds (Phylum MYXOMYCOPHYTA)

In one stage of their life cycle these plants are so much like animals that they are frequently classified in the animal kingdom. During this stage, a given species exists as a **plasmodium** (Fig. 5.15)—a multinucleate mass of protoplasm that is not divided into cells. A plasmodium, which may be several square feet in area, is capable of moving over the ground. As it travels, it consumes bacteria and bits of debris in true phagotrophic fashion. Eventually, however, the entire mass comes to rest (often on a decaying log) and forms **spore-**producing structures. Spores are reproductive structures consisting of one to several cells, and are characteristic of a great many plants. In most cases, they give rise directly or indirectly to a new plant body like the one which produced them. However, the spores of slime molds develop into gametes. When these gametes are released, two of them may fuse, thus forming a new plasmodium.

Fig. 5.16 A mycelium of a mold, *Penicillium*. Grown in culture for its production of the antibiotic penicillin, this mold is an ascomycete (sac fungus), not an algal fungus. It is shown here to illustrate the appearance of a typical mold mycelium. For a representative algal fungus, see Fig. 5.20.

Courtesy General Biological Supply House

The True Fungi (Phylum EUMYCOPHYTA)

Both the bacteria and the slime molds bear a somewhat obscure evolutionary relationship to those plants which are unquestionably fungi, some of which are discussed below. For this reason, the phylum Eumycophyta ("true fungi") is made to include all fungi other than the bacteria and slime molds. The three most important groups within the Eumycophyta are the algal fungi, the sac fungi, and the club fungi.

The **algal fungi** form a group that includes several common molds, such as the fuzzy mold you sometimes see on bread, cheese, or on the surface of preserves that have been standing about for a long time. Like certain other fungi, the algal fungi form filamentous masses called **mycelia** (Fig. 5.16). But unlike other fungi, the filaments of algal fungi are not divided into cells by cross walls. The algal fungi were given their common name because they resemble certain filamentous algae.

Fig. 5.17 This view of a sac fungus shows ascus (sac) formations, the oval structures near the outer edge.

Courtesy Carolina Biological Supply Company

Fig. 5.18 The dead trees shown here were killed by parasitic fungi. After they have died, trees are usually attacked by a variety of decomposing fungi.

Courtesy of The U.S. Forest Service

The **sac fungi** (Fig. 5.17) got their name because they bear their spores within a sac-like enclosure, the **ascus**. The major common types are the **yeasts**, several **molds**, the **mildews**, and fleshy forms known as **cup fungi**.

The yeasts have become extremely valuable to man because of their ability to produce alcohol and carbon dioxide, which makes them particularly important to the beverage and baking industries. Furthermore, yeasts are an important source of the B vitamins, and are either processed into medicines or food, or consumed directly for their nutritive value by humans and livestock. Several molds of this group have come into prominence within recent years because they are capable of producing important drugs. The antibiotic penicillin is produced by a blue-green mold. Certain molds are also valuable in biological research. Study of the pink mold *Neurospora* has led to discovery of some of the fundamental principles of genetics and biochemistry.

The **club fungi** received their common name because they produce spores on club-like structures, known as **basidia**. As a group, they are probably more familiar to most people than any other fungi. Some of the major types are the **rusts** and **smuts**. They are parasites that attack higher plants, particularly cereal grains such as corn and wheat.

The rusts and smuts are of considerable economic importance. They often seriously damage their host plants, thus reducing yield to the farmer. Although parasitic **shelf fungi** may damage timber, most of them and their near relatives, including the **puffballs** (Fig. 5.19) and **mushrooms** (Fig. 5.13), are important decomposers. In this capacity they are beneficial to man. Many types of fungi, including club fungi, grow on both living and dead trees (Fig. 5.18). A few species, including certain mushrooms, are edible and can be grown as a marketable food item.

As we have seen from this short survey, the fungi show remarkable diversity in form. They range in size and complexity from the unicellular bacteria to the larger club fungi. Compared with many of the higher plants, and even with the larger algae, they are all relatively small and simple in structure. Even the mushrooms are essentially masses of interwoven filaments. Only those cells controlling reproduction show much specialization. Like the algae, very few fungi have structures resembling the roots, stems, or leaves of higher plants.

Almost all species of fungi reproduce sexually, but their life cycles vary a great deal. At the simplest level, some bacteria are known to carry on a form of sexual reproduction. At the other end of the spectrum, some parasitic fungi (rusts) have extremely complicated life cycles. The reproductive cycles of

Fig. 5.19 Geaster, the earth star, a common puffball.

Courtesy Carolina Biological Supply Company

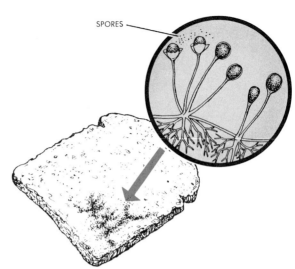

Fig. 5.20 *Rhizopus*, the black bread mold, is an algal fungus. In the photograph shown at the left, a single stalk with its anchoring filaments is shown enlarged many times. The knob on the end of the stalk is the sporangium, in which hundreds of spores are produced. The drawing above shows the relationship between the filaments (which compose a mycelium) to the material upon which they grow.

Fig. 5.21 A representative lichen growing on a tree limb.

Courtesy Carolina Biological Supply Company

most fungi are somewhat intermediate between these extremes; *Rhizopus*, the bread mold pictured in Fig. 5.20, is a good example. Filaments that originate from a single spore are either "plus" or "minus." If plus and minus strains are grown close together, their filaments fuse at various points and form spiny zygotes. Each zygote has several pairs of fused nuclei (zygote nuclei). Only one nucleus within a zygote, however, eventually germinates and begins a new growth of filaments.

Many fleshy fungi, such as the mushrooms, have elaborate filamentous growth beneath the soil. After sexual fusion occurs between opposite mating types, the large "fruiting body" (the mushroom itself) appears above the soil.

The greatest importance of fungi is their ability to decompose organic materials. When they do so, elements and compounds are released into the environment to be used again by higher plants and animals. Furthermore, fungi prevent an accumulation of dead bodies, animal wastes, and various other materials that would gather on the surface of the earth. For these and other reasons, the fungi are one of the most important plant groups.

The Lichens

You have probably noticed grey-green patches of growth on rocks and tree trunks, especially on the more shaded side. These are **lichens** (Fig. 5.21), a composite of an algal species and a fungal species growing together in a close relationship (mutualism). Neither the alga nor the fungus could survive separately under most natural conditions, but the two thrive in biological partnership.

Lichens are very widespread in nature. Frequently they grow where other plants exist only with great difficulty, or not at all. Lichens thrive on the surface of bare rocks, on high mountain peaks, in the Arctic and in deserts alike. They are of little economic importance, although some are processed for dyestuffs or other chemicals, and certain forms are useful in biochemical research. They help break down rocks and so play a part in the formation of soil. They also support animal life in regions where they are virtually the only available vegetation.

Since a given lichen reproduces consistently as a lichen—and not as either an alga or a fungus—it is considered a species in its own right. Hence, the lichens are given generic-specific names in spite of the fact that their algal and fungal components have generic-specific names of their own. This creates an awkward situation taxonomically. All lichens are classified in the phylum Eumycophyta, because the fungal component (a "true fungus") belongs to this group. The algal component is always either green or blue-green.

The Liverworts (Phylum BRYOPHYTA)

These plants are not widespread in nature. Their environments usually are moist, shaded areas. A few are aquatic (fresh water, never marine), but most species grow on soil, where they may form a flat, green carpet. Some, but not all, exhibit a habit of growth that makes individual plants branch out and resemble the lobed human liver. For this reason they have long been known as liverworts (pronounced *wurt*, an old name meaning "plant"). See Fig. 5.22.

Although liverworts are of no economic importance and play a very minor role in nature, they are of tremendous significance to botanists. For one thing, the liverworts have a life cycle (Fig. 5.23) that is typical of all other plants studied in the remainder of this chapter.

The liverwort plant itself consists of cells whose nuclei contain only one set of chromosomes. Any organism in which this condition exists is said to be **haploid**. This term is also applied to any single cell or nucleus, and frequently, the haploid condition is symbolized by the letter **n**. In contrast, the body cells of animals and those of most higher plants contain two sets of chromosomes, and their bodies or cells are said to be **diploid**. The symbol **2n** is used to describe this condition. In the case of liverworts, the specialization of certain cells results in haploid

Fig. 5.22 A typical "carpet" of liverworts, shown here somewhat smaller than natural size. The "heads," which superficially resemble tiny flowers, are reproductive structures. Can you make out the lobes of individual plants?

Courtesy Carolina Biological Supply Company

88   *Animal and Plant Diversity*

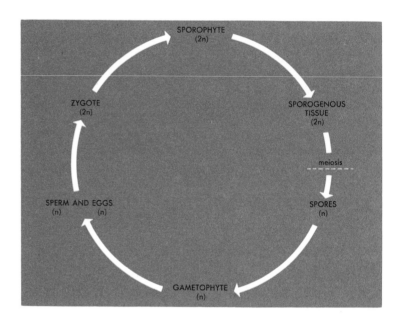

Fig. 5.23 This is a life cycle diagram of plants exhibiting both the gametophyte and the sporophyte stages. The designation "n" means that the cells at that particular stage have a single "set" of chromosomes; "2n" means that cells have paired chromosomes; that is, there are two "sets." Meiosis reduces the chromosome number from 2n to n; fertilization doubles the chromosome number from n to 2n.

eggs and sperm. Zygotes produced by the union of eggs and sperm develop into many-celled, diploid **sporophytes** that lie within the tissues of the liverwort. Eventually, these sporophytes produce cells that undergo meiosis and form haploid **spores.** When these are released from the plant and germinate, they grow into liverwort plants.

The significant thing about the liverwort life cycle is that there are two distinct phases. The plant itself is called a **gametophyte,** because it produces **gametes,** that is, eggs and sperm. Although sporophytes produced by the union of eggs and sperm are not really different plants, their diploid condition makes them a separate **phase** of the life cycle. Hence, the life cycle of liverworts consists of two phases that alternate with each other, the haploid gametophyte and the diploid sporophyte.

With some important variations that we shall note, this type of life cycle is characteristic of the vast majority of plants. This is in contrast to animals, where the gametes are the only haploid cells in the life cycle. Although the gametophyte is the outstanding phase in liverworts, the sporophyte gains in prominence as we proceed up the scale of complexity in the plant kingdom. When we reach the flowering plants, which are the most advanced of all, we find that the sporophyte is the main body of the plant. The gametophyte has been greatly reduced in relative size and importance.

In emphasizing the importance of liverworts as the simplest of a long line of green plants whose life cycles bear a general resemblance to each other, we have not meant to imply that liverworts were the original ancestors of these plants. The fossil record does not indicate that this was the case. In fact, many algae have a sporophyte-gametophyte life cycle of this type, and it is generally supposed that the algae, not liverworts, gave rise to the higher plants, including liverworts. As you saw earlier, the algae have other types of life cycles as well. However, the liverwort type of life cycle has been generally more successful than other types in evolution. As a result, it is characteristic of those plants that are more complex than algae and fungi.

The Mosses (Phylum BRYOPHYTA)

Like liverworts, most mosses (Fig. 5.24) are relatively small plants usually found in moist surroundings. However, mosses are more abundant than liverworts, and, as a group, they require less water. Most mosses are terrestrial, although some forms grow in bogs or marshes, and a few are submerged aquatics. In addition, certain species may grow on wood or rock. The mosses are of limited direct economic value to man, although some are used as packing materials. Peat moss (*Sphagnum*) is widely used by gardeners to increase the acidity and water-holding capacity of soil. The sponge-like quality of mosses has also made them useful in certain types of surgical dressings. Their indirect economic value can be great, since they frequently prevent soil erosion.

Fig. 5.24 The two phases of the moss life cycle: At left is a dense growth of moss gametophytes. At right are several sporophytes, each growing from a gametophyte.

R. H. Noailles

R. H. Noailles

The life cycle of mosses (see Fig. 5.25) is essentially that of liverworts. In both groups the gametophyte is more obvious than the sporophyte. However, the moss sporophyte is far more prominent than that of liverworts. Typically, it is a knobbed stalk borne by the female plant (or a female branch in the case of plants that are bisexual). The physical relationship between the sporophyte (stalk) and gametophyte (leafy plant) is thus a very close one. Spores are produced in the **sporangium** (plural, **sporangia**), the enlarged tip of the sporophyte. If they are carried to a moist surface, the spores germinate and form gametophytes. The mature gametophytes, in turn, produce eggs and sperm, either on separate plants or on separate branches. Rain or heavy dew carries the sperm from the male plant (or from the male branches of a male-female plant) to

Fig. 5.25 Life cycle of a moss. Compare this diagram with the schematic life cycle shown in Fig. 5.23.

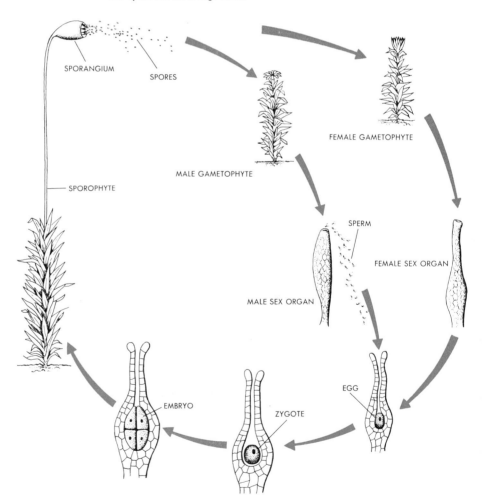

the several eggs produced by a gametophytic stalk. Fertilization takes place and several zygotes are produced at the tip of a female stalk, but, for some reason, only one survives and develops into a sporophyte.

The moss sporophyte achieves some degree of independence from the gametophyte through the development of chlorophyll, yet throughout its life it remains attached to the gametophyte. Thus it depends on the gametophyte for water, minerals, and perhaps more complex materials that regulate its growth and function. Even though the moss sporophyte is more independent of the gametophyte than is the case in liverworts, there is still a close physical relationship between the two.

The Ferns (Class FILICINEAE, Phylum TRACHEOPHYTA)

As a rule, the plants forming the groups you have studied so far are rather small. With few exceptions they are either aquatic or they are restricted to moist habitats. Both characteristics—small size and dependence on moist habitats—reflect a structural principle common to virtually all of them. Except for certain large brown algae, not one of the plants has specialized tissues for carrying water, minerals, and food throughout the body of the plant. As a result, these substances usually must diffuse slowly from one part of the plant to another. This inferior transportation system limits the size a plant can reach. Since aquatic forms are surrounded by water and dissolved minerals, the handicap is not as great as it is for land plants. This explains in part why certain algae grow to a large size.

The ferns (Fig. 5.26) are a major step upward in the plant kingdom. For the first time we meet plants with a transport system of specialized tissues that carry fluids throughout the plant body. These are called **vascular tissues,** and those plants that have them (ferns, fern-like plants, and seed plants) are called **vascular plants,** or **tracheophytes** (see Fig. 5.27). By virtue of their ability to transport water and minerals relatively long distances in a relatively short time, vascular plants are able to reach far greater sizes than nonvascular plants. Although the ferns most of us see do not attain great size, certain tropical tree ferns sometimes reach a height of 80 feet. Even the common ferns of the United States are much larger than mosses.

There are two types of vascular tissues, **xylem** and **phloem** (Fig. 5.28), and each has a special function in plants. Xylem tissues carry water and dissolved minerals from the roots upward to other parts of the plant. Phloem transports soluble foods upward whenever the plant releases them from storage

Courtesy of The U.S. Forest Service

Fig. 5.26 Above: typical ferns growing on the forest floor (temperate zone). Below: tree ferns growing in Java.

W. H. Hodge

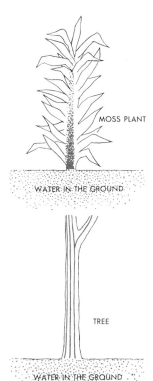

Fig. 5.27 Because nonvascular plants such as mosses have no specialized tissues for water conduction, there is a mechanical limit to the size they can reach. In contrast, vascular plants are frequently quite large.

Fig. 5.29 This drawing of a typical fern shows the plant in relation to the soil. Notice that the stem grows beneath the surface; only the leaves (fronds) are visible.

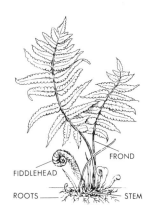

for use in the manufacture of new plant parts, and it transports newly-manufactured foods downward from the leaves. These tissues usually form a highly organized system that is spread throughout the roots, stems, and leaves. The system illustrated in Fig. 5.28 belongs to a typical young vascular plant, and as it matures, the system becomes more complex. Typically, the xylem and phloem tissues are grouped into **vascular bundles** in the stem (see drawing of cross section). If one of these bundles is magnified in longitudinal section, the xylem and phloem appear as slender tubes.

The typical fern is a plant whose leaves (called **fronds**) are often the only visible part. The stem and roots grow underground. In almost all species, the stem and roots survive for many years, while new leaves are produced each season. The fact that they grow year after year makes the ferns **perennial** plants. The fronds may project upward to a height of several feet. As they develop, they unroll in a curious fashion that makes a young leaf resemble the scroll of a violin. For this reason, developing fronds are sometimes called **fiddleheads.** In many species, the fronds are divided into **leaflets** that project outward from a central **rachis,** or axis. Fern structure is depicted in Fig. 5.29.

Fig. 5.28 A cross section of corn stem is shown below, highly magnified. Note the relative position and size of xylem and phloem tubes. The accompanying drawing shows how regions of vascular tissue fan out to all parts of a plant.

Courtesy Carolina Biological Supply Company

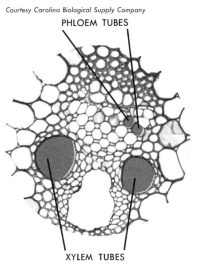

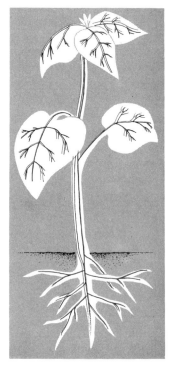

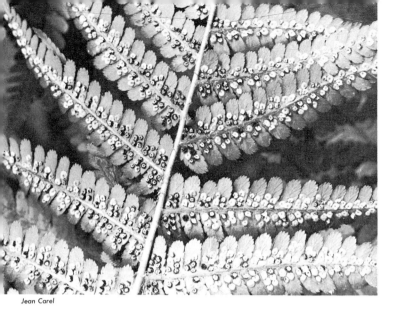
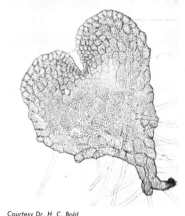

Jean Carel

Courtesy Dr. H. C. Bold

Fig. 5.30 Above: the underside of a fern leaf, showing numerous *sori*, which are groups of sporangia (spore-producing structures). Right: a fern gametophyte, enlarged.

We have seen that in liverworts and mosses the gametophytic phase is more obvious than the sporophytic phase. The situation is reversed in ferns. The leafy part we see above ground is the sporophyte. The gametophyte is a small structure from which the sporophyte develops.

In the typical fern, spores are produced on specialized areas of the leaf, usually on the underside. The sporangia—spore cases—are produced in groups known as **sori** (singular, **sorus**). The sori (Fig. 5.30 left) are arranged in an orderly pattern on the fern leaf. When the plant is mature, the spores are released from their cases, fall to the ground, germinate, and develop into gametophytes. The gametophytes (Fig. 5.30 right) are flat, green structures that grow closely pressed to the soil. Seldom are they more than a quarter-inch in diameter. Gametophytes of most species of ferns are bisexual. Each gametophyte produces both sperm and eggs in specialized structures that develop on the under side, next to the soil. Usually, it takes several weeks from the time the spore germinates until the gametes reach maturity. Water from a rain or even a heavy dew enables the sperm to swim to the eggs. Although several eggs may be fertilized and form zygotes, only one zygote of a gametophyte succeeds in developing into a sporophyte. Once the zygote is formed, development is rapid. A root projects downward, and stem and leaf tissues form above the root. From this point on, growth and development produce the mature sporophyte, and thus complete the two-phase cycle (Fig. 5.31).

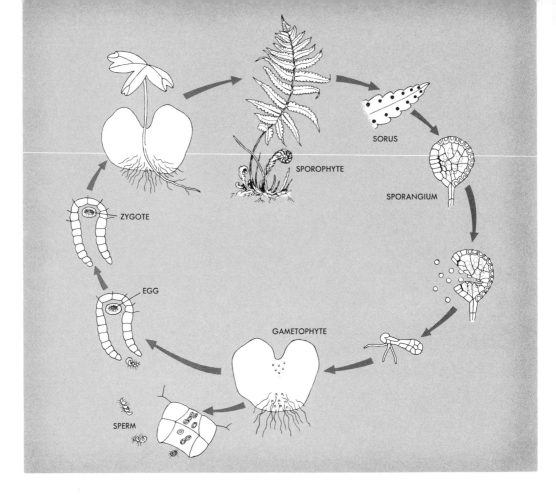

Fig. 5.31 Life cycle of a fern. Compare these stages with those of the schematic life cycle shown in Fig. 5.23. Where does meiosis occur in the fern life cycle? See the text for further clarification of this diagram.

Although some of the larger ferns are grown for ornamental purposes, the group is not of great economic importance to man, at least not the ferns that are alive today. About 300 million years ago, however, ferns and fern-like plants were the dominant vegetation on earth. During that time they formed the major basis for coal deposits, which have been of great economic importance to us.

The Seed Plants
(Classes GYMNOSPERMAE and ANGIOSPERMAE, Phylum TRACHEOPHYTA)

During the "Age of Ferns," some 300 million years ago, a new development in the reproductive structure of plants evolved. It has proved to be the most successful mechanism for the propagation of plants that nature has ever produced. That new development was the seed. It appeared first among the ferns. We have abundant fossil evidence of a group of extinct plants called **seed ferns.** It is interesting that modern ferns do not bear seeds; however, the seed plants form the most widely distributed and numerous plant group on Earth today. Apparently, this large and successful group of plants evolved from the seed ferns.

What is a seed, and why is it an especially efficient reproductive structure? The body of a seed plant, like that of a fern, is the sporophyte. The spores that it produces develop into gametophytes. However, the seed plant sporophyte produces two kinds of spores. One of these, the **microspore,** develops into a male gametophyte, while the **megaspore** gives rise to a female gametophyte. A seed plant's life cycle is shown in Fig. 5.32.

In seed plants, the gametophyte is even smaller than in ferns. The male gametophyte is simply the microspore (sometimes referred to as the **pollen grain**) and a tubular outgrowth containing several nuclei. The female gametophyte is only slightly more complex than this, and it remains imbedded within the tissues of the sporophyte throughout its functional existence. The female gametophyte develops within a structure called an **ovule,** which eventually matures and forms a seed. The ovule is composed of tissues specialized for the storage of food. Its outer cells form a protective layer. When the male gametophyte produces sperm, they reach the egg of the female through the pollen tube, and fertilization occurs. The zygote produces an immature, many-celled plant (the sporophyte), which reaches a limited stage of development. At this time, the stored food and protective tissues of the ovule surround the young sporophyte during a period of dormancy, or relative inactivity. In some species, the young sporophyte may take the storage materials into its tissues before dormancy.

A seed (Fig. 5.33), then, is a three-part reproductive structure. It has: 1. a young (embryonic) sporophyte, 2. stored food and protective tissue that the embryo will need when it begins to grow, and 3. outer tissues that surround and protect everything within.

Seeds may be rather small, such as seeds produced by mustard and carrot plants, and a variety of "weeds," or they may be quite large, as coconuts are. The size and construction of a seed are results of adaptation by natural selection. Small, light seeds are carried great distances by the wind. The coconut is adapted to float in water. Because the young sporophyte is protected from dryness, heat, cold, and other adverse conditions until conditions are favorable for growth, the species has an excellent chance of survival. Many seeds are so well constructed

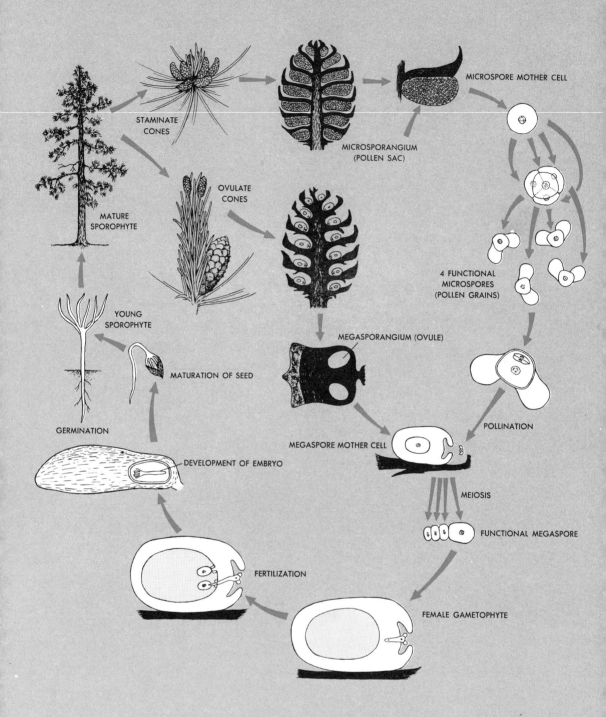

Fig. 5.32 Life cycle of a pine tree: Although this life cycle appears complex, it is essentially the same kind of cycle you have been following since we discussed the liverworts.

that they can pass unharmed through the digestive systems of animals. Considering this superior reproductive device, it is not surprising that the vast majority of plant species and individual plants are seed plants (Table 5-1).

There are two clear-cut groups of seed plants—the **gymnosperms** (from the Greek words *gymnos*, naked, and *sperma*, seed) and the **angiosperms** (from the Greek word *angeion*, vessel). Among the gymnosperms are such familiar trees as pine, spruce, and fir, as well as many evergreen ornamental shrubs. The angiosperms include all flowering plants. The angiosperms are far more numerous than the gymnosperms. Of the approximately 250,000 known species of seed plants, fewer than 1,000 are gymnosperms.

Of the several types of gymnosperms, the **conifers,** cone-bearing trees, are by far the most outstanding (Fig. 5.34). Common evergreen trees and shrubs belong to this group. Pine, fir, and spruce are representative evergreen conifers. A few species, such as larch and cypress, are **deciduous,** meaning that all their leaves are shed at one season, instead of gradually, as in the case of evergreen leaves.

There are more than 500 known species of conifers, many of which are widely distributed. The great pine forests of the northern latitudes of the United States attest to the successful adaptation of this particular genus. Coniferous plants are of great economic importance to man. Many of them are sources of lumber and other products. The pine tree, for example,

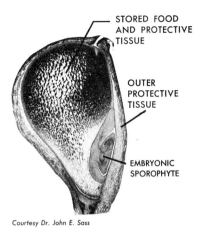

Courtesy Dr. John E. Sass

Fig. 5.33 Section of a corn seed, showing its three general areas.

Fig. 5.34 Stem tip of pine showing staminate cones. The needle-shaped leaves are characteristic of pines and many other conifers.

Courtesy of The American Museum of Natural History

Fig. 5.35 *Zamia floridana*, a cycad. Can you see why these plants are often called "palms?"

Julia Morton

Fig. 5.36 With so many species of flowering plants, there is great diversity of floral structure. The three types shown here are representative of the larger flowers. Top: oriental poppy. Middle: cactus flower. Bottom: day lily.

Courtesy of The American Museum of Natural History

Lynwood M. Chace

Leonard Lee Rue III

produces not only valuable lumber but wood pulp, turpentine, and various resins. In addition, conifers are widely grown and sold by nurserymen as ornamental shrubs or trees.

Palm-like trees called **cycads** form another group of gymnosperms, particularly numerous in the tropics. The cycads (see Fig. 5.35) are plants with thick, fleshy stems which, if above ground, are armored with leaf bases, bearing a crown of large fern- or palm-like leaves. For this reason, they are sometimes confused with ferns and palms, which are not gymnosperms. Only one genus of cycad (*Zamia*) is native to the United States, and it is restricted to Florida. However, several other cycads are widely grown indoors as ornamental plants, and outdoors in subtropical and tropical climates.

There are more angiosperms—flowering plants—than all other plants combined (Fig. 5.36). The reason for this seems obvious. The flower is a highly efficient device for producing seeds. Essentially, a flower is a group of modified leaves that have specialized for various functions, including the production of spores. The typical flower exhibits four types of modified leaves: **sepals, petals, pistils** and **stamens,** as shown in Fig. 5.37.

While the stamen bears microspores (pollen grains), the pistil bears one or more ovules in the **ovary,** which is the lower portion of the pistil. Each ovule develops a megaspore within

Fig. 5.37 This generalized diagram of a complete flower shows the various floral parts. The ovary persists far longer than the others, and it develops into the fruit.

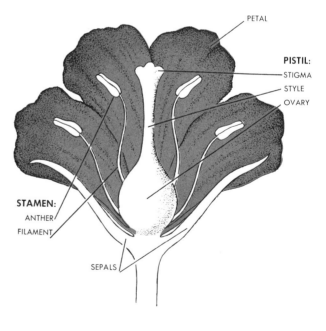

its tissues. The pistil and stamen are called the **essential** parts of the flower. Because the sepals and petals play no direct role in reproduction, they are called **accessory** parts. A flower that has these four parts is called a **complete** flower.

Although most angiosperms bear complete flowers, there are some species that do not. Grasses, for example, bear flowers that are generally without definite sepals and petals. Other plants bear flowers that have either stamens or pistils, but not both. Corn (Fig. 5.38), for example, has staminate flowers on the "tassel," and the pistillate flowers on the "ear." Still other species, such as the pussy willow and mulberry, have staminate and pistillate flowers on separate plants. Sometimes these are called "male" and "female" plants, but since the plant bodies are sporophytes, some botanists take issue with the "male" and "female" designations.

Even though the flowers of various angiosperms differ widely, all species reproduce sexually in a similar way. When the microspores (pollen grains) are produced by stamens, they are carried to pistils by a process called **pollination**. A variety of agents may be the pollinators—man, wind, water, insects. When pollen is carried from a flower of one plant to a flower of another plant belonging to the same species, the process is called **cross-pollination**. However, some flowers are self-pollinated.

Fig. 5.38 An experimental field of corn. The small staminate flowers are located on the "tassel," or terminal portion of the stalk, and the small pistillate flowers are on the "ear."

Courtesy Allied Chemical Corporation

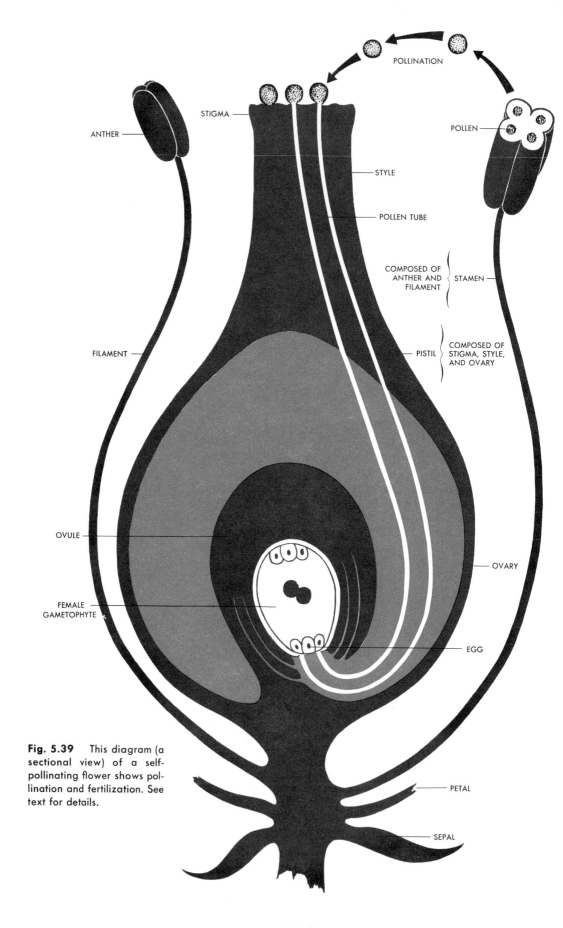

Fig. 5.39 This diagram (a sectional view) of a self-pollinating flower shows pollination and fertilization. See text for details.

In such plants the pollen transfer may take place by direct contact of a stamen with a pistil. In each situation, the pollen grain (microspore) nucleus divides and forms two nuclei. One of these two then divides, giving each pollen grain a total of three nuclei. This may take place before the pollen grain leaves the stamen, or it may be completed later. When a pollen grain reaches a particular part of the pistil (the **stigma**), it germinates and forms a **pollen tube.** One of the three nuclei is typically associated with growth of the tube and is called the **tube nucleus.** The other two, which will function as gametes, are called **sperm nuclei.** Note the diagram in Fig. 5.39.

One of the cells that lies deeply within the tissues of the developing ovule specializes as the megaspore and begins development into the female gametophyte. In most angiosperms, the megaspore nucleus undergoes three successive divisions and produces eight haploid nuclei, six of which become surrounded by cell walls. One of these six develops into an egg cell. The other five play no important part in the further development of the ovule. The two remaining nuclei are called **polar nuclei,** and they occupy a common cytoplasm. The mature female gametophyte, then, consists of seven cells, one of which has two nuclei.

By this time, a pollen tube has grown to the ovule, discharging its two sperm nuclei into the female gametophyte. One of these unites with the egg and forms a zygote. The other one joins the two polar nuclei (which may unite with each other first) and forms the **endosperm nucleus.** This, together with the surrounding cytoplasm develops into storage tissue (endosperm). The zygote develops into the embryonic sporophyte. When the embryo, endosperm, and outer coverings of the ovule have completed their development, the seed is ready to be released from the plant.

While the ovules are maturing into seeds, the ovary undergoes further development, forming the **fruit.** This may be a fleshy structure, as in the case of a tomato or watermelon, or it may be merely a dry pod, as in the case of bean, mustard, and many wild plants. Regardless of the many differences in details that exist in seed and fruit development, the seeds of flowering plants are *always* produced within a floral ovary. It is this characteristic that gives rise to the term "angiosperm." Gymnosperms do not bear flowers or fruits, and their seeds are borne *upon* modified leaves rather than *within* an enclosure such as the floral ovary. Apparently, the fruit is a highly successful adaptation resulting in wide distribution of the seeds of flowering plants. The flower-seed-fruit mechanism of reproduction goes far in explaining why the flowering plants are the dominant vegetation on the Earth.

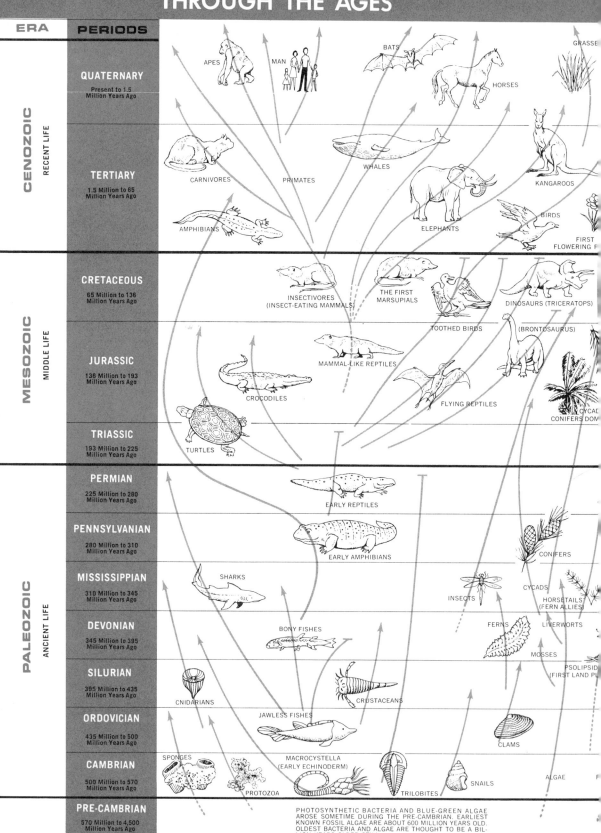

Fig. 5.40

Flowering plants have far more economic importance than all other plants combined. Among the flowering plants, the large deciduous trees, such as oak and hickory, are a source of wood for lumber and for a great variety of wood products. The textile industry uses the fibers of cotton and other flowering plants for the weaving of cloth. Still other flowering plants are valuable as sources of drugs or oils. Cereal grains, such as corn, wheat, and rice have been cultivated by man for thousands of years. Many other plants have been developed for their fruits—for example, apples, bananas, the citrus fruits, tomatoes, grapes, melons, and many common garden "vegetables" such as squash (the edible part of which is really a fruit, since it is a matured ovary of a flower). The roots of many plants, such as carrots, beets, and sweet potatoes, are also an important food source. Still others of this group are grown for their stems—the Irish potato (the "potato" is a modified stem, not a root), asparagus, and sugar cane. This list is hardly a complete one, but it serves to remind us of the extent to which we depend on the flowering plants every day.

The Evolution of Plants

When we study evolution, we must use the approach of the historian. First, we must gather evidence. When it is inconclusive, then we must do the best we can to reconstruct events. In plant evolution, the major sources of evidence are the fossil record, comparative morphology, biochemistry, cytogenetics, and geographical distribution of the plants now living.

In spite of the abundance of data that we have from these fields of study, there are large gaps in our knowledge of plant evolution. One of the major difficulties is that many of the smaller plants did not lend themselves well to fossilization. But even if the fossil record were complete, it would not necessarily tell us any more about the **origins** of specific plant groups than we now know. Any account of plant evolution has to be based to a certain extent on some intelligent guesswork. The modern biologist is in the position of a detective who is faced with the obvious fact that a safe has been blown open and yet may never find out exactly how it was done.

Many botanists believe that the first living organisms were heterotrophs, not unlike modern bacteria, and that photosynthetic bacteria and the blue-green algae developed very early from these organisms. This view is upheld by the fact that the earliest known fossil plants (about 600 million years old) are those of simple algae that are much like modern blue-green algae. Also, there is indirect evidence that both algae and bacteria had existed for nearly a billion years before that time (see the time chart in Fig. 5.40).

The origins of the fungi are rather obscure, but they probably arose from the algae rather than from original heterotrophic cells such as bacteria. There is strong evidence that land plants evolved from green algae perhaps 500 million years ago, after which there was great diversification. But the exact origins of the different groups are obscure, and the evidence can be interpreted in different ways. Vascular, fern-like plants are among the earliest land plants known; fossil mosses and liverworts appear in later fossil deposits, which may mean that they evolved more recently than did vascular plants.

Ferns and fern-like plants, some of which were gigantic, made up the major vegetation of the Earth until some 150 million years ago. At that time the cycads and conifers became dominant. The earliest known fossil angiosperms are at least 125 million years old. Within a few million years they had succeeded in replacing gymnosperms as the major plants. This trend has continued to the present time, with the result that the vast majority of modern plants are angiosperms. Figure 5.40 shows something of these general evolutionary trends and relates the plants to the animals in their evolution.

SUMMARY

We have presented the plant kingdom in seven groups of plants—algae, fungi, lichens, liverworts, mosses, ferns, and seed plants. Although certain other plant types are known, these seven groups represent the plants we most commonly see.

Of the approximately 350,000 known species of plants, about 250,000 are flowering plants. This indicates the successful adaptation of this group.

In their reproductive habits, the vast majority of plants produce spores by a process of meiosis, and these spores give rise to haploid plants. The union of gametes produces the sporophyte, thus completing a cycle that shows alternation of two generations—the gametophyte and the sporophyte. Seed plants, and particularly flowering plants, have reproductive structures that are highly complex and specialized.

The algae probably evolved from ancestral forms at least a billion years ago. Land plants evidently arose from the algae about 500 million years ago. Since that time, ferns and fern-like plants, gymnosperms, and angiosperms have succeeded each other in that order as the dominant plants on Earth.

FOR THOUGHT AND DISCUSSION

1 What good are algae?
2 A student once wrote on a test, "Bacteria are very bad. They cause diseases and eat up the good plants and animals." Evaluate this statement. Can we really speak of organisms as "bad" and "good"? Assuming that we can call some bacteria "bad," or, perhaps, harmful to man's interests, does this apply to all bacteria? Incidentally, do bacteria "eat?"
3 Generally speaking, vascular plants reach much larger size than nonvascular plants. How do you account for this?
4 Why do we consider the lichens to be a major plant group, and yet they are not given their own phylum in Table 5-1?
5 Distinguish between sexual and asexual reproduction.
6 Distinguish between a fruit and a vegetable.
7 The cells of endosperm tissue in most flowering plants are triploid (3n). How do the cells of this tissue come to have this particular chromosomal constitution?

SELECTED READINGS

Bold, H. C. *Morphology of Plants*. New York: Harper and Brothers, 1957.
Bold gives an exhaustive and authoritative presentation of the world of plants.

——— *The Plant Kingdom* (2nd ed.). Englewood Cliffs, N.J.: Prentice-Hall, Inc., 1964.
The author of the rather comprehensive book cited above surveys the plant kingdom in this small volume.

Cronquist, A. *Introductory Botany*. New York: Harper and Brothers, 1960.
This is one of several excellent textbooks of general botany. It is mentioned especially because it presents the plant kingdom in an evolutionary perspective.

Schery, R. W. *Plants for Man*. Englewood Cliffs, N.J.: Prentice-Hall, Inc., 1952.
The viewpoint of this book is the economic importance of various plants.

6 DIVERSITY, SIMILARITY, AND THE HUMAN VIEWPOINT

In the introductory chapters we said that certain principles govern the world of living things. We did not say that those principles apply to only one kingdom of organisms and not to the other. If you have followed the arguments in this book carefully, you should now be able to talk quite specifically about similarities *and* differences that exist between plants and animals (see Fig. 6.1). You should also be able to describe diversity among the various categories of animals and the various categories of plants as you move upward or downward among the phyla.

In describing diversity, we have not meant to establish a hard-and-fast cleavage between plants and animals. To do so would destroy the concepts of basic similarities among all organisms. In this concluding chapter we shall attempt to show something of the similarities that exist in the very presence of diversity.

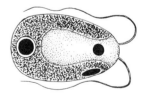

Fig. 6.1 What are the *basic* similarities between an elephant and a maple tree? What are their *basic* differences? What biological characteristics does *Chlamydomonas* share with each?

Order in Diversity

In the 1800's, when descriptive biology was making its greatest gains, major emphasis was placed on differences existing among organisms. Actually, this period of activity extended well back into the 1700's, before Linnaeus began his work, and continued into the 1900's. While there is still a need for this viewpoint in biological research, the emphasis of modern biology on cellular and molecular aspects of life is revealing striking similarities among organisms. These similarities help us to see order in diversity and, by so doing, enable us to view the world of living things as a meaningful whole.

Let us consider a few examples. So far as we know, all organisms (except for certain viruses, if these are to be considered organisms) have within their cells control systems consisting of the substance DNA. As we have seen, this substance has remarkable properties. For one, it is self-duplicating. For another, DNA molecules control the formation of new protein molecules. These, in turn, dictate exactly how much and what kind of new materials will be built within the cell. If the organism in which this occurs is very complex, as in a human being or a tree, the ultimate shape, size, color, and texture of mature parts will be dictated indirectly by cellular DNA. It seems incredible that so much in nature is dependent upon so little; also, that in spite of the differences between a man and an oak tree, each develops from a cell whose nucleus contains a substance very much like that contained by the other.

The cellular metabolism of all organisms is remarkably similar. For example, water is the common solvent of all living matter; it makes up about two-thirds of the mass of virtually all cells. The relative proportions of the fuel substances (carbohydrate and fat) and of the structural and enzymatic material (protein) are about the same in most cells, no matter where they are found. The same is true of vitamins and minerals. Many of the enzyme systems in cells of organisms quite low on the evolutionary scale are also found in the cells of organisms which are only distantly related to them. What does this mean? It is a reflection of at least some basic genetic similarities. Enzyme (protein) molecules are formed as a result of "instructions" that are "dictated" by genes. Certain genes, then, are apparently widespread in genetic systems.

Finally, to conclude this brief list of similarities in cell metabolism, organic molecules are broken down for their potential energy in much the same way by all organisms. Certain compounds within cells function in the transformation of this energy to a more readily expendable form. These compounds are found in all living matter.

Mechanisms of reproduction are also very similar among all organisms. Although certain differences exist, the basic processes are very much the same. Remember that reproduction occurs at **molecular, cellular,** and **organismal** levels. We have already mentioned that only the nucleic acids are capable of reproduction at the molecular level, and that this process is characteristic of all cells. At the cellular level, cell division is the reproductive process. There are some cells of *all species* that have the ability to divide and thus produce daughter cells that have an equal distribution of vital materials.

At the organismal level, there are many differences in reproductive processes, but these differences are simply variations of either asexual or sexual reproduction. There are very few basic types of asexual reproduction—the body of an animal may divide in some fashion and form two or more individuals; a plant may produce special cells or clusters of cells called spores, or a part of its body may develop into a complete individual. The essence of sexual reproduction is the union of two specialized gametes which form a zygote. The zygote then develops into a new organism. This phenomenon is widespread in both the plant and the animal kingdoms. Finally, the process of meiosis, described in an earlier chapter, is a remarkably similar process wherever it occurs.

These few examples of fundamental similarity among organisms point up the common origin of organisms from primitive life forms existing many millions of years ago. In the long processes of adaptation and natural selection, many of these molecular and cellular characteristics have remained virtually unchanged ever since the great proliferation of species began. If life began some two billion years ago, and evidence indicates that it did, then evolutionary history divides itself into two time periods of roughly equal length.

During the first period, the cell established itself as the basic unit of life. It was during this time that the characteristics we have just discussed, and others, developed as a result of natural selection. The cells that were able to evolve these mechanisms were the "fittest," survived, and played a further role in evolution. During the second period, natural selection moved to a new level. It gave rise to diversity at the organism level, that is, diversity of external form. But the most efficient mechanisms were already firmly established at the molecular and cellular levels, and they have persisted to the present time.

The vast array of organisms around us today are all closely interrelated. To study them as a whole, we must adopt the viewpoint of the ecologist. The complicated food webs are but one example of the many relationships between plants and animals. In the water, algae and other small plants are the producers, both of energy-yielding organic molecules and oxygen, while land plants serve this function in terrestrial environments. Fungi and other decomposers provide the mineral substances for these producers, while various animals consume plants and each other. At the same time these animals and the decomposers produce carbon dioxide and other materials that are useful in the metabolism of the producers.

In addition to these nutritional relationships, there are those which have to do with maintaining the temperature, light, and moisture needs of various species. For example, most animals depend on a plant habitat in which they can live and reproduce successfully. Various plants have a characteristic fauna associated with them; remove the plants, and the animals also disappear. Finally, there are such relationships as parasitism and mutualism through which plants and animals are associated. These examples point up the fact that we cannot dissect nature into species and kingdoms without distorting it to some degree. The great web of nature is an endless series of relationships among all organisms inhabiting the Earth.

Man's Place in Nature

As the oak tree and the variety of animals that live among its branches and that burrow among its roots are a part of nature, so is man. Considering the long line of success of other species, man is a relatively recent arrival on Earth, yet the place he has made for himself, as a species, has been unparalleled in the great succession of life. As they have in the past, his fortunes depend largely on the wisdom with which he exploits his environment.

Unfortunately, man tends to be enormously wasteful of natural resources, both living and nonliving (Fig. 6.2). It is only within recent decades that much organized effort—conservation programs—has been made to curb this wastefulness. As a result, wanton destruction of natural resources has been reduced somewhat. However, more effective conservation measures are still needed at the national and international levels. In addition, many of the laws governing the use of natural resources were passed when certain aspects of conservation were poorly understood; consequently, they need revision. This is particularly true of laws concerning game and fish. Many laws are not reasonable in the light of recent ecological findings, but it is difficult for wildlife biologists to have them changed as rapidly as our knowledge advances.

We have been more successful with conservation of plant

Courtesy Rhodesia National Tourist Office

Fig. 6.2 When man alters the face of the earth, plant and animal life are usually affected. The photograph above shows the plight of displaced animals during the building of a reservoir in Rhodesia. In some such cases, game rescue operations, as shown in the photograph below, were effective.

Courtesy Rhodesia National Tourist Office

resources than we have with animals (Fig. 6.3). For example, many years ago it became apparent that our forests were being depleted. Such safeguards as selective cutting of timber, reseeding, and the prevention of forest fires helped enormously. In general, these measures are our best defense against forest depletion.

One of the greatest problems in conservation is the constant danger of upsetting what is often called the "balance of nature." For example, an area may be sprayed with insecticide to kill an insect pest. In killing the insects, however, its natural enemies may also be killed, including other insects and birds. Quite frequently, spraying brings about a situation worse than the situation it was intended to cure, or relief may be only temporary. This upset of "balance" sometimes has such far reaching effects that a farmer's crops, or even his domesticated animals, are harmed. Ranchers of the western United States once conducted an intensive campaign against coyotes, which were killing livestock. The ranchers' efforts were quite successful, but with the decline of the coyote population, rabbits began to overrun the country. The rabbits were more than a mere nuisance because they ate vegetation that the livestock needed. When people realized that the coyotes were necessary to keep the rabbits in check, the ranchers began to consider the coyotes the lesser of two evils and relaxed their efforts.

Fig. 6.3 Man's influence over nature is not always destructive. At left conservationists make a study of a pine forest in British Honduras. At right a fish ladder at Bonneville Dam makes it possible for salmon to swim past the dam to spawn. Through the institution of such conservation measures, our natural resources can be preserved.

Man's relation to nature, however, reaches beyond those organisms that are useful or harmful to him. What of man's relation to man? Aside from political, social, and ethical considerations, which are beyond the scope of this book, two outstanding dangers exist. The first is **pollution** (Fig. 6.4), the accumulation of waste materials. The body wastes of man are not the major problem, but even so, they sometimes spread disease which kills plants and animals. Every large city has its sewage problems, especially where there are rivers or harbors. Industrial wastes, insecticide residues, and a variety of such products enter the air, water, and soil. They remain there, or are deposited elsewhere, and damage plants and animal life, and man himself. Also, there are radioactive wastes from industrial and military operations. It is extremely difficult to dispose of them, and possibly they constitute a greater long-range threat to man than nuclear warfare.

The second danger, and one that compounds the pollution problem, is **overpopulation** (note Fig. 6.5). We are now seeing a "population explosion" in the world. Within the past century the population of the world has more than doubled, and with longer life expectancies and lower mortality rates among infants, it is climbing at an alarming rate. Just how long this trend can continue is difficult to say, but it is a problem that demands the attention of all of us. It is not simply a matter of food. It is more a problem of pollution and of sheer living space. Of all of the social and biological problems facing modern man, the world population problem is the greatest, if not the most urgent.

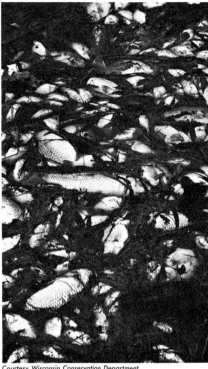

Courtesy Wisconsin Conservation Department

Fig. 6.4 These fish were killed by pollution that was allowed to enter their waters. With increased human populations, it is becoming more difficult to control the pollution of water, soil, and air.

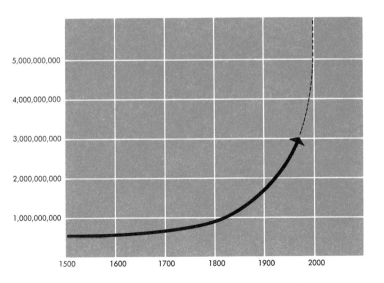

Fig. 6.5 World population figures and rate of increase are shown since the year 1500. The dotted line represents a projection of human population increase based on past and present trends. You might ponder Question 4 at the end of this chapter in the light of this chart.

FOR THOUGHT AND DISCUSSION

1 Do plant and animal kingdoms actually exist in nature? If you feel that this question might logically be answered by either yes or no, defend each answer.
2 In external form and function, a human being and an oak tree would seem to have little, if anything, in common. What basic similarities can you list between them?
3 What do we mean by "upsetting the balance of nature?" Is there really a "balance" in nature?
4 At the present rate of increase in the human population, it has been estimated that there will be one human being per square yard of Earth's land surface by the year 2500! Obviously, something will be done to alter the rate of increase before such a population is reached. What do you think this "something" will be?

SELECTED READINGS

Bates, M. *Man In Nature* (2nd ed.). Englewood Cliffs, N.J.: Prentice-Hall, Inc., 1964.
  A distinguished student of human biology considers man's relationship to his surroundings.

Bonner, J. T. *The Ideas of Biology*. New York: Harper and Brothers, 1962.
  This excellent short book is a synthesis of broad principles of biology. Although the book is well worth reading in its entirety, Chapter 5 is of special interest. It is entitled "Simple to Complex," and discusses certain issues closely related to the first part of Chapter 6 of this book.

Carson, R. *Silent Spring*. Boston: Houghton Mifflin Co., 1962.
  The author of this book urges caution in our attempts to control nature, and presents an excellent picture of ecological relationships.

Thomas, W. L., Jr. (ed.). *Man's Role In Changing the Face of the Earth*. Chicago: University of Chicago Press, 1956.
  An informative presentation of man's various activities in exploiting nature.

PART TWO
Modern Cell Biology

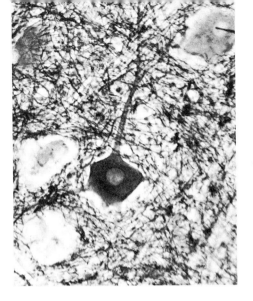

ABOUT THE PHOTOGRAPH

Section through spinal cord (yellow-red-brown). This section shows a large ventral horn cell of the spinal cord. The cell body is clearly seen with an axon extending from it. This axon leaves the spinal cord and carries impulses to muscles. The field is interlaced with great numbers of branches and twigs from other nerve axons. The stain is Cajal-Silver.

Section by E. Evans   Photo by A. Blaker

ABOUT PART TWO

In Part 1 we considered those aspects of biology that could be explored without an extensive use of instruments. Emphasis was given to the wide variety of plants and animals that abound on our planet, some of them were familiar, others were not. We also emphasized the importance of classifying the array of diverse organisms, including man. The concept of adaptation was also presented—what is meant by adaptation to an environment, and our current thinking about how adaptation was achieved, maintained, and sometimes lost. You have, we hope, gained the impression that the present population of organisms is not the same as others that have existed in the past, and probably is not the same as still others that will exist at some future time. The world of life is a constantly changing one.

In Part 2 we deal with the interplay of matter, energy, and life, with the intimate details of structure and function, and with the major concepts that form the basis of present-day biology. Specifically, we shall be concerned with how an organism, having a particular structure, acquires and manipulates matter, energy and information. Matter, of course, consists of the atoms and molecules required by the organism for maintenance, growth and reproduction; the organism *uses* this matter first by acquiring it, then by transforming it through chemical reactions. Information, on the other hand, is inherited *and* acquired by experience. You are a human being because of the information inherited by the cells of your body, but the information you use as a functioning human being is that which you have acquired through experience as well as that which your cells have inherited.

Neither matter nor information can be manipulated by an organism without a source of energy. These relationships are shown in the diagram, on page 116 and you should make frequent references to this diagram as you progress through Parts 2, 3, and 4.

There is a limit to the energy upon which organisms can draw, and a limit to their ability to use that which is available. But an organism, to survive, must have the means for acquiring, converting, storing, and using

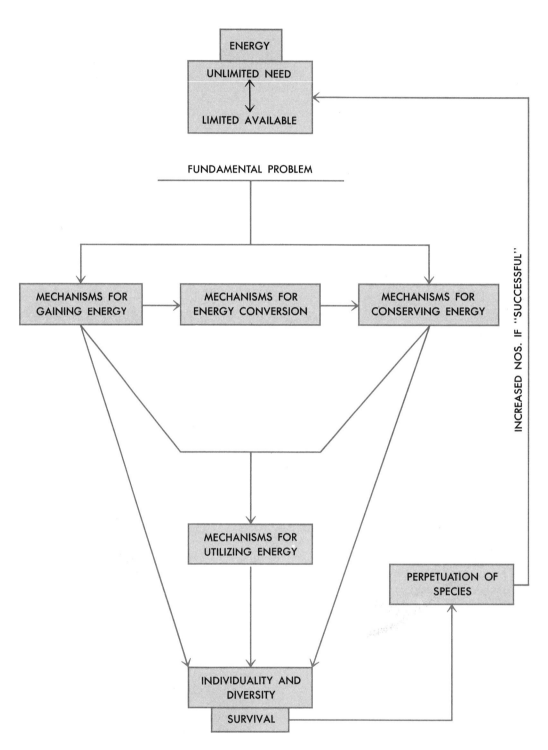

Courtesy of Dr. H. H. Hagerman

energy. An automobile engine cannot use the energy available in gasoline unless its parts are organized and work in a particular way. Species must be similarly organized to use their energy sources. In Part 2, we have chosen to discuss this problem at the cellular level, but the problem could be equally well discussed at the organism or population levels.

Just as engines differ in the way that they extract energy from gasoline, so organisms differ in their management of energy. This is, in part, a source of individuality and diversity, and it leads to competition for energy sources. Successful acquisition and management of energy determines survival and the perpetuation of species. Such success leads to an increase in the numbers of individuals, and this factor, in turn, leads to greater demands on the environment for more matter and more energy. Since both matter and energy are not limitless, the competition for them limits both the numbers and kinds of organisms. Evolution is the inevitable result.

Since the cell is the basic unit of organization of living things, our focus in Part 2 will be on cells: their basic structure, how their various parts contribute to the life of the cell, and how cells grow, divide, and differentiate. Since the cell consists of organized matter and requires energy for existence, we will deal with atoms, molecules, and chemical reactions. When an ameba divides, it produces two ameba; when an oak tree reproduces, it produces other oak trees like itself. Information, matter, and energy have obviously been manipulated to achieve these results. These facts lead us to a study of inheritance, to its physical and molecular basis, and to an understanding of how a cell contains and transmits the instruction needed for the development of an ameba, or an oak tree, or a human being. And finally, we will return to the topic of evolution again, and examine it in greater detail so that we can appreciate the origin of diversity and the course of evolution through time. In doing so, we will examine the special case of man himself. An animal, man is an animal who, during the course of evolution, has achieved enormous success in manipulating information, managing energy, and using matter.

7 BIOLOGY IN A MODERN WORLD

Nature is difficult to define in a simple scientific way. This is so because it includes so many different things—atoms as well as mountains; a single drop of water as well as the vast oceans; our own Earth and the other planets of our Solar System, and countless stars and galaxies comprising a universe so large that we do not know its outer limits in space. Nature includes all living things as well.

These are things that we can see, either with our eyes or by means of detecting instruments. But there is another aspect of nature that is equally important. It is always in motion. Often, in the cases of ourselves and of the planets, we can detect motion with our unaided senses. But often it is difficult to detect motion, even when our senses are aided, yet nature is never at rest. It is

Fig. 7.1 The Crab Nebula. The gaseous remains of a star that exploded in the year 1054. It is a particularly strong source of radio signals.

Courtesy of Mount Wilson and Palomar Observatories

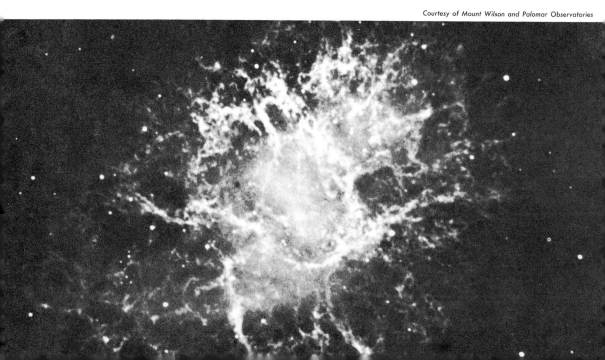

constantly changing. The mountains seem eternal to us, but we know they are thrust up from the Earth's crust, and after millions of years will be worn away. Water falls in the form of rain or snow, flows in our streams and rivers, evaporates, and begins the cycle again. Living things arise, grow, age, and die. Their remains decay, blend with the soil or water, and are used anew by other organisms in a seemingly endless process.

Perhaps the simplest scientific description of nature is that it is **matter in motion.** This may convey to you the idea that nature is disorganized. When we try to comprehend nature in all of its aspects, this is the first impression we get. As we learn more about the natural world, however, we find in it a remarkable degree of order in both matter and energy. Matter, from the smallest atom to the largest star, has organization that gives it shape and motion of particular kinds. One of the particular functions of science is to perceive and describe the order that exists in this vast amount of seeming disorder.

By matter we mean anything having mass and, consequently, weight. Matter is in motion because some form of energy is acting on it. In trying to understand nature, science therefore deals with matter and energy. The sciences of physics, chemistry, and biology, for example, differ only in that they are concerned with different aspects of matter and energy.

The Growth of Science

Before getting to the heart of biology, let us first take a brief look at science in general. We live in an *Age of Science,* an age during which we are gaining knowledge of ourselves and the world around us more rapidly than at any other time in our history. Think of the commonplace things which are a part of your everyday world and about which your parents, at your age, had no knowledge whatsoever: the jet airplane, transistor radio, synthetic fibers of many kinds, packaged frozen foods, "wonder" drugs, electronic computers, hydrogen bombs, artificial satellites and space ships. Each year it becomes possible to add many items to this list. Often the new replaces the old even before we become fully acquainted with the old. Just as you live differently from the way your parents did at your age, so your children will grow up in a world which will be vastly different from yours.

The things we have listed are products of the **applied sciences** —engineering, medicine, and agriculture. But behind each lie ideas developed by the **basic sciences.** What is an applied science as contrasted with a basic science? How does one depend upon the other? The discovery and use of the antibiotic penicillin may help to illustrate this point. In 1928, Sir Alexander Fleming,

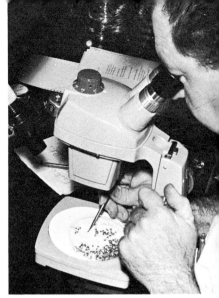

Paul Knipping

Fig. 7.2A An entomologist separating species of mosquitoes in an effort to determine which species transmits disease-producing viruses. He will crush the mosquitoes and then add the "brei" to growing chick embryos to search for the presence of viruses.

Fig. 7.2B A field biologist searching a stream for organisms that transmit disease-producing viruses, in this case a virus that causes vesicular stomatitis. The disease is characterized by blisters on the tongue, lips, and membranes of the mouth in horses, cattle, and swine, and can be transmitted to man.

Paul Knipping

the English bacteriologist, discovered that a mold (fungus), *Penicillium notatum,* had contaminated some of his bacterial cultures. Not only did the mold kill the bacteria it was touching, but it was killing bacteria at some distance away also. From this observation, it was possible to conclude that a product formed by the mold diffused, or spread outward, killing the bacteria with which it came into contact. After more than a decade of further research and testing, medical men were able to isolate, purify, and use penicillin to cure many bacterial infections in man. In this sense, they *applied* a *basic* idea to solve a particular problem confronting them.

Although this illustration is clear cut, it is not always so easy to distinguish between these two aspects of science. In the treatment of cancer, as in so many instances, the applied precedes the basic information. We know how to control some forms of cancer, if caught in time, but we still lack a basic knowledge of *why* cancer develops and behaves as it does. Ultimately, this knowledge may come from basic research, or from applied research, or from a combination of the two.

One thing we can be certain of is that knowledge of ourselves and the world around us advances as basic scientific discoveries are made. This has been proven time and again. And use of this knowledge is a form of power: to improve and increase our food supply, to control and eliminate disease, to educate the peoples of the world, to communicate faster and more precisely. Fleming's work with antibiotics changed the whole complexion of medical science just as the molecular aspects of structure, growth, and behavior have changed the complexion of biology over the past few years. The unique power of science is that man can now transform the world if he chooses to do so. What is needed is planning, co-operation, and a vision of what he wishes the world to be for himself and his descendants.

One reason for the rapidity of change today may be gained by considering the recent increase in the number of scientists. We think of modern science as having its origins sometime around the year 1500, when men of the Renaissance, in their effort to understand the world around them, turned from a life of contemplation to one of direct observation and experimentation. This time was the beginning of the scientific revolution, which is still going on. Yet of all the scientists who have ever lived, nearly 90 per cent are alive and active today. In recent decades the scientific population has also been given greater financial support by individual donors, governments, industry, and foundations. No wonder, then, that the volume of information and ideas produced by scientists today is proportionately greater. It has been estimated that our store of scientific knowledge doubles every 10 years or less. Scientific discoveries made during the last 20 years alone far exceed all that man had discovered up to that time. Most important, the information obtained has not been haphazard, but directed to the solution of specific problems chosen because of their importance. No person can possibly read all of this printed information. Even the devoted scholar finds it difficult to keep himself informed about all the new findings in his own field. Arnold Toynbee, the English historian, has pointed out that we are in danger of being drowned in an ocean of information.

The situation reminds one of Alice's conversation with the Queen in *Through the Looking Glass.* "Well, in our country," said Alice, "you'd generally get to somewhere else, if you ran very fast for a long time as we've been doing."

"A slow sort of country," said the Queen. "Now here, you see, it takes all the running you can do to keep in the same place. If you want to get somewhere else, you must run at least twice as fast as that."

The Ways of Science

The aim and purpose of science is to gain an understanding of the natural world, living and nonliving. By "understanding" we really mean *the ability to predict* events or relationships in nature to a more or less correct degree. You readily *understand* that $1 + 1 = 2$, but in reality you have sufficient familiarity with numbers to *predict* that the same result will always be obtained. The more exact a science becomes, the more precise will be the power of prediction. We can, for example, predict the time of an eclipse of the Sun by the Moon to a fraction of a second, but you are equally aware that the science of weather forecasting is much less accurate.

Science, therefore, seeks to discover order in what often appears to be disorder. Science explores new ways of knowing, just as the arts explore new ways of seeing and expressing. The domain of science is the structure, behavior, and history of matter and energy. Unlike the arts, science deals best with those things that can be observed, described, measured, tested, and verified—things that we can directly or indirectly detect with our senses, and that enable us to project our thoughts beyond our senses. In the final analysis our senses are our *only* authority. They are our means of contact with the world around us. However, it is the mind that assembles these observations, puts them into meaningful relations to give us our ideas of an atom, a gene, or a universe (Fig. 7.3).

This does not mean that science must deal only with those phenomena which can be directly sensed; after all, we cannot see x-rays and we do not hear radio waves. Yet we know of their

Courtesy of Mount Wilson and Palomar Observatories

Fig. 7.3 Without the aid of instruments man's view of the universe is severely limited. The Hale telescope on Mount Wilson, California enables us to probe outward into space and backward in time. In the photograph above, notice the observer in the focus cage and the reflecting surface of the 200-inch mirror ahead of him. Microscopes, like the student research microscope below, which can magnify about 1200 times, enable us to view the micro-universe.

Courtesy of Leitz-Labolux

Courtesy of the American Museum of Natural History

Fig. 7.4 Above, the beautiful symmetry of a snow crystal, no two of which are exactly alike. Below, the equally symmetrical siliceous shell of a diatom, a single-celled alga.

Courtesy of the American Museum of Natural History

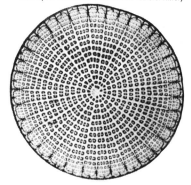

existence for we have instruments to detect them. These instruments become our extended senses: the telescope, microscope, and spectrometer are examples of such instruments. And in the hands of an imaginative scientist they can be used to enrich our lives every bit as much as the painter's brush or the writer's pen. The great scientist must be as much a creative artist as a Shakespeare or a Rembrandt. The great writings of science are permeated with imagination, enthusiasm, mystery, and humility.

We do not mean to imply that there is a clear and sharp separation between science and the arts. It is true that we generally think of science as dealing with phenomena lying outside of us, phenomena which we can measure, test, and experiment with. But we must not forget that each of us is an organism, and our emotions, sensations, desires and impulses, thoughts, and purposes arise out of our animal attributes. In their totality, these behavioral phenomena comprise in each of us an inner, private world from which we view and interpret the outside, measurable world.

We can think of science in two ways. It is, first of all, *a body of knowledge organized by men and reflecting the order and disorder we see in nature.* Gradually we have come to sense that there is a rule of order instead of chaos, a pattern of law instead of anarchy. We can see this strikingly revealed in the symmetry of crystals and diatoms. It can be just as beautifully demonstrated in the realm of ideas and discoveries. Our image of the origin of the universe, the Solar System, the Earth, and life itself has been laboriously and gradually built from ideas and discoveries of astronomers, geologists, physicists, chemists, and biologists over a period of many years. We can think of this approach—the gradual building up of a body of knowledge—as the conservative side of science, but there is another and equally important side.

Science is also an attitude. When no answers are immediately forthcoming, the scientist devises means of wresting answers from the unknown. This is the progressive side of science; it is what gives science its vigorous spirit of inquiry. It includes the priceless quality of wonder and curiosity that has understanding as its goal. In a way, the endless "what" questions of the wondering child are not so very different from the "what" and "how" questions of the scientist. It includes also the restless, endless pursuit of knowledge, even though it is difficult to know in advance whether a bit of knowledge will be "useful" or enlightening. Understanding is also a means to power. That is, power to control and use our environment to our advantage. Much of the financial support given to science is for the extension of this power: control of disease, use of space, improvement in the quality and quantity of food, and an understanding of ourselves are examples.

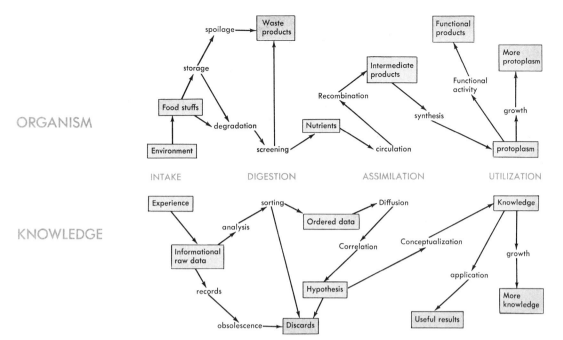

Fig. 7.5 A comparison, by means of analogy, between the growth of an organism and the growth of knowledge. An organism takes in food from its environment and converts part of it into the performance of its activities. A portion of the organism's food will also end up as waste products. Knowledge is "mental food," it is generated from experience, absorbed, digested, assimilated, and used. But some will never be used, and some will be incorrect. The organism—you—must decide what will be used and what will be discarded.

Why do we apply the term "progressive" to this aspect of science? There are two reasons. The quest for new ways of knowing is an advance against the unknown. Here man finds his greatest challenge, and his sight is ever forward. His searching ground is at the boundary of the known and the unknown. The second reason is that science is the only field of human activity in which progress is certain and inevitable. Civilizations rise, decay, and fall. But the high point of science is today, and tomorrow it will be still higher.

When we combine the conservative and progressive aspects of science, we find that science emerges as a process of growth similar to that taking place in an organism. This has been depicted by Paul Weiss, biologist at the Rockefeller University. On the organism side of our illustration (Fig. 7.5) the key to the chain of events is the organism. It takes from the environment what it needs and transforms matter and energy into more of itself. In this way it metabolizes, grows, and reproduces. On the side of knowledge, the key agent is not shown. It is the human mind, constantly observing, questioning, testing, and devising new tools and new techniques as it seeks new ways of knowing.

The Science of Biology

Biology is that part of science dealing with the matter and energy which is, or was, a part of a living system. The word comes from the Greek *bios*, which means "life," and *logos*, which means "word" or "thought." In short, biology is a body of knowledge of living systems. We often add adjectives or the prefix *bios* to describe narrower segments of the larger subject of biology: molecular biology, developmental biology, biochemistry, symbiosis, and so on.

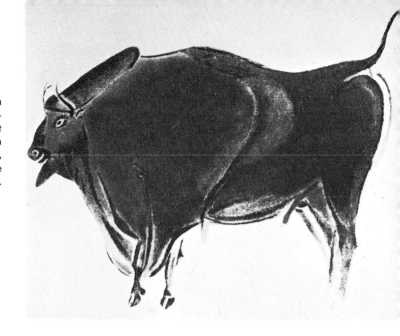

Fig. 7.6 A portion of a cave painting—a bison—made by prehistoric man at Altamira, Spain. The bison belongs to an extinct species, and the history of this race of artists is not fully known.

Biology is both an old and a vast science. Our earliest recorded biology comes from the ancient Greeks, but even prehistoric man left behind him in his caves beautiful drawings of animals. He seems to have had a sensitive awareness of proportion, anatomy, and motion. When man existed principally as a hunter, he practiced biology of a sort as he sought out his food, for he had to know the ways of the hunted animal, and the source of edible plants. When he turned from his nomad life to a more stable, agricultural existence, he had to have a greater knowledge of plants and animals before he could domesticate them sufficiently well to provide himself with a ready source of food. It is from these early beginnings that biology had its start.

The vastness of biology today can be seen by the various subdivisions into which the science has been broken. A look at any natural habitat will impress you with the great diversity of living forms that range from trees and large animals to organisms of microscopic size. Classification of organisms into convenient groups showing common similarities is **taxonomy**. Investigation of structure is **anatomy;** of function, **physiology;** of the details of cells, **cytology;** of inheritance from one generation to the next, **genetics;** of the relation of organisms to their environment, **ecology;** of disease, **pathology**. Biology, too, can be divided according to the particular group of organisms being studied: **entomology** (insects), **malacology** (shells), **mycology** (molds and mushrooms), **bacteriology, virology,** and **botany** (plants), and **zoology** (animals). The list can be greatly lengthened, depending on how narrowly one wishes to limit a given area of study.

Table 7-1 will give you some idea about subdivisions as one goes from the molecules of biological importance to the total living world, and as one views the particular subject in terms of

TABLE 7-1 The Science of Biology—Disciplines, Methods, Tools, Concepts

Level of Investigation	Disciplines	Methods	Tools—Equipment	Concepts
Molecule	Biochemistry Biochemical genetics Organic chemistry Immunology Enzymology Polymer chemistry Pharmacology	Biologically important, molecules—isolation, purification, analysis, synthesis, function. Use of radioactive tracers Determination of reaction rates X-ray diffraction Electron microscopy Fluorescent microscopy	Analytical balance Centrifuges Spectrometers Chromatographs Electrophoresis Counting devices for radioactivity X-ray sources Electron microscope Fluorescent microscope	Mechanism of enzyme action Source, storage and transfer of energy Source and transfer of information Control and inheritance of biosynthesis Antibody-antigen relations Molecular structure and function Action of drugs
Organelle	Cell biology Plant physiology Animal physiology	As above Cell fractionation Microdissection Staining procedures	As above Microscopes (various)	Organelles—formation, structure, function, interaction
Cell	Cell Biology Embryology Protozoology Microbiology Bacteriology Pathology	Fixation, sectioning, staining of cells Cell culture techniques Autoradiography Microdissection	Microscopes Microtomes Incubators Tracer counters	Mitosis, meiosis Haploidy vs diploidy Sex determination Differentiation, growth Host—parasite relations Cell death Classification
Organ	Anatomy Histology Physiology Pathology Embryology Pharmacology	Sectioning Staining Dissection Surgery Transplantation Transfusion Drug action	Microscopes Microtome Electrocardiograph Electroencephalograph Blood, bone banks Heart-lung machines	Comparative anatomy and physiology Action of drugs Action of hormones Replacement of organs Regeneration Problems of disease
Individual	Embryology Physiology Pathology Psychology Animal behavior Genetics Medicine (various) Taxonomy Physical Anthropology	As above Conditioning Psychoanalysis Inheritance tests Psychological tests Diets Hypnosis Learning	Psychological tests—IQ, aptitude, personality Breeding plots Statistics Twin studies	Homeostasis Consciousness Ideas of learning Individual differences Gene action; limits of variation Genotype vs phenotype Goals, purposes, free will Aging
Species	As above Evolution Paleontology Cultural anthropology Climatology Geography	Methods of classification Cytotaxonomy Statistical methods Population sampling	Herbaria Museums Microscopes Breeding plots	Natural selection Mutation rates Breeding structure Species formation Race concepts Ecotypes Fossil formation
Community	Ecology Population genetics Sociology Social Anthropology Group Psychology Evolution Environmental medicine Epidemiology Meteorology Geology	As above Sampling techniques Field observations Isolation techniques Determination of environmental variables	Control of environment—physical and biological Introduction of new variables	Cultural evolution Social structure and evolution Group reaction Environmental variables Population dynamics Species succession

structure, function, or history. We can also look upon biology as being subdivided by ideas (concepts) or by methods (tools and techniques). Try doing this as you proceed through this book. It will help you organize your thoughts and enable you to visualize the relation of one part of biology to all of the other parts.

Many of the terms in the table are unfamiliar to you now, but by referring to them as your knowledge of biology grows, you will begin to see that structure always has a history or an evolution, and that structure and function are intimately related to each other.

How does a science grow from the humble beginnings of casual or studied observation stemming from a need to know something? Two things are required to develop a science—facts and ideas. Men of prehistoric times had to rely on their unaided senses. They witnessed the slow passages of the seasons and associated with them the migrations of birds, the coming of plants into flower and fruit, and the cycle of birth, growth, death, and decay. Men gradually recognized that the great variety of life about them exhibited similarities as well as differences. All of these things had to be named and sorted out so that they could be handled conveniently and discussed intelligently. Imagine trying to keep track of your friends or enemies if they had no names. This would be the **taxonomic,** or classification, stage of biology. No experimentation is needed here, for the purpose is to group similar things into similar categories in order to form broad generalizations or patterns which reveal basic similarities in a mass of different phenomena. This activity is based on the assumption that there is an orderliness and a pattern in nature.

At this stage the question generally asked is "What is it?" or "What is it like?" Having answered the "what" question, it is then natural to ask, "What is it made of?" or "How is it put together?" Sometimes the "how" of the situation requires instruments (the microscope is a good example) and at this point the **structural stage** of biology begins. Here again, experimentation is not a necessary part of the science, for good description is often enough.

Proceeding onward, the more curious will inevitably ask, "What do the parts do?" This is a question of function (for which the biological term is **physiology**), and we can say that we have arrived at an **analytical stage** of scientific inquiry. Description alone is inadequate at this stage, and some kind of experimentation becomes necessary. If we are keen-minded, we begin to notice that structure and function go hand-in-hand. It is also important to recognize that there can be no appeal to authority at this stage or at any other stage of science. We must find out for ourselves. The mind must be open, curious, and questioning, and the experimenter is judge for the moment. History and the test of time will be the final judge of what is good or bad observation or experimentation.

Nowhere is this better illustrated than in one of the first, great experiments that ushered in a new era in biology. In 1628, William Harvey demonstrated the circulation of the blood and the function of the heart as a pump. His ideas, although correct, contradicted the great Greek physician, Galen (A.D. 131–201), whose ideas had prevailed for hundreds of years. Harvey rebelled and voiced his annoyance in these words:

> The blood is supposed to ooze through tiny pores in the septum of the heart from the right to the left ventricle, while the air is drawn from the lungs by the large pulmonary vein. According to this many little openings exist in the septum of the heart suited to the passage of blood. But, damn it, no such pores exist, nor can they be demonstrated.

Like most scientists, Harvey felt strongly about his work. But more important, he trusted his own careful observations.

The final stage in the solution of a problem generally involves the question: "What is the relation of this information to other bits of knowledge?" **Synthesis** must follow analysis. Although description and experimentation are not involved, synthesis, which attempts to tie things together, depends on them for basic information. It is also the most difficult stage of science and the least certain to produce answers. But when an answer emerges, synthesis is a truly creative step. The theory of evolution put forth by Charles Darwin is an example of this aspect of scientific study. It is not fundamentally different from the struggles of the poet who, in his way, also recreates experience. Synthesis is a tying together of information in order to answer questions such as "What are the functional requirements that demand this type of structure?" or "How is this structure related to its function?"

Probably science proceeded along such lines. Yet it would be wrong to suppose that each stage developed independently of the others, and that all stages developed in a 1, 2, 3 fashion. Furthermore, no stage is ever complete. Each generation tends to center its interests on some broad aspect of a science, and each generation reinterprets the past in order to make the past a useful part of the present. The Swedish botanist, Carolus Linnaeus (1707–78) established the rules of taxonomy in 1735 with his great book, *Systema Naturae,* and taxonomy continued to be a most important part of biology throughout the 18th and 19th Centuries. But taxonomy as a dynamic, rather than as a descriptive, part of biology had to wait for the 20th Century, a

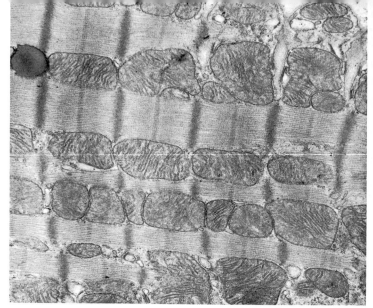

Courtesy of Emory University

Fig. 7.7 This electron micrograph of a section of heart muscle of a rabbit reveals the great amount of detail resolved by the electron microscope. Magnification here is 50,000 times.

time when the sciences of genetics, evolution, and biochemistry enriched it. The first half of the 1900's witnessed the beginnings of tissue and cell analysis, although the latter half of the 19th Century was known as the "Golden Age of Cytology" because of the many discoveries made then, but, once again, biochemistry and the development of the electron microscope have changed our science of cellular biology (Fig. 7.7). By the beginning of this century, the organ systems of plants and animals were generally understood in a structural and functional way. At present biologists tend to concentrate their attention on the structure and function of the many parts making up a cell— **molecular biology** as it is popularly called.

Scientific Method

Let us now turn the whole problem of biology around, assuming that we know a bit about a particular organism. Let us suppose that we wish to add to the present store of information. How do we do this? You have probably heard of something called the **scientific method.**

Unfortunately, this term is like a battered football; it has been kicked around and abused by a good many people. Actually, the definition is simple: it is what scientists do when they try to solve a problem. It can be a simple or a complex process; it varies with the scientist and with the problem. And its procedures are not rigidly set, despite attempts of many people to make it into a neat formula.

Let us consider a number of scientists and what they have done so that we may gain some idea of how they have contributed to our fund of knowledge. We shall find that these men have several things in common, but they followed no set formula in attempting to understand the world (Figs. 7.8, 7.9, 7.10).

Today we accept the idea that the Sun is the center of the Solar System, and that the Earth, together with the other planets, revolves around the Sun. This hypothesis was put forth in 1543 by Copernicus, a Polish astronomer, in a book published as he lay on his death bed. Certain Greek scientists had thought similarly, but a point of view, the Ptolemaic hypothesis, persisted and had dominated men's minds up to the time of Copernicus: namely, that the Earth was the center of the universe, and that the Earth and all that lived on it were created for Man's use and delight. It was a tight, small, comfortable, and orderly world. Each planet was thought to revolve in a perfect circle, and the fixed stars were in crystalline heavens, the home of the many gods of the Greeks—Mars, Jupiter, Venus, and others—who oversaw all things. Some of the words in our language come from the world of the Greeks: *jovial* (Jove) and *saturnine* (Saturn) tell us something of the character of these gods. *Disaster* originally meant "against the stars," and *exorbitant* meant being "out of regular orbit."

However, some of the astronomical events that could be observed, and that were needed for navigation and the prediction of the seasons, did not fit the neat and tidy picture of Ptolemy. Copernicus spent 30 years recalculating astronomical data and concluded that these data would equally well fit the idea that the Sun, not the Earth, was the center of things. Copernicus had not set out to change man's position in the scheme of things, but he did, and the world has not been the same since.

Charles Darwin was an equally tireless collector of information. An English country gentleman, he started out to be a doctor, then a clergyman, and finally became a naturalist. A trip on HMS *Beagle*, as ship's naturalist, took him over the world and brought him face to face with the enormous diversity of living things and their irregular patterns of distribution. He knew a good deal of biology and geology, and of the character and origin of fossil-bearing rocks. He also knew the practices of plant and animal breeders and the value of selection in changing the character of domesticated plants and animals. But all of this diversity made sense to him only after reading an essay by Thomas Malthus on how human populations were controlled in number by food supply, disease, famine, and war. Darwin's theory of evolution through natural selection came, therefore, only after a long period of observations and the tedious collection of vast amounts of information. Alfred Wallace, an animal

Courtesy of Burndy Library

Fig. 7.8 Above, Ptolemy (Claudius Ptolemaeus), Greek mathematician, astronomer, and geographer, whose idea of an Earth-centered universe, advanced in A.D. 127, was widely accepted until it was displaced by the idea of a Sun-centered Solar System, advanced in 1543 by Nicolaus Copernicus, a Polish astronomer.

Courtesy of Burndy Library

Courtesy of Burndy Library

Fig. 7.9 Charles Robert Darwin (1809–1882), an English naturalist, developed the theory of evolution that is accepted today.

Camera Press-Pix

Fig. 7.10 In 1953 James Watson (left), an American biologist, and Francis Crick (right), an English physical chemist, with Maurice Wilkins, an English crystallographer, discovered the structure of DNA, the crucial molecule of inheritance. A model of the molecule is in the background. These scientists were awarded the Nobel Prize for their discovery.

geographer, had virtually the same idea at the same time; but instead of putting his theory in the form of a book, as Darwin did, he sent Darwin a short note which contained the essence of Darwin's theory.

Our world today is a safer one to live in because of our knowledge of antibiotics, but the discovery of these substances was a matter of chance. As mentioned earlier, some of the bacterial cultures of Sir Alexander Fleming, the English microbiologist, became contaminated with mold, and some of the bacteria were killed. But instead of just considering this bad luck or poor technique, Fleming investigated the matter further, and the science of antibiotics was born. He treasured his exceptions and made the most of them.

The most important biological discovery in the last 25 years has been the discovery of the structure of DNA, deoxyribose nucleic acid, the molecule of inheritance in the vast majority of organisms. This knowledge has changed the whole character of biology. The significance of DNA, which we will deal with later, was recognized in 1944. Through the efforts of three men working in Cambridge, England—J. D. Watson (an American virologist), Francis Crick (an English chemist), and Maurice Wilkins (an English crystallographer)—a model of the DNA molecule was painstakingly built, which corresponded to all of the information gathered by them and many other scientists. They were awarded the Nobel Prize for their brilliant efforts.

These men had several things in common: they knew their science and the tools of their trade thoroughly, they valued sound data, they trusted their observations, they were not afraid of new ideas, and they had the courage to believe in their convictions. The great idea that ties odds and ends of miscellaneous information together into a meaningful whole can come in a flash, or only after a lifetime of work. Whenever it comes, it is the product of an alert and disciplined mind, a mind that can take advantage of ideas when they do appear.

The final step in the treatment of a problem is the process of making this information public. The scientist does this, usually, in writing, though he may on occasion simply present his ideas in lectures or discussions. The scientist has an obligation to communicate his findings; he cannot rightly shirk this responsibility, for he is, in a very real sense, a public servant. What is bad science is usually forgotten, but what is good becomes part of the great heritage of knowledge and understanding passed on to other generations.

Unfortunately, the more specialized the science the more technical and difficult the language, and the proportion of information that gets back into our common language is correspondingly small. Even so, you might be surprised at the number of technical words that have crept into common usage from the field of rocketry alone in recent years. How many can you think of? Where did you first hear of them? A word such as **feedback,** arising out of the fields of electronics and communications, often provides us with a new way of seeing a problem in our everyday experience. The extent to which such words and concepts find their way into common usage, thereby enriching our language, is one measure of public response to science.

Why Study Biology?

As human animals, all of us, individually and collectively, are part of the world, related directly to, or in association with, all other things, living and nonliving. A study of biology is a study of ourselves and of certain of those associations. Its goal is to help us know ourselves and the world we live in. We also want to know where we stand in the stream of time and in the immensities of space. Some 250 years ago, Blaise Pascal, the great French philosopher, said:

> "When I consider the short duration of my life, swallowed up in the eternity before and after, the little space which I fill, and even can see—engulfed in an infinite immensity of spaces of which I am ignorant, and which know me not, I am frightened, and am astonished at being here rather than there; for there is no reason why here rather than there, why now rather than then. The eternal silence of these infinite spaces frightens me."

But Pascal's despair was counterbalanced by a knowledge of his ability to think, and by a knowledge of his own limitations.

> "It is not in space that I should look to find my dignity, but rather in the ordering of my thought. I would gain nothing by owning territories: in point of space the universe embraces me and swallows me up like a mere point; in thought, I embrace the universe."

You, too, can embrace the universe in your thoughts.

SUMMARY

Science is an attempt by men to understand, and thereby to predict, the structure and behavior of themselves and the physical, chemical, and biological world in which we live. It does so through controlled and orderly observations, and through experiments designed to test our ideas about energy and matter—what it is, how it is put together, how it changes, and how simple forms of matter can be built up into complex forms. When we study matter that is living, or has been part of a living organism, we are studying biology.

Biology is a vast subject, ranging from the study of molecules to the study of populations composed of millions of individuals. It has grown from simple descriptions of living things to the present day emphasis on biochemistry, cell biology, genetics, growth and development, evolution, and population studies. This does not mean that other aspects of biology are unimportant. They are, and they will continue to be; but a concentration on specific aspects of biology reflect today's interests in biology.

FOR THOUGHT AND DISCUSSION

1 Develop a problem that you can solve; for example, a light switch fails to turn on your desk lamp, or the rear wheel of a bicycle is jammed. How do you go about finding out what is wrong? What hypotheses do you develop, and how do you test them? Outline every step of your thinking and action.

2 Think of an organism of some kind, say a pet animal. In scientific terms, how many ways can you think of describing this animal? Into what subsciences of biology would these descriptions fit?

3 Assume that a scientist has an idea that can drastically alter our way of life, if the idea proves to be correct. When does consideration of this problem cease to be scientific and become nonscientific? What do we mean by "nonscientific?"

4 Read a biography of some famous scientist, such as Pasteur, Galileo, or Darwin. From your knowledge of history and of science, consider the time in which he lived in relation to the problems he worked on, how he worked, and his contribution to our present knowledge. What advantages are available to a scientist today compared with the advantages of a scientist of Galileo's time, or a scientist who lived in the 19th Century, when Pasteur and Darwin lived?

5 Why did Copernicus' work change the world?

6 How would you decide if astrology, history, and economics were "scientific"?

7 Make a list of words common in everyday use which were not in the vocabulary of your parents when they were your age. From what sciences have these words come?

8 We have not tried to define "life" in this chapter. How would you define this term?

9 Consider some activity in which you are engaged; for example, riding a bicycle. You now want to explain the physical

and biological events and processes that make riding a bicycle possible. What activities can you name? Are they physical or biological? When, in this process of examination, can you say with assurance, "I can now understand what is involved in riding a bicycle"?

10 You are planning a week-end camping trip; you want to have a good time, new experiences, and return safely, and on time. What facts do you need to know, how do you plan for the unexpected, and how do you plan for your well-being and safety? How does such a trip differ from planning for a journey to the Moon?

11 We have stated that "understanding" is a form of power. What does this mean to you? Can you give concrete examples?

SELECTED READINGS

DuBos, R. *Pasteur and Modern Science.* New York: Doubleday, Science Study Series, 1960.
   A popular and readable account of the impact Pasteur has had on science today.

Fermi, Laura, and G. Bernardini. *Galileo and the Scientific Revolution.* Greenwich, Conn.: Premier Books, 1961.
   An account of Galileo's life and the times in which he worked.

Gourlie, Norah. *The Prince of Botanists: Carl Linnaeus.* London: H. F. & G. Witherby, 1953.
   A popular account of the life of one of the world's great botanists and scientists.

Johnson, W. H., and W. C. Steere. *This is Life: Essays in Modern Biology.* New York: Holt, Rinehart and Winston, 1962.
   A group of essays by eminent biologists on aspects of today's biology.

Mees, C. E. K. *The Path of Science.* New York: John Wiley & Sons, 1946.
   An account of the gradual growth of ideas in the physical and biological sciences, and how ideas changed with the passage of time.

Moment, G. (ed.). *Frontiers of Modern Biology.* Boston: Houghton Mifflin, 1962.
   Twenty essays on how each scientist views his own subject matter.

Poole, L., and G. Poole. *Scientists Who Changed the World.* Dodd, Mead, 1960.
   An account of scientists of the world, from the Greek Hippocrates to Einstein, and how their ideas have changed our thinking about ourselves and the world around us.

8 THE CELL

The world around us consists of matter and energy in many different forms. To deal with them in an orderly fashion we need to find those things that are common to all matter, and common to all energy. We need, in other words, to get down to *basic units,* and to reach some general agreement about what these units are.

When we deal with matter we say that the basic unit is the **atom,** and that atoms group together into **molecules.** Molecules, in turn, group together and form the objects and living things around us. Energy, too, has its own basic units. As our knowledge of such basic units increases, we can describe them more precisely. It is in this way that scientific language, common to all scientists, develops. Science makes use of two kinds of units. Those used to define time, mass, and distance are arbitrarily defined. Kilometers may be used instead of miles to measure distance, for example, or centimeters instead of inches. Other units, however, are world-wide in use. All scientists agree that atoms are the units of matter and of molecular construction even though no one has ever seen an atom. Mathematicians do not dispute the value of *pi* (3.1416). These units, like the others, have a physical reality, but they are not arbitrarily defined. Anyone having the proper instruments and knowledge can verify their existence, structure, and behavior.

The **cell** is such a unit (Fig. 8.1). It is the basic unit of all living things, except viruses, which may or may not be "living." We can break cells apart and extract fragments of them much as a physicist might break up an atom in a cyclotron. We find that these fragments, when separated from the cell, can carry on many of their activities for a time. They may consume oxygen, ferment sugars, and form new molecules. But these activities

Fig. 8.1 The cell is the basic unit of all living things. This electromicrograph shows a human leucocyte, white blood cell. The large nucleus, with its smaller, dark nucleolus, occupies the middle of the cell, with the less structured cytoplasm surrounding it. *Courtesy of Imperial Cancer Research Foundation*

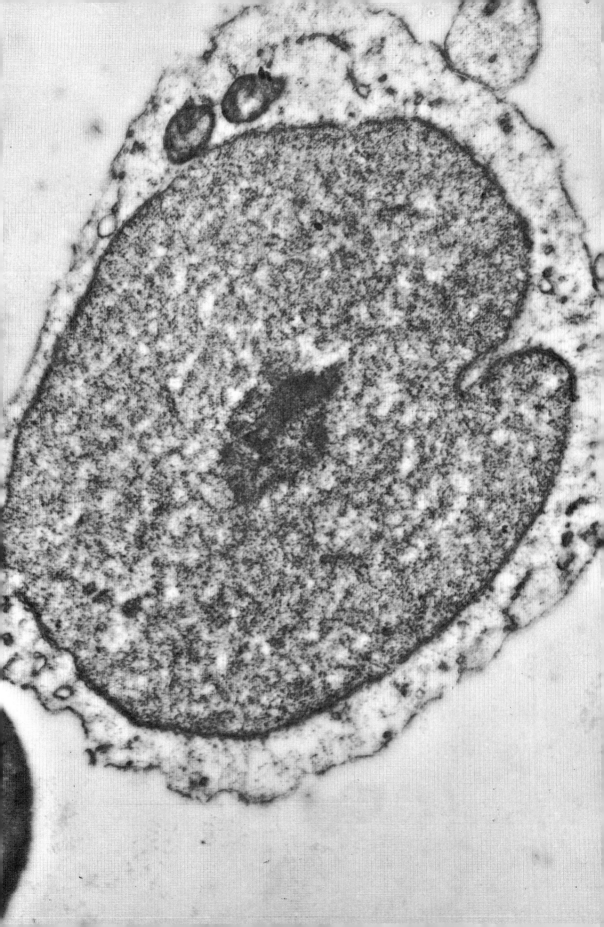

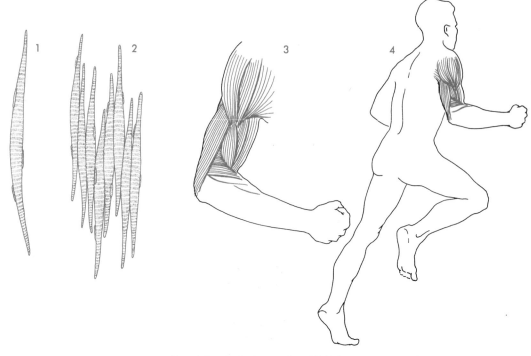

Fig. 8.2 A single muscle cell (1) becomes grouped with other similar cells into a cellular mass (2), which is part of the muscle tissue of the arm (3), which in turn is but one of many parts of the entire organism (4).

individually do not constitute "life" any more than a lone electron or a proton typifies the behavior of a whole atom. A disrupted cell can no longer continue life indefinitely, so we conclude that the cell is the simplest unit that can carry on life.

Compared to an atom or a molecule, a cell is very large and complex. It has a definite boundary within which chemical activity is going on constantly. In the area of **cytology** (the science of cells) one learns to recognize the many kinds of cells, and to understand their organization and structure in terms of their function. One also learns to visualize the cell not only as an individual unit, but also as a working member of a more elaborate organ or system of an entire plant or animal (Fig. 8.2).

How the Cell Concept Was Formed

The now familiar idea that the cell is the basic unit of life is known as the **cell theory.** The theory was formulated in 1839 by two German scientists, the botanist M. J. Schleiden and the zoologist Theodor Schwann. Today we regard the statement as one of the basic ideas of modern biology. We understand life itself only to the extent that we understand the structure and function of cells.

The general acceptance of a great truth such as the cell concept is a slow process. Every great thought has its origin in the mind of one man, but even when clearly stated, it is rarely accepted by everyone. The date 1839 and the names Schleiden and Schwann are not significant because these men discovered cells. They did not. Cells had been known since 1665, when the English scientist, Robert Hooke, first saw them in a piece of cork under his microscope (Fig. 8.3). It was Hooke who coined the word "cell" to describe the tiny structures he had observed. He thought they looked like the cells of a monastery in which the monks lived.

Also, Schleiden and Schwann were not the first biologists to believe in the idea that plants and animals were composed of cells and cell products. Many others had stated this clearly some years earlier. What they did was to associate a number of ideas and observations, and make a whole structure out of them. Theirs was an act of synthesis rather than of original discovery. By viewing the cell as both the structural and functional unit of organization, Schleiden and Schwann defined the basic unit of life. Because they were both prominent scientists of their time, they were more readily believed than were their predecessors.

Some 20 years later (1859), Rudolf Virchow, the great German physician, made another important statement: *cells come*

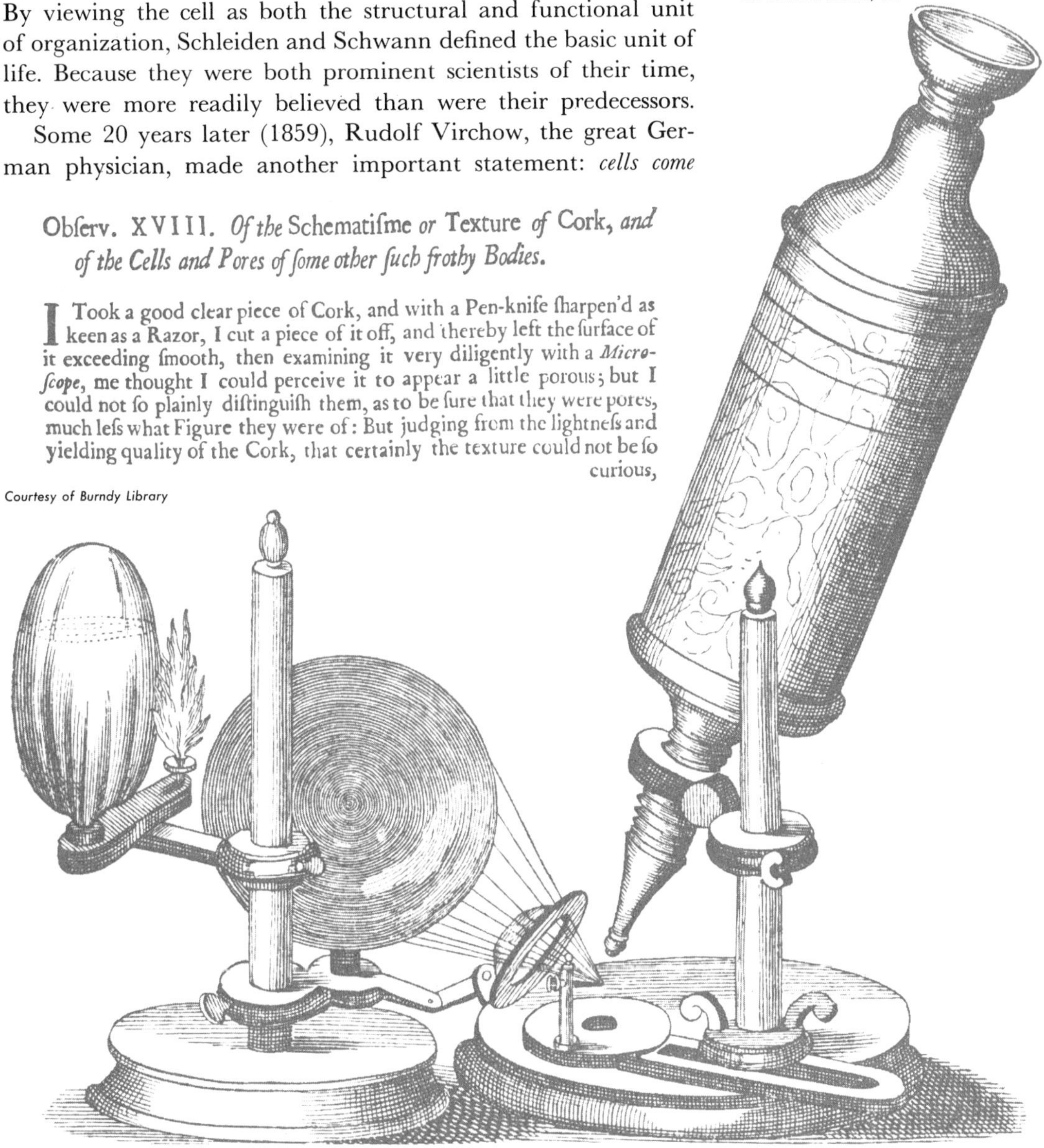

Fig. 8.3 With this primitive microscope and illuminating system, Robert Hooke observed cells in a plant tissue. Above the illuminating system is a portion of a page from his book *Micrographia*, in which he described his observations.

The Bettmann Archive, Inc.

Obſerv. XVIII. *Of the* Schematiſme *or* Texture *of* Cork, *and of the Cells and Pores of ſome other ſuch frothy Bodies.*

I Took a good clear piece of Cork, and with a Pen-knife ſharpen'd as keen as a Razor, I cut a piece of it off, and thereby left the ſurface of it exceeding ſmooth, then examining it very diligently with a *Microſcope*, me thought I could perceive it to appear a little porous; but I could not ſo plainly diſtinguiſh them, as to be ſure that they were pores, much leſs what Figure they were of: But judging from the lightneſs and yielding quality of the Cork, that certainly the texture could not be ſo curious,

Courtesy of Burndy Library

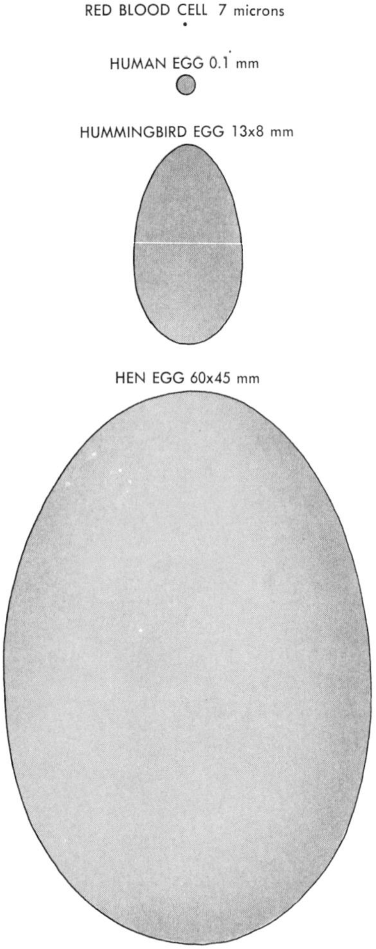

RED BLOOD CELL 7 microns

HUMAN EGG 0.1 mm

HUMMINGBIRD EGG 13x8 mm

HEN EGG 60x45 mm

Fig. 8.4 Relative sizes of four cells: The eggs of the humming bird and hen are about natural size; human egg and red blood cell are enlarged.

only from pre-existing cells. When biologists further recognized that sperm and eggs are also cells which unite with each other in the act of fertilization, it became clear that life from its earliest beginnings was a continuous and uninterrupted succession of cells. From the time, some two billion years ago, when life first developed on Earth, there has been an unbroken chain of cells leading up to the organisms alive today.

Cells Under the Microscope

The world of cells, except in a gross sense, lies beyond the limits of ordinary vision. Most cells are far too small to be seen with the naked eye. For example, the period at the end of this sentence is about 0.5 mm in diameter. It would cover about 25 to 50 cells of average diameter. Of those we can see, none of their details can be made out. We must, therefore, use instruments to extend our vision into the minute world of cells.

The 200-inch Hale telescope on Mount Palomar, California reaches across billions upon billions of miles of space, bringing distant galaxies of the universe into view. The cell biologist, however, needs to overcome the problem of small size rather

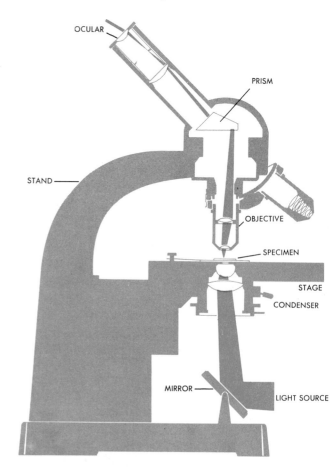

Fig. 8.5 The diagram at the right shows the optical system of a light microscope and the passage of light through the condenser, objective, prism, and ocular.

than that of great distance. Light microscopes and electron microscopes do this for him, thereby revealing otherwise invisible worlds. Magnification, therefore, is as much of a problem for the biologist as it is for the astronomer.

The problem of magnification is a problem of **resolving power**. An optical system has good resolving power if it enables us to distinguish objects—stars or parts of a cell—lying very close together. In reading these pages, for example, some people see each letter and word distinctly. Proper image formation is part of the problem, but others, with poorer resolving power, will see only a blur, and their resolving power and image formation have to be improved through the use of eyeglasses. Some people can see that the middle star in the handle of the Big Dipper is really a double star; others can see the two stars only with binoculars or a telescope. In a microscope, the resolving power of the **magnifying lens** is the critical factor. As Fig. 8.5 shows, the lens nearest the specimen being examined—the **objective**—is the key element of the compound microscope. The uppermost lens—the **ocular**—enlarges only what the objective lens has resolved.

The human eye is, of course, a crucial factor in making use of instruments. The unaided human eye with 20/20 vision has a resolving power of about 0.1 mm (Fig. 8.6). Lines closer than this appear as a single line. Objects that have a diameter smaller than about 0.1 mm are invisible or appear only as blurred images. Unlike the human eye, microscopes both resolve *and magnify*, but they cannot magnify that which they cannot resolve. We have the same problem in photography. A 35 mm negative can be printed into a huge picture, but if the negative is a blur because it did not resolve the object in question, the positive will also be a blur, no matter how we try to correct it.

The resolving power of a microscope is limited by the kind of illumination used. Objects that are less than one-half the wavelength of light apart cannot be distinguished in a light microscope. This is a fundamental law of optical physics. Since light has an average wavelength of 5,500 **Angstroms (A)**, even with the most perfectly ground lenses, the objective cannot resolve objects with a diameter less than 2,750 A (which is 0.275 **micron**, or 275 **millimicrons**. Micron is abbreviated μ, and a millimicron is abbreviated mμ.) Since many parts of biological systems are much smaller than 2,750 A, we could not detect them unless a means of greater resolution were available.

The **electron microscope** (Fig. 8.7) provides increased resolving power by making use of "illumination" of a different sort. High-speed electrons, which are parts of atoms, are used instead of light. As the electrons pass through a specimen being viewed,

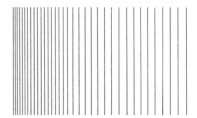

Fig. 8.6 Anyone with good vision can resolve each of these into individual lines. If you have poor resolution, the lines at the left begin to merge and become indistinct.

Fig. 8.7 In the optical system of the electron microscope, electrons instead of visible light provide a means of resolution and magnification.

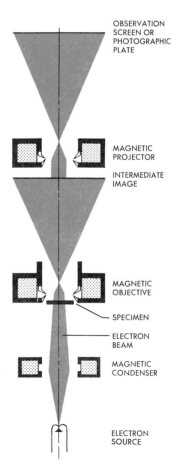

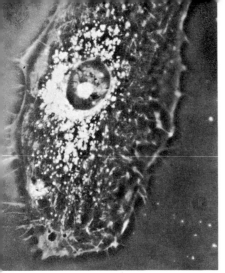

Courtesy Dr. George O. Gey

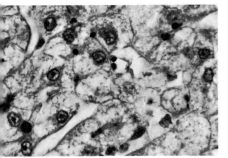

Courtesy E. Leitz

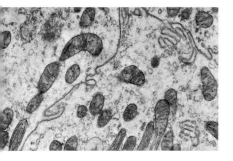

Fig. 8.8 Top, a human cancer cell growing in tissue culture and photographed in a living state by means of a phase-constant microscope, X100. Center, cells from the liver of a dog. The tissues were fixed, sectioned, and stained before being photographed—at about X1,000. Bottom, portions of three cells from the kidneys of a frog, revealing a richness of detail when photographed through an electron microscope, X38,000.

the more dense parts of the specimen deflect or absorb more electrons than less dense parts. This contrast forms an image of the specimen on an electron-sensitive photographic plate or fluorescent screen. The human eye, of course, is not stimulated by electrons, hence the need for plates or screens. The "optical" system of an electron microscope is similar to that in the light microscope, except that the "illumination" is focused by electromagnetic lenses instead of conventional glass lenses.

When electrons are propelled through the microscope by a charge of 50,000 volts, they have a wavelength of about 0.05 A. This is 100,000 times shorter than the average wavelength of light. An electron microscope can thus theoretically resolve objects with a diameter of one-half of 0.05 A, or 0.025 A. This dimension is far less than the diameter of an atom (the smallest atom—hydrogen—has a diameter of 1.06 A). However, due to difficulties in lens construction, the actual resolving power of a good modern instrument is about 10 A. In approximate figures, then, the human eye can resolve down to 100 μ, the light microscopes to 0.2 μ, and the electron microscope to 0.001 μ. Or, to put it another way, if the human eye has a resolving power of 1, that of the light microscope is 500, and that of the electron microscope 100,000. The electron microscope has opened up a whole new domain to the biologist.

Sometimes biologists need more than just clear resolution and high magnification. They also must be able to distinguish clearly between one small part of a cell and its immediate surroundings. The electron microscope can do this much better than the light microscope. To overcome this problem with the light microscope, we use killing agents **(fixatives)** and stains to bring out the parts we want to examine. Literally, hundreds of fixing and staining procedures are known (Fig. 8.8).

Another way biologists study a living substance is to grind it up and examine it. This is done with a special mortar and pestle so as to burst the cells and release their contents in solution. This solution is then centrifuged at carefully regulated speeds. The heavier material settles out at lower speeds, the lighter material at higher speeds. Most of the individual parts of cells can be collected in this way. Once separated, each portion can be analyzed for chemical content or tested for activity, since in the test tube some parts continue to function for a while as they did in the intact cell.

Biologists, then, have several ways of making a cell give up its secrets. Any one tool or technique, however, is not enough by itself. Usually, several methods must be used before an answer to a particular problem can be found.

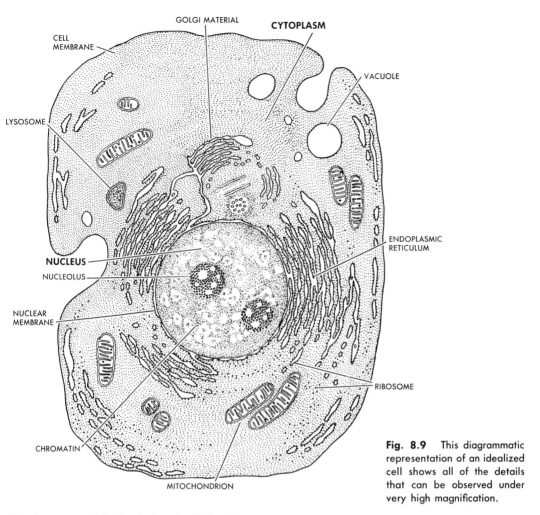

Fig. 8.9 This diagrammatic representation of an idealized cell shows all of the details that can be observed under very high magnification.

The Structure of Cells Under the Light Microscope

Let us now take a look at a cell under the light microscope so that we can become familiar with its parts. Later we will deal with the finer structure of the cell as revealed by the electron microscope. The cell we will describe now does not really exist, for there is no *typical cell*. Our cell is idealized for purposes of discussion (Fig. 8.9). We should remember that any cell in a living organism has a particular structure necessary for its activities, and it is, therefore, unique, not typical.

As the diagram shows, our idealized cell has two major components: a rounded central body, the **nucleus,** and a surrounding mass, the **cytoplasm.** Each plays an important role in the life of the cell.

The nucleus can best be seen in cells that have been stained. It is bounded on the outside by the **nuclear membrane.** The membrane is invisible in the light microscope, but the electron microscope shows it. Within the nucleus is the **chromatin,** which appears as a network of fine threads. There is also the **nucleolus,**

141

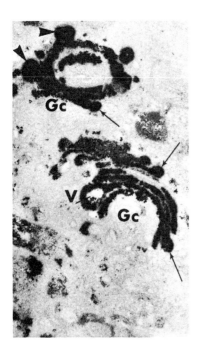

Fig. 8.10 The Golgi material of an animal shows up clearly when stained with a silver-containing dye and photographed through a light microscope. (See also Fig. 8.9.)

a more solid appearing body. The nuclear membrane does two things: (1) it separates the nucleus from the cytoplasm so that each is partially independent of the other; and (2) it permits the passage of some materials in and out of the nucleus while keeping other substances out. Such a membrane is said to be **selectively permeable,** a term which can be applied to any living membrane wherever it is found. The nuclear membrane, unlike the **plasma membrane,** has pores, but there is some doubt that they are free passageways.

The chromatin of the nucleus is a very special substance made up of **proteins** and **nucleic acids.** These are very large molecules which we will come to know intimately and which are responsible for determining the unique structure and function of each cell. We shall return to the nucleus time and again in this book. At the moment we need remember only that the nucleus controls the cell, and that the chromatin and the nucleolus are its essential materials.

If we punctured a cell with a microneedle, we would find that the cytoplasm would leak out. This suggests two things: (1) that cytoplasm is a watery substance; and (2) that it is bounded on the outside by a membrane—this is the **cell membrane** (also called the **plasma membrane**), which separates the cell contents from the outside environment. Like the nuclear membrane, the cell membrane is selectively permeable.

The fluid nature of the cytoplasm is somewhat misleading. Although it contains water, it contains other materials as well. When properly stained, or when viewed in the living condition in a phase-contrast microscope, it presents a varied appearance. We can see many granules in the form of rods, long filaments, or spheres. These are the **mitochondria** (singular, **mitochondrion**). Bacteria, blue-green algae, and a few other primitive organisms are the only cellular organisms lacking them. No human cell, except the red blood cell, is without them. A liver cell may have 1,000 or more mitochondria in its cytoplasm. The more active a cell is, the more mitochondria it is likely to have. This suggests that they play an important role in the life of the cell. This role, which we shall discuss in detail later, is concerned with the energy requirements of the cell.

If you wish to stain a cell to show the mitochondria particularly well, use **haematoxylin** on cells that have been previously fixed (killed in a fixative), or **Janus Green B** on living cells.

If a cell is stained with a silver-containing dye, a network of black substance becomes visible. This is the **Golgi material** (Fig. 8.10). Like mitochondria, Golgi materials (named after their discoverer) are found in most cells; it appears that they play a role in the aggregation and secretion of cell products.

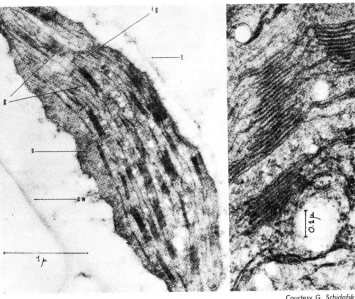

Fig. 8.11 At left is an electron micrograph of a chloroplast from a tobacco leaf cell, with the stacked lamellae (layers) of the grana, and the more open spaces of the stroma. At right, the layered grana are shown at a higher magnification. (g = grana; ig = intergrana space; s = stroma; t = membrane surrounding a vacuole; cw = cell wall.)

Cells, particularly plant cells, occasionally show large clear areas when they are stained. These are the **vacuoles.** As Fig. 9.4 shows, they force the cytoplasm into a thin layer against the cell membrane and keep the cell from collapsing. It appears that certain waste products are dumped into the vacuoles for indefinite storage. More recently they have been found to contain enzymes. A membrane enclosing the vacuolar material keeps the waste products away from the remainder of the cell.

Plant cells have two features that are generally lacking in animal cells. One is the rather thick **cell wall** which lies outside of the plasma membrane, and which serves as a supporting skeleton for the plant cell. The cell wall is made up largely of a substance called **cellulose,** which is of tremendous economic importance to man. It determines the strength and character of wood and paper. In the cotton plant, it provides us with the thread for our cotton fabrics, and it is the major ingredient in cellophane.

In the cytoplasm of green plants, and particularly in the cells of leaves, we find another structure—the **plastids.** They come in many shapes and sizes, each characteristic of a particular kind of plant. The number of plastids per cell can vary from one to a great many. Figure 8.11 (left) shows one kind of **chloroplast,** those plastids that contain the green pigment **chlorophyll.** Also, there are several kinds of storage plastids which may contain starch, fat, or protein, and chromoplasts containing pigments other than chlorophyll. The chloroplast is of particular importance. It is this structure which can absorb light energy from the Sun and convert it into chemical energy which the cell can then use.

The rest of the cytoplasm seems to be without visible structure, but this is merely because the light microscope cannot resolve the fine structure of membranes and particles which are present. When a growing cell is stained with **basic dyes,** the cytoplasm often shows a deep, rich color, indicating the presence of definite substances. This organization remained unknown until the greater resolving power of the electron microscope revealed a wealth of details. We now know that the cytoplasm is the main synthesizing part, or manufacturing plant, of the cell. Its fine structure will be discussed in later chapters when we can relate the cellular structure to particular chemical activity.

Possible Exceptions to the Cell Theory

For nearly every rule there is an exception. The cell theory, which is a rule of a sort, also has its exceptions. The fact is that the viruses do not fit into the cell theory scheme.

More than 300 different kinds of viruses are known. Many of them are the infective agents in such diseases as yellow fever, rabies, poliomyelitis, small pox, mumps, and measles in humans; and peach yellows and tobacco mosaic disease in plants. Some viruses appear to be the cause of some kinds of cancer. Those that infect plants tend to be elongated structures. The viruses

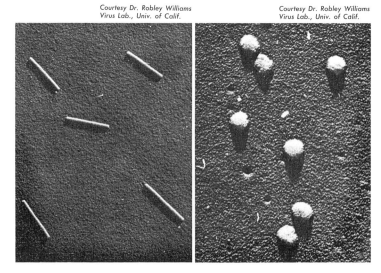

Fig. 8.12 Two different kinds of viruses are shown here: at right, the sphere-shaped influenza virus; at left, the rod-shaped tobacco mosaic virus.

Courtesy Dr. Robley Williams Virus Lab., Univ. of Calif.

Courtesy Dr. Robley Williams Virus Lab., Univ. of Calif.

that infect animals tend to be spherical (Fig. 8.12). If we now apply our usual definition of a cell, viruses do not qualify as living organisms. They lack the internal organization normally considered indispensable to a functioning cell. A distinction cannot be made between nucleus and cytoplasm. Although they contain the nucleic acids characteristic of nuclei, they seem to be totally lacking in cytoplasm.

When they exist outside a living cell, viruses behave as inactive molecules, although very elaborate and complex ones. Inside a cell, however, they act as parasites and display the usual characteristics of life: they multiply and produce exact replicas of themselves. They also have a type of inheritance not too different from our own. Viruses also contain the key molecules of protein and nucleic acid invariably found in every living organism.

Their doubtful nature has led biologists to describe them in various ways: as living chemicals, cellular forms that have degenerated by adopting a life of parasitism, or primitive organisms that have not reached a cellular state. Fortunately, we are not forced to decide whether a virus is or is not a cell, or even whether it is living or nonliving. We generally treat them as if they were individual cells. Their extreme simplicity of structure, when compared with a normal cell, makes them ideal objects for certain types of biological research.

Certain of the less complex forms of life, such as the protozoa, algae, and fungi are also difficult to fit into the plan that the cell is the basic unit of life. The protozoan *Paramecium* is seemingly a single cell, but it has a mouth, or gullet, vacuoles for the elimination of water and waste, other vacuoles for digestion, and many **cilia** (fine surface hairs) for moving about. Is *Paramecium* a true cell, or not? It is a difficult question to answer.

The same thing is true for certain algae such as *Valonia*, or for fungi such as the black bread mold (Fig. 8.13). They are simply a mass of cytoplasm containing many nuclei, and are bound by a continuous outer retaining wall. It would be difficult to define the basic unit of such living bodies. Are they single-celled, or multi-celled? These organisms, however, are related to more conventional cellular forms so we can speculate that they have simply lost the usual type of cellular organization and have acquired one that is mechanically better suited to their mode of existence.

The Structure of Cells Under the Electron Microscope

Let us now examine an animal cell with an electron microscope, remembering that with this instrument we can see details well below the limits of resolution of the light

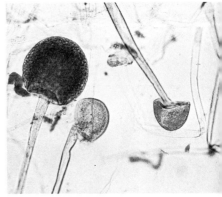

The Bergman Associates

Fig. 8.13 There are no typical cells making up the body of this fungus, the black bread mold, *Aspergillus niger*.

Fig. 8.14 A high magnification electron micrograph of an animal cell reveals detail that is not visible in a light microscope.

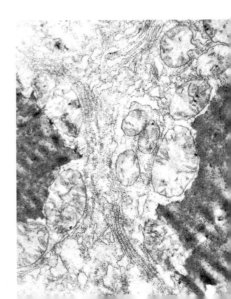

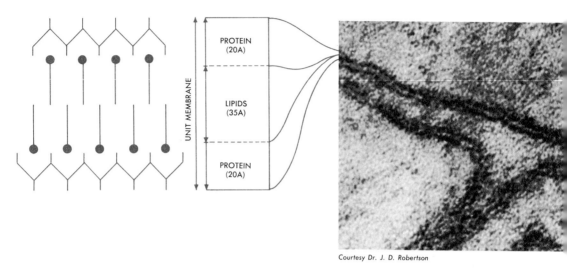

Courtesy Dr. J. D. Robertson

Fig. 8.15 At right is an electron micrograph of cell membranes at the juncture of three cells. At left is an interpretation of the cell membrane in molecular terms. The paired dark layers in the electron micrograph consist of proteins; the light layer sandwiched in between is made up of lipids (fats). The three layers together constitute a unit membrane.

microscope. The photograph of the animal cell (Fig. 8.14) shows such a cell greatly magnified. Our first impression is that of a great richness of structures. We will want to examine each one, but before we do we need to become acquainted with the fact that these structures are composed of atoms and molecules.

The atoms are mainly carbon, oxygen, hydrogen, and nitrogen, along with sulfur and phosphorus. These are the principal atoms commonly found in living structures. The molecules formed from these atoms are of four major classes, and all are distinguished by the fact that they are **macromolecules.** That is, compared with ordinary small molecules such as water (H_2O), carbon dioxide (CO_2), and common table salt (NaCl), they are enormous structures. They are usually elongated molecules, although some are globular, and are composed of thousands of atoms hooked together, or **bonded,** in a particular way. The four molecules are as follows: (1) **proteins,** each made up of amino acids strung together like beads on a chain; (2) **polysaccharides,** constructed similarly but of smaller sugar molecules; (3) **lipids,** made up in large part of fatty acids; and (4) **nucleic acids** constructed out of several different kinds of smaller molecules. We need not be concerned now with their chemical structures, but we need to be familiar with their names and where they are found.

As we make our way from the outside of the cell toward its middle, the first structure we encounter is the cell membrane. This was invisible in the light microscope, but the electron microscope reveals its structure clearly. When we view it under high magnification, we can see three distinct layers. The outer and inner dark layers are separated by a light region in a sandwich-like arrangement. From a variety of experiments, we now believe that the dark layers are protein, while the central, light area consists of two layers of lipid molecules. Figure 8.15 shows how the layers are arranged.

Why do biologists believe that this is the structure of a cell membrane? If we suspend red blood cells in water, they swell and soon release their contents (hemoglobin) into solution. We can in this way obtain a mass of pure cell membranes and analyze them. They are called "ghosts." We can then separate the proteins and lipids. When digested, or broken down, the proteins yield amino acids, the lipids fatty acids. There are small amounts of other molecules but they are unimportant to our present consideration.

But how do we know that there are two layers of lipids between the outer and inner protein coats? Suppose that we have a cell whose surface area on the outside is $100\ \mu^2$. We now extract the lipids from the cell membrane and allow them to spread out on the surface of water in a dish. Like a drop of oil, these molecules will form a thin film one molecule thick. If there is one layer of lipids, an area $100\ \mu^2$ will be covered. If there are two layers, $200\ \mu^2$ of water will be covered; three layers, $300\ \mu^2$, and so on. It turns out that the lipids from one cell will cover

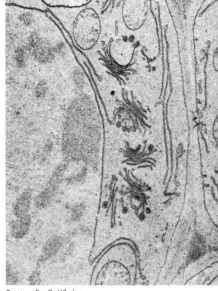

Courtesy Dr. G. Whaley

Fig. 8.16 An electron micrograph of plant cells shows the heavy cell wall, the long membranes of the endoplasmic reticulum, the stacked membranes of the Golgi materials, and several mitochondria.

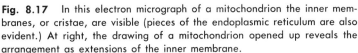

Fig. 8.17 In this electron micrograph of a mitochondrion the inner membranes, or cristae, are visible (pieces of the endoplasmic reticulum are also evident.) At right, the drawing of a mitochondrion opened up reveals the arrangement as extensions of the inner membrane.

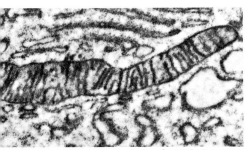

Courtesy Dr. B. L. Munger

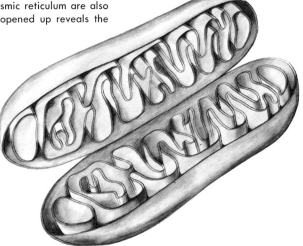

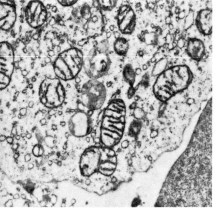

Courtesy Imperial Cancer Research Fund

Fig. 8.18 Many ribosomes, free in the cytoplasm, are revealed in this electron micrograph of part of a human white blood cell. Several rounded mitochondria are also visible.

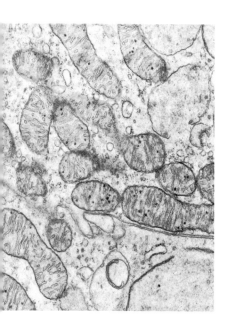

Fig. 8.19 This electron micrograph of part of a kidney cell of a frog shows the elongated mitochondria characteristic of this kind of tissue.

twice the area of the cell. Two layers of lipids are, therefore, sandwiched between the inner and outer protein layers.

We have spent this much time on the cell membrane because all membranes of the cell, and there are many kinds, are of a similar nature. Furthermore, knowledge of the cell membrane permits us to understand the passage of substances in and out of the cell, a topic of much importance to be discussed later.

As we continue our exploration of the cell, we come to the cytoplasm and find that it is not just a watery fluid. It is filled with membranes and particles. The long membranes running through the cytoplasm are conspicuous. In some cells they are quite numerous, filling the cell with many channels. In others, membranes are sparse, or even absent. These are the membranes of the **endoplasmic reticulum** (inside network). In some cells, these membranes are studded with small bodies called **ribosomes** (Fig. 8.18); in other cells the membranes are smooth (Fig. 8.20).

The endoplasmic reticulum is believed to be the manufacturing, packaging, and transporting portion of the cytoplasm; it is the assembly line of the cell. The nature of the endoplasmic reticulum determines in part what the cell does. When ribosomes are present, as they are in virtually all cells, they are there for the purpose of making proteins. When the endoplasmic reticulum is free of ribosomes, some other substance is being made by the cell.

Another series of membranes, similar to the endoplasmic reticulum, but recognizably different, is the **Golgi material.** Compare the photograph of Golgi material on page 142, taken with the light microscope, with Fig. 8.16, taken with the electron microscope. The difference in detail is striking. Actually, the Golgi material is like a group of balloons flattened together. They do not have ribosomes attached to them, and consequently are not concerned with protein formation. They seem, rather, to accumulate certain materials such as fats or mucus and then move them out of the cell. One might think of Golgi material as a place in the cell for packaging molecules for export.

The ribosomes, whether numerous or few in number, are about 250 A in diameter. A bacterial cell of 1 μ diameter may have as many as 30,000 ribosomes. Often they are grouped in clusters of five to eight. In this state they are known as **polyribosomes,** or simply **polysomes.** Again, they are concerned with the synthesis of proteins.

The **mitochondria** appear complex when viewed in the electron microscope. They have both an outer and inner membrane. The inner membrane is thrown into folds called **cristae,** which give a greatly increased surface area (Fig. 8.17). A great amount of research has been done on these structures, for the mitochondrion is of vital importance to the cell. It is the chief source of usable energy. Whenever the cell does work of any kind, the mitochondrion (and, in plant cells, the chloroplast) supplies it with the necessary energy.

A structure similar to the mitochondrion is the **lysosome.** It is different in appearance in that it does not have cristae. It is also different in function. Its main task in the economy of the cell is to break large molecules of food materials into smaller ones, which are then passed on to the mitochondria to serve as energy sources. However, if the lysosomes are injured, their destructive action can cause digestion and death of the cell.

The nucleus is the last of the structures we shall view in our cell. Surrounding it is a double membrane, its two parts often separated from each other by a **perinuclear** space. The outer membrane often has pores and is connected with the endoplasmic reticulum (Fig. 8.20). Just as the cell membrane is the barrier separating the cell from the outside environment, so the nuclear membrane separates the nuclear contents from the cytoplasm. It differs from the plasma membrane, however, particularly in behavior, for it can disappear and reform during cell division. All substances passing in and out of the nucleus must pass through the nuclear membrane. Except for the nucleolus, the nuclear contents show very little defined structure. This is because the chromatin, the major nuclear component, is widely dispersed.

A plant cell differs from an animal cell in three major ways. First, there is usually a heavy wall just outside of the cell membrane. This is a major supporting structure, and it is that part of a cell that gives "character" to the different kinds of wood. Such heavy walls are lacking in animal cells. Second, the center of a plant cell is generally occupied by a water-filled sac, or **vacuole.** Covered by a membrane called the **tonoplast,** the vacuole pushes the cytoplasm toward the outer edges of the cell where a ready exchange of gases can take place. The vacuole may be filled with colored pigments—for example, the red pigment, anthocyanin, of the beet cell. Third, the plant cell may contain structures known as **plastids.** These, as indicated, are of great importance, for they capture the energy of sunlight for use in the manufacture of sugars. The structure of such a plastid is very complex (Fig. 8.11). The many layers of its membranes are often stacked into **grana;** and the green pigment chlorophyll, which traps the energy of sunlight, is layered on these membranes. Each plastid, therefore, has a tremendous amount of surface area, and is a most efficient energy trap. Other kinds of plastids are used for storage. Some store starch, others protein, and others fats. Each differs slightly from the other in appearance.

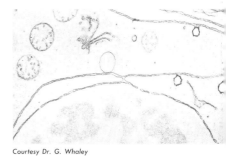

Courtesy Dr. G. Whaley

Fig. 8.20 The double membrane (with its pores) of the nucleus, the endoplasmic reticulum connected (connected to the membrane as well as free in the cytoplasm), the stacked membranes of the Golgi materials, and several round mitochondria are visible in this electron micrograph showing part of a plant cell.

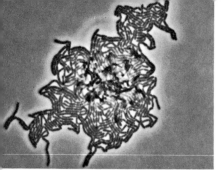

E. Leitz, Inc.

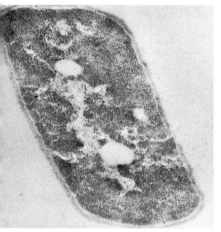

Courtesy Dr. C. Robinow

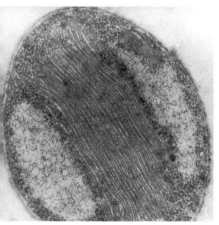

Courtesy Dr. S. Watson

Fig. 8.21 Top: a low-power photomicrograph of a cluster of bacterial cells. Middle: a *Bacillus subtilis*, showing its dense cytoplasm (which contains the bacterium ribosomes), and the less dense nuclear areas. Below: another electron micrograph shows the elaborate membrane system, in *Nitrosocystis Oceanus*, a marine bacterium.

The Bacterial Cell

Bacteria of many kinds abound in the world. Some are disease-producing organisms that give rise to a variety of infections, such as tuberculosis. Others are useful and, indeed, necessary, such as those that break down waste organic materials or those that make vinegar. All bacteria are cellular organisms, but their structure is different from that of ordinary plant or animal cells. As Fig. 8.21 shows, a bacterial cell has a cell membrane, and a cell wall may lie outside the membrane. Although ribosomes are present, there are no plastids, no mitochondria, and no nuclear membrane. Some bacteria, however, have elaborate membrane systems. The functions carried on by these structures in more complex cells are not missing in the bacterial cell. They are performed in the cytoplasm, or at the surface of the plasma membrane, with less highly organized structures.

SUMMARY

Except for viruses, all organisms are composed of **cells,** or cell products. Cells arise only by the division of another cell. All organisms are, therefore, related to each other through **cellular descent.**

Cells are so small that we can see their details only with the aid of a microscope. With a light microscope, we can see that cells consist of three main parts: (1) a **nucleus;** (2) **cytoplasm;** and (3) a **cell membrane.**

The nucleus contains **chromatin** and a **nucleolus,** which are enclosed within a nuclear membrane. The cytoplasm contains **mitochondria, Golgi material,** and **vacuoles.** These individual parts may be seen well only when they are fixed and stained before being viewed through a microscope. An electron microscope reveals many more details than a light microscope does, particularly in the cytoplasm. Ribosomes, endoplasmic reticulum, and lysosomes show up well. Also, the detailed structure of membranes, the mitochondria, and Golgi material can be seen.

Plant cells differ from animal cells in that plant cells have a heavy cell wall and plastids in addition to the structures found in animal cells. Bacteria differ from both plant and animal cells by lacking a nuclear membrane, mitochondria, and plastids.

FOR THOUGHT AND DISCUSSION

1 Imagine a spherical cell having a radius of 30 microns. What is the radius in millimeters? Millimicrons? Angstroms? What is its surface area and volume in these units of measure?

2 Cells have often been compared to the bricks in a house. Do you think this is a reasonable comparison? What are your reasons for your conclusions?

3 A cell is a miniature factory: it either performs a task or makes a product. From what you know of a cell, and of a factory, draw up a list of comparisons for sources of energy, steps carried out by particular parts, and tasks performed, or products made. Can you point out where you need more information in order to understand the functioning of a cell?

4 Animal cells, in contrast to plant cells, have no cell walls. How does an insect, or crab, or dog keep its shape and move in a purposeful manner? Why doesn't it collapse?

5 Two spherical cells, one having twice the diameter of the other, have the same metabolic rate (metabolism per unit mass). Let us assume that oxygen limits the rate of metabolism. If the concentration of oxygen outside of the smaller cell is 20 per cent, what must the concentration be outside the larger cell in order to maintain the same metabolic rate?

6 You want to examine two similar cells, one by light microscopy and the other by electron microscopy. Both are spherical and 30 μ in diameter, and both are to be sectioned for study, with a 5 μ thickness for light microscopy, and a 150 A thickness for electron microscopy. How many sections of each do you need to include the entire cell?

SELECTED READINGS

Butler, J. A. V. *Inside the Living Cell.* New York: Basic Books, 1959.
   A nontechnical account of cell structure and function, and the cellular aspects of radiation, chemicals, cancer and aging.

Butler, J. A. V. *The Life of the Cell.* New York: Basic Books, 1964.
   A nontechnical account of how we have come to understand the secrets of life at the cellular level.

Jensen, W. *The Plant Cell.* Belmont, California: Wadsworth, Foundations of Botany Series, 1964.
   A college level treatment of the plant cell in terms of structure and function.

Kennedy, D., (ed.). *The Living Cell.* San Francisco: W. H. Freeman and Sons, 1965.
   A collection of *Scientific American* articles on cell structure and function, with the editor providing a preface to each article.

Swanson, C. P. *The Cell* (2nd ed.). Englewood Cliffs, N.J.: Prentice-Hall, Inc., Foundations of Modern Biology Series, 1964.
   Chapters 1 through 4 deal with cell structure and function in a general way, and consider in some detail the problems of dimension in the light and electron microscopes.

9 GENERAL CELL FEATURES

It must be apparent to you that all cells do not perform the same kinds of work. Muscle cells contract and relax, nerve cells send messages to various parts of the body, cells of the eye receive impressions of the outside world and transmit them to the brain, while cells of the stomach and intestine play a role in digestion. Cells become specialized and do different kinds of work, just as individual members of our society do particular jobs and so keep a community as a whole functioning in a coordinated way.

When an organism consists only of a single cell, that cell must carry on all of the functions required by the organism to live, grow, and reproduce. We find such single-celled, or **unicellular,** organisms among the bacteria, protozoa, and algae. They differ from each other in appearance, fine structure, and behavior because each lives a somewhat different existence. Other organisms are multicellular, some consisting of relatively few cells.

Fig. 9.1 A single fertilized egg cell gives rise to countless billions of other cells, most of them specialized in the performance of certain functions. Shown here are five stages in the development of a human being: fertilized egg; two-cell stage; 16-cell stage; embryo with the three cell layers that give rise, respectively, to the brain, heart, and lungs. These three layers will also form all of the other tissues of the body.

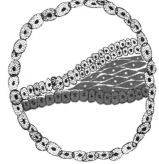

Some, such as man, have billions of cells. Each cell, or group of similar cells, has a special function and a special structure which enables it to perform its duties.

In this chapter we want to discuss general cellular features: size, shape, number, and the length of time each cell lives. We want also to determine, as well as we can, what it is that determines these features of the cell. As we shall see, this is not an easy task. Definite answers are not always readily available.

Cell Size

By examining cells of plant and animal origin, we can see that their sizes vary widely. The smallest cells we know of belong to a group (called **pathogens**) that causes respiratory diseases. These are the pleuro-pneumonia-like organisms (abbreviated to **PPLO**). Their diameter is about 0.1 μ, which means that they can be seen only with the aid of an electron microscope. Louis Pasteur, the great French bacteriologist, knew of them nearly a hundred years ago, but it is only recently that biologists have studied them intensively. Pasteur simply could not find them, even though he recognized their disease-causing properties. Their size places the PPLO's among the viruses. But unlike the viruses, which can grow only within another cell, the PPLO's grow readily in a test tube when given the proper nutrition. Their structure is simple, and not unlike that of the bacteria. Compare the photograph of the bacterial cell on page 150 with the drawing of a PPLO in Fig. 9.2. Within this tiny and simply-constructed cell, all of the functions of life are carried out.

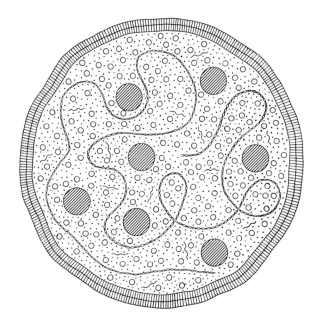

Fig. 9.2 The smallest cell we know of is a pleuro-pneumonia-like organism or PPLO. A membrane surrounds the cell, which contains the hereditary material (long thread structure), ribosomes (large spheres), and molecules in solution (smaller structures in the cytoplasm).

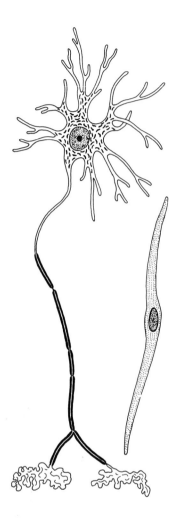

Fig. 9.3 Two relatively long animal cells. Left: a nerve cell with its many-branched projections from the cell body, and the long axon terminating at points where it could make contact with a muscle. The axon may reach a meter in length. Right: a smooth muscle cell such as might be found in the wall of the intestine.

How small can a cell be? Clearly, it must have a diameter greater than 0.02 μ, or 200 A, because the membrane surrounding the cell must have a width of at least 100 A. But how much room is needed for the cytoplasm? (You will notice that the PPLO's, like the bacteria, have no definitive nucleus.) We do not really know. About all we can say is that the smallest cell we know of is about 0.1 μ in diameter. Taking away 200 A for the cell membrane, this leaves a sphere 800 A in diameter for the remaining contents. Within the material filling this remarkably tiny space life expresses itself. Here must be contained all of the matter and forms of energy which we call "life." This minute amount of cellular material grows, reproduces more of its own kind, and undergoes variation or change.

The many forms of bacteria range in size from 0.2 to 5.0 μ in diameter. The smallest, therefore, are just at the limits of resolution of a good light microscope. The largest cell known is the ostrich egg, if we consider volume alone. It measures about 15 cm. around the outside, and about 7.5 cm. when the shell is removed. If we compare such a cell with the smallest bacterium, the ratio of linear dimensions would be about 75,000:1. The ratio of their volumes would be $75,000^3:1$. To put this in terms of easy reference, the order of difference is about the same as that between a sphere one inch in diameter and one that is more than a mile wide.

The range of cell size in the human body extends from the small leucocyte (white blood cell), which has a diameter of 5 to 6 μ, to a nerve cell, which may be more than a meter in length. This is a difference of about 300,000:1. Such a comparison is misleading, however, for the *main body* of the nerve cell is not nearly so different in size (100 μ) from that of the small leucocyte. When computed on this basis, the ratio is only about 20:1. Even this comparison has little meaning without a consideration of other factors governing size. A hen's egg, for instance, has outside dimensions of 60 × 45 mm. Like the ostrich egg, it is large because it contains food stored in the form of an enormous yolk, food which will be used for the growth of the developing embryo over a three-week period. A human embryo, on the other hand, draws its nutrition from its mother during development, so the human egg (about 0.1 mm. in diameter) need not be adapted for the storage of a large amount of food.

The different sizes of nerve, muscle, and blood cells reflect the particular task they do. With cells of similar function, we find that three things tend to govern size: (1) the ratio of the amount of nuclear material to the amount of cytoplasmic material; (2) the ratio of cell surface area to cell volume; and (3) the rate at which the cell carries on its many chemical activities. Although all three are related to each other, we shall consider them separately.

Let us first accept the fact, which we shall prove later, that the nucleus is the control center of the cell. In cooperation with the cytoplasm, the nucleus regulates the growth, development, and continued existence of the cell. Although a cell can function for a time without a nucleus, it eventually runs down and ceases to operate. The nucleus, however, cannot extend its control over too large an amount of cytoplasm. As the cell enlarges, the surface area of the nucleus, across which an interchange of materials must pass, increases only as the square of the nuclear radius (s.a. $= 4\pi r^2$) while the volume of the cell increases as the cube of the cell radius ($v = 4/3\pi r^3$). Too large an amount of cytoplasm would soon put some part of the cell far removed from nuclear control. The nucleus can increase its surface area by changing its shape, or by doubling its amount of chromatin, the main component of the nucleus. Adjustments can, therefore, be made. Most mature cells, however, seem to maintain a relatively constant nucleo-cytoplasmic ratio, and we find that growing cells do not vary greatly in size.

The amount of surface area tends also to limit a cell's size. As we saw earlier, chemical activity occurs continuously throughout the cell mass, and the various substances the cell needs in order to carry on this activity must pass through the surface membranes. Oxygen, for example, is required by most cells. If sufficient oxygen is to reach those parts of the cell where it is needed, its concentration outside of the cell must be at or above a certain level. The particular concentration is related to the rate at which oxygen moves into the cell, the rate at which oxygen is being used, and the dimensions of the cell. A cell that is too large has difficulty in getting oxygen to its more inaccessible parts.

The surface-volume limitations can be overcome in a variety of ways. Cells can change their spherical shape by being flattened, folded, or elongated. These changes either increase the surface area without increasing the bulk of the cell, or they bring the cellular contents nearer to the membrane surfaces. If the surface area is increased, the flow of materials in and out of the cell is also increased, and the cell can consequently enlarge without changing the rate of its chemical activities, called **metabolism.** This enlargement can go on as long as the expansion does not become so great that the nucleus can no longer exert control over the cytoplasm.

The surface-volume problem appears over and over again in living organisms, and it is met and solved in various ways. A nerve cell, for example, may reach a meter in length, but the

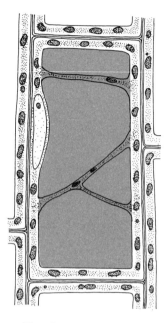

Fig. 9.4 Diagram of a plant cell with the heavy cell wall, the nucleus and chloroplasts in the cytoplasm, and the center occupied by a large vacuole. Strands of cytoplasm cross the vacuole.

Fig. 9.5 Here is a section of an intestinal wall with its convoluted surface, or *villi* (left), and with two of the absorbing cells at the right.

main body of the cell is not especially large. The **axons** and **dendrites** of the nerve cell are of such small diameter that an exchange of materials across their membranes can readily take place.

The individual plant cell, with its rather rigid cell wall, has solved the surface-volume problem in yet another way. It has a large liquid-filled vacuole in the center (Fig. 9.4). This pushes the cytoplasm to the outside so that a rapid exchange of materials is possible. In addition, the contents of the cytoplasm are kept in constant motion, thus preventing any portion of it from becoming stagnant through lack of oxygen or nutrients.

For cells that cannot exceed a certain size, an increase in number rather than size is the only solution. An organ of the body (as opposed to a single cell) has more elaborate means to meet the demands made on it. In mammals, for example, the digestive system is a long coiled tube, and its function is to digest and absorb the food that is eaten. To increase its ability to absorb, the lining of the intestine is composed of many folded surfaces arranged much as the piling of a bath towel (Fig. 9.5). Also, the surface of each of the absorbing cells is similarly folded so that its absorbing surface is greatly enlarged (Fig. 9.6). The inner surface of the lungs is also designed to make the rapid exchange of oxygen and carbon dioxide possible. Each cell is in contact with a blood vessel on one side and air on the other. Oxygen can readily pass into the blood, and carbon dioxide can easily move out.

The third condition that affects the size of a cell is the rate of chemical activity carried on inside it. It is tempting to say that the smaller a cell or an organism is, the higher its rate of chemical activity. Although cell size is not absolutely correlated with the rate of metabolism, the rapidly metabolizing cells of such

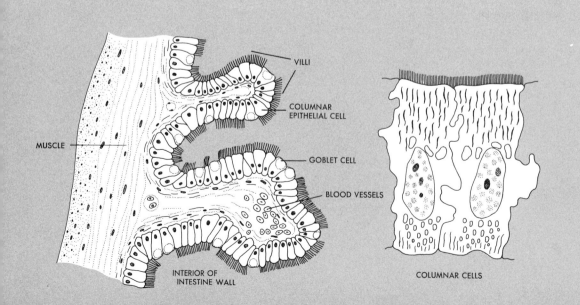

organisms as bacteria, hummingbirds, shrews, bees, flies, and mosquitoes are generally small. Those of the more slowly metabolizing animals such as man, elephants, amphibians (frogs, toads, and so on), and grasshoppers are considerably larger. The surface exchanges of materials in the cells of the smaller and more active animals must be more rapid because of their greater need for a constant supply of energy. Also, the cells must be smaller so that the amount of surface area relative to volume can be at a maximum. If this were not the case, the movement of substances into the interior of the cell would be insufficient to maintain the rate of metabolism; the metabolic processes would bog down. Also, heat is released from the breakdown of food substances, and the heat must escape. For example, if the elephant metabolized at the same rate as the hummingbird, it would roast itself. The tremendous amount of energy, some of it liberated as heat, could not escape. The same result would occur if a human cell metabolized at the same rate as a small bacterial cell.

Cell size can also be viewed as a structural problem. The cell contents need support of some kind so that they can be held together and function efficiently, hence the cell membrane. Such an elastic but firm support would seem wasted on cells of minute size, but it is necessary as a boundary between the cell and its environment. On the other hand, oversized cells without the aid of a membrane, would burst like punctured balloons because of internal pressure. A balance must be maintained between the need for support and the need for a proper surface-volume relationship.

What, then, is the ideal size at which a cell performs most efficiently? Obviously there is no simple answer. Since cells of various sizes exist, we must assume that each is efficient. If we exclude eggs from our consideration, since they really belong in a category by themselves, we find that maximum cell diameter is about 100 μ, while the minimum is approximately 0.1 μ, a ratio of 1,000:1. (We have also excluded viruses because, whether they are cells or not, their energy for metabolism comes from the cell they have parasitized.)

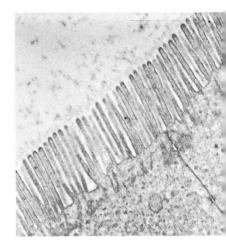

Fig. 9.6 The surface of an absorbing cell (Fig. 9.5) has many projections, or *microvilli*, as shown by this electron micrograph. Each absorbing cell may have as many as 3,000 microvilli, so that its surface area is increased enormously.

Cell Shape

Let us now look at organisms that are unicellular. When we remember that protoplasm is a somewhat fluid substance bounded by an elastic cell membrane, we might well assume that the shape of these cells would be spherical. Their surface tension, particularly in those that are free-floating, should shape them in the same way that surface tension shapes airborne soap bubbles or rubber balloons. Many cells, indeed, do have a

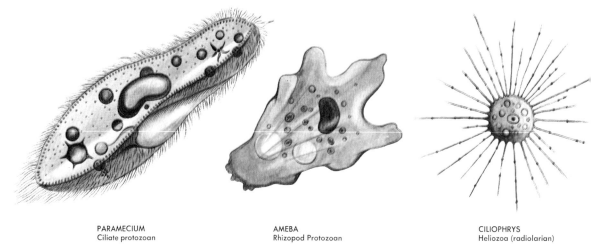

PARAMECIUM
Ciliate protozoan

AMEBA
Rhizopod Protozoan

CILIOPHRYS
Heliozoa (radiolarian)

Fig. 9.7 Three species of protozoans: *Paramecium* and *Ameba* are commonly found in waters all over the world. Paramecia may range in length from 200 to 350 microns. Amebas may be around 600 microns in diameter. Both are considered to be rather large protozoans.

Fig. 9.8 *Acetabularia* is a stalked one-cell plant living in warm marine waters. The plant may be a centimeter or more in height, but it remains as one cell until the nucleus in the rhizoid (base) rises to the cap and divides, thus forming many cells.

General Biological Supply House, Inc., Chicago

spherical shape: the eggs of many marine animals, many yeasts and bacteria, and a variety of unicellular algae. But from the different shapes attained by other forms of unicellular life, it appears certain that the organism itself, through its inheritance, determines its own shape. Some bacteria are shaped like rods, spirals, or commas. Among the algae, the diatoms, desmids, and dinoflagellates, with their unusual contours and outer skeletons, take on a bizarre appearance. Even the ameba, familiar to most of you, is not normally a sphere. Generally flattened because it rests on a surface, it has no particular shape, but rather is a fluid glob of protoplasm that can flow this way or that. Only at rest or in death does it become spherical.

One of the most remarkable single-celled organisms, and one widely used in biological research, is *Acetabularia*, an alga found in warm marine waters (Fig. 9.8). Some species are 9 to 10 cm in height. A distinctive cap, characteristic of each species, tops the whole structure. Until it begins its fruiting stage, it is essentially a single cell with the nucleus located at the base of the stalk, and just above the root-like structure **(rhizoid)**.

When we turn to a consideration of the cell shapes found in multicellular organisms, we find that mechanical forces still determine shape to some extent, but that the function a cell performs can also help determine the shaping process. As pointed out above, free-floating cells with thin membranes tend to be spherical. This is the most economical (i.e., the most compact) shape a given mass of protoplasm can assume. A physicist would say that the cell is obeying the "law of minimal surfaces." When spherical cells are packed together, however, they tend to become faceted as they come in contact with neighboring cells, much as the sides of soap bubbles become flattened when the bubbles

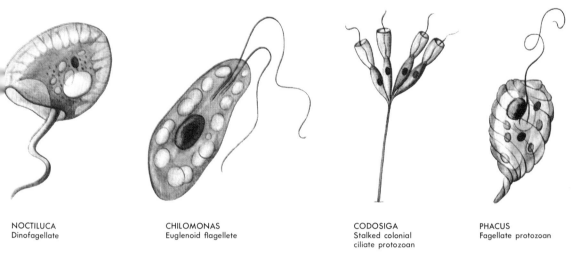

| NOCTILUCA | CHILOMONAS | CODOSIGA | PHACUS |
| Dinoflagellate | Euglenoid flagellete | Stalked colonial ciliate protozoan | Fagellate protozoan |

Fig. 9.9 Four species of protozoans: three of which are free-swimming and one lives fastened by a stalk and is colonial in habit. *Noctiluca* is the protozoan responsible for the luminescence in warm marine waters.

are jammed together in a confining space. In animals this can be seen in the early stages of embryo development (Fig. 9.10). For a brief period the cell mass still retains the rounded shape and size of the original egg, but in adjusting to the available space the cells shape themselves accordingly. Similar arrangements of plant cells can be seen in the growing tips of roots or stems.

Cells are not always packed in the same way. Some are layered in flat sheets, as in the linings of blood vessels, or skin. Such cells tend to be longer and wider than they are thick (Fig. 9.10). Presumably, they are forced into this shape by tension. However, it would be a mistake to overemphasize the role of tension in shaping cells. The function a cell performs is also related to its shape, but we cannot say that function *determines* shape, or that shape *determines* function. In any particular organism, the two go together. Let us consider the human red blood cell as an example.

Viewed face on, a red blood cell from a vertebrate has a circular shape. From the side, it may appear flattened and concave (Fig. 9.11). Its function is to carry oxygen from the lungs to various tissues of the body, and to carry carbon dioxide from the tissues to the lungs. While its flattened shape allows for the exchange of gases, its rounded contours and small size permit it to slide easily through the smallest blood vessels **(capillaries)** without clogging them. A spherical cell of the same size would be inefficient, because the rate of gas exchange between the exterior of the cell and its center would be very slow in proportion to its size.

The shape of muscle and nerve cells further emphasizes the relation of shape to function. Both kinds of cells are elongated. The muscle cell can alternately contract and relax. The nerve cell is part of the communication system of the body. Why would

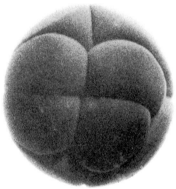

Courtesy John A. Moore

Fig. 9.10 Above: the cells of an amphibian embryo become faceted as the number of cells (eight here) increases through division. With additional divisions they become more and more crowded together. Below: layered cells, such as those found in the skin or in the walls of arteries. They become flattened, and are greater in width and length than they are in depth.

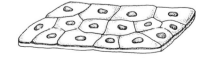

Fig. 9.11 Left: red blood cells appear rounded in face view, but dumbbell-shaped in profile. This shape prevents clogging in the capillaries and allows for an easy and rapid exchange of gases. Right: the melanocyte, containing the black pigment melanin, changes its shape as it matures. These from a human being do not alter their shape once they are mature, but those in the flounder (Fig. 9.12) can expand or contract, depending on the environment, in which the flounder finds itself.

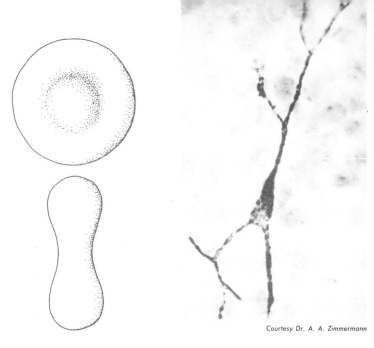

Courtesy Dr. A. A. Zimmermann

R. H. Noailles

R. H. Noailles

Fig. 9.12 These photographs show a flounder on rocky (above) and sandy (below) bottoms. Its melanocytes are expanded above and contracted below so that it can blend better with the environment, and so escape detection.

these cells perform their function less well if they had a spherical or a flattened shape?

Some cells can alter their shape quite rapidly. An example of this kind of cell is the **melanocyte,** so-called because it contains a pigment named **melanin,** which is responsible for the dark color in the skin of animals. As these melanocytes mature in man, they go through a series of changes that transform them from a spherical into an elongated and branched shape. Once they reach maturity, they do not change shape. However, in the flounder or the chameleon, the shape varies with the background of the environment in which the animal finds itself. When the melanocytes are contracted, the animal appears light. When expanded, with its cell branches extended, the animal is darker (Fig. 9.12). The melanocyte, therefore, not only gives color to the skin but also provides protective coloration: it enables the animal to blend into either a light or a dark background. No other shape of cell could perform this function so well.

Fig. 9.13 shows a number of cell shapes from various parts of a plant, each type having a special function. The particular shape adopted reflects the particular position and function of each cell. We find, therefore, that the shapes of cells are related both to the body plan of the organism and to the varied activities the organism performs.

Plants and animals differ in their structure, mobility, and mode of nutrition. Except for certain free-floating plants, plants generally have devices for anchorage and for the absorption of water and mineral salts (roots), for the conduction of substances through the system (stem and veins of leaves), for the manufacture of food through photosynthesis (leaves), and for reproduc-

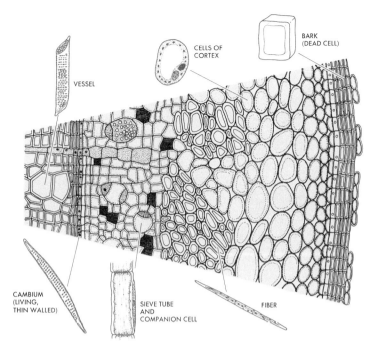

Fig. 9.13 A cross-section of a woody stem shows different kinds of cells: vessels for water conduction; cambial cells for the production of new cells through division; sieve tubes and companion cells for the conduction of food materials; fibers for protection and tensile strength; cortical cells for photosynthesis; and cells of the bark for protection.

tion (flowers or other reproductive organs). Animals, however, must search for their food, and must have a bodily construction designed for this purpose. They need structures for support (bones, cartilage, or **exo-skeleton,** as in the lobster), for movement (muscles), for communication (nerves), and for digestion, excretion, secretion, circulation, and reproduction. Since the cell is the unit of construction in all these structures, we must conclude that it is an extraordinarily flexible unit. Each cell in a body has the same inherited constitution, yet it is capable of adapting to the situation it finds itself in, and to the function it performs.

Cell Number

Let us consider the *number* of cells found within any given organism. This number may seem impossible to calculate in so large an organism as man, but since cells are roughly the same size, simple arithmetic can give us a fairly good estimate. The difference between a human dwarf and a giant is primarily a matter of cell number rather than of cell size. Generally speaking, within any one species the larger the organism the greater the number of cells in its body.

Ewing Galloway

Fig. 9.14 The difference in size of a giant and a dwarf is due to numbers of cells, not to a difference in cell size.

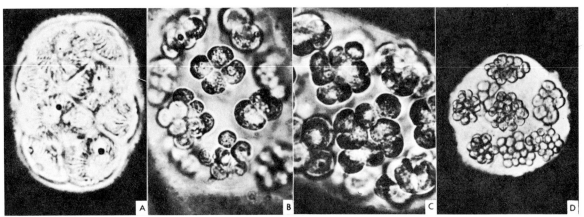

Courtesy of Annette Coleman

Fig. 9.15 *Pandorina morum* is an alga that has a definite number of cells (16) per colony. When new colonies are to be formed, each cell in a colony undergoes four divisions, forming 16 new 16-celled colonies.

Among unicellular forms, the cell and the organism are one and the same. Most multicellular organisms consist of an indefinite number of cells, but in a few the number is fixed. The green alga, *Pandorina morum* (Fig. 9.15), found in fresh-water ponds, is a plate made up of eight or 16 cells. A similar and related species, *P. charkowiensis*, has either 16 or 32 cells. Intermediate numbers of cells are not found in these species. When new colonies are formed in *P. morum*, each of the 16 cells divides four times, producing 16 new colonies, each with 16 cells apiece. Such definite cell numbers are found in only a few of the animal forms.

Most organisms have an indefinite number of cells. The number is limited only by the ultimate size the organism reaches at maturity. Suppose we take man as an example. During the first 285 days of prenatal life, a human being grows from a single fertilized egg that divides again and again, forming a newborn infant weighing about seven pounds. The egg cell originally weighed about one-millionth of an ounce, but at birth the infant consists of approximately 2,000,000,000,000 (2×10^{12}) cells. At maturity the average human male, weighing about 160 pounds, consists of about 60 thousand billion (6×10^{13}) cells. If all cells divided simultaneously and at the same rate, how many cell divisions would have occurred by the time of birth? How many by the time of maturity?

Not all cells, however, divide at the same rate and for the same duration of time. Some reach maturity and stop dividing long before others; some continue to divide through old age and until death. The development of the human being from a single fertilized egg involves many things; an increase in cell number is but one of them. For example, at birth all of the nerve and the great majority of muscle cells are formed; growth after birth is a matter of cell enlargement, not an increase in number. The destruction of a nerve cell would be an irretrievable loss, for nerves, once formed, cannot multiply. Muscle cells can replace themselves to a limited extent, as the repair of cut or torn muscles indicate, but the enlargement of a muscle through exercise is due to enlargement of cells, not to an increase in the number of cells. On the other hand, cells in the blood stream are continually being produced to replace those that die. Because the number of new blood cells formed is about the same as the number of old blood cells that die, the number of blood cells in your body stays fairly constant.

Cell Death

Any organism has a life span that is characteristic of the species to which it belongs. Usually, we think of life span as an *average* figure: a few days for certain insects, a few months for annual plants, 60 to 70 years for man, 250 to 300 years for an oak tree. The sequoia (red wood) of our West Coast and the bristle-cone pine of California's White Mountains are probably the longest-lived organisms. Some of these trees reach an age of 3,000 to 4,000 years.

Cells, too, have a life span that is characteristically long or short. Yet some cells can be considered to be immortal. When a unicellular organism divides, the life of the single mother cell becomes part of the life of two new daughter cells. So long as a unicellular species continues to live, so too does the life of the original cells that began the species sometime in the long-distant past. Among sexually reproducing organisms, only the cells of the germ line (those cells producing eggs or sperm) can lay claim to immortality. They are the only ones that span the generations, thus keeping the species alive. But among the cells of the body, death is a very necessary process. If its role were altered, the functioning of the organism would be drastically affected.

Some parts of your body are continuously renewed as new cells replace old ones that die. These parts are characterized by rapid rates of cell division. Other tissues cannot renew themselves. The original cells live on and on; they die only through disease, accident, or when the individual dies. Cell division does not occur in such parts of your body. As pointed out above, all nerve cells are formed by the time of birth. They continue to mature and function as long as the individual lives (barring injury, of course). Yet nerve cells have been estimated by pathologists to die at a rate of 12,000 to 14,000 per day. Using a brain capacity of 1,350 cc, and a nerve cell diameter of 100 μ, can you calculate how many brain cells you had at birth, and what percentage you will have at age 50 or 100 years? How long would you have to live to have no brain at all?

The fact that an organ remains constant in size tells us nothing about its rate of cell replacement. The constancy of size merely indicates that there is no *net* gain or loss of cells. The death of the old cells can be equalled by the production of new cells.

Some biologists estimate that one to two per cent of the cells of the human body die each day. But if a person's weight remains constant, these old cells must be replaced by new ones—by the billions every day. Since muscles and nerve tissue do not produce new cells, there must be some very active centers of death and replacement elsewhere. These include the protective layers of skin, the blood-forming centers, the lining of the digestive tract, and the reproductive system. The other organs of the body have much slower replacement rates. A cell in the liver, for example, is thought to have an average life span of about 18 months. Consequently, if we look at a liver slice under the microscope, we expect to find very few cells in division.

The outer surface of the human body is covered with a protective layer of skin. This includes the cornea of the eye and such modified skin derivatives as nails and hair. The cells of these

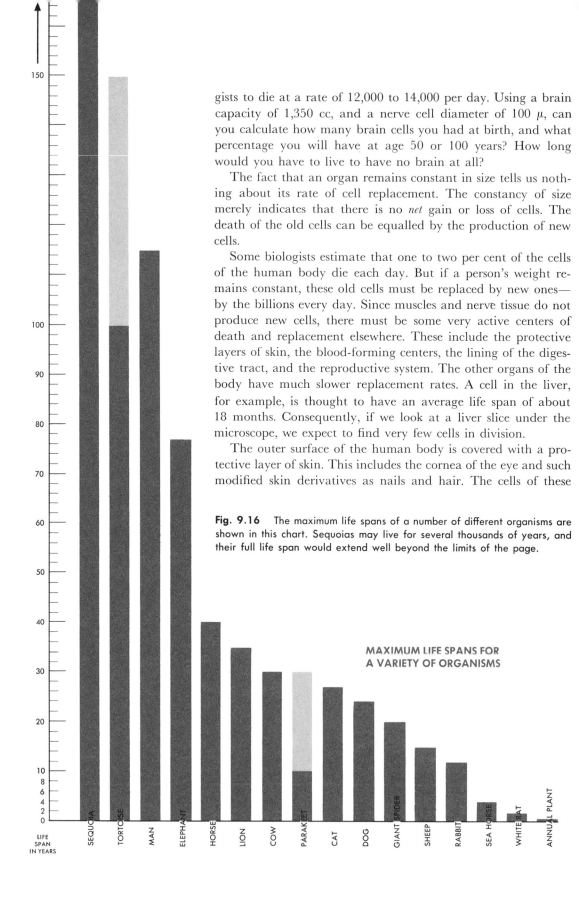

Fig. 9.16 The maximum life spans of a number of different organisms are shown in this chart. Sequoias may live for several thousands of years, and their full life span would extend well beyond the limits of the page.

MAXIMUM LIFE SPANS FOR A VARIETY OF ORGANISMS

structures are constantly being lost through death. You can easily scrape the inside of your mouth and pick up living cells that can be viewed in the microscope. The skin constantly sloughs off, and the growing nails and hair are composed of cells that have died. The process of replacement, then, must be a relatively rapid one. The underlying cells are constantly dividing, and are pushed outward toward the skin surface, while the outermost cells harden, or become **cornified,** as they die (Fig. 9.17). It takes about 12 to 14 days for a cell in the skin of the forearm to move from the dividing to the outermost layer of the skin. Callouses on the hands are thickened areas of dead cells.

The cornea of the eye is a special type of skin in which the rate of cell death and replacement is high. The cornea, in fact, is an excellent type of tissue to examine for active cell division since it is only a few cell-layers thick. When it is fixed, stained, and mounted intact on a microscope slide, the dying cells can be seen at the outer surface, while the underneath cells can be seen in active division.

The cells of the blood are not formed in the blood. The red blood cells are formed in the marrow of the long bones; the white cells form in the lymph nodes, spleen, and thymus gland. Together, these cells and a clear fluid, the **plasma,** constitute the blood, which has an average ratio of one white cell to 400–500 red cells. The blood-forming areas usually manage to maintain this ratio.

Let us consider the red blood cell. Its life span is about 120 days. Mature red blood cells do not have a nucleus, which means

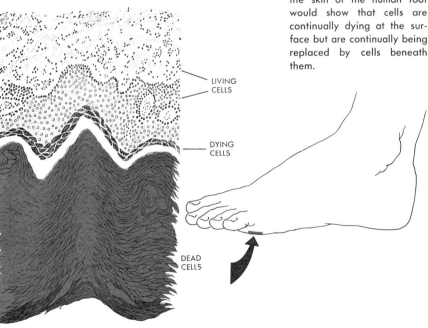

Fig. 9.17 A section through the skin of the human foot would show that cells are continually dying at the surface but are continually being replaced by cells beneath them.

that the damage done to them as they pass through the vessels cannot be repaired. The cells grow fragile and eventually burst. Certain types of illness may shorten the life span of red blood cells. In a person afflicted with pernicious anemia, the life span is reduced to about 85 days; with sickle-cell anemia, to 42 days. The rate of replacement in persons afflicted with these diseases cannot keep pace with the loss of cells, and the red-cell count falls below normal and results in an anemic state. The cause of the shortened life span of these abnormal cells is not known. The digestive system consists of some organs in which the cell death rate is very high. Figures for the human are not known, but it has been estimated that the cells lining the intestine, of the rat are replaced every 38 hours.

In the plant kingdom, we find that the lower plants—the algae and fungi in particular—have a rather low loss of cells through death. In the higher plants, however, the rate is enormous. In herbaceous plants, all the cells above ground are lost every season. But consider a large tree. The annual loss of cells in the leaves, flowers, and fruits is high enough, but when all the cells forming dead wood and bark are added, the loss of cells through death in animals is small by comparison. A high rate of cell death, therefore, is as much a pattern of existence as the continuation of living cells. Unfortunately, we know very little about why cells die. Until we advance our knowledge in this field we cannot hope to understand the whole pattern of aging.

SUMMARY

Cells from various organisms and from the several parts of a single multicellular organism differ greatly in size, shape, length of life, rate of metabolic activity, and function performed or product produced. What a cell does and what kind of morphology it possesses is determined by several things: its heredity, which is largely located in the nucleus; the environment in which the cell finds itself and which determines its nutrition, metabolic rate, and imposing stresses; the part of the body in which it is located; and the function which the cell performs. All cells have a common group of organelles, and some have special structures, but the organelles shared by most cells can vary in kind and number. We must, therefore, view the cell as a very flexible structure, capable of doing many things in a variety of environments. Every cell is an efficient unit, otherwise it would not persist.

FOR THOUGHT AND DISCUSSION

1. We have about 5 liters of blood in our body, and there are 5,400,000 red blood cells per milliliter in the average human male (females have about 600,000 fewer). These cells have an average life span of 120 days. How many RBC's must die and be formed each day to keep the number of RBC's constant?
2. Wood of various kinds consists only of dead cells. How do you suppose the different kinds of wood—pine, oak, cherry, walnut, balsa—get their distinctive characteristics?
3. Plan a menu consisting of soup, salad, a fish course and meat course, vegetables, dessert, and something to drink. What cells and cell products do you consume?
4. Metabolic needs are proportional to the volume of cells; exchanges are proportional to the surface area. What does this statement mean? What does it have to do with the limitation of the size of cells? If the metabolic rate increases, what can the cell do to meet the increased needs (assuming that the volume does not change)?

SELECTED READINGS

(All the books listed in the previous chapter can serve also for this chapter.)

Gerard, R. *Unresting Cells.* New York: Harper, 1940.
   One of the finest books available on cells. Although somewhat out of date, it should be read by all students of biology.

Hoffman, J. *The Life and Death of Cells.* New York: Hanover House, 1957.
   A nontechnical account of the world of cells, examined in the light of modern theories of matter and energy.

10 CELL DIVISION

Cells increase their numbers by **division.** In unicellular species, such as an ameba or a bacterium, division results in an increase in the number of individuals in a population. In multicellular forms, such as a human being, the cells do not separate after division. This means that an increase in cell number is part of the processes of growth and maintenance of the individual. As we saw in the last chapter, an adult man has about 6×10^{13} cells. Among these are cells of many kinds, each kind having a different structure, function, and life span. The cells that die must be constantly replaced if the body is to maintain itself. Within any organism that grows and requires routine repair, the process of cell division must produce new cells at varying rates and for varying periods of time.

The process of cell division is essentially the same in all orga-

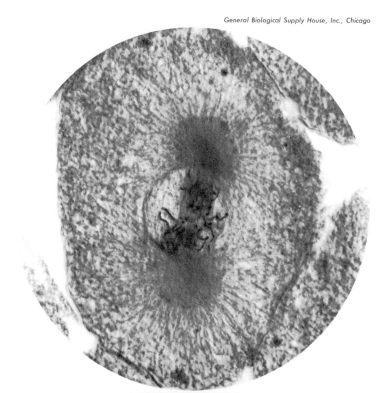

General Biological Supply House, Inc., Chicago

Fig. 10.1 This cell, part of a whitefish embryo, is dividing. The nuclear membrane is still intact, the spindle is being formed between the poles, and the chromosomes are readily visible.

nisms. If we describe this process as it takes place in one or two kinds of cells, we gain a reasonably clear picture of how it operates in all organisms. Most of the activity of cell division is centered in the nucleus. However, the cytoplasm also undergoes a significant series of changes, so the whole process of division involves both nucleus and cytoplasm.

William Bateson, the great English biologist of an earlier generation, once wrote: "When I look at a dividing cell I feel as an astronomer might do if he beheld the formation of a double star: that an original act of creation is taking place before me." Other biologists who have also watched a cell divide, particularly through motion pictures, have been as fascinated as Bateson was.

Each of us developed from a single cell. This cell came from preceding sex cells, and so on back to the beginnings of cellular life. Our most important heritage from the past, and our most precious gift to the future, is the unbroken chain of perfectly formed, individual cells. They are "individual" in the sense that eggs and sperm are individual cells, and the fertilized egg is an individual being, even if not yet fully formed. They are "perfect" in the sense that they are able to live, reproduce, and give rise to new cells, and hence to new individuals. Every time you wash your hands thousands of dead cells making up the outer layers of skin are removed. So far as these cells are concerned, the chain of cellular life has been broken. Death is the inevitable and inescapable fate of most cells, just as it is for us as individuals. Among sexually reproducing organisms, only the eggs and sperm maintain the unbroken chain that links one generation to another.

Ever since biologists came to understand the cellular nature of organisms, they have been investigating and debating the ways in which new cells originate. Not until the middle of the 19th century was it generally accepted that cells originate through the division of pre-existing cells. We attribute this idea to the German, Rudolf Virchow, who stated in 1858:

> Where a cell exists there must have been a pre-existing cell just as the animal arises only from an animal and the plant only from a plant. The principle is thus established, even though the strict proof has not yet been produced for every detail, that through the whole series of living forms, whether entire animal or plant organisms or the component parts, there are rules of eternal law and continuous development, that is, of continuous reproduction.

Although we cannot reconstruct the beginnings of life from what we know today, presumably it did not begin in the form of the cells that we now see under our microscopes. They are the products of ages of evolution. Nor can life be originated anew, at least with the techniques and knowledge we possess today. It

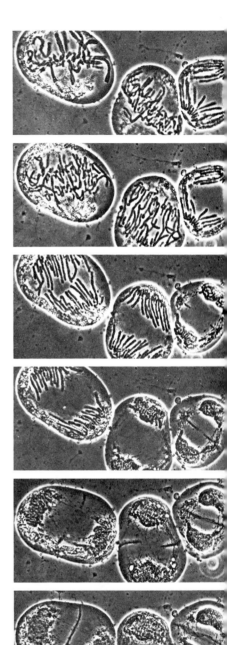

Courtesy W. T. Jackson

Fig. 10.2 These living plant cells are also undergoing division, as shown in the series of time-lapse photographs. The difference in time from the top view to that at the bottom is 81 minutes.

can come only from pre-existing life. This theory, known as **biogenesis,** we credit to the French biologist Louis Pasteur, although he was not the first scientist to demonstrate this fact. Pasteur showed that life could not arise anew under the conditions existing in his laboratory. He took two flasks of broth, a good culture medium for micro-organisms, and boiled the broth in such a way as to kill all the organisms in them. One flask he left open; the other he made airtight by sealing it. Within a few days the open flask contained "germs" of various sorts—bacteria, yeast, or molds—which, as we now know, were present in the air. The sealed flask, however, contained no life whatsoever, and could not acquire any until air (a carrier of germs) was once more admitted.

Roottip Cells in Division

The growing roottips of plants have long provided a good source of actively dividing cells. Because the cells do not all divide at the same time, a single roottip has cells in all of the stages of division. Here we will examine the process as it takes place in the roottips of the broad bean, *Vicia faba,* and the onion, *Allium cepa,* although almost any kind of an actively growing roottip will serve for this purpose.

Roottip cells present a variable appearance, depending on how the cells are prepared and stained. Study the low-magnification

Fig. 10.3 A number of different stages of cell division are visible in this low-power view of a longitudinal section through the roottip of an onion.

Courtesy M. S. Fuller

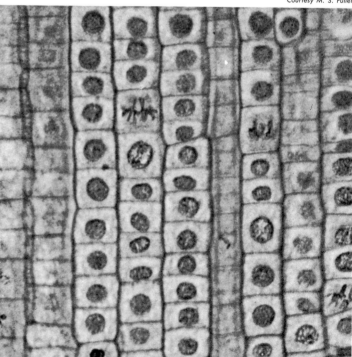

photograph of an onion roottip, sectioned in a longitudinal direction and stained with haematoxylin (Fig. 10.3). Nuclei are in various stages of division, and nucleoli are prominent. Now compare these cells with those of the broad bean (Fig. 10.5), stained with Feulgen and then squashed on a slide. Feulgen stains only the chromatin, leaving all other cellular structures virtually invisible; it is a highly specific stain. Haematoxylin, on the other hand, is a general stain. When we use it we can see other structures in addition to the chromatin.

Our knowledge of why a cell divides at a particular time is not very complete. However, the cell goes through an active stage of synthesis prior to division, doubling the amount of nucleic acid in the nucleus. This activity occurs during **interphase** (Fig. 10.4), a stage in which the nucleus shows little definable structure, except for the lightly stained chromatin and the more deeply stained nucleoli. The **chromosomes,** into which the chromatin will form itself, are not individually distinguishable during interphase.

Prophase begins when the chromosomes first become visible as long, slender threads which are longitudinally double. Each longitudinal half is called a **chromatid,** and the two chromatids of each chromosome are twisted around each other like two wires of an electrical cord. During prophase the two chromatids shorten by becoming coiled. The progress of coiling is shown in Fig. 10.4. The mechanism of coiling is not yet known, but it is obviously a

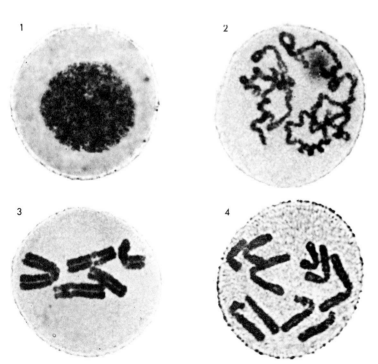

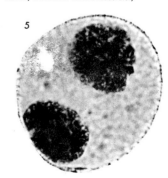

Fig. 10.4 Cell division in the microspores (immature pollen grains) of the wake robin, *Trillium erectum*: (1) interphase; (2) mid-prophase, during which the chromatids shorten by becoming coiled; (3) metaphase the spindle is not stained, hence does not show; (4) anaphase; and (5) telophase.

Courtesy Brookhaven National Laboratory

Fig. 10.5 Top: Feulgen stain was used to show the prophase, metaphase, and anaphase stages of division in the roottip of the broad bean, *Vicia faba*. The stain reveals only the structure of the chromosomes, leaving the spindle, cytoplasm and cell walls unstained. Bottom: The spindle is clearly revealed when iron haemotoxylin stain is used on dividing cells of the onion roottip.

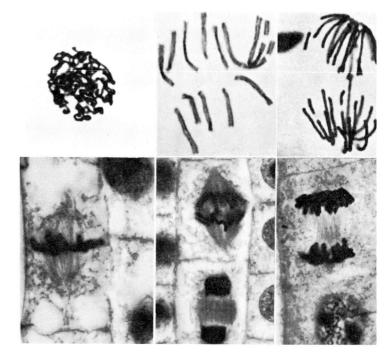

Courtesy S. Incue

Fig. 10.6 Here is the spindle of a dividing animal cell with the chromosomes on the metaphase plate. The fibers that stretch from chromosomes to poles, and from pole to pole, are actually fine tubules. The poles show astral rays radiating outwardly into the cytoplasm.

means of converting a long and unmanageable strand into a more compact, maneuverable one.

During prophase the nucleoli are large at first, but gradually they become smaller. They also free themselves from the chromosome to which they are attached, and finally disappear. The nuclear membrane breaks down and disappears in late prophase, and the chromosomes become attached to a new structure called the **spindle** (Fig. 10.6). The cell is now in **metaphase.** At this stage, it is evident that the number of chromosomes per cell is constant, and that each chromosome has a particular shape and size which is maintained from cell to cell. It becomes possible, therefore, to identify the chromosomes individually. In the broad bean, the shapes and sizes of the chromosomes are of two major classes; in the human, they are grouped differently.

The spindle consists of chains of protein molecules arranged longitudinally between the two poles of the spindle. These proteins are formed in the cytoplasm during interphase and prophase. The mechanism which causes the proteins to aggregate into a spindle structure is unknown. However, by placing roottips in a solution containing the drug colchicine, spindle formation can be prevented. The chromosomes shorten normally, but they fail to aggregate. They lie free in the cytoplasm where they are easy to count and where their shapes and sizes can be seen clearly.

When the spindle appears, the chromosomes move to a position midway between the poles. Here they become attached to the spindle at their **centromere** regions. The centromere of the chromosome is its organelle of movement. Without it, the chromosome fails to orient properly on the spindle, and later will fail to separate its two chromatids into the new daughter cells. The position of the centromere is clearly marked by a constriction in the chromosome. It consequently divides the chromosome into two arms of varying length. Very few chromosomes have centromeres at their ends. Since the position of the centromere is constant, this serves as an additional feature to aid in the identification of particular chromosomes.

The two chromatids of each chromosome now move apart from each other and migrate to the poles of the spindle. This period of movement is called the **anaphase** stage. Protein fibers running from the poles to the centromere of each chromosome shorten during anaphase and bring about the movement. Anaphase ends when the two groups of chromatids (which can now be called chromosomes) reach the poles.

Telophase begins when chromosome movement has stopped. At this stage a new nuclear membrane forms, the nucleus enlarges, the spindle disappears, the chromosomes uncoil and become long and slender, and nucleoli appear. The events are essentially the reverse of what took place in prophase. Across the middle of the spindle a new cell membrane is formed. Eventually, the **cell plate,** as it is called, grows outward until it reaches the side walls and cuts the cell in half. Two new cells are now fully formed.

Cell Division in Animal Cells

Cells from the embryo of the whitefish illustrate very beautifully the division process. They also reveal the differ-

Fig. 10.7 Stages of cell division in the whitefish embryo. Left to right: prophase, metaphase, anaphase, and telophase.

General Biological Supply House, Inc., Chicago

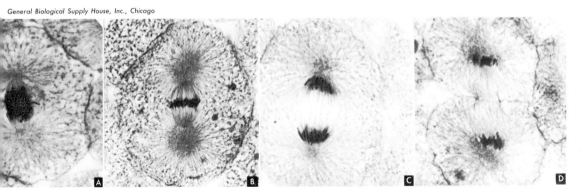

ences that distinguish animal from plant cell division. The behavior of the chromosomes is the same as in plant cells, although the chromosomes are so small and numerous in the whitefish that they cannot be individually identified. The most immediate difference is in the process of spindle formation. As Fig. 10.7 shows, the whitefish cell in prophase has a radiating structure adjacent to the nuclear membrane. This is the **centrosome,** with **astral rays** radiating from it, and with a central body, or **centriole,** within it (but not visible in the illustration). In early prophase, the centrioles, which had previously divided, migrate along the nuclear membrane until they lie opposite each other. As the centrioles migrate, they organize the spindle between them. The nuclear membrane disappears and the chromosomes line up in the middle of the spindle (Fig. 10.7, metaphase stage). Anaphase movement, and the telophase reorganization of the two new nuclei take place as in plant cells.

The division of the cell into two daughter cells is another point of difference. A process of *furrowing,* beginning at the outer edges of the cell, cleaves the cell in two. Plant cells, with their rigid cell walls, cannot do this, but cell plate formation accom-

Fig. 10.8 The cycle of cell division: interphase is the longest stage because during this time the chromosomes must replicate and the cell prepare itself for division. Metaphase and anaphase, the most dramatic of the stages when the chromosomes are clearly visible, are of shortest duration.

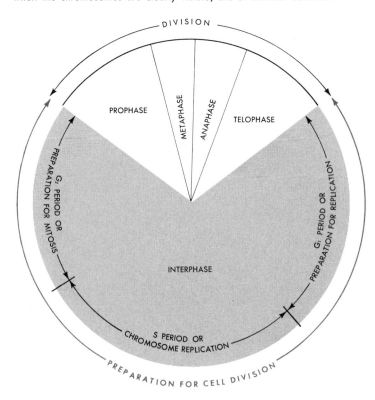

plishes the same thing.

Figure 10.7 shows the complete sequence of events that take place during the course of cell division. The several processes and structures involved must be coordinated in time and place if the cell is to divide successfully. One mistake, and the daughter cells will be abnormal. They may die.

Time Sequence of Cell Division

The time required to complete an entire cycle of cell division varies quite widely. Bacterial cells, for example, accomplish a cycle in 15 to 20 minutes. Roottip cells in the broad bean require about 20 hours at room temperature. Higher temperatures can shorten the time span, lower temperatures lengthen it.

The over-all time for cell division can be readily determined by finding the time necessary for doubling the cell number. This can be done with cells in tissue culture, or bacterial cells in a test tube; one simply counts them. Such knowledge, however, does not tell us much about the timing of the various stages of division. Let us consider this in terms of a human cell, easily grown in tissue culture. At 37°C, the temperature of the human body, a division cycle takes about 18 hours. Yet the period from the beginning of prophase to the end of telophase is only 45 minutes. More than 17 hours is spent in interphase preparing for division. The dramatic events of cell division take place, therefore, in a rather explosive fashion.

Figure 10.8 shows the cycle of division of a human cell. It is possible to show that interphase has three distinct stages: the G_1, S, and G_2 stages (S stands for synthesis, and G for the interphase gaps before and after synthesis). The important stage is the S period when the nucleic acids and some of the proteins of the chromosome are being synthesized. Some six to seven hours are required for the process, and the end result is that the chromosomes become longitudinally double. Some of the chromosomes double, or replicate, early in the S period, others later. **Histone,** a protein component of the chromosome, is also synthesized at this time. The stages from prophase to telophase occur in rapid fashion, with metaphase and anaphase being the shortest stages.

Cell division is an act of survival—survival of life itself, survival of a species, and survival of each individual organism. A cell that does not, or cannot, divide will ultimately die. Cell division is also part of the process of growth and maintenance, providing more cells for the developing organism and replacing cells that die. As we saw earlier, the control center of a cell is the nucleus. It is here where chemical "decisions" of great importance are made, decisions which not only affect all division, but the day to day general well-being of the cell. In the next chapter we will look at this control center in detail.

SUMMARY

Cell division is the source of new cells. Each cell, as it divides, goes through a number of recognizable stages: **interphase,** when the chromosomes in the nucleus replicate; **prophase,** during which the chromosomes shorten and the chromatids become visible, the nuclear membrane breaks down, and the nucleoli disappear; **metaphase,** when the spindle forms and the chromosomes orient on the spindle; **anaphase,** when the chromatids separate and move to the poles; and **telophase,** when the nuclei reorganize and the cytoplasm is divided for the formation of two new cells.

The time required for division can vary from several minutes to many hours. Although plant and animal cells differ somewhat in the way their cytoplasm divides, cell division is basically similar in all organisms.

FOR THOUGHT AND DISCUSSION

1 Construct a chromosome out of pipe cleaners or wire so that the centromere and nucleolar organizer are visible features. Show how this chromosome changes as it goes through the several stages of division, from interphase to telophase.

2 Why do you suppose that it is necessary for the chromosome to shorten before the two chromatids separate at anaphase?

3 Every daughter cell is like the mother cell from which it originated. How does the behavior of chromosomes during division ensure that this will be so? What would happen if it were not so?

4 Why would it be difficult for a plant cell to divide its cytoplasm in the same manner as an animal cell?

5 Asexually reproducing organisms are essentially immortal. Why is this so? The introduction of sexual reproduction into the plant and animal kingdoms made sexually reproducing individuals mortal; each individual must eventually die. What does this statement mean? What keeps the species going if the individuals die?

6 What advantage is there to an organism being made up of many cells instead of just one large cell?

SELECTED READINGS

Kennedy, D., (ed.). *The Living Cell.* San Francisco, Calif.: W. H. Freeman & Co. 1965.

A collection of articles from *Scientific American* covering most aspects of cell structure, behavior, division, and metabolism.

Mazia, D. *Cell Division.* D. C. Heath, 1960.

A small pamphlet written as a supplement to high school biology texts, and containing a clear account of the processes of cell division.

McLeish, J. and B. Snoad. *Looking at Chromosomes.* New York: St. Martin Press, 1958.

This little book contains one of the finest collections of photographs illustrating mitosis and meiosis.

Swanson, C. P. *The Cell* (2nd ed.). Englewood Cliffs, N.J.: Prentice-Hall, Inc., Foundations of Modern Biology Series, 1964.

Chapters 5 and 6 in particular give a clear description of the processes of cell division in both plants and animals.

11 THE NUCLEUS — CONTROL CENTER OF THE CELL

We have mentioned several times that the nucleus exercises a control over the activities of the cell. What we are saying is that a cell carries out its functions because it has a nucleus that in some way provides it with a blueprint for action. In stating this, we do not mean that the nucleus controls every immediate action of the cell. With a change in temperature or nutrition, for example, the cell may change its metabolic action or even its appearance. It is possible for a cell without a nucleus to function for a fair length of time. The human red blood cell, for example, loses its nucleus at about the time it enters the blood stream; it then operates very well for about four months without a nucleus. It may perhaps be more accurate to say that the nucleus in all kinds of cells provides a kind of long-range guidance over the affairs of the cytoplasm. But before accepting this statement, let us see what evidence can be found to support it.

Fig. 11.1 The dark objects in this electron micrograph are chromosomes in a dividing cell at the metaphase stage. The chromosomes are so compacted at this stage, and so dense to the electrons, that internal structure cannot be seen.

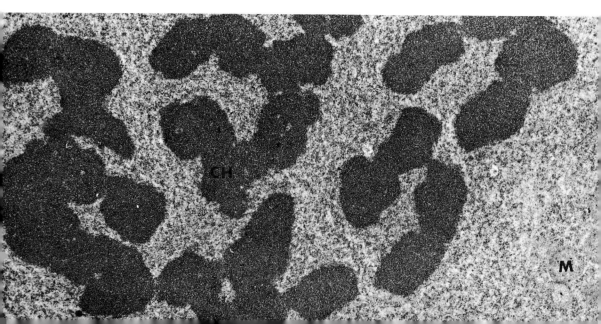

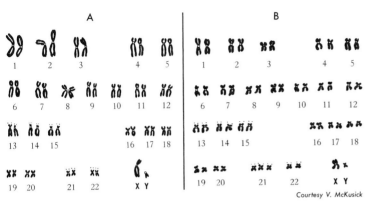

Fig. 11.2 Chromosomes of a variety of organisms are shown on this page. (A) normal human male, with the chromosomes arranged as 22 pairs plus an X and a Y chromosome. (B) chromosomes from a human male characterized by Down's syndrome (Mongolian idiocy), which is caused by the individual having three instead of two chromosome 21's.

(C) Metaphase stage from the testes of a salamander (normal body cells would show 28 individual chromosomes instead of 14 pairs).

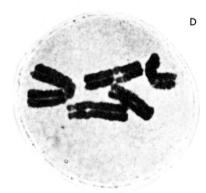

(D) The five chromosomes of *Trillium erectum*, each one different in shape and in size.

First of all, observation tells us that the usual state of cellular affairs is one nucleus per cell. A departure from this rule is rather unusual in the cells of higher organisms, although among some bacteria and fungi, multinucleate cells may be the usual state of affairs. Living material, therefore, is generally so subdivided as to provide a single nucleus for a given amount of cytoplasm. There must be a good reason for this. From many observations, we also know that the character of the cytoplasm can change, the cell functions may differ, and the size and shape of the cell can be altered drastically. Yet the nucleus remains stable in what is often a changing cellular picture. This is not proof of a nuclear control, but it suggests that we should probably expect a control center to have a degree of stability such as that observed.

Another fact that suggests nuclear stability is the remarkably constant number and kind of chromosomes in each nucleus. The illustrations in Fig. 11.2 show the chromosomes of several organisms, including man. If you compare figures A and B, you will notice that B shows 47 chromosomes instead of the usual 46, the number characteristic of normal human individuals. Notice that the extra chromosome in B is one of the very smallest. Nevertheless, its addition is sufficient to alter a normal child into one decidedly abnormal. This is a dramatic, if tragic, example of the nuclear influence of one extra chromosome on the character of the individual possessing it.

We can, therefore, adopt the hypothesis that for normal development and behavior, a normal number of chromosomes (23 pairs in man) is necessary. Since the nucleus consists very largely of chromatin that, during division, changes into distinct chromosomes, we add evidence to our idea that the nucleus is the control center of the cell.

We still, however, lack definite proof. There are many kinds of proof, but let us use that provided by an ingenious set of experiments carried out by the German biologist, Max Hämmerling, on the single-celled alga, *Acetabularia*. The technique and the results are illustrated (see Fig. 11.3). The two species of *Acetabularia*, *A. mediterranea* and *A. crenata*, differ principally in the shape of their caps. If the cap of either is cut off, it will form again as before and without change in shape. But it is also possible to remove the cap and graft a portion of the stalk of one species onto the nucleus-containing rhizoid of the other species. This means then that the new stalk contains both *med.* and *cren.* cytoplasm and cell wall but has either a *med.* or a *cren.* nucleus in the rhizoid. When the cap forms again, the cap is always characteristic of the species contributing the nucleus and not that of the grafted stalk.

The proof can be made even more secure. It is possible to graft two rhizoids together, one containing a *med.* nucleus, the other a *cren.* nucleus. Again, a cap will form, but in this instance the cap will be intermediate in shape, reflecting the influence of both nuclei. We can now say with a good deal more assurance that it is the nucleus that instructs the cell to do as it commands. In other words, the nucleus contains the hereditary elements of the cell. This is an important fact to remember when one recalls that when a sperm fertilizes an egg, the major contribution of the sperm in the act of fertilization is a nucleus, but virtually no cytoplasm. Yet the contribution of the egg and sperm to the inheritance of the newly developing offspring is equal.

The Controlling Element in the Nucleus

We have now reached the point where we can say with some degree of scientific support that the chromosomes are the principal controlling elements of the cell. There is, in fact, very little else in the nucleus except chromosomes. It is possible to extract the chromosomes from burst cells and to analyze them chemically to find out what kinds of molecules they are made of.

Before looking into this, however, we want to consider a series of experiments which proved in a most conclusive way the nature of the *molecular basis of heredity*. *Pneumococcus*, the bacterium responsible for causing some forms of pneumonia, is a small, somewhat elongated cell covered with a thick coating, or capsule, of **polysaccharide,** a complex sugar. When plated out on a solid agar medium, this cell multiplies rapidly and forms a smooth, shiny colony. Occasionally, a colony is rough and dull appearing. The bacteria in these colonies are **avirulent** (Fig. 11.4); they

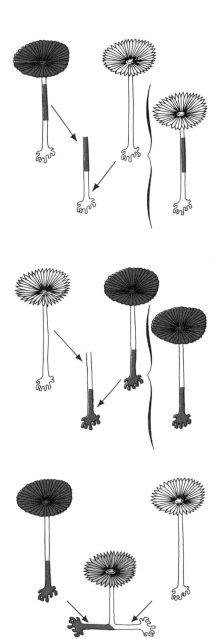

Fig. 11.3 Hämmerling's grafting experiments demonstrated that the character of the cap of *Acetabularia* is determined by the kind of nucleus present. (See text for explanation.)

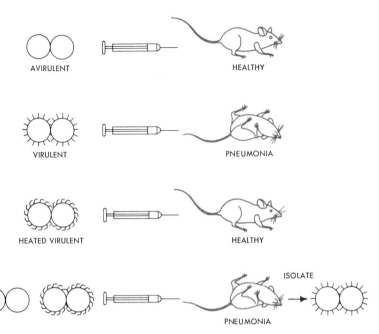

Fig. 11.4 Griffith's experiment showed that heat-killed virulent bacteria could contribute "something" to avirulent strains and cause them to be changed to a virulent form. Avery and his colleagues eventually showed that this "something" was DNA (see Fig. 11.5).

are incapable of inducing pneumonia if injected into mice. They produce rough colonies because they lack the polysaccharide capsule, and they are unable to manufacture it. The rough trait, therefore, is heritable. Frederick Griffiths, an English bacteriologist, injected into mice a mixture of two strains of pneumococci. One strain was living, but rough and avirulent. This was called Type II, and it would not induce infection. The second strain, Type III, was virulent, but its virulence had been destroyed by killing the bacteria with heat. One would naturally assume that the mice should not have been bothered by these injections. Strangely enough, however, the mice died of pneumonia, and from them a living virulent Type III pneumococcus was recovered.

The results of the experiment suggest that some interaction took place between the two strains of bacteria in such a manner that the heat-killed cells changed a living, rough, avirulent Type II strain into a virulent Type III strain. This is basically a change in a heritable trait, because the cells remained Type III after repeated divisions. The phenomenon is called **bacterial transformation.** It was later demonstrated that the same phenomenon could take place in a test tube as well as in mice. Further, it was also shown that smashed dead Type III cells could do the same thing as intact, but heat-killed, cells. The most likely explanation was that some molecule in the Type III cells was incorporated into the avirulent Type II cells, and subsequently changed its heritable characteristics.

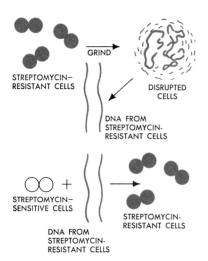

Fig. 11.5 Avery's experiment demonstrated that DNA from one strain of bacteria possessing a known inherited trait could enter another strain lacking the trait, be incorporated, and thereby alter the host bacterium.

Oswald Avery and his colleagues at the Rockefeller Institute set out to track down this *transforming principle*. By bursting bacterial cells open, and by laboriously isolating one kind of molecule after another for testing, Avery finally demonstrated that the critical molecule was **deoxyribose nucleic acid (DNA)** (Fig. 11.5). By adding purified DNA, extracted from virulent Type III cells, to a living culture of avirulent Type II cells, the latter can be permanently transformed into Type III cells. The frequency of transformation is not high—about one in every million cells was transformed at first—but the fact that it could be done at all was an immensely important discovery.

Since Avery's work in 1944, scientists have shown that a number of other heritable traits—penicillin-resistance, ability to ferment sugars, and so on—can be similarly transformed.

The Chemistry of Chromosomes

The importance of Avery's discovery, when considered in relation to our hypothesis that the nucleus is the control center of the cell, becomes obvious: DNA is found *only* in structures capable of self-replication—mitochondria, plastids and, most importantly, in chromosomes. DNA is not, therefore, a unique nuclear substance, although most of it is found in the nucleus, and specifically in the chromosomes. In bacterial transformation, DNA from killed Type III cells must be incorporated into the DNA of Type II cells, thereby changing them into virulent Type III cells. Interestingly enough, however, DNA was discovered nearly 100 years ago. From that time it has been known to be a nuclear constituent. One can see that it often takes a long time for all of the facts to fall into a pattern that makes scientific sense.

Avery's discovery prompted a closer scrutiny of the chemical nature of chromosomes. From a mass analysis of isolated chromosomes it appears that they contain from 26 to 40 per cent DNA; a small amount of **ribose nucleic acid (RNA)**, which is related to but different from DNA; an amount of **histone** (a low molecular weight protein), and a complex protein called **residual protein**. Some calcium, magnesium, and **lipids** (fats) are also present.

The important question that has been raised is whether molecules other than DNA occupy the same central role in governing cellular activity. The answer at the moment appears to be that DNA is the master molecule—the chemical basis of heredity—and that other molecules play supporting roles.

In this regard, however, it is important to point out that some viruses—tobacco mosaic virus (TMV) is an example—contain

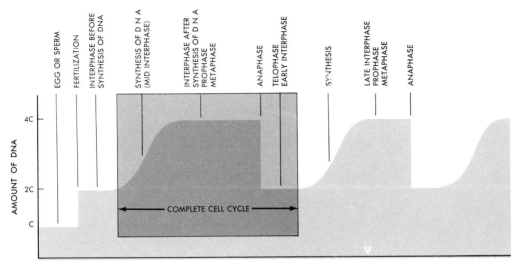

Fig. 11.6 This diagram shows the sequence of cellular events and their relation to the amount of DNA per cell. The amount of DNA in an egg or sperm equals C. When these cells unite at the time of fertilization and their nuclei fuse, the DNA value is 2C. (In a human cell, the 23 chromosomes equal C; the 46 chromosomes, 2C). When a cell prepares to divide, the DNA is doubled to a 4C value at interphase, and then reduced to a 2C value at anaphase, when the chromosomes separate into their chromatids. Successive cell divisions continue the 4C-2C-4C-2C alternation of DNA change.

only RNA and protein, but no DNA. It is possible to separate the TMV protein from the RNA, and then to test each for its ability to produce the mosaic disease. The RNA is infective, the protein is not. The RNA of the virus is its hereditary material, but the answer is not quite so clear as in the bacterial transformation experiment. It is known that the TMV requires the cooperation of the DNA of the tobacco cell for the formation of additional viral particles.

But let us return to the chemistry of chromosomes. If we postulate that the DNA is the hereditary molecule of the cell, we would also expect it to be exceedingly stable in character and amount. This turns out to be true. By "tagging" a DNA molecule with radioactive atoms, we can show that, once formed, it remains intact and does not break down and reform. The histone is also quite stable and seems always associated with DNA. But only DNA has the capacity for transformation. On the other hand, RNA and residual protein in the nucleus and in most molecules in the cytoplasm continually break down and reform. They do not have the stability we would expect of the critical molecule of heredity.

The amount of DNA in nondividing cells is constant. In dividing cells it goes through a regular cycle that is coordinated with the stages of division. At anaphase, when the chromosomes move to the poles, the amount of DNA in each daughter nucleus is one-half that of the original mother cell. During interphase DNA is again synthesized, and the normal amount is soon attained (Fig. 11.6).

The amount of DNA per human cell is approximately 5.6×10^{-12} grams, or about two hundred billionths (2×10^{-11}) of an ounce. This is the amount contained in the fertilized egg that developed into you, and that was responsible for determining that you are what you are. This remarkable molecule somehow has built into itself the exceedingly complex store of information needed to direct the growth and activities of the simplest and most complex organisms. To understand how this is accomplished, we need to know something of the chemistry of DNA—what kind of a molecule it is, how it can store information, and how it can form more DNA. The next few chapters will consider the chemistry of cells, after which we will return to DNA again, and its role in cellular behavior.

SUMMARY

Usually, cells consist of a single nucleus surrounded by cytoplasm. Through grafting experiments performed on the alga *Acetabularia*, it could be shown that the character of a cell is determined by the kind of nucleus it contains. It can also be shown that a particular species is characterized by a given number of chromosomes, and that if this number is changed, the individual organism is usually abnormal in some way. The nucleus, therefore, is the control center of the cell, and the chromosomes are the key structures governing cellular morphology and function.

Chemical analyses of chromosomes show that they contain four kinds of macromolecules: deoxyribose nucleic acid (DNA), ribose nucleic acid (RNA), and two kinds of protein. Transformation studies in bacteria have shown that DNA is the crucial molecule governing heredity. DNA is also stable in amount from one cell to another in the same organism, as we would expect of a controlling element. We can, therefore, trace the control of cellular characteristics to the nucleus, then to the chromosomes, and finally to the DNA molecule within the chromosome.

FOR THOUGHT AND DISCUSSION

1. What advantage is there in having only one nucleus per cell? Why not have an indefinite number?
2. What does the expression "the molecular basis of heredity" mean to you?
3. Why do you think stability in the amount of DNA is a necessary condition for exercising cellular control?
4. Why do you think it important that we know the chemical basis of heredity?
5. It was once thought that our destiny, as individuals, was controlled by the stars. How would you classify such a belief? How does science help to alter our thought structure?
6. DNA extracted from a human cell has not been shown to be capable of transforming another human cell, even in cell cultures. What do you see as difficulties in this kind of experiment?
7. Without knowing in advance the structure of DNA as a molecule, how can you visualize this chemical as a controlling agent? Can you think of any analogies which would help you in your thinking?

SELECTED READINGS

Asimov, I. *The Genetic Code*. New York: Signet Science Library, 1962.
   An elementary discussion of how decoding the gene has aided us in understanding the meaning of inheritance.

Barry, J. M. *Molecular Biology: Genes and the Chemical Control of Cells*. Englewood Cliffs, N.J.: Prentice-Hall, Inc., Foundation of Modern Biology Series, 1964.
   An excellent discussion of how we have come to understand the gene from Mendel to today.

Maddox, J. *Revolution in Biology*. New York: MacMillan Co., 1964.
   A nontechnical discussion of how our understanding of the genetic code has changed the entire field of biology.

Watson, J. A. *Molecular Biology of the Gene*. New York: W. A. Benjamin, Inc., 1965.
   The most up-to-date treatment of the structure and function of genes.

12 ATOMS AND MOLECULES

In Chapter 7 we defined nature as "matter in motion," including living and nonliving things. In view of the highly organized nature of cells, we can define living things as organized systems (of matter) in motion. One attribute of all organisms, plant and animal alike, is that life impresses special kinds of organization on all of the matter that becomes part of a living system. As atoms and molecules are brought into a cell, they enter into organized chemical reactions and become part of an organized structure.

To remain alive, grow, and reproduce, all living things, from microorganisms to man, require certain substances and undergo certain chemical changes. A cell, therefore, must be able to take up certain substances from the surrounding fluid. It must also contain the necessary machinery for making use of the substances in one way or another. Some of the absorbed **nutrients,** as these substances are called, become part of the cell itself. But this building process requires energy. The cell obtains energy by

Fig. 12.1 Radioactive elements can be used to trace the distribution of nutrients taken up by plants. In this case sulfur-35 is fed to a fern frond. The sulfur accompanies the food to various parts of the plant. The frond is then placed against a photographic film. The radiation emitted by the radioactive sulfur leaves light traces against the dark outline of the frond.

Courtesy Brookhaven National Laboratory

breaking down the molecules it takes in, or by manufacturing new molecules and then breaking them down.

Basically, the primary purpose of the machinery of a cell is to convert nutrients into useful energy, and at the same time to make new chemical compounds which the cell needs for a variety of functions. We speak of these chemical transformations as **metabolism.** For the past 75 years research into the metabolism of cells has been a primary activity in the field of biochemistry. One of our purposes of including chemistry chapters in this book is to pave the way for a detailed discussion of cell metabolism later. Our task now, however, is to examine the composition and structure of a variety of important substances of biological origin.

All matter in the universe is made up of specific combinations of a limited number of substances called **elements.** Today we know of 103 elements—among them oxygen, gold, hydrogen, uranium, and so on. An element is matter composed of identical atoms. We classify the atoms of different elements by the differences in atomic mass, that is, the quantity of matter in, say, a hydrogen atom compared with the quantity of matter in a uranium atom. The smallest amount of an element that can be involved in a chemical change is an atom. But there are many subatomic particles, building blocks of the atoms themselves. More than 30 have been discovered so far; however, only three need concern us in this book.

The Structure of Atoms

Atoms are made up of three primary particles: **electrons, protons,** and **neutrons.** The electron was the first particle to be identified as a constituent of all atoms; it is a negatively charged particle which moves about a positively charged

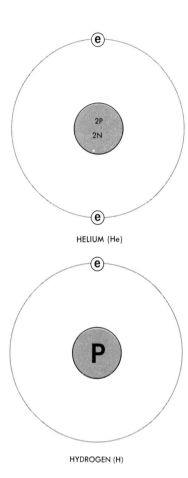

HELIUM (He)

HYDROGEN (H)

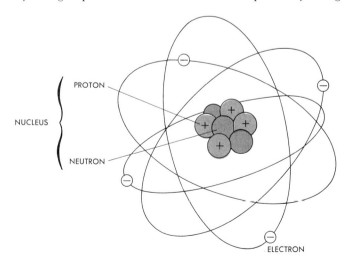

Fig. 12.2 Diagrammatic sketch of an atom (left). This is not a picture of the atom but rather a working model. The electrons move about the dense nucleus at varying distances and at high velocity so that one speaks of a "cloud" or "haze" of electrons about the necleus. The representation of a hydrogen and a helium atom are shown above.

central mass, called the **nucleus.** (The term nucleus is the Latin word for "nut" or "kernel." Just as a nucleus is the central part of a cell, so is a nucleus the central part of an atom.) The nucleus is usually composed of two different kinds of particles—**protons** and **neutrons.** A proton is a positively charged particle with a mass 1,845 times that of the electron. Neutrons have about the same mass as protons, but are neutral; that is, they do not have an electric charge. Except in certain cases, which we shall discuss later, atoms are electrically neutral. This means that each must contain an equal number of negative electons and positive protons. The electrons are the particles directly involved in chemical reactions.

The hydrogen atom, having one electron and one proton, is the simplest of all the atoms. The proton accounts for most of the mass of a hydrogen atom. As a unit of measure, we use the mass of the proton (or neutron), calling it **one mass unit,** or **one atomic mass unit.** Accordingly, the weight of hydrogen is approximately one because of the single proton forming its nucleus. The electron mass is so small ($1/1{,}845$ of a proton) that we consider its mass to be negligible. Helium, the next largest atom, has a nucleus composed of two protons and two neutrons (giving it an atomic mass of 4) plus two electrons. As we work our way up the atomic scale toward heavier and heavier atoms, we find some, such as uranium, with dozens of protons. One form of uranium, in fact, has 92 protons and 146 neutrons, hence an atomic weight of 238.

Fig. 12.3 An atom of sodium and carbon are represented here. Notice the distribution of electrons at different energy levels.

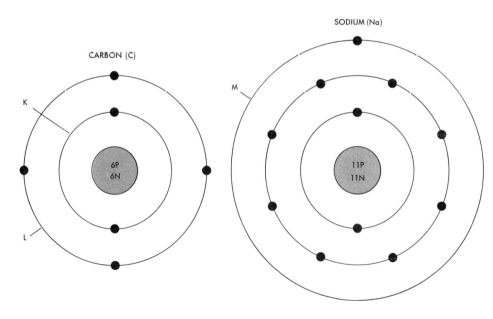

In general, electrons travel in specific **orbitals,** or shells, around the nucleus of an atom. Although an atom can have a large number of orbitals, all atoms seem to have at least seven in which electrons can be located. They are called the **K, L, M, N, O, P,** and **Q** levels. There is a maximum number of electrons which each orbital can accommodate. This does not mean, however,

TABLE 12-1 Energy Levels and Orbitals

Atomic Number	Element	Energy Levels						
		K	L	M	N	O	P	Q
1	Hydrogen	1						
2	Helium	2						
3	Lithium	2	1					
4	Beryllium	2	2					
5	Boron	2	3					
6	Carbon	2	4					
7	Nitrogen	2	5					
8	Oxygen	2	6					
9	Fluorine	2	7					
10	Neon	2	8					
11	Sodium	2	8	1				
12	Magnesium	2	8	2				
13	Aluminum	2	8	3				
14	Silicon	2	8	4				
15	Phosphorus	2	8	5				
16	Sulfur	2	8	6				
17	Chlorine	2	8	7				
18	Argon	2	8	8				
19	Potassium	2	8	8	1			
20	Calcium	2	8	8	2			
21	Scandium	2	8	9	2			
22	Titanium	2	8	10	2			
23	Vanadium	2	8	11	2			
24	Chromium	2	8	13	1			
25	Manganese	2	8	13	2			
26	Iron	2	8	14	2			
27	Cobalt	2	8	15	2			
28	Nickel	2	8	16	2			
29	Copper	2	8	18	1			
30	Zinc	2	8	18	2			
36	Krypton	2	8	18	8			
47	Silver	2	8	18	18	1		
53	Iodine	2	8	18	18	7		
56	Barium	2	8	18	18	8	2	
79	Gold	2	8	18	32	18	1	
92	Uranium	2	8	18	32	21	9	2

that an electron is bound to any one orbital. The electrons circling the nucleus of an atom are free to jump back and forth from one orbital, or energy level, to another.

However, energy uptake or release is necessary for this movement. For example, when an intense high-energy ultraviolet light ("black light") is directed at certain chemicals, the electrons in the outer energy orbital are "pushed" into a higher energy orbital for a brief period of time. This is usually an unstable situation and subsequently the electrons drop back spontaneously to the lower energy level. Energy is released in the process and may appear as visible light. The shining of safety badges and of a television screen are examples of this process. We will discuss how light can move electrons to higher energy levels when we consider how green plants convert sunlight into useful chemical energy.

The distribution of the electrons in the various orbitals gives each element its particular chemical property. Their electron distribution of several atoms is shown on this page. Notice that the outermost orbital, or energy level, of an atom never contains more than eight electrons. This is the stable electron pattern; atoms containing this arrangement are usually inactive, or **inert.** Neon and argon are two examples of inert gaseous elements.

How Atoms Combine

The number of electrons in the outermost energy level of an atom is what determines the ability of one atom to combine with another and form a molecule. These outer orbital electrons are called **valence** electrons. Understanding how atoms combine and are held together by a **chemical bond** is important to an understanding of the chemical changes that take place in living organisms. When two atoms collide or come close to one another, their electron clouds may overlap. The overlapping may lead to a sharing of electrons by the two atoms, or it may lead to an actual transfer of an electron from one atom to the other. If an electron is transferred, the atom losing it will be less negatively charged; that is, it becomes positive. The atom gaining the electron will be more negatively charged. This creates a situation in which the two atoms are held together because of their opposite charge. In such cases, **electromagnetic attraction** is the chemical bond. Thus, the chemical bond is an energy relationship between atoms; so in any chemical reaction we can expect energy changes. When a chemical bond is broken, potential energy is converted into **kinetic** energy, or energy of motion.

Gases such as neon and argon are chemically inert because they have eight electrons in their outermost shell. Atoms which are **reactive,** or combine with other atoms, do so because they

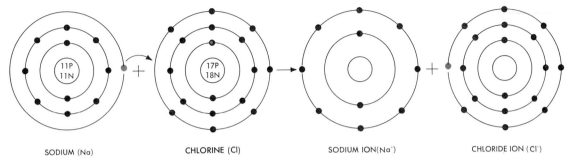

Fig. 12.4 This diagram shows the formation of sodium and chloride ions by the transfer of an electron from sodium to chlorine.

do not have a full stable of eight outer-level electrons. In general, atoms with fewer than four electrons in the outer orbital tend to give up electrons; those with more than four tend to gain electrons.

Let us see how this works by examining a common kitchen chemical—sodium chloride, or table salt. During the process where sodium (Na) and chlorine (Cl) atoms combine and form the compound sodium chloride (NaCl), one electron seems to be transferred from the sodium to the chlorine atom (Fig. 12.4). Sodium has 11 protons and 12 neutrons in its nucleus, giving it an atomic mass of 23. Because of its 11 protons, it has 11 electrons in the various orbitals. Two of these electrons are at the K energy level, eight at the L energy level, and one in the M. The nucleus of chlorine, on the other hand, has 17 protons and 18 neutrons, giving it an atomic mass of 35. Two electrons are at the K level, eight at the L level, and seven at the M level. Neither atom is stable (eight electrons in the outer orbital are required for stability or chemical inertness).

When sodium and chlorine react, the sodium atom acquires a net positive charge by losing an electron; it is designated Na^+. The chlorine atom acquires a net negative charge by gaining an electron and is designated Cl^- (Fig. 12.4). Both atoms are now **ions**. An ion is an atom with an unbalanced electrostatic charge; or, putting it another way, an ion is a charged atom. For this reason, we say that sodium and chlorine are joined by **ionic** bonding. A crystal of NaCl is held together, therefore, by the electrostatic forces that act between oppositely charged ions.

Another type of bonding between atoms is one in which electrons are not transferred, but shared. In effect, a shared electron fills an orbital for each atom, giving stability to each atom by completing an electron pair. A shared pair of electrons is called a **covalent** bond. Atoms of a number of elements form covalent bonds, and can also occur as two-atom, or **diatomic**,

Fig. 12.5 A shared pair of electrons is called a *covalent* bond. Atoms of a number of elements form covalent bonds. Hydrogen and chlorine are shown here.

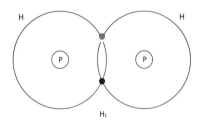

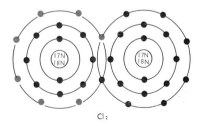

molecules. Hydrogen and chlorine gases are good examples. The two nuclei of chlorine atoms are held together because both electrons of the electron pair are attracted to both positively charged nuclei. Knowing that chlorine exists as a diatomic molecule, and that sodium does not, we can now write a chemical equation showing that the reaction of chlorine with sodium requires two molecules of sodium, but only one of chlorine gas:

$$2\,Na + Cl_2 \longrightarrow 2\,Na^+ + 2\,Cl^- \longrightarrow 2\,NaCl$$

A number of elements of biological interest combine by forming covalent bonds. Oxygen and hydrogen are two. For all organisms, the most important compound these two elements form is water. In the next chapter we shall take a detailed look at this substance and other compounds that are of particular importance to life processes.

SUMMARY

All matter in the universe is made up of specific combinations of substances called **elements,** such as hydrogen, oxygen, and nitrogen. The basic unit of an element is called an **atom.** Atoms are composed of three primary particles: (1) the heavy positively charged **protons;** (2) **neutrons,** which are electrically neutral and which are packed with the protons in the nucleus; and (3) **electrons,** negatively charged particles which surround and circle the nucleus in definite energy levels (orbits).

The number of electrons in a given atom always equals the number of protons in the nucleus, so that the atom is electrically neutral. Atoms combine to form specific chemicals by sharing electrons **(covalent bond),** or by transferring electrons **(electromagnetic attraction,** or **ionic bonding).** When table salt, NaCl, is dissolved in water, the sodium and chlorine exist as ions in solution (Na^+, Cl^-).

FOR THOUGHT AND DISCUSSION

1 Look at the table of elements and determine the number of electrons, protons, and neutrons in some of the common elements that are familiar to you.

2 How many of the following terms do you understand?

atoms	nucleus	ionic bonds
electrons	atomic mass	covalent bonds
neutrons	orbitals	electromagnetic attraction
protons	valence electrons	ions

SELECTED READINGS

Baker, J. J. W., and G. E. Allen. *Matter, Energy and Life.* Reading, Mass.: Addison-Wesley Publishing Co., Inc., 1965.

An excellent small paperback written for the purpose to provide students with a background in the chemistry and physics essential to understanding modern biology.

Bush, G. L. and A. A. Siluidi. *The Atom: A Simplified Description.* New York: Barnes and Noble, 1961.

An elementary description of atom structure.

Drummond, A. H., *Atoms, Crystals and Molecules* (Part 2). Columbus, Ohio: American Education Publication, 1964.

Written for secondary school students this booklet discusses in detail ionic covalent bonds, hydrogen bonds, and electronegativity.

13 THE CHEMISTRY OF BIOLOGICAL COMPOUNDS

Life as we know it on this planet is intimately associated with a water environment. Even those organisms that live in deserts are made up of large quantities of water. It is only in this aqueous environment that cells can function and maintain normal life. When cells or organisms lose too much water, all of their life processes cease. Yet certain cells, particularly among the microorganisms, can be completely dehydrated and later restored to activity by submersion in water. To gain some appreciation of the importance of water to metabolic function, let us now look into the structure and properties of this most important of all biological compounds.

What Is Water?

The union of two hydrogen atoms and one oxygen atom produces the remarkable molecule we call water. As you saw in the previous chapter, a hydrogen atom consists of one proton and one electron. An oxygen atom, on the other hand, has eight protons and eight neutrons in its nucleus. This makes it 16 times as massive as a hydrogen atom. The eight protons give the nucleus of the atom a positive charge of eight. An oxygen atom also has eight electrons, two of which are in the K orbital close to the nucleus, while the remaining six are in the outer (L) orbital. As we said earlier, atoms with more than four electrons in the outer orbital are electron gainers. Oxygen atoms tend to gain two electrons from some other source, thus filling the outer orbital (to a total of eight electrons) and becoming a stable electronic structure. One source of two electrons is two hydrogen atoms. When one oxygen atom and two hydrogen atoms combine by covalent bonding, a stable molecular union results. So much for the composition of a water molecule. What about its structure?

Fig. 13.1 To function and maintain normal life, the cells of all organisms require a water environment. How does the desert cactus plant obtain water?

The American Museum of Natural History

The Chemistry of Biological Compounds 195

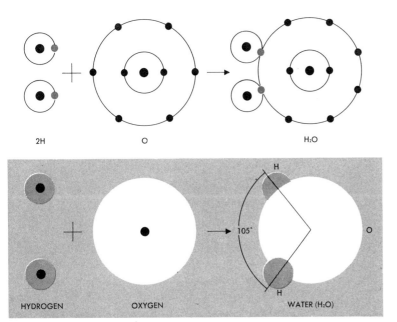

Fig. 13.2 The formation and geometry of the water molecule.

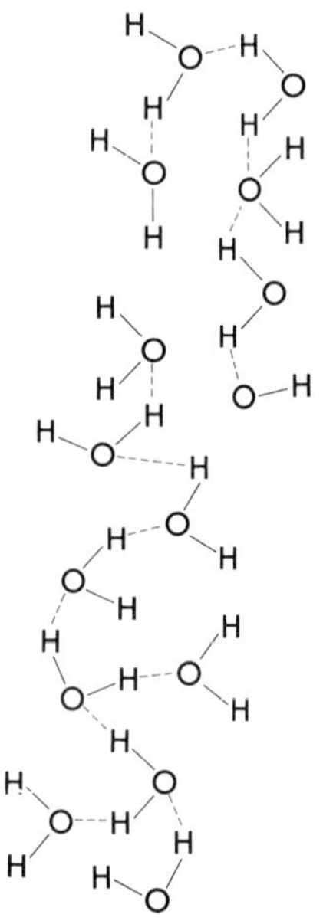

Because of their like (positive) charges, the two hydrogen nuclei tend to repel each other. This results in their taking the position shown in the diagram. The fact that the electrons of the hydrogen are shared with the oxygen atom in a covalent bond means that each hydrogen atom electron no longer spends as much time on one side of its nucleus as on the other side (see Fig. 13.2). As a consequence, the hydrogen "end" of a water molecule has a positive charge, while the oxygen "end" has a negative charge. Such a molecule is called a **polar** molecule.

This polarity of a water molecule makes the molecule an electron seeker. It will attach itself to any other atom or molecule having an available electron. This results in a **hydrogen** bond, a very important bond in biochemical reactions. Thus the hydrogen bond results from the tendency of a hydrogen atom to share electrons with two other atoms, usually oxygen atoms—for example, with other water molecules. Formic acid is another molecule which tends to form hydrogen bonds with itself (Fig. 13.4).

Although hydrogen bonds are weak compared with covalent bonds, they are strong enough to give rise to some of the unique properties of water. Water, for example, is the only substance that is commonly present in all three states as a solid, liquid, and gas in the range of temperatures found at the Earth's surface. It is true that various other atmospheric gases can also be

Fig. 13.3 Hydrogen bonding between water molecules is indicated by the colored lines. The tendency of hydrogen atoms in one water molecule to combine with oxygen in a second water molecule gives water an organized structure.

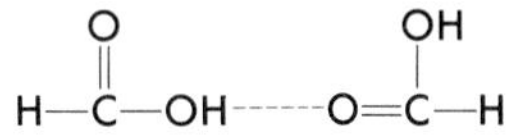

Fig. 13.4 Formic acid is another molecule tending to form hydrogen bonds with itself.

converted into a liquid or solid, but only in the laboratory where we can make use of extreme pressures or temperatures.

If we compare the boiling and freezing points of water with other chemicals similar to water, the variations we see in other compounds are surprising. For example, hydrogen sulfide (H_2S) has a freezing point at $-83°$ centigrade and a boiling point at $-62°C$. Water, on the other hand, has a freezing temperature that is 83 centigrade degrees higher ($0°C$), and a boiling temperature that is $162°$ higher ($100°C$). This difference is accounted for if we consider the relatively greater thermal energy required to break the hydrogen bonds joining water molecules. Thus, water molecules tend to attract one another into an organized structure because of the hydrogen bonding. Energy is required to break these water molecules apart before the individual molecules can escape from solution as a vapor. This is what we mean by "boiling"—the escape of individual water molecules.

The great stability in the water molecule structure was important for the origin and evolution of life as we know it on this planet. At room temperature, the thermal energy in a water molecule is so great that hydrogen bonds between molecules are being broken continuously and are reforming continuously. When water flows, we can visualize the molecules as tumbling over each other, breaking and re-establishing hydrogen bonds as they go. When the thermal energy of water molecules drops below the level necessary to break these bonds, ice forms. Later we shall return to the role water and hydrogen bonds play in relation to the function of enzymes.

Carbon Compounds

The carbon atom is of special interest to students of chemistry, biochemistry, and biology. It plays a key role in the structure of molecules which are essential to living things. A carbon atom has six protons and six neutrons forming the nucleus, giving the atom a mass of 12. It also has a total of six electrons; two at the K level and four at the L level (Fig. 13.5). Carbon is neither strongly electronegative (electron attracting), nor strongly electropositive (electron repelling). For the most part, therefore, carbon enters into chemical combination by sharing electrons—covalent bonding—with other carbon atoms, or with atoms of other elements.

From a biological standpoint, the most outstanding property of carbon is the fact that it can share electrons with other carbon atoms and form long straight or branched chains to which atoms of other elements can attach themselves. There are so many different carbon compounds (more than half a million)

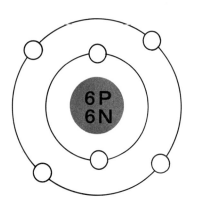

Fig. 13.5 Schematic representation of the carbon atom

that a special branch of chemistry—**organic** chemistry—is devoted to their study.

When carbon atoms combine and form a chain, and when all the left-over electrons are shared with hydrogen atoms, substances known as **hydrocarbons** are formed. Figure 13.6 shows some of the simplest hydrocarbons: methane CH_4, ethane C_2H_6, and propane C_3H_8. Notice that the electronic structure is shown for methane. The open dots represent the carbon's electrons originally in the L orbital; the solid dots represent the hydrogens' electrons in the K orbital. Neither atom has *lost* its electrons. Instead, they *share* their electrons with each other, and the result is a stable configuration of eight electrons in the outer orbital.

The shared electrons form a moving cloud and may circle the nucleus of any atom in the molecule. This makes a very stable covalent link between carbon and hydrogen. Since hydrogen has only one electron to share, and carbon needs four to complete the stable configuration, it should be clear why four hydrogens combine with one carbon, forming methane. Because carbon will accept four electrons from another source, and because it can share its own four electrons with another atom, we say that carbon has a valence of four.

Earlier you saw that two hydrogen atoms each contributes one electron when they combine with oxygen and form water. Oxygen, then, has a valence of two. Oxygen also combines with carbon, forming carbon dioxide (CO_2). However, in this case two oxygen atoms are involved, each one sharing a pair of electrons with the carbon atom. This type of covalent bond is called a **double bond,** and is represented by the bar in Fig. 13.7. Both the electronic and bond formulae are shown. Note that the sharing of four electrons by each oxygen and the carbon help complete the stable electronic configuration of both atoms.

As in the case of carbon dioxide above, one carbon atom can form a double bond with another carbon atom, and the remaining valence electron can combine with hydrogen to form the **unsaturated** hydrocarbons. They are called "unsaturated" because the *carbon* atoms are sharing the electrons, rather than some other atom, such as hydrogen. Some examples are shown in Fig. 13.8.

Fig. 13.6 Saturated hydrocarbons.

METHANE

ETHANE

PROPANE

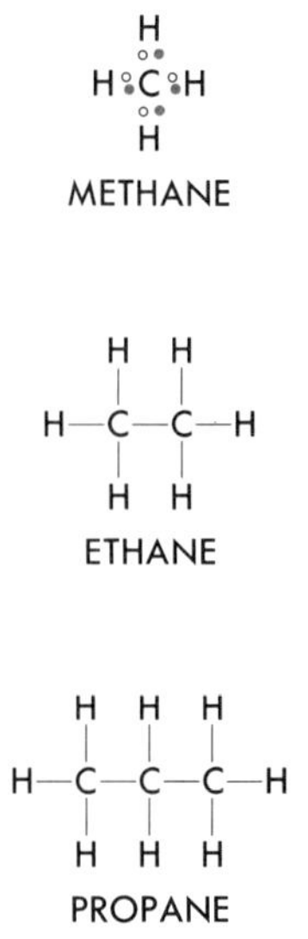

Fig. 13.7 The atoms forming carbon dioxide are held together by a type of covalent bond called a *double bond*.

Fig. 13.8 Unsaturated hydrocarbons are shown below.

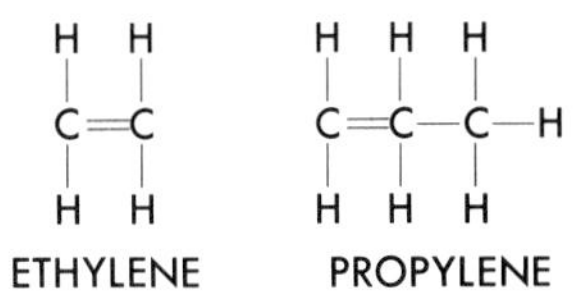

ETHYLENE   PROPYLENE

Fig. 13.9 The removal of water from alcohol results in the formation of a double bond in the ethylene.

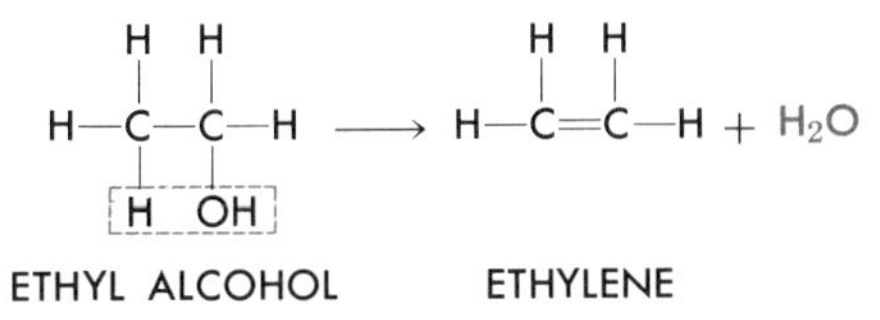

ETHYL ALCOHOL → ETHYLENE + H_2O

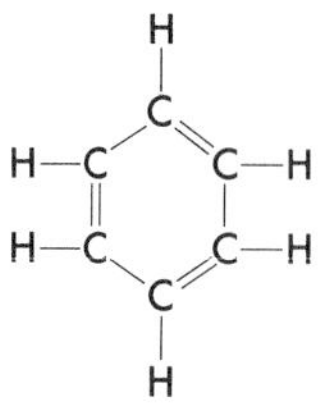

Fig. 13.10 The benzene ring.

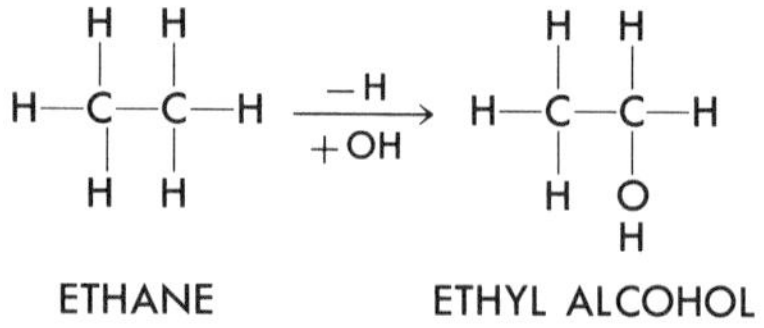

Fig. 13.12 Phenol.

Fig. 13.14 Acetone.

ETHANE   ETHYL ALCOHOL

Fig. 13.11 The formation of an alcohol can be brought about by adding an OH (hydroxyl) group to a hydrocarbon.

Important Chemical Groups

In addition to a straight-line arrangement, carbon atoms can link together and form ring structures. Because many of these **ring compounds** have rather fragrant odors, they are called **aromatic**. The parent compound of all these aromatic substances is **benzene**, the structure of which is shown in Fig. 13.10. Many important biological compounds have a ring structure.

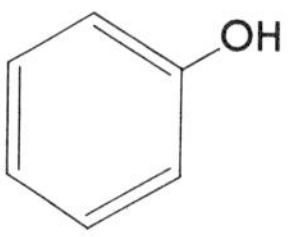

If we replace one of the hydrogen atoms of a hydrocarbon with a unit of oxygen and hydrogen (OH), called a **hydroxyl group**, we produce a compound belonging to the class of **alcohols**. If the hydroxyl group replaces a hydrogen of an aromatic ring, the compound formed is known as a **phenol**. The chemistry of alcohols and phenols is primarily that involving the properties of a hydroxyl group attached to a carbon atom.

When an alcohol such as ethyl alcohol is **oxidized** (see Fig. 13.13), that is, when it loses electrons, it is converted to an **aldehyde**. In the example illustrated, it is **acet**aldehyde. Compounds containing a **carbonyl** group (C=O) are known as aldehydes or **ketones**. If one of the valences of the carbonyl compound is used in bonding with a hydrogen atom, the compounds are aldehydes. The remaining valences may bond with the hydrogen atom or a number of other groups. If one is a **methyl** group (CH_3), as in Fig. 13.13, then the compound becomes acetaldehyde. If both valences of the carbonyl carbon are used to bind other carbon atoms, then a ketone is formed. **Acetone** is a good example.

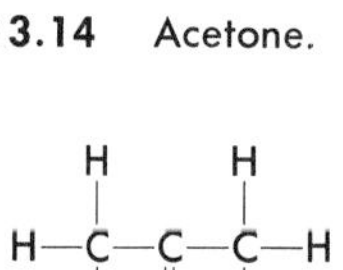

Another group of organic substances of biological importance is comprised by those compounds containing the **carboxyl** group.

Fig. 13.13 The oxidation (removal of hydrogen) of an alcohol produces an aldehyde, as shown below.

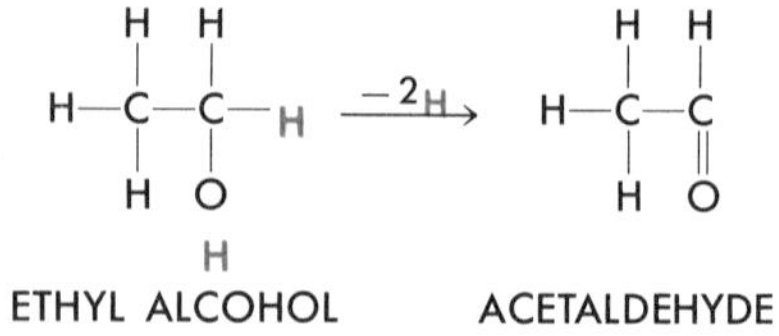

ETHYL ALCOHOL   ACETALDEHYDE

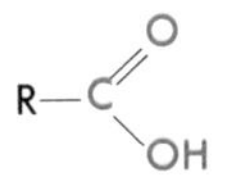

Fig. 13.15 The R group shown attached to the carbon atom of the carboxyl group represents any number of groups. If R=H we have the structure of formic acid. If R=CH$_3$— we have acetic acid.

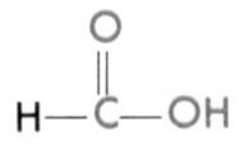

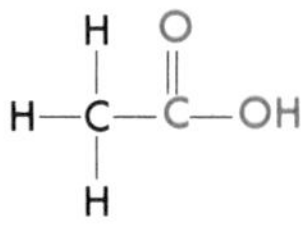

This is a name coming from "**carb**onyl" and "hydr**oxyl**." The structure of a carboxyl group is shown in Fig. 13.15. Compounds of this general class are known as **carboxylic acids.** In the ordinary straight chain series, those compounds containing the carboxyl group are known as **fatty acids,** two of which are shown in Fig. 13.15. There are many other general classes of acids which contain the carboxyl group.

Chemical Reactions

When a substance such as hydrochloric acid (HCl) is dissolved in water, the hydrogen separates from the chlorine. When this happens, however, the electron normally associated with the hydrogen atom remains with the chlorine. This dissociation in water, therefore, leads to the formation of ions—a hydrogen ion (a proton with a net charge of +1), and a chloride ion with a net charge of −1. The ionization of HCl in water can be written as follows:

$$HCl \rightleftarrows H^+ + Cl^-$$

Normally, H^+ and Cl^- would immediately recombine and form HCl again; however, in water this charge separation is partly maintained by water molecules located between the two ions. Not all compounds ionize when they are dissolved in water. Molecules held together by ionic bonds, such as NaCl, tend to dissociate. Those held together by covalent bonds do not. In the case of the ionic bond, one atom has *given up* an electron to another, so the only forces holding the two atoms together is an electromagnetic attraction which water can overcome. In covalent bonds, the electrons in the outer energy levels are *shared* by the atoms involved; this means that more energy is needed to separate them.

We discuss this problem of charge on ions because of its importance to the combination of atoms in chemical reactions. In all chemical reactions, atoms combine in very definite proportions. Knowing whether an atom tends to give up or accept electrons, and knowing how many, helps us to predict the nature of the reaction. For example, NaCl dissociates into Na$^+$ and Cl$^-$. When calcium reacts with chlorine, it forms calcium ($CaCl_2$). Calcium has two electrons in the N energy level, and that chlorine needs only one electron to complete the octet at the M level. Thus, for each calcium atom we need two chlorine atoms as follows:

$$Ca + Cl_2 \longrightarrow CaCl_2$$

If we then add water to the calcium chloride, the $CaCl_2$ ionizes and forms two chloride ions for each calcium ion. The calcium ion carries a $+2$ net charge:

$$CaCl_2 \longrightarrow Ca^{++} + 2Cl^-$$

Oxidation-Reduction

When an atom loses an electron, we say that it is **oxidized;** when an atom gains an electron we say it is **reduced.** We emphasize electron transfer at this time for the following reason: When electrons move to different energy levels (orbitals), there is a release of energy that can be used by organisms. When sodium and chlorine combine as sodium chloride, there is a transfer of an electron from the sodium to the chlorine. The sodium is oxidized; the chlorine is reduced. Another way of saying the same thing is that sodium is a **reducing agent,** and chlorine is an **oxidizing agent.** When oxidation-reduction reactions occur in a cell, we need to know if *useful* energy is released, and if so, whether it is available for use by the cell. As we shall see later, the oxidation of carbohydrates, fats, and other food stuffs is the primary source of energy for cell function.

Acids and Bases

To describe chemical reactions that take place in cells and organisms, we must know something about the chemical nature of **acids** and **bases.** There are many types of acidic and basic substances in organisms. Some play major roles in cell structure; others are important in metabolism. Acids are substances that can donate a proton (hydrogen ion). The concentration of the hydrogen ions determines the degree of acidity of a solution. Bases are substances that combine with hydrogen ions. In aqueous solution, for example, hydrochloric acid dissociates into hydrogen ions and chloride ions. The chloride ions comprise a base (because they can combine with hydrogen ions).

$$\underset{\text{Acid}}{HCl} \rightleftharpoons \underset{\substack{\text{Hydrogen} \\ \text{ion}}}{H^+} + \underset{\substack{\text{Chloride ion} \\ \text{(a base)}}}{Cl^-}$$

Sodium hydroxide (NaOH), on the other hand, is a base. It dissociates into a sodium ion (Na⁺) and a base (OH⁻).

$$\underset{\text{Base}}{\text{NaOH}} \longrightarrow \underset{\substack{\text{Sodium} \\ \text{ion}}}{\text{Na}^+} + \underset{\substack{\text{Hydroxyl ion} \\ \text{(a base)}}}{\text{OH}^-}$$

Consequently, NaOH can neutralize the acid HCl, forming water and NaCl.

$$\text{HCl} + \text{NaOH} \longrightarrow \text{H}_2\text{O} + \text{NaCl}$$

The carboxyl groups of fatty acids, such as acetic acid, can also donate protons.

$$\underset{\text{Acetic acid}}{\text{CH}_3-\text{C}\overset{\text{O}}{\underset{\text{OH}}{\diagup}}} \rightleftharpoons \text{H}^+ + \underset{\substack{\text{Acetate ion} \\ \text{(a base)}}}{\text{CH}_3-\text{C}\overset{\text{O}}{\underset{\text{O}^-}{\diagup}}}$$

Notice that when an acetate ion takes up a proton it is converted to an acid. Thus, when a base accepts a proton it is converted into a molecule which acts as an acid. Hydrogen ion concentration, then, is very important in determining the net charge on molecules inside cells. The negative and positive charges on very large molecules are important in determining their biological reactivity.

Fig. 13.16 The hydrogen ion concentration determines the acidity of a solution: high concentration produces high acidity, low concentration, low acidity.

Concentration of H⁺

One way of expressing the hydrogen ion concentration of a substance is to speak of the **pH** of the substance. On the pH scale, 7 is the neutral point. Any substance with a pH of 7 is neither acid nor base; it is neutral. Water, for example, is a substance that is neutral, so its pH value is 7. A small amount of HCl in water, on the other hand, has a much higher concentration of hydrogen ions; its pH value is less than 7.0. NaOH in water gives pH values much greater than 7.0. (The definition of pH is "the negative log of the hydrogen ion concentration." Thus, if the hydrogen ion concentration is 10^{-3} molar, the pH is 3; if the concentration is 10^{-8} molar, the pH is 8, and so on.

```
←acidic                                              basic→
                        ─── Neutral ───
0   1   2   3   4   5   6   7   8   9   10   11   12   13   14
```

The pH is very important in cell metabolism. Big changes in acidity of the cells of an organism can greatly affect metabolism. The pH of blood in mammals, for example, is kept very close to

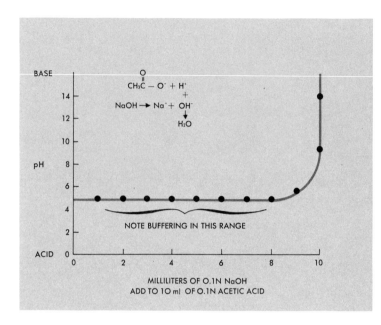

Fig. 13.17 Titration of acetic acid with the base, sodium hydroxide.

7.35. A shift as small as 0.2 can result in death. Organisms have evolved chemical systems called **buffers,** which tend to keep the pH relatively constant. Buffer systems resist changes in pH when acids or bases are added. There are many buffers in the human body, but in principal they all work the same way. We can use acetic acid (CH_3COOH) as an example. When the acid is dissolved in water, it dissociates into the negative acetate ion and hydrogen ion to give a pH of about 4.0. If we add a small amount of base, such as NaOH, to this acetic acid solution, OH⁻ ions

$$CH_3\overset{O}{\overset{\|}{C}}-OH \rightleftharpoons CH_3-\overset{O}{\overset{\|}{C}}-O- + H^+$$

neutralize the H⁺ and water is formed; however, additional acetic acid will dissociate and contribute more H⁺ to the solution, making it acid again. In other words, acetic acid acts as a reservoir for H⁺ so that the pH of the solution tends to remain constant. The buffer role of acetic acid is shown in Fig. 13.17.

You have now seen how certain compounds vital to cells are built up from elements. In Chapter 14 we turn our attention to several major compounds present in cells.

SUMMARY

Life on this planet is intimately associated with a water environment. Water molecules tend to associate with each other because of their **hydrogen bonding** properties. The tendency of hydrogen in one water molecule to associate with an oxygen in a second water molecule leads to the formation of a structure involving multiple combinations of water molecules. Energy is required to break these water molecules away from one another, which suggests why water has a high boiling point.

The carbon atom is of special interest because it is part of so many compounds that are essential to life. Carbon atoms can combine with each other and form long-chain hydrocarbons. They can also form ring structures, such as benzene.

There are a number of important chemical groups such as the **hydroxyl** (OH), **aldehyde** ($-C\overset{O}{\underset{H}{\diagdown}}$), and **carboxyl** ($-C\overset{O}{\diagdown}-OH$).

Substances which contain the carboxyl group are **acids** because they can dissociate, which enables them to contribute a hydrogen ion (H^+). The concentration of the hydrogen ions determines the degree of acidity of a solution. **Bases** are substances that combine with hydrogen ions. We use **pH** to express the concentration of hydrogen ions, or acidity, of a solution—pH 7.0 is neutral; values less than 7.0 are acid; those greater than 7.0 are alkaline, or basic.

FOR THOUGHT AND DISCUSSION

1 When ice is thawing and forming liquid water can you visualize the state of the water molecules? Why can H_2O exist as a solid *and* a liquid? Remember hydrogen bonding and thermal energy.

2 What are **oxidation** and **reduction?** Can you oxidize something without reducing something in an ordinary chemical reaction?

3 What is a **hydrocarbon, hydroxyl group, carbonyl group, acid, base?** What is **pH?**

4 Can you think why the pH, or hydrogen ion concentration (H^+), is important in determining the electronic charge of chemical substances? Write down an example and explain.

SELECTED READINGS

See the books listed at the end of the previous chapter.

14 MAJOR COMPOUNDS OF CELLS

Carbohydrates, fats, proteins, and nucleic acids are the major large molecules that occur inside of cells. These molecules, in turn, can associate with each other and make even larger cellular structures.

THE ROLE OF CARBOHYDRATES

Carbohydrates play a key role in the energy requirements of organisms. They are the principal products formed when a green plant captures light energy in the process of **photosynthesis.** In this process, the energy of sunlight is used to convert CO_2 and H_2O to the energy-rich bonds of carbohydrates:

Fig. 14.1 The sunlight reaching these green plants underwater provides energy needed by the plants in the production of carbohydrates.

Richard F. Trump

$$\text{CO}_2 + \text{H}_2\text{O} \xrightarrow{\text{Light energy}} \text{C(H}_2\text{O)} + \text{O}_2$$
Carbon dioxide / Water / Carbohydrate

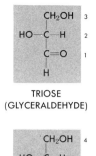

TRIOSE
(GLYCERALDEHYDE)

Sugars and starches are the principal sources of energy in the ordinary diet of most organisms. The carbohydrates, although used primarily as an energy source, also supply important carbon "skeletons"—carbon atoms linked together and associated with hydrogen and oxygen—that are necessary in the manufacture of the basic components of the living material of all cells.

The basic units in carbohydrates are carbon, hydrogen, and oxygen. The term "carbohydrate" means *hydrate of carbon*. This name was used because it included many compounds that contain atoms of hydrogen and oxygen in the same proportion occurring in water—two hydrogens to one oxygen. So a carbohydrate can be described by the general formula $\text{C(H}_2\text{O)}$. Carbohydrates range from relatively simple molecules called sugars to the complex molecules of starches and cellulose. The simplest class of carbohydrates is represented by the **monosaccharides.** They are further classified according to the lengths of their carbon chains: three-carbon compounds **(trioses)**; four-carbon sugars **(tetroses)**; five-carbon sugars **(pentoses)**; six-carbon sugars **(hexoses)**, and so on to the 10-carbon sugars (Fig. 14.2).

Glucose and **fructose** are the two hexoses which serve as the principal source of energy and building material for most cells. Both can be represented by the same formula, $\text{C}_6\text{H}_{12}\text{O}_6$. Glucose contains five hydroxyl groups, each attached to a different carbon atom (see Fig. 14.3). But many sugars can have the same chemical composition. The differences among them depend on the positioning of the hydrogen and hydroxyl groups around the carbon. Different groupings produce different chemical properties. Those sugars of the same chemical composition, but different chemical properties, are called **isomers.** Cells are able to distinguish one isomer from another because of the different chemical groupings of the H and OH around the carbon atoms. When we talk about isomers, therefore, we must consider their molecular geometry.

Other important monosaccharides are the two pentoses, **ribose** and **deoxyribose.** They are components of the nucleic acids comprising RNA and DNA, which are discussed in Chapters 18 and 19. Figure 14.4 shows the difference between ribose and deoxyribose. Notice that carbon atom number 2 of deoxyribose lacks an OH group. The OH has been replaced with a hydrogen.

Monosaccharides can be linked to form larger units. The **disaccharides,** those complex sugars containing two monosaccharides, are the most common. Disaccharides are formed by combining two monosaccharides. Energy is required to do this—an

TETROSE
(ERYTHROSE)

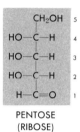

PENTOSE
(RIBOSE)

Fig. 14.2 Carbohydrates with different numbers of carbon atoms. Names of specific examples are given in parentheses.

Fig. 14.3 Structure of the hexose sugars: glucose and fructose. The carbon atoms of glucose are numbered as shown. The positioning of the H and OH groups are important. For example, if we exchange the positions of H and OH groups on carbon 4 of glucose we have the formula for the sugar galactose.

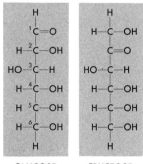

GLUCOSE   FRUCTOSE

Fig. 14.4 Ribose and deoxyribose. Notice that carbon atom number 2 of deoxyribose lacks an OH group. It has been replaced by an H.

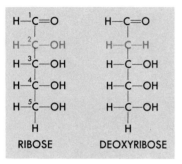

Fig. 14.5 The formation of a disaccharide from two monosaccharides by the loss of water. For simplicity, only the carbon skeleton and the two OH groups are shown.

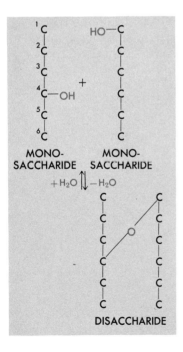

OH from one sugar combines with a H from a second sugar and forms water. This leads to the formation of a carbon-oxygen-carbon bond between the two sugars (see Fig. 14.5). When two glucose units are linked through the number 1 and number 4 carbons, the disaccharide **maltose** is formed.

One of the most important disaccharides is **sucrose,** or cane sugar. Sucrose is made up of equal quantities of glucose and fructose. Very complex molecules, such as starch and glycogen, are called **polysaccharides** because they are made up of a large number of monosaccharides joined in a long chain. **Starch** is the reserve carbohydrate in most plants and is formed by green plants in the process of photosynthesis. Starch is made up of glucose units linked in much the same way as maltose. **Glycogen** is the reserve carbohydrate of animals. It is found in high concentration in liver and muscles. Disaccharides and polysaccharides are readily broken down to monosaccharides by splitting the C—O—C bond with water.

Cellulose is another polysaccharide of great importance. It is the chief ingredient of the cell walls of plants. Cellulose is the major polysaccharide occurring in nature and is probably the most abundant organic compound found on our planet. It com-

Fig. 14.6 The glycogen molecule is made up of many units of glucose. In Fig. 14.3 the basic chemical unit (glucose) is shown. Glucose units are held together by a bond between the number 1 carbon of one glucose molecule and the number 4 carbon of a second glucose molecule. At intervals a glucose molecule is attached to the number 6 carbon. This leads to a branching point. Both arms of the branch grow by the addition of glucose making use of the 1-4 bond. A limited part of a glycogen molecule is shown schematically at the right in this diagram. The circles represent glucose molecules.

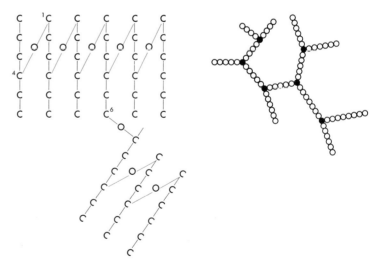

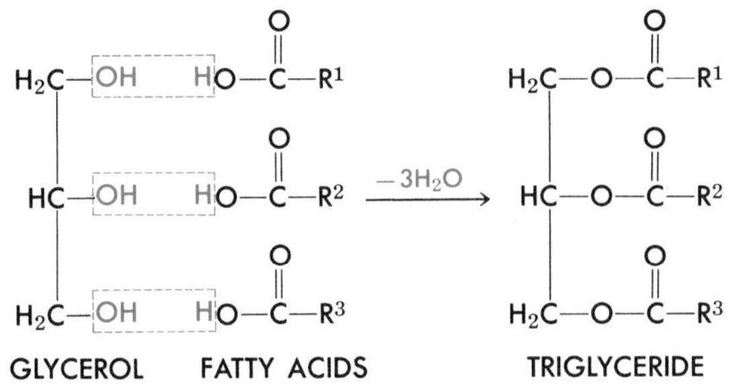

```
H₂C—OH   HO—C—R¹           H₂C—O—C—R¹
          ‖                      ‖
          O                      O
          
HC—OH    HO—C—R²  —3H₂O→   HC—O—C—R²
          ‖                      ‖
          O                      O
          
H₂C—OH   HO—C—R³           H₂C—O—C—R³
          ‖                      ‖
          O                      O
GLYCEROL  FATTY ACIDS          TRIGLYCERIDE
```

Fig. 14.7 The synthesis of a triglyceride. The fats contain a mixture of what are called *triglycerides*, which are links between glycerol and fatty acids. As shown in this figure, the hydroxyl group (OH) of glycerol reacts with the carboxyl group (—C(=O)—OH), forming the bond. The loss of water and the formation of this new bond requires metabolic energy. The R^1, R^2, and R^3 compounds shown in the figure can be identical long-chain fatty acids or they can be a mixture. The hydrophobic ("water-repelling") region of the molecule is due to the long-chain fatty acids while the hydrophilic ("water-loving") region is due to attached compounds that have OH, COOH, H_3PO_4, or other such water-soluble groups. See the structure of lecithin, for example.

prises at least 50 per cent of all the carbon in the plant world. Cellulose is the main constituent of cotton, wood pulp, linen, straw, and many other substances of plant origin.

Cellulose is insoluble in water and most organic solvents. Yet it is made up of glucose units linked together in very much the way maltose is, but with a slightly different configuration. These differences in shape are very important in stabilizing the molecule and making it resistant to biochemical attack by organisms. For example, the digestive systems of man and most animals do not contain the necessary chemical agents to break cellulose down. This is in contrast to their ability to break down compounds such as maltose, which has a very similar linkage. However, many microorganisms can decompose cellulose and use it as a food. Although termites eat wood, they cannot themselves break the cellulose down, or digest it, as a source of nourishment. However, microorganisms living in the intestines of termites do the job of digesting. They are able to split the cellulose into the simple nonsaccharides which the termites can use. This is an example of **symbiotic** relationship. The microorganisms live in a sheltered environment and in turn provide the host (termite) with sugar.

Thus, a very subtle difference in the chemical makeup of the compound can determine whether it can be metabolized by an organism, and hence be useful or useless as a source of food.

LIPIDS

The **lipids**, or fats, are a group of organic substances originating in the living cell. They are classified into several subdivisions on the basis of their chemical and physical properties. All fats, such as butter, the fat on meat, olive oil, are built up from carbon, hydrogen, and oxygen atoms, and consist of two major components—fatty acids and **glycerol** (see Fig. 14.7).

In general, animal fats are in a solid or semisolid state at room temperature; vegetable fats or oils are in a liquid state

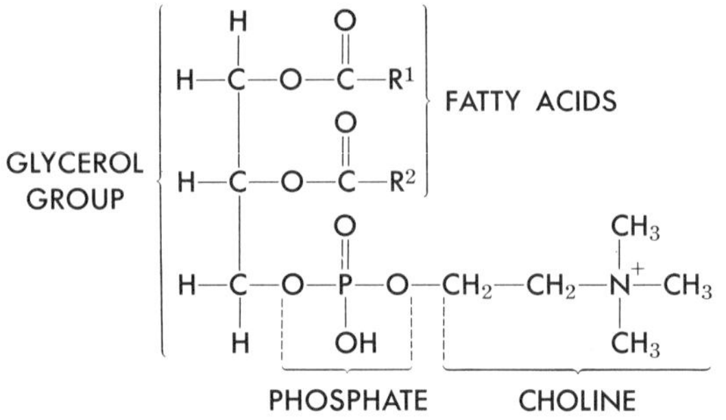

$$\text{GLYCEROL GROUP} \begin{cases} H-\overset{H}{\underset{|}{C}}-O-\overset{O}{\overset{\|}{C}}-R^1 \\ H-\overset{|}{\underset{|}{C}}-O-\overset{O}{\overset{\|}{C}}-R^2 \\ H-\overset{|}{\underset{H}{C}}-O-\overset{}{\underset{OH}{\overset{\|}{P}}}-O-CH_2-CH_2-\overset{CH_3}{\underset{CH_3}{\overset{|}{N^+}}}-CH_3 \end{cases}$$
 FATTY ACIDS

PHOSPHATE CHOLINE

Fig. 14.8 Structure of lecithin.

largely because their fatty acid components have a greater number of unsaturated bonds (double bonds). Waxes, such as beeswax and the cuticle waxes of fruits, leaves, and flower petals, are principally esters of a fatty acid with a long chain alcohol rather than with glycerol. Bees wax, for example, is an ester of palmitic acid (a 16-carbon saturated fatty acid) and myricyl alcohol (a 30-carbon chain saturated alcohol).

In the simple lipid, one of the fatty acids may be replaced by compounds containing phosphorous and nitrogen. When this happens, **lecithin** and **cephalin** are formed. Called **phosphotides**, they frequently represent the major portion of cellular lipids. These two compounds are soluble in both water and fats and, therefore, serve a vital role in the cell by binding water-soluble compounds, such as proteins, to lipid-soluble compounds. Lecithin is a key structural material in the cell membrane. It maintains continuity between the aqueous and lipid faces of the inside and the outside of the cell. In the diagram showing the structure of lecithin, the R groups represent a long-chain fatty acid.

PROTEINS

The structure, function, and metabolic activity of a cell or tissue all depend on a class of molecules called **proteins**. Proteins make up a significant portion of the protoplasm of plant and animal cells alike. Proteins also are an important part of the structure of the chromosomes, the nucleoplasm, and the nuclear membrane. All the structures in the cytoplasm—including the mitochondria, the ribosomes, the spindle fibers, and the flagella structures which are used for motion—are made up in part of protein molecules.

In addition, there are various proteins that operate as **catalysts**—agents that speed up chemical reactions. These catalytic proteins are called **enzymes**. Enzymes are intimately associated with all chemical processes occurring in the cell—muscle contrac

Fig. 14.9 Electron micrograph of a fragment of subcutaneous connective tissue, air dried and shadowed with chromium. The photograph probably represents the normal arrangement of the fibers in the tissue.

The Upjohn Company—Jerome Gross, M.D.

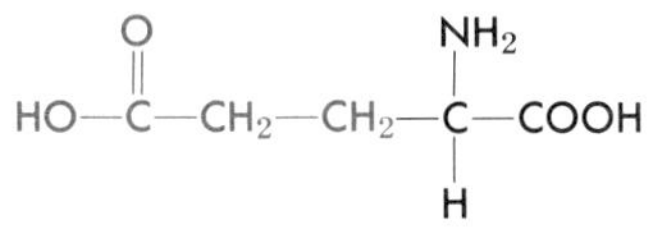

$$\text{HO}-\overset{\overset{\displaystyle O}{\|}}{C}-CH_2-CH_2-\overset{\overset{\displaystyle NH_2}{|}}{\underset{\underset{\displaystyle H}{|}}{C}}-COOH$$

Fig. 14.10 Structure of glutamic acid.

tion, nerve conduction, excretion, absorption, and general metabolic reactions. Many proteins, such as **keratin** of the skin, finger nails, wool, and hair, and the **collagen** of connective tissue and bone, serve in a structural capacity. The **antibodies,** which combine with disease-producing agents and destroy them, are also proteins of a very specific kind. There are important protein hormones that regulate cellular and tissue functions. Insulin is one. The amount of insulin in the blood determines whether we are normal or diabetic.

The Structure of Proteins

Proteins are molecules of gigantic size, sometimes containing tens of thousands of atoms. They are tremendously complex and have no competitors in the diversity of roles they play. In addition to containing carbon, hydrogen, and oxygen, proteins also contain nitrogen. They range in molecular weight from about 5,000 (insulin) to 40,000,000 (tobacco mosaic virus protein). Although we have much more to learn about the chemical and physical properties of protein molecules, we do know that their complex and diverse properties are related to at least three major aspects of their structure.

Primary Structure: When proteins are broken down in the presence of water **(hydrolyzed),** they yield a mixtuxe of simple organic molecules that contain nitrogen. These molecules are called **amino acids** (see Fig. 14.10). The $-NH_2$ group is the amino group while $-COOH$ is the carboxyl or acidic group since it can donate hydrogen ions to the solution. All amino acids contain these two groups. Thus, in general we can write the structure of an amino acid as follows:

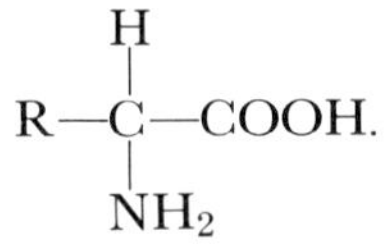

$$\text{R}-\overset{\overset{\displaystyle H}{|}}{\underset{\underset{\displaystyle NH_2}{|}}{C}}-COOH.$$

The chemical nature of the R group in amino acids can vary considerably. In fact there are about 20 different R groups in nature; therefore, there are 20 different amino acids. For instance, glutamic acid shown here has an R group that is made up of two CH_2 groups and one carboxyl group.

In a protein molecule the amino acids are joined by a carbon-nitrogen bond between the carboxyl group of one amino acid

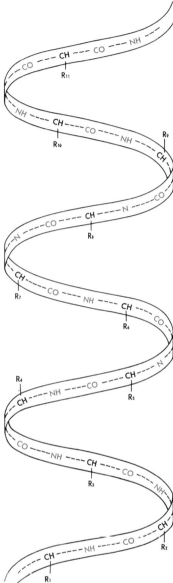

Fig. 14.12 The spiral pattern of a polypeptide chain. The peptide bonds are shown in color. The R groups represent the remainder of the amino acids. Hydrogen bonds along the axis of this helical structure stabilize it and give rise to what is called the secondary structure. The amino acid sequence ($R_1 \rightarrow R_{11}$) is called the primary structure.

$$\underset{\text{AMINO ACID 1}}{R-\underset{NH_2}{\underset{|}{C}}-\underset{\|}{\overset{O}{C}}-OH} + \underset{\text{AMINO ACID 2}}{H-\underset{COOH}{\underset{|}{N}}-\underset{|}{\overset{H}{C}}-R^1} \longrightarrow \underset{\text{A DIPEPTIDE}}{R-\underset{NH_2}{\underset{|}{C}}-\underset{\|}{\overset{O}{C}}-\underset{|}{\overset{H}{N}}-\underset{|}{\overset{H}{C}}-R^1}$$

Fig. 14.11 Synthesis of a peptide bond by joining two amino acids.

and the amino group of another by the removal of water. The reaction is shown in Fig. 14.11. This bond between two amino acids ($-\overset{O}{\underset{\|}{C}}-\overset{H}{\underset{|}{N}}-$) is called the **peptide** bond. If only two amino acids are linked, it is called a **di**peptide; if three, a **tri**peptide, and so on. If a large number are connected, it is called a **poly**peptide (Fig. 14.12).

Proteins are made up of long polypeptide chains consisting of hundreds upon hundreds of amino acid units connected to one another by peptide bonds. Since there are approximately 20 different amino acids in nature, it follows that the amino acids in the long polypeptide chains of proteins must be present in many different combinations. The possible number of different combinations, in fact, is enormous. The thousands of different proteins are in part ascribed to an almost countless variety of possible combinations. Thus, what we call the *primary* structure of the protein is determined by the number, kind, and sequence of the amino acids in the polypeptide.

Hydrogen bonding, which we discussed earlier, is important in maintaining the structure of proteins. In the protein molecule, the hydrogen bonds generally occur between the oxygen of one peptide group and the amino nitrogen of a neighboring group. Although the forces of hydrogen bonding are relatively weak compared to most covalent bonds, the fact that a large number of hydrogen bonds can occur in a polypeptide makes them collectively a major force in protein structure. However, they are weak enough so they can form and break with relative ease, thus giving the protein molecule a mobile structure.

Secondary and Tertiary Structure: We now know that the long polypeptide chains of proteins are not perfectly straight chains. Instead, they exist as spirals held together by hydrogen bonding. The spiral nature of the polypeptide chains of proteins is spoken of as its secondary structure.

Many proteins also have a tertiary structure. The tertiary structure is due to the spiral polypeptide chain folding forward and backward on itself, forming a globular, rather than a long fibrous, molecule. The folding of most proteins is not a chance arrangement. It is a definite spatial configuration specific for each particular protein, and it is maintained, in part, by hydrogen

Fig. 14.13 A schematic representation of a protein molecule showing the spiral polypeptide chain folding in a number of ways to give a *tertiary* structure to the proteins. Hydrogen bonding and the formation of disulfide (S—S) cross links stabilize the structure.

bonding. Thus, some of the very important properties of proteins are related to their secondary and tertiary structure.

Among other important properties of a protein is its electrical charge. Electrical charge depends on the acidic and basic properties of the individual amino acids. Thus, the *p*H of a solution in which a protein occurs determines many of the protein's chemical properties. At neutral *p*H many of the free carboxyl groups are dissociated and give a negative charge (COO^-) to the protein. As the hydrogen ion (H^+) concentration is increased this negative charge is neutralized. The ability of various small molecules to combine with the protein often depends on the electric charge.

GROWTH REQUIREMENTS OF ORGANISMS

Although we have mentioned some of the organic chemicals which are either found in cells or are produced by them, we have not said anything about the growth requirements of organisms. Many different types of molecules are necessary for the formation and maintenance of protoplasm, the living material of cells. Organisms vary widely in their ability to manufacture certain molecules with their own metabolic machines. Some cells, for example, cannot make certain of the amino acids; consequently, if the organism is to survive, these substances must be supplied in its diet. They are, therefore, **essential growth factors** for the organism. A study of growth requirements is called **nutrition,** and it encompasses all the ways in which food is used.

Fig. 14.14 This five-month-old tung seedling shows symptoms of manganese deficiency. The lower leaves are normal because they developed before the manganese in the seed was exhausted.

Vitamins are essential for the normal functioning and growth of all organisms. While some simple microorganisms and most green plants can manufacture all the vitamins they need, other organisms cannot; those that cannot must have another source of vitamins—their diet. If a required vitamin is omitted from the diet of a higher organism, a vitamin deficiency disease will occur. If the organism is a single cell organism, it will not grow or multiply.

When vitamin deficiency diseases were first studied, the chemical structure of vitamins was not known; therefore, the vitamins were referred to simply as vitamins A, B, C, and so forth. In the early study of vitamin nutrition it was found that certain vitamins were soluble in fat solvents (such as ether or alcohol), while other vitamins were soluble in water. It turns out that many of the water-soluble vitamins occur in what is now known as the **B vitamin complex.** Some of the B vitamins include pyridoxine, nicotinic acid (niacin), pantothenic acid, biotin, folic acid, vitamin B_{12}, and others.

The fat-soluble vitamins include vitamins A, D, E, and K. Vitamin A exists only in animal products, but there is a yellowish substance in plants, called *carotene,* (because it is found in carrots), which can easily be changed into vitamin A by animal cells. Vitamin A is essential for the growth of higher organisms: for the maintenance of nerve tissue and the growth of bone and tooth enamel. A deficiency of vitamin A produces "night blindness," the inability to see in dim light.

Vitamin D is another fat-soluble vitamin. Since it is made in the body under the influence of sunlight, it is sometimes called the "sunshine vitamin." It is essential for the absorption of calcium from the intestinal tract. Vitamin E appears to be necessary to prevent sterility in male animals. In cellular metabolism it is involved in the process of oxidation of complex molecules such as carbohydrates and fats. Vitamin K also seems to play a role in a general oxidative process, and in adult mammals it is essential for the normal coagulation of blood.

Autotrophic Organisms

Autotrophic organisms are able to grow and multiply in a purely inorganic medium. In other words, they do not depend on an outside source for vitamins, amino acids, or other complex organic molecules. From a nutritional viewpoint, autotrophs are the least exacting group of organisms, for they produce their own sugars, fats, amino acids, and so forth from carbon dioxide and ammonia (NH_3). They obtain their energy in one of two ways and are subclassified on the basis of this energy source.

Chemosynthetic Autotrophs: Organisms of this group make their own protoplasm from carbon dioxide, ammonia, or nitrate (NO_3^-) and obtain the energy for the synthesis by the oxidation of inorganic substances. For example, one chemosynthetic autotroph present in the soil is the bacterium *Nitrosomonas*. It is capable of oxidizing ammonia to nitrate, thus generating useful energy:

$$2NH_3 + 3O_2 \longrightarrow 2HNO_2 + 2H_2O + \text{energy (79,000 calories)}$$

Fig. 14.15 *Bacilli Nitrosomonas is a chemosynthetic bacterium present in soil.*

Walter Dawn

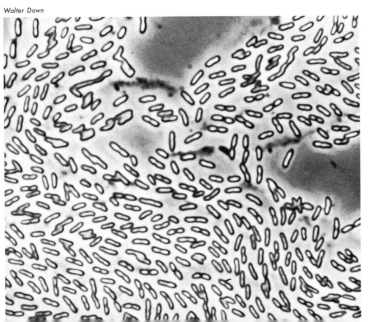

Fig. 14.16 *Chlorella* is a photosynthetic unicellular organism.

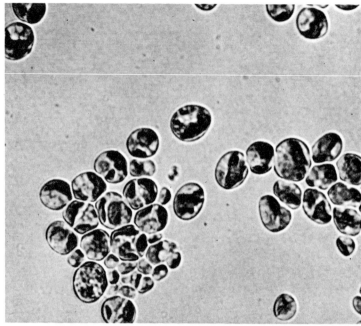

Walter Dawn

This organism is a typical chemosynthetic autotroph. Since its cells contain all the complicated carbohydrates, fats, proteins, nucleic acids, and vitamins, it represents a magnificent synthetic factory for making protoplasm.

Photosynthetic Autotrophs: These organisms obtain the energy for their synthetic activities by converting light energy into chemical energy by photosynthesis. They obtain their nitrogen from ammonia or nitrate, and their carbon from carbon dioxide in the air. Some of the organisms in this group are colored sulphur bacteria, diatoms, the blue-green, red, brown, and green algae, and complex green plants. The colors in the plant result from mixtures of pigments, including the crucial one, chlorophyll, which is capable of trapping light energy. Of all the biochemical processes in nature, photosynthesis is of paramount importance. In the green plant, the end product of photosynthesis is a reserve of chemical energy (carbohydrates) that serves as a sole source of energy for most living things.

Heterotrophic Organisms

The other major category of organisms is the heterotrophic group. It consists of organisms that get their energy mainly from organic sources, such as carbohydrates. Hetero-

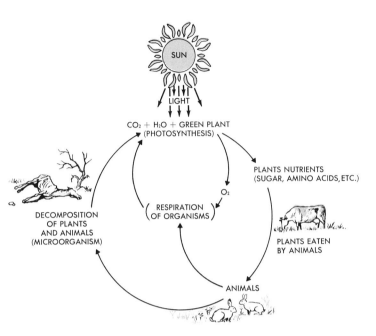

Fig. 14.17 The Carbon cycle: Green plants use the energy of sunlight, carbon dioxide, and water to make all the organic chemicals necessary for life. During this photosynthetic reaction oxygen is also liberated. Thus, animals depend on plants for all of their food as well as oxygen. The end products of animal respiration are water and carbon dioxide. When an animal dies it rapidly decays to carbon dioxide and water —except for the skeleton. The over-all effect of these various activities leads to a carbon cycle.

trophic organisms, therefore, are related to animals in their general metabolism; autotrophic organisms are related primarily to plants.

Man is a very complex heterotroph. He must eat other organisms in order to obtain the growth factors he himself cannot make. He requires not only certain kinds of amino acids in his diet, but other required growth factors include a number of vitamins as well as certain fatty acids. Compared with many other organisms, the synthetic machinery of man is less versatile, hence less complex. The photosynthetic green plants, on the other hand, are remarkable in their ability to use the energy of sunlight to make essentially all the known vitamins and amino acids. It is for this reason that many higher vertebrates, such as man, depend on plants as their primary supply of food—if not directly, then indirectly by eating other animals that eat the plants (Fig. 14.17).

Aerobic and Anaerobic Organisms—Oxygen as a Nutrient

Some organisms cannot live and grow without molecular oxygen (O_2); such organisms are said to be **aerobic**. Those organisms which can grow in the complete absence of oxygen are said to be **anaerobic**. For anaerobic organisms, oxygen is often toxic; it inhibits growth.

There are other organisms, such as the yeast cell, that can

live either anaerobically or aerobically. These organisms offer excellent material for studying the transitional changes that occur when the organism is changed from one environment to another. It appears that during embryonic development of vertebrates, for example, that cells and tissues may shift from an anaerobic to an aerobic way of life. The mechanisms involved in this shift are not understood, although a great deal of interesting experimental work is now being done on this important problem.

INORGANIC SALTS

In addition to gases such as oxygen, carbon dioxide, and, in some cases, nitrogen, there are a large number of inorganic minerals that are essential for growth. The need for these minerals varies from cell to cell, particularly from plant cells to animal cells. The mineral nutrients can be classified into two broad categories: **macro**nutrients, which are required in large quantities; and **micro**nutrients, which are required only in trace amounts.

For both plants and animals, the major macronutrients are sodium, chlorine, potassium, calcium, phosphorus, and magnesium. The principal micronutrients are iron, copper, manganese, and zinc. In addition to these, animals need cobalt, iodine, and possibly vanadium and selenium. Plants require, in addition, boron, molybdenum, and vanadium. But vanadium is essential for only certain forms of plants.

Most of the 103 elements in the periodic table have been found in living cells, but this does not mean that all that have been found are *essential* to life. If an organism is burned, about 95 per cent of the ash is made up of potassium, phosphorous, calcium, magnesium, silicon, aluminum, sulphur, chlorine, and sodium. The remaining 5 per cent or less is accounted for by the micronutrients and other elements.

There is one important class of organic compounds that we have not mentioned in this chapter—the nucleic acids. Their role in the manufacture of protein is the subject of Chapter 18, for we need first to discuss the chemical reactions that take place within cells, and how energy is handled by the cell.

SUMMARY

In this chapter we have briefly reviewed most of the major chemical substances that are associated with living organisms, such as carbohydrates, fats, and proteins. We also saw that the source of chemical energy preserved in the carbohydrates is photosynthesis, which is carried out by green plants. Simple sugars such as glucose can combine with one another to form complex polysaccharides, such as glycogen (in animals) or starch (in plants). One of the chief ingredients of the cell walls of plants is cellulose, a complex polysaccharide. It can be broken down by microorganisms and be made available to other organisms as a useful nutrient.

Proteins are also complex polymers made up of varying combinations of simple amino acids. The primary structure of proteins is determined by the kind and number of amino acids linked together by the peptide bond. This polypeptide can form spirals and fold back and forth on itself, giving a secondary and tertiary structure to this complex macromolecule. The secondary and tertiary structures are important for certain catalytic and functional roles for the protein, and the structures are maintained, in part, by hydrogen bonds.

In order for organisms to make these complex structures internally—and thus grow and reproduce—certain nutrients are required. An energy source is essential (carbohydrates), and a nitrogen source (for the synthesis of amino acids and other nitrogen-containing compounds). If an organism cannot make the vitamins it requires, then it must consume other organisms that can make the vitamins, or obtain the vitamins from other sources in the diet.

Certain autotrophic organisms can live and multiply on a diet that contains only CO_2, ammonia, minerals, and water. They obtain energy from the oxidation of ammonia (NH_3). These are called **chemosynthetic autotrophic** organisms. The green plants obtain their energy from sunlight and are called **photosynthetic autotrophs.** Man and most mammals are complex **heterotrophic** organisms. They must obtain essential amino acids, fatty acids, and vitamins from other organisms that are capable of making them.

Oxygen is essential for a large number of organisms, but there are some that can live and reproduce without this gas (**anaerobic** organisms). All organisms require certain major elements, such as sodium, chlorine, potassium, calcium, phosphorous, and magnesium. In addition, trace amounts of iron, copper, manganese, and zinc are also essential. Cobalt and iodine (and possibly vanadium and selenium) are required by animals; boron, molybdenum, and vanadium are required by plants. Because these latter elements are needed in trace amounts only does not mean that the trace elements are less important than the major elements (which are required in larger amounts). The trace elements function, as we shall see later, as catalysts, thus only small amounts are required.

FOR THOUGHT AND DISCUSSION

This and the previous chapter attempt to give you the necessary chemical background for understanding some of the metabolic processes that we will be discussing in the next few chapters.

1 Make sure that you understand what a carbohydrate is.
2 Can you write the straight-chain general formula for glucose?

It is also important to remember that there are six-carbon sugars, five-carbon sugars, and so on.

Proteins and their structure are also very important. We will discuss them in greater detail later.

3 Make sure that you understand what an amino acid is composed of.
4 How is a peptide bond formed?
5 What is a **polypeptide?**

You should have a general understanding of the "important" chemicals that are associated with living organisms. If an organism cannot make essential chemicals, then it must obtain them in its diet.

6 What is a vitamin?
7 A trace metal?
8 An autotrophic organism?
9 A heterotrophic organism?
10 Is oxygen essential to all organisms?

SELECTED READINGS

Baldwin, E. *Dynamic Aspects of Biochemistry* (3rd ed.). New York: Cambridge University Press, 1957.
  An outstanding introduction to the concepts of cellular metabolism.

Bennett, T. P., and E. Frieden. *Modern Topics in Biochemistry.* New York: Macmillan Co., 1966.
  An excellent discussion in greater detail of the various subjects discussed in our last three chapters. Written for the beginning college student.

Loewy, A. C. and P. Siekevitz. *Cell Structure and Function.* New York: Holt, Rinehart, and Winston, 1963.
  A detailed elementary discussion of various aspects of cell structure and biochemistry.

McElroy, W. D. *Cell Physiology and Biochemistry* (2nd ed.). Englewood Cliffs, N.J.: Prentice Hall, Inc., 1964.
  A small paperback intended for beginning college students.

Neilands, J. B. and P. K. Stumpf, *Outlines of Enzyme Chemistry* (2nd ed.). New York: John Wiley & Sons, 1958.

An excellent elementary introduction to enzyme chemistry.

Readings from Scientific American

Allen, R. D., "Amoeboid Movement," February 1962.
Allfrey, V. G. and A. E. Mirsky, "How Cells Make Molecules," September 1961.
Brachet, J., "The Living Cell," September 1961.
Doty, Paul, "Proteins," September 1957.
Frieden, E., "The Enzyme—Substrate Complex," August 1959.
Fruton, Joseph S., "Proteins," June 1950.
Green, D., "The Synthesis of Fat," February 1960.
Hayashi, T. and G. A. W. Boehm, "Artificial Muscle," December 1952.
Holter, Heinz, "How Things Get into Cells," September 1961.
Huxley, H. E., "The Contraction of Muscle," November 1958.
Kendrew, J. C., "Three-Dimensional Structure of a Protein," December 1961.
Lehninger, A., "Energy Transformation in the Cell," May 1960.
Lehninger, A., "How Cells Transform Energy," September 1961.
McElroy, W. D. and C. P. Swanson, "Trace Elements," January 1953.
Moore, S. and W. H. Stein, "The Chemical Structure of Proteins," February 1961.
Pfeiffer, John E., "Enzymes," December 1948.
Robertson, David J., "The Membrane of the Living Cell," April 1962.
Siekevitz, P., "Powerhouse of the Cell," July 1957.
Solomon, A. K., "Pores in the Cell Membrane," December 1960.
Solomon, A. K., "Pumps in the Living Membrane," August 1962.
Stein, W. H. and S. Moore, "The Chemical Structure of Proteins," February 1961.
Stumpf, P. K., "ATP," April 1953.

15 METABOLIC PROPERTIES OF CELLS

To observe some of the properties of a single cell, we have only to peer through a small hand lens or a microscope. The shifting of the protoplasm inside the cell, the changing size and shape of the cell, and the movements of the cell toward or away from light are only some of the cell's dramatic properties. However, there is a realm of smaller things—atoms and molecules—that we cannot see through an ordinary microscope. Our task in this chapter is to find out how a cell maintains itself as a living chemical unit.

THE PHYSICAL AND CHEMICAL NATURE OF PROTOPLASM

Cells of *all* animals are almost identical to the ones in your own body and are all basically like the cells of

Fig. 15.1 The granular material visible in this cell is part of the protoplasm. The protoplasm is the "living" material in all cells.

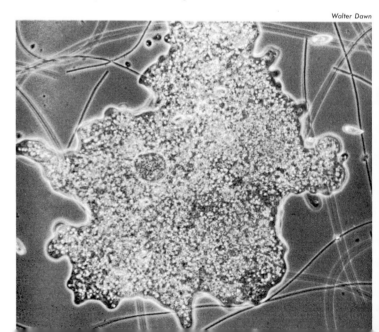

Walter Dawn

plants. In an earlier chapter, you found that the cell is made up of three main parts: (1) the **cell membrane** (also called the **plasma** membrane), which surrounds the total contents of the cell; (2) **cytoplasm,** which contains a large number of important structures such as the mitochondria, the endoplasmic reticulum, lysosomes, and ribosomes; and (3) the **nucleus,** which contains the hereditary material.

The cytoplasm, nucleus, and cell membrane are all living parts of a cell and collectively are called **protoplasm.** It is the chemical and physical organization of protoplasm with which we are immediately concerned at this point. We have talked about the chemical elements and how they join to make compounds. Protoplasm is a mixture of a large number of both simple and complex chemical compounds. Compounds made up of carbon, hydrogen, oxygen, nitrogen, and many other elements, are found in protoplasm in many different combinations. Not only are these elements united in a variety of compounds, but the compounds themselves are often combined in very definite proportions, and with a very definite organizational pattern. A cell, it should be emphasized, is not an indiscriminate mixture of compounds, but an intricate and organized arrangement.

Earlier, we found that cells contain a watery fluid. When we have such a fluid system containing molecules of various kinds, we can classify them in a number of ways. If the particles are small, such as sugar molecules, they can dissolve in the fluid and become a **true solution.** If, on the other hand, the particles are very large, such as grains of sand which can sink to the bottom, the system is called a **coarse suspension.** If particles are of intermediate size, and neither settle out nor form a true solution, we call the fluid system a **colloid.**

Protoplasm and Colloidal Systems

There are many different types of colloidal systems. Probably the one best known to you is mayonnaise—a mixture of fat and protein in water. Fog also is a colloidal system, one made up of small water particles suspended in air. Most biological systems, however, are liquid-liquid combinations. Protoplasm is in part a true solution and in part a colloidal system. There are several important properties of protoplasm that make us classify it as a colloid. For example, the dispersed particles in the cytoplasm of the cells—large protein molecules and large oily fat molecules—are of such size and charge that they do not settle out under ordinary gravitational force. Instead, they keep moving about constantly. This is called **Brownian motion,** and its effect is easily observed under the microscope. Since the

Aerofilms, Ltd. From Ewing Galloway, N.Y.

Fig. 15.2 Fog is a colloidal suspension. Liquid water droplets are suspended in a gas (the air).

Fig. 15.3 Small particles suspended in water tend to move about with a haphazard motion. This is due to the random bombardment of the suspended particles by the water molecules and is called Brownian motion. Molecules of all kinds are in constant motion, and when they collide with another object some of their kinetic energy imparts motion to that object.

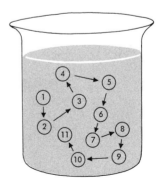

221

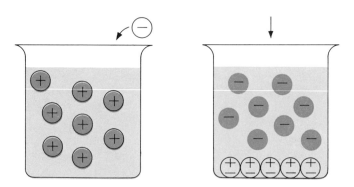

Fig. 15.4 Large particles tend to stay dispersed if they have a positive or negative charge (colored circles). Neutralized particles tend to settle out.

speed at which molecules move about depends on the amount of thermal or kinetic energy they have, Brownian motion becomes more vigorous as the temperature is raised. You can easily see this for yourself in a pot of water as it is heated to a boiling temperature.

One of the more important properties of a colloid is the electric charge of the dispersed particles. Protein molecules, fat droplets, and other large molecules of biological interest usually carry a net positive or negative charge. Because of these differences in charge, many of the molecules are prevented from aggregating. If two molecules carry a net positive charge, they repel one another; thus, the particles are kept apart. If the charges are neutralized by a negative charge, they settle out (Fig. 15.4).

Another interesting property of colloids is that they can form **gel**-like substances. For example, if we greatly increase the concentration of protein in a water solution, the protein molecules line up parallel to one another, making a compact rod-shaped

Fig. 15.5 Examples of living and nonliving gel-like structures: human skin (left) jello (center) and an ameba.

particle. When many such particles interlock and form a network with water droplets throughout, the gel state of the colloid is formed. Your skin, jelly fishes, jello, and gelatin are colloids in this gel condition (Fig. 15.5). High temperatures tend to dissolve the gel state and convert it into what appears to be a viscous solution. This change from a gel to a **sol** is very important in a number of biological processes. The movement of an amoeba or the ability of a jellyfish to maintain its structure depends on the gel-like property of protoplasm.

Protoplasm, then, is a colloid with gel-like properties, and it is composed largely of protein, lipids, and water droplets. Such is the internal environment of a cell. Outside the cell membrane, there is usually a water environment. Let us now find out what exchanges take place between these two environments, and the role of the cell membrane in permitting or preventing an exchange of material. (For a review of the structure and composition of the cell membrane, see Chapter 8.)

THE CELL MEMBRANE

Permeability

When a cell is placed in a suitable nutrient medium, it takes up the nutrients, and thrives. Just because a nutrient is close at hand outside the cell, however, does not mean that the cell is capable of using it. The substance must first pass through the cell membrane.

All the food products a cell takes in must be soluble, to a certain degree, in the water outside the cell in order to pass through the cell membrane. Likewise, waste products within the cell must be soluble in the protoplasm in order to pass out of the cell. Since not all dissolved substances can penetrate the membrane with equal ease, the membrane is said to be **selectively permeable.** This selectivity is vital in maintaining the life of a cell. Although the cell membrane is the major structure safeguarding the cell's internal environment, other parts of the cell, such as the nucleus and mitochondria, are also bounded by selective membranes that control their internal environments.

Osmosis

Whenever two different solutions are separated by a selectively permeable membrane, an **osmotic system** is established. Each cell, therefore, represents an osmotic unit. **Osmosis** may be defined as the movement of water molecules through a selectively permeable membrane. In cells, water is exchanged between the protoplasm and the solution surrounding each cell.

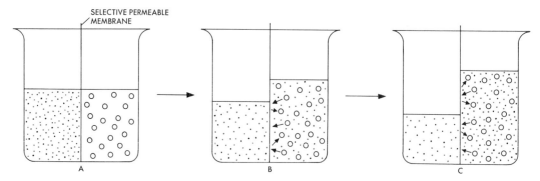

Fig. 15.6 A schematic representation of osmosis. Initially, we have a vessel divided into two chambers by a selectively permeable membrane (A). On the left side we have a solution containing small molecules that are capable of passing through the membrane. On the right side we have a solution of larger molecules that cannot pass through the membrane. In time—(B) and (C)—the small molecules will be distributed throughout both chambers. However, since the larger molecules cannot pass through the membrane, there is a tendency for the water molecules to move from left to right because of the differential bombardment of the two sides of the membrane by the water molecules. The pressure that must be applied on the right-hand column to prevent the increase in volume is known as the osmotic pressure.

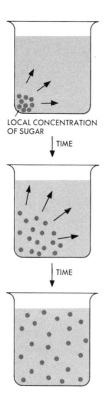

Fig. 15.7 A high concentration of sugar in water diffuses. In time the concentration of sugar will be the same throughout the water.

Take a simple osmotic system in which water is separated by a selectively permeable membrane from a solution of sugar (Fig. 15.6). In this case, both the **solute** (sugar) and **solvent** (water) tend to diffuse, because of Brownian movement, from an area of higher concentration to one of lower concentration (Fig. 15.7). In a perfect system, where the membrane keeps the sugar isolated, only the water is able to penetrate. The two solutions can reach equilibrium, that is, a state in which further change does not occur, only by the transfer of water into the sugar solution.

Since the water concentration on one side of the membrane is higher than on the other side, water molecules will pass into the sugar solution and increase the volume of water. Eventually, the pressure of the water on the sugar solution side of the membrane will stop the entrance of more water. This pressure is called **osmotic pressure.**

The cell is not a perfect osmotic system. Not only water, but many dissolved substances commonly present in and around the protoplasm are able to penetrate the cell membrane. This enables the exchange of water between the cell and its surroundings to be accompanied by the exchange of other substances as well. Oxygen and carbon dioxide, for instance, pass readily into and out of the cell. The permeability of the complex cell membrane depends not only on the nature of the surrounding particles, but also on the changing conditions inside and outside the cell.

Although permeability varies in different cells, and sometimes on different sides of the same cell membrane, we can make certain generalizations about osmosis. For example, we know that water rapidly penetrates most cells. Gases such as carbon dioxide, oxygen, and nitrogen, and fat solvent compounds such as alcohol, ether, and chloroform easily penetrate all cell membranes. Somewhat slower to penetrate are such organic substances as glucose, amino acids, and fatty acids. Inorganic salts, acids, and large disaccharide molecules such as sucrose, maltose, and lactose penetrate even more slowly by the process of simple diffusion.

The bulk of all solutes present in protoplasm (including proteins and most sugars and inorganic salts) penetrate the cell membrane very slowly, if at all. The solvent water, on the other hand, enters and leaves the cell very quickly. This, coupled with the fact that water is more abundant than all other components combined, means that the water must bear the main burden of establishing osmotic equilibrium between the cell and surrounding solutions. If the cell is placed in a solution with a water concentration drastically different from that in the protoplasm, so much water enters or leaves the cell that the cell may be destroyed. The cell wall in plants and most microorganisms is usually rigid enough to prevent the swelling of the cell when water rushes in, but an animal cell, bounded only by the cell membrane, can be easily ruptured in this way.

In an **iso-osmotic** solution, the concentration of the water outside the cell is the same as that in the protoplasm. This concentration occurs only when the total concentration of solute particles in the external solution equals that of solute particles in the protoplasm (Fig. 15.8). The water balance is achieved because the water molecules that are continuously escaping from the cells are matched by an equal number of water molecules entering the cells. A true iso-osmotic solution, therefore, contains a concentration of nonpenetrating or very slowly penetrating solute molecules that approximate the total concentration of nonpenetrating solutes in the protoplasm. An equal water concen-

Fig. 15.8 The effect of the osmotic environment on cell structure. When a plant cell (A) is placed in a hypertonic solution of a slowly penetrating solute (B), water is rapidly lost from the cell and the cell membrane shrinks away from the cell wall. As the solute slowly penetrates—(C) and (D)—water re-enters and the cell swells, resuming its original size.

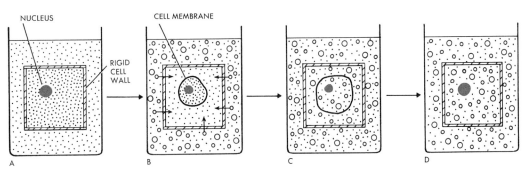

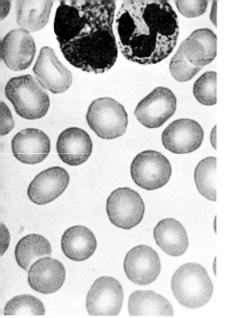

tration inside and outside the cell could not otherwise exist.

A **hypotonic** solution has a relatively low concentration of non-penetrating solutes compared to the protoplasm of the cell it surrounds. Thus the water concentration in it is relatively high. Animal cells placed in a hypotonic solution tend to take in water and swell. If the solute on the outside is of very low concentration, the swelling will continue until the cell membrane ruptures. When human red blood cells are placed in a solution containing only 0.2 per cent sodium chloride, instead of the usual 0.9 per cent, the corpuscles swell and burst, even before they can be observed under a microscope.

Cells placed in a **hypertonic** solution, on the other hand, tend to shrink. This happens because the solution contains a higher concentration of solute molecules than the protoplasm does. When animal cells are placed in a hypertonic solution, they shrivel up so much that we cannot detect their original shape. When a red blood cell, for example, is placed in a concentrated sucrose solution, it shrinks from loss of water. If this process is not carried too far, it is reversible (Fig. 15.9).

If we place a plant cell in a hypertonic solution whose solute particles slowly penetrate the cell, the plant membrane contracts away from the cell wall (Fig. 15.8). This process is called **plasmolysis**. As the solute particles slowly enter, water returns and the cell returns to its original size. This reverse process is called **deplasmolysis**. By observing deplasmolysis we can determine the rate of penetration of solute particles into the cells.

Fig. 15.9 Human red blood cells in a hypertonic solution (below) are shown here compared with normal red cells.

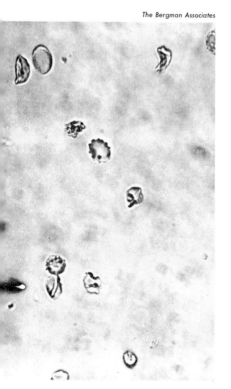

Active Transport

The selective cell membrane is not a *passive* organ. While simple diffusion and osmosis account for much of the exchange of material between cells and surrounding fluids, the cell membrane also plays an active part in the exchange process. Many marine algae, for instance, accumulate iodine to a concentration more than a million times greater than that of the sea. Such situations cannot be accounted for by the simple laws of diffusion or osmosis alone. Cells, therefore, have a means of forcing molecules of a particular substance to move in a direction opposite to that dictated by the laws of diffusion. This is the process of **active transport.**

There seem to be specific proteins and other molecules in the cell membrane that act as catalysts for moving substances across the membrane and into the cell. We call this transport machinery "metabolic pumps," and speak of the "sodium pump," the "potassium pump," and so on.

One of the best ways to show how the metabolic pumps work is to use **protoplasts** of bacteria. Protoplasts are bacteria with

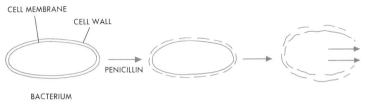

Fig. 15.10 Normally when bacteria are placed in a solution that contains fewer osmotically active molecules (hypotonic) than the interior of the cell, the bacterium tends to swell because of the inward movement of water. The strong cell wall, however, prevents the cell membrane from expanding. If we add penicillin to the medium the synthesis of the cell wall is inhibited and the wall begins to break down. When this happens the cell is enlarged by the inward movement of water. If the molar concentration of molecules outside the cell is too low, the bacterium bursts and dies.

their cell walls removed. Certain bacteria accumulate so much sugar lactose, for example, that the protoplasts swell and burst due to the internal increase of osmotically active particles. The point is that the concentration of lactose within the cell is higher than it is outside the cell, and it continues to get higher still due to the active transport process. Thus, the cell does not depend entirely upon the external concentration of a substance.

In normal conditions, active transport makes it possible to regulate the entry of substances at a rate which satisfies the requirements of the cell. Let us assume, for example, that an organism is in an environment where the concentration of an essential nutrient is extremely low. Because of active transport carried out by the cell membrane, the organism does not have to depend upon the process of free diffusion to obtain an adequate internal concentration of the essential nutrient. It is obvious that the process of active transport has tremendous survival value for an organism.

Pinocytosis and Phagocytosis

Cells such as white blood cells, epithelial cells of the intestines, and others can bring substances into themselves by forming pockets, or **invaginations,** in the cell membrane. The invagination is pinched off and the captured material floats free in the cytoplasm. Molecules that are too large to pass through the cell membrane can be brought into the cell in this way. The small free-floating vacuoles are called **pinosomes,** and the general process is called **pinocytosis. Phagocytosis** is a similar process in which amoeba and white cells are able to engulf large particles or bacteria by extending "arms," called **pseudopods,** of the cell's surface out and around the particle to be taken up (Fig. 15.11).

Pinocytosis seems to take place only under certain conditions. For example, if proteins or salts are added to water a cell begins the process of pinocytosis, but continues for a limited time only. Then, after a lull, the process is begun again. There appear to be specific sites on the cell surface capable of forming the pinosomes. We do not yet understand how the metabolic machinery of the cell begins and stops pinocytosis according to its needs.

Eric V. Grave

Fig. 15.11 An amoeba engulfing a unicellular microorganism.

The material in the pinosomes is not inside the cell in a metabolic sense. The membrane around the engulfed material persists and behaves very much like the cell membrane itself. Small molecules can diffuse into the cytoplasm, but larger ones appear to be left behind. Gradually, the pinosome decreases in size and eventually becomes part of the cytoplasmic granules.

ENZYMES

Cells must be able to take up nutrients from the surrounding fluid and must contain the machinery for creating new parts of themselves from the food material. As pointed out earlier, while some of the nutrients become part of the cell itself, others are broken down and provide the energy needed to synthesize new molecules. In general, all these nutrients are called "food" and include the carbohydrates, fat, proteins, minerals, vitamins, and water. Taken as a whole, these cellular processes are rather complicated, but if we isolate and examine selected parts of the elaborate machinery of the cell, we can study the individual steps that lead to the synthesis of complex molecules. This is best done by following certain reactions as they take place in a test tube.

One group of molecules particularly important to all organisms are the large complex protein molecules called **enzymes**. For every essential chemical reaction that occurs in a living cell, there is a specific enzyme capable of speeding up that step. Without enzymes, these chemical reactions would not take place fast enough at normal temperatures to sustain life. Enzymes, then, are very efficient catalysts. They are also effective in very small amounts, and are usually unchanged by the chemical reaction they promote. Cells can duplicate themselves in a few minutes or hours because enzymes are capable of catalyzing the chemical changes associated with life processes. Enzymes are intimately associated with all life processes such as muscle contraction, nerve conduction, excretion, and absorption.

In some instances, for an enzyme to function it must be associated with a small molecule called a **coenzyme**. Parts of these coenzymes are often vitamins such as **riboflavin** (vitamin B_2) and **thiamin** (vitamin B_1). In addition to the organic **cofactors** (coenzymes) like the vitamin-containing compounds, several inorganic cofactors are required. For example, iron is necessary for the transfer of electrons; copper, too, apparently plays a role in electron transport. Magnesium is essential for the transfer of phosphate, and so on. In some cases, then, the enzyme protein alone is not enough to speed a chemical reaction. One or more coenzymes may be needed.

Naming enzymes is very easy once we know the type of reaction, or the substance acted on by an enzyme. Enzymes are denoted by the suffix *ase*. For example, an enzyme that catalyzes the breakdown of proteins is called a prote*inase*. One that catalyzes an oxidation is called an ox*idase*, and so on.

Effect of Temperature on an Enzyme Reaction

The individual molecules of a cell are in ceaseless motion. Occasionally, they react with one another when they collide. If we deprived a cell of its enzymes, the chance that a molecular collision would result in chemical reaction is very small. However, add the right enzyme and the chance of a collision resulting in a chemical reaction is greatly increased. The major question is how enzymes perform this catalytic activity.

All molecules in a given "population" do not have the same kinetic energy, or speed. Some, through collision, acquire more energy than others. These energy-rich molecules are more likely to react with energy-poor ones than are other energy-poor ones. In other words, there is an "energy barrier" to reaction. The energy required to hurdle molecules over this barrier is called the **energy of activation.**

Figure 15.12 shows a hypothetical reaction in which compound A is converted into the product compound B. This diagram in general holds true for all chemical reactions, although the height of the energy barrier varies from one reaction to another. Note that for A to be converted into B it must first acquire the necessary energy to form the activated molecule A*. The rate of a chemical reaction must depend, therefore, on the number of A* molecules existing at any one moment, and the speed with which they break down and form product B. Thus, the rate of a chemical reaction can be given by the following simple equation: **rate = (A*) k,** where k is a constant (which is essentially the same for all chemical reactions) and (A*) means concentration of A*.

The number of molecules activated depends on the temperature. As the temperature is raised, the number of A* also increases, thus the rate of the reaction increases. One of the significant features of enzymes is their ability to lower the energy of activation, thus *increasing the rate of chemical reaction.* We do not know exactly how this is done, but we are certain that reactions that take place only at boiling temperatures in a test tube can take place with relative ease at body temperature in the cell when enzymes are present. See Fig. 15.13.

High temperatures tend to destroy enzymes so that no further reaction can take place. Thus, biological processes that

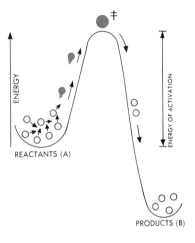

Fig. 15.12 Energy of activation for a chemical reaction.

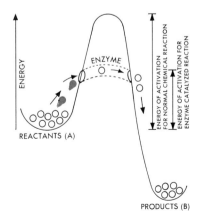

Fig. 15.13 The effect of enzymes on a chemical reaction are shown in this diagram. Enzymes, in some unknown way, lower the energy of activation for reactions so that the rate is greatly accelerated. It is as if the enzyme makes a tunnel in the energy barrier.

Fig. 15.14 The effect of temperature on an enzyme-catalyzed reaction.

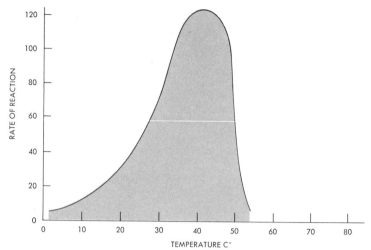

depend on enzyme reactions have an optimal temperature at which they function best. Above certain critical temperature enzymes do not perform as shown in Fig. 15.14. In the example shown, although the reaction rate increases rather rapidly from 0°C up to about 25°C, above this temperature the rate begins to slow; and at approximately 35°C it starts to decrease. If the damaging temperature is maintained for too many minutes, the enzyme will be completely inactivated. Other enzymes have different temperature optima.

This thermal behavior of enzymes imposes serious limitations on organisms. Most cells lose their capacity to carry on metabolism at temperatures above 40°C. Only the few organisms that have heat-resistant enzymes are able to survive in exceptionally hot places, such as in hot springs. In most cells, the rate of metabolism, and hence the intensity of the life processes, changes as the temperature varies from day to day and from season to season. The winter metabolism of most organisms declines con-

Fig. 15.15 A chipmunk coiled up in its nest underground in winter hibernation. In the state of hibernation the heart beat slows down extensively; in general, the over-all metabolic rate is greatly depressed. In contrast, the high metabolic rate of the Alaskan Brown Bear, which regularly fishes for salmon in cold water, maintains the body temperature at constant value.

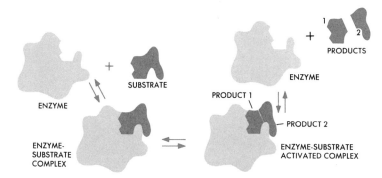

Fig. 15.16 The formation of an enzyme-substrate complex, followed by catalysis.

siderably. In general, plants in the colder winter climates cease to grow. We refer to this state of low metabolism as **dormancy**. Warm-blooded organisms, mammals and birds, are the only creatures that have evolved a method of controlling their body temperature.

END

How an Enzyme Works

The single molecule which an enzyme breaks down into separate new molecules, or molecules which an enzyme combines into a single new molecule, are called **substrate** molecules, or simply the substrate. An enzyme and a substrate fit together like pieces of a jig-saw puzzle. By combining with the enzyme, some of the bonds of the substrate molecule lose their form, thus producing a condition that favors chemical reaction.

While Fig. 15.16 shows something of the geometry of enzyme-substrate activity, Fig. 15.17 shows on a chemical level how a single maltose molecule is split into two glucose molecules (in the presence of water) by the enzyme maltase. Since the enzyme molecule is so large compared with the substrate maltose molecules, only an outlined section of the maltase molecule is shown. The first step in the reaction is the bonding of the maltose molecule to a proper position **(active site)** on part of the maltase

Fig. 15.17 Maltose combines at a specific site (the catalytic site) on the enzyme molecule. When bound to the site, water can readily split the linkage between the two sugar molecules to give free glucose.

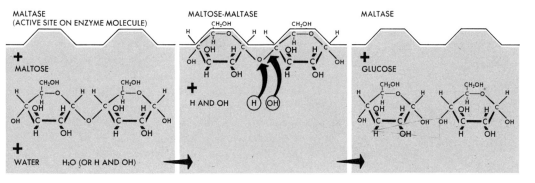

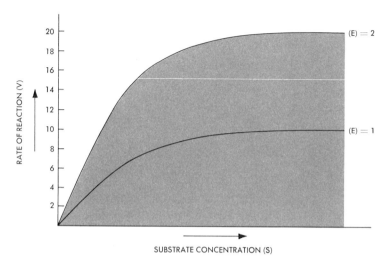

Fig. 15.18 The effect of substrate concentration on the rate of enzyme-catalyzed reactions.

molecule. Next, the oxygen bond is broken by hydrolysis (the addition of a molecule of water). One hydrogen and one oxygen atom take up a position on the end carbon atom of the glucose subunit, while the second hydrogen atom from the water molecule joins the old oxygen of the other glucose subunit. The final step is for the two new glucose molecules to break away from the enzyme maltase, which has in no way been permanently altered during the reaction. The enzyme is now ready to be used again.

If we isolate an enzyme and dissolve it in a water solution, we can study the rate of the chemical reactions it catalyzes. In general, if we add only a few substrate molecules, so that our solution is of low concentration, the reaction rate will be relatively low; but if we add many substrate molecules, so that the solution is of high concentration, the reaction rate will be relatively high. This variation of rate is over a limited range of concentration. As illustrated in Fig. 15.18, the rate of enzymatic reaction at first rises as we increase the substrate concentration. Then there is a slowing of the rate, and eventually the rate becomes constant—no matter how much more substrate we add. When the rate of the reaction becomes constant, we assume that the surfaces of all of the enzyme molecules are completely covered, or saturated, with the substrate molecules. Two different enzyme concentrations are shown in Figure 15.18.

The combination of a substrate with an enzyme is a very specific process, and each enzyme has a unique chemical arrangement containing precise sites where the substrate molecules join

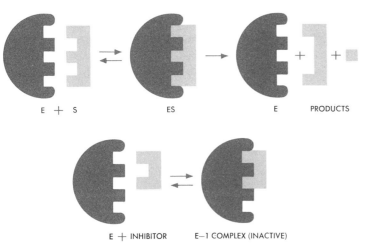

Fig. 15.19 When a substance has the same shape or geometry as the normal substitute, it can combine at the active site of the enzyme, thus inhibiting enzyme activity. This is called *competitive inhibition* because the inhibitor competes with normal substrate for the active site.

it. Because of the unusual folded surface of proteins, only very specifically shaped substrate molecules can gain access to the active region of the enzyme.

The specificity and the geometry of an enzyme reaction are beautifully illustrated by the use of enzyme inhibitors, "intruder" molecules that compete with the substrate molecules for the active site on the enzyme. Let us consider **succinic dehydrogenase,** an enzyme that catalyzes the removal of hydrogen from succinic acid. If malonic acid, whose molecule is very similar in shape to a succinic acid molecule, is added to the solution, the enzyme's effectiveness is considerably reduced. Malonic acid apparently attaches itself to the enzyme at a position that would normally be filled by succinic acid. Therefore, malonic acid competes with succinic acid for the active sites of the enzyme molecules. (See Fig. 15.19.) By so doing, it prevents the enzyme from acting as a catalyst. Since substrate and inhibitor compete for the same active sites as the enzyme molecules, the degree of inhibition depends on the relative amounts of substrate and inhibitor. If there is more substrate, the inhibitor might not be able to compete.

There are many different types of **competitive inhibitors.** Some of them are very effective in killing bacteria. One of these is **sulfanilamide** (see Fig. 15.20). The success of sulfanilamide and other bacterial killers opens the exciting possibility that all disease-producing organisms, as well as the abnormal growth in cancer, might be susceptible to the antimetabolite approach, that is, to inhibitors that make an enzyme inactive. Unfortunately,

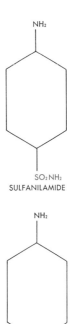

Fig. 15.20 Sulfanilamide is a competitive inhibitor. P-aminobenzoic acid (PABA) is one of the B vitamins and is essential for normal metabolic reactions. Sulfanilamide is very similar to PABA in structure and can interfere with the metabolic processes that depend upon PABA.

the host and disease-producing organism usually have very similar enzyme systems. This means that the inhibitor might be just as toxic to the host organism as it is to the disease-producing organism. However, by careful screening we may be able to find an appropriate chemical that will kill the disease-producing organisms without harming the host organism.

SUMMARY

Cells of both plants and animals are all basically alike. The cell membrane surrounds a **colloid** with gel-like properties. The colloid is composed largely of proteins, lipids, and water droplets. Within this gel-like structure are subcellular particles such as the nucleus, mitochondria, and ribosomes. The cell membrane is **selectively permeable.** Water moves back and forth across the membrane with great ease while other molecules have difficulty penetrating this barrier. The separation of two solutions (one inside the cell, the other outside) by a selectively permeable membrane creates an **osmotic system.** Molecules can often move through the cell by free diffusion. However, the cell membrane is not a passive barrier. In some cases metabolic energy is used to **actively transport** nutrients into the cell. Thus, cells may accumulate a high concentration of a particular substance. In such a system water also moves in and maintains osmotic equilibrium. Consequently, the cell may increase in volume due to the increased osmotic pressure inside. Cells may take in larger particles by the processes known as **pinocytosis** and **phagocytosis.**

Cells are capable of carrying out chemical reactions at high speeds and at room temperature because of the presence of specific catalysts called **enzymes.** Enzymes are complex proteins that often require **coenzymes** in order to function. Parts of these coenzymes are often vitamins such as riboflavin (Vitamin B_2). Inorganic metals such as copper and iron are essential cofactors for some specific enzymes. As in ordinary chemical reactions, enzyme-catalyzed reactions increase as the temperature is raised. Because very high temperatures destroy enzymes, biological processes have an optimal temperature at which they function best. The enzyme molecule must combine with the substrate molecule in a specific way in order for catalysis to occur. Each enzyme has a specific shape at the site where the substrate combines. Only substrate molecules with this specific geometry can combine at the **active site.** Certain antibiotics inhibit enzyme reactions because they have shapes similar but not identical to the substrate. They compete with the substrate for the active site on the enzyme, thus inhibiting the catalytic process. These are called **competitive inhibitors.**

FOR THOUGHT AND DISCUSSION

1 What happens to the colloidal properties of protoplasm when cells are heated at high temperature?
2 If a cell that contains the equivalent of 0.5 molar osmotically active particles is placed in a 1 molar solution of sucrose or glucose, what will ultimately happen to the water content of the cell?
3 Draw a sequence of rough pictures showing the process of pinocytosis.
4 Draw a scheme depicting the mechanism of enzyme reaction.

SELECTED READINGS

See the books listed at the end of the previous chaper.

16 METABOLISM AND ENERGY

Fig. 16.1 The metabolic machinery of this short-tail weasel (*Mustela erminea*) enables the animal to convert nutrients into useful energy. The animal's energy needs in winter are greater than in summer, because in winter the weasel loses more energy (in this case heat) to the environment.

When nutrients pass through the cell membrane, they enter a new environment which contains the metabolic machinery of the cell. One of the machine's chief purposes is to convert these nutrients into *useful* energy. When an organism breaks or splits certain atomic links during metabolism, we want to know whether the energy released is free to do useful biological work. If it is, we call it **free energy**. In biochemical systems, free energy is not literally liberated into the environment. It is conserved as a special **bond energy** (special bonds between elements) that is passed on to other molecules and is used to form new chemical bonds.

The liberation of energy in a biological system is really energy distribution and the formation of new chemical bonds of different **potential** energy. Potential energy is stored energy that can be released to do work. During oxidation and reduction reactions, large amounts of free energy are transferred and used to form energy-rich bonds (potential energy). In oxidation, you will recall, electrons are removed; and in reduction electrons are

Courtesy Charles J. Ott from National Audubon Society

gained. (Oxidation is not merely the use of molecular oxygen, a special case of oxidation.) Oxidation and reduction, then, involve the movement of electrons from one place to another, with a resulting transfer of energy that can be used to create new bonds of biological interest. Some compounds act as strong reducing agents (giving up electrons); others act as strong oxidizing agents (taking up electrons). Let's take a hypothetical situation in which a reducing substance (AH) combines with an enzyme (E) and is subsequently oxidized by another substance (X), according to the following equations:

$$E + AH \longrightarrow E—AH$$
$$E—AH + X \longrightarrow E{\sim}A + XH$$

Normally, when AH reacts with X and forms A plus XH, the energy in the A—H bond is lost as heat. However, in the illustration here, when hydrogen is removed [removal of an electron (e) plus a proton (H^+)] the energy of the oxidation is conserved in the bond between A and the enzyme (E$\sim$A), indicated by $\sim$. The energy of this bond can be used to combine A with other molecules to make new compounds, as illustrated here:

$$E{\sim}A + B \longrightarrow E + AB$$

This is what we mean by a *special bond energy* that is conserved in oxidation reactions. **It is one of the most important concepts in chemical energy transformations in organisms.** It is essential in the synthesis of all known substances in the cell.

Oxidation and Energy

The oxidation of carbohydrates or related compounds is the main source of energy for most organisms. During oxidation, *energy-rich* bonds are formed. The organism may then use these bonds to make other bonds, as discussed above. Energy-yielding reactions, then, can be coupled with energy-consuming reactions, with the result that energy is not actually liberated, but is redistributed in the reacting molecules.

The simplest type of oxidation is called **dehydrogenation,** in which hydrogen (an electron plus a proton) is removed from a compound. Every substance undergoing oxidation must be accompanied by a substance undergoing reduction, and vice versa. Hydrogen is not released as a gas in the process. Thus, in the reaction below, AH_2 is the reductant and is being oxidized by B.

$$AH_2 + B \rightleftharpoons A + BH_2$$

If the reaction is reversible, we can start with the chemicals on the right, then BH_2 is the reductant and A is the oxidant. Usually, when electrons and protons flow from a reducing system to

Fig. 16.2 The oxidation of an aldehyde to an acid and the generation of an energy-rich group. Note that in the oxidation of the aldehyde, an energy rich ($\sim$) intermediate is formed on the enzyme. If water splits this intermediate, the energy is lost as heat. If phosphate splits the intermediate, the energy is conserved in the phosphate bond and can be stored in the form of ATP (see text for details).

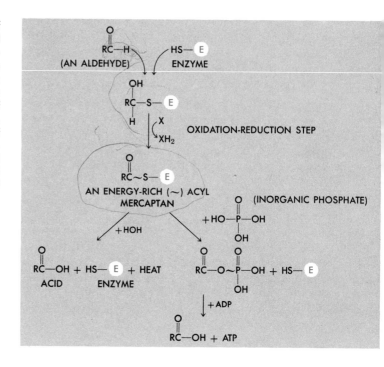

an oxidizing system; energy is liberated. We now want to find out how the energy is liberated, trapped, and used for a typical biological oxidation.

First let us consider a typical reaction, the oxidation of an aldehyde. You'll recall from Chapter 13 that an aldehyde is a compound having a $-C\begin{smallmatrix}O\\H\end{smallmatrix}$ group. You will also recall from the previous chapter that enzymes are essential for most reactions and combine with a substrate. Enzymes often contain an SH group **(sulfhydryl),** as indicated in Fig. 16.2. In this case, the aldehyde first combines with the enzyme. During the combining process, the double bond oxygen in the aldehyde part of the molecule reacts with the SH group. This produces a compound that contains an OH and an H on the end carbon atom. One of the bonds that was formerly connected to the oxygen reacts with the sulfur on the enzyme molecule. The hydrogen normally associated with the sulfur moves to the oxygen. This can be described, then, as the enzyme substrate complex which is the important intermediate which must be formed before catalytic oxidation can occur. In the next step, the oxidation process occurs and energy in the molecule is redistributed to form the energy-rich bond, as indicated in Fig. 16.2.

If this energy-rich bond between the carbon and sulfur is

broken by water (hydrolysis), a large amount of energy in the form of heat will be liberated. If the molecule is hydrolyzed, it forms an acid (carboxyl group) and the SH group is restored on the enzyme. In short, the aldehyde is oxidized in the presence of water, forming an acid and heat. During oxidation, therefore, an energy-rich group is formed on the enzyme. Ultimately, this bond energy is lost as heat if it is split by water. On the other hand, if the organism is to use this bond energy, it must transfer the energy-rich group to some other molecule, or use the energy to make a new and useful compound. When this occurs, a new compound is synthesized by drawing on the energy of the carbon-sulfur link. If the organism does not immediately need the reacting group for synthesis, it can store the energy in some other chemical form as a chemical energy reservoir. The enzyme is then freed for further work.

Coupling of bond energy in the reservoir to other systems is the key reaction in all biosynthetic processes. In the illustration used above, we can conserve the oxidation energy by splitting the high-energy bond with phosphoric acid (H_3PO_4) instead of water (HOH). Perhaps now you can begin to see why phosphate (PO_4^{-3}) is extremely important in biological reactions, particularly in the conservation of energy. For some reason which we do not completely understand at the present time, the phosphate group can split the carbon-sulfur link and prevent the loss of energy of the link. This results in the formation of an energy-rich link between the carbon and the phosphorus compound. The energy of the link is conserved, and at the same time the enzyme is freed to catalyze the oxidation of another substrate molecule (an aldehyde) and thus form more energy-rich groups.

As more and more such energy-rich groups are formed, they are transferred to another molecule, called **adenosine diphosphate (ADP)**, whose structure is shown in Fig. 16.3; and from

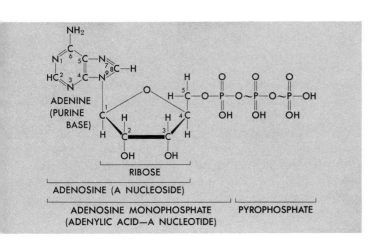

Fig. 16.3 Adenosine triphosphate

Fig. 16.4 All biological reactions depend on the energy derived from converting ATP into ADP.

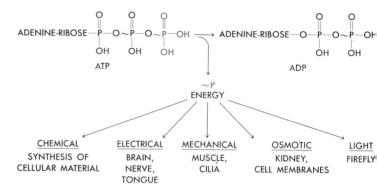

ADP a compound called adenosine **triphosphate (ATP)** is formed (Fig. 16.3). Notice that on the right-hand side of the ATP molecule there are three phosphate atoms. The last two are energy-rich bonds. When one is broken by water, large amounts of heat are liberated, showing its energy-rich character.

ATP is found in all plants, microorganisms, and animals. It serves as the initial storehouse for the energy generated by oxidation reactions taking place in the cell. This bond energy of the phosphate groups directly or indirectly drives all the energy-requiring processes of life—talking, walking, and building cellular material (Fig. 16.4).

Alcoholic Fermentation

Although man has known about alcoholic fermentation since prehistoric times, its cause was not discovered until about 1860. A short time before this, however, the French chemist Gay-Lussac had described the production of alcohol from sugar by the following equation:

$$\text{Glucose} \longrightarrow 2 \text{ Carbon dioxide} + 2 \text{ Ethyl alcohol}$$
$$(C_6H_{12}O_6) \qquad (2CO_2) \qquad (2CH_3CH_2OH)$$

In the middle of the 19th Century, there were two conflicting ideas about alcoholic fermentation. One group thought the process to be strictly chemical; the other claimed that the process was intimately associated with living organisms. The great French chemist, biochemist, and physiologist Louis Pasteur concluded that fermentation takes place in the absence of oxygen **(anaerobic)** and only in the presence of certain microorganisms. Without these organisms, fermentation would not occur. Pasteur's experiments indicated that fermentation is a physiological process closely bound up with the life of cells. This contradicted the idea that all living things needed oxygen. Pasteur felt that there were substitutes for molecular oxygen. In short, he felt that fermentation is the result of life activities being carried on in the absence of air.

Courtesy Standard Brands, Inc.

Fig. 16.5 Yeast being grown in a large commercial vat.

More than 20 years passed before the next major breakthrough in the investigation of fermentation occurred. In 1897, the German physiologist Eduard Buchner accidentally stumbled onto the discovery that opened the door to the secrets of fermentation, and the whole field of modern enzyme chemistry as well. He was primarily interested in making what he called "protoplasmic extracts" from yeast which were to be injected into animals. Hopefully, the injections might make the animals live longer. He ground yeast with sand, mixed it with certain other compounds, and finally squeezed out the juice with a hydraulic press. Since it was difficult to prepare this material daily, he made various attempts to preserve the cell-free extract. Because it was to be injected into animals, ordinary antiseptics, such as chloroform, could not be used, so he tried the usual kitchen chemistry method of preserving fruit by adding large amounts of sugar.

To Buchner's surprise, the sugar was rapidly decomposed into carbon dioxide and alcohol, or fermented, by the yeast juice! So far as we know, he was the first person to observe fermentation in the complete absence of living cells. At last it was possible to study the process of alcoholic fermentation outside the living cell. Buchner's work was soon followed by intensive studies of the properties of yeast juice. Yeast juice was found to ferment many sugars such as glucose, fructose, mannose, sucrose, and maltose.

Glucose was converted by the juice into ethyl alcohol and carbon dioxide, according to the Gay-Lussac equation. The next question to be answered was *how* yeast juice brought about the fermentation of sugar.

Phosphorylation of Glucose

The first important analysis of the activity of yeast juice was made by two English scientists, Harden and Young, when in 1905 they added fresh yeast juice to a solution of glucose. At *p*H 5 fermentation began almost at once. Although the rate of carbon dioxide production soon fell off, they could restore fermentation by adding inorganic phosphate. The recovery was only temporary, however, because the phosphate was soon used up; the rate of fermentation dropped as the phosphate concentration declined. But by adding more phosphate they sparked another burst of fermentation.

The work of Harden, Young, and others has shown that phosphate is essential in the metabolism of carbohydrates. Phosphate aids in breaking down carbohydrates to smaller molecules, and is also essential for the conservation of energy released during the breakdown process.

Subsequent investigations have led to the conclusion that when

Fig. 16.6 Phosphorylation of glucose and the formation of aldehyde phosphates. The hydrogen and oxygen atoms are not shown on all of the phosphorus and carbon atoms.

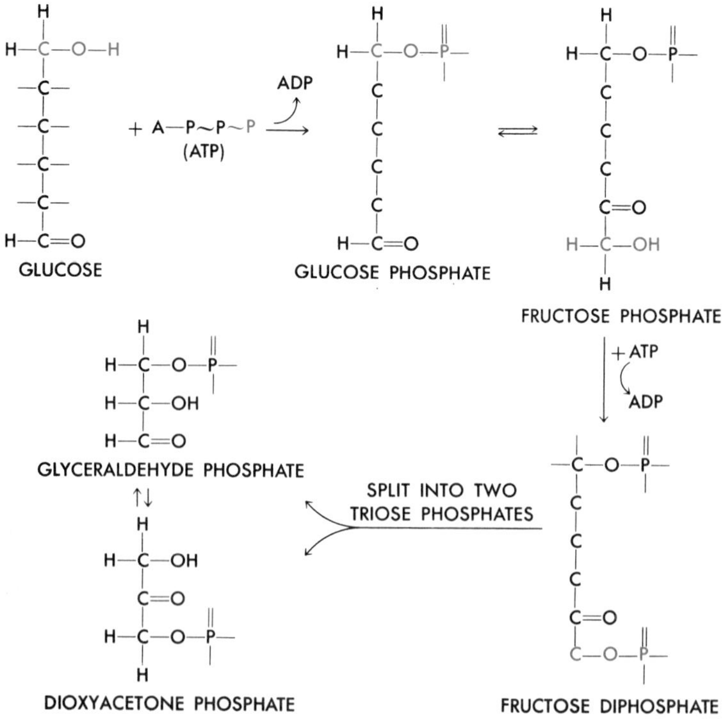

yeast juice metabolizes glucose, ATP energy is required for the initial reaction. As Fig. 16.6 shows, a phosphate from ATP is transferred to glucose, making a glucose phosphate compound plus ADP. Through a series of reactions, glucose phosphate is converted to the hexose sugar **fructose phosphate.** Fructose phosphate accepts a second phosphate from another ATP molecule, forming a hexose sugar (a C_6 sugar) with two phosphate groups —fructose **di**phosphate. The first product of the phosphorylation reactions which prepare glucose for metabolic breakdown, then, leads to the formation of a hexose diphosphate, the Harden and Young ester.

Next, the fructose diphosphate is split into two compounds, each containing three carbons **(triose)** and one phosphate. These triose phosphate compounds are interchangeable and are, in effect, identical. The important one for fermentative metabolism is an aldehyde compound called **glyceraldehyde-3-phosphate** (see Fig. 16.7). This compound is oxidized in exactly the same

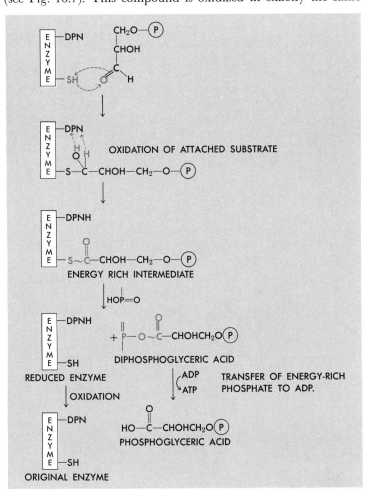

Fig. 16.7 The formation of ATP from the oxidation of glyceraldehyde phosphate.

Fig. 16.8 The formation of ATP and pyruvic acid from phosphoglyceric acid.

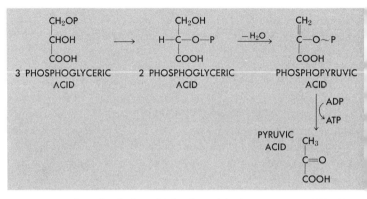

way as we described the aldehyde oxidation on page 238. During the process, energy is liberated in the form of phosphate bond energy, which leads to the synthesis of a molecule of ATP. The other product of the reaction is **phosphoglyceric** acid (Fig. 16.7). It is this reaction that requires the inorganic phosphate which makes the ATP that is essential to the introduction of glucose into the metabolic pathway.

In additional reactions, water is removed from phosphoglyceric acid in such a way that the phosphate in the phosphoglyceric acid also becomes energy-rich. This group is transferred to ADP and forms ATP. We end up also with a molecule of **pyruvic** acid (see Fig. 16.8). Thus, if we start with a glucose molecule, we put in two ATP molecules to make two molecules of glyceraldehyde-3-phosphate. During the oxidation and dehydration reactions we recover two energy-rich groups for each triose. Consequently, from the two trioses we recover four ATP molecules. The net gain in terms of bond energy in the fermentative metabolism of glucose is two ATP molecules.

During the oxidation of the glyceraldehyde-3-phosphate, an enzyme is essential for catalysis. In addition, the enzyme requires a coenzyme capable of accepting the hydrogens during the oxidation of the aldehyde. The coenzyme required for the oxidation of an aldehyde is a derivative of the B vitamin, **niacin.** This finding provided the first revealing clue to the function of vitamins in cellular metabolism. The particular coenzyme that is involved in the dehydrogenation of the aldehyde is known by its initials as **DPN (diphosphopyridine nucleotide).** Recently it has been suggested that the name be changed to nicotinamide adenine dinucleotide (NAD). The name of the enzyme requiring DPN as a coenzyme is **glyceraldehyde-3-phosphate dehydrogenase.** Sometimes it is referred to as **triose phosphate dehydrogenase.** Thus, in the oxidation of the aldehyde DPN is reduced. It is during this oxidation process that the energy-rich intermediate is formed.

For each molecule of glucose fermented, two molecules of alcohol and two of CO_2 are formed. The final key step in the

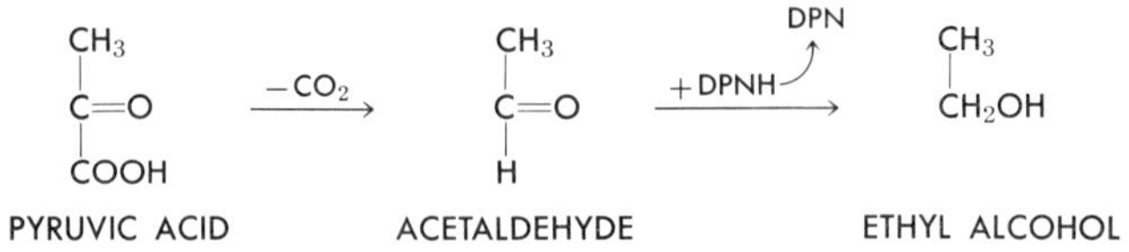

$$\underset{\text{PYRUVIC ACID}}{\begin{array}{c}CH_3\\|\\C=O\\|\\COOH\end{array}} \xrightarrow{-CO_2} \underset{\text{ACETALDEHYDE}}{\begin{array}{c}CH_3\\|\\C=O\\|\\H\end{array}} \xrightarrow[+DPNH]{DPN} \underset{\text{ETHYL ALCOHOL}}{\begin{array}{c}CH_3\\|\\CH_2OH\end{array}}$$

Fig. 16.9 The formation of alcohol from pyruvic acid.

formation of alcohol is the removal of CO_2 from pyruvic acid (Fig. 16.9). The B vitamin, thiamine, is required for this reaction. Thus, when pyruvic acid loses the carboxyl group by **decarboxylation** it forms a new compound, acetaldehyde. Certain organisms can oxidize acetaldehyde and form acetic acid. Yeast cells, however, do not have the enzyme capable of oxidizing this short-chain aldehyde. But they do have an enzyme capable of catalyzing the transfer of hydrogen from reduced DPN to aldehyde (see Fig. 16.10). The reduction of acetaldehyde leads to the formation of **ethyl alcohol;** the enzyme which catalyzes the reaction is called **alcohol dehydrogenase** (that is, the enzyme catalyzes the reverse reaction and has been named as indicated.)

In the final steps of fermentation, we have decarboxylation and reduction—forming one molecule of alcohol and one molecule of CO_2 from each pyruvic acid molecule. Since two pyruvic acid molecules are formed during the breakdown of one of glucose, we end up with two molecules of alcohol and two of CO_2, thus satisfying the Gay-Lussac equation for this process.

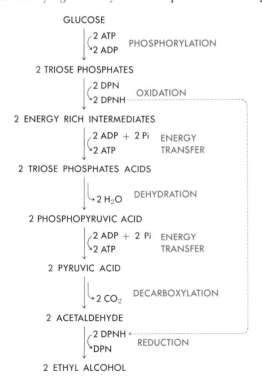

Fig. 16.10 Summary of key reactions in alcoholic fermentation.

OVER-ALL REACTION:
GLUCOSE + 2 ADP + 2 H_3PO_4 ⟶ 2 ALCOHOL + 2 CO_2 + 2 ATP

Muscle Metabolism—Glycolysis

For a long time after Harden and Young's discovery of the phosphorylation of sugars in alcoholic fermentation, the process was not considered significant, except as a means of shaping the hexose molecule for fermentative breakdown. However, similar studies of other cells, particularly muscle cells, revealed that phosphate plays a dominant role in energy transformation. This is especially true in muscle contraction.

Among the questions asked by the earlier investigators of muscle contraction were these: What is the chemical source of energy that powers the muscle machine? And, how is chemical energy of the cells transformed into mechanical energy of muscle contraction? The investigators soon discovered that muscle can contract in a normal manner in the complete absence of oxygen. They also discovered that **lactic** acid is produced during the contraction process and is associated with muscle fatigue. If a fatigued muscle, however, is supplied with oxygen, it soon recovers its ability to contract, and the lactic acid disappears.

By the late 1920's, it became evident that most of the energy spent by muscle contraction came from the metabolism of reserve **glycogen,** which is stored in muscle. Investigators also found that glycogen was converted to lactic acid in a process very similar to that of alcoholic fermentation. We call this process **glycolysis,** that is, the breakdown of glycogen. Eventually, other investigators pointed to ATP as the immediate energy source for muscle contraction. The breakdown of glycogen is essential for the formation of ATP, and the reactions are the same as in the breakdown of glucose by the yeast cell.

An additional compound, however, was isolated from muscle, and it proved to be a very important storehouse of energy-rich phosphate. This compound is **creatine phosphate** (Fig. 16.11). Creatine, it turned out, could react with ATP and form creatine phosphate, which accumulates in muscle cells in very high concentrations. When phosphate from ATP is transferred to creatine, ADP is produced; and ADP is essential for the continued metabolism of glucose.

When a muscle contracts, it uses ATP vigorously, but the mechanism that converts the chemical energy of ATP into mechanical work remains obscure. When we study muscle for its chemical composition, we find two important proteins. One is called **actin,** and the other **myosin.** When actin and myosin are mixed they form **actomyosin,** which can be made into threads that rapidly decompose ATP. At the same time, the threads contract. Actin and myosin are the major proteins of contractile muscle fiber (Fig. 16.12).

Fig. 16.11 Creatine phosphate and the relationship to ATP. Creatine phosphate acts as an important storehouse of energy-rich phosphate groups. When ATP reacts with creatine, ADP and creatine phosphate (CP) is formed. The ADP is now free to "pick up" another energy-rich phosphate. The process continues until essentially all of the creatine is converted into creatine phosphate. When ATP is used rapidly the ADP formed can be rapidly converted back into ATP by using the CP reserve.

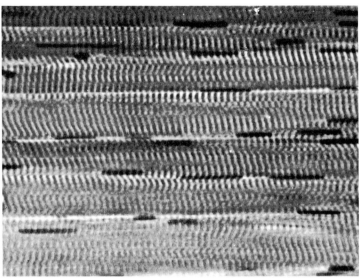

Fig. 16.12 Skeletal muscle is made up of a number of parallel bundles of very large multinucleated cells. These muscle cells contain a number of parallel myofibrils which are the functional units of the contractile machinery. The cross-striated appearance is due to the differences in the optical properties of the myofibril substances.

Eric V. Grave

Gradually, the mechanical arrangements of these fibers in the muscle, and the function of ATP in causing contraction, are being cleared up. We now think that a muscle contracts when its actin and myosin fibers slide over one another. ATP can dissociate, or separate, actin and myosin. It is possible that when a nerve impulse stimulates a muscle to contract, the first thing that happens may be the activation of the enzyme **ATPase,** which splits ATP. The destruction of the ATP would allow the actin and myosin to combine. The actomyosin threads may then contract (Fig. 16.13).

One important difference exists between the breakdown of glycogen and carbohydrate by muscle and yeast. When a muscle breaks down carbohydrate, alcohol and CO_2 are not produced! Muscle does not have the enzyme which is capable of catalyzing the removal of CO_2 from pyruvic acid, thus, the latter substance must accept the hydrogens from reduced DPN. For this reaction to occur, the muscle has an enzyme called **lactic acid dehydrogenase.** It is this enzyme which catalyzes the transfer of hydrogens from reduced DPN to pyruvic acid, making lactic acid. As the name indicates, it will catalyze the reverse reaction. This reaction is shown in Fig. 16.14. Thus, pyruvic acid in muscle acts very much like acetaldehyde in alcoholic fermentation. The net effect of the anaerobic reaction sequence in muscle is as follows: For every glucose unit derived from glycogen, two molecules of lactic acid are produced. The coenzyme DPN is alternatively reduced and oxidized. In yeast juice, as in muscle extract, the sequence generates four new energy-rich phosphate bonds for each glucose molecule metabolized.

Fig. 16.13 This schematic representation shows the relationship between actin and myosin filaments from muscle and what is thought to occur when ATP energy induces contraction.

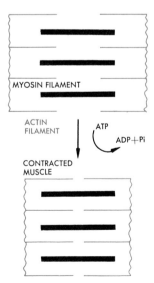

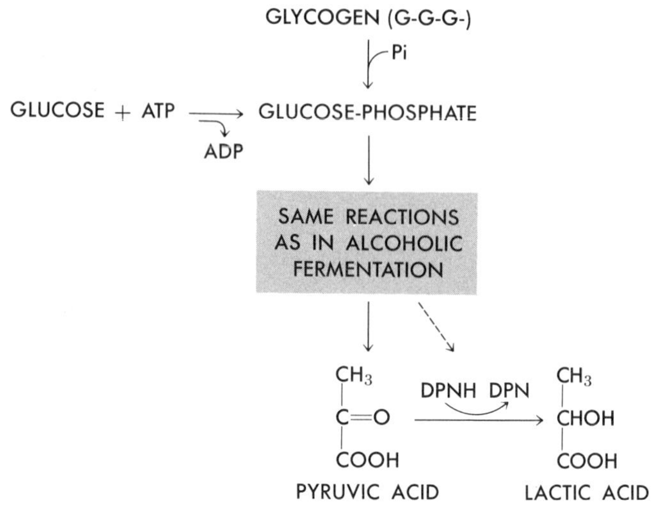

Fig. 16.14 Glycolysis: the formation of lactic acid from glycogen. The energy in the glucose-glucose links in glycogen is such that inorganic phosphate (Pi) can be used to form the glucose phosphate. When free glucose is used, ATP energy is necessary to add phosphate to the sugar molecule. In the synthesis of glycogen from free glucose, phosphate bond energy must be used.

GLYCOGEN (G-G-G-)
↓ Pi
GLUCOSE + ATP → GLUCOSE-PHOSPHATE
ADP
↓
SAME REACTIONS AS IN ALCOHOLIC FERMENTATION
↓

$$\begin{matrix} CH_3 \\ | \\ C=O \\ | \\ COOH \\ \text{PYRUVIC ACID} \end{matrix} \xrightarrow{DPNH \;\; DPN} \begin{matrix} CH_3 \\ | \\ CHOH \\ | \\ COOH \\ \text{LACTIC ACID} \end{matrix}$$

The control of glucose metabolism in muscle, then, is very similar to the process described for yeast. The reoxidation of reduced DPN, the action of inorganic phosphate, and the use of ATP are the essential processes. Thus, in muscle glycolysis, as well as in alcoholic fermentation, the mechanisms of oxidation, reduction, dehydration, and phosphorylation are at work. These

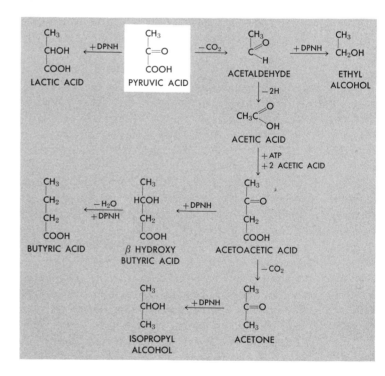

Fig. 16.15 Some products of fermentative metabolism.

reactions are the basic ones involved in energy and carbon transformation in fermentation. In the past 25 years we have found that in the metabolism of various organisms these steps account for almost all of the fermentation products formed from carbohydrates by various organisms. For example, lactic acid bacteria produce lactic acid by a sequence of reactions identical to those observed in muscle. Outlined in Fig. 16.15 are a number of reactions carried out by different organisms and products of fermentative metabolism. Figure 16.15 shows some products of fermentative metabolism.

Oxidation of Reduced DPN by Other Substances

So far, we have discussed the oxidation reactions in which DPN acts as the initial hydrogen or electron acceptor. This step is the important initial dehydrogenation in the metabolism of several substrates. In a series of brilliant studies beginning in the early 1920's, the German biochemist Otto Warburg found traces of two other important coenzymes in cells grown under aerobic conditions.

He was intrigued by the ability of iron-containing compounds to catalyze the oxidation of many different organic substances by using molecular oxygen. Warburg suspected that the iron contained in these compounds was responsible for their catalytic activity. He reasoned that an iron-containing substance is needed in the cell if oxygen is to be activated and used. A search led to the discovery of several iron-containing compounds, the function of which is to carry electrons to molecular oxygen. The compounds are called **cytochromes.**

It turned out that a second coenzyme was needed to transport electrons (plus the protons H^+) from reduced DPN to molecular oxygen—**flavin,** a derivative of the vitamin **riboflavin.** It soon became clear that when hydrogen atoms (electrons plus protons H^+) were removed from various substances they were transported by DPN flavin, and iron-containing cytochromes to molecular oxygen, with a resulting production of H_2O. In addition to the compounds mentioned so far, there are other electron acceptors that are important in the metabolism of carbohydrates.

[It should be noted here that we often use the terms *electron transport* and *hydrogen transport* interchangeably. We do so because, although both processes occur, the net effect of each one is the same. Thus, a compound such as AH may dissociate into A^- and H^+ before oxidation (electron removal). Thus,

$$AH \longrightarrow A^- + H^+$$

$$A^- + B \longrightarrow A + B^-$$

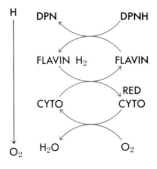

Fig. 16.16 The chain of electron transport from DPNH to oxygen.

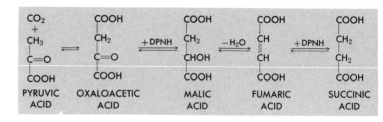

Fig. 16.17 During the fixation of carbon dioxide by pyruvic acid adenosine triphosphate is needed. The oxaloacetic acid which is formed can be reduced and in several steps, as indicated, can form succinic acid.

B^- may now take up a proton from the environment as follows:

$$B^- + H^+ \longrightarrow BH$$

The net effect is to transfer a hydrogen atom:

$$AH + B \longrightarrow A + BH$$

In some cases, this actually happens without the proton dissociation; that is, a hydrogen transport instead of electron transport.]

Carbon Dioxide Fixation

Green plants take up CO_2 in the presence of water; then they make use of light energy and form carbon-carbon links which, eventually, are reduced to a carbohydrate. The reaction is called CO_2 fixation and reduction. This is the ultimate source of all complex organic molecules on the surface of the Earth. However, it is not the only way in which CO_2 is **fixed.** In the presence of ATP, CO_2 can be added to several compounds. Pyruvic acid is one such compound. When pyruvic acid takes up CO_2, it forms a new four-carbon compound capable of accepting hydrogens, which leads to the formation of other types of acids, in particular, succinic acid (Fig. 16.17). These acids are very important in stimulating cell respiration, and, as we shall find in a moment, are very important in converting carbohydrates completely into CO_2 and H_2O. Large amounts of energy are released in this process.

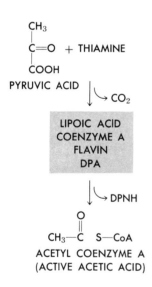

Fig. 16.18 Oxidative decarboxylation of pyruvic acid forms the energy-rich acetic acid molecule.

Oxidative Metabolism

One of the additional key reactions in the metabolism of pyruvic acid is the removal of the carboxyl (—COOH) group, forming CO_2. The two-carbon aldehyde molecules remain attached to a coenzyme and are oxidized; the result is the

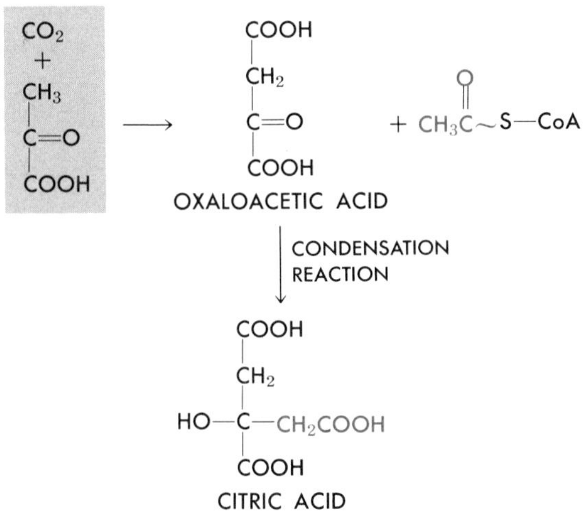

Fig. 16.19 The synthesis of citric acid from active acetic acid and oxaloacetic acid.

$$\begin{array}{c} CO_2 \\ + \\ CH_3 \\ | \\ C{=}O \\ | \\ COOH \end{array} \longrightarrow \begin{array}{c} COOH \\ | \\ CH_2 \\ | \\ C{=}O \\ | \\ COOH \end{array} + CH_3\overset{O}{\overset{\|}{C}}{\sim}S{-}CoA$$

OXALOACETIC ACID

↓ CONDENSATION REACTION

$$\begin{array}{c} COOH \\ | \\ CH_2 \\ | \\ HO{-}C{-}CH_2COOH \\ | \\ COOH \end{array}$$

CITRIC ACID

formation of an energy-rich acetic acid (Fig. 16.18). The activated acetic acid combines with the four-carbon compound (**oxaloacetic** acid), which is formed from another pyruvic acid by the CO_2 fixation reaction discussed above. This leads to the formation of the complex molecule called **citric** acid (Fig. 16.19). Eventually, citric acid is broken down step by step until two of its carbons are converted into CO_2, and oxaloacetic acid is reformed. The electrons removed are transported to oxygen, and water is formed. The oxaloacetic acid regenerated accepts another two-carbon unit from the decarboxylation of another pyruvic acid molecule. The cyclic process of adding two-carbon units to oxaloacetic acid to form citric acid, and then breaking it down to reform oxaloacetic acid, is called the **citric acid cycle** (Fig. 16.20).

This integrated group of enzymes—the dehydrogenases and the decarboxylases—which engineer the complex series of reactions in citric acid cycle are located in the mitochondria of the cell. Bound to this complicated structure, in some unknown way, are the various enzymes and coenzymes. The mitochondria from several different kinds of cells are known to be capable of carrying on this final oxidative stage of cell metabolism by themselves.

The significance of these oxidative reactions is that energy-rich phosphate groups are generated during the complete combustion of pyruvic acid. These oxidative processes generate ATP much in the same way that the oxidation of the aldehyde does. The process is called **oxidative phosphorylation** and is one of the crucial events in oxidative metabolism. The energy liberated in the oxidation of these coenzymes, is thus used to synthesize ATP. A mitochondrion seems to have three sites for the formation

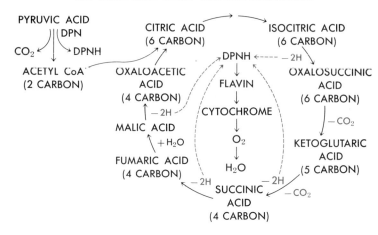

Fig. 16.20 When active acetic acid condenses with oxaloacetic acid, citric acid is formed. The citric acid then goes through a series of oxidation and decarboxylation reactions leading ultimately to the reformation of oxaloacetic acid. The net effect is the complete combustion of pyruvic acid (active acetic acid) to carbon dioxide and water. The hydrogens removed are transported to oxygen by means of the electron transport chain. All of these reactions occur in the mitochondria.

of ATP during the oxidation of reduced DPN by molecular oxygen (Fig. 16.21). The oxidation of one molecule of reduced DPN by the mitochondrial enzyme complex produces three molecules of ATP. The total combustion of pyruvic acid to CO_2 and water leads to the net synthesis of 15 ATP molecules. It is no wonder that the mitochondria have been called the "powerhouses" of the cell.

Since 15 ATP molecules are formed from the metabolism of pyruvic acid, it is clear that the majority of the energy in a compound becomes available as ATP. But this occurs only when a compound is oxidized all the way to CO_2 and water. Thus, aerobic organisms are capable of obtaining large amounts of energy from very small amounts of food. This is in contrast to those organisms that must live anaerobically. For example, a yeast cell that must grow and multiply under anaerobic conditions uses more than 30 times as much sugar as it would use if it grew and metabolized under aerobic conditions. In other words, there is still a large amount of energy in the alcohol that is formed by the yeast cell.

Thus, in the aerobic metabolism of glucose the initial pathways of breakdown are the same as those we discussed for the fermentation or anaerobic system. Under aerobic conditions the DPNH that is formed in the triose phosphate dehydrogenase

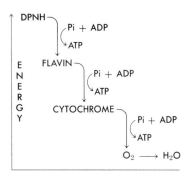

Fig. 16.21 Oxidative phosphorylation, the formation of ATP during the transfer of electrons over the electron-transport chain.

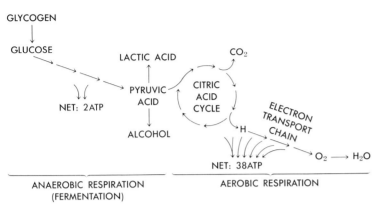

Fig. 16.22 Comparison of the anaerobic and the aerobic phase of metabolism. Note the large amount of ATP that is synthesized under aerobic conditions.

reaction is not used to form alcohol (in yeast) or lactic acid (in muscle); instead it is oxidized by molecular oxygen by way of the flavin and cytochrome systems. When the oxygen concentration is varied it is possible to get some fermentation products as well as aerobic products (Fig. 16.22).

A muscle under vigorous activity will consume oxygen so fast that some of the hydrogens of DPNH may be diverted to pyruvic acid and form lactic acid. Thus, there may be competition between two systems for the hydrogen derived in the dehydrogenase reactions. When the muscle accumulates lactic acid because of the exhaustion of oxygen, it acquires what is called an **oxygen debt.** When resting, the muscle will then use more oxygen than it normally does in order to metabolize the accumulated lactic acid.

Although we have talked about the citric acid cycle as one which liberates large amounts of energy, it is also valuable in the synthesis of carbon skeletons which can be used for making many other compounds. For many years the breakdown of carbohydrates was thought to be primarily a "tearing-down" process **(catabolism),** whose sole purpose was the liberation of energy. This is true so far as energy utilization is concerned, but we know that the formation of key intermediates in the breakdown of carbohydrates is also significant in the "building" (anabolism) of compounds of biological interest. During the rapid growth of cells, the principal function of the citric acid cycle may be to supply the cellular carbon skeletons for biosynthesis.

Fatty Acid Oxidation

The mitochondrion also functions as the cellular furnace for the combustion of fatty acids, amino acids, and other fuels. A typical fatty acid, **butyric** acid, is shown in Fig. 16.23. The metabolism of this acid is very much like the

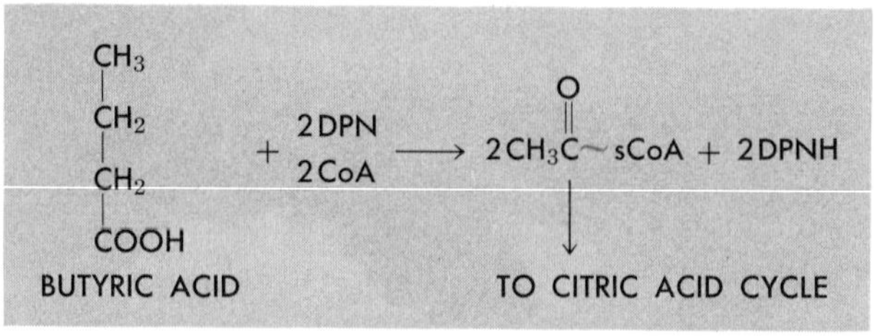

Fig. 16.23 Metabolism of a fatty acid (butyric acid).

$$\begin{array}{c} CH_3 \\ | \\ CH_2 \\ | \\ CH_2 \\ | \\ COOH \end{array} + \begin{array}{c} 2\,DPN \\ + \\ 2\,CoA \end{array} \longrightarrow 2\,CH_3C\!\sim\!sCoA + 2\,DPNH$$

BUTYRIC ACID → TO CITRIC ACID CYCLE

metabolism of pyruvic acid. It is first oxidized and eventually split into two carbon fragments which can add to oxaloacetic acid in the citric acid cycle (Fig. 16.20).

Because there are more hydrogens on fatty acids than there are on compounds such as glucose or pyruvic acid, the metabolism of fatty acids liberates much more energy than the consumption of an equivalent length of carbohydrate. For example, from butyric acid we obtain two active C_2 units; on entering the citric acid cycle they lead to the generation of 24 ATP molecules. Five additional oxidative steps lead to the formation of five more ATP molecules, giving a total of 29, which come from a single C_4 unit. Fatty acids, then, are a much better source of energy than carbohydrates are.

Amino Acid Metabolism

Proteins are the most important macromolecular structures of the cells. Proteins, as we have mentioned before, are composed of amino acids. A supply of amino acids—either from the diet or from the biosynthetic machine—is essential, since we cannot take ready-made proteins from the environment and use them. Rather, the plant or animal proteins we eat are broken down and digested into their individual amino acids. The amino acids are then absorbed by our cells. In the cells the amino acids are reassembled to form the right kind of protein required by a particular cell.

For man there are about 10 essential amino acids which must be supplied in the diet, since animal cells do not have the

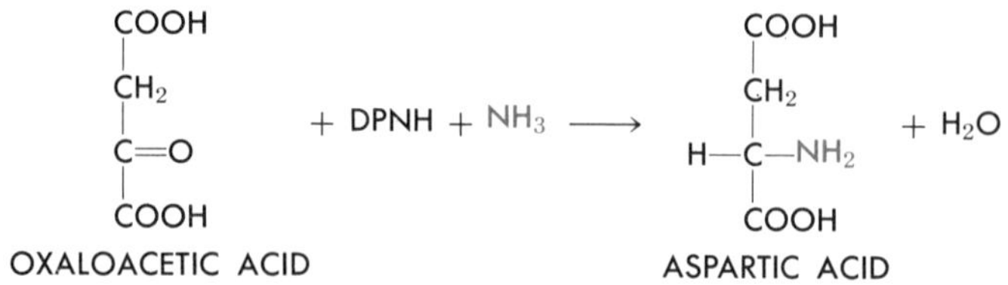

Fig. 16.24 The synthesis of an amino acid (aspartic acid) from oxaloacetic acid, DPNH, and ammonia.

$$\begin{array}{c} COOH \\ | \\ CH_2 \\ | \\ C\!=\!O \\ | \\ COOH \end{array} + DPNH + NH_3 \longrightarrow \begin{array}{c} COOH \\ | \\ CH_2 \\ | \\ H\!-\!C\!-\!NH_2 \\ | \\ COOH \end{array} + H_2O$$

OXALOACETIC ACID → ASPARTIC ACID

machinery (enzymes) to make them. In order to make the other amino acids, an additional supply of nitrogen is needed. The nitrogen can come from the amino acids in the diet or can be supplied in the form of **ammonium salts.** A crucial reaction in the uptake and incorporation of ammonia into amino acids—and eventually into proteins—involves one of the intermediates in the citric acid cycle (Fig. 16.24). Once ammonia is converted into amino nitrogen, it can be transferred to other carbon skeletons and form different amino acids, the raw materials for protein syntheses.

SUMMARY

When organisms break down **(metabolize)** sugars and other food substances, energy is trapped in special energy-rich bonds. The energy in these "rich" bonds can be used to make new compounds that are important to the organism. This useful energy that can be used in synthetic reactions is called **free energy.** This special form of chemical energy can be stored in the terminal phosphate bonds of a compound called **adenosine triphosphate (ATP).**

Large amounts of ATP are formed when compounds are oxidized aerobically by the mitochondria of the cell (citric acid cycle). During the cycle, molecular oxygen is used as the final hydrogen acceptor, thus leading to the complete combustion of foodstuff to CO_2 and water. The formation of ATP from inorganic phosphate and ADP in this process is called **oxidative phosphorylation.** This does not occur under anaerobic conditions. Yeast without oxygen forms alcohol and CO_2, while muscle cells make lactic acid. Even under these conditions, the limited amount of ATP formed is due to oxidative reactions.

Coenzymes that contain the B vitamins niacin and riboflavin are essential for oxidation (electron transport), and iron-containing cytochromes are necessary for activating molecular oxygen so that it can accept electrons (e) and protons (H^+) to form water.

Fatty acids, amino acids, and other simple organic compounds can also be metabolized by the mitochondria to form large amounts of ATP. In addition, the intermediates formed from carbohydrate breakdown can be used as carbon skeletons for the synthesis of specific amino acids or fatty acids. By the addition of ammonia to one of the intermediates in the citric acid cycle, the cell can make the amino acid, aspartic; or by addition to another intermediate, it can make glutamic acid.

Thus, in the breakdown of carbohydrates energy liberation is important. In addition, key intermediates are formed. The intermediates are essential for building new compounds of biological interest.

FOR THOUGHT AND DISCUSSION

1 Vitamins are important in the functioning of enzymes. Describe the function of niacin in alcoholic fermentation by yeast, and lactic acid formation in muscle.
2 What do we mean by the **phosphorylation** of glucose? Describe the role of ATP.
3 What is **fermentation?** How does it differ from aerobic metabolism?
4 Define **oxidation-reduction,** and show the process in a typical reaction.
5 Describe a hypothetical oxidation reaction and show how an **energy-rich** bond can be formed.
6 Do you get more ATP energy from glucose or ethyl alcohol when both are metabolized completely to CO_2 and water? Explain.

SELECTED READINGS

Baldwin, E. *Dynamic Aspects of Biochemistry* (3rd ed.). New York: Cambridge University Press, 1957.

An outstanding introduction to the concepts of cellular metabolism.

Bennett, T. P., and E. Frieden. *Modern Topics in Biochemistry.* New York: Macmillan Co., 1966.

An excellent discussion in greater detail of the various subjects discussed in our last three chapters. Written for the beginning college student.

Loewy, A. G. and P. Siekevitz. *Cell Structure and Function.* New York: Holt, Rinehart, and Winston, 1963.

A detailed elementary discussion of various aspects of cell structure and biochemistry.

McElroy, W. D. *Cell Physiology and Biochemistry* (2nd ed.). Englewood Cliffs, N.J.: Prentice Hall, Inc., 1964.

A small paperback intended for beginning college students.

Neilands, J. B. and P. K. Stumpf. *Outlines of Enzyme Chemistry* (2nd ed.). New York: John Wiley & Sons, 1958.

An excellent elementary introduction to enzyme chemistry.

Readings from Scientific American

Allen, R. D., "Amoeboid Movement," February 1962.
Allfrey, V. G. and A. E. Mirsky, "How Cells Make Molecules," September 1961.
Brachet, J., "The Living Cell," September 1961.
Doty, Paul, "Proteins," September 1957.
Frieden, E., "The Enzyme—Substrate Complex," August 1959.
Fruton, Joseph S., "Proteins," June 1950.
Green, D., "The Synthesis of Fat," February 1960.
Hayashi, T. and G. A. W. Boehm, "Artificial Muscle," December 1952.
Holter, Heinz, "How Things Get into Cells," September 1961.
Huxley, H. E., "The Contraction of Muscle," November 1958.
Kendrew, J. C., "Three-Dimensional Structure of a Protein," December 1961.
Lehninger, A., "Energy Transformation in the Cell," May 1960.
Lehninger, A., "How Cells Transform Energy," September 1961.
McElroy, W. D. and C. P. Swanson, "Trace Elements," January 1953.
Moore, S. and W. H. Stein, "The Chemical Structure of Proteins," February 1961.
Pfeiffer, John E., "Enzymes," December 1948.
Robertson, David J., "The Membrane of the Living Cell," April 1962.
Siekevitz, P., "Powerhouse of the Cell," July 1957.
Solomon, A. K., "Pores in the Cell Membrane," December 1960.
Solomon, A. K., "Pumps in the Living Membrane," August 1962.
Stein, W. H. and S. Moore, "The Chemical Structure of Proteins," February 1961.
Stumpf, P. K., "ATP," April 1953.

17 LIGHT AND LIFE

The oxygen we breathe and most of the food we eat are formed by plants through **photosynthesis.** The power to drive the photosynthetic machinery of plant cells comes from sunlight, which is absorbed by **chlorophyll,** the green pigment in the plant (Fig. 17.2). All of life that we know on Earth depends directly or indirectly on photosynthesis. We know a great deal about the nature of the machinery which traps the sunlight and converts carbon dioxide into carbohydrate; however, our knowledge is still incomplete. So far, man has not been able to devise a chemical system that can serve as a substitute for photosynthesis. Only the green plant is able to use light energy to convert carbon dioxide into organic matter, and at the same time produce free oxygen from water.

Fig. 17.1 Sunlight provides the power needed to drive the photosynthetic machinery of plant cells. Light and life are intimately related.

Walter Dawn

One of the earliest experiments in photosynthesis was performed by the Belgian scientist Jean Baptiste van Helmont in the latter part of the 16th Century. He planted a small willow shoot in a tub of soil, and for a few years added only water to the growing plant. After five years he removed the plant and weighed it. He found that it had gained well over 150 pounds; meanwhile, the soil had lost only a few ounces. He concluded that the extra weight of the tree must have come from the water, the only thing he had added. Although his conclusion was not entirely right, it opened the way to a more careful study of photosynthesis.

In 1771, the British scientist Joseph Priestley showed that green plants could regenerate "good air" which had been converted to "bad air." Priestley placed a mouse in a closed container of air. After a short time the mouse died. He then placed a lighted candle in the same container of air and found that the flame quickly went out. He said that this air was "bad," that some life-giving part of it had been taken away. When he put a sprig of mint in a similar closed container in which a candle had been extinguished, he found that the bad air became good air, and that a candle could be burned in it. We now know that Priestley's experiment demonstrated that the mouse required oxygen and produced CO_2; also, that the plant used the CO_2 and some water, and released oxygen.

The real quantitative aspects of the use of light by plants were not considered seriously until about 1845. At that time the German physician Robert Mayer formulated the law of conservation of energy. The idea that energy could be changed from one form to another—that it does not just disappear—was important to an understanding of photosynthesis in plants. Mayer realized that photosynthesis was a conversion of light energy to a form of chemical energy which was stored by the plant cell. We can now summarize photosynthesis as being the process in which CO_2 is taken up by green plants in the presence of water. Sunlight is then used to form carbohydrate and molecular oxygen, as shown in the following equation:

$$6CO_2 + 6H_2O + 672{,}000 \text{ calories (light)} \longrightarrow C_6H_{12}O_6 + 6O_2$$

What this equation says is that sunlight energy is absorbed by the plant and converts six units of CO_2 and water into one unit of carbohydrate and six units of oxygen. But this general equation tells us only what goes in and what comes out. The problem is a more difficult one when we ask what takes place inside the cell, and *how* CO_2 is taken up and *how* light energy is used.

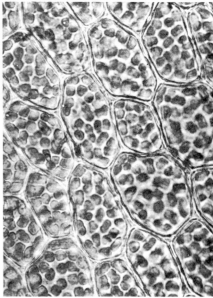

Photo by Hugh Spencer

Fig. 17.2 Chloroplasts, the cell components that give green plants their green color, are clearly visible in the cells of this moss (*Minium*) leaf.

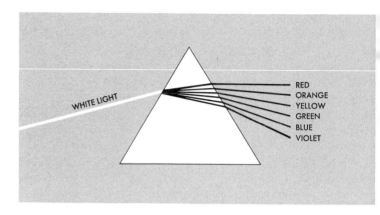

Fig. 17.3 A spectrum of colors is formed when light passes through a glass prism. Light of all wavelengths is slowed down when it passes from the air into the glass of the prism. Violet light is refracted the greatest amount and travels slower than red light which is refracted least.

PHOTOSYNTHESIS AND THE ATMOSPHERE

The oxidation of organic compounds during respiration, and during fermentation, gives off CO_2 to the air. In addition to this source, the burning of oil and coal also releases large amounts of CO_2 into the air. It is remarkable that the concentration of CO_2 within the air remains very nearly constant. Evidently the total rate of CO_2 production is exactly balanced by CO_2 photosynthetic consumption. The exact mechanism of the regulation of the atmospheric CO_2 by photosynthesis is not clearly understood. However, we do know that it is of great importance in affecting the Earth's temperature.

The Role of Light

Before describing photosynthesis and the general effect of light on biological processes, we should consider some of the properties of light. When a fine beam of sunlight is passed through a prism, it is separated into its component colors: violet, blue, green, yellow, orange, and red. If the different colors are then passed through a second (reversed) prism, the colors recombine as white light. But if only a single color is selected from the spectrum, no treatment can change it in any way. The individual colors can be regarded as resulting from the behavior of discrete units, or particles, called **photons.** The energy of the photons determines the color of the light. The photons of violet light have more energy than the photons of blue light; blue has more energy than green; green more than yellow, and so on to red, whose photons are least energetic of all.

The word "light" is usually applied to that range of radiation that can be detected by the eye. Scientists, however, speak of ultraviolet light and are able to "see" other types of radiation by means of photocells. Then there would seem to be more to light than meets the eye. While light behaves as particles (photons), it also behaves as waves. When the wavelength of violet light is measured, it is found to be shorter than the wavelength of red (Fig. 17.4). From violet to red along the spectrum, the wavelengths become progressively longer; and the longer the wavelength of a particular color of light, the less energy its photons have. Or, the shorter the wavelengths, the greater the energy. Visible light represents only a small part of the total range of radiation of the **electromagnetic spectrum.**

Ultraviolet and x-rays, beyond the violet end of the spectrum, have very short wavelengths and extremely high energies. Infrared and radio waves, beyond the red end of the spectrum, have very long wavelengths and low energies.

During photosynthesis, only the light in the visible part of the spectrum is used as an energy source. Let us now examine in some detail what happens when light energy falls on a green plant. In the process of photosynthesis, light energy of various wavelengths excites a chlorophyll molecule. As Fig. 17.5 shows, a photon strikes an atom of the chlorophyll molecule, causing an electron to jump to a higher energy level, called the **excited state.** This electron, which is far removed from the positive nucleus of the atom, is very active in reducing other substances. For example, it can reduce DPN which, in turn, can reduce CO_2 to carbohydrate. In addition, the reduced coenzyme DPN can be oxidized by the various enzymes previously described and, in the process, generate phosphate bond energy in the form of ATP is made. The oxidized chlorophyll that is generated in this process is reduced by electrons obtained from water (Fig. 17.6). An additional photon seems to be required for this reduction process. One hydrogen (electron) is moved from water by light and so

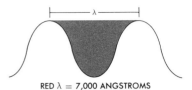

RED λ = 7,000 ANGSTROMS

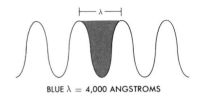

BLUE λ = 4,000 ANGSTROMS

Fig. 17.4 Comparison of wavelengths of red and blue light.

Fig. 17.5 Formation of reduced DPN by chlorophyll and light energy.

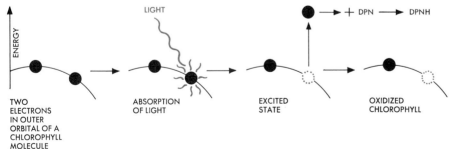

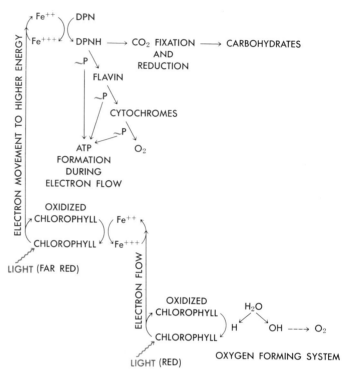

Fig. 17.6 Recent evidence indicates there are probably two types of chlorophyll in green plants and that both must be excited by light in order for photosynthesis to take place. One type of chlorophyll absorbs light in the far-red and creates an excited state that is capable of reducing DPN. The second type absorbs light in the red region and uses electrons from water to reduce the chlorophyll of the first type. In some unknown way the OH groups that are generated interact and form water.

reduces the oxidized chlorophyll. This leaves an OH which in some way interacts with other OH groups and forms water and molecular oxygen (Fig. 17.6).

Although we do not fully understand how ATP is made in the chloroplasts of green plants, we do know that it is a major product of the photosynthetic reaction—that is, light energy is converted into the chemical energy of ATP. ATP synthesis generated by light energy is called **photophosphorylation.**

One particularly fascinating thing about photosynthesis is that only about two per cent of the total energy of sunlight falling on a plant is used. A large amount of the energy is never absorbed. About 30 per cent of the light goes right through the leaf, while about 20 per cent is radiated away as heat (longwave radiation). Close to 48 per cent is used up as heat in the evaporation of water.

The two per cent of sunlight that is used in photosynthesis provides enough energy to maintain the whole plant, which, in turn, directly or indirectly is the energy source for all other living things. Considering the plant itself, only some parts of it are useful for human consumption; a large fraction of a typical plant consists of inedible fibers. If a crop is eaten, only a very small amount of it is converted into animal tissues: Possibly 90 per cent of the energy in a crop fed to animals is lost as heat or waste. This means that when we eat meat, we are obtaining only about 10 per cent of the energy originally fed into the animal. Using plants to feed farm animals in order to produce food for humans is a very wasteful and inefficient process. It would be much more efficient—if not quite so pleasant—to omit the animal stage and eat only plants.

The Rate of Photosynthesis

Several things affect the rate of photosynthesis. An adequate supply of CO_2 and water is essential. In bright sunlight, the relative rate of photosynthesis increases as the CO_2 concentration increases up to a given level (Fig. 17.7). The same can be said for water. In addition to the CO_2 and water concentration, temperature influences the rate of photosynthesis. When the light intensity is high and the CO_2 concentration is great, the effect of temperature becomes very apparent. The rate of photosynthesis at different temperatures increases up to about 30° or 35° and then starts to decline because of the destruction of the plant enzymes by heat.

The Production of Chlorophyll

Light is essential not only for photosynthesis, but also for the production of chlorophyll. Chlorophyll begins as a substance called **protochlorophyll.** Protochlorophyll is different from chlorophyll in that it needs two additional hydrogens to make it into chlorophyll. Only in the presence of light can protochlorophyll change into chlorophyll. You can see that this is so by growing bean seedlings in the dark and then exposing them to light until various amounts of chlorophyll are formed. If you

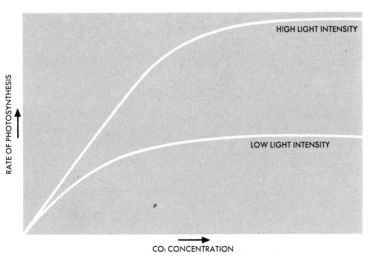

Fig. 17.7 Effect of light intensity and carbon dioxide concentration on the rate of photosynthesis.

use light bulbs of different colors, you can study the effectiveness of different wavelengths of light in making the leaves become green.

There are other pigments in green plants capable of absorbing light and transferring its energy to chlorophyll. Called **accessory pigments,** they are very important in that they allow the plant to use light of different wavelengths for photosynthesis. Were it not for accessory pigments, certain wavelengths of light would not be absorbed and used by the chlorophyll system. The blue pigment, phycocyanin, in blue-green algae is an accessory pigment.

Photosynthesis as an Energy Source

Not only is photosynthesis a supplier of food for living organisms, it is also a supplier of energy for other processes. Our industrial civilization has depended on a reservoir of coal and oil, which we have been removing from the ground at a rapid rate during the past century. Both coal and oil are derived from plants which grew in past geologic ages. Although we cannot say exactly how long these reservoirs will last, at the present we need not worry about what seems to be an inexhaustible supply of fossil fuel. However, the day will come when this stored supply of coal and oil will be used up, and a continuing source of energy will have to be found.

We could, of course, make use of the abundant energy coming from the Sun. It has been calculated, for example, that $1\frac{1}{2}$ square miles of the Earth's surface receives from the Sun during one day approximately the same amount of energy released by the explosion of one small atomic bomb. The difficulty at the present time is that man does not understand how to make really good use of the sunlight directly. About all we can do now is convert

Courtesy NASA Photo

Fig. 17.8 Space crafts are equipped with solar panels which collect solar energy and convert it into electrical power.

sunlight into electricity by using solar batteries. Possibly in the future we may learn how to store this energy in some other more stable form; it could then be released as we needed it.

The need for power sources is one of the reasons for our great interest in photosynthesis. A process which, on a large scale, captures light energy and stores it in a chemical form is of considerable economic interest. Photosynthesis takes place over the surface of our planet and fixes CO_2 at the rate of about a million tons of carbon per minute. Unfortunately, we do not yet know if we can duplicate these feats in the laboratory in an economical way. If we do manage to build a machine capable of converting solar energy to power on a large scale, the machine will probably be quite different from the green plant. Considering that the Earth's energy sources are limited, a photosynthesizing machine, or something akin to it, must one day become a necessity. As one investigator has pointed out, unless more progress is made in solar energy conversion during the next hundred years, we may again go back to the horse and buggy method of transportation.

VISION

Photosynthesis is but one aspect of light and life. Although there are many, many others, we shall consider only two—vision and bioluminescence. The visual process is another striking biological example where electrons in excited states must be involved. In spite of a great deal of outstanding work on the eye, we know very little about the basic mechanisms underlying vision. By attaching electrodes to the optic nerve, which connects the eye to the brain, we can tell that a nerve impulse is produced when photons strike the retina. A man's eye contains about four million units called **cones,** and an additional 125 million units called **rods.** The rods lead to about one million optic nerve fibers. They control our dim light vision, which is mostly black and white vision. The cones control bright light, or color, vision.

We assume that the triggering mechanisms of the nerve impulse must lie in the rods and cones because most of the visual pigments are found in the outer segment of them. One of the visual pigments, **rhodopsin,** has been extracted and has been shown to consist of a protein, called **opsin,** attached to a vitamin A derivative called **retinene.** Vitamin A is converted into retinene by the reduction of the aldehyde group to the alcohol by reduced DPN. Alcohol dehydrogenase catalyzes this reaction. When light strikes the visual pigment rhodopsin, the pigment

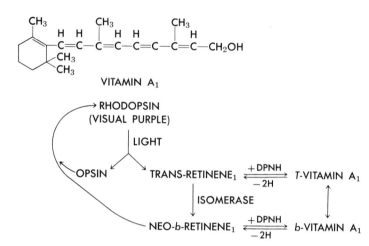

Fig. 17.9 When light strikes the visual pigment the rhodospin immediately dissociates into opsin (a protein) and retinene (the pigment). It is during this time that the nerve impulse is triggered in some unknown way and registered in the brain as a visual impulse.

immediately dissociates into opsin and retinene (Fig. 17.9). It is during this time that the nerve impulse is triggered.

The relationship between vitamin A and vision was first noticed when a vitamin A deficiency was associated with the eye's ability to adapt to the dark. If you do not have enough Vitamin A, the reconstruction of the visual pigment is retarded, and you become night-blind. The **absorption spectrum** of rhodopsin is identical to the sensitivity of the eye to different wavelengths of light. The peak absorption and sensitivity is around 500 millimicrons. How the photoexcitation in the pigment's molecules produce an impulse in the nerve fibers is not understood. Rhodopsin dissociates into opsin and retinene when exposed to light and recombines in the dark. These facts led to the following conclusion.

The photocurrent must be associated with the chemical changes in the retinene that follow after dissociation. Recent studies, however, show that these changes probably are quite slow compared to the rate of initiation and conduction of the nerve impulse. We now believe that the small chemical changes that might have occurred in the unfolding of the protein during the photochemical event may be sufficient to initiate the electrical impulse which is conducted by the nerve to the brain. The impulse is registered in the brain as a picture and gives us vision. The method of decoding this electrical message to give a visual sensation still remains a mystery.

Research into brain function is rapidly becoming one of the most interesting and challenging fields of research in biology. It seems possible, by using special electrical and computer techniques, that we can come to understand these and other coded messages transmitted by nerves to the brain.

Considerable progress has been made recently in understanding certain aspects of color vision in man, goldfish, and other animals. There seem to be three distinct types of cones, each of which is sensitive to three different colors of light: blue, yellow-green, and red. The difference in the sensitivity is due to the presence of pigments which absorb photons primarily at the three different wavelengths. The combination of electrical impulses from these three basic color receptors gives us the color spectrum that we can see. But much work remains to be done on this fundamental problem before we can be certain about the basic mechanisms involved.

BIOLUMINESCENCE

There are many organisms which are capable of giving out light, a process called **bioluminescence.** These are a variety of luminous forms, ranging from the bacteria that give off a blue light to the South American railroad worm (a larva of a beetle) that gives off a red light. The light of decaying wood is produced by luminous fungi. The luminescence of the sea is caused by a variety of forms including various protozoans, sponges, jellyfish, brittle stars, snails, clams, squid, shrimp, small crustaceans, fish, and many other forms (Fig. 17.10). Probably the best known luminous forms on land are the fireflies and the glow worms. In addition to these forms, there are luminous spring tails, flies, centipedes, millipedes, earthworms, and snails. The mechanism of this bioluminescence is not, at present, too well understood.

Walter Dawn

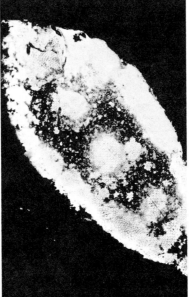

Fig. 17.10A Photobacterium fischeri beginning to develop in colonies on flounder. Photo taken under subdued light.

Fig. 17.10B Photobacterium fischeri in colonies of high luminescence. Photo taken by luminescence only.

In 1887, the French physiologist Raphial DuBois suggested that the light from a luminous clam, *Pholas dactylus,* is caused by the oxidization of a substance he called **luciferin**. In the presence of oxygen and an enzyme called **luciferase**, luciferin is destroyed by an oxidative reaction which gives off light. Although a great deal of progress has been made in recent years in probing the nature of the chemical substances required for light emission, the basic mechanism remains obscure. In some respects bioluminescence is very similar to photosynthesis and vision. In the latter two cases, the excited states are generated by light, while in the case of luminescence the excited state is created by chemical reactions and the energy is lost as light.

How do those organisms displaying this remarkable ability to emit light use the light? There is no clear answer yet. There are many examples in which bioluminescence has been adapted to good biological use. The reproduction cycle of many organisms in the sea is intimately tied to light emission. The flash of the firefly is used by some species as a sex signal. Light emission by deep-sea organisms living at great depths provides the only light available to them. In some cases, luminous bacteria are known to grow in special glands in fish and to provide a regular source of light. Whether light emission in the depths of the ocean triggers other photobiological processes is not known, but it is safe to guess that it does.

EFFECT OF LIGHT ON BIOLOGICAL PROCESSES

In addition to the photosynthetic reactions in green plants that are dependent on light there are other striking effects of light. Most green plants, for example, display striking growth responses to light, and their response may be independent of photosynthesis. The bending and growth of a green plant toward light is known as **phototropism** and is attributed to the effect of light on the metabolism of one of the plant growth hormones, **auxin**. The duration of light and dark periods also regulates the flowering and reproduction of plants, an effect known as **photoperiodism**. The phenomenon of photoperiodism is not limited to plants. Most organisms that have been studied carefully show a day-night periodicity for a number of physiological processes. Thus both animals and plants exhibit rhythms in a manner indicating that a "biological clock" plays an important role in controlling behavior and physiological function.

Plants and animals are able to "recognize" the seasons of the year by measuring the changing length of night and day. For example, the migration of Canadian geese in the autumn is initiated by the change in the length of the nights and is one of the many important and interesting displays of photoperiodic responses in animals. The length of the day or the shortness of the night have been demonstrated to have important effects in the sexual cycle of numerous fishes, reptiles, and in the reproductive cycle and migration of birds.

The flowering of the poinsetta at the Christmas season is another example of photoperiodism. Other plants grow, flower, and bear fruit at different times of the year. Thus, we come to recognize that some clocks in plants cause flowering in the spring, while other periodic clocks cause flowering in the summer, and still others in the autumn.

Unfortunately, we know very little about the details of the biochemistry or physiology of the biological clock systems that control these rhythmic responses. There must be specific pigments in the cells that are affected by light and, in turn, regulate cellular responses. One such pigment in plants, **phytochrome,** has been studied in detail. We know from studies on flowering response at various wavelengths of light that the interruption of a long dark period by red light inhibits the flowering response in the so-called short-day plants (that is, those that require long night). If the plants are subsequently exposed to far-red light, flowering is stimulated. The pigment which absorbs the light is phytochrome. One form absorbs in the red region and is then converted into a form that absorbs in the far-red region. For flowering to occur in short-day plants (for example, soybean) the period of darkness (or far red) must be long enough to decrease the concentration of the far-red-absorbing pigment to a low level by conversion to the red-absorbing form. This state must be maintained for several hours if flowering is to be initiated. If the dark period is interrupted with a brief flash of red light the far-red-absorbing pigment produced suppresses flowering.

Unfortunately, we do not know the mechanisms involved in the flowering response. One important lead has been obtained by grafting experiments. By such techniques we can show that the phytochrome is in some way associated with the production of a plant hormone that initiates flowering. Basic research in this area is of considerable importance since it could lead to important agricultural advances.

Flowering in plants is not the only response to the red-far-red light system. Leaf expansion, stem elongation, seed germination, and other processes also demonstrate a red-far-red antagonistic response.

SUMMARY

Light is essential for all life. Through the process of photosynthesis in green plants, light energy is used to convert carbon dioxide into complex organic matter (food). At the same time molecular oxygen is produced from water. The green pigment, chlorophyll, is the primary trapping agent for the light quanta. When light is absorbed an electron in the chlorophyll molecule is raised to an excited state. The electron in the high-energy state can be used to reduce DPN and to make ATP in a process called **photophosphorylation.** The ATP energy and the reducing power of DPNH is essential for the fixation and reduction of CO_2 to carbohydrate. The oxidized chlorophyll is reduced in the light by an electron from water and oxygen is formed.

The eyes of animals are a very special organ for trapping light quanta. The eye of man contains both rods and cones. The rods are associated with black and white vision while the cones control color vision. When light quanta strike the rods and cones the quanta are absorbed by a special pigment called **rhodopsin.** Rhodopsin consists of a protein called **opsin** and a Vitamin A derivative called **retinene.** In some way the absorption of light triggers a nerve impulse from the eye along the optic nerve to the visual center in the brain. Unfortunately, we do not understand the mechanism of this aspect of the visual process.

There are many organisms that produce light by special chemical reactions. This process of **bioluminescence** is particularly noticeable among beetles (fireflies) and in the ocean. There are many examples in which bioluminescence has been adapted to biological use. However, we are not certain how this unique ability arose in nature.

In addition to photosynthesis and vision, light has other effects on biological processes. The bending of plants toward light **(phototrophism)** is an excellent example. The length of the day or the shortness of the night has been demonstrated to have important effects in the sexual cycle of numerous marine invertebrates, fish, reptiles and in the reproductive cycle and migration of birds. Duration of light and dark periods also regulate flowering and reproduction in plants. This is called **photoperiodism.** Numerous studies of this type have lead to the general notion that there are "biological clocks" which regulate specific functions. Light seems to have a pronounced effect on the rhythm or period of these clocks.

FOR THOUGHT AND DISCUSSION

1 Draw a graph indicating the approximate relationship between the energy in a light quantum and the wavelength.
2 If a green plant absorbs light in the red and far-red region of

the spectrum does this give you an idea why the leaf looks green?
3. What is an **excited state?** How is it used by the green plant?
4. What would happen to the CO_2 and oxygen content of the air if photosynthesis by plants was stopped?
5. What is **photophosphorylation?**
6. In human vision which wavelength of light is the eye most sensitive to? What determines this sensitivity?
7. Can you think of any good reasons why organisms give off light?
8. Can you describe any example where photoperiodism in either plants or animals may be of economic importance?

SELECTED READINGS

Scientific American articles

Arnon, D. I. "The Role of Light in Photosynthesis," November 1960.
Bassham, J. "The Path of Carbon in Photosynthesis," June 1962.
Brown, F. A., Jr. "Biological Clocks and the Fiddler Crab," April 1954.
Evans, Ralph M. "Seeing Light and Color," August 1949.
Land, E. H. "Experiments in Color Vision," May 1959.
McElroy, W. D. and H. H. Seliger. "Biological Luminescence," December 1962.
Rushton, W. A. H. "Visual Pigments in Man," November 1962.
Wald, G. "Life and Light," October 1959.
Waterman, T. H. "Polarized Light and Animal Navigation," July 1955.

18 DNA—THE MOLECULE OF LIFE

Up to this point we have considered the nature of some molecules of crucial biological importance. In particular, we have dealt in some detail with the carbohydrates, fats, and proteins: their molecular structure, their involvement in cellular structures and reactions, and the flow of energy in and out of these molecules. We recognize that these molecules and the role they play in the life of a cell form an organized pattern, that a cell is an organized structure in which organized reactions take place. A bacterial cell does what is proper for a bacterial cell to do, and it does not do those things that are proper for an elephant, grass, or a human cell. We need, therefore, to consider the control mechanisms that operate within cells.

Two major controlling systems are recognized. The first governs the general character of cells. That is, a system exists which determines whether a cell is part of a man or part of a man-eating shark. If the cell is human, its control system determines whether it is part of Mary Smith or Tom Jones. Both of these aspects are part of the problem of inheritance. If the cell is part of Mary Smith, the system also determines whether the cell will be part of her liver, muscles, or skin—a problem of cellular differentiation. This system has already been discussed briefly in Chapter 11, and we now know that the crucial molecule is deoxyribose nucleic acid, or DNA. Its structure and function will be discussed in this chapter.

The second controlling system is that which governs cellular metabolism. Consideration of it will be the subject of the next chapter.

We want now to do three things: (1) inquire into the chemical and physical nature of the DNA molecule; (2) consider how such a molecule can be a source of information which determines the character and, in part, the activities of the cell; and (3) explore how this information passes from DNA to other parts of the cell where it is put to use.

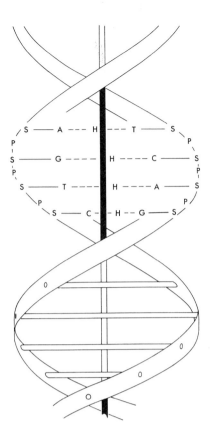
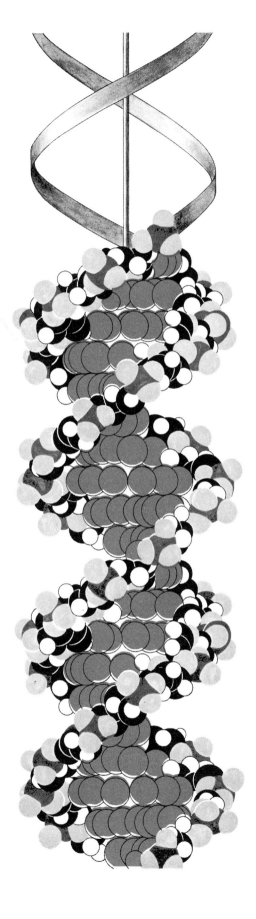

Fig. 18.1 DNA molecule. In the model above, the double helix is shown with the base pairs extending across the molecule from the sugars on either side. The sugar residues (S) are linked by phosphates (P) on both sides, forming the two continuous "backbones" of the long molecule. Only the base pairs A—T, T—A, C—G, and G—C are possible, but they can vary, giving internal variety to the molecule. The vertical rod running through the center of the molecule is an imaginary axis around which the helix entwines. At right, the same molecule is shown, but with the atoms indicated.

COLOR KEY

BASE PAIRS   H   O   P   C

The story of the inquiry, which has taken place in the last 10 to 15 years, is one of the most exciting chapters in the history of the biological sciences. Although the existence of nucleic acids has been known for nearly 100 years, only recently have their structure and role been fitted into general biological theory. An understanding of the role of these molecules has revolutionized biology.

DNA is unique in three respects. First, it is a very large molecule, having a certain outward uniformity of size, rigidity, and shape. Despite this uniformity, however, it has infinite internal variety. Its varied nature gives it the complexity required for information-carrying purposes. One can, indeed, think of the molecule as if it had a chemical alphabet somehow grouped into words which the cell can understand and to which it can respond.

The second characteristic of DNA is its capacity to make copies of itself almost endlessly, and with remarkable exactness. The biologist or chemist would say that such a molecule can **replicate,** or make a carbon copy of itself, time and again with a very small margin of error.

The third characteristic is its ability to transmit information to other parts of the cell. Depending upon the information transmitted, the behavior of the cell reflects this direction. As we shall see, other molecules play the role of messenger, so that DNA exercises its control of the cell in an indirect manner.

THE STRUCTURE OF DNA

DNA can be isolated from nearly every organism—from viruses, bacteria, and fungi to man and the plants and animals with which we are all familiar. Only certain of the viruses lack DNA; in such viruses DNA is replaced by a comparable molecule, RNA. No matter where it is found, all DNA has much the same chemical and physical properties.

DNA is a **polymer**—a very large molecule made up of repeating units. In this sense it is much like rubber or many plastics, but with the single, important exception that the repeating units of DNA can vary. To give you some idea of the size of the molecule, Fig. 18.2 shows the length of a bacterial virus DNA molecule in relation to the length of the virus itself. For such a large molecule to be compacted into so small a space, DNA must be capable of considerable folding.

The repeating units of DNA are called **nucleotides;** these are rather complicated molecules. All DNA nucleotides contain three units: **phosphoric acid,** a 5-carbon sugar called **deoxyribose,** and a **base,** in this case a ring structure which can take up hydro-

Fig. 18.2 A drawing of the entire DNA molecule contained within a bacterial virus. The dimensions of the virus and the length of the molecule are drawn to scale, so the molecule must be highly compacted to fit within the hollow head of the virus.

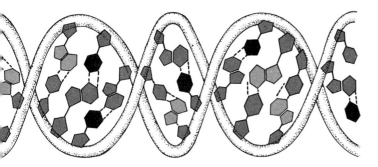

Fig. 18.3 In this diagram of a DNA molecule, the base pairs are stretched in flat planes perpendicular to the sugar-phosphate backbones. Actually there are 10 base pairs for every full turn of the double helix. Only five are shown here.

gen ions. The way these components are arranged to form repeating units is shown in Fig. 18.3. The bases project from one end of each of the sugars, while the phosphoric acid serves as a link binding the successive sugars.

The bases are of four major types. Two of the four are **purines (adenine** and **guanine),** and two **pyrimidines (cytosine** and **thymine).** Four nucleotides can consequently be formed from these four bases by the addition of phosphoric acid and a sugar to each one. The two purine nucleotides are **deoxyadenylic acid** and **deoxyguanilic acid.** The pyrimidine nucleotides are **deoxycytidylic acid** and **deoxythymidilic acid.**

It is the number, type, and arrangement of the bases of DNA that determine what kind of an organism will develop. Since these bases can vary in sequence along the length of the DNA molecule, much as the amino acid sequences of protein can vary, the DNA's of different organisms are composed differently. We can, therefore, characterize the various DNA's by their base composition, that is, by the relative proportions of the four nucleotides to each other. As yet we know little about the ordered arrangement of the bases in the DNA from any given organism. The discovery that DNA from a variety of organisms can vary was an important step in our understanding of this molecule as a source of hereditary information.

The biochemist Irving Chargaff and his collaborators at Columbia University, however, made two additional important discoveries. First, when they analyzed DNA for its base composition, they found that the number of purines was always equal to the number of pyrimidines; second, they found that the number of adenines was equal to the number of thymines, and that the number of guanines equals the number of cytosines. We can now discuss DNA in terms of its base ratios. Using capital letters to represent the bases, we can say the following: $A + G = T + C$. Also, $A = T$ and $C = G$, but A does not have to equal

TABLE 18.1 Base Ratios of Several Well-known Organisms
(values are arbitrary but accurate as ratios)

	Adenine	Thymine	Guanine	Cytosine	Ratio of $\frac{A+T}{C+G}$
Man	29.2	29.4	21.0	20.4	1.53
Sheep	28.0	28.6	22.3	21.1	1.38
Calf	28.0	27.8	20.9	21.4	1.36
Salmon	29.7	29.1	20.8	20.4	1.43
Yeast (fungus)	31.3	32.9	18.7	18.1	1.19
Staphylococcus	31.0	33.9	17.5	17.6	1.85
Pseudomonas	16.2	16.4	33.7	33.7	0.48
Colon bacterium	25.6	25.5	25.0	24.9	1.00
Vaccinia virus	29.5	29.9	20.6	20.0	1.46
Pneumococcus	29.8	31.6	20.5	18.0	1.88
Clostridum	36.9	36.3	14.0	12.8	2.70
Wheat	27.3	27.1	22.7	22.8	1.19

C or G. The base ratios for several well known organisms are shown above (Table 18.1).

The importance of these discoveries is that we can now visualize how each organism can pass a specific kind of information to its cells and to its offspring. For example, in a given small section of DNA, the base sequence of ATCCGATT may mean something very different from ACCGTTAT, even though the number and kinds of letters are the same. In a sense, this is not very different from the words *heat* and *hate,* which have the same letters, but a different arrangement and, consequently, a different meaning.

A Molecular Model of DNA

With the discovery that DNA is the key molecule of heredity, scientists asked the question: "What is the structure of DNA?" The answer came in the early 1950's through the efforts of an English crystallographer, M. H. F. Wilkins, an American biologist, James Watson, and an English chemist, Francis Crick.

The findings of these men suggested that a molecule of DNA was a double-strand structure. The two polynucleotide strands are intertwined in the form of a long helix, or spiral. The outside edges of the helix have alternating molecules of phosphoric acid and sugar. These bases project to the inside of the helix and form base pairs, A pairing with T, and C pairing with G. One very important aspect of the model, therefore, is that the sequence of bases in one strand of the helix determines, or is complementary to, the sequence of bases in the other strand. Thus, if strand X has the sequence shown above, strand Y must have the complementary sequence. That is, if X strand has the sequence of ATGGC, then Y strand must be TACCG. The base pairs are connected to each other by hydrogen bonds.

The usefulness of this model of DNA has been demonstrated several times. For example, biologists have long known that the hereditary materials in the cell—that is, the chromosomes in the nucleus—replicate themselves exactly at each cell division. In a gross way, we can see the result of this replication in the microscope. The doubleness of each chromosome at metaphase, and the separation of the two chromatids at anaphase, tell us that the chromosome has replicated itself. The Watson-Crick model demonstrates how this can be accomplished chemically. A double-stranded piece of DNA (Fig. 18.4) is gradually separated into two single polynucleotide strands. Each of these strands directs the synthesis of a new strand, producing two double-stranded structures. If we were to label the two original strands X and Y, it would become evident, because of the manner of accurate base pairing, that X would direct the formation of a new Y strand. At the same time, the original Y strand would direct the formation of a new X strand.

This synthesis of new DNA, like other cellular reactions, is under the direction of a nuclear enzyme called **polymerase.** (The enzyme is also known as the **Kornberg enzyme,** after Arthur Kornberg, of Stanford University, who discovered it and who was awarded the Nobel Prize for this discovery.) Two double helices, apparently identical in all respects to the original double helix, can thus be formed, and the piece of DNA replicates itself. It is through this mechanism of chemical synthesis that new chromosomes, which are exact replicas of previous chromosomes, are formed. Hereditary information is passed from one cell to another in division, and from one generation to the next through reproduction.

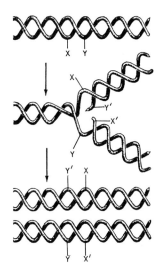

Fig. 18.4 As a DNA molecule separates into its two polynucleotide strands (X and Y), each strand directs the formation of a complementary strand to pair with it; that is, X directs the formation of Y', and Y the formation of X'. The entire strand is progressively replicated in this way, and the two new helices are identical to each other in nucleotide sequence.

omit

Proof of DNA Replication

The DNA story is an attractive and exciting one. It is attractive because it ties together a lot of information about cell behavior; exciting because it provides an opening wedge into some of the mysteries of the cell.

Let us now play the role of the skeptic, a role that comes naturally to scientists. Two experiments will be described, both cleverly conceived and beautifully executed. One deals with the chromosomes of higher plants since these chromosomes are large enough to be seen and they contain the DNA that interests us.

The other deals with the DNA of a bacterium which can be seen only in the electron microscope.

You will recall that we stated that DNA is newly synthesized during the interphase stage of cell division. Let us assume, as we have indicated in the diagram (Fig. 18.4), that the original (X and Y) strands of DNA will be preserved intact and that the X' and Y' strands will be the newly synthesized ones. If this is so, it should then be possible to identify the old from the new if the new one can be "tagged" in some way. This was accomplished by J. H. Taylor of Florida State University by using the following procedure.

At the time the DNA was being synthesized, he fed the root-tip cells of a broad bean plant a solution containing radioactive thymidine. This molecule is incorporated *only* into DNA, where it becomes the deoxythymidilic nucleotide. Since DNA is a very stable molecule, the old strand would not pick up any of the radioactive thymidine. During formation, however, the new strand would. When the chromosomes reach metaphase and anaphase, at which time they are large and distinct enough to see, the distribution of the radioactivity can be determined by the use of **autoradiographic** techniques.

If you understand how a photographic negative is produced, you can understand **autoradiography.** A root tip is squashed on a glass slide so that the cells are flattened out. Then, a thin photographic emulsion is placed over the flattened cells. As the radioactive atoms in the thymidine decay, atomic particles or rays are released and pass through the emulsion. These darken the emulsion much as ordinary light affects a photographic film.

Taylor showed that cells fed radioactive thymidine during interphase would show radioactivity in both chromatids during metaphase stage. If, however, he allowed these cells to pass through another division, then the chromosomes in the next metaphase would show one chromatid radioactive and the other "cold," or nonradioactive. These events are shown in Fig. 18.5.

Let us now relate these results to the model of DNA we have discussed. Assume that the chromosome in interphase, before DNA synthesis, consists *only* of a double helix. (Actually it has other molecules—protein and RNA—but we can disregard these for the moment.) Assume, also, that when the DNA replicates itself the old strands are preserved intact, but when the new ones are synthesized they take up radioactive thymidine. The diagram shows what will be expected at the first metaphase, and also after another division. The results of the Taylor experiment are, therefore, in agreement with what we might have predicted on the basis of the Watson-Crick model and its mode of replication.

Let us now take stock of what we *know* as opposed to what

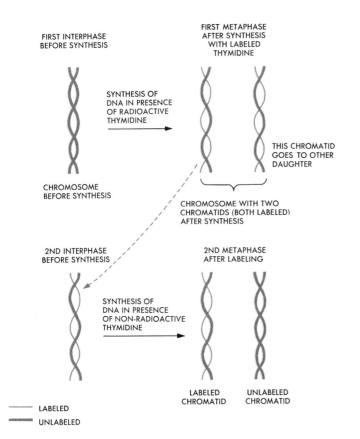

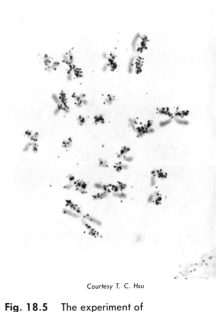

Courtesy T. C. Hsu

Fig. 18.5 The experiment of H. J. Taylor demonstrated that the method of replication shown in Fig. 18.4 is correct for chromosomes as well as for the double helices of DNA. The newly synthesized strands showed that both chromatids of each chromosome were radioactive at the first metaphase following labeling. At the second metaphase, each chromatid had replicated again, but in the presence of nonradioactive thymidine, and one chromatid showed itself radioactive. The other was not. The results at the second metaphase can be seen in the photograph (above) of hamster chromosomes.

we have *assumed*. Taylor showed by means of autoradiography that radioactive thymidine is incorporated into DNA. He also demonstrated that only the newly synthesized strands pick up the radioactive thymidine; the old strands are unaffected. In order to prove the correctness of this, the old strand of DNA had to be identified with certainty. This problem was attacked by Matthew Meselson and Frank Stahl, working at the California Institute of Technology, in a beautiful experiment with bacteria (Fig. 18.6).

The most common nitrogen has an atomic weight of about 14. Some nitrogen atoms, however, have an extra neutron, which gives the atom a weight of 15. Ordinary nitrogen is written N^{14}, heavy nitrogen, N^{15}. If nitrate, NO_3^-, is used as a food source, the bacteria can grow equally well on $N^{15}O_3^-$ as on $N^{14}O_3^-$. Also, if the nitrate used is the only source of nitrogen available to the bacteria, the N atoms from the NO_3^- will become incorporated into the cells; in particular, for our discussion, the N atoms will go into the nucleotides of DNA.

Therefore, if one culture of bacteria is fed $N^{15}O_3^-$, and another culture is fed $N^{14}O_3^-$, the DNA extracted from the first should be heavier, or denser, per unit of volume, than that from the second. If a mixture of "heavy" and ordinary DNA is put into a centrifuge, the heavy DNA can be separated from the light DNA, since the movement of a molecule in a centrifugal field (that is, its sedimentation rate) depends upon its weight, or density. Stated in another way, heavy DNA will move a greater distance before coming to rest, or equilibrium, in a centrifugal field, than will ordinary DNA. A molecule of DNA made up of one-half N^{14} and one-half N^{15} will come to rest halfway between DNA^{14} and DNA^{15}.

Meselson and Stahl grew bacteria for several generations in $N^{15}O_3^-$, until virtually all of the nitrogen of the DNA was N^{15}. Cells were then removed from the culture medium containing $N^{15}O_3^-$, and placed in a fresh medium containing $N^{14}O_3^-$ *but only long enough to allow each cell to divide only once.* During this time the number of cells doubled, and each molecule of DNA had replicated itself. Of the total amount of DNA, one-half should have been DNA^{15}, the other half DNA^{14}. When the DNA from these cells was removed, its sedimentation rate was determined to be between that of DNA^{14} and DNA^{15}. So far, so good. But let us now consider the results in relation to our prediction: that the old strands of DNA^{15} are preserved intact, and that the new ones will contain only N^{14}. Essentially, this is what Taylor found when he used radioactive thymidine.

However, let us suppose, as a contradictory hypothesis, that the old DNA breaks down completely instead of being preserved intact. The new DNA will then be assembled at random from bases containing N^{15} (from the old DNA) and from newly synthesized bases containing N^{14}. Both strands of the double helix

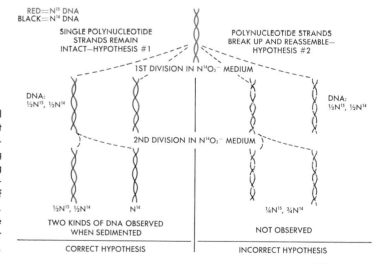

Fig. 18.6 The Meselson-Stahl experiment demonstrated that the polynucleotide strands separate but remain intact during synthesis rather than breaking down and reassembling. Compare these results with that of the Taylor experiment in Fig. 18.5. The results are the same whether whole chromosomes or only DNA molecules are used.

will have a mixture of bases of N^{15} and N^{14}, and we could label this as $DNA^{1/2n15,1/2n14}$. This kind of DNA would be indistinguishable, insofar as sedimentation rate is concerned, from DNA in which one strand had only N^{15} and the other only N^{14}. It is the amount of N^{15} and N^{14} bases—not their distribution in the strands—that determines sedimentation rate. The experimental results described so far, then, cannot distinguish between the two hypotheses.

However, Meselson and Stahl carried the experiment one step further. They allowed some of the bacteria to divide once more in $N^{14}O_3^-$, harvested the cells, and extracted the DNA. Let us now see what would be predicted on the basis of our two hypotheses. If the DNA breaks down, as our second hypothesis supposes, each strand of DNA would have $\frac{1}{4}N^{15}$ and $\frac{3}{4}N^{14}$, with the N^{15} and N^{14} randomly distributed. *There would be only one kind of DNA.*

On the other hand, if the old N^{15} strand had been preserved intact, two kinds of DNA would be found: $DNA^{1/2n15,1/2n14}$ and DNA^{n14} (see diagram). Also the heavy DNA should contain one-half of the original N^{15}. These two kinds were found, and we can, therefore, state that our first hypothesis has been supported.

We are not certain, at this moment, that the answers are final. The DNA molecule we have been discussing is a **macromolecule,** meaning that it is a huge molecule with a molecular weight of several million or more. It has many thousands of turns in its spiral configuration, and the base sequence can be arranged in any order. The possible variations, therefore, are astronomical in number and give an infinite variety to the DNA's from various sources. It has been estimated, for example, that every cell in the human body has approximately eight billion nucleotide pairs making up the DNA of its 46 chromosomes. If we look upon the base pairs as the letters in a genetic alphabet, which when put together in a particular sequence form a "word" having meaning to the cell, then we can readily grasp the idea of how DNA carries information.

DNA-RNA Protein Chain of Relationships

The question we must ultimately ask is "How does DNA carry out its cellular role of command?" Before trying to answer this question, we need to remember that what a cell does and how it is constructed depend largely on the kinds of proteins it contains. These include all of the enzymes and membranes of the cell, and are found as well in ribosomes, spindle, plastids, and chromosomes. Therefore, if we maintain our hypothesis that DNA is the controlling agent of the cell, then we must also hypothesize further that DNA in some way controls the production of the various kinds of proteins.

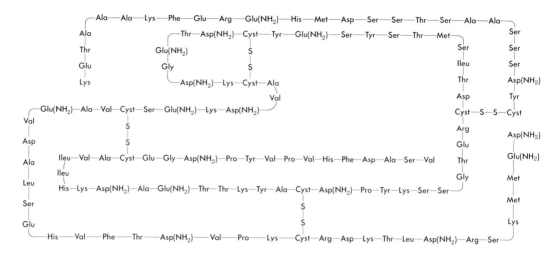

Fig. 18.7 The entire molecule of ribonuclease (or insulin) showing the sequence of amino acids, and the manner by which cross-links cause the molecule to fold.

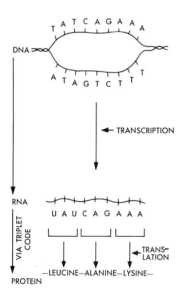

Fig. 18.8 Here is an outline sequence of events from DNA to RNA to protein. The determination of the amino acid sequence by means of the triplet code is also shown.

However, we know that DNA is mainly in the nucleus, while protein is formed largely on the ribosomes of the cytoplasm. Some substance must consequently be a messenger, carrying the information from DNA to the ribosomes so that they will "know" just what protein to make. We now know that this messenger is **RNA,** a nucleic acid molecule which differs from DNA in having a substance called **uracil** in place of thymine, and in having a **ribose sugar** in place of the deoxyribose sugar. In a very much over-simplified way we can then say that DNA makes RNA, and that RNA makes protein.

If this is so, then each RNA molecule produced by DNA must contain enough information to spell out the order, kind, and number of amino acids in each protein, and the features which give each protein its uniqueness of structure and function. (Such a protein is illustrated in Fig. 18.8.) Three different RNA's participate in the formation of protein, and each plays a special role. These are **ribosomal RNA (r-RNA), messenger RNA (m-RNA),** and **transfer,** or **soluble, RNA (t-RNA or s-RNA).** All of these are formed by DNA, much in the same manner that DNA forms more DNA. This being so, each piece of RNA must have a sequence of nucleotides *complementary* to the strand of DNA which formed it—but with one exception: that the thymine of DNA will be replaced by uracil in the RNA. Thus, if a piece of DNA having the sequence ATAGTCTTT makes RNA, the RNA would have the sequence UAUCAGAAA (see Fig. 18.8).

The solution to the problem of how the chemical information in DNA is translated into the formation of a protein is one of the most significant biological discoveries of the 20th Century. The problem was basically one of unscrambling a code. What we need is a four-letter code that can form a dictionary of 20 words. The four letters are the four nucleotides of DNA: A, T, C, and G. The 20 words are the 20 amino acids found in virtually all proteins. Immediately we know that one nucleotide cannot be responsible for one amino acid (see Fig. 18.9). If it were, only four of the 20 amino acids would be accounted for.

DNA—The Molecule of Life

SINGLET CODE (4 WORDS)	DOUBLET CODE (16 WORDS)				TRIPLET CODE (64 WORDS)			
A	AA	AG	AC	AU	AAA	AAG	AAC	AAU
G	GA	GG	GC	GU	AGA	AGG	AGC	AGU
C	CA	CG	CC	CU	ACA	ACG	ACC	ACU
U	UA	UG	UC	UU	AUA	AUG	AUC	AUU
					GAA	GAG	GAC	GAU
					GGA	GGG	GGC	GGU
					GCA	GCG	GCC	GCU
					GUA	GUG	GUC	GUU
					CAA	CAG	CAC	CAU
					CGA	CGG	CGC	CGU
					CCA	CCG	CCC	CCU
					CUA	CUG	CUC	CUU
					UAA	UAG	UAC	UAU
					UGA	UGG	UGC	UGU
					UCA	UCG	UCC	UCU
					UUA	UUG	UUC	UUU

Fig. 18.9 The possible numbers of code words based on singlet, doublet, and triplet codes are shown here. We now believe that the triplet code is correct.

Two nucleotides for each amino acid are also not enough. The four letters, arranged in groups of two, would produce only 16 words, accounting for only 16 amino acids. What about a three-letter code: ATT, ATG, ATC, AAT, AAG, AAC, and so on? This would give us $4 \times 4 \times 4 = 64$. It would now appear that we have too many words. Possibly some of the words make sense, while others do not. For example, take the letters N, O, and W. These three letters can form six possible groups: NOW, OWN, WON, ONW, WNO, and NWO. The first three are sense words while the last three are nonsense words, at least in the English language.

We now believe that the three-letter, or **triplet, code** is the correct one for DNA. But is nature so wasteful that it makes use of only 20 out of 64 possibilities of making amino acids into protein molecules? It now appears that several amino acids have more than one triplet code (see Table 18.2). Much still needs to be done to clarify the coding system. For example, the first two letters of the code may stand for the particular amino acid while the third letter may have something to do with the efficiency of the process. These and other questions about the coding system remain to be worked out.

Figure 18.10 shows how the over-all system works and summarizes our present state of knowledge of the DNA-RNA protein story. The three RNA's are made in the cell nucleus. The r-RNA makes up the ribosomes, the s-RNA is attached to the ribosomes after becoming activated, and the m-RNA is wrapped around one or more ribosomes. The way the m-RNA and the s-RNA's interact determines the kinds and order of amino acids forming a protein molecule.

TABLE 18.2 The Amino Acid Code*

Code Triplets	Amino Acid	Code Triplets	Amino Acid
AAA	lysine	CAA	glutamine
AAG	lysine	CAG	glutamine
AAC	asparagine	CAC	histidine
AAU	asparagine	CAU	histidine
AGA	arginine	CGA	arginine
AGG	arginine	CGG	arginine?
AGC	serine	CGC	arginine
AGU	serine	CGU	arginine?
ACA	threonine	CCA	proline
ACG	threonine	CCG	proline?
ACC	threonine	CCC	proline
ACU	threonine	CCU	proline
AUA	isoleucine?	CUA	leucine
AUG	methionine	CUG	leucine
AUC	isoleucine	CUC	leucine
AUU	isoleucine	CUU	leucine
GAA	glutamic acid	UAA	gap (comma)
GAG	glutamic acid	UAG	gap (comma)
GAC	aspartic acid	UAC	tyrosine
GAU	aspartic acid	UAU	tyrosine
GGA	glycine?	UGA	tryptophan
GGG	glycine?	UGG	tryptophan
GGC	glycine?	UGC	cysteine
GGU	glycine	UGU	cysteine
GCA	alanine?	UCA	serine
GCG	alanine?	UCG	serine
GCC	alanine?	UCC	serine
GCU	alanine	UCU	serine
GUA	valine?	UUA	leucine
GUG	valine	UUG	leucine
GUC	valine?	UUC	phenylalanine
GUU	valine	UUU	phenylalanine

A most elegant proof that the DNA-RNA-protein system operates in the manner suggested was provided by Marshall Nirenberg of the National Institutes of Health. It is possible to make an artificial m-RNA containing only the base uracil. It is, therefore, polyuridylic acid, or **polyU**. When polyU was added to a test tube containing a mixture of ribosomes, amino acids, and all of the other components of the cell necessary for protein synthesis, Nirenberg found that polypeptides were formed. But the newly formed polypeptide contained *only* the amino acid

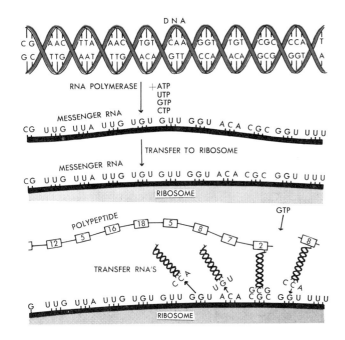

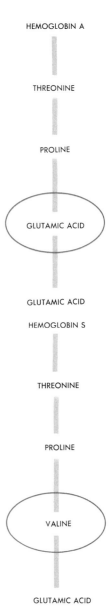

Fig. 18.10 The sequence of events from DNA to protein is shown here. Only one of the two polynucleotide strands of DNA transcribes and forms messenger RNA. The m-RNA becomes attached to the ribosomes where the transfer RNA's, each carrying a particular amino acid, match up in complementary fashion with the successive triplet codes of m-RNA. As this occurs, the amino acids are linked and form a polypeptide chain, and the t-RNA's float free.

phenylalanine. The triplet code for this amino acid was therefore UUU. By using other artificial m-RNA's, the triplet code for other amino acids can be determined. Thus, polyC (CCC) codes for the amino acid proline, CUU for leucine, GUG for valine, and so on (Fig. 18.10).

DNA, by means of the several RNA's is, therefore, the controlling agent for the formation of proteins of the cell. If a protein consists of several subunits, or polypeptides, as many of them do, then several genes are necessary for the formation of the protein. It is more correct, then, to say that one gene, or one sequence of DNA, forms one polypeptide chain. If the chain is made up of 200 amino acids in sequence, then the block of DNA must consist of 600 nucleotides. One can readily see, therefore, that a change in the nucleotide sequence of DNA will lead to a change in the sequence of nucleotides in m-RNA. This in turn will lead to a change in the sequence of amino acids in a protein or a polypeptide. A changed function or structure then results, since the structure of a protein determines how it behaves in a cell (either as an enzyme or as a structural part of an organelle). Such changes in a DNA molecule are called **mutations**. We can illustrate how mutations affect a protein by considering the hemoglobin molecule.

Fig. 18.11 A small part of the polypeptide chain of hemoglobin A and hemoglobin S is shown here. A change in the nucleotide sequence of the DNA responsible for forming hemoglobin A led to the insertion of valine instead of glutamic acid, and resulted in hemoglobin S. This change of one amino acid out of several hundred greatly affects the function of the hemoglobin molecule.

Museo Del Prado

The Granger Collection

Fig. 18.12 These portraits of two members of the Hapsburg royal family of Austria (top: Philip IV; bottom: his son, Charles II) show the typical long jaw and protruding lower lip. These inherited traits occurred in most members of this family.

Hemoglobin is an iron-containing red pigment, a protein, contained in the red blood cells. Its function is to carry oxygen and carbon dioxide in the blood stream. The kind of hemoglobin in most individuals is **hemoglobin A,** and it is made up of a total of about 600 amino acids. (Since there are only 20 different amino acids, each one is represented many times in the hemoglobin molecule.) **Hemoglobin S** is an abnormal type, and anyone having it develops a disease known as **sickle-cell anemia,** the name being derived from the sickle shape of the red blood cells. The difference between hemoglobin A and hemoglobin S is caused by a mutation and affects only one amino acid out of the several hundred in the whole molecule (see Fig. 18.11), yet the change profoundly affects the individual and his health. The actual nucleotide change that took place in the DNA at the time of mutation is not known, but once we know how m-RNA codes for the hemoglobin molecule, we can then work backward and say what actually happened in the DNA molecule. It is apparent, then, that our future knowledge of how cellular activity is controlled lies in a better understanding of the interactions and interrelations of DNA, RNA, and protein. Until these are fully known, our insight into the behavior of cells will continue to be fragmentary.

GENES AND ENZYMES

DNA is a molecule having certain physical and chemical properties. It also has an hereditary function. An hereditary unit, or gene, can therefore be defined in a preliminary way as a segment of DNA ultimately responsible for the formation and character of a protein. Only rarely do we recognize the change that takes place in the protein.

More commonly we recognize a change, or mutation, in DNA by a change in a **trait.** Brown vs. blue eyes is one such variable trait that is inherited. In a short-hand way of speaking, we say that the gene is responsible for determining the trait. We should be aware, however, that heredity is an inherited pattern of chemical reactions and that a process of development is required before a trait can be recognized. A change in a gene through mutation leads to an altered pattern of chemical reactions. If we now recall a statement made earlier that specific reactions in the cell are controlled by specific enzymes, then it is logical and correct to assume that genes control enzymes, since enzymes are proteins. In fact, one of the very fruitful hypotheses, made in 1941 by G. W. Beadle and E. L. Tatum, then at Stanford University, is the "one-gene-one-enzyme" hypothesis. It states that a single gene acts by determining the specificity of a particular enzyme (or one polypeptide). In turn the enzyme governs a particular chemical reaction.

A change in the gene controlling an enzyme can have two consequences. Either the enzyme is no longer formed, or it is altered structurally so that it no longer functions in a normal way. In either case, the chemical reaction stops or is changed. We already know that the product of a chemical reaction can change, for we have seen how a mutation can alter both the structure and function of normal hemoglobin to give a trait known as sickle-cell anemia. The first can also occur for it can be shown in some instances that when a chemical reaction is missing, so too is the enzyme.

To illustrate the control of single chemical reactions by genes, let us consider the synthesis of **arginine,** one of the amino acids. Our knowledge originally came from a study of the red bread mold, *Neurospora,* but the results are generally applicable to virtually all organisms.

Ordinarily, *Neurospora* can manufacture its own arginine from a simple medium containing nitrate, a vitamin (biotin), and glucose (one of the sugars), the latter providing a source of energy as well as of carbon. By exposing spores of *Neurospora* to X-rays or ultraviolet light, we can cause various genes to mutate. These mutated spores will grow, and among them will be some that are unable to synthesize their own arginine. To keep such strains alive, it is necessary to supply them with arginine since it is an essential amino acid.

There are many steps in the synthesis of arginine from glucose, but we need be concerned only with the three end steps (see Fig. 18.13), each governed by an enzyme. If any one of these

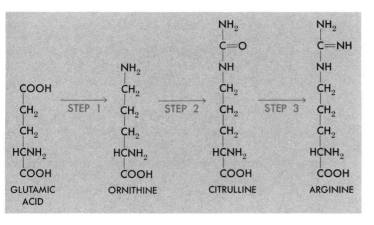

Fig. 18.13 This diagram shows the last three steps in the sequence of reactions leading to the formation of arginine. Each step is enzyme-controlled and would be blocked if the enzyme were missing or altered to a nonfunctional state.

steps is blocked by lack of a functional enzyme, arginine will not be synthesized. It must be added to the growth medium as an essential nutrient. Let us now test our original hypothesis. We are assuming that genes control specific chemical reactions through their control of the specific enzymes. But we must make a further assumption. We must also assume that an arginine-less mutant (a strain that cannot synthesize arginine) can result from a loss or impairment of three different enzyme systems. As a consequence, not all arginine-less mutants are similar in character. We can now prove that this is the correct assumption.

Refer again to Fig. 18.13. Assume that Step 1 has been blocked, and that no arginine has been formed. We know that the strain will grow if supplied with arginine, but will it also grow if it is supplied with the amino acids **ornithine** or **citrulline?** It turns out that it will, since Steps 2 and 3 are not impaired, and the block is earlier in the sequence of reactions. Similarly, if Step 2 is blocked, the strain can use either citrulline or arginine, but the addition of only ornithine would be inadequate for growth since ornithine cannot be converted to citrulline. If Step 3 is blocked, the organism can grow only if arginine is added. Further work has also revealed that the particular enzymes are either missing or are altered so that they no longer function properly.

In order to demonstrate that each step is controlled by a single gene, it is necessary to use the inheritance test, which involves controlled matings and careful observation of the offspring. This is the subject of Chapter 20.

SUMMARY

DNA is a macromolecule consisting of two complementary strands of repeating units wrapped around each other in the form of a double helix. The repeating units are of four kinds, each consisting of a base, a sugar, and a phosphoric acid. Each one is a nucleotide, and the four bases—thymine, adenine, guanine, and cytosine—give infinite variety to the DNA molecule. They also supply the genetic information to the cell and the organism.

DNA does two things: (1) it makes copies of itself by replication; and (2) it makes several kinds of RNA. These RNA's pass to the cytoplasm where they participate in the formation of proteins. These proteins form structures such as membranes, spindles, or ribosomes, act as enzymes which control the chemical reactions of the cell, or perform both kinds of functions. The uniqueness of a cell or an organism, therefore, results from the DNA it contains and the kinds of proteins formed as a result of the action of DNA and its derivative RNA molecules.

FOR THOUGHT AND DISCUSSION

1 The triplet code of the m-RNA is called the **codon,** the complementary on the t-RNA's the **anticodon.** Suppose that a piece of transcribing DNA has the following arrangement of nucleotides:

   A T C| G G A| C T T |A C A| C C T| A G G

   What will be the nucleotide sequence of the m-RNA? What t-RNAs will be necessary for protein formation? Using Table 18.2 determine the nature of the polypeptide to be formed.

2 In the nucleotide sequence of DNA given in Prob. 1, suppose that the second nucleotide from the left (T) is lost, but that the process of transcription is unimpaired. What change will be found in the m-RNA? Will the t-RNAs be the same as before, and used in the same sequence? What will be the nature of the polypeptide that is formed?

3 DNA is a double helix, but all evidence indicates that for any given gene only one of the two polypeptide strands can transcribe. Can you suggest why? Can you suggest a mechanism that permits one to be transcribed, but not the other?

confuses cell

SELECTED READINGS

Beadle, G., and M. Beadle. *The Language of Life.* Garden City, N.Y.: Doubleday and Co., 1966.

   An introduction to the science of genetics, including an account of the role of DNA.

Kendrew, J. *The Thread of Life.* Cambridge, Mass.: Harvard University Press, 1966.

   An introduction to molecular biology, with an emphasis of proteins and nucleic acids.

19 CONTROL OF CELLULAR METABOLISM

The DNA-RNA-polypeptide chain of synthesis provides proteins for the building of cellular structures. It also forms enzymes, which catalyze the chemical reactions taking place within a cell. As we found earlier, there are many different kinds of cellular structures; also, each cell is a complex, but orderly, mixture of hundreds of different enzymes. Although the cells of all living organisms are remarkably alike in some ways, in other ways they are different. A liver cell, for example, looks and behaves differently from a cell in a muscle, the brain, or skin. We also know that each kind of cell is stable chemically, even though it is chemically active. It performs its own special task regularly, doing a kind of work not done by other cells. Each cell is a closely regulated system; and because it is regulated, some kind of control is constantly guiding its activity. We do not yet have answers to many of the problems relating to cellular control, but as we gain more and more insight into the cell, we find that control usually involves enzyme synthesis and action.

Fig. 19.1 The Tiger Salamander is an animal that is unable to keep up the production of enough heat to remain active in cold weather. It survives by hibernating in winter.

Richard F. Trump

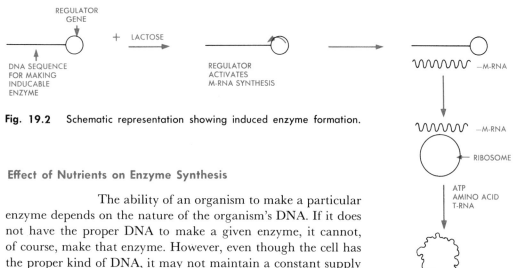

Fig. 19.2 Schematic representation showing induced enzyme formation.

Effect of Nutrients on Enzyme Synthesis

The ability of an organism to make a particular enzyme depends on the nature of the organism's DNA. If it does not have the proper DNA to make a given enzyme, it cannot, of course, make that enzyme. However, even though the cell has the proper kind of DNA, it may not maintain a constant supply of the enzyme. The nutritional environment of a cell seems to control the kind and amount of enzyme a cell has.

If a culture of bacteria is being grown on a medium containing glucose as the only source of carbon and energy, the bacteria will have the enzymes necessary to metabolize glucose. The bacteria can also grow on lactose, a disaccharide sugar found in milk. Lactose contains two hexoses—glucose and galactose. The enzymes in the cells which have been grown on glucose cannot break (hydrolyse) the lactose into the two single sugars. If we replace glucose with lactose as a nutrient source, we find that there is a delay—a lag period—before the cells begin to metabolize lactose. After a while, however, the cells make use of lactose with no difficulty. During the lag period the cells manufacture the enzyme necessary for the hydrolysis of lactose, which makes the simple sugars available. When enough of the enzyme is formed, the cells grow as well on the disaccharide lactose as they did on glucose. In this instance, the lactose *induced* the formation of the enzyme needed to metabolize it. Such an enzyme is said to be **inducible** in contrast to the **constitutive** enzyme normally present in cells grown on glucose.

Clearly, the bacteria must have the proper DNA for making the inducible enzyme. In the absence of lactose, however, this DNA is inactive. With the addition of lactose, the DNA is activated and the proper enzyme is made. This means, of course, that the formation of a specific enzyme requires the prior formation of a specific messenger RNA. This takes time, and accounts for the lag period; m-RNA is being formed, and it, in turn,

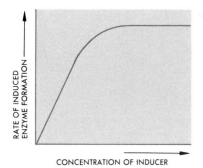

Fig. 19.3 Rate of induced enzyme formation depends upon the concentration of the inducer.

forms the enzyme. This information, on the other hand, does not tell us how lactose can induce the formation of a specific m-RNA. Although we are still not certain, some experiments have given us a clue. It appears that there are at least two kinds of DNA. One kind is concerned directly with the formation of m-RNA, and consequently with the synthesis of the enzyme. This kind of DNA is called the **structural** gene. The other kind of DNA, which controls the structural gene by turning it on or off, is called the **regulator** gene. The situation is similar to the difference between a light bulb and a light switch. It is the light bulb that does the work, but the switch determines whether it will be on or off. It appears that the regulator gene produces a repressor substance which can inhibit the action of the structural gene. The inducer—lactose in this instance—combines either with the regulator gene and prevents the formation of repressor substance, or it combines with the repressor substance itself. In either case, the structural gene proceeds with the formation of m-RNA, and the enzyme can be made.

The rate of synthesis, as well as the kind of enzyme, is also controlled in part by the inducer substance. The greater the amount of inducer added, the faster is the rate of enzyme induction. At very high concentrations of substrate, the rate tends to level off (Fig. 19.3), so the relation holds only over a limited range of concentrations. Also, the induction of enzymes is quite specific. In fact, some inducers cause the formation of enzymes even though the enzyme which is formed cannot metabolize the inducer. This indicates that DNA can be activated—or the regulator gene repressed—thus stimulating enzyme formation, but the enzyme is not used. The cell, therefore, can make "mistakes," just as an organism can.

Enzyme Repression and Feedback Inhibition

If an enzyme and its substrate are present in a cell, a chemical reaction will take place. The reaction can be controlled in two ways: (1) by controlling the activity of the enzyme; or (2) by controlling the amount of enzyme formed. For example, if product A is converted to B by the action of an enzyme, the rate at which A goes to B can be shown to be controlled by the amount of B present in the cell. The pathway of synthesis of the amino acid arginine shown in Fig. 19.4 illustrates this point. Bacterial cells grown in a medium containing a high arginine concentration have low amounts of the enzyme converting ornithine to citrulline. As the arginine is used up during growth, the amount of enzyme rises, and the rate of formation of arginine will consequently increase. A careful analysis of this and similar systems reveals that arginine actually represses,

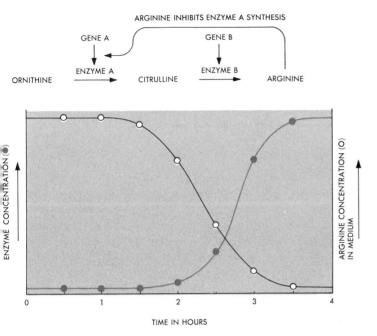

Fig. 19.4 Enzyme repression —see text for details.

and thereby controls, the formation of this enzyme. The enzyme converting citrulline to arginine, in this system, at least, is not subject to the same control. Only one enzyme in the system is, therefore, affected, but by this the whole system is controlled. The whole system illustrates what is called **enzyme repression.**

At the present time we know of many instances in which enzyme activity can be demonstrated in a cell-free extract only after an inhibitor has been removed. Some of these inhibitors are proteins. The products of an enzyme-catalyzed reaction can also act as inhibitors if they are tightly bound to the enzyme; in this case the enzyme is inactive until the product is removed. In addition, the product may regulate a metabolic pathway by inhibiting an enzyme associated with an earlier step. This mechanism of control, called **feedback inhibition,** may be of considerable importance in regulating the amount of a particular product formed in a biosynthetic pathway (Fig. 19.5). For example, in the biosynthesis of histidine in certain bacteria there are at least 10 different steps requiring eight different enzymes. If we add histidine to the culture in which the cells are growing, they stop making the particular amino acid and use the outside source of histidine for protein synthesis. This blockage of histidine synthesis by the cell is due to the inhibition by histidine of the enzyme which catalyzes the first step in the biosynthetic pathway. Thus, by this mechanism of feedback inhibition the cell's machinery and energy supply are relieved of an additional duty. Since the first step in the pathway is inhibited, wasteful intermediates are not accumulated.

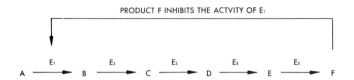

Fig. 19.5 Feedback inhibition of enzyme activity.

Positive feedback systems also operate in a cell. For example, energy is provided for the cell when ATP is broken down and ADP is formed. ADP stimulates the enzymes involved in energy release. When ATP is reconstituted from ADP, it depresses enzyme action as it increases in amount. By maintaining this positive and negative feedback relation, the energy metabolism of a cell can be effectively regulated.

Induction, repression, and feedback controls are, therefore, important factors in the regulation of cell metabolism. They allow a cell to be efficient, yet economical. Enzymes are formed when needed, and their amount and rate of activity are controlled according to the needs of the cell. The advantages to a cell are obvious. We can only assume that these control mechanisms made their appearance during the course of evolution and were retained because of their selective value to the organism.

Protein Structure and Enzyme Activity

The special function of an enzyme is due not only to the sequence of amino acids in the molecule, but to the shape or configuration of the molecule. Lactic acid dehydrogenase (LDH) is not a single polypeptide chain, but is made up of four associated chains, or monomers. It is, therefore, a tetramer, and the monomers are of two kinds, called A and B. The number of A and B monomers in a tetramer can vary, and as a result there are five different LDH's. LDH^1 consists only of B monomers, and is given the symbol A^0B^4, which means there are no A and 4 B monomers in the active enzyme; $LDH^2 = A^1B^3$; $LDH^3 = A^2B^2$; and $LDH^4 = A^3B^1$; and $LDH^5 = A^4B^0$. It is possible to isolate LDH^1 and LDH^5 in a pure state and to break them down into their monomers. A solution of A monomers and B monomers can then be mixed in a test tube. If the conditions are right, they will reaggregate and form tetramers. All five tetramers are formed, the proportions of each depending on the initial concentration of A and B monomers.

The LDH tetramers are called **isozymes**. Interest in them arises from the fact that different isozymes occur in different kinds of cells and act in slightly different ways. LDH^1, for example, is found predominately in heart muscle, while LDH^5 is more prominent in skeletal muscle. LDH^5 functions best when the oxygen concentration is low and lactic acid tends to accumulate, as happens in skeletal muscles. LDH^1, on the other hand, functions best under high oxygen conditions, as is characteristic for heart muscle. The various cells of the body have different internal environments, and respond to varying conditions by modifying the number and kinds of enzyme present. This is a more efficient practice than one requiring a different enzyme for each change in the cellular environment. We do not yet know how this enzyme modification takes place.

The monomers of LDH do not act as enzymes. However, glutamic acid dehydrogenase (GDH), a tetramer similar to LDH, behaves differently. As a tetramer, it acts on glutamic acid, but when it is dissociated into its monomers, the monomers act on the amino acid alanine. The monomers, therefore, are molecules of alanine dehydrogenase. We can see, as a result, that the enzymes' function varies, depending upon the state of aggregation. This is under the control of steroid hormones. The presence or absence of a hormone provides a means for a rapid shift in function without the necessity of forming new enzymes. The gain in efficiency by this kind of a control mechanism gives a cell the capability of meeting new situations rapidly.

Not all hormones act in this manner. The hormone adrenalin, for example, increases the activity of an organism. When you are excited or frightened you can often run faster than you can normally. During such times of excitement you use up more energy than you do when at rest. What happens is that the adrenal gland produces more adrenalin under stress, and pours it into the blood stream. The adrenalin stimulates an enzyme that produces a cofactor, which, in turn, is necessary for the activity of the enzyme phosphorylase. The latter enzyme stimulates the breakdown of glycogen, with the release of energy which is necessary for muscular exertion.

Enzyme Complexes

Fatty acids are long-chain molecules formed by linking smaller molecules, such as those of acetic acid, in sequence. The fatty acid synthetase system is responsible for this biosynthesis, and at least seven different chemical reactions are known to be involved. According to our usual definition, each chemical step is catalyzed by a different enzyme, but attempts to separate the synthetase complex into distinct enzymes have been unsuccessful. It would appear that the enzymes are inactive alone, perhaps because they do not have the proper shape, but when they are aggregated into a **multienzyme complex** they can carry out all of the necessary reactions in sequence. Not all single enzymes in a multienzyme complex are inactive when isolated by themselves, but it is clear that their close association in a complex makes for a more efficient system.

There are obvious advantages to such a system. If we are dealing with a chemical series:

$$A \xrightarrow{e_1} B \xrightarrow{e_2} C \xrightarrow{e_3} D \xrightarrow{e_4} E,$$

and the product of one reaction, say B, is the substrate for the next enzyme, e_2, then the enzyme complex becomes a sort of cellular assembly line. This would be immensely more efficient than a system in which all of the enzymes and substrates are in solution, and must haphazardly find each other. Also, control is more effectively exercised, since blockage of one step blocks the entire system.

The mitochondrion and the chloroplast in the cell are examples of multienzyme complexes. Here one can visualize how such enzymes may be structured. A large number of enzymes, for example, are known to be associated with mitochondria, and they fall into two major groups: (1) the respiratory, or citric acid cycle, enzymes; and (2) those enzymes engaged in oxidative phosphorylation. ATP is one of the principal end products formed by the mitochondrion, and both enzyme groups join together in converting the energy in carbohydrates, fats, and proteins into the useful energy of ATP. Some of the mitochondrial enzymes appear to be in solution in the matrix (interior); others are definitely bound to the walls of the mitochondrion.

It seems that even the soluble enzymes are bound in an organized array, but free themselves from the membranes

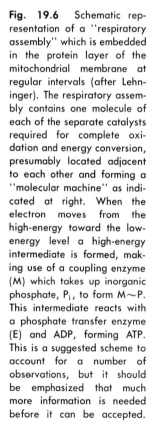

Fig. 19.6 Schematic representation of a "respiratory assembly" which is embedded in the protein layer of the mitochondrial membrane at regular intervals (after Lehninger). The respiratory assembly contains one molecule of each of the separate catalysts required for complete oxidation and energy conversion, presumably located adjacent to each other and forming a "molecular machine" as indicated at right. When the electron moves from the high-energy toward the low-energy level a high-energy intermediate is formed, making use of a coupling enzyme (M) which takes up inorganic phosphate, P_i, to form M~P. This intermediate reacts with a phosphate transfer enzyme (E) and ADP, forming ATP. This is a suggested scheme to account for a number of observations, but it should be emphasized that much more information is needed before it can be accepted.

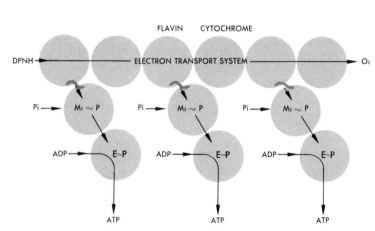

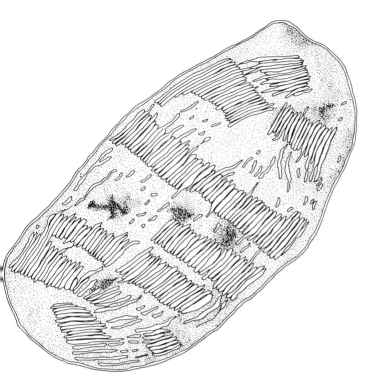

Fig. 19.7 This chloroplast from the leaf cell of tobacco shows the stacked lamellae of the grana and the more open regions of the stroma.

during the course of extraction. The structure of a chloroplast (Fig. 19.7) also leads us to believe that the enzymes for the conversion of CO_2 and H_2O into carbohydrates, and for photophosphorylation, are grouped into an assembly line array for greater economy of action. The difficulty is that while we can see, by means of electron microscopy, the elaborate structure of the mitochondrion or the chloroplast, we still cannot distinguish structural protein from enzyme protein in the membranes.

Biological Membranes and Control Systems

Membranes are an important structural feature of cells. They enclose the cell and the nucleus; they separate mitochondria, Golgi materials, lysosomes, and plastids from the rest of the cytoplasm; and they form a rich network as the endoplasmic reticulum in the cytoplasm. Membranes, therefore, are intracellular partitions. When we realize that most enzymes are also membrane-bound within organelles, we have some concept of how reactions in part of a cell can go on independently of reactions in another part. Thus, the breakdown of food materials occurs in lysosomes and mitochondria with the conversion and conservation of energy in the form of ATP, while the build-up

Fig. 19.8 Proposed bimolecular lipid-protein structure for membranes. The basic structure of cell membranes appears to be composed of fatty material sandwiched between inner and outer layers of protein. This structure is not completely uniform, however, for we know that there are special proteins and other molecules attached to the surface of membranes.

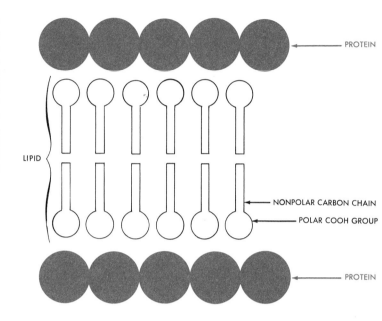

of other chemicals, utilizing ATP, can occur in the endoplasmic reticulum and nucleus. Since membranes are made up of protein and lipid materials (Fig. 19.8), water in the cell is also controlled in its movement. It is also possible to immobilize substances in the cell by tying these materials into membranes. For example, cholesterol, a fatty substance associated with hardening of the arteries and with aging, is often found in the smooth endoplasmic reticulum of liver cells. The cell is, therefore, a compartmentalized system, both structurally and functionally, and membranes are the compartmentalizing agents.

One of the most interesting phenomena in cellular studies is how membranes act as "pumps"—how they govern the flow of substances in a given direction. The process is called **active transport,** and requires energy. The result is the movement of soluble substances *against a concentration gradient,* quite the opposite of what occurs during diffusion or osmosis.

Let us consider active transport in terms of the cell membrane. This membrane has an ATPase associated with it. When this enzyme breaks down ATP, the energy released forces sodium ions to move in one direction, potassium ions in the other (Fig. 19.9). When these ions are absent, the ATPase is inactive, so their movement appears to be definitely related to the splitting of ATP and the release of energy. The enzyme acts as a traffic policeman as well, giving direction to the particular ions. We can visualize the ATPase as an active site, or crossroads, in the

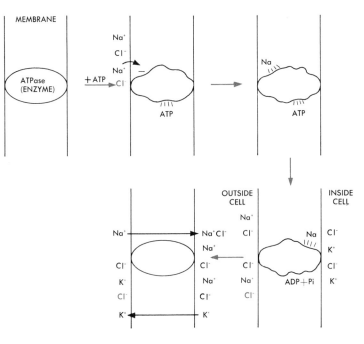

Fig. 19.9 Proposed mechanism for the sodium and potassium pump in the cell membrane. A special membrane bound ATPase makes use of ATP to transport sodium ion across the membrane. As suggested above when ATP is bound to the enzyme a negative charge is exposed on the protein which attracts a sodium ion from outside the cell. When ATP is hydrolyzed the energy release allows the movement of sodium to the inner surface of the membrane. The unbalanced negative charge on the outside attracts a potassium ion from the inside. When an unbalanced charge is maintained by the pumps, there develops an electrical potential difference between the inside and the outside of the cell.

membrane, with the enzyme being able to "look" in two directions at once and directing the flow of ionic traffic. The ATPase in this manner acts as a **pump.** Other pumps at other active sites direct the flow of other molecules. Again, this means that the cell, through its enzymes, governs and compartmentalizes the activities going on within it.

Other metabolic pumps are also known. One in the mitochondria governs the flow of calcium and magnesium ions. One in muscle cells determines the direction of flow and the amount of calcium, an important aspect of muscular contraction and relaxation. In all such systems, the protein-lipid structure of the cellular membranes play a key role. It is the boundary between one environment and another, and permits the development of little "islands" of activity both within and between cells. It is unlikely that effective control systems could develop in the absence of such membranes.

SUMMARY

Each cell is a highly controlled and regulated metabolic machine. The regulation in most cases involves enzyme synthesis and function. The nutrients surrounding the cell have an effect on the enzyme composition of the cell. In some cases synthesis of enzymes can be induced by an appropriate substrate provided the cell has the right kind of DNA. **Induced enzyme formation,** therefore, is an important factor that determines the enzymatic activity of cells.

Enzyme synthesis can also be inhibited or repressed by certain nutrients. For example, if a cell is capable of making an essential nutrient (such as an amino acid) it will stop doing so if the nutrient is added to the medium in which the cells are growing. Analysis of this situation indicates that the nutrient inhibits the synthesis of the enzyme which catalyses the first step in the biosynthetic pathway. This enzyme repression conserves cell energy and specific chemicals.

The product of a biosynthetic pathway can also inhibit an enzyme that catalyses an earlier step. This is called **feedback inhibition.**

The structure of enzymes is an important factor in their control and regulation. For example, some enzymes are made up of several different polypeptides or a multiple combination of two or more polypeptides. Dissociation of the polymers either by specific hormones or other agents will alter their catalytic activity.

Enzyme activity is also affected by being bound to membrane-like structures in the cell. The **multienzyme complexes** are important in the mitochondria where the aerobic metabolism of fatty acids and carbohydrates occurs. Enzymes bound to the cell membrane are also important in that they are capable of making use of metabolic energy (ATP) to move substances across the membrane against a concentration gradient. These "metabolic pumps" are essential for regulating the concentration of nutrients inside the cell.

FOR THOUGHT AND DISCUSSION

1 Can you think of various mechanisms that would lead to a different enzymatic make-up for a liver cell and a heart cell during the time a fertilized egg is developing into an adult organism?

2 In studying the scheme for aerobic metabolism of carbohydrates can you think of a way that enzyme repression or feedback inhibition could function to regulate oxygen consumption?

3 Why do you think there are different types of isozymes in different cells? What are some of the factors leading to the synthesis of one type over another?

4 What is the value of having enzymes associated with one overall chemical process bound onto a structure in a multi-enzyme complex?

SELECTED READINGS

Baldwin, E., *Dynamic Aspects of Biochemistry* (3rd ed.). New York: Cambridge University Press, 1957.

An outstanding introduction to the concepts of cellular metabolism.

Bennett, T. P., and E. Frieden, *Modern Topics in Biochemistry*. New York: The Macmillan Co., 1966.

An excellent discussion in greater detail of the various subjects discussed in our last three chapters. Written for the beginning college student.

Loewy, A. G. and P. Siekevitz, *Cell Structure and Function*. New York: Holt, Rinehart, and Winston, 1963.

A detailed elementary discussion of various aspects of cell structure and biochemistry.

McElroy, W. D., *Cell Physiology and Biochemistry* (2nd ed.). Englewood Cliffs, N.J.: Prentice Hall, Inc., 1964.

A small paperback intended for beginning college students.

Neilands, J. B. and P. K. Stumpf, *Outlines of Enzyme Chemistry* (2nd ed.). New York: John Wiley & Sons, 1958.

An excellent elementary introduction to enzyme chemistry.

Readings from Scientific American

Allen, R. D., "Amoeboid Movement," February 1962.
Allfrey, V. G. and A. E. Mirsky, "How Cells Make Molecules," September 1961.
Brachet, J., "The Living Cell," September 1961.
Doty, Paul, "Proteins," September 1957.
Frieden, E., "The Enzyme—Substrate Complex," August 1959.
Fruton, Joseph S., "Proteins," June 1950.
Green, D., "The Synthesis of Fat," February 1960.
Hayashi, T. and G. A. W. Boehm, "Artificial Muscle," December 1952.
Holter, Heinz, "How Things Get into Cells," September 1961.
Huxley, H. E., "The Contraction of Muscle," November 1958.
Kendrew, J. C., "Three-Dimensional Structure of a Protein," December 1961.
Lehninger, A., "Energy Transformation in the Cell," May 1960.
Lehninger, A., "How Cells Transform Energy," September 1961.
McElroy, W. D. and C. P. Swanson, "Trace Elements," January 1953.
Moore, S. and W. H. Stein, "The Chemical Structure of Proteins," February 1961.
Pfeiffer, John E., "Enzymes," December 1948.
Robertson, David J., "The membrane of the Living Cell," April 1962.
Siekevitz, P., "Powerhouse of the Cell," July 1957.
Solomon, A. K., "Pores in the Cell Membrane," December 1960.
Solomon, A. K., "Pumps in the Living Membrane," August 1962.
Stein, W. H. and S. Moore, "The Chemical Structure of Proteins," February 1961.
Stumpf, P. K., "ATP," April 1953.

20 INHERITANCE OF A TRAIT

In an earlier chapter we discussed DNA as the principal molecule of heredity. Of the three billion human beings on Earth, no two are exactly alike (except, possibly, identical twins); it follows that their DNA's are also dissimilar, which, in turn, means that their proteins must be dissimilar. It is impossible to pinpoint this dissimilarity by determining the number and distribution of the eight billion nucleotide pairs in the DNA of a human cell. But, if we want to know something of the details involved in the inheritance of similarities and dissimilarities from one generation to another, we must have some way of tracing the passage of DNA from one generation to another. The simplest way is to concentrate on a character or trait which is determined by DNA.

We should make it clear that patterns of inheritance can be studied *only* when variation exists. We would not learn very much

Fig. 20.1 These identical twins have strikingly similar facial features. What do we mean when we say that identical twins are "identical?"

Ewing Galloway

about inheritance if we studied a population of organisms all alike. Without "fingerprints" of some sort, there would be no way of distinguishing one individual organism from another. In a way, genetics has developed by focusing attention on differences between individuals rather than on their similarities. Fortunately, variations of all kinds exist. The problem is to select that variation best suited to the situation being investigated. Whenever possible, the variation should consist of a pair of easily distinguishable, sharply contrasting, and mutually exclusive characters. For example, red versus white flower color; the presence or absence of an enzyme; black as opposed to white spores; straight versus curly hair, and so on. When studying a pair of contrasting characters we disregard all other characters so that we do not complicate our results.

The basic tool in studying the inheritance of a trait is the **inheritance test.** It was used long before the role of DNA was understood, and it is still used today. The inheritance test is relatively simple. It consists of controlled matings and the study and tabulation of the offspring. Let us examine a particular inheritance test.

The Life Cycle of Neurospora

Neurospora, which is easy to grow, handle, and observe, is a fungus (mold) and an **ascomycete.** It gets this name from one of its main fruiting structures, the **ascus.** The diagram outlines the life cycle of this plant.

The normal growth of *Neurospora* is by means of a spreading mat of fine threads, the **mycelium.** The individual threads, or **hyphae,** are not broken up into cells with single nuclei; rather there are many nuclei in a common mass of cytoplasm. When the mycelium is broken up, each hyphal piece forms a new mycelium (Fig. 20.2). This represents a form of asexual reproduction. The mycelium can also form small asexual spores called **conidia** (singular, **conidium**), each having one or more nuclei in them. These germinate and form mycelial masses similar to the parental mycelium.

Neurospora also has a form of sexual reproduction, even though it is not possible visually to distinguish male from female mycelia. The different kinds of mycelia are classified into mating types A and a. A will not mate with A, nor a with a. A will mate only with a.

When the two mating types A and a are brought together, the mycelial strands fuse, or a conidium from one type unites with a **protoperithecium** from the other, and two nuclei—one of each type—move into a single cell that is cut off by a wall. The

Fig. 20.2 The life cycle of the ascomycete, *Neurospora*, indicates both the asexual and sexual aspects.

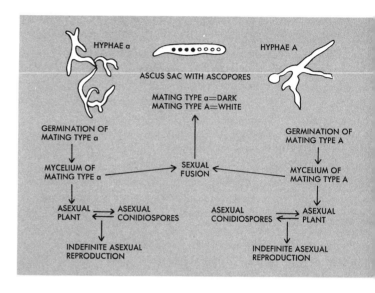

two nuclei fuse and then undergo a series of three divisions producing eight nuclei. A wall forms around the mass of cytoplasm surrounding each nucleus, and eight ascospores are formed. When the ascospores germinate, each one produces a mycelium. The mycelium from any single ascospore will be of mating type *A* or *a*, never a mixture of the two. Of the eight ascospores, four will produce *A* mycelia, and the other four, *a* mycelia. That is, the ratio of *A* to *a* is 1 : 1.

The Inheritance Test

Let us now consider in detail the meaning of the 1 : 1 ratio. A portion of the DNA—how much we do not know—is responsible for determining traits *A* and *a*. This portion of DNA is a gene, and we can speak of genes *A* and *a* which have the function of determining mating types. How they do this is unknown, but it is probably through some kind of a protein structure or reaction. Since genes *A* and *a* are also in chromosomes, the behavior of chromosomes should tell us something about the behavior of genes; since genes determine traits, the inheritance of a trait should follow the inheritance of a chromosome.

Genes *A* and *a* are never found in the same mycelium. They are, therefore, mutually exclusive and we can think of them as alternate states of a single character. The term **allele** is used to designate such contrasting genes. If we assume that *A* represented the original state, then *a* is a mutation of *A*. (It could, however, be the other way around, with *A* a mutation of *a*.) If *A* were the original allele, then *a* probably arose by some change in the portion of DNA controlling mating type. Our present guess is that this change reflects an alteration of nucleotide pairs.

With the information above, let us now consider chromosome behavior in our inheritance test. The nuclei in the mycelium of *Neurospora* contain seven chromosomes. These can be numbered 1 through 7. From other evidence we know that gene A (or a) is located on chromosome 1. This group of seven chromosomes, each of which is different from all others in gene content, is known as a **haploid set** (for a more detailed explanation of "haploid" see Chapter 21). Since we tend to characterize cells, tissues, or organisms by their chromosome content, *Neurospora* is essentially haploid; only the fusion nucleus is diploid. We shall see that other organisms may differ in the degree of haploidy and diploidy of their cells, and their features of inheritance also differ. All sexually reproducing organisms, however, show the alternation of haploid and diploid phases.

The diagram (Fig. 20.3) shows that the mating of the two types of mycelia is soon followed by fusion of the two nuclei. This is the principal act of sexual reproduction. It is the act of fertilization. The single fusion nucleus now contains two haploid sets of chromosomes, and is said to be **diploid** (see Chapter 21). Instead of 14 different chromosomes, however, we can think of the nucleus as containing seven pairs of chromosomes. Thus, mating type A contributed chromosome 1 containing gene A; mating type a contributed its chromosome 1 containing gene a. For our purposes, we can consider that these two chromosomes are identical, except for the particular allele carried at the mating type position or **locus**. The two chromosomes are said to be **homologous** with each other.

The fusion diploid nucleus now goes through a series of divisions called **meiosis.** The behavior of the chromosomes differs significantly from that occurring in prophase of the earlier mitotic division. In mitosis each chromosome entered prophase with the chromosomes longitudinally double. At anaphase the chromatids separated from each other and formed two nuclei having the same chromosome number as the mother cell from which they arose. In prophase of meiosis the homologous chromosomes pair with each other—number 1 from type A pairs with number 1 from type a, and so on. They arrive at metaphase as pairs of chromosomes, called **bivalents,** instead of single chromosomes longitudinally double. The homologues separate from each other at anaphase, and the newly formed nuclei now have seven instead of 14 chromosomes. The diploid fusion nucleus has now been reduced to two haploid nuclei. In terms of chromosomes, therefore, the first division is a **reduction division.**

Fig. 20.3 The sequence of events taking place in the sexual cycle of *Neurospora*. Chromosome I from each strain, or mating type, joins with its homologue soon after formation of the fusion nucleus, and then separates from it at the first meiotic anaphase. The two chromatids of each chromosome separate at the second meiotic anaphase, and each again divides at the post-meiotic mitosis. Each of the eight chromosomes is now enclosed in a separate ascospore in the mature ascus. The behavior of the chromosomes, like the behavior of genes, illustrates the law of segregation discovered by Mendel.

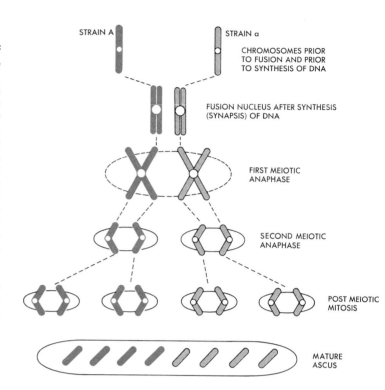

The second division merely separates the two chromatids of each chromosome, producing four haploid nuclei. The third division is a normal mitotic division, so the ascus now has eight haploid nuclei within it. These now are enclosed with a mass of cytoplasm inside of a spore wall, and eight ascospores result.

A summary diagram of these events is shown in Fig. 20.3 for one pair of chromosomes. It is clear that sexual reproduction involves the union of haploid nuclei to give a diploid state, and that meiosis is the reverse of this process. The life cycle of a sexually reproducing organism is, therefore, an alternation of haploid and diploid states. In *Neurospora*, however, the diploid state is a brief one and is confined to the fusion nucleus. *Neurospora*, consequently, is primarily a haploid organism.

The events we have just described follow what is known as **Mendel's first law of inheritance.** Gregor Mendel, an Austrian monk, formulated this law in 1865 when he was studying inheritance in garden peas. It is also called the **law of segregation.** In terms of the *Neurospora* life cycle, the law states that within sexual organisms there are pairs of factors (we now call them genes) which unite at the time of fertilization (when the two nuclei fuse), and which segregate at the time of meiosis. Stated another way, genes of sexually reproducing organisms come together in a single cell and then segregate during any given life cycle. The genes do this without losing their identity.

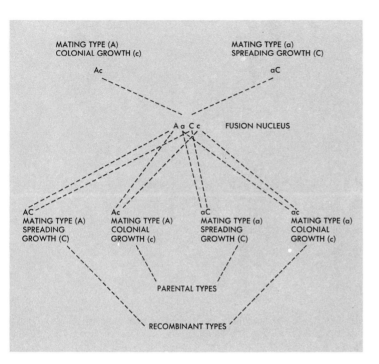

Fig. 20.4 Random assortment at meiosis of two pairs of genes found on two different and nonhomologous pairs of chromosomes. Their behavior illustrates the Mendelian law of independent assortment.

Random Assortment of Genes

The remaining six haploid chromosomes of Neurospora, numbered 2 to 7, also contain genes. On chromosome 4 is a mutant gene c which causes a colonial type of growth. The normal gene C, of which c is an allele, governs the spreading type of growth. Suppose we cross a colonial strain of mating type A with a spreading strain of mating type a. These strains can be designated respectively, as Ac and aC (Fig. 20.4). From such a cross, ascospores will be produced. These can be isolated, grown, and the mycelial colonies produced can be examined for traits. We would find that 25 per cent of them would be Ac; 25 per cent aC; 25 per cent AC; and 25 per cent ac. However, what if we took individual asci, isolated the ascospores one by one, grew them into colonies, and then examined their genes? We would find that half of the asci produced four Ac spores, and four aC spores, while the other half produced four AC spores and four ac spores.

Let us consider this in terms of chromosomes 1 and 4 which contain these genes (see Fig. 20.5). The fusion nucleus would contain two chromosomes of number 1 type and two chromosomes of number 4 type, each with their respective genes. At meiosis the homologous chromosomes would pair and then segregate.

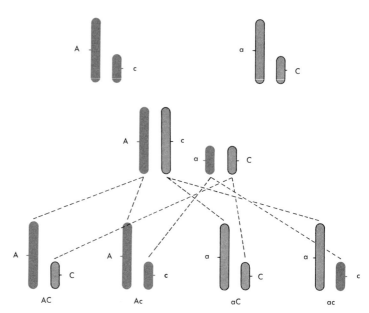

Fig. 20.5 The independent assortment of two pairs of nonhomologous chromosomes produces the same results as those obtained by following the genes in Fig. 20.4.

At the first meiotic division, chromosome 1 (*A*) goes to one pole, chromosome 1 (*a*) to the other. There is an equal chance of chromosome 4 (*C*) going with 1 (*A*) or with 1 (*a*). If the former, then four spores in the ascus will be *A, C*, the other four *a, c*. If 4 (*C*) goes to the same pole with 1 (*a*), then four spores will be *a, C*, the other four *A, c*.

The combinations *a, C* and *A, c* are parental types; that is, they are the same as the original strains with which the cross was started. The *A, C* and *a, c* spores are **recombinant** types. The results obtained indicate that the chromosomes assort at random during meiosis, and so do the individual genes.

Mendel also discovered this principle, now known as the **law of independent assortment** of genes. We shall see in a later chapter, however, that when two genes are on the same chromosome, the law does not always run true.

SUMMARY

The **inheritance test** is the basic tool of genetics. Its use requires pairs of contrasting characters or traits, which are controlled by contrasting genes or alleles. The inheritance test also requires controlled matings and a quantitative record of the offspring produced from such matings. In a haploid organism, using trait A in one parent as contrasted to trait a in the other, the inheritance test shows that the ratio of A to a among the offspring is $1:1$. When two traits are studied simultaneously, they are inherited independently of each other to give ratios of $1AB: 1Ab$; and $1aB: 1ab$.

These two ratios provide us with information needed to formulate the two basic laws of genetics: 1. the **law of segregation of genes;** and 2. the **law of independent assortment of genes.** The first law is used when only one pair of contrasting traits is being studied, the second law when two or more pairs of traits are being studied. The physical basis of genetic inheritance lies in the behavior of chromosomes at meiosis.

FOR THOUGHT AND DISCUSSION

1 List a group of contrasting characters in man. Select characters that you think might be relatively easy to follow in inheritance studies. Discuss them and try to decide which ones would serve your purpose and which would not.

2 Some genetically determined characters are said to show "qualitative" differences, other "quantitative" differences. What do these terms mean? Which would be the easier to study by the inheritance test?

3 Why is the use of the inheritance test in man impractical? Can you think of any way to get around this problem so that genetic studies in man can be made?

4 Can you study the inheritance of a trait in an asexually reproducing organism? Give reasons for your answer.

5 It is thought that the average size of a gene is about 1,000 nucleotide pairs long. If this is so, how many genes can man have in a haploid set of chromosomes? Have you any reason for thinking this number to be reasonable or not?

6 If a gene of average size controls mating types A and a in *Neurospora*, and the difference between the two alleles is due to differences in nucleotide pairs, would you expect more than two kinds of mating types? If only two are present in nature, can you think of any reason why this should be so?

7 Genes X and x are found on the members of one pair of homologous chromosomes, and genes Y and y on another homologous pair. A cross is made between an XY strain and an

xy strain, and the ascospores are removed in serial order from a number of asci. How many different combinations can you expect to get from individual asci? Diagram your results to show the distribution of chromosomes in each case.

8 Yeast is like *Neurospora* in that it is haploid and is an ascomycete, but it is unicellular rather than forming a mycelium. After the fusion of two haploid yeast cells, meiosis occurs, but only four spores instead of eight are formed (the final mitotic division is missing). One strain of cells cannot ferment glucose (glu$^-$), the other strain cannot ferment the sugar maltose (mal$^-$) because of defective genes. These genes segregate independently of each other. The two strains are crossed, and the spores are isolated from individual asci. What kinds of spores will you get from each ascus? Will all asci be alike in this respect? How are you to test for the genotype of each spore? If you group the results from many asci, what kinds of ratios would you obtain?

SELECTED READINGS

Auerbach, Charlotte. *Genetics in the Atomic Age.* New York: Essential Books, 1956.
   An elementary treatment of genetics and the effects of radiation on genetic systems.

Auerbach, Charlotte. *The Science of Genetics.* New York: Harper & Row, 1961.
   An elementary text covering the various aspects of inheritance in plants and animals.

Bonner, D. M., and S. E. Mills. *Heredity.* Englewood Cliffs, N.J.: Prentice-Hall, Inc., Foundations of Modern Biology Series, 1964.
   The emphasis in this small book is on the microbial and biochemical aspects of inheritance.

Cook, S. A. *Reproduction, Heredity and Sexuality.* Belmont, California: Wadsworth Publishing Co., Foundations of Botany Series, 1964.
   Deals broadly with the many variations of reproduction and inheritance in the plant world.

Herskowitz, I. H. *Genetics.* Boston: Little, Brown, & Co., 1965.
   A college textbook that includes as a supplement some of the classical papers of genetics.

Levine, R. P. *Genetics.* New York: Holt, Rinehart and Winston, Inc., 1962.
   Covers most aspects of genetics at a college level, including haploid and diploid inheritance and its chromosomal relationships.

Peters, J. A., (ed.). *Classic Papers in Genetics*. Englewood Cliffs, N.J.: Prentice-Hall, Inc., 1959.

Contains 28 classic papers that laid the foundations of genetics from its early beginnings to today.

Srb, A. M., R. D. Owen, and R. Edgar. *General Genetics*. San Francisco: W. H. Freeman & Co., 1965.

One of the most broadly comprehensive textbooks on genetics at the college level, including both classical and modern aspects of the subject.

21 MEIOSIS AND ITS RELATION TO SEXUAL REPRODUCTION

In the life history of a sexual organism such as *Neurospora*, two events are of particular significance. One is fertilization. In *Neurospora*, fertilization occurs when the haploid nuclei of the two mating types fuse and form a diploid nucleus. In higher plants and animals, fertilization occurs when the sperm penetrates the egg and their two nuclei fuse, giving rise to a **zygote,** or new individual, which is diploid.

If you stop for a moment and think of the consequences of fertilization, it should soon become apparent that there always must be some other process as well, an accompanying counterpart. Remember that when two nuclei fuse, the chromosome number doubles. In each succeeding generation, therefore, the chromosome number would continue to double. If no event occurred to offset this trend, it would not take long to arrive at unmanageable numbers of chromosomes. Furthermore, we know that the haploid number of chromosomes in *Neurospora* remains at seven, generation after generation, just as the diploid number of man remains 46.

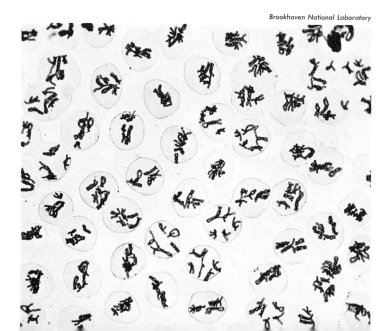

Brookhaven National Laboratory

Fig. 21.1 These cells at the first meiotic metaphase in the anther of the wake robin, *Trillium erectum,* were taken at low magnification. The cells are synchronized in division so that most of them arrive simultaneously at metaphase. A single anther contains hundreds of such cells.

Brookhaven National Laboratory

Fig. 21.2 In this meiotic prophase cell from the testes of an amphibian the chromosomes have formed homologous pairs. Can you identify all 15 of the pairs?

The answer, of course, is that meiosis is the opposite of fertilization; it is a form of nuclear division which halves the number of chromosomes at some stage in the life cycle. In *Neurospora,* meiosis follows fertilization immediately. In man, however, the zygote is formed when fertilization occurs. Afterwards the zygote develops first into an embryo, and then gradually over the years into an adult. Meiosis occurs when eggs and sperm are produced. Only these two kinds of human cells are haploid. Man, like all higher animals and plants, is essentially a diploid organism during most of his life cycle. Here we will consider meiosis as a variation of mitosis in which a number of interesting innovations appear.

The Stages of Meiosis

Neither *Neurospora* nor man is a favorable organism in which to study meiosis. Maize (corn), grasshoppers, amphibians, and lilies are much more suitable because of their large chromosomes and the ease with which the cells can be fixed and stained. Prophase is the first stage of meiosis, as it is in mitosis, but in meiosis it lasts longer and is more complicated.

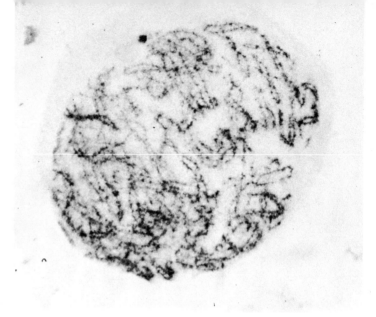

Brookhaven National Laboratory

Fig. 21.3 An early meiotic prophase stage in a testicular cell of an amphibian. The nucleolus is not visible because the stain used is absorbed only by chromatin, but the chromomeres are visible along the lengths of the elongated chromosomes. Some indications of pairing are evident.

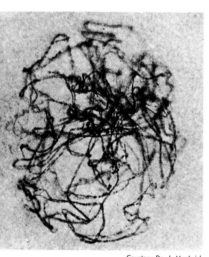

Courtesy Dr. J. MacLeish

Fig. 21.4 A meiotic prophase cell from the anther of the regal lily, *Lilium regale*. The pairing of homologous chromosomes can be seen at various points; at other regions pairing is not yet complete.

In the early stage of meiotic prophase the chromosomes of the nucleus are seen as delicate threads within the nuclear membrane. Because of their great length, the chromosomes are not individually identifiable. The nucleoli are quite large and easily visible. This stage is not strikingly different from the early prophase of mitosis, except that along the extended length of the chromosomes are dense granules of irregular size. These are the **chromomeres.** Any given chromosome has chromomeres of characteristic size and position. It is thought that the chromomeres arise as a result of the chromosome contracting unevenly along its length. However they arise, the number, size, and position of chromomeres are constant features of each chromosome.

The chromomeres gradually become more visible, and the individual chromosomes become thicker as the result of contraction. The nucleoli can be seen to be attached to particular chromosomes at special points called the **nucleolar organizers.** This is the stage when the homologous chromosomes begin to pair. The beginning of this stage is shown in Fig. 21.2. The homologues contact each other at one or more points. Then, like a zipper, the two chromosomes pair, chromomere by chromomere. When pairing is complete, it appears as if only a haploid number of chromosomes were present in the nucleus. With the aid of a high-power microscope, however, each closely associated pair can be seen as two. Each pair of homologues is called a **bivalent.**

Contraction continues to shorten the chromosomes and makes them more readily identifiable. Figure 21.6 shows a few of the chromosomes of maize (the haploid number is 10). Remember that what appears to be a single chromosome consists of two

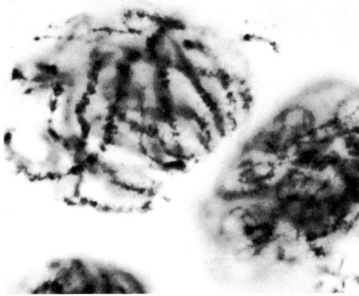

Brookhaven National Laboratory

Fig. 21.5 Mid-prophase stage of meiosis in an amphibian: The chromosomes are nearly paired along their entire lengths, the chromomeres are large and coarse, and considerable shortening has taken place (compare with Fig. 21.3).

homologues closely paired. This is shown clearly in Figure 21.6. Here chromosome number 6 of maize has the nucleolus attached to it, the individual chromomeres are visible, and the paired nature of the strands can be seen in several places.

An abrupt change in the appearance of the bivalents now takes place. The paired homologues fall apart, showing that each consists of two chromatids. The homologues do not separate completely, however. They are held together at various points called **chiasmata** (singular, **chiasma**). The chiasmata can be seen as points where chromatids cross over from one homologue to another. We shall refer to these chiasmata in a later chapter.

The chromosomes are more contracted in this stage, and in some organisms small coils can be seen. It is by means of a coiling process that a long thin chromosome is converted into a shorter, more maneuverable one.

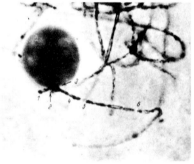

B. McClintlock

Fig. 21.6 A mid-prophase stage in meiosis of maize, showing the paired homologues, and the attachment of one pair to the nucleolus. Recognizable regions of the bivalent are numbered. This figure is comparable to Fig. 21.2 from an amphibian, but the chromosomes are clearer as to intimate details of pairing and structure. There are 10 pairs of chromosomes in maize, but only Chromosome 6 is shown in its entirety.

Fig. 21.7 A photograph of a single bivalent from a male salamander revealing the individual chromatids of the homologous chromosomes, and the two chiasmata where the chromatids cross over from one homologue to the other. The centromeres of the two homologues are seen as the dark circular areas at the left of the bivalent.

At the end of prophase the chromosomes are shorter, the nucleoli disappear, and the nuclear membrane disintegrates. As these events occur, the spindle forms and the bivalents orient themselves on it halfway between the poles. The **first meiotic metaphase** stage has thus begun.

You will recall that at *mitotic* metaphase, the centromere of each chromosome occupied a position on the metaphase plate. At *meiotic* metaphase, however, each bivalent has two centromeres, one for each homologue. Instead of both being at the metaphase plate, they come to lie on either side of, and equidistant from, the metaphase plate. This appears to be a position of equilibrium.

The **first anaphase** of meiosis begins with the movement of the chromosomes to the pole (see Fig. 21.8). The two centromeres of each bivalent remain undivided, and their movement causes the chiasmata to be undone and to free the homologues from each other. When movement ceases, a reduced or haploid number of chromosomes is located at each pole. Unlike mitotic anaphase, in which the chromosomes appeared longitudinally single, each chromosome now consists of two distinctly separate chromatids held together only at their centromeres.

A nuclear membrane forms around the chromosomes, and the chromosomes uncoil. A wall or membrane divides the cell in two in some kinds of meiotic cells; in others it does not. This is the **first telophase** of meiosis.

The second meiotic division follows the first without appreciable delay (Fig. 21.9). It appears to be quite similar to a mitotic division, but there is one important distinction. The second meiotic prophase chromosomes have the same structure as those in the previous first meiotic anaphase. In other words, *no chromosomal replication took place during meiotic interphase,* and the centromere of each chromosome remains undivided.

A spindle forms in each cell, initiating the second meiotic metaphase, and at the second anaphase the centromeres divide and the chromatids separate and pass to the poles. The four haploid nuclei are reorganized in the second telophase. Depending on the species and the kind of cells, the nuclei may or may not be segregated into individual cells by segmentation of the cytoplasm.

Neurospora is representative of those organisms which include many algae and fungi in which the two meiotic divisions are followed by a mitotic division resulting in eight haploid nuclei and then eight ascospores. This mitotic division does not occur in animal species.

If you look back over the events of meiosis, you will find that the structure of each chromosome remained unchanged.

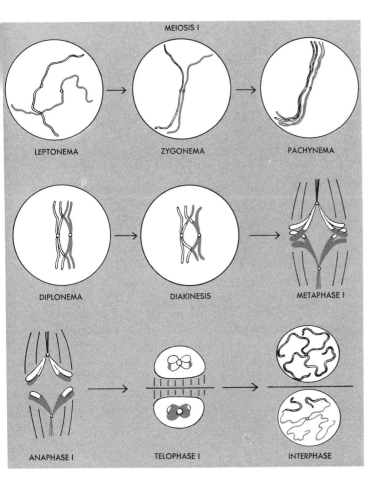

Fig. 21.8 The sequence of stages that occur in the first meiotic division of both plants and animals are shown here. Only a single pair of chromosomes is represented. Prophase includes the first five stages shown in the diagram, and distinctive terms have been given to each recognizable stage. The progression of prophase, however, is continuous, and one stage leads directly into the following one without delay. The line drawn between the telophase and interphase nuclei represents the cell membrane, which cuts the cell in two. The important thing to remember is that *chromosomes*, not chromatids, segregate in the first meiotic division.

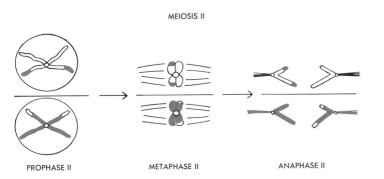

Fig. 21.9 The second meiotic division follows the first with little delay. Each chromosome now divides and its two chromatids segregate at anaphase II. As a result, the four chromatids of a first-division bivalent become segregated into four separate cells.

Fig. 21.10 This diagram demonstrates the relation of meiosis and fertilization to the haploid and diploid states of a sexually reproducing species such as man. The only haploid cells are the eggs and sperm cells after meiosis has been completed. All other cells of the body at all other stages of the life cycle are diploid.

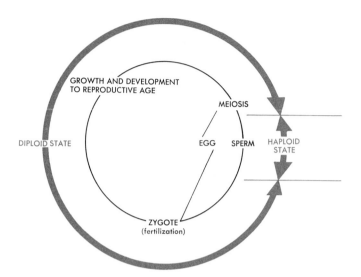

From mid-prophase to the second anaphase, the chromosomes maintained their longitudinal subdivisions. The replication of each chromosome took place in the interphase preceding meiosis. Then two divisions occurred: during the first the homologues separated, thus reducing the chromosome number; during the second the two chromatids of each chromosome were separated. It might be argued that a reduction in chromosome number could have been accomplished by only one division. But when viewed in terms of gene recombination, the two divisions make good genetic sense. This point will be discussed in the following chapter.

Reproduction in Animals

Sexual reproduction in most animals differs from that in *Neurospora* in two major respects. First, the sexual cells, or gametes, in animals are produced in specialized structures: **eggs** are produced in **ovaries,** and **sperm** are produced in **testes.** The gametes are visibly different, both in size and in character. In *Neurospora* there are no special organs in which sexual cells are produced, and the nuclei differ only in mating type genes. Secondly, the body of an animal is a diploid structure in contrast to the haploid mycelium of *Neurospora.* Let us consider what takes place in man (Fig. 21.10).

The Egg

The ovaries of a female begin development early in fetal life, and all the eggs which an individual will produce are formed by about the sixth month of fetal life; that is, three months before the child is born. Figure 21.11 shows a section of

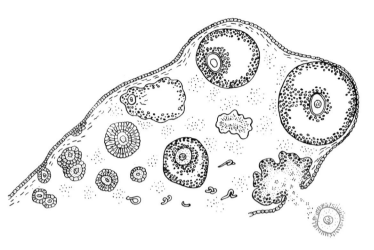

Fig. 21.11 A section through a mammalian ovary would show the progressive enlargement of the eggs, and the eventual shedding of one egg or ovum (lower right). The human female, after the age of puberty, sheds one egg per month, usually.

an ovary with eggs in the course of development. On the outside is the **germinal epithelium**, which gives rise to **oogonial** cells. These cells then develop into **oocytes**. The oocytes are those cells in which meiosis takes place. The first stages of meiosis in all oocytes begin before birth, stopping, however, at a stage just prior to the first metaphase. Since a female does not shed eggs until 12 to 14 years of age, and may continue to do so up to 50 years of age, the oocytes remain retarded in division for many years.

During this period the oocyte prepares itself for future activity. Each one is surrounded by **follicle cells** which provide protection and nutrients, and the whole structure is known as a **Graafian follicle**. Yolk materials, which the young embryo will use for food, are stored inside. In a chicken egg, the yolk can be seen as a yellow mass, but in the human egg, the yolk is much less prominent and is scattered throughout the cytoplasm of the egg.

When the egg is released from the ovary, and then fertilized by a sperm, meiosis starts up again. The first meiotic division gives rise to two cells. One, very minute, is called a **polar body**. This is difficult to detect in humans, but the results can be seen in a fish egg (see Fig. 21.12). The division, however, brings about a reduction in chromosome number.

General Biological Supply House, Inc.

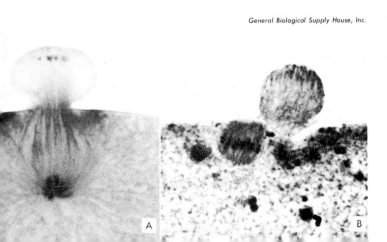

Fig. 21.12 Polar body formation in the egg of a fish: (left) late anaphase of the first meiotic division, with the first polar body being pinched off; (right) metaphase of the second meiotic division, with the first polar body resting on the cell membrane above and to the right. This division will eventually pinch off a second polar body. In the meantime, the first polar body may divide again, producing a total of three polar bodies and one functional egg.

Fig. 21.13 A section through a mammalian testes would show the development of mature sperm cells: (A) outer membrane of the testis tubule; (B) accessory cell; (C) early prophase in a spermatocyte; (D) later prophase stage (metaphase I and anaphase I stages are also present but not labeled); (E) second spermatocyte ready to begin the second meiotic division; (F) spermatids at two stages of development; (G) mature sperm cells.

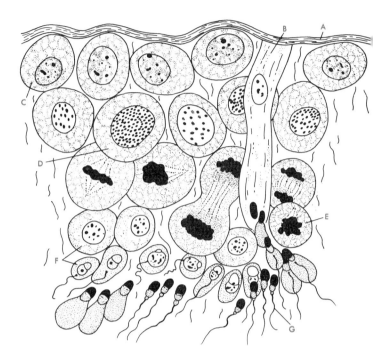

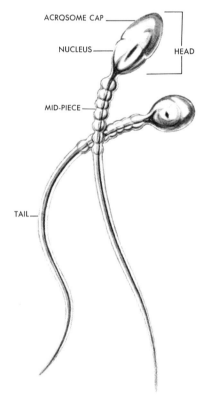

Fig. 21.14 Mature human sperm cells. The acrosome caps the nucleus, with the two parts making up the head of the sperm cell. The mid-piece contains the mitochondria, and the tail extends beyond the mid-piece, providing a means of locomotion.

The second meiotic division is similar, and another polar body is produced. In the meantime, the first polar body has divided. Thus, as a result of meiosis, four haploid cells are produced: three small polar bodies which eventually disintegrate, and one large egg which will be functional.

The Sperm

The testes in a male consists of a tissue made up of many tiny tubules, all lined with germinal epithelium. The epithelium contains a group of dividing cells which produce **spermatocytes** when a male reaches about 14 years of age, and continues to produce them by mitosis until late in life.

The spermatocytes undergo meiosis, and each one produces four haploid cells. These then undergo a series of changes (see Fig. 21.13) during which: 1. most of the cytoplasm is lost; 2. the nucleus becomes compacted into the head of the sperm; 3. the Golgi materials form an **acrosome** which assists the sperm in penetrating the egg; 4. a long tail is produced from one of the centrioles; and 5. the mitochondria wrap themselves around the tail just below the attachment to the nucleus, presumably supplying the sperm with energy. By the whipping action of its tail, the sperm swims to the egg, which has no power of movement. The sperm, therefore, can be thought of as a haploid nucleus with an outboard motor.

Fertilization

The mature egg and sperm must unite in a limited period of time, since their life span is limited. Fertilization begins when the sperm enters the egg. This occurs in the **fallopian tube,** a structure which connects the ovary to the uterus, where the embryo will develop. Once one sperm has entered, other sperm are blocked from entering by the quick development of an impenetrable membrane. The sperm loses its tail, and the nucleus fuses with the nucleus of the egg. The sperm also contributes a **centriole** to the egg. The centriole aids in the formation of a spindle, and the fertilized egg, or zygote, prepares to divide and become an embryo.

In summary, then, the sperm does several things: 1. it adds a haploid set of chromosomes to form a diploid zygote; 2. it contributes a centriole, which enables the zygote to divide; and 3. it provides a stimulus for division of the egg.

Reproduction in Plants

If we were to consider reproduction in all kinds of plants—algae, fungi, mosses, ferns, pine trees, and flowering plants—we would find a bewildering array of life cycles. Here we will consider only one, that in the lily.

The simplicity of the lily flower makes it ideal for an examination of the reproductive organs and cells. The lily plant is a diploid structure, having developed from a fertilized egg. The lily flower is part of the diploid plant and contains both male and female organs in the same flower.

The female part of the flower is the **pistil,** consisting of **ovary, style,** and **stigma.** The seeds develop in the ovary and each seed develops from a single egg. The egg develops from a cell which enlarges and becomes the **megasporocyte.** The megasporocyte goes through two meiotic divisions forming four haploid nuclei (Fig. 21.17), after which three of the nuclei fuse and form a triploid nucleus (it contains three sets of chromosomes). The megasporocyte, therefore, contains only two nuclei, each different from the other. These two nuclei now go through two successive divisions, giving a megasporocyte with eight nuclei, four of which are haploid, and four triploid. These nuclei now rearrange themselves. Three triploid nuclei go to the basal part of the megasporocyte; three haploid nuclei cluster at the micropylar end (with the middle one becoming the egg); and the remaining two nuclei (one triploid and one haploid) fuse in the center, forming a tetraploid **polar nucleus.** The whole structure is now called an **embryo sac** and is ready to be fertilized.

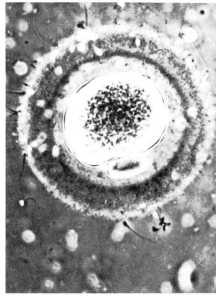

Courtesy Dr. L. B. Shettles

Fig. 21.15 Photograph of a human egg at the moment of fertilization by a sperm cell.

Fig. 21.16 A lily flower at a point in its life cycle when the ovule has just been fertilized. The stages leading up to this point are shown in Fig. 21.17

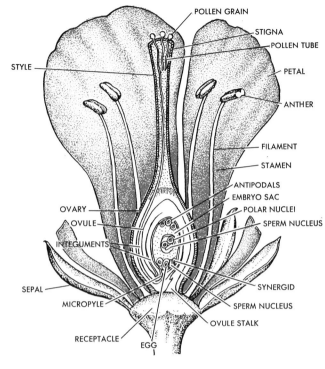

Fig. 21.17 Formation of an embryo sac of lily: (A) the megasporocyte. (B) 4-nuclei stage at the end of the two meiotic divisions. (C) 2-nuclei state, one nucleus (the smaller one) being haploid, and the other triploid. (D) 8-nuclei stage resulting from two successive mitotic divisions; 4 nuclei are haploid and 4 triploid. (E) rearrangement of the eight nuclei in the embryo sac. (F) fertilization, with one sperm (shown in solid color) uniting with the egg and forming a zygote. The other sperm unites with the polar nucleus and forms a pentaploid endosperm nucleus which will produce a nutritive tissue for the developing zygote.

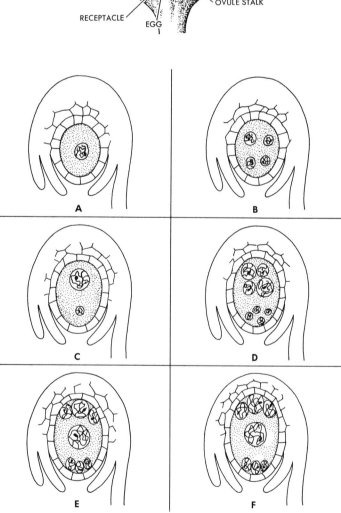

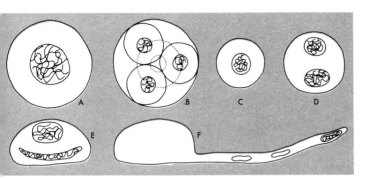

Fig. 21.18 Stages in the formation of sperm cells in the lily: (A) microsporocyte entering meiosis. (B) the four microspores formed as a result of the two meiotic divisions. (C) a single haploid microspore (immature pollen grain). D, microspore with two nuclei, one of which will govern the growth of the pollen tube, the other of which will divide and form two sperm cells. (E) a mature pollen grain at the time the flower opens, containing the long nucleus which will divide in the pollen tube after germination and form two sperm cells. (F) a germinating pollen tube with its two sperm cells (in color).

The male portion of the flower is the **anther,** in the center of which are many cells called **microsporocytes.** These undergo two meiotic divisions, producing four haploid cells which will develop into mature pollen grains. First, however, the single nucleus in each of these cells divides by mitosis, providing a cell with two nuclei. One will control the growth of the **pollen tube.** The other will divide in the pollen tube and produce two sperm nuclei.

Fertilization

When a pollen grain is carried by an insect to the stigma, a pollen tube grows down the style and empties its contents, including the two sperm nuclei, into the embryo sac. One of the sperm nuclei fuses with the egg, producing a diploid zygote. The other fuses with the diploid fusion nucleus, giving rise to a **triploid** (three sets of chromosomes) **endosperm nucleus.** Each of these nuclei now divides. The zygote develops into an embryo lily plant while the endosperm nucleus forms a rich nutritive tissue upon which the embryo feeds.

A seed, therefore, is a mosaic of tissues. The seed coats are maternal tissue and diploid like the mother plant. The embryo is also diploid, but since the sperm nucleus may have come from another plant, its genetic constitution may be different from the seed coat. The endosperm is triploid tissue, and is used up when the seed germinates and draws upon it for nutrient.

SUMMARY

Fertilization and **meiosis** are of particular significance in the life history of a sexually reproducing organism. Fertilization doubles the chromosome number; meiosis halves it. In the two processes, genes are recombined and segregated, giving a physical basis for the inheritance of traits, and a means for producing offspring with varying combinations of traits.

The key to an understanding of meiosis is the fact that chromosomes occur in pairs. The two members of each pair come together, and then segregate from each other during the first meiotic division. At the second meiotic division, the chromatids of each chromosome separate from each other, resulting in four haploid cells from each diploid cell entering meiosis.

The cells resulting from meiosis can be eggs, sperm, or spores depending on the kind of organism and the kind of tissue in which meiosis takes place.

FOR THOUGHT AND DISCUSSION

1 Does a haploid gamete contain a full set of genes? Explain. What is meant by a "full set" of genes? A "full set" of chromosomes?

2 What are the basic differences that distinguish meiosis from mitosis? In what ways are they similar? Consider your answers in terms of both genes and chromosomes.

3 Can you think of any reason why it is of advantage to form only one functional egg instead of four from each oocyte?

4 What would be the result if an oocyte failed to go through meiosis but was fertilized by a sperm?

5 What would a zygote be like if the second division of meiosis did not take place, but fertilization occurred?

6 What do you think would happen if no chiasmata formed between homologues?

7 Why should the replication of DNA be lacking between the two meiotic divisions? Is the lack simply an economical measure, or is the lack necessary for the success of meiosis?

8 Why, in both plants and animals, should more male than female gametes be produced?

9 Can you think of any reason why it is of advantage to have the endosperm of corn triploid?

10 Let us assume that the maternal member of a pair of homologous chromosomes is white, the male member red. If a particular meiotic cell has two pairs of homologues, how many kinds of gametes are possible in terms of red and white chromosomes? How many kinds if three pairs of homologues are present? Four pairs? Twenty-three pairs, as in man? Can you diagram this for four pairs? Can you think of a mathematical formula to describe the number possible for any given number of pairs of homologues? What law of Mendel's does this behavior of chromosomes support?

1 By combining knowledge of the inheritance of genes and the behavior of chromosomes in meiosis, the *chromosome theory of inheritance* was developed. How would you state this theory in your own words?
2 In what cells does meiosis take place in a lily? In man?
3 What would happen if meiosis took place in other cells of the body?

SELECTED READINGS

See the books listed at the end of the previous chapter.

22 INHERITANCE IN A DIPLOID ORGANISM

The Granger Collection

Fig. 22.1 Gregor Mendel, an Austrian monk and science teacher, carried out his studies on the inheritance of characters in the garden pea in the monastery garden. From the results obtained, he discovered the basic laws of biparental inheritance operating in diploid organisms.

In our discussion of *Neurospora* we made the point that it is a haploid organism; that is, each nucleus has but a single set of unpaired chromosomes. The only diploid phase (each nucleus having paired chromosomes) is that brief period of time after the two nuclei fuse and each contributes its single set of chromosomes. For any single pair of segregating alleles, ratios of $1A:1a$ are to be expected. With two pairs of independently assorting genes, ratios of $1AB:1Ab:1aB:1ab$ are the rule.

As we found earlier, many organisms are diploid. The fertilized egg divides by mitosis, and continued cell division brings about a mass of cells which differentiates into an adult organism. The genetic constitution of an individual, therefore, is expressed by cells containing two sets of chromosomes and, consequently, two sets of genes. For example, a given gene could exist in cells or in organisms such as *AA, Aa,* or *aa*. When both *A* and *a* are present, do both genes express themselves? Does *A* mask *a*? Or does some new expression make itself evident? In the haploid mycelium of *Neurospora*, any gene generally expresses itself because no other allele is present. As we shall see, several different conditions are possible in diploid organisms. We can consider them all under the general term of **diploid inheritance**.

The 3:1 Ratio

Let us describe here the classical experiments performed by Mendel, the results of which enabled him to formulate his laws of inheritance (Fig. 22.1). Mendel studied inheritance in the garden pea, and he selected a number of pairs of contrasting, mutually exclusive characters. One of these was *tall* vs. *dwarf* plants (Fig. 22.2).

The garden pea is ordinarily a self-fertilized plant. That is, the eggs it produces in its ovaries are fertilized by sperm brought in by its own pollen. With care, however, the anthers in a given

Fig. 22.2 Two varieties of peas which were planted at the same time. They have about the same number of leaves, but the internodes (that portion of the stem between successive leaves) of the taller variety grow much faster than those of the dwarf variety. At maturity, the dwarf variety will reach a height of only 20 inches. The taller variety will reach a height of about 60 inches.

Richard F. Trump

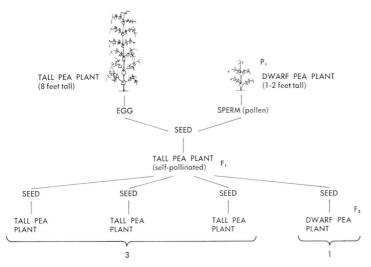

Fig. 22.3 This diagram shows the results of Mendel's experiment, in which he crossed two distinct strains of peas—one tall, the other dwarf. The F_1 generation were all tall, but when these plants were self-pollinated, the seeds developed into both tall and dwarf plants, in a ratio of 3:1.

flower can be removed, and pollen from another plant can be substituted. It is thus possible to control all matings, and to make certain which sperm fertilizes each egg.

The tall and dwarf plants used by Mendel bred true when they were allowed to self-pollinate. They showed no segregation of the tall vs. dwarf character in that tall plants produced tall offspring, dwarf plants dwarf offspring. It can be assumed that the genes controlling height were in a pure allelic state, even though it can be further assumed that the genes responsible for the tall character were different from those governing the dwarf character. Mendel then crossed the two strains of plants. The cross is shown in Fig. 22.3, with the parent generation labeled P_1. The seeds were harvested, sown, and produced the F_1 **(first filial)** generation. *The F_1 plants were all tall; the dwarf character was not expressed in any way.*

Mendel then allowed the F_1 plants to self-pollinate themselves. Seeds again were harvested and sown, and the F_2 generation plants grew. The population consisted of 1,064 plants. Of these, 787 were tall, and 277 were dwarf. This is essentially a 3:1 ratio of tall to dwarf progeny.

Mendel next carried the experiment further by allowing each F_2 plant to self-pollinate and produce seed. From these seeds he discovered that the dwarf plants bred true; that is, they produced only dwarf progeny. However, only one-third of the tall plants bred true for the tall character. The other two-thirds segregated for tall vs. dwarf, again in a 3 tall:1 dwarf ratio. These segregating tall plants behaved, therefore, exactly like the F_1 tall plants.

Remember that Mendel published his work in 1865. He knew about the necessity of pollen as an agent for fertilizing the egg, and he had a good mathematical training, but he knew nothing about chromosomes, meiosis, or the concepts of diploidy and haploidy. Mendel's analysis of his data is all the more remarkable when we consider the meager foundation of knowledge he had to build on. Let us now list the facts that emerge from his experiments, and then list the assumptions he made to explain the facts.

HERE ARE THE FACTS:

1 The characters **tall** (6 to 7 feet) and **dwarf** ($3/4$ to $1 1/2$ feet) are constant. There are no plants of intermediate height.
2 The dwarf character does not appear in the F_1 generation. But it appears unchanged in the F_2 generation; and when it does, it reappears with a predictable frequency.
3 Part (one-third) of the tall F_2 generation breeds true for tallness. The remainder (two-thirds) continues to show segregation in the F_3 generation.
4 The dwarf F_2 plants always produce dwarf progeny when self-pollinated.

The fact that predictable ratios were achieved suggested to Mendel that factors were being passed from generation to generation in a constant manner; also that the factor for dwarfness was present in the F_1 generation even when not expressed.

HERE ARE MENDEL'S ASSUMPTIONS:

1 Assume that T stands for the tall factor (we would now call this a gene), t for dwarfness.

2 Assume that one factor is contributed by the egg and one by the sperm (pollen). The zygote and the plant itself would have two factors existing together in each cell. These would segregate from each other when the eggs and pollen were formed. (Meiosis was discovered and understood in the 1880's so Mendel formed his assumptions without this knowledge.)
3 Assume that T dominates t in such a way that when the two are together only T is expressed. These assumptions can be diagrammed as follows:

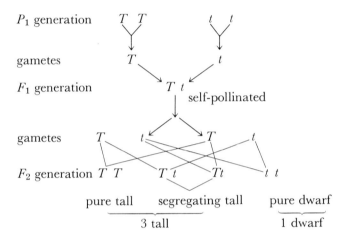

The assumptions made by Mendel fitted his data very well, and they have been fully supported since by hundreds of similar crosses in a wide variety of organisms. The law of segregation, therefore, applies not only to garden peas, but to all sexually reproducing diploid organisms.

Mendel, however, was not fully satisfied. He devised another way of testing his hypothesis that the F_2 tall plants were of two kinds. He made what is known as the **backcross** or **testcross**. The testcross involves the crossing of the individual being tested to another individual so that its genetic constitution will be revealed. The testcross made by Mendel was to cross the tall F_2 plants to any dwarf plant. When TT was testcrossed to tt, only Tt tall progeny resulted. If Tt were testcrossed to tt, then two kinds of progeny resulted in a 1:1 ratio.

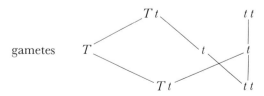

The testcross, therefore, fully supported his contention that dwarf factor could be obscured when in the presence of T, but that it could be recovered unchanged in the next generation when combined with another t gene.

Some terms, commonly used, should now be defined: T is said to be **dominant** over the **recessive** t; TT and tt are, respectively, **homozygous dominant** and **homozygous recessive;** Tt is **heterozygous** state. The terms tall and dwarf describe the **phenotype** (appearance) of the pea plant. However, because of dominance and recessiveness, the phenotype tall is the same even when the **genotype** (genetic constitution) is different, that is, Tt or TT.

9:3:3:1 Ratio

The breeding experiment just described is known as a **monohybrid cross.** It involves only one pair of characters. All other characters are disregarded. Mendel also made **dihybrid crosses,** in which two pairs of independent characters were tested simultaneously. For example, in one of his crosses he used seed characters: yellow (Y) vs. green (y) and round (R) vs. wrinkled (r). Mendel knew from previous studies that yellow was dominant over green, and round dominant over wrinkled. Using the symbols as before, the cross can be expressed as follows:

$$YYRR \times yyrr$$

By using the Punnett square method of demonstration (see Fig. 22.4) to show the F_1 gametes and the F_2 individuals, we can demonstrate that the expected phenotypic ratio will be 9 yellow round : 3 yellow wrinkled : 3 green round : 1 green wrinkled. In the legend of the diagram the exact values obtained by Mendel are shown. The pairs of genes behaved independently of each other. The **law of independent assortment** of genes, therefore applies to diploid inheritance as well as to haploid inheritance. The difference between the $1AB:1Ab:1aB:1ab$ ratio in haploid organisms such as *Neurospora*, and the $9AB:3Ab:3aB:1ab$ ratio in the diploid garden pea is due to single vs. double doses of genes in the individual organisms, and to the phenomena of dominance and recessiveness. Otherwise, the same rules of inheritance apply. One needs only to remember that in a haploid organism *every* gene has the opportunity of expressing itself since there is no other allele to interfere with it. Its phenotype directly reveals its genotype. This, of course, is not necessarily true in a diploid organism.

In Chapter 20 the segregation of genes during meiosis in *Neurospora* was explained in terms of chromosome segregation. Those same rules of chromosome behavior operate in the garden pea, man, or any other diploid organism having normal meiosis.

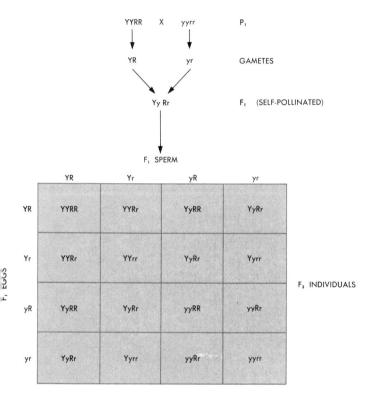

Fig. 22.4 The Punnett square method for demonstrating the F_1 gametes and the F_2 progeny resulting from a dihybrid cross. Among the F_2 progeny, Mendel obtained 315 plants that had yellow, round seeds (Y-R-); 108 plants that had yellow, wrinkled seeds (Y-rr); 101 plants that had green, round seeds (yyR-); and 32 plants that had green, wrinkled seeds (yyrr). The dash (-) indicates that the missing symbol can be either dominant or recessive. For example, Y- can be either YY or Yy, but the phenotype will be yellow. The ratio obtained is basically a $9:3:3:1$.

Chance and Probability in Inheritance

Imagine a married couple who want to have children. The mother displays a recessive trait governed by gene c (her genotype will be cc), and the father is heterozygous (Cc) for the genes involved. All eggs, of course, will be c, and the sperm produced by the father will be C or c. Suppose that they have only one child. Will it be Cc or cc? If they have two children, will they both be Cc? Or cc? Or will one be Cc and the other cc? If cc leads to the expression of a trait that will affect the future of the child or children, it is important that the parents have some idea of the probabilities of inheritance.

Mendelian law tells us that a $1:1$ ratio of Cc to cc is to be expected among the offspring if the parents are Cc and cc. But what is expected is not always realized; so we must recognize that a predicted ratio, such as $1:1$, or $3:1$, or $9:3:3:1$ represents an ideal situation that will hold true only if certain conditions are met: 1. that the number of offspring is large; 2. that there is an equal chance of any sperm fertilizing any egg; and 3. that all offspring survive.

If condition 1 is not present—that is, if small numbers of

offspring are produced—the predicted ratios are not always realized because of the element of chance. You can readily determine this for yourself by flipping a coin, counting the number of times heads or tails turns up, and comparing the ratio when the number of flips is small as opposed to when they are large. If conditions two and three are not realized, even the expected ratios are bound to be distorted.

If condition two holds in our $Cc \times cc$ mating above, the probability, or chance, that a C sperm will fertilize the c egg is one half. The probability of the c sperm fertilizing the egg is also one-half. There is, in other words, a 50:50 chance that the first offspring will be cc. Let us assume that this is so. If the family has two children, what, then, is the probability that both will have the cc genotype? Here we must make one assumption— namely, that the occurrence of the first event does not influence the probability of the second event. In terms of coins, if the first flip shows heads, it does not influence the next flip so that tails is any more likely to turn up than heads. Each flip is still a 50:50 heads or tails.

Therefore, if the probability of the first child being cc is one half, the same probability holds for the second child. The answer to our question, then, is $\frac{1}{2} \times \frac{1}{2}$, or $\frac{1}{4}$; or to phrase it differently, there is one chance out of four that both children will be cc. On the same basis, there is an equal probability that both children will be Cc, or that one will be Cc, the other cc. The probability of three cc children, will be $\frac{1}{2} \times \frac{1}{2} \times \frac{1}{2}$, or $\frac{1}{8}$.

The Punnett square previously described is a pictorial representation of Mendelian segregation and recombination to give ideal ratios. When dealing with small numbers, however, it is often easier to make use of a simple algebraic representation of the genetic information. This is done by expanding the binomial $(a + b)^n$, where a is the probability of one genotype, b the probability of the other, and n the number of offspring. In our $Cc \times cc$ family above, and with three children, the binomial would be written $(\frac{1}{2}Cc \times \frac{1}{2}cc)^3$, which becomes

$$\tfrac{1}{8}(Cc)^3 + \tfrac{3}{8}(Cc)^2 \times cc + \tfrac{3}{8}Cc \times (cc)^2 + \tfrac{1}{8}(cc)^3$$

The exponents assigned to the genotypes indicate the character of the three children: $(Cc)^3$ means 3 Cc children, $(Cc)^2 \times c$ means two Cc children and one cc child, etc. The fractions indicate the probabilities for the occurrence of the particular families; that is, in many families of three children, arising from a $Cc \times cc$ mating, you would expect all children to have a cc only one-eighth of the time.

If one is dealing with a 3:1 ratio, the same binomial can also be used, but in this instance, the formula would be $(\tfrac{3}{4}a +$

$_4b)^n$, with a being the dominant phenotype and b the recessive phenotype. With two pairs of alleles, we can write the usual 9:3:3:1 ratio as $9/16 A-B- + 3/16 aaB- + 3/16 A-bb + 1/16 aabb$, the dash (−) indicating that the second allele can be either dominant or recessive without altering the phenotype. Let us now ask: What is the probability in a family of three that the offspring will be two $aaB-$ and one $aabb$? The probability is $3/16 \times 3/16 \times 1/16$, or $9/4,096$. Therefore, in families of three children whose parents were heterozygous for both pairs of alleles, we should expect the above genotypes to appear once in about 455 times.

If we assume in the problem above that A is normal and a is albino, and that B is normal and b is dwarf, we can tabulate all aspects of the problem in the following way:

Genotype	Phenotype		Probability	Phenotypic frequency
	Color	Height		
A-B-	normal	normal	$(3/4)(3/4)$	$9/16$
aaB-	albino	normal	$(1/4)(3/4)$	$3/16$
A-bb	normal	dwarf	$(3/4)(1/4)$	$3/16$
aabb	albino	dwarf	$(1/4)(1/4)$	$1/16$

SUMMARY

In a **haploid** organism, all genes express themselves without hindrance. In a **diploid** organism, each gene is represented twice in every cell, and problems of gene expression (dominant vs. recessive) are raised. The ratios resulting among the offspring differ accordingly, and monohybrid ratios of 3:1 and dihybrid ratios of 9:3:3:1 are encountered. The laws of inheritance and the relation of meiosis to gene segregation, however, are the same for diploid and haploid organisms; only gene dosage and expression of genes differ.

FOR THOUGHT AND DISCUSSION

1 In a cross between pure breeding red-flowered and white-flowered plants, the F_1 offspring are pink-flowered. In the F_2 generation, the offspring appear in a ratio of 1 red : 2 pink : 1 white. From these data what can you conclude about the action of the genes concerned with color formation? What ratios would you expect if you crossed an F_2 red with an F_2 pink? F_2 pink with F_2 white? Pink with pink? In the original

cross, diagram the genotypes of all individuals from parental types to F_2 offspring.

2. If two yellow mice are crossed, they will produce offspring with an average ratio of 1 black:2 yellow. Can you explain these results? How would you test your conclusions?

3. The major blood types in human beings, due to a chemical substance in the blood, and determined by several alleles of a given gene, are of four kinds: A, B, AB and O. The pattern of inheritance is indicated at left:

Parents	Offspring
A × A	A or O
B × B	B or O
A × B	A, B, AB or O
A × O	A or O
B × O	B or O
AB × O	A or B
AB × A	A, B or AB
AB × AB	A, B or AB

Can you give an explanation for these patterns of inheritance? What combinations of alleles show dominance and recessiveness, and which reveal a lack of dominance? If a child has an AB blood type, what possible genotypes can its parents have? What use is the above information in determining the parentage of a disputed child?

4. In chickens, assume that the gene I inhibits the formation of feather color; i permits color formation. Assume also that gene C, which is independent of I, governs the formation of black pigment; gene c inhibits its formation. In a cross of $IiCc \times IiCc$, what is the expected phenotypic ratio of black to white offspring? Explain your results.

5. In rabbits, the genotypes C^hC^h or C^hc produce the Himalayan phenotype in which the extremities (nose, ears, feet) are black, and the remainder of the body white. The black pigment, melanin, is formed by enzyme action. From this information and your knowledge of enzyme action, what explanations can you offer for the Himalayan phenotype? How could you test your hypothesis?

6. The parents of two offspring are heterozygous for the genes A and B. What are the probabilities that both offspring will show only the dominant phenotype? One with the double recessive phenotype and one with the double dominant phenotype? Both with an aaB^- phenotype? One with aaB^- and the other with A-bb? If there are four offspring, what is the probability that all will have the double recessive phenotype? That all will have the dominant phenotype and all will be girls? (Assume that being boy or girl has a 50:50 probability.)

7. The ability to taste PTC (phenylthiocarbamide) is inherited as a dominant trait. Using the tester materials provided by your instructor, test yourself and your family. Do the results agree with the hypothesis that it is a dominant trait?

8. Perform the following experiment yourself, using two pennies and two nickels. Put them in a container, shake them, and then toss them onto a table. Score the results for 100 such tosses as follows: H = heads, T = tails.

Class	Pennies	Nickels	Observed	Expected
1	HH	HH		
2	HH	HT		
3	HH	TT		
4	HT	HH		
5	HT	HT		
6	HT	TT		
7	TT	HH		
8	TT	HT		
9	TT	TT		
		Total	100	100

9 Before beginning the experiment, calculate your expected results for 100 tosses. How do your observed results agree with the expected? If there is a difference, explain it. Compare your observed results with those of another student, and then with the pooled results of all students. Is the difference between observed and expected greater for a single comparison or for the pooled results? Why?

10 Assume heads to be dominant, tails recessive, and pennies and nickels to be two different traits. What is the expected genotypic ratio? Phenotypic ratio? What kind of a genetic experiment is this comparable to? How do your observed results agree with what you expect? What accounts for the difference, if it exists?

SELECTED READINGS

Auerbach, C. *Genetics in the Atomic Age.* New York: Essential Books, Inc., 1956.
An elementary discussion of chromosomes and genes, and how radiation can affect them.

Auerbach, C. *The Science of Genetics.* New York: Harper & Row, Publishers, Inc., 1961.
An excellent and clear description of the basic principles of genetics.

Baker, W. K. *Genetic Analysis.* Boston: Houghton Mifflin Co., 1965.
A difficult but excellent book concerned with the analysis of genetic results.

Bonner, D. M., and S. E. Mills. *Heredity.* Englewood Cliffs, N.J.: Prentice-Hall, Inc., Foundations of Modern Biology Series, 1964.
More concerned with the physical nature of the gene, but containing much good material on inheritance.

Brewbaker, J. L. *Agricultural Genetics.* Englewood Cliffs, N.J.: Foundations of Modern Genetics Series, Prentice-Hall, Inc., 1964.
Contains a good deal of information on basic Mendelian inheritance as well as data relating to inheritance in domesticated plants and animals.

McKusick, V. A. *Human Genetics.* Englewood Cliffs, N.J.: Foundations of Modern Genetic Series, Prentice-Hall, Inc., 1964.
A readable volume on the methods and results of genetic studies in humans.

23 LINKAGE, CROSSING-OVER AND GENE MAPS

Our understanding of genetic systems has been built up largely from information gained through inheritance tests. Controlled matings and accurate recording of offspring are required in such tests (Fig. 23.1). Fortunately, the behavior of those genes which we cannot see is reflected in the behavior of the chromosomes which are visible. Therefore, when we find genetic ratios unlike those we expect, we can often turn to the chromosomes for possible explanations.

We have already pointed out that the law of segregation applies to most organisms. There are exceptions, however, to the law of independent assortment. We can understand the reasons for many of the exceptions if we relate them to the behavior of chromosomes.

The garden pea, for example, has seven pairs of homologous chromosomes. When two different pairs of genes are on different

Fig. 23.1 Photographs of two females of the fruit fly *Drosophila melanogaster*. The one on the right is a normal individual, commonly referred to as *wild type*. That on the left exhibits two mutant characters: white eye instead of the wild type (red eye), and miniature wing instead of the longer wing of the wild type.

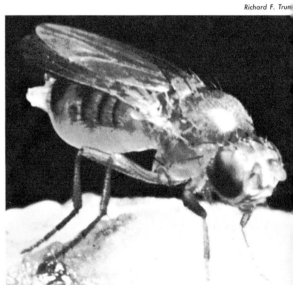

Fig. 23.2 The genes in a single chromosome can be represented by the different symbols shown here. It is not known, however, that genes are separated from each other by nongenic material, as the diagram would suggest. The chromosome is probably a continuous piece of DNA, in which case the genes would be adjacent to each other and distinguishable only by variations in the nucleotide sequence.

pairs of chromosomes they segregate independently of each other, giving rise to the familiar 9:3:3:1 dihybrid ratio characteristic of a diploid organism. *Neurospora* also has seven chromosomes, but being haploid the dihybrid ratio is 1:1:1:1. But both the garden pea and *Neurospora* have many more than seven pairs of genes. It follows, then, that any given pair of chromosomes contains many genes. If the two pairs of genes being studied happen to be on the same chromosome, the chances are great that these genes will not show independence of assortment. This tendency for the genes in a given chromosome to segregate together is called **linkage**.

Let us think of genes as beads on a string. Each bead may be separate from the others and is individually distinct (Fig. 23.2). But if one bead is moved the others follow, simply because of the string that holds them together. In the same way, genes show linkage because they are on the same chromosome. If the genes on any chromosome are always segregated as a group, these are said to show **complete linkage**. Complete linkage, however, is rare and groups of linked genes can recombine with other groups within homologous chromosomes by a process of **crossing-over**. A study of meiotic chromosomes indicates this at the cell level by the formation of chiasmata.

Linkage and Crossing-Over in Drosophila

Two easily recognized variations in *Drosophila melanogaster*, the fruit fly, are black body (*b*) and vestigial wings (*vg*). Both are recessive to the normal gray body color and to normal wing size. If a cross is made between a normal fly homozygous for these traits (*BBVgVg*) and a black one with vestigial wings (*bbvgvg*), the F_1 is normal in appearance, but heterozygous for the two pairs of genes (*BbVgvg*). A testcross of an F_1 female with a double recessive (*bbvgvg*) male would be expected to produce offspring in a ratio of 1*BVg*:1*Bvg*:1*bVg*:1*bvg* if the genes were on different chromosomes. However, the actual results are quite different in two ways.

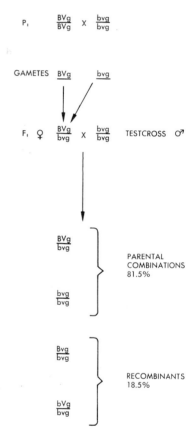

Fig. 23.3 Diagram of the results of a dihybrid cross: The genes being studied are part of the same chromosome rather than being on separate chromosomes. In this particular cross, the F_1 female was the heterozygous parent, and the testcross male was homozygous recessive. Compare these results with those in Fig. 23.4

Notice in Fig. 23.3 that the genes are designated somewhat differently than before—$\frac{BVg}{BVg}$ instead of $BBVgVg$. This is the geneticist's way of indicating that the genes are linked. The genes BVg being located on one chromosome, while the other set of BVg genes, indicated below the line, are located on its homologue. The F_1 individuals are, of course, heterozygous—$\frac{BVg}{bvg}$

After the F_1 females are mated to the double recessive male in the testcross, and 1,000 of the offspring are singled out, the results might be as indicated in Fig. 23.3. The same results would be obtained if the experiment were repeated, although the actual numerical figures might vary somewhat from one experiment to another.

The results clearly indicate that a marked deviation from a 1:1:1:1 phenotypic ratio has been obtained. The parental phenotypic combinations—$\frac{BVg}{bvg}$ and $\frac{bvg}{bvg}$—account for 81.5 per cent of the total population, with the two parental combinations being represented almost equally. The new combinations—$\frac{Bvg}{bvg}$ and $\frac{bVg}{bvg}$—add up to 18.5 per cent of the population. Again each new combination of genes is represented equally. From these data we can say that these two pairs of genes do not segregate independently of each other. The BVg and bvg parental combinations show a strong tendency to be inherited together. The new arrangements of genes, Bvg and bVg, called **recombinations,** show up relatively infrequently. The strength of the tendency to be inherited as a unit, calculated as a per cent $\left(\frac{407 + 408}{1,000}\right)$ of the total population, is a measure of the strength of linkage. Conversely, the frequency of recombinations again calculated as a per cent $\left(\frac{92 + 93}{1,000}\right)$ of the total population, is a measure of the frequency of crossing-over between these two genes.

If we do the same experiment, but cross F_1 males to double recessive females this time, we obtain the results indicated in the Fig. 23.4. Again, if 1,000 flies were counted, we could find that they would be about equally divided into $\frac{BVg}{bvg}$ and $\frac{bvg}{bvg}$ individuals. Therefore, only the parental combinations were recovered, and we must assume that linkage is complete.

The percentages obtained from these two experiments are not haphazard. When *Drosophila* F_1 males are used, complete linkage is the rule. When F_1 females are used, parental combinations will be found in about 80 per cent of the population, recombinations in about 20 per cent. Clearly there must be some physical reason for this. Since we have already discussed the fact that gene segregation is mirrored in the behavior of chromosomes, it is natural that we should turn to the behavior of meiotic chromosomes. It is in meiosis that segregation occurs.

If you look back to Fig. 21.6, you will notice that in meiotic prophase the homologous chromosomes, which were paired, fall apart except at those points along their length where they are held together by chiasmata. *Notice carefully that a chiasma is an exchange of chromatin between only two of the four chromatids,* one from each of the two homologous chromosomes. This exchange results from the process of crossing-over, with the chiasma being visible evidence of it at the four-strand stage.

Assume now that the genes for body color and wing size are located on this pair of chromosomes. Gene B and its allele b occupy corresponding places in the homologues, and the same is true for gene Vg and its allele vg. A chiasma forms between these two pairs of genes, and a cross-over results (see Fig. 23.5). Two of the chromatids are unaffected; they retain the parental gene combinations, BVg or bvg. The other two have new gene combinations, Bvg and bVg. These are **cross-over chromatids.** Therefore, every cross-over yields only 50 per cent cross-over chromatids, and 50 per cent parental, or non-cross-over, chromatids. If a cross-over occurred in every meiotic cell between these two genes, no more than 50 per cent recombination of genes could result. When crossing-over occurs less frequently, more non-cross-over chromatids are found.

Let us now view our data in terms of crossing over. In the experiment where the F_1 individual was a female, 18.5 per cent of the offspring had new gene combinations, 81.5 per cent had the parental combinations. With the knowledge that only two of the four chromatids are involved in a single cross-over, we can state that 37 per cent (18.5 × 2) of all eggs had a cross-over taking place between the genes b and vg.

Fig. 23.4 This cross is similar to the one shown in Fig. 23.3, but here the F_1 male is heterozygous for the two genes being studied, and the testcross female is homozygous recessive. Drosophila males do not exhibit crossing-over; therefore, recombinants do not appear among the progeny.

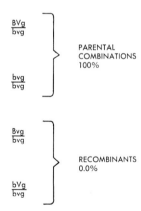

Fig. 23.5 This diagram shows a cross-over that has taken place between the chromatids of homologous chromosomes. The chromosomes are marked with the mutant genes shown in Figs. 23.3 and 23.4.

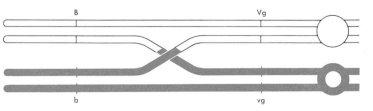

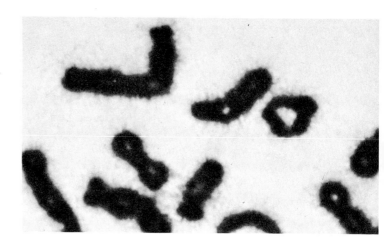

Fig. 23.6 In this meiotic cell of a male amphibian you can see the chiasmata holding the homologous chromosomes together. The shorter bivalents have two chiasmata while several of the longer ones have as many as four, possibly five.

In the experiment where the F_1 individual was a male, no cross-over chromatids were recovered in the offspring. Only the parental combinations appeared. Therefore, no cross-overs take place in *Drosophila* males. This is not true for males of all species. In fact, *Drosophila* males are exceptional in this regard, but they are useful in demonstrating complete as opposed to partial linkage.

If we look at meiotic chromosomes in a number of species we find in general that the longer the chromosomes the more numerous are the chiasmata. Small pairs of homologues generally have one chiasma holding them together. Larger ones may have two or more (see Fig. 23.6). Furthermore, the position of a chiasma is not constant. It may be any place along the length of a chromosome. A general rule can therefore be stated that the farther apart two genes are on a chromosome, the greater is the opportunity for a cross-over to occur between them. The frequency of recombination will, of course, also be greater. When the percentage of recombination is low, the genes can be physically close to each other, or crossing-over for some reason does not occur, as in male *Drosophila*.

Proofs of Crossing-Over

Those of you who have a skeptical turn probably have been asking a few questions as you have been reading this chapter: How do we *know* that crossing-over involves an exchange of chromatin? How do we *know* that crossing-over takes place in the four-strand stage and between only *two* of the four chromatids? How can we be *certain* that a chiasma is visible evidence of a cross-over between genes? These questions

need answering if we are to take the position that the behavior
of genes is reflected in the behavior of chromosomes, because
this assumption lies at the heart of genetics.

The first question was answered by a simple yet elegant
experiment carried out by Harriett Creighton and Barbara
McClintock, then at Cornell University. They took advantage
of the fact that some chromosomes in maize have visibly distinct
knobs while others with which they are homologous do not.
They crossed two strains of maize, producing a hybrid with
the chromosomal composition shown in Fig. 23.7. The knobbed
chromosome also carried an additional cell marker indicated
by the dashed portion in the diagram. In addition, the chromosomes were marked genetically with two pairs of heterozygous
alleles: C, colored kernel: c, colorless kernel: Wx, starchy kernel,
wx, waxy kernel. This hybrid plant, heterozygous now for the
two genes, was crossed with one carrying knobless chromosomes,
and with c and wx in a homozygous state. The testcross progeny was then examined genetically and cytologically.

Since C was very close to the knob, little or no crossing-over
took place in this region. When a cross-over does occur between
the two genes, however, and if an exchange of chromatin accompanies crossing-over, then all CWx plants should have the
knobbed chromosome. The cwx plants should not have them.
However, the cwx chromosomes should have the dashed portion
at the other end of the chromosome. In all cases studied there
was complete agreement between the genetical and cytological
information. Whenever a cross-over occurred, an exchange of
chromatin resulted. A comparable study by Curt Stern in *Drosophila* confirmed the maize work in all respects.

To answer our second question about crossing-over in the
four-strand stage, we turn to *Neurospora* because all of the chromatids from a single meiotic cell can be recovered. Each one is
contained within a single ascospore. Assume that genes A and B

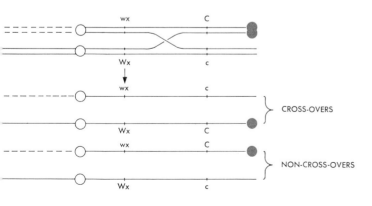

Fig. 23.7 An experiment carried out in maize demonstrates that when a crossover occurs there is also an exchange of a block of chromatin. See text for further explanation.

are linked, and that strain *AB* is crossed with *ab*. If crossing-over occurs, two alternatives exist: either crossing-over occurs at the two-strand or the four-strand stage. Figure 23.8 shows these two possibilities. If crossing-over occurs at the two-strand stage the ascospores will contain *only* cross-over chromatids. If no crossing-over occurs, only the parental chromatids will be found. *No ascus will contain both parental and cross-over chromatids.* If crossing-over occurs at the four-strand stage, however, both parental and cross-over chromatids are to be expected in the same ascus. The fact that the latter situation prevails is proof of the fact that crossing-over does take place at the four-strand stage, and that only two of the four chromatids are involved in any given crossover.

Proof that all chiasmata represent points of crossing-over has not been established with certainty. The kinds of cross-overs we have been discussing should be visible in good preparations of meiotic cells as chiasmata. However, we are faced at the present time with the fact that meiotic cells in male *Drosophila*, the homologous chromosomes, are sometimes held together by chiasmata but yield no evidence of having had any recombination of genes. The best we can say, therefore, is that in the majority of organisms, a chiasma is very probably the cytological equivalent of a genetic cross-over. To go beyond this is to go beyond the evidence at hand.

Chromosome Maps

You will recall that linked genes are on the same chromosome; and that the frequency of crossing-over between two genes is constant and characteristic. For example

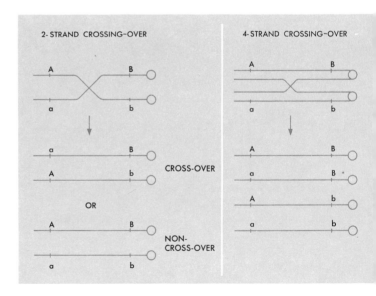

Fig. 23.8 Crossing-over can occur in *Neurospora* in two ways. The spores from a single ascus show that crossing-over occurs at the two-strand, not at the four-strand, stage. Crossing-over, therefore, is between chromatids, and not between whole chromosomes.

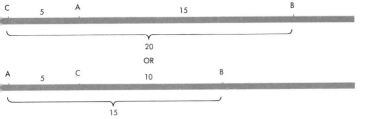

Fig. 23.9 Gene C has two possible positions when genes A and B are 15 map units apart, and when gene C is five map units from gene A. The only way to resolve the position of gene C in respect to the other two genes is by use of the three-point testcross described in the text.

In *Drosophila*, *b* and *vg* always show about 18.5 per cent recombination in females if the environment is kept constant. Another pair of genes might show a different rate of crossing-over, but it, too, would be constant for these two genes.

Genes, therefore, must occupy particular positions on a chromosome. This being so, it should be possible to construct a map showing their relationship one to another. This would, in essence, be a road map of a chromosome, with genes being the "towns" along the route. We have no precise way yet, however, of determining the exact physical distance between genes, particularly in higher organisms. This would require a full knowledge of the nucleotide sequence of the chromosome. At the moment, we can do this only in terms of the percentages of recombination. Using this system, *one map unit of genetic distance is defined as that length of chromosome within which one per cent of crossing-over takes place*. For example, since genes *b* and *vg* show 18.5 per cent recombination, they are 18.5 map units apart.

Let us now suppose that genes A and B are 15 map units apart on the chromosome. Suppose also that gene C is five map units from A. What is the position of C on our gene map? Clearly, there are two possibilities. As Fig. 23.9 indicates, C could be between A and B, and 10 units from B, or to the left of A and 20 units from B. Therefore, we need more information than has been given before we can decide whether the order of the genes is $A\ C\ B$ or $C\ A\ B$.

A standard test procedure, developed by A. H. Sturtevant, then at Columbia University, and known as the **three-point test cross**, is now commonly used in determining gene order. The basis of the test is as follows: suppose that the correct serial order of three genes is *a b c*. Suppose also that the distances between genes is reflected in the frequency with which crossing-over takes place between them. If these two assumptions are correct, then a testcross of the heterozygote $\frac{A\ B\ C}{a\ b\ c}$ to the triple recessive $\frac{a\ b\ c}{a\ b\ c}$ can yield the following phenotypes, the complementary cross-over and non-cross-over types being grouped: (In all instances, each individual would also have the $\frac{}{abc}$ chromosome, but it is omitted for the sake of convenience.)

parental or non-cross-overs	A B C
	a b c
single cross-overs	A b c
Type I	a B c
single cross-overs	A B c
Type II	a b C
double cross-overs	A b C
	a B c

On the basis of the mechanism of crossing-over described in the diagram on page 341, the non-cross-over types are the most frequent. Type I single cross-overs result from a single cross-over between genes a and b. Type II single cross-overs result from a cross-over between genes b and c.

If the Type I cross-overs show a frequency of 10 per cent, and the Type II cross-overs a frequency of 20 per cent, then *the double cross-overs would be expected to occur at a rate much lower than either alone.* This is because the simultaneous occurrence of both Type I and Type II in the same cell is the product of the two frequencies (10 per cent of 20 per cent), and the double cross-overs should occur in approximately 2 per cent of the cells.

An experiment carried out with maize will illustrate the use of the three-point cross. The genes brown midrib (*bm*), red seed color (*pr*), and virescent (light green) seedling (*v*) are located in one chromosome, and, of course, are linked. The crosses and the data obtained are as follows:

$$P_1: \frac{Bm\ Pr\ V}{Bm\ Pr\ V} \times \frac{bm\ pr\ v}{bm\ pr\ v}$$

$$F_1: \frac{Bm\ Pr\ V}{bm\ pr\ v}$$

$$\text{Test cross: } F_1 \times \frac{bm\ pr\ v}{bm\ pr\ v}$$

Testcross progeny (The $\frac{}{bm\ pr\ v}$ is omitted for all progeny but assumed to be present):

Bm Pr V	232	non-cross-overs = 42.1 per cent
bm pr v	235	
Bm pr v	84	single cross-overs between *bm* and *pr* = 14.5
bm Pr V	77	per cent

Bm Pr v	201	single cross-overs between *pr* and *v* = 35.6
bm pr V	194	per cent
Bm pr V	40	double cross-overs between *bm* and *pr*, and
bm Pr v	46	*pr* and *v* = 7.8 per cent

Among the progeny above we can recognize the non-crossover parental types. They are the most numerous. We can also recognize the double cross-overs. They are the two least frequent classes of progeny. The double cross-overs also tell us the order of the genes. That is, *pr* has shifted position with *Pr*, while *Bm* and *V*, and *bm* and *v* are still linked. Therefore, the *Pr* (or *pr*) gene lies between the other two genes.

In calculating map distances, all cross-overs that have taken place between any two genes must be considered. Thus, the distance between *bm* and *pr* is not 14.5 map units, but 14.5 plus 7.8, or 22.3. In other words, the single cross-over frequency between *bm* and *pr* does not represent the total frequency. The double cross-overs must be added since they also represent cross-overs in this same region. Similarly, the total cross-over frequency between *pr* and *v* is 35.6 plus 7.8, or 43.4. Therefore, the total distance from *bm* to *v* is 65.7 map units.

There is one more point to consider. If the cross-over frequency between *bm* and *pr* is 22.3 per cent, and between *pr* and *v* is 43.4 per cent, then we should expect the double cross-over frequency to be 22.3 per cent of 43.4 per cent, or 9.7 per cent. Actually, however, we find that the frequency is only 7.8 per cent, less than expected. Many studies have shown that this is due to something called **interference.** It turns out that two adjacent cross-overs in a single pair of chromosomes are not generally independent of each other; one tends to interfere with the other and reduce the frequency. In general, and in such organisms as maize and *Drosophila,* double cross-overs do not occur between two genes if the genes being tested are 10 map units or less apart. Interference is complete. If the genes are more than 45 map units apart, interference is difficult to detect, or is nonoperative. For reasons that we do not yet understand, the cross-overs on one side of the centromere of a chromosome do not interfere with those on the other side.

On the basis of many three-point tests, it is possible to build up the gene maps of organisms. Figure 23.10 shows the gene maps of the four chromosomes of *Drosophila melanogaster.* The left end of the chromosome is indicated as zero (0) map distance. Successive distances are added to give the location of individual genes.

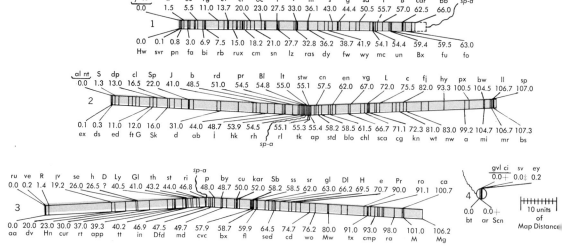

Fig. 23.10 Genetic maps of the four chromosomes of Drosophila melanogaster: Each gene has a lettered symbol, with small letters indicating recessiveness, capital letters indicating dominance in relation to the normal, or wild type, gene. Each gene also has a given position on the chromosome. For example, in chromosome 1 at the top of the illustration, the gene yellow body color (y) is at the extreme left at position 0.0; the gene white eye (w) is close to it at position 1.5, or 1.5 map units distance from yellow. Bobbed bristle (bb) is at the other end of the chromosome, 66.0 map units from yellow. Chromosomes 2 and 3 are the longest in size and in map distance. Chromosome 4 is very small, rarely undergoing crossing-over.

SUMMARY

When two genes are on the same chromosome, the law of independent assortment does not hold. The genes are linked to each other and tend to be inherited together. **Crossing-over** breaks up linkage groups, and the frequency of crossing-over gives a measure of the distance one gene is from a neighboring gene. A chiasma formed between homologous chromosomes in the four-strand state is physical evidence of crossing-over between genes.

Since crossing-over is a measure of distance between genes, gene maps of chromosomes can be, and have been, constructed. In such maps, one per cent crossing-over is equal to one map unit. Map distances calculated between genes close together are more accurate than map distances for distant genes, because of the occurrence of double cross-overs.

The position of one gene in respect to two other genes on the same chromosome can be determined through the **three-point testcross**. In such a cross, the double cross-overs form the least frequent class and provide a clue about the order of the three genes.

FOR THOUGHT AND DISCUSSION

1 Why do we use testcross offspring rather than F_2 offspring in calculating cross-over frequencies? What do we mean by the term "frequency"?

2 If a double cross-over occurred between the linked genes A and B, could it always be detected? In considering this question, remember that at any position of crossing-over, any chromatid can cross-over with any nonsister chromatid. Can crossing-over be detected if it occurs between sister chromatids? Explain.

3 An organism has five pairs of chromosomes as a diploid number, and an examination of many meiotic cells reveals that these pairs, from longest to shortest, have the following chiasma frequencies: 3.72, 2.60, 2.10, 1.50 and 1.00. Assuming that a chiasma always represents a cross-over between 2 of the 4 chromatids, what is the maximum genetic map length of these chromosomes?

4 If two genes are more than 50 map units apart on a chromosome, what do we need to know in order to determine their true map distance accurately?

5 Genes a, b, and c are linked in that order; a is 40 map units from b, and b is 20 map units from c. In a test cross, only 5.5 per cent double cross-overs were detected among the offspring. How much interference was there?

6 Two homozygous parents were mated, and the F_1 offspring were testcrossed to the double recessive. The testcross progeny were counted as follows: AB–22; Ab–29; aB–31; ab–18. Determine the genotypes of the parents, and the amount of crossing-over that has occurred?

7 Consider meiosis in *Neurospora*, and suppose that a gene m is not too far from the centromere on one of the pairs of chromosomes. You have mated a strain with M to one with m. How can you determine the genetic map distance from the centromere to the gene? Diagram your results.

8 The genes x, y, and z are linked on the same chromosome, but not necessarily in that order. The homozygous parents were mated to produce an F_1 offspring, and this F_1 was then testcrossed to a homozygous triple recessive. The following progeny were obtained:

XYZ–230   What are the genotypes of the parents?
XYz–26   What is the order of the genes on the chromosome?

XyZ–145   What are the map distances between the genes?
Xyz–98   How much interference was there?
xYZ–102   Why do we list only part of the genotype of the testcross progeny?

xYz–155
xyZ–24
xyz–220

SELECTED READINGS

See the books listed at the end of the previous chapter.

24 SEX AS AN INHERITED TRAIT

The most obvious difference among human beings is that of sex. Maleness and femaleness represent a pair of contrasting mutually exclusive characters (Fig. 24.1), and they serve well to illustrate the basic mode of inheritance and its physical, cellular basis.

Before the principles of inheritance were firmly established, biologists generally viewed heredity as some mechanism that operated in the offspring and blended the characters of the two parents. Common experience tells us that this is true for such characters as weight, height, and skin color. For example, the child of a white-Negro parentage is colored (intermediate shade), and it makes no difference which parent is white and which is Negro. Sex, however, is not a "blended" character. Except in rare instances, human offspring are either male or female.

Fig. 24.1 The markings that distinguish a male drake (lower left) and the female duck (upper right) are obvious in this photograph of a pair of mallards. Not all animals show such clear-cut sexual differences. The horns of a ram, the tail of the peacock, and the mane of a lion are distinguishing features of maleness; but bears, rabbits, and horses, for example, do not show such striking differences.

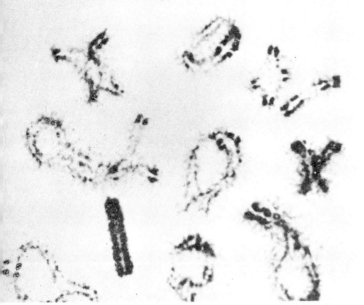

Fig. 24.2 The single X chromosome in this photograph of a mid-prophase meiotic cell from the testis of a male grasshopper can be seen as the dark-staining rod in the lower left. Since there is only one X chromosome, the grasshopper males have an XO constitution, the females an XX constitution.

J. J. Tijo and A. Levan

The hereditary basis of sex was experimentally established about 1900. By this time microscopic techniques had been developed, and the behavior of chromosomes was well understood. It was around this time that biologists noticed that the males and females of certain insects differed in their chromosome number. The male generally had one less chromosome in its somatic cells than did the female. This odd chromosome in the male became known as the **X-chromosome** simply because its role was initially poorly understood. An illustration on this page shows the X-chromosome as it appears in a grasshopper (Fig. 24.2).

During meiosis the paired chromosomes are equally distributed to the gametes—except for the X-chromosome. Because it is without a pairing partner, it is found in only half of the total number of sperm cells. When this was first worked out, biologists thought that here, obviously, was a mechanism that could account for the equality of numbers of males and females in sexually reproducing populations. If so, then the female must have a pair of X-chromosomes, while the male has but a single one. It also follows that all eggs must have an X-chromosome in their haploid set, and that sex is determined at the time of fertilization: an X-bearing sperm combining with an X-bearing egg produces a female zygote; a sperm lacking an X-chromosome combines with an X-bearing egg and produces a male zygote.

These assumptions have been fully borne out. Female organisms bearing the X-chromosome can be designated **XX**, while the males are labeled **XO** (the **O** stands for the lack of an X). Figure 24.3 shows the inheritance of X-chromosomes. Notice that the male offspring always receive their X-chromosomes from their mother. Female offspring receive an X-chromosome from each parent.

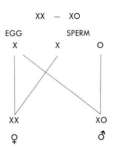

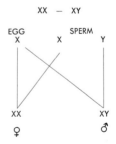

Fig. 24.3 Diagrams of the inheritance of XX-XO and XX-XY species.

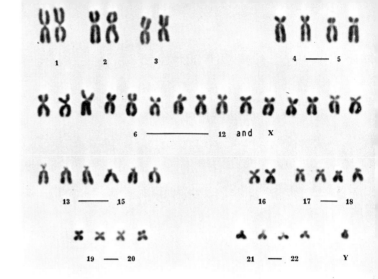

Fig. 24.4 The chromosomes of a normal human male: A cell was photographed and the chromosomes were cut out, arranged in homologous pairs, and numbered according to size. The human male has an XY sex-determining system. The tiny Y chromosome is shown at the bottom right. The X chromosome, which is difficult to identify positively, is one of those in the second row.

In many species of organisms, including man, both males and females have the same number of somatic chromosomes. In these males, the X-chromosome has a pairing partner which may be either similar to it in shape and size, or strikingly dissimilar. This is known as the **Y-chromosome** (see Fig. 24.4). In the male the chromosomal composition is designated **XY** instead of **XO**, while the female remains XX as before. The Y-chromosome may or may not exert a sex-determining influence; in either event, two types of sperm are produced—those bearing an X- and those bearing a Y-chromosome.

Proof that sperm of two kinds are produced in equal numbers is provided by *Protentor*, an insect of the XX-XO type. At the end of meiosis, the **spermatids** (those cells which will be transformed into motile sperm) are of two kinds. One shows very little nuclear structure; the other has a deep staining body within it (see Fig. 24.5). This body has been identified as the X-chromosome, which has a tendency to stain darkly during interphase.

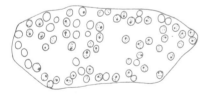

Fig. 24.5 This section cut through the testis of the bug, *Protentor*, shows the haploid spermatid cells before they have differentiated into motile spermatozoa. About one-half of the cells show a dark-staining body which has been identified as the X chromosome, the other half do not.

The remainder of the chromosomes in a cell are known as **autosomes.** Of the 46 chromosomes in humans, females have 22 pairs of autosomes plus one pair of X-chromosomes. Males have the same number of autosomes plus one X-chromosome and one Y-chromosome. We shall see that the autosomes may also exert a sex-determining influence. Not all sexually reproducing species show the female to be XX, and the male to be XY or XO. Birds, moths, butterflies, fishes, and some reptiles and amphibia are the reverse; that is, the female is XY, and the male is XX.

Sex Differentiation

The *X* and *Y* chromosomes provide us with only part of the genetic basis for the development which transforms a fertilized egg into either a male or a female. Sex is the **phenotypic** character, and like all phenotypes it is dependent upon a genotype. Sex, as a distinct character, becomes a reality only after a long and rather complex period of development. The early embryo in human beings, which arises from the zygote through cell division is neither male nor female. Essentially it is neutral, and it remains neutral even after the early development of **gonads,** or sex organs. This is so because the embryonic gonad consists of two parts: 1. an external layer of tissue, the **cortex,** which can develop into an ovary; and 2. an internal mass, the **medulla,** which can develop into a testis. In addition, two pairs of sex ducts are present in the neutral embryo. One of these is the **Müllerian duct,** which persists if the individual becomes a female. The other, the **Wolffian duct,** persists if the individual becomes a male (Fig. 24.6). There are additional embryonic parts, which are later transformed into the external genitalia. But at an early stage in embryogeny these parts are in such a primitive state that they can be transformed into either male or female organs, depending on genotypic influences.

It is only after the neutral stage of gonadal development that the specific sex determiners take over and stimulate the development of sex organs that are distinctly male or female (Fig. 24.7).

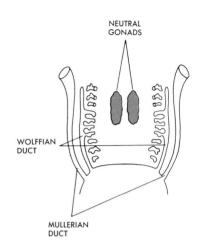

Fig. 24.6 In early embryonic development the gonads appear the same in both sexes. Males and females alike have comparable ducts. In females, the Müllerian duct will persist, and the Wolffian duct degenerate. The reverse occurs if the individual is to be a male.

Fig. 24.7 This diagram shows the developmental change that takes place in each gonad as sexual differentiation occurs.

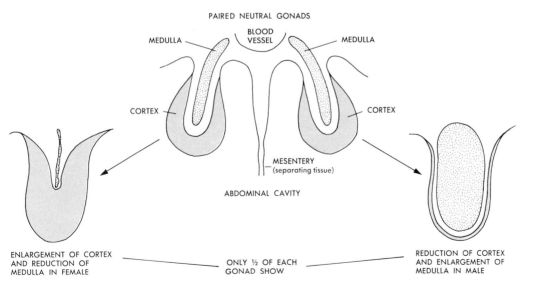

In an embryo whose chromosomal constitution is *XX,* and, therefore, potentially female, the cortical part of the gonad begins to develop, while the medullary part becomes inconspicuous and may even degenerate. In this way the neutral gonad is transformed into an ovary. If the embryo, on the other hand, happens to be *XY,* the reverse situation takes place. The medullary part of the gonad enlarges and begins to differentiate, while the cortical portion disappears; the neutral gonad is transformed into a testis.

It is at a particular point in normal development, therefore, that the genetic constitution of the individual becomes decisive. In a morphological sense this is the time when either the cortex or the medulla enlarges and the other part degenerates. In a genetic sense the "decision" was probably made much earlier, since a whole chain of biochemical changes must precede any obvious structural change. The chromosomes, consequently, act as a trigger mechanism to start off one or the other of a set of developmental reactions that leads either to maleness or femaleness. Any upset or delay in these reactions leads to an individual who is undeveloped sexually, or who falls into a "twilight" zone of intersexuality. Later on, other influences tend to reinforce the established sexual pattern so that maleness and femaleness can often be recognized visually as well as by the type of gonad and gamete produced. The plumage of some birds, which differs in the two sexes, is an example of a secondary sex character (Fig. 24.1).

In human beings, sexual differentiation is by no means complete at birth. Secondary sex characters develop as beginning adulthood is reached, that is, during the period of puberty. These may involve anatomical differences which change the larynx and consequently change the character of the voice; differences in general body conformation, which influence the structure of the pelvic region, and the mammary glands; and differences in hair growth and distribution, the male generally being more hairy than the female. All of these are influenced by the **gonadal hormones,** chemical substances that are produced as development proceeds. The important thing to remember, however, is that these influences begin in the individual with the initial genetic constitution present when the fertilized egg begins to grow.

Sex Linkage

The distribution of the X and Y chromosomes in meiosis, and their combinations in the fertilized zygote, account for the fact that there are approximately as many males as females. Sex is but one of many inherited characters, however. Do we need to assume that the sex chromosomes are concerned solely with maleness and femaleness? In many organisms, including man, the answer is "no." They may govern other characters, unrelated to sex. If the factors governing the appearance of other characters are carried on the X-chromosome, then these particular characters should, however, show a relationship to sex. We refer to such a relationship as **sex linkage.** Those factors carried on the autosomes should show transmission relationships unrelated to the sex of the individual.

Let us use an example from *Drosophila melanogaster*. In 1910 the American geneticist T. H. Morgan, then at Columbia University, found that one of the fruit flies arising in a bottle culture had white instead of the usual red eyes (Fig. 23.1). Morgan isolated this fly and eventually obtained a strain of true-breeding white-eyed flies. When a white-eyed fly was crossed with a conventional red-eyed fly, the eye color of the offspring depended on whether the white-eyed parent was male or female. The results suggest that eye color is transmitted in much the same way that sex is. The details of these crosses are shown in Figs. 24.8 and 24.9.

From the cross of a white-eyed male with a homozygous red-eyed female, the first generation (F_1) cross shows all red-eyed individuals. From this we can assume that whatever is determining red eyes prevails over that factor controlling the development of white eyes—in other words, red is dominant over the

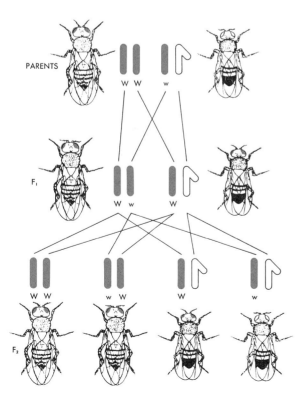

Fig. 24.8 A cross in which the mutant gene white (w) is in the male parent, and hence in only one X-chromosome. Of the F_2 progeny, only one-half of the males, and none of the females, are white-eyed. The Y chromosome, although partially homologous with the X-chromosome, lacks the genes found in the X-chromosome, hence does not contribute to sex linkage.

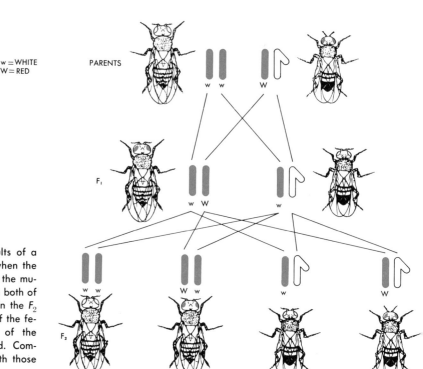

w = WHITE
W = RED

Fig. 24.9 The results of a cross in *Drosophila* when the female parent carries the mutant gene *white* (w) in both of her X-chromosomes. In the F_2 generation one-half of the females and one-half of the males are white-eyed. Compare these results with those of Fig. 24.8.

recessive white. When these F_1 individuals are bred, white reappears in the F_2 generation, but only in one-quarter of the flies. The diagram also shows that all of the white-eyed individuals are males. White-eyed females are absent from both F_1 and F_2 generations. If we diagram the cross in such a way (Fig. 24.8) that the red-eye character is designated by a large W, and the white-eye character by a small w, we find that in the F_2 generation the females must be of two different kinds—one homozygous WW, the other heterozygous (Ww). We can further support our assumption by more breeding. One type of F_2 female, when crossed with a white-eyed male, produces only red-eyed offspring. The other type of female produces half white and half red offspring. The latter female must, therefore, carry white in a recessive heterozygous condition, but its heterozygosity is revealed only by further breeding trials. The white-eye character is not lost when heterozygous; it is merely suppressed in its action.

If we cross a red-eyed male with a white-eyed female, we get different results (see Fig. 24.9). Among the F_1 offspring, all of the females are red-eyed and all of the males are white-eyed. When these are bred, the F_2 offspring show red and white-eyed characters in equal proportions, and with an equal distribution among the sexes. If the F_2 females are back-crossed once again to a white-eyed male, the white-eyed female will not produce red offspring but the red-eyed female produces half red and half white offspring. It therefore appears that the white-eyed F_2 female is pure in respect to the characters shown, but that the red-eyed female, again, is heterozygous. The red- and white-eyed males bred back to a white-eyed female would produce either white or red offspring, respectively. Therefore, it appears that these F_2 males are pure for the character they display.

Let us analyze what we have just been through. The male transmits the eye character to his grandsons through his daughters, but never directly to his sons. The mother transmits the character to both sons and daughters. A glance at Figs. 24.8 and 24.9 will reveal this, if we assume that the factor for eye color is carried on the X-chromosome, and that the Y-chromosome plays no part in eye color. Only the daughters get an X-chromosome from the father, but both sons and daughters receive an X-chromosome from the mother. Therefore, we can conclude, as Morgan did, that the factor for white-eye color, as well as red-eye color, is carried on the X-chromosome, and that red and white are allelic to each other in a genetic sense. The factor, or gene, exists in two forms, and a single X-chromosome carries only one at a time. A particular genetic factor, then, must be located on a particular chromosome. It remained for C. B. Bridges of Columbia University to prove that this is so.

Bridges made a series of crosses between white-eyed females and red-eyed males of *Drosophila*. Ordinarily, if everything goes

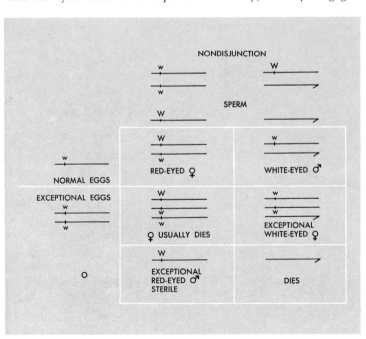

Fig. 24.10 This diagram shows what happens when nondisjunction of the X-chromosomes occurs during egg formation in females of *Drosophila* and forms exceptional eggs. The discovery, made by C. B. Bridges, proved that a particular gene is carried on a particular chromosome.

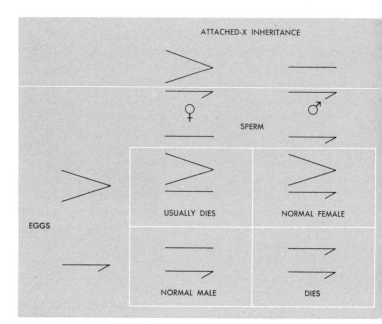

Fig. 24.11 Inheritance in attached-X females of *Drosophila*: These females also contain a Y-chromosome, but this has no influence on sex determination or sex linkage.

as it should, the offspring would be white-eyed males and red-eyed females, as indicated in Fig. 24.9. Out of a large number of offspring, however, he found an occasional red-eyed male or white-eyed female in the F_1 generation. This would seem to suggest that the original hypothesis (that the factors for red and white eye color are carried on the X-chromosome) was incorrect. Instead, however, it turned out that these flies provided proof of the conclusiveness of Morgan's hypothesis. Bridges suspected that what had happened was that the X-chromosomes in the female had failed to segregate from each other during meiosis. This failure resulted in some eggs having two X-chromosomes instead of only one, and other eggs having no X-chromosomes. He called this abnormal process **nondisjunction** (see Fig. 24.10).

Bridges could now predict that if a 2-X-egg with both of the X-chromosomes carrying the factor for white were fertilized by a Y-bearing sperm, which does not carry any eye-color factors, a white-eyed female should be produced. Similarly, a no-X-egg fertilized by an X-bearing sperm carrying the factor for red should produce a red-eyed male. The transmission in this instance was directly from father to son, whereas in the first case it was the transmission of the white factor from mother to daughter. If this were true, then each of the exceptional white-eyed F_1 females should have, in addition to the 2-X-chromosomes, a Y-chromosome in each cell of her body. The red-eyed F_1 exceptional males should not have any Y-chromosomes in their cells. To test this, Bridges made a microscopic examination of the cells. The chromosomal composition of these exceptional flies was exactly as he had predicted. Every time there was an irregularity in the pattern of heredity, there was a corresponding irregularity in the distribution of chromosomes.

Bridges' study provided a very convincing demonstration of the close relationship of genes to chromosomes. Additional evidence was supplied by L. V. Morgan, wife of T. H. Morgan. She had experimented with a strain of *Drosophila* in which the two X-chromosomes of the mother were always passed on to the daughters, never to the sons. It is difficult to imagine how this could happen unless the two X-chromosomes were somehow attached to each other, and always remained together in the egg nucleus, or passed together into the polar body. It is now known that they are actually attached to each other. Females bearing this type of chromosome are known as **attached-X females;** in addition, they usually carry a Y-chromosome. Figure 24.11 shows the type of transmission that takes place when an attached-X female is mated with a normal male.

Of the four kinds of zygotes produced, those with 3-X-chromosomes die, as do those lacking an X-chromosome. Thus, the two surviving types are exactly like their parents. Attached-X transmission, therefore, stands in contrast to the usual pattern of sex-linked inheritance. Once again the close relationship between gene and chromosome transmission is convincingly demonstrated.

The attached-X studies also provide evidence of the nonimportance of the Y-chromosome in *Drosophila* sex determination. The Y-chromosome can be found in either males or females without affecting sex. However, we cannot generalize and say that the Y-chromosome is *always* without such effect. Also, males lacking a Y-chromosome are sterile, so the Y-chromosome has a function to perform. As we shall see later, it appears that mammals, including man, have a sex-determining mechanism in which Y-chromosomes play a prominent role.

Let us now examine one case of sex-linked inheritance in humans—color-blindness. Say that a woman who is heterozygous for color-blindness marries a normal man (see Fig. 24.12). Their daughters would all be normal, although there is a 50–50 chance that each daughter is heterozygous, and thus able to transmit the color-blindness character to sons in a later generation. On the other hand, one-half of all the male children of the marriage are likely to be color-blind since one of the X-chromosomes carries the factor for color-blindness. The Y-chromosome does not play any part in color-blindness.

The number of known characters found on the X-chromosome of man that show a sex-linked inheritance of the type just described are relatively few. In *Drosophila* many more

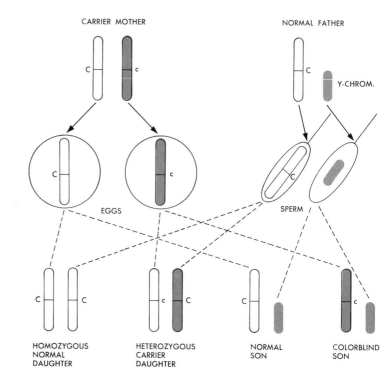

Fig. 24.12 Sex-linked inheritance of color-blindness in man: Normal vision is designated by C, color-blindness by c. The Y-chromosome carries no comparable gene, and so is without influence in this respect.

characters are known. At the present time several hundred are known, involving the wings, bristles, eye-color, as well as more subtle characters that can be determined only by careful experimentation. Any character, regardless of whether it is involved in the determination of sex, is always sex-linked if it is borne on the X-chromosome.

In birds, butterflies, moths, reptiles, and some fish, and amphibians, the sex-determining mechanism is the reverse of that found in *Drosophila*. The female is the heterozygous sex (*XY*) whereas the male carries two *X*-chromosomes. The study of certain sex-linked genes has borne this out.

How Chromosomes Determine Sex

We have said that species can have an *XX-XO*, or an *XX-XY* sex-determining mechanism. The real situation, however, is not quite so simple. In Fig. 24.13 are the chromosomal conditions observed in *Drosophila* and man. Each haploid set of autosomes is designated by the letter *A*. A glance at the

Drosophila data indicates that while the principal action of the X-chromosomes is to push development in a female direction, the autosomes push development in a male direction. The sex of an individual is determined, therefore, by a balance between the number of X-chromosomes and the number of sets of autosomes. If the $X:A$ ratio is 1 or greater, femaleness results. If the ratio is 0.5 or less, the individual is a male. However, a ratio of $2X:3A$, or 0.67, is an **intersex**. This is an individual who shows an intermediate kind of maleness or femaleness.

In man the role of the X-chromosome in sex determination is not clear. An XO individual develops into a female, not into a male as in *Drosophila*. In addition, the XO individual is an abnormal female. The ovaries are abortive or missing, the individual is sterile, intelligence may be subnormal, and there are a number of structural abnormalities particularly of the heart. In the mouse, an XO animal is a fertile female and indistinguishable from an XX animal. The X-chromosome, at least in the mouse, and possibly in man, does not seem to have any pronounced male- or female-determining tendencies.

No YO individuals are known to exist in man or in other mammals. The X-chromosome is necessary for survival. Any mammal with a Y-chromosome, however, is a male. This again is different from the situation in *Drosophila* where the Y is without any sex-determining influence. In *Drosophila*, an XXY individual is a normal, fertile female. In man, the same chromosomal situation results in a sterile, abnormally underdeveloped male. It is of interest to point out that the silkworm, *Bombyx mori*, carries sex-determining factors in its Y-chromosome. In regard to sex determination, man is therefore more like a moth than a fly.

CHROMOSOME CONSTITUTION AND SEX IN DROSOPHILA AND MAN (A = HAPLOID SET OF AUTOSOMES)

DROSOPHILA			MAN		
CHROMOSOME CONSTITUTION	SEX	X/A RATIO	CHROMOSOME CONSTITUTION	SEX	X/A RATIO
2A XXX	SUPERFEMALE	1.5	2A X	FEMALE	0.5
2A XX	FEMALE	1.0	2A XX		1.0
2A XXY			2A XXX		1.50
3A XXX			2A XXXX		2.00
3A XX	INTERSEX	0.67	2A XY	MALE	0.5
3A XXY			2A XYY		0.5
			2A XXY		1.0
			2A XXYY		1.0
2A X	MALE	0.50			
2A XY					
2A XYY					
3A X	SUPERMALE	0.33			

Fig. 24.13 The relation of chromosomal constitution to sex in Drosophila and man: A haploid set of autosomes is designated by the letter A. In *Drosophila*, the X/A ratio determines maleness and femaleness; in humans, maleness and femaleness are determined by the presence or absence of a Y-chromosome. It is probable that human femaleness is influenced by the autosomes, but we do not know at the present time.

It is, of course, tragic that *XO* and *XXY* individuals exist in a human population. Nondisjunction of chromosomes at meiosis is undoubtedly the cause, but nothing as yet can be done to alter this condition, once it arises. It is through our understanding of these cases, however, that we gain some of our knowledge of inheritance in the human race.

What we have just described points out the fact that while sex is an inherited character, it is not determined by a single gene in the same way that eye color is. Sex is determined by whole chromosomes rather than by single genes, at least in diploid organisms. No distinct sex gene has been found in *Drosophila*, man, or mouse. However, the *A* and *a* mating types found in *Neurospora* can be thought of as a primitive kind of sex, not yet differentiated into maleness or femaleness, and these are determined by single genes. Sex probably arose in some early organism as a mutation, and its inheritance later became a chromosomal rather than a genic feature of inheritance.

SUMMARY

Sex is an inherited trait, but rather than being determined by a single gene, it is determined by a balance between the sex chromosomes and the autosomes. In *Drosophila*, the *X*-chromosome is female-determining, the autosomes male-determining; the *Y*-chromosome is without influence on sex and is concerned only with fertility of the male. In man, the *Y*-chromosome is male-determining, and it would appear as if the autosomes rather than the *X*-chromosome were female-determining.

Genes located on the sex chromosomes follow a type of inheritance linked to sex, and are called **sex-linked.**

FOR THOUGHT AND DISCUSSION

1 The sex ratio in human populations is not equal; there are 105 boys born to every 100 girls. What reasons can you give to explain this difference? At about age 21, males and females are about equal in number. What genetic explanation can you offer for the shift in ratio?

2 Does the inheritance of *X*-chromosomes suggest that sons or daughters are likely to be more similar to their mothers than their fathers in certain characteristics? Why?

3 The attached-*X* chromosome in *Drosophila* is peculiar in that the order of genes in the two arms is in reverse; that is, ABCDEFG–centromere–GFEDCBA. By folding at the centro-

mere, the two arms can synapse and cross-over in meiosis. Yellow (yy) bodied offspring can be obtained from a female known to be heterozygous for this character (Yy). Can you show by means of a diagram how this can happen?

4 Color-blindness for red-green is due to a sex-linked recessive gene in man. A father is color-blind, and the mother is normal with no evidence of color-blindness in her ancestry. What is the probability that their son will be color-blind? Their daughter? Their grandson, if the grandson is a son of the daughter?

5 In cats, which are XX-XY, the sex-linked gene Y determines black when homozygous, yellow when homozygous recessive, and tortoise-shell or calico (mixture of black and yellow) when heterozygous. The Y-chromosome lacks the gene.
   (a) A calico mother has a litter of six: one yellow male, two black males, one yellow female, and two calico females. What is the genotype of the father?
   (b) A calico mother has a litter of three black females. If the father was black, how often would you expect the same result to happen again?
   (c) A yellow mother has a litter consisting of two yellow and three calico offspring. What is the genotype of the father, and what is the sex of the offspring in relation to color?

6 A convenient way of determining the rate at which genes in *Drosophila* mutate when exposed to x-rays or chemicals is to treat males and mate them to attached-X females. What advantage is there to such a technique? What kinds of mutations would be missed in this sort of experiment? If the treated males were mated to normal XX females, what procedures would have to be followed to obtain the same information? Can you think of any advantage of the second technique over the first?

7 In chickens, the barred pattern of feathers of the Plymouth Rock breed is due to a dominant sex-linked gene, B. An autosomal gene F affects the feathers such that FF has brittle, curly feathers, Ff has feathers that are slightly curly but otherwise normal, and ff has feathers that are fully normal. A number of barred, normal feathered hens are mated to a non-barred, slightly curly cock, and 200 hatchable eggs are produced. What is the expected frequency of offspring in terms of feather character and sex?

SELECTED READINGS

See books listed at the end of Chapter 16.

25 HEREDITY AND ENVIRONMENT

The inheritance test is the major tool of the geneticist in his study of the transmission of characters from parents to offspring. The sharper the contrast between a particular pair of characters, the easier it is to determine the pattern of inheritance. We have used, for example, such contrasting characters as maleness and femaleness, and red and white eyes in *Drosophila*. These characters are remarkably constant. Because they do not have a wide range of variation, no one has difficulty in identifying the phenotype.

Fig. 25.1 Three members of the cat family: the leopard, Canada lynx, and house, or tabby, cat. Among other characteristics, they are readily distinguishable by size, fur color and markings, disposition and native home.

Frank Stevens
From National Audubon Society

We can extend this idea of contrast in individual characters to groups of characters (Fig. 25.1). An oak tree has an array of characters—leaf, bark, wood, and growth habit—that enables us to distinguish it from a maple or pine tree. Even if the oak tree grew in such widely different environments as a shady forest, a dry open hillside, or a swamp, the characters remain constant enough for us to identify it as an oak. Similarly, a cocker spaniel can be identified not only as a vertebrate, a mammal, and a dog, but a dog different from a setter, a foxhound, or a wire-haired terrier. No change in diet or environment will alter the general character of a cocker so that it will be mistaken for another breed.

These facts stress the importance of the role heredity plays in determining individual characteristics. We often remark on this in everyday speech by saying, "He has his mother's eyes," or "She has her father's chin," or "She favors the Smith rather than the Jones side of her family." Rarely do we say that the environment plays an important role in determining the phenotype of an individual. Yet all of us are aware of it. What about the relation of your diet to your weight? We know that some individuals readily gain or lose weight. Others change hardly at all, no matter what they eat.

Leonard Lee Rue III

Joyce R. Wilson
From National Audubon Society

Every individual organism, of course, inherits certain things from its parents. These things determine its genotype. Every organism also grows and develops *in* an environment, never in the absence of one. The two influences, genotype plus environment, give rise to the individual's phenotype. Beethoven was a musical genius. Would he have been had he not had access to a piano? Here are some examples from biology of how we can separate, and then determine the role of each of these two influences.

Some people are born with a kind of diabetes that is hereditary. They lack **insulin,** a hormone which regulates the level of sugar in the blood. Without insulin, which is formed in the pancreas, such individuals would go into diabetic shock and die. The internal environment of these people can be corrected and controlled by periodic injections of insulin, but the genetic deficiency is not altered by this treatment. It is the internal environment—not the genetic deficiency—that determines whether the individual lives or dies.

An equally dramatic example occurs in human females. A single abnormal gene leads to the absence of a certain chemical **(21-keto steroid)** needed for normal development and feminization of the individual. Without medical treatment, a young girl with this genetic deficiency would have a tendency toward masculinity: the ovaries would fail to develop, the mammary glands would not enlarge, facial hair would become prominent. If this deficiency is recognized and treated before adulthood, there can be striking results. The ovaries become fully functional, development of the mammary glands proceeds, facial hair disappears, and the individual can lead a normal life. The treatment, however, must continue for as long as the individual wishes to preserve her feminine traits. Here is another example of how the environment can drastically alter a phenotype. The phenotypic alteration in this instance involves both appearance and reproductive capabilities.

The same problem has been approached experimentally by using plants which can be divided. The subdivisions are then grown in different environments. Remember that the original plant always comes from a single fertilized egg, so all of its cells have the same set of genes. This, of course, is true also for the subdivisions. How do the subdivisions behave in different environments? Which characters are altered, and which remain unchanged?

We will use the cinquefoil *Potentilla glandulosa* as an example. This species is widespread, and in California it grows from sea level to alpine heights (10,000 feet) in the Sierra Nevada Mountains. At sea level the species is active throughout the year, but its primary growth begins in late January. The seeds are mature

Fig. 25.2 Shown here are four altitudinal types of the cinquefoil, *Potentilla glandulosa*. They were taken from four different locations in the Sierra Nevada Mountains and were subdivided into three parts, each part grown in a different environment. The origin and altitude of the four types are as follows: top row, 10,000 feet at Timberline; second row, 4,600 feet at Mather; third row, 2,500 feet in the foothills; bottom row, 900 feet near the coast. The contrasting environments are at Stanford (1,100 feet), Mather (4,600 feet), and Timberline (10,000 feet). Each type, in general, grows best in the region from which it was originally collected, and less well at other locations. This indicates that each type is adapted to a particular locality.

Jens Clausen, Carnegie Institution of Washington

by mid-July (about 175 days). At sea-level, therefore, the plant is timed to a slow rhythm of seasonal development and a long season. At Mather the season is from mid-April to the end of July (about 105 days). At Timberline the season is from July into September (less than 90 days). Plants from each station were selected. Each plant was then subdivided into three parts, and then transplanted at each of the three stations. Figure 25.2 shows growth at each station.

The environment clearly altered the growth habit. The length of the flowering period was also changed, being lengthened at low altitudes, and hastened at high altitudes. However, each of the plants tended to retain certain characters that are constant, regardless of the environment; for instance, arrangement of the leaflets, openness or compactness of the plant, and the shape of the petals. This is also shown for two different species of yarrow growing at different altitudes (Fig. 25.3).

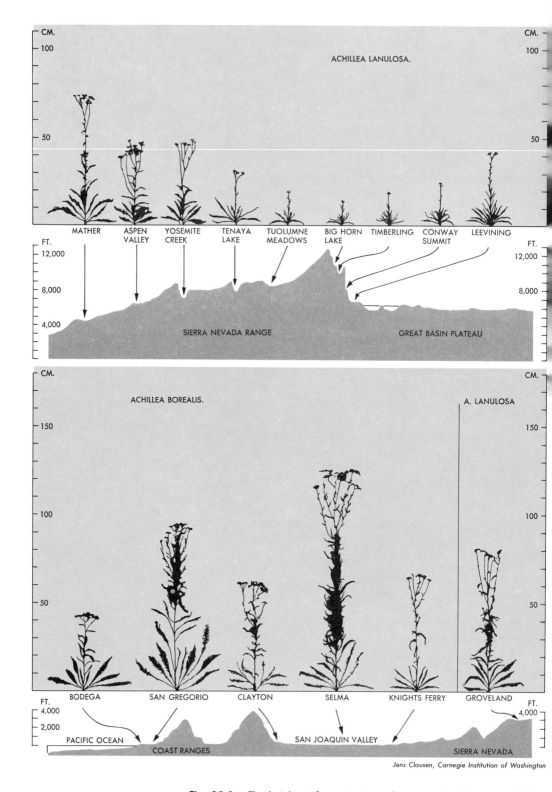

Fig. 25.3 The height at flowering time of two species of yarrow, *Achillea lanulosa* and *A. borealis*, indicates how altitude affects the habit of growth.

The examples we have considered demonstrate that a single genotype can exhibit different phenotypes when environmental conditions are varied. But, you may wonder, "Can two genotypes exhibit the same phenotype?" The answer is "Yes" when environmental conditions are varied. In *Drosophila*, for example, a mutant gene *y* (yellow) causes the body of the fly to be yellow instead of gray. But a wild-type gray fly, with the dominant *Y* gene, turns yellow if silver nitrate is included in its food. Such a fly is called a **phenocopy,** since it copies the phenotype of a mutant strain. The yellow color in this instance would not be transmitted to any offspring, because the genes are unaffected by the silver nitrate.

The same kind of thing happens in rabbits. Figure 25.4 shows three rabbits of different coat color. Inheritance tests show that the fully colored rabbit (bottom) has a dominant phenotype. Its genotype may be CC, Cc or C^h. The solid white individual is cc, while the Himalayan is either $C^h C^h$ or C^h. The order of dominance of these three alleles is $C \rightarrow C^h \rightarrow c$.

If we vary the temperature as a means of changing the environment, the phenotypes of the white and colored rabbits stay constant. Temperature seems to have no influence whatever on the genetic expression of C or c. In the Himalayan form, however, coloring pattern is temperature sensitive. Notice that the black hairs are located at the extremities—feet, nose, and ears. This pattern of distribution of black hair by itself does not necessarily prove anything about temperature sensitivity. Even though the body extremities—feet and ears—have a lower temperature (because of rapid heat loss) than the rest of the body does, we cannot accept this alone as proof of temperature sensitivity. However, if hair is plucked from the back, the new hair coming in is white if the temperature is high, but it is black if the temperature is low. That is the kind of proof needed in order for us to make a definite statement about temperature sensitivity, which in this instance can be related to the formation of **melanin,** a black pigment. The chemical steps leading to the production of melanin are catalyzed by a series of enzymes. One of the enzymes is temperature sensitive and cannot function at high temperatures. Melanin, therefore, can be produced only when the temperature is lowered, artificially in the case of the plucked hair, naturally in the extremities.

Can we always be so certain of how much of a role the genotype or environment plays in determining the phenotype? For example, the average American soldier of World War II was several inches taller and somewhat heavier than his counterpart in World War I. Amherst College has recorded the weight and height of its students since it was founded. Today's freshman

Charlie Ott From National Audubon Society

Walter Dawn

Walter Dawn

Fig. 25.4 Three color patterns in rabbits are shown here. Top: the white phenotype has a cc genotype. Bottom: the black phenotype has one of three possible genotypes, CC C^h or Cc. Middle, the Himalayan phenotype has either a $C^h C^h$ or C^hc genotype.

averages four inches taller and twenty pounds heavier than the freshman of a previous generation. Is this due to a change in the genotypes of American men, so that there are more genes favoring tallness? Or is it the result of better nutritional standards? How do we find out? We might suspect that height is genetically, not environmentally, determined, but can we be certain? Weight, on the other hand, is a character easily changed by diet. One of the best ways we have of answering such questions is to make use of twin studies.

Twin Studies

Twins are either **fraternal** or **identical**. If they are fraternal, they are **dizygotic (DZ)** arising from two separate eggs fertilized by two separate sperm. Despite their birth relationships, they are no more like each other than they would be like their other brothers and sisters born singly, except, of course, that they are of the same age. Identical, or **monozygotic (MZ)** twins, on the other hand, develop from a single fertilized egg. They are, consequently, of the same sex and have the same genotype (Fig. 25.6). Division of the single fertilized egg into two units does not occur at the first cleavage division, as was once believed, but only after the embryo consists of many thousands of cells. Complete separation of the cellular mass into two parts is necessary for twin formation. Incomplete separation leads to Siamese twin formation, with the union between the two being of varying degrees. As Fig. 25.5 shows, the degree of union often

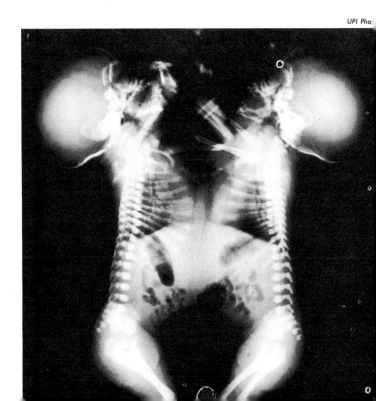

Fig. 25.5 An x-ray photograph of Siamese twins at birth. Joined along the chest and abdominal regions of the body, they exhibit various degrees of shared and separate internal organs and bone structures. An operation, in this instance, would not permit complete separation and survival of the twins.

interferes with full development, and a variety of abnormalities can result.

Among human twin births, which occur at a rate of about one in 80, the ratio of identicals to fraternals is about 1:2. Higher multiple births of triplets, quadruplets, and quintuplets also occur, but their occurrence is too rare to provide a ready source of individuals for study.

The important thing for us to remember here is that identical twins provide the geneticist with a measure of control over two important variables—inheritance and environment. Consequently, the most important source of information has been studies of identical twins who have been separated from each other early in life, and who have developed in quite different environments. Similarities exhibited by such identical twins would point to the independence of heredity from the environment in determining specific characteristics. Dissimilarities, on the other hand, do not minimize the importance of heredity. Instead, they indicate how the environment and heredity interact. They show us that certain inherited potentialities may be expressed one way in one environment, but quite another way in a different environment.

If you have ever seen identical twins, you know how very much alike they are physically. They are, of course, always of the same sex. Their height, weight, and build are strikingly similar (see Table 25.1). So, too, are other physical features: hair and eye color, texture and pigmentation of skin, structure and placement of teeth, and the shape and size of nose, eyes, ears, and mouth. These similarities also extend to less obvious

TABLE 25.1 Differences, expressed as percentages, between MZ twins reared together, MZ twins reared apart, and DZ twins (from J. Shields, Monozygotic Twins, Oxford Univ. Press).

Trait and intra-pair difference	MZ twins reared together (44 pairs)	MZ twins reared apart (44 pairs)	DZ twins (39 pairs)
Height			
under 1 inch	82	58	24
over 1 inch	18	42	76
Weight			
under 14 lbs.	68	64	40
over 14 lbs.	32	36	60
Intelligence			
0–4 points	44	32	14
5–9 points	21	30	14
10–14 points	23	14	14
15– points	12	24	57

Fig. 25.6 This diagram illustrates the origin of monozygotic (identical) and dizygotic (fraternal) twins at the time of fertilization.

DIZYGOTIC TWINS	MONOZYGOTIC TWINS
OCCUR WHEN 2 DIFFERENT SPERM FERTILIZE 2 DIFFERENT EGGS BY INDEPENDENT BUT NEARLY SIMULTANEOUS FERTILIZATIONS. THE ZYGOTES ARE GENETICALLY DIFFERENT FROM EACH OTHER.	OCCUR WHEN A SINGLE SPERM FERTILIZES A SINGLE EGG...
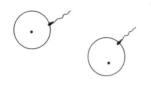	
	WHICH THEN DIVIDES SOME TIME DURING EARLY DEVELOPMENT,...
EACH FERTILIZED EGG DEVELOPS INTO A SEPARATE INDIVIDUAL, BUT WITHIN THE SAME UTERUS.	GIVING RISE TO TWO GENETICALLY SIMILAR INDIVIDUALS IN THE SAME UTERUS.
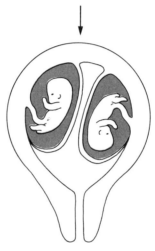	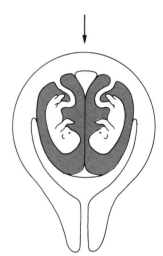

features: blood type, fingerprint pattern, pulse and respiration rates, and blood pressure.

You are also well aware of the fact that fraternal twins often look much alike, but no more so than their brothers or sisters born singly. Their likeness, of course, is due to the fact that they have the same parents, and consequently have many of the same genes. However, the segregation of genes during meiosis, in preparation of eggs and sperm formation, makes it highly unlikely that fraternal twins would have the same genotype.

Table 25.1, however, indicates that MZ twins (monozygotic) reared together are more alike than are MZ twins reared apart. Environmental influences, therefore, are exerted on virtually all characteristics. We must remember that the environment is never the same for any two individuals, however similar their genotypes may be.

If we now consider such hard-to-define characteristics as intelligence, ability, or personality, the influence of the environment appears to be greater. Let us concentrate on intelligence since we all are aware of the existence of I.Q. tests. Intelligence is not a single, clearly defined character. It is a composite of many attributes, all of which provide a basis for rational behavior. But we recognize that differences do exist between individuals, and I.Q. tests, while not wholly reliable (since it is difficult to rule out the role of experience), can provide usable estimates of these differences.

The table provides information about both MZ and DZ twins. It is clear that MZ twins reared together have less variation in their I.Q. scores than similar twins reared apart. Fraternal twins and paired siblings, however, are still wider apart in score than either of the MZ groups. Additional tests designed to measure personality factors and several kinds of ability show the same trends. It is clear, however, that experience, education and the general environment play a greater role in determining intellectual and social traits than they do physical characteristics.

Of some interest is a study made by Johannes Lange, a German professor, on heredity and criminal tendencies in MZ and DZ twins. He selected those twins, one member of which had a criminal record. Among the MZ twins, most of whom had been reared together, 10 of the 13 pairs showed both members with criminal records, and generally for the same kind of crime. Of the DZ twins, only 2 of the 15 pairs revealed tendencies in both members. It is, of course, difficult to determine the relative weight to be given to heredity and environment in these cases. However, it seems clear that certain social and behavioral responses have an inherited basis, and that the environment will determine how these responses will be expressed.

SUMMARY

The **phenotype** of an organism is the result of a given **genotype** developing in a given environment. The contribution of the environment vs. that of heredity can be assessed through the study of monozygotic and dizygotic twins. The results show that many traits—physical, behavioral, and intellectual—have a strong genetic basis, but that the environment does play a role. The genotype determines the limits of expression of a given trait, but the environment also determines it to a certain extent.

FOR THOUGHT AND DISCUSSION

1 Assume that each of the 23 chromosomes a man inherited from his mother are different from the 23 he inherited from his father. How many genetically different kinds of sperm

can he form? What is the probability that a sperm will contain all of the chromosomes derived from the paternal parent? What does this information tell us about the possible variability found among children in the same family? If striking likeness shows up among all of the children, what conclusions can be drawn?

2 Baldness in human beings is genetically determined. What possible mechanisms or differences in internal or external environment can be responsible for the relative absence of baldness in women?

3 Tuberculosis is an infectious disease caused by a bacterial organism, yet it tends to characterize certain families as though it were inherited. What can you say about the possible environmental and hereditary factors responsible for the expression and distribution of the disease?

4 Some kinds of cancer are hereditary. Does this mean that a gene is directly responsible for the disease?

5 Houseflies and mosquitoes, which were once easily controlled by an insecticide such as DDT, are now often found to be resistant to the insecticide. Would you think that this change in susceptibility is hereditary or environmental? Explain.

6 Parents of a set of girl twins have the following genotypes: MmNNOoPpRR and MmNnooPPRr. Both twins show the same phenotype for these genes and are very much alike in all other respects. Assuming that the genes are inherited independently of each other, what is the probability that the twins are dizygotic?

7 Diabetes is a lethal trait if left untreated. Is the gene responsible for the lack of insulin a lethal gene? Explain.

8 Inbred strains of animals such as mice have long been used in testing the effects of new drugs. What is meant by "inbred," and what are the genetic consequences of inbreeding? Why should the investigator insist on inbred strains instead of any random group of mice?

9 Do you suppose that Mendel could have derived his laws of segregation from a study of characters which varied widely in different environments? Explain.

10 If a trait is expressed in both members of a pair of monozygotic twins, can one be certain that the trait is genetically determined? Explain.

11 If the changing height and weight of Amherst College freshmen is not due to a changed pattern of inheritance, how can you account for the differences environmentally?

12 The Napoleonic Wars were said to have had a marked effect on the average height of Frenchmen. From your knowledge

of history can you suggest a possible reason for this? In time of war and with heavy loss of life among members of the armed forces, do you think that the present selective service draft procedure—which is an environmental factor of a sort—can alter the genetic structure of a population? If so, in what way?

SELECTED READINGS

Auerbach, C. *Genetics in the Atomic Age.* New York: Essential Books, Inc., 1956.
An elementary discussion of chromosomes and genes, and how radiation can affect them.

Auerbach, C. *The Science of Genetics.* New York: Harper & Row, Publishers, Inc., 1961.
An excellent and clear description of the basic principles of genetics.

Baker, W. K. *Genetic Analysis.* Boston: Houghton Mifflin Co., 1965.
A difficult but excellent book concerned with the analysis of genetic results.

Bonner, D. M., and S. E. Mills. *Heredity.* Englewood Cliffs, N.J.: Prentice-Hall, Inc., Foundations of Modern Biology Series, 1964.
More concerned with the physical nature of the gene, but containing much good material on inheritance.

Brewbaker, J. L. *Agricultural Genetics.* Englewood Cliffs, N.J.: Prentice Hall, Inc., Foundations of Modern Genetics Series, 1964.
Contains a good deal of information on basic Mendelian inheritance as well as data relating to inheritance in domesticated plants and animals.

McKusick, V. A. *Human Genetics.* Englewood Cliffs, N.J.: Prentice Hall, Inc., Foundations of Modern Genetic Series, 1964.
A readable volume on the methods and results of genetic studies in humans.

26 DEVELOPMENT— AN INHERITED PATTERN

Our study of identical twins has shown that two individuals with the same genotype, developing in the same environment, will be strikingly similar to each other in appearance and behavior. If you stop to think about this, you should soon realize that it is an extraordinary thing. Consider what it means.

Identical twins are separated from each other at a very early stage in their development. At this time each twin consisted of only a few hundred or a few thousand cells, and all of the cells were very much alike. From then on, the two individuals developed independently: from prenatal life, through childhood and adolescence to adulthood. Yet throughout the course of development, a striking similarity was retained. This must mean that both twins follow a precise pattern of development, step by step, as they proceed from the egg to adulthood; the fertilized egg must contain the information that determines this parallel behavior. Development, therefore, is the realization of an inherited pattern: it brings to fulfillment the potentiality present in the egg at the time of fertilization.

There is no resemblance between an adult and the egg that gave rise to the adult. In size the human egg is no larger than a speck of dust. It weighs about one-millionth of a gram; and the sperm, when fertilizing it, adds only another five-billionths of a gram. If you examine an egg under the light microscope, you will see little visible structure: a clear membrane surrounding the egg, fat droplets (yolk), and a nucleus within. There is no hint among these details of what the egg can become.

The fertilized egg, in addition to being a single cell, is also an organism. Even though it is relatively simple and undeveloped, it is very much alive and completely coordinated. The adult, on the other hand, is one of the most complex structures known, consisting of billions of cells. If the pattern of cell division were one cell to two cells, two cells to four, four to eight, and so on,

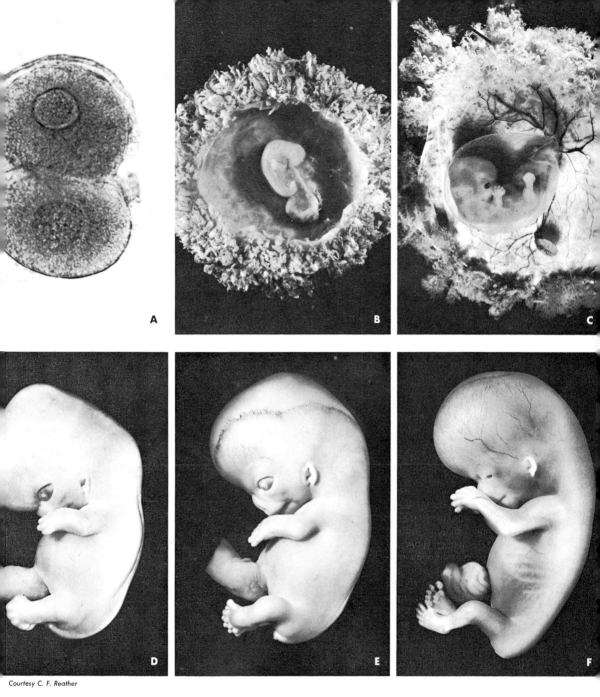

Courtesy C. F. Reather

Fig. 26.1 This sequence of photographs shows the development of a human embryo from the two-celled stage to an embryo 56 days old. A, two-celled stage; B, 28-day embryo with chorion removed; C, 39-day embryo with chorion laid open but with amnion (inner membrane) intact; D, 40-day embryo, with eyes, ears and limbs evident; E, 44-day embryo; F, 56-day embryo. The umbilical cord, or placenta, which attaches the embryo to the wall of the mother's uterus, and through which the embryo obtains its nutrition, is seen in figures I and J.

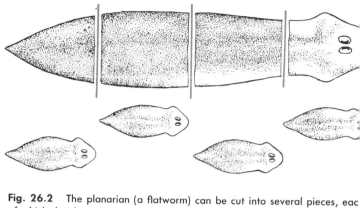

Fig. 26.2 The planarian (a flatworm) can be cut into several pieces, each of which develops into a complete individual. This is an example of asexual reproduction.

Richard F. Trump

Fig. 26.3 Examples of asexual reproduction. The geranium stem (above) which has been rooted in damp sand and the strawberry plant (at right) are examples of asexual reproduction in the plant world. The strawberry plant develops new individuals by means of "runners." The parent plant (1) forms a lateral stem, or runner, which roots when it comes in contact with the ground, at which points new plants develop (2 and 3). These can be detached and grown as separate individuals.

it would take about 45 cellular divisions to produce a human adult. But a human adult could never be formed by a group of similar cells, no matter how they were organized. As our body cells form, they also become different. This is the process which gives character and uniqueness to us as individuals. Some cells form bone, others muscle, still others skin, kidney, liver and other organs and tissues of the body. Development is the result of many coordinated processes: **growth,** which leads to an increase in mass and is generally accompanied by cell division, **differentiation,** a process by which similar cells become different in structure and function; and **integration,** or **regulation,** which is really growth and differentiation coordinated in time and space (Fig. 26.1).

The problems of development are, therefore, cellular problems. Before we look into this, however, we might ask a simple question. Why should development begin with a single cell, an egg? The animal could release a highly developed portion of itself which could then grow into a new adult form. As Figs. 26.2 and 26.3 show, this happens in certain flatworms, and it is a chief means of reproduction in many plants. But the fact that sexual reproduction is so widespread would suggest that it has

Richard F. Trump

certain advantages over asexual and vegetative reproduction.

One advantage of sexual reproduction is that, through fertilization, the characters of the two parents can be combined in the offspring. New combinations of genes can be tried. Such a trial would be much more difficult if asexual reproduction were the rule. Furthermore, fertilization could not occur in an organized way if a multicellular body were involved. A single cell (an egg), therefore, is needed. The sperm cell brings to the egg not only a set of genes, but also an activating mechanism which stimulates and initiates the whole process of development.

Growth

Cells can do one of three things: 1. They can grow and divide; 2. They can differentiate; or 3. They can die. All of these aspects are included in the general term "development." We can think of growth as an increase in mass. Since we are dealing with cells, growth can occur by an enlargement of cells, but is usually accompanied by cell division as well. Sometimes, however, early development may not involve growth as an increase in mass although cell division is very much a part of the pattern of change. During the early **cleavage** stages in the development of an embryo, there is active cell division, but the mass of the embryo remains the same. This is why it is called "cleavage;" the egg is cleaved into smaller and smaller units by cell division. The individual cells must, therefore, become smaller as their number is increased (Fig. 26.4).

As the number of cells increases, the mass of cells takes the form of a hollow sphere. This is the **blastula.** No particular change in shape or size has occurred, and there is no obvious advance toward an adult form. If we were to cut the blastula of a frog in half, however, we would find several features not visible before. There are cells of many different sizes. The **dorsal** or upper cells, for example, are larger than the **ventral** cells. Also there is a mass of still smaller cells (see Fig. 26.5) at the ventral side. The region between the large and small cells will be the point of change, or **gastrulation,** which eventually will lead to the sphere changing into a tadpole, then into a frog.

Gastrulation is a period of cell movement. The embryo becomes completely reorganized. By marking the blastula so that selected regions can be recognized later, it is possible to see where certain cells take up their final residence, and what they eventually give rise to (see Fig. 26.6). In a general way, all of the small cells of the ventral pole go to the interior of the embryo, while the larger cells of the dorsal pole spread over the outside of the embryo. This is accomplished by a movement of cells.

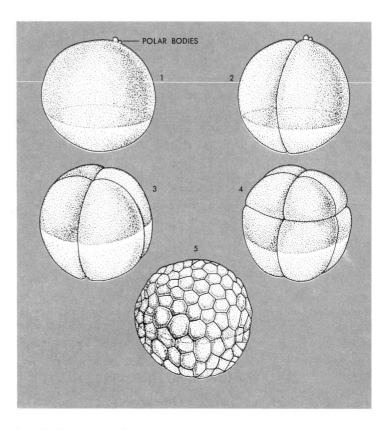

Fig. 26.4 These early cleavage stages in the development of the frog show the one-, two-, four-, and eight-cell stages, and the external appearance of the early blastula.

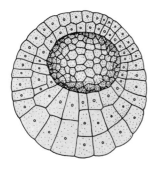

Fig. 26.5 A section through the blastula of a frog shows the central cavity, the larger ventral cells, which are filled with yolky materials for nutrition, and the smaller dorsal cells.

An opening forms in the ventral region, and the cells turn in at this point, moving internally. The opening which began as a slit eventually becomes a full circle, with the cells which were once external being rolled inwardly. The opening surrounds the mass of the **yolk**, which is finally enveloped, leaving a tiny slit. This becomes the anus of the developing frog. In the meantime, the yolk is being absorbed by the growing organism.

Gastrulation, therefore, has given the embryo shape and organization by a rearrangement of cells. Cell division continues at the same time. The reorganization continues, and the gastrula becomes transformed into a **neurula** (see Fig. 26.7). This is characterized by the formation of the **neural folds** on the top of the embryo. Continued growth and unfolding, followed by a fusion of the two sides of the fold, produces the **neural tube.** This structure later develops into the brain and spinal cord.

At the end of about four days, the single fertilized frog egg has become a recognizable tadpole. As Fig. 26.8 shows, the main internal organs are beginning to form while externally the tadpole has gills with blood running through them, a tail, an olfactory (smelling) organ, and identifiable regions for eyes and mouth. In this condition it is released from the egg membranes and becomes a free-swimming tadpole. The obvious dramatic changes are over. The remainder of the growth period is that of filling in the details. To achieve the same stage of development as that described for the frog, a human embryo requires about a month.

The filling in of details is, itself, equally interesting, but is of a more localized nature as one or another organ is formed. One of the facets of this period of development is that some parts grow faster or slower than others; furthermore, the relative rates of growth of different parts and regions differ from time to time. At two months of fetal age, the human head is equal in length to the rest of the body. Arms grow faster at an early period than do legs. The body, on the other hand, grows at a steadier rate until maturity is reached. In this way the organism reaches its adult form (Fig. 26.9).

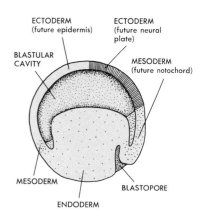

Fig. 26.6 This section through an embryo at the beginning of *gastrulation* shows the cells beginning to move inward by a process of *invagination*. The regions of the gastrula are labeled to indicate that particular groups of cells will eventually contribute to the formation of particular parts of the body as development continues. Compare this with Fig. 26.10.

Fig. 26.7 The formation of the neural tube and the outward appearance of the neurula of the frog are shown here. Above, the folding of the ectoderm and eventual pinching off of the neural tube; below, successive stages of neurula formation as viewed externally.

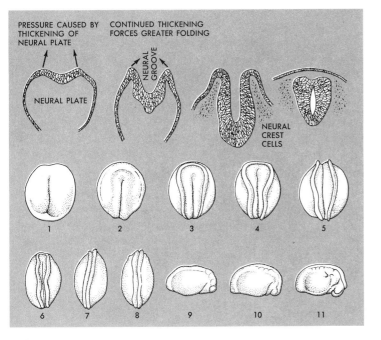

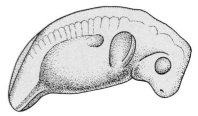

Fig. 26.8 External appearance of a frog embryo shortly before it is released from the egg membranes and becomes a swimming tadpole.

Fig. 26.9 Relative sizes of different portions of the human body are shown in this diagram as development proceeds from the early fetal stage to adulthood.

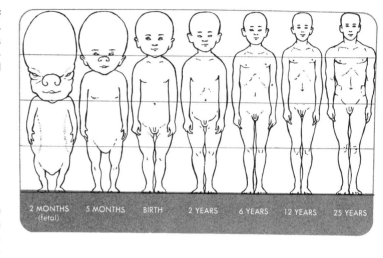

Fig. 26.10 A single cell in an early embryonic stage is capable of becoming a nerve, muscle, liver, or kidney cell. What it becomes depends on its place of origin in the embryo and the local environment among other things. The chart shows the time course of development, during which a single cell (the egg) becomes an organized mass of different kinds of cells as the result of cell division and differentiation.

Differentiation

As the embryo grows, cells are increasing in number, invaginating, and regrouping. Each change in the embryo takes place at a certain time, gradually altering a formless mass of cells into an organism of distinctive characteristics. In the frog, some of these changes can be followed with the naked eye. But other, less obvious events are also occurring and giving the embryo character and individuality. Internal as well as external organs are being formed. Within each, individual cells are beginning to assume the shapes and structures related to their adult function. This is the process of **differentiation**. We can think of this process occurring at the level of the organism itself, at the level of an organ, or at the cellular level. No matter what the level, differentiation places a stamp of uniqueness on the whole organism or its parts.

Basically, differentiation is genetically controlled, but environment also plays a role. For example, the cells destined to be-

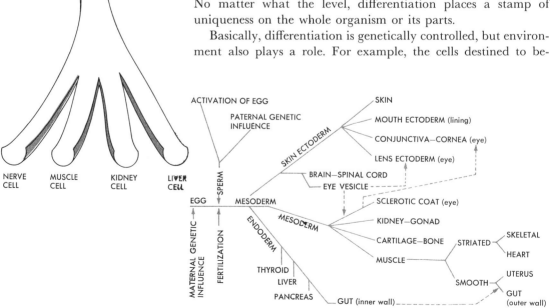

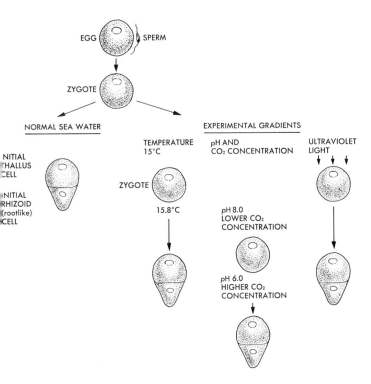

Fig. 26.11 The *Fucus* (brown alga) egg has its polarity or direction of development determined by a number of environmental variables. The rhizoidal cell forms in the warmer region, where the pH is lower and the CO_2 concentration higher, and on the side opposite from a source of ultraviolet light.

come part of the liver, heart, or kidney are determined by origin, position, and local environment, even though each cell has the same set of genes as all other cells (Fig. 26.10). The effect of specific environments can be seen acting on the development and differentiation of a fertilized *Fucus* egg (see Fig. 26.11). After the first division, one of the two cells will form the thallus, the other the rhizoid. The rhizoidal end of the undivided egg is indicated early by the formation of a small protuberance. But as the illustration indicates, the position of the protuberance is determined by light, temperature, and *p*H. Once the protuberance develops, the developmental pattern of the embryo is set.

At the cellular level, differentiation causes cells to lose their generalized form. Muscle cells begin to elongate and acquire contractile fibers (Figs. 26.12 and 26.13). Their tendency to do so is established chemically before the eye can detect it visually, and is beautifully demonstrated by tissue culture techniques. Heart cells change internally and begin a rhythmic beating that will continue throughout the life of the organism. A simple cell thus becomes a specialized cell. In doing so, the differentiated cell loses its ability to do other things as its specialized ability increases. For example, the more specialized the cell, the less likely it is to divide. In any individual cell, therefore, growth and differentiation appear to be mutually exclusive states of

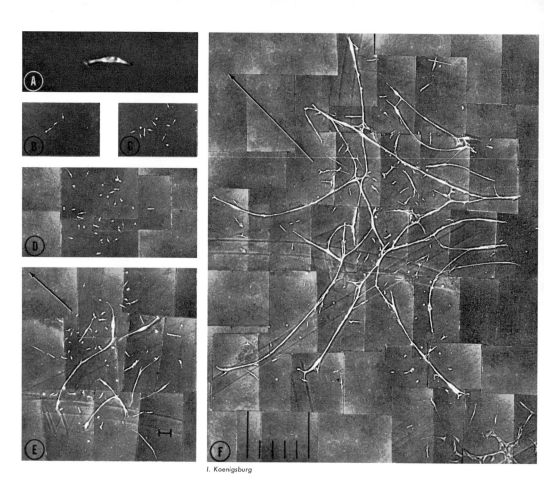

I. Koenigsburg

Fig. 26.12 Muscle cells (myoblasts) can be grown in tissue culture. A through G show the same culture at various time intervals. A shows the single cell before division, which gives rise to all other cells in the culture. B, C, and D show an increase in the number of cells through division. E, F, and G show the same culture at later stages, after the cells have aggregated into bundles and differentiated into striated muscle cells (see Fig. 26.13).

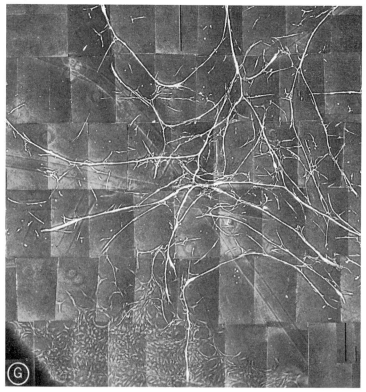

E. Leitz, Inc.

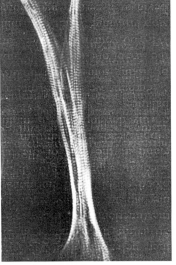
I. Koenigsburg

Fig. 26.13 When shown at higher magnifications, muscle cells from the culture in Fig. 26.12 reveal the cross-striations that develop and enable the cells to contract.

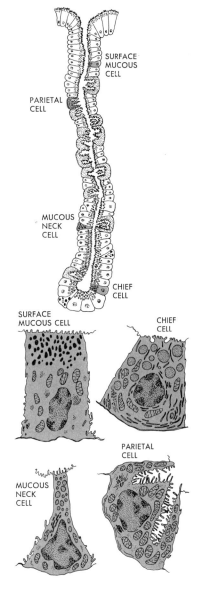

Fig. 26.14 The diagram below shows four kinds of cells found in the lining (gastric mucosa) of the stomach of a bat. Two of the cells form mucus (polysaccharide). The parietal cell forms hydrochloric acid, and the chief cell forms enzymes (protein). All of these substances participate in the digestive processes of the stomach.

activity. Growth involves the ceaseless formation of new cells, while differentiation makes specialists out of various cells of this general mass. Figure 26.14 shows the result at a cellular level. Each of the cells of the gut has a different function, and each has the necessary internal structures needed to perform that function.

We are now faced with the problem of trying to explain *why* differentiation occurs. The problem in its simplest form is this: the exactness of DNA replication and of chromosomal segregation in mitosis makes it reasonably certain that all somatic cells in an organism will have the same genotype. Then why do they come to have different phenotypes? Let us take three cells in the gut concerned with digestion: **mucous** cells, **parietal** cells, and **chief** cells of the **gastric mucosa** (stomach lining). Fig. 26.14 shows that they are quite different in internal structure, even though they have a common origin from a group of deeper cells. Their functions are also different. They produce, respectively, mucus, hydrochloric acid, and pepsin as their major product. Their enzymes must, consequently, be different.

As we have already learned, enzymes are proteins, and genes are responsible for the production of proteins. Therefore, if cells of the same genotype have different kinds of proteins (enzymes are only one kind of protein that can differ), cellular differentiation must be due to *differential gene action*. In other words, not all genes are active all of the time; those that are active determine the phenotype of the cell. We do not know what determines when a gene will be active or inactive, but we do know that genes respond differently to different environmental conditions.

In the bacterium, *Escherichia coli,* for example, there is a gene that controls the formation of an enzyme called **tryptophan synthetase.** The function of this enzyme is to catalyze a reaction which combines two chemicals, **indole** and **serine,** into **tryptophan,** an essential amino acid. If the colon bacterium is grown in a medium containing tryptophan, no enzyme is formed. If tryptophan is removed from the medium, the enzyme is immediately synthesized. From these observations we can conclude that the gene controlling enzyme formation is inactivated in the presence of tryptophan, and is active in its absence.

Another gene in *E. coli* controls the formation of a different enzyme, **beta-galactosidase.** Its function is to break down a sugar, **galactose,** so that it can be used as a carbon source within the cell. When galactose is present, the enzyme is being formed; when galactose is absent, so, too, is the enzyme.

Here, then, are two genes whose activity depends on the presence or absence of a particular chemical. But their activities are opposed to each other. In one instance, tryptophan inhibits enzyme formation; in the other, galactose promotes enzyme formation. The two examples, on the other hand, show that environmental differences can act on a uniform genotype to evoke different phenotypic responses.

Are the gene inhibitions of a temporary nature as the ones discussed appear to be? Or can they be of a more permanent type? At least a partial answer can be given by certain experiments performed in the frog. It is possible surgically to remove a nucleus from one cell and transplant it to another cell. If the cells are of the same sort, say from a blastula, nothing unusual happens. However, let us now transplant nuclei from progressively more differentiated cells and put them into an egg which has had its own nucleus removed. Will these nuclei permit the egg to continue its development? If nuclei from blastula cells are so transplanted to an egg, the egg proceeds with development and everything is normal. If nuclei from cells of the gastrula are similarly transplanted, the egg will develop through the blastula stage, but stop development abruptly at the onset of gastrulation. Nuclei from more differentiated cells, such as from adult liver, will cause a similar change at gastrulation. It would appear, then, that nuclei as well as cytoplasm can become differentiated.

This is very beautifully demonstrated in the larvae of *Drosophila.* As Fig. 26.15 shows, the chromosomes in the cells of the salivary gland are very different in character from those of ordinary somatic cells, but they are also found in other portions of the larval body—in the gut and rectum, for example. In each type of cell, and at different times during larval develop-

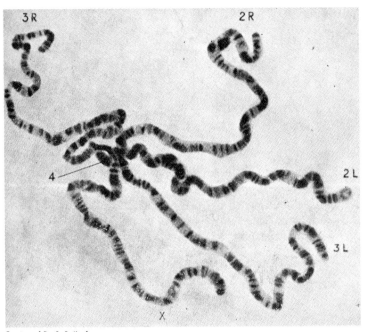

Fig. 26.15 The chromosomes of *Drosophila melanogaster*, as they appear in the cells of the salivary gland. Each chromosome can be recognized by a distinctive banded structure. The chromosomes are united to each other at their centromeric regions. Chromosomes 2 and 3 have median centromeres, and their right and left arms are identified. The X chromosome and the tiny 4 do not have two arms.

ment, the chromosomes have a characteristic appearance. Puffs (see Fig. 26.16) appear or disappear. It is now believed that the appearance of a puff means that the gene at that location is active, and is producing RNA. Genes, therefore, can be active or inactive at different times, and the environment as well as the stage of development determines how they behave. As a very rough guess, it is likely that in a given cell at a given time, only one-tenth of the genes are active. The remainder will be active in other cells and at other stages of development or activity.

Fig. 26.16 A portion of a single giant chromosome from a midge, *Rhyncosciara angelae*, appears differently at different stages of larval development. The arrows indicate comparable bands in each of the figures. When a particular region puffs out and becomes diffuse in appearance, it is believed that the gene or genes in that region are active in making messenger RNA.

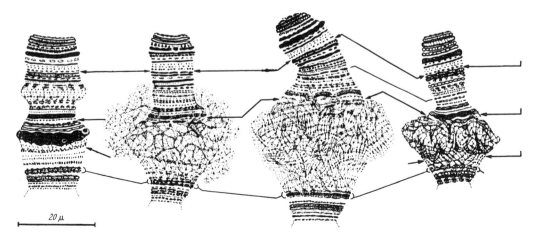

Integration

The problems of growth and differentiation at the cellular level are among the most intriguing facets of modern biology; development in its total aspect, however, is still more challenging, but also more perplexing. The whole course of development is characterized by a unity and a harmony of structure and behavior and maintenance. The egg, in its own way, is a complete and total organism. So, too, are seeds, embryos, larvae, and pupae. They are as complete as the mature individual to which they will give rise in the sense that they are fully functional entities developing as a whole and not simply as a collection of parts.

For lack of a better word, we can refer to this harmony of existence as **integration** even though we do not understand how it is maintained. So far as we know, life is a series of chemical reactions taking place within an organized structure. Integration therefore, must begin at the molecular and cellular level, and then be expressed at the level of tissue, organ, and organism. Structure and behavior at all levels must be compatible with the functioning of the whole organism. Among the integrative mechanisms are those relating to our sense organs, nervous system, blood supply, and hormonal condition. For example, an inadequate supply of growth hormone produced by the pituitary gland at the base of the brain, leads to dwarfism; an oversupply leads to gigantism. The body, at maturity as well as during development, is maintained by a series of checks and balances.

Perhaps we can best point out the significance of integration by asking some as yet unanswered questions. What determines the size of one organ as compared with the size of other organs? Why do your nose, fingers, and toes reach a certain length and stop growing? What determines "mature" size? Why does aging begin at a certain age? Some of these aspects of development may be answered at a cellular level, but others may involve higher levels of integration. If some day we can comprehend these aspects of regulative control, we may be well on our way to an understanding of the biology of cancer and aging.

SUMMARY

An organism inherits from its parent or parents a set of cellular structures capable of performing or directing certain chemicals reactions. **Development** is the process whereby a fertilized egg or a group of cells is transformed into a recognizable organism. Both the genes and the environment participate in determining the direction of development.

Development consists of three major phases: 1. **growth,** 2. **differentiation,** and 3. **integration.** Growth is an increase in mass. It includes cell enlargement or cell division, or more commonly, it includes both proceeding together. Differentiation is a process which transforms an unspecialized cell into a cell having a special structure and function. The more complex the organism, the greater is its need for specialized cells and organs, and the greater is the degree of differentiation. Integration, or regulation, is a term which describes all of the control mechanisms which keep the organism functioning as a unit.

FOR THOUGHT AND DISCUSSION

1 Suppose that fertilization, instead of uniting single cells and forming a zygote, involved a mass of cells each of which had to be fertilized by a single sperm. What difficulties would be encountered? What would be the genetic consequences?

2 As you found in an earlier chapter, immortality was lost when sexual reproduction became a way of life for most organisms. What does this statement mean to you? Think deeply. What are the philosophical implications?

3 A chicken is said to be an egg's way of making more eggs. Does this statement make biological sense? Does it make sense if it is applied to you as an individual?

4 Be certain that you understand the difference between "growth" and "differentiation." What would happen if a zygote of a frog or a daisy underwent one process but not the other? Can you think of any reasons why the cells of an organism

should not divide continuously until the adult number of cells were attained, and then go through the processes of differentiation all at once?

5 Cancer may develop in only one organ of the body, but the cancerous cells may spread and invade other organs. Why don't normal cells do this?

6 You fall and skin your knee. The skin grows back and covers the wound, but only just covers it. What biological processes are involved in the repair job?

7 We have implied that chemical differentiation must precede visible morphological differentiation. Why?

8 Think back to the chapter on chromosome structure and ask yourself, "Can the proteins of the chromosome play some role in determining whether genes are active or not in a differentiated cell?" Can you think of some way to test this question?

9 There is an insect hormone, "ecdysone," that controls molting. Remembering that *Drosophila* salivary gland chromosomes undergo puffing when particular genes are active, can you devise an experimental plan to test for the effects of the hormone on genes?

10 The developing limb bud of a frog contains recognizable cells of different kinds: those that will eventually form skeleton, muscle, inner skin (dermis), and outer skin (epidermis). If the limb bud is removed and the cells are separated from each other (this can be done with the enzyme trypsin) and randomly dispersed in a cell culture, they will reaggregate and form something like a limb bud, with each particular cell in its right place—skeletal cells inside the mass, then muscle, dermis, and, finally, epidermis on the outside. Cell position is an example of a regulatory process. From the way the cells behave, regulation is cellular and is not lost by dissociating the cells. Can you give any reasons for this regulation? Use this problem to illustrate how you would go about understanding this kind of regulation in a scientific manner. What are your facts? What hypotheses can you develop? How can you test them?

SELECTED READINGS

Corner, G. W. *Ourselves Unborn.* New Haven: Yale Univ. Press, 1944.
 A classic book dealing with the development of the human individual.

Medawar, P. B. *The Uniqueness of the Individual.* New York: Basic Books, Inc., 1957.

A discussion of the interaction of heredity and development in the determination of individual uniqueness.

Moore, J. A. *Heredity and Development.* New York: Oxford Univ. Press, 1963.

A readable account of the interrelations of hereditary and developmental processes.

Sussman, M. *Growth and Development.* Englewood Cliffs, N.J.: Prentice-Hall, Inc. Foundations of Modern Biology Series, 1964.

An elementary account of the processes of growth and development in a wide variety of organisms.

27 THE EVOLUTION OF INHERITED PATTERNS

So far in this book, we have discussed a number of basic concepts in biology: the cell theory, cellular structure and behavior; the chemistry of biological reactions and their control; modes of inheritance and the physical basis of heredity; and the realization of inherited patterns of development. Let us now try to examine them from the points of view of **unity** among organisms, **diversity** among organisms, and **continuity** (Fig. 27.1).

As we stressed earlier in this book, the cell is the basic unit of organization. The individual cells of a tree, a cow, and a person are more alike than unlike. Only the viruses represent a major departure if we consider them as organisms. Nearly all cells fall within definite size limits, and as a rule they contain a single nucleus in a mass of cytoplasm bounded by a plasma membrane. In most cells the nucleus also is bounded by a membrane. Only the bacteria and the blue-green algae depart from this general arrangement.

Fig. 27.1 An example of continuity through time is shown by a comparison of a fossil lobster claw (left), which is millions of years old, and the claws of the common lobster, *Homarus vulgaris*. The claws are not greatly different from each other.

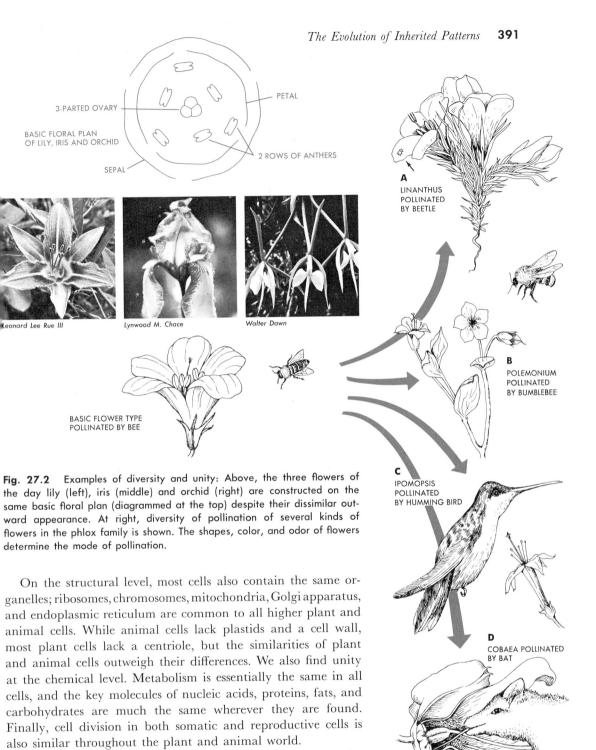

Fig. 27.2 Examples of diversity and unity: Above, the three flowers of the day lily (left), iris (middle) and orchid (right) are constructed on the same basic floral plan (diagrammed at the top) despite their dissimilar outward appearance. At right, diversity of pollination of several kinds of flowers in the phlox family is shown. The shapes, color, and odor of flowers determine the mode of pollination.

On the structural level, most cells also contain the same organelles; ribosomes, chromosomes, mitochondria, Golgi apparatus, and endoplasmic reticulum are common to all higher plant and animal cells. While animal cells lack plastids and a cell wall, most plant cells lack a centriole, but the similarities of plant and animal cells outweigh their differences. We also find unity at the chemical level. Metabolism is essentially the same in all cells, and the key molecules of nucleic acids, proteins, fats, and carbohydrates are much the same wherever they are found. Finally, cell division in both somatic and reproductive cells is also similar throughout the plant and animal world.

Unity also exists at higher levels of organization. Despite their differences in color, shape, and size, flowers are built on the same basic plan (Fig. 27.2); unity also characterizes the structure of their stems, roots, and leaves. The arm of a man, the fore leg of a dog, and the wing of a bird are again the

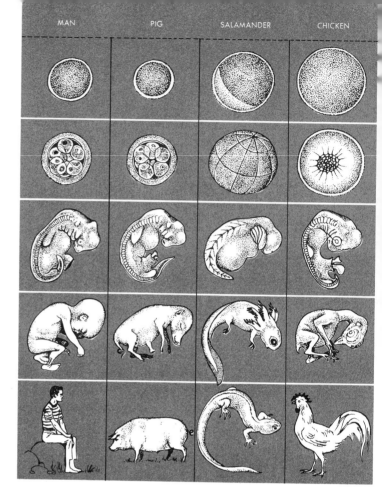

Fig. 27.3 There are striking similarities, as well as dissimilarities, in the development of the four vertebrates shown here. The diversities become more apparent as the embryo progresses toward adult form.

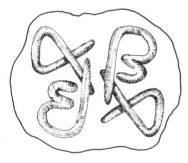

Fig. 27.4 Diagrammatic representation of the hemoglobin molecule of man which consists of four polypeptide chains linked together in the center. The Greek letters alpha (α) and beta (β) indicate the positioning of the *alpha* and *beta* polypeptide chains. The heme and iron (Fe) portion of the molecule is represented by the central dot.

same, despite their differences. We emphasize this by referring to them as **homologous** structures. The science of comparative anatomy, founded by the great French naturalist, Georges L. Cuvier (1769–1832), deals with this kind of information, and reveals essentially how these structures are related. Furthermore, we can see by examining the developmental stages of these animals that these structures arise from similar sites within the embryo, and in the same sequence. It is for this reason that the embryos of related animals show a marked similarity (Fig. 27.3)

Diversity

If we now shift our view from unity to diversity, we are confronted by a bewildering array of differences. Diversity exists at all levels. Table 27.1 shows that the nucleotide composition of DNA from various organisms varies widely. On another scale, this can be demonstrated within a particular molecule, hemoglobin, which is made up of two identical halves, each half consisting of an alpha (α) and a beta (β) peptide chain. These four polypeptide chains make up the protein portion of the molecule called **globin** (Fig. 27.4). In addition, the

TABLE 27.1 Base Ratios of Several Well-known Organisms
(values are arbitrary but accurate as ratios)

	Adenine	Thymine	Guanine	Cytosine	Ratio of $\dfrac{A+T}{C+G}$
Man	29.2	29.4	21.0	20.4	1.53
Sheep	28.0	28.6	22.3	21.1	1.38
Calf	28.0	27.8	20.9	21.4	1.36
Salmon	29.7	29.1	20.8	20.4	1.43
Yeast (fungus)	31.3	32.9	18.7	18.1	1.19
Staphylococcus	31.0	33.9	17.5	17.6	1.85
Pseudomonas	16.2	16.4	33.7	33.7	0.48
Colon bacterium	25.6	25.5	25.0	24.9	1.00
Vaccinia virus	29.5	29.9	20.6	20.0	1.46
Pneumococcus	29.8	31.6	20.5	18.0	1.88
Clostridum	36.9	36.3	14.0	12.8	2.70
Wheat	27.3	27.1	22.7	22.8	1.19

center of the molecule consists of four **heme** groups linked together around an iron (Fe) atom.

Hemoglobin A is the most prevalent type in humans. Its general structure may be written as

$$\frac{_\alpha A}{_\beta A} \cdot \frac{_\beta A}{_\alpha A}$$

Variations, however, can occur in the polypeptide chains. One such variation is **hemoglobin S**. The change is in the β polypeptide, and the homozygous and heterozygous types, respectively, are

$$\frac{_\alpha A}{_\beta S} \cdot \frac{_\beta S}{_\alpha A} \quad \text{and} \quad \frac{_\alpha A}{_\beta A} \cdot \frac{_\beta S}{_\alpha A}$$

Individuals possessing these hemoglobins are anemic (**sickle cell anemia**, so called because the red blood cells are sickle-shaped), and the homozygous individuals die early. Only one amino acid has been changed to alter hemoglobin A to hemoglobin S. There are about 560 amino acids in the hemoglobin molecule, yet the change of one amino acid can mean the difference between a normal and an anemic condition.

On the level of organisms, diversity is more obvious. We can readily tell plants from animals. We can also distinguish various kinds of mammals, and various breeds of dogs. Diversity enables us to distinguish not only among races of human beings, but also among individual human beings. In fact, of the three billion humans inhabiting the planet, no two are *exactly* alike (Fig. 27.5). Even identical twins have their minor dissimilarities of structure, behavior, and ability. If we now realize that there are about two

Fig. 27.5 Of the many people in this crowd, no two are exactly alike.

Ewing Galloway

million known different species of plants and animals, the magnitude of the diversity comes home to us.

Organic life is, therefore, characterized by both unity and diversity. Which is the more prominent depends on our point of view. At times these two aspects of life seem to oppose each other, but many things said in this book tell us that unity and diversity have a common basis. Heredity and environment, for example, are the sources of both unity and diversity in identical twins. Although unity implies common ancestry and, in particular, sets of similar genes, superficial unity may arise because of a particular environment. Diversity, on the other hand, implies different genes, different ancestry, and diverse environments.

Continuity

Life does not begin anew with each generation. Life, in fact, does not cease between generations, but rather, is passed on without interruption. When reproduction is asexual a living fragment of one individual develops into a complete organism. This can be readily demonstrated in the flatworm *Planaria* (see Fig. 26.2). Each piece from a single animal has the ability to form a complete organism. In sexual reproduction, the egg and sperm are the living cellular bridges that span the gap between generations.

The basis of continuity resides in the ability of a cell to replicate itself. As the German physician Rudolf Virchow stated in 1858, every cell comes from a pre-existing cell. The living cells now existing are only the temporary ends of long chains of cells that extend backward in time to that period in the history of life when the first cell, or cells, arose. On a molecular level, the basis of continuity resides in the ability of the cellular organelles to replicate themselves exactly. We have already discussed how DNA can do this, but centrioles, mitochondria, and plastids also have this ability to replicate themselves. Perhaps the most remarkable thing about the whole process of replication is that life at various levels of organization—molecules, organelles, cells, organisms, species, and communities of species—is characterized by this power of replication. Life is immortal even if individuals are not.

The Theory of Evolution

Although life is continuous, it is also continuously changing. It is characterized by unity, but also by diversity. The theory of evolution was advanced to account for these

The American Museum of Natural History

Fig. 27.6 An artist's conception of the Milky Way, our home galaxy, which consists of about 100 billion stars. The Sun is located near the edge of the galaxy.

three aspects of life. By embracing unity, diversity, and continuity, the theory states that all living organisms have arisen from common ancestors by a gradual process of change that leads to diversification. The theory also denies the validity of the Doctrine of **Special Creation,** which advanced the belief that organisms were immutable, that is, they were created as they now appear, and that no change has taken place.

The Doctrine of Special Creation dominated the thinking of men during medieval times and up to the 19th Century. This was, in part, a religious belief, but it was also due to misconceptions that man had about the age of the Earth. Before the 15th and 16th Centuries, man had a limited idea of the enormity of the universe. The Ptolemaic theory of the universe placed the Earth at the center of a limited universe. Man thought of himself as the creature for whom the Earth and all plants and animals were created. It was natural, then, to place man's home, the Earth, at the center of the universe. Copernicus and Galileo, who lived in the 1500's and 1600's, destroyed this idea. We know now that our Solar System is out near the edge of a galaxy (the Milky Way), which is composed of 100 billion stars (Fig. 27.6), and that there are a billion or more galaxies in addition to our own. Space, in fact, is so vast we have difficulty comprehending its immensity.

Hal H. Harrison from National Audubon Society

Joyce R. Wilson from National Audubon Society

Jeanne White from National Audubon Society

Fig. 27.7 The English Setter (top), Scottish Terrier (middle), and Boxer (bottom) represent three breeds that have been developed by man through artificial selection.

It was the geologists who demonstrated that the Earth did not spring into being in the year 4004 B.C., as Archbishop Ussher of Ireland once calculated the Earth's age to be. Today it is calculated not in thousands of years, but in billions; it is possibly five billion years old. The oldest known fossils are about two billion years old. Life, therefore, has had a long time to change, a long time during which the great chain has never been disrupted from the beginning to the present.

What evidence do we have that life has evolved? It is so overwhelming that evolution cannot be rationally disbelieved. Evidence can be found in the physiology, biochemistry and anatomy of organisms but let us choose examples that are familiar, say domesticated plants and animals. The dog was probably the first animal domesticated by man. As Fig. 27.7 shows, the various breeds of dogs are quite different from each other, in size, body conformation, facial shape and expression, hair, the way they bark, instinctive behavior, and so on. Yet all breeds arose from a wolf-like animal that man domesticated early in his rise to civilization. By selective breeding man has been able to isolate certain variations and produce pure breeds. Mongrels, showing a mixture of various breeds, result from a random, not a selective, pattern of breeding.

The development of the many breeds of dogs illustrates the course of evolution. It is, however, an artificial evolution in the sense that man directed it for his own purposes; man instead of the environment becomes the selective agent. Then, again, man is part of the environment, and his purposefulness is as natural to him as a trunk to an elephant. The basic principles are the same, however, whether evolution is artificial or natural. For one thing, there must be variation among the breeding population in order for selection to take place. If all organisms were alike, selection could have no effect. There must also be a pattern of breeding, that is, the parents must be selected by the breeder rather than by random picking. And once a particular variation is achieved, and breeds true in succeeding generations, the parents of one breed are kept reproductively isolated from other breeds. On a larger scale and with a longer time period, the same events occur naturally and give rise to new forms of plants and animals.

What man has done through selective breeding of dogs he has also done with horses, cattle, swine, and poultry. One needs only to open a seed catalog to see the variations man has selected to produce the great diversity in our vegetable and cereal crops, and in our garden flowers.

What proof do we have that organisms have changed as one period of time succeeds another? Although early evidence came from fossil remains in rock formations (Fig. 27.8), we can visualize

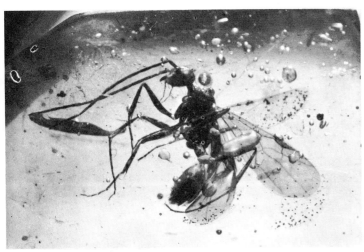

Fig. 27.8 This winged ant was fossilized in amber. Amber itself is fossilized pitch of coniferous trees.

The American Museum of Natural History

this more clearly by observing alterations of a more subtle nature as they take place in the laboratory. The bacterium, *Staphylococcus aureus,* is a **pathogen** (disease-inducing organism) that affects humans. It can be grown in a test tube. If we add a minute amount of streptomycin (an antibiotic) to the medium in which this organism is growing, we find that the great majority of cells are killed. About one in a million cells survive, however, and the survivors continue to grow. They are resistant to the antibiotic. If we continue to add streptomycin to the media, thus gradually increasing the total amount, we find that each additional amount of streptomycin kills the great majority of remaining cells, but that each time some survive.

If we now test the cells surviving high concentrations of streptomycin, we may encounter one or both of two situations. One kind of cell will be highly resistant to streptomycin; another kind may actually require streptomycin as a nutrient. We can separate these two kinds of cells by placing them in a medium lacking streptomycin. Those that are streptomycin-dependent die. But both types of cells—the resistant and the dependent—are different from those in the original culture. By breeding techniques it can be shown that some of their genes are different. They have **evolved.**

The situation in the laboratory is an "artificial" one controlled by the investigator. The fact that the same situation has occurred without the purposeful intention of man, however, is indicated by the fact that some of our hospitals are infected by a staphylococcus that is antibiotic-resistant. Since antibiotics came into general use in the mid- and late 1940's, the changed nature of some staphylococci is a relatively recent phenomenon.

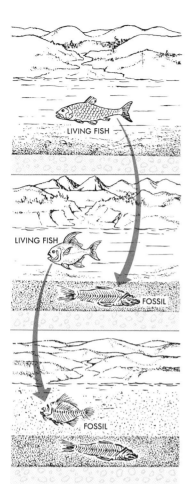

Fig. 27.9 In this schematic representation, a fish dies (top) and is covered over by sediments (middle) and fossilized. Thousands of years later living fish in the middle panel dies and its remains are covered by a second layer of sediments. Both layers are compressed and hardened into rock. A fossil may be simply the impression of an organism in rock, or it may be a portion of the organism (generally the hard parts, such as bone) that is preserved.

When we wish to reconstruct the changing nature of organisms over long spans of time, we must turn to the fossil record. Our source of information is from sedimentary rocks. The surface of the Earth is continually being eroded by wind and water, and the fragments are carried away by rivers as sand and silt (Fig. 27.9). When deposited as sediment in quiet waters, the sand and silt are often compressed and hardened into sedimentary rock. An animal or plant that dies and becomes embedded in the sediment may be preserved well enough to be recognizable millions of years later when the rocks are eroded and exposed (Figs. 27.9 and 27.10). The Grand Canyon is such an exposed mass of sedimentary rock.

In general, the bottom layers of sedimentary rock should contain earlier forms of life than the top layers do. This is what has been found. Scientists can thereby follow a succession of life. By determining the age of the rocks, it is then possible to calculate the time required to bring about such changes and to visualize the lines of succession of life.

Determination of lines of succession requires one major assumption—namely, that later organisms show a relation to the earlier ones that gave rise to them. We can see this beautifully demonstrated in the evolutionary history of the horse. The Cenozoic Era has been called the Age of Mammals, and it was during the Eocene period, approximately 60 million years ago, that *eohippus* (*Hyracotherium*), the "dawn-horse," made its appearance. Fossil remains show that *eohippus* was not very much larger than a modern cat, 10 inches high at the shoulder and weighing eight or nine pounds. Its teeth reveal that it was a brows-

Fig. 27.10 The cliffs of the Grand Canyon were formed by the erosion of rock by the river. The oldest rocks and most ancient fossils are found at the lower levels of the canyon.

American Airlines

ing animal, that is, it fed on the leaves and twigs of trees. It walked on three toes, even though the forefoot had four toes. It had disappeared in Europe by mid-Eocene, and in North America by late-Eocene. Before doing so, it left descendants, and these in turn left other descendants, with the modern horse the living member of a long line of ancestors. During the 60 million years, a change in tooth structure took place, showing that the horse became a grazing animal, that is, it fed primarily on grass. Also during this period the number of toes was being reduced, resulting in a single toe, or hoof. Other lines of horses were also evolved. Some were even smaller than *eohippus*, others larger; tooth structure varied widely and not all became one-toed. Only the line which led to the present-day horse persisted. All other lines became extinct. Only the fossil record remains.

The Course of Evolution

The theory of evolution states that organisms arose through a process of gradual change from more primitive ancestors. We cannot now and never will be able to reconstruct the entire evolutionary history of all living things. The fossil record is not complete. Some organisms were not suitable for fossilization, and often the conditions for fossilization were not right. There are, as a consequence, great gaps in our knowledge of past life on this planet. However, by making use of what fossils have been found, and by assuming that similarities of structure provide evidence of relationships, the major sequences of life are known (Table 27.2).

The table shown on the next page is a geological timetable, also providing information about changes in environment and the rise and fall of plant and animal groups. The oldest rocks of which we have knowledge do not contain fossil remains. These were formed in the Archeozoic era. Life apparently began in the Proteozoic. We make this assumption not because of fossil evidence, of which there is very little, but because the Cambrian rocks of the Paleozoic era contain abundant fossil remains. In fact, by the end of the Cambrian period most modern groups except the chordates were already established.

The first organisms were unicellular plants and animals: the **protists.** These had their origin in the seas. There were few, if any, fresh-water forms, and the land masses were barren. Multicellular algae and fungi among the plants, and the coelenterates (jelly fishes), sponges, mollusks, arthropods, and echinoderms among the animals soon followed. All of these are **invertebrates,** lacking a vertebral column. The first chordates (fishes), which lead eventually to man, appeared in the middle of the Ordovi-

TABLE 27.2 GEOLOGICAL TIME SCALE

Eras (Years of duration)	Major Divisions	Periods (Years from present)	Epochs	Dominant Organisms	Events of Biological Significance	Geological and Climactic Phenomena
Cenozoic (60 million)	Quaternary	2 million	Recent	Age of man and herbs	Rise of civilized man	
			Pleistocene		Extinction of great mammals and many trees	Periodic glaciation
		Late Tertiary	Pliocene		Rise of herbs; restriction of forests; appearance of man	Climactic cooling; temperate zones appear; rise of Cascades, Andes.
	Tertiary		Miocene	Age of Flowering Plants, Mammals, and Birds	Culmination of mammals; retreat of polar floras; restriction of forests	Cool and semi-arid climate; rise of Himalayas, Alps.
		Early Tertiary (60 million)	Oligocene		World-wide tropical forests; first anthropoid apes; primitive mammals disappear.	Climate warm and humid; rise of Pyrenees.
			Eocene		Modernization of flowering plants; tropical forests extensive; modern mammals and birds appear	Climate fluctuating.
Mesozoic (125 million)	Late Mesozoic	Cretaceous (125 million)		Age of Higher Gymnosperms and Reptiles	Rise and rapid development of flowering plants; gymnosperms dominant but beginning to disappear; rise of primitive mammals.	Rise of Rockies and Andes; great continental seas in N. America, Europe climate fluctuating.
					Extinction of great reptiles.	Climate very warm.
	Early Mesozoic	Jurassic (157 million)			First known flowering plants; gymnosperms prominent but primitive ones disappear; dinosaurs and higher insects numerous; primitive birds and flying reptiles.	Great continental seas; rise of Sierras; climate warm.
		Triassic			Gymnosperms increase; first mam-	Climate warm and semi-arid.

400

Era	Period	Age	Life	Geology
Paleozoic (368 million)	Late Paleozoic	(223 million)		land vertebrates.	Appalachians; Urals.
		Pennsylvanian (271 million)	Age of Lycopods, Seed Ferns, and Amphibians	Primitive gymnosperms dominant; extensive coal formation in swamps.	
		Mississippian (309 million)		Lycopods, horsetails and seed ferns dominant; some coal formation; rise of primitive reptiles and insects.	Shallow seas in N. America.
	Middle Paleozoic	Devonian (354 million)	Age of Early Land Plants and Fishes	Rise of early land plants; rise of amphibians; fishes dominant.	Shallow seas in N. America.
		Silurian (381 million)	Age of Algae and Higher Invertebrates	First known land plants; algae dominant; first air-breathing animals (lungfish and scorpions).	
	Early Paleozoic	Ordovician (448 million)		Marine algae dominant; corals, star fishes, bivalves; first vertebrates (fishes).	Shallow seas in N. America.
		Cambrian (553 million)		Algae dominant; many invertebrates.	Shallow seas in N. America.
Proterozoic (900 million)		(1,500 million)	Age of Primitive Marine Invertebrates.	Bacteria, algae, worms, crustaceans prominent.	Formation of Grand Canyon, Laurentians. Sedimentary rocks.
Archeozoic (550+ million)		(2,000 million)	Age of Unicellular Forms	No fossils; organisms probably unicellular; origin of first life.	Rock mostly igneous.
		(10 billion) ?			Beginning of present universe and the Solar System

Fig. 27.11 The present-day lycopods and horsetails are remnants of a once vast tree-like flora that contributed heavily to the formation of the coal beds of the world.

Fig. 27.12 A fossil of *Archaeopteryx*, the oldest known bird. It lived during the Jurassic Period, some 157 million years ago.

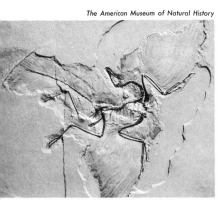

cean, and there is some evidence that they arose in fresh water. The fishes increased greatly in the Silurian and Devonian periods. At about the same time, the spiders and insects took their place on the evolutionary stage. Invasion of the land by plants occurred in the Silurian, and by the Devonian, the gymnosperms (ancestors of the present-day pines and spruces) had become established. Amphibians, lung fish, and sharks also arose during the Devonian.

During the Carboniferous and Permian periods great forests of gymnosperms and seed ferns covered the land. These, together with the lycopods and horsetails, led to the formation of the great coal beds of the world. The fallen plants piled up in great layers, were compressed into strata, and were carbonized into coal as the volatile substances evaporated. By the end of the Permian, the lycopods and horsetails declined, but they persist today as miniature remnants of a once great flora (Fig. 27.11).

The reptiles appeared in the upper Carboniferous, or Pennsylvanian, period when insects and amphibians abounded. The rise of the Appalachian Mountains signaled the end of the Paleozoic and the beginning of the Mesozoic era. The Mesozoic is the Age of Reptiles, the time of the rise and fall of the dinosaurs. It is also the time when the first birds appeared, the earliest ones bearing teeth and being decidedly reptilian in character (Fig. 27.12). Mammals, too, came into existence, the earliest of them being egg-laying, like their reptilian ancestors.

The Mesozoic also witnessed the rise of the flowering plants. The modern cone-bearing gymnosperms were present, the seed ferns were disappearing, and first the dicotyledonous and then the monocotyledonous plants emerged. The latter two groups were to become dominant members of our flora.

The mammals came into prominence in the Cenozoic. Of the egg-laying type, only the duck-billed Platypus and the echidnas remain (Fig. 27.13). The others were placental, giving rise to

Fig. 27.13 The duck-billed Platypus has many features of the mammal group, yet it lays eggs which are hatched outside of the body. Consequently it lacks the mammary glands characteristic of mammals.

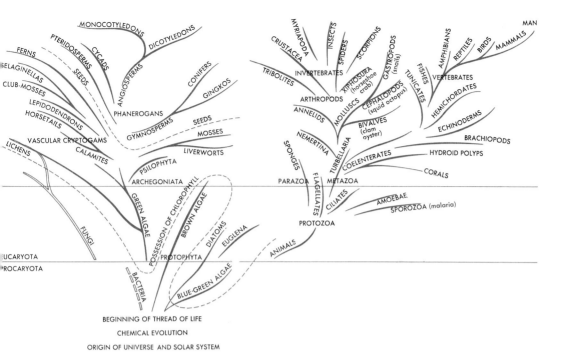

Fig. 27.14 The thread of life has continued unbroken, so far as we know, from the beginnings of life to the present time.

living young. The rise of the anthropoid apes and then of man (toward the end of the Cenozoic) began the present period. Meanwhile the great deciduous forests developed, later to give way to herbs and grasses. The world of today was then ushered in.

The fate of most species of living things over the great sweep of time since life began has been extinction, without descendants. This is as much a part of evolution as is persistence. The present flora and fauna, therefore, represent those species that managed somehow to keep the thread of life going. This thread of life, from its earliest beginnings to the present time is illustrated by Fig. 27.14. The main avenues are reasonably clear, although not every connection is known for certain. For example, the ancestors of both the chordates and the flowering plants remain in doubt. Both groups appear suddenly in the fossil record, with transitional forms absent, not yet discovered, or possibly misinterpreted.

SUMMARY

Organisms differ from each other, but they also show a common unity of metabolism, cell structure, and developmental patterns. Both **unity** and **diversity** result from inheritance; and since life from one generation to another is passed on by way of cells, there is also **continuity**. All three aspects of life are understandable if we assume that life has evolved through millions of years, giving rise to the present forms of plants and animals.

Evolution is change. For the process to be effective, variation and selection are necessary. Man has imposed evolution artificially among domestic plants and animals by acting as the selecting agent. Nature does it more slowly; we now realize that life has existed on Earth for more than two billion years. Present species of plants and animals are the present ends of long chains of continuous life that extend backward in time to the moment when life first began on our planet. The most striking evidence of this is found in sedimentary rocks, where the remains of former life appear as fossils.

Past ages of the Earth show that there is a progression from one form of life to another. Older rocks have more primitive organisms than younger rocks. Through comparative anatomy, a history of life can be reconstructed, but gaps in the fossil record prevent us from reconstructing a *complete* history of all organisms.

FOR THOUGHT AND DISCUSSION

1 How do you suppose Archibishop Ussher arrived at the date 4004 B.C. as the beginning of all things?

2 Consider two points of view: 1. that the Earth is the center of the universe and that all plants and animals are on Earth for man's use (a view prevalent up to the 19th Century); and 2. that we exist near the edge of a galaxy and revolve around a star in a universe of millions of other galaxies; further, that all life is related to us through evolution. Does acceptance of one or the other of these points of view affect your thoughts about yourself: who you are, where you came from, what is to become of mankind?

3 Why can selection have no effect if there is no variation?

4 The Earth is about 5 billion years old. How is this age determined? Look up information on the carbon-14 dating process. Why is the process reliable only to about 30,000 years in the past? What is carbon-14? How does it differ from the more common carbon-12?

5 Why are there no fossils in igneous rocks? If fossils are not found in a certain layer of sedimentary rocks (where they normally occur), can we conclude that there was no life when the layer of sedimentary rock was formed? Explain.

6 How can the teeth of a fossil animal reveal anything about its dietary habits?

7 Clover is a plant that shows variation in the height of individual plants: some are tall, some intermediate, some creeping. Imagine two fields of clover, one closely grazed for several years by sheep, the other ungrazed. At the end of this time would you expect the genotypes of the surviving plants to be the same or different in the two fields? Why? How would you test your hypothesis?

8 The American horses of today originated in the Old World, but abundant horse fossils are found in North America. Why do you suppose that only one kind of horse survived to the present time? What could have caused their disappearance in America, but not in the Old World?

9 Does the fact that cells are the basic unit of organization through most of the plant and animal kingdoms, and that cells of all kinds have quite similar modes of metabolism, have any evolutionary significance?

10 Consider your classmates and yourself. What features do you all share in common? What features show a great deal of diversity? What does this kind of comparison tell us about development, ancestry, and evolution?

11 Would you expect to find as much variation among a group of sexually breeding organisms as among a group arising through asexual reproduction? Explain.

12 If extinction is a more likely evolutionary fate than persistence, can man escape extinction? What aspects of man's structure and/or behavior could promote and hasten extinction? What aspects could favor persistence?

13 Make a list of common plant or animal species that display considerable diversity. Make a list in which no diversity is present.

14 "The purpose of every unit—organism, cell, or chemical compound—is to be itself for a short period and then to become something else. . . ." G. Ehrensvärd. What does this statement mean for the units mentioned? What does it mean when you apply it to yourself?

SELECTED READINGS

deBeer, Sir Gavin. *Atlas of Evolution.* Camden, N.J.: Thomas Nelson & Son, 1964.

An expensive volume, but one of the finest and most beautifully illustrated books on evolution. It covers all of the plant and animal kingdom, including man.

White, J. F., (ed.). *Study of the Earth.* Englewood Cliffs, N.J.: Prentice-Hall, Inc., 1962.

Includes discussions of fossils and how one can determine the age of rocks.

Simpson, G. G. *Horses.* New York: Doubleday, The Natural History Library, 1951.

An excellent presentation of one of the most completely known fossil histories of an organism.

28 CAUSES AND RESULTS OF EVOLUTION

One of the most common observations we could make, but one so obvious that we rarely give it thought, is that organisms in their native habitat seem to belong there. The organism and the environment fit each other: a cactus in the desert, a polar bear in the frozen north, an ameba in a wayside pool, an earthworm in rich garden soil. Indeed, our everyday language reflects our thinking about this: we say that someone is "in his element" when the situation is such as to give the individual an opportunity for full expression of his talents. Or, we say that someone is like "a fish out of water" when the individual and the situation are mismatched.

While some organisms exist only in a restricted kind of environment, others seem relatively indifferent to their surroundings. A fish cannot exist out of water. Yet water is not the only limiting factor. Temperature, salinity, available food, and amount of oxygen must also be considered. Man, on the other hand, lives in a wide variety of environments: in the dry and treeless desert, the hot and humid tropics, and in the frozen Arctic. Man's ability to invade and live in difficult environments depends on his ability to protect himself from harmful aspects of the environment—housing and clothing, for example—or to change the environment, as he does in high-flying aircraft, space ships, in submarines, and in deep mines (Fig. 28.1). Air conditioning and irrigation are but two means which enable him to make an environment more suitable to his needs as an organism.

The simple fact is that continued existence of a species requires that it be adapted to an environment. Failure to adapt can lead only to extinction of the species. Individuals, of course, can make temporary adjustments to changes in the environment, and can live for a time in an unsuitable environment. But if the species is to persist, its members must grow, maintain themselves, and reproduce generation after generation. We have already discussed this problem in another context. We showed that when

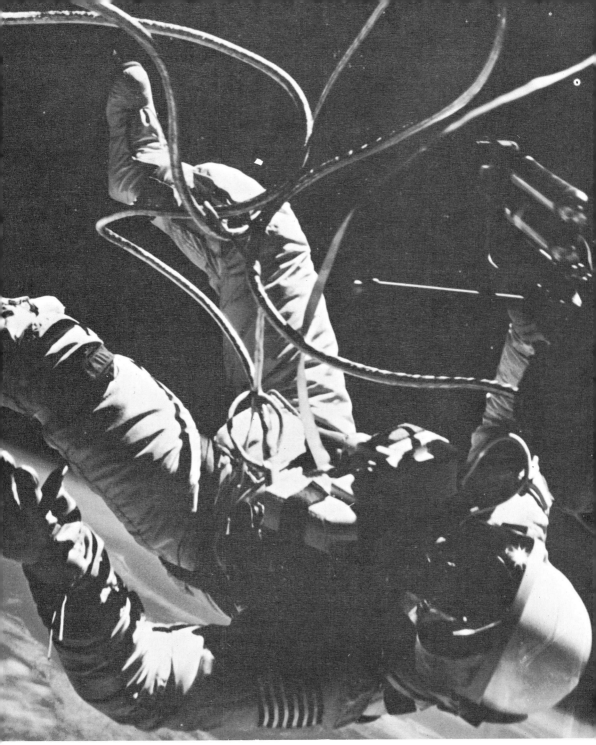

NASA

Fig. 28.1 The American astronaut E. H. White, II "walks" in space at a rate of about 17,000 miles an hour. A line tethers him to the spaceship so that he will not wander off and be lost. His ability to survive in this airless, frigid environment depends upon his space suit, within which a controlled environment is maintained.

Fig. 28.2 The rock Ptarmigan is shown here in its brown summer plumage (left), and as it is changing from its white winter to its brown summer plumage (right).

certain plants are moved to lower or higher altitudes, they fail to reproduce and sometimes die. Again, we emphasize a most important fact: the species and the environment must be in harmony with each other if the species is to persist.

An environment itself is seldom constant. There are daily and seasonal changes in temperature, precipitation and other forms of weather, and many destructive forces such as hurricanes, floods and prolonged dry spells. The daily rhythm of light and darkness is one of the most important aspects of our environment on the Earth. Why do chrysanthemums bloom in late Summer and Fall rather than in the Spring? Why do birds produce their young in the Spring? The answer, in part, is found in the changing length of day and night. We know that light or darkness changes the behavior of organisms profoundly. But temperature and the amounts of food, water, and oxygen are also important.

In response to seasonal aspects of the environment, seeds and spores carry plants over periods of cold or dryness; some animals hibernate, others migrate, while still others may change their coat color as winter approaches (Fig. 28.2). We also know that an environment may be altered "permanently." Although the bread fruit tree now grows only in the tropics, fossils of this plant are found beyond the Arctic Circle. At one time in the past, the climate must have been warm in this region. Mountains form and, as they do, the environment around them is altered. If organisms cannot adapt to the new conditions, they must migrate to more suitable climates, change so that they are better adapted, or die.

Environmental changes are not only physical in nature. Some changes are biological, such as groups of organisms competing for living space. Fossil-bearing rocks record the failure of many organisms to survive the competition for "elbow room." The dinosaurs appear to be a notable example. We still do not understand fully the reason for their disappearance, but the blunt fact is that extinction is a more likely fate of a species than is continued survival over long periods of time. In the continuing fight for living space, man is an important biological factor. As he hunts, farms, and in other ways alters the land to suit his way of life, he destroys much around him: the dodo bird, the auk, and the ivory-billed woodpecker have disappeared within the last century, and conservation is required to protect other animals and plants.

Over the span of many hundreds, thousands, and millions of years, many species have died out, but some have continued to survive. These have given rise to the species living today. The fact that we can write these words, and you can read them, means that we are just the present-day members of an unbroken chain of organisms extending far back in time. *Homo sapiens* has existed for a million years or more, but for how much longer he will continue to inhabit the planet, no one can say. Environments change, and so do organisms. Those that have survived have been adaptable, and our understanding of evolution rests upon our understanding of the *causes of adaptation.*

Charles Darwin's Theory of Natural Selection

In the previous chapter we stated that life was continuous, but also that it was constantly changing. We also said that within the continuity of life we find both unity and diversity. The transmission of genes determines unity in that a cat always produces kittens, not some other kind of animal. But since genes can undergo change (mutate) and also recombine through meiosis, diversity is continuously introduced into the life of a species. It is this diversity which, through a gradual process of change, leads to the evolution of species on a grand scale, and to the adaptation of the species in a more restricted sense. If we phrase this thought somewhat differently, we can state that all adaptive changes that are inherited are evolutionary changes. The reverse, however, is not true; all evolutionary changes are not adaptive. It is important to make this distinction.

Theories attempting to account for the evolution of organisms go back to the Greeks, but our current point of view was developed principally by Charles Darwin, the great English naturalist. His book, *The Origin of Species,* which appeared in

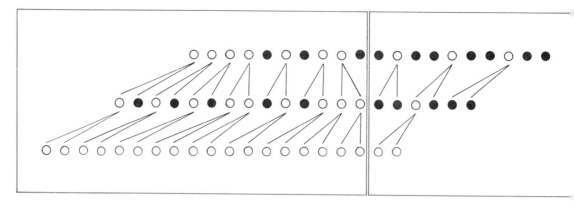

Fig. 28.3 In three generations of individuals, the number of individuals per generation remains the same. (The nonreproducers are shown in black.) If there is a heritable difference between the reproducers and the nonreproducers, a gradual change will take place in the population.

1859, is probably the single most influential volume of biology ever written. He termed his ideas on evolution the **theory of natural selection.** We can understand this best by following Darwin's line of reasoning.

The first point, or assumption, made by Darwin was that the number of individuals in a species tends to increase in geometric ratio, generation after generation. This means that each generation has more offspring than there are parents. Thus if a single bacterium divides and produces two bacteria, that generation of two will produce four bacteria; the four will give rise to eight, the eight to 16, the 16 to 32, and so on. A single maple tree produces thousands of seeds, each of which is capable of producing more maple trees. A single female codfish annually sheds several million eggs, each of which can hatch into a young codfish.

The fact remains, however, as Darwin clearly knew, that the number of individuals in each generation remains fairly constant. Great fluctuations in numbers do not generally occur. Some individuals are eaten by predators, and there is competition among those that survive, competition for the basic essentials of life: water, light, food, and "elbow room." Some organisms thrive amid competition, others do not. The individuals that survive and become the parents of the next generation are, therefore, only a small number of the original population.

Darwin made a third assumption: that among the individuals belonging to a single species there is diversity, or **variation.** This is an observation you can make for yourself by comparing the cats and dogs of your neighborhood, or the plants in a garden or flower bed. It is more difficult to make the same observation among wild plants and animals, but the variations are equally numerous.

On the basis of these assumptions, Darwin developed his theory of natural selection. He argued that if some variations were more favorable for survival than others, these gave certain individuals a better chance of passing these variations on to the next generation. That is, these individuals are more likely to be the parents of the next generation. Individuals with less favorable variations stand a lesser chance. Gradually, over a period of time, the favorable variations would tend to accumulate, and the species would undergo a gradual evolution. Species with short generation times could evolve faster than those with longer times, since over a given period of time the number of generations would be greater.

Only those variations which are heritable are of any significance in evolution. Many variations are caused by environment, but these play no role in the evolution of species. Furthermore, favorable variations do not guarantee survival to a reproductive age. Survival or death of an individual is often a matter of chance, and many favorable variations are lost in the course of evolution. Also, a favorable variation in one environment may be unfavorable in another, or be of no use at all. The ability to swim is of no value to an individual who lives in a desert. We must, therefore, think of variations in terms of the organism in a given environment, and define as "favorable" those changes which in the long run increase the individual's chance of producing more and better adapted offspring. If the individual fails to participate in production of the next generation, any unique variation it has will be lost when the individual dies.

Let us take a closer look at what Darwin meant by the term *natural selection*. We have just stated that all of the individuals of one generation do not contribute their genes to the next generation. In terms of human populations, some individuals are sterile, some choose not to produce offspring, and some die before they reach reproductive age. The parents of one generation are, therefore, a selected part of that generation, and their hereditary materials are a selected part of the hereditary material of that generation (Fig. 28.3). As Bruce Wallace and Adrian Srb of Cornell University have stated:

"The disparity between parents as one group of individuals and the rest of the population as another *is* natural selection. . . . Thus, when we say that natural selection results in the adaptation of organisms to their environment and evolutionary changes in populations, we are simply saying that the continual contrast, generation after generation, between reproducing individuals as one group and the remainder of the population as another results in adaptation and evolution."

Natural selection, therefore, is a random process in the sense that it has no predetermined goals. Variations are produced in low numbers and environmental changes occur gradually. But if both changes occur, then evolutionary change is unavoidable; it will take place, as, indeed, we believe it has been doing since life first arose on our planet.

We need now to inquire in more detail into the source of variation, its fate in succeeding generations, and its effect on the adaptability of populations.

Source of Variation

The heritable material of an individual is its DNA (RNA in some viruses). It is found primarily in the nuclei of cells, where it forms the chromosomes. Short segments of DNA form the genes. As yet we do not know, except in a few instances, how many nucleotide pairs make a gene, but we must assume that genes differ among themselves in number of nucleotide pairs if only because proteins differ in the number of amino acids they contain. We do know, however, that a gene is made up of a particular sequence of nucleotide pairs, and that this sequence is the genetic alphabet. It spells out the message through RNA, which, in turn, takes part in the formation of a particular protein. The uniqueness of every individual, plant or animal, is determined by the proteins it contains and also, of course, by its genes.

As indicated in an earlier chapter, DNA is a large molecule so constructed that it can replicate itself in a most exact fashion, one cell generation after another. This is the basis of biological continuity and biological unity. DNA is also responsible for biological diversity. If this is so, diversity, or variation, has its origin in the fact that changes can occur in the sequence of nucleotide pairs within a gene. Gene A, for example, then becomes a, or A_1, A_2, A_3, and so on, depending on where within the gene a change occurred in the sequence of nucleotide pairs. A gene, therefore, can change into many forms. Each one is a mutation, and the several mutants arising from a particular gene are called **multiple alleles.** Each mutation alters the action of a gene by changing the gene product, or protein. This, in turn, changes the organism in some manner, allowing us to detect that a mutation occurred. From these facts, you will recognize that the existence of a gene can be detected and studied only when at least two alleles of the same gene can be compared.

How often do genes mutate? Genes are very stable structures, but we now recognize that each gene has its own rate of mutation. This can be seen in the table on the facing page showing the

TABLE 28.1 Spontaneous mutation rates of specific genes in several organisms.

Organism	Gene	Rate
MAIZE	R	492 per 10^6 gametes
	I	106 per 10^6 gametes
	S	1 per 10^6 gametes
	Wx	0 per 10^6 gametes
E. COLI	leucine-1	0.07 per 10^9 cells
	leucine-2	1.42 per 10^9 cells
	arginine-2	0.37 per 10^9 cells
	tryptophan-6	5.61 per 10^9 cells
MAN	Achondroplasia (dwarfism)	41 per 10^6 gametes
	hemophilia	32 per 10^6 gametes
	albinism	28 per 10^6 gametes
	total color blindness	28 per 10^6 gametes
	infantile amaurotic idiocy	11 per 10^6 gametes

rates of mutation of a number of genes in corn (*Zea mays*). The seed color gene, A, mutates to the a allele at a rate of 492 per million (or 10^6) cells, or gametes. *Wx*, on the other hand, is exceedingly stable, and no mutations to *wx* were found in this particular experiment. The rate of change of the other genes tested was intermediate.

Bacteria or viruses provide a convenient way for us to study mutations, since millions of cells or virus particles can be grown in a matter of hours or days. Some mutation rates are as low as one in a billion (10^9) cells. In human beings, mutation rates are difficult to determine. We can recognize them only by knowing how frequently particular mutations make their appearance in a population. The mutation that gives rise to hemophilia, a disease in which the blood fails to clot readily, occurs once in about every 50,000 persons. This is a fairly high rate of mutation.

If gene A mutates to a, and this change is an alteration in the nucleotide sequence, then we should expect that a can mutate to A. This is known to occur, and can be expressed as $A \rightleftarrows a$, with the change of $A \rightarrow a$ being a forward mutation rate, and $a \rightarrow A$ being a backward mutation rate. The two rates need not be the same, and, in fact, generally are not.

We do not know how many genes there are in any particular organism. The number in human beings has been estimated to be between 10,000 and 50,000. But if each gene mutates at a

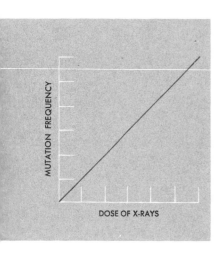

Fig. 28.4 If a given dose of x-rays induces X mutations, twice the dose will produce 2X mutations. This relationship is indicated by the straight line.

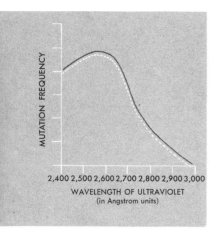

Fig. 28.5 The dose of ultraviolet light remains the same for all wavelengths. The solid line represents the degree of absorption of UV of different wavelengths by the DNA of the cell, the broken line represents the frequency of induced mutations. The similarity of the two curves indicates that the UV must be absorbed in order to be effective as a mutation-inducing agent.

given rate, it means that every cell has a good chance of containing a mutant gene, newly arisen. If these arise in somatic cells, it may be detected, but it is not passed on to future generations. If arising in eggs or sperm, however, they can be passed on. A reasonable guess would be that one out of every 100 gametes has a new mutation in it. Since these are being passed on to future generations, the individuals of each generation have a wealth of genetic diversity.

Each individual, therefore, is likely to be different from others in the population. This fact reinforces three points already made: 1. Gene mutations provide the initial diversity which makes evolution possible. If there were no diversity, clearly there would be no evolution. Furthermore, since genes recombine in meiosis, the number of possible combinations of genes increases the amount of diversity to an enormous degree. For example, if a species has 1,000 heterozygous genes in its population (a conservative estimate) the number of possible combinations is $2^{1,000}$, a number so large that it exceeds the number of atoms in the universe. 2. Since individuals differ genetically, they will also differ in their degree of adaptability to a given environment. Some will be more, others less, adaptable. 3. Since only a selected number of individuals in a population give rise to the next generation, the chances are good that the genetic nature of the population as a whole will change with each generation. If the change goes continuously in a given direction, the population will evolve in that direction. We will return to these points later in the chapter.

The mutations just discussed arise naturally in populations. We speak of them as **spontaneous mutations** to distinguish them from those artificially induced. X-rays, radioactive materials released by a nuclear bomb, ultraviolet light, and a wide variety of chemicals can induce mutations. The greater the amount of x-rays received by the cells, the greater is the number of mutations induced (see Fig. 28.4). We can think of x-rays as small rifle bullets that pass through the genes and alter them during passage. Ultraviolet radiation, on the other hand, behaves differently. It is absorbed (captured) by DNA, and the absorbed energy of ultraviolet makes the DNA unstable, thus bringing about change. Some regions of ultraviolet are more effective than others in inducing mutations. The region around 2,600 Å is particularly effective. The relation of wavelength to mutation induction is shown in Fig. 28.5.

X-rays are very energetic rays, and readily pass through cellular substances. They can induce mutations in any organism. Ultraviolet, however, is quickly absorbed by cellular substances, and it is a good mutation-inducing agent for viruses, bacteria, and single layers of cells.

Many chemicals cause mutations. Some are merely destructive substances, damaging any part of the cell including DNA. Others are more selective in their action. One of these (BUDR, 5-bromouridylic deoxyriboside) is closely related to the nucleotides in DNA and can replace thymine specifically. When it does so, the DNA becomes unstable and tends to mutate.

Interestingly enough, the mutations induced by these agents are the same as those which arise spontaneously. No agent has been found which gives rise to a particular mutation, so we are unable as yet to mutate one particular gene and leave all other genes unaffected. Certain genes, however, can cause other genes to mutate in a specific way. One such mutator gene is found in corn, and is called **Dotted (Dt).** When Dt is in a cell in the presence of gene *a*, which governs color, gene *a* will mutate to *A*, at a given frequency. As Fig. 28.6 shows, the mutations show up as colored spots on an uncolored seed. Table 28.2 shows how the number of Dt genes in a cell influence the mutation rate $a \rightarrow A$.

Other changes can occur in cells and produce genetic changes which are inherited like mutations, but which are really changes in chromosome structure rather than changes in nucleotide sequences. Some of these are losses of genes, called **deletions.** In *Drosophila* a group of *Notch* mutants, which produce a nick in the wings, and *Minute* mutants, which reduce the over-all size of the fly, are in most instances due to losses (Fig. 28.7). The *Notch* effect appears when a particular piece of chromatin on the *X*-chromosome is missing, but the *Minute* mutants are scattered over all of the chromosomes, and appear to be due to losses of chromatin forming t-RNA.

The *Bar* gene, so-called because it affects eye shape in *Drosophila*, is the result of a gain in genetic material. The piece **duplicated** is in the *X*-chromosome, and by manipulation it is possible to increase the number of times the region is duplicated. Each added piece reduces still more the size of the eye (Fig. 28.9).

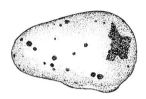

Fig. 28.6 In these two kernels of corn, gene *a* has mutated to *A* (dark areas). In the top kernel the mutations occurred late in the development of the seed and the spots of *A* are, consequently, small. In the bottom kernel one or more mutations occurred early, giving a large patch of dark color.

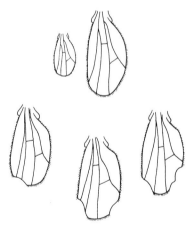

Fig. 28.7 Top: The normal wing of *Drosophila melanogaster* (right) is compared with the reduced wing size found in the *Minute* phenotype (left). Bottom: Several types of wing alterations result from *Notch* "mutations."

TABLE 28.2 Mutations of $a \rightarrow A$ occurring in maize seeds (triploid tissue) when the number of Dt and *a* genes are varied (after Rhoades, M.M.)

Genetic composition of plant	Number of mutations per seed
aaa/Dt dt dt	7.2
aaa/Dt Dt dt	22.2
aaa/Dt Dt Dt	121.9
aa^{p}a^p/Dt dt dt	3.20 (calculated)
aaa^p/Dt dt dt	5.64
aaa/Dt dt dt	8.16

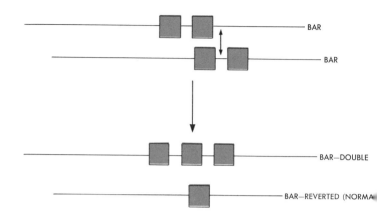

Fig. 28.8 The *Bar* "mutation" is a duplication of a segment of chromatin. It causes a change in the shape of the eye and reduces the number of facets per eye (see Fig. 28.9). Occasionally, crossing-over occurs (as diagrammed here), giving a normal chromosome without duplication, and a chromosome in which the Bar region (represented by the dark blocks) is triplicated. This results in the *Bar-double* phenotype in which the eye is even more severely altered.

Fig. 28.9 The eye of *Drosophila melanogaster* is shown in the normal form (left) and when the *Bar* region is duplicated or triplicated. The diploid chromosome situation for each one is represented above. The homozygous *Bar* and heterozygous *Double Bar*/normal have the same number of *Bar* regions, but the arrangement of 2/2 and 3/1 have a different effect on eye structure and facet number.

The gain or loss of whole chromosomes also produces a mutation-like effect. **Down's syndrome** (mongolian idiocy) in human beings appears when the small 21 chromosome is present three times in each cell, rather than in the normal diploid condition. Loss of an X-chromosome, to give an XO instead of an XX condition, leads to the defective physical development of female individuals **(Turner's syndrome)**. An added X in males, to give an XXY instead of an XY situation, also leads to a comparable defective development (Klinefelter's syndrome).

Gains or losses of chromosomes, like deletions and duplications, produce their effects by upsetting genetic balance. Think of a

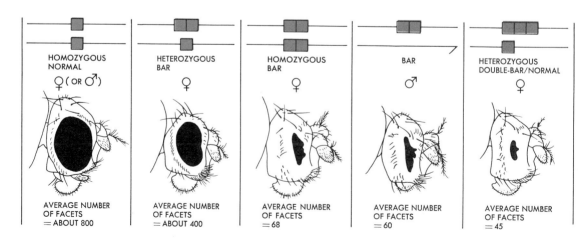

normal set of genes, single in the haploid state and double in the diploid state, as that array of genes needed for normal expression of development. Gains or losses will shift expression away from normality. Their role in evolution is still being investigated. Duplications at least provide a possible source of added genetic material which, over a period of time and through mutation, might come to serve some useful genetic purpose.

When whole sets of chromosomes are added, to give triploids ($3n$), tetraploids ($4n$), and so on, the changed appearance of the organism is slight (generally of an increased size). Such organisms are called **polyploids,** in contrast to normal haploids and diploids. Since one-third of the flowering plants are polyploid, polyploidy has played a significant role in their evolution. This phenomenon is rare among animals, however.

The Fate of Variation

Let us assume that gene A and its allele a exist in a population. Our knowledge of Mendelian inheritance tells us that all of the individuals in the population will be AA, Aa, or aa. If A or a do not mutate to other allelic forms of the gene, no other genotypes for this gene are possible. If these individuals breed together in random fashion, and no other events change the proportion of A to a, then the frequencies of A and a will remain the same from one generation to the next.

This is known as the **Hardy-Weinberg principle,** and we can demonstrate its validity in the following way. The frequency of A in a population is equal to the frequency of AA individuals plus one-half of the Aa individuals. Let us designate this by the letter p. Similarly, the frequency of a is equal to the aa individuals plus one-half of the Aa individuals. This frequency is designated by the letter q. Since all of the AA, Aa, and aa individuals in a population equals 100 per cent, or 1.00, then $p + q = 1.00$. In a breeding population, A and a sperm and A and a eggs will be produced, and random fertilization will occur as indicated below:

Eggs	Sperm	Resulting Individuals	Frequency
A	A	AA	$p \times p$ or p^2
A	a	Aa	$p \times q$
a	A	Aa	$p \times q$ } $2pq$
a	a	aa	$q \times q$ or q^2

Fig. 28.10 Cultures of the bacterium *Staphylococcus aureus*. Plate 1, without penicillin; the colonies are so numerous they merge to form a solid layer over the surface of the dish; Plate 2, medium contains 0.016 mcg/ml of penicillin, which has killed most of the bacteria; most of these colonies are probably not strongly penicillin-resistant due to mutation, but some may be; Plate 3, contains 0.032 mcg/ml and all bacteria have been killed except two colonies (top and bottom of dish) which are probably penicillin-resistant; Plate 4 contains 0.063 mcg/ml and all bacteria have been killed. Resistance to penicillin can result from the mutation of several genes, causing different degrees of resistance.

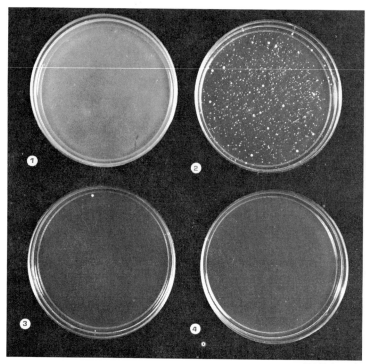

Bristol Laboratories

The equation $p^2 + 2pq + q^2$ is another expression of the use of the binomial. It represents the proportion of AA, Aa, and aa individuals in a population, and for all succeeding generations. Let us assume that there are nine times more A genes than a alleles; p^2 is then equal to 0.81 (0.9 × 0.9); $2pq = 0.18$ (0.9 × 0.1 × 2); and $q^2 = 0.01$ (0.1 × 0.1). If these individuals breed together randomly, they will give rise to the next generation, in which the proportion of genes remains the same.

The Hardy-Weinberg principle is a theoretical one, but it shows that variation—once arisen—will persist in a population if it is found in the breeding individuals. It assumes, however, random breeding, equal survival of all genotypes to reproductive age, a large population, no additional mutations, and absence of migration of individuals in and out of the population. Few populations exist in this ideal state. Mutations do occur, although many are lost because the individuals possessing them are not among the selected parents of the next generation. But even if lost, they can occur again and again (Fig. 28.10). Over long periods of time, measured by geologic time, a mutation rate as low as one in 10^6 can insure continued variability upon which natural selection can act. Those mutations that are clearly harmful—that is, harmful in the sense that the individual possessing it has a lowered reproductive capacity—tend to be eliminated. Beneficial ones tend to be retained, and gradually increase in the population.

The retention of variability in a population is often aided by the fact that heterozygous Aa individuals are favored over the homozygous AA or aa. Why this is so is not always clear, but in one well-known instance it is clearly demonstrated. Human hemoglobin A, the common type, is found in all populations, and is controlled by gene A. In Africa, among certain tribes, a mutant type, A_s, causes the appearance of a changed hemoglobin. It is recognizable because the red blood cells collapse into a sickle-shaped configuration. As we saw earlier, hemoglobin S can cause anemia. It has no effect in AA_s individuals because their blood is a mixture of A and S hemoglobins; but it brings on severe and lethal anemia in A_sA_s individuals. Malaria is also a prevalent disease in Africa, and an AA genotype stands a good chance of contracting the disease and dying early in life. The A_sA_s will die of anemia. Only the AA_s heterozygotes have good prospects of being the selected parents of the next generation. They do not suffer from anemia, and the malarial parasite cannot exist in AA_s blood. Yet both genes must be present in order for the heterozygote to be formed. Should malaria be wiped out as a disease, the environment would be drastically altered, and the AA homozygote would then be at a greater advantage.

This kind of variability is called **polymorphism** (many forms). The type described above is of a kind restricted to one gene, but many instances are known where polymorphism involves many genes. The advantages are obvious. Where the environment over a wide area is highly uniform, polymorphism is likely to be absent, for one kind of genotype would be selected. But this is rare. Most environments are a mixture of innumerable microenvironments, differing from each other in temperature, light, moisture, amount and kind of food. Each one favors the persistence of a given genotype, and a wide-ranging population of organisms is almost certain to consist of as many subpopulations as there are microenvironments. These populations overlap and interbreed, but they tend to possess a large amount of variability. Only when small populations become isolated, and a certain type of variation becomes fixed throughout the population, do we find the development of new species.

Results of Variability as Adaptations

The great diversity of life around us, ranging in size and form from viruses and bacteria to redwoods, whales, and man, is the result of heritable variations that have survived. Just being alive is a measure of evolutionary success. Being alive and able to reproduce carries such success to another generation further in time. You are here today because when life first arose on Earth, some combination of variations continuously met the challenge of the environment successfully.

During the past, possibly, two billion years, the combination of variations underwent continuous change, sometimes slowly, sometimes quite rapidly as measured by geologic time. Different forms of life, exhibiting different modes of existence, came into being. Some have persisted, others have fallen by the wayside. The fossil record in the rocks tells us a partial story of these evolutionary "failures." But these are failures only in the sense that they are not alive today. In their time, before their bodily remains were fossilized, they were a success; they were alive. Yet it is a mistake to think of all fossils as evolutionary failures or as temporary successes. In our discussion of the evolution of the horse, we saw that some lines of evolution came to a dead end. However, the line that gave rise to the modern horse was an evolutionary success even though the modern horse differs from its earlier ancestors. The chain of life was never broken at the same time that the horse was undergoing change.

Consider the huge dinosaurs. So far as we know, they disappeared from the Earth without giving rise to any present-day species. As such they were temporary evolutionary successes. But the reptiles as a whole, the class to which the dinosaurs belonged, were more successful. Some forms gave rise to the birds, others to the mammals, and still others to the reptiles of today. Evolutionary or adaptive success is, therefore, to be measured and judged only in terms of reproductive success. The failure of parents to produce offspring represents the end of the line for the genetic material found in those individuals.

It is difficult, if not impossible, to assess the results of genetic variability in terms of adaptation, except in very simple cases. Adaptation is concerned with the whole business of living, from the egg through birth, reproductive age, and death. All of the structures of an individual and all of its physiological processes are part of and contribute to adaptability, and these are controlled by many genes. It is true, of course, that human beings can lose an arm or eye, or lack the ability to produce insulin (a diabetic), and still live and reproduce, but we are less adaptable when these organs or processes are missing.

We have already discussed a simple case of adaptation, that of *Staphylococcus aureus,* to an environment suddenly changed by the introduction of antibiotics. Let us examine this somewhat further. This time we will use the common colon bacillus, *Escherichia coli,* a nonpathogenic form that you can experiment with if you choose. It will respond to antibiotics, much as does *Staphylococcus,* by displaying resistance.

Bacteria such as *E. coli* multiply by simple division. One cell becomes two, and the two daughter cells will, in general, be genetically identical. If we now spread a large number of bacteria onto an agar medium containing streptomycin, most of the cells will die. These are nonresistant cells. An occasional colony will form, however, each colony being the result of repeated divisions of a single cell when the plating was done. Cells from this colony can be repeatedly transferred to a streptomycin-containing medium, and they will continue to grow. These cells are resistant. Genetic tests show that resistance is determined by one or, possibly, several genes.

A question then arises. Was the genetic variation in the form of streptomycin resistance induced by the streptomycin? Or did the mutation occur in the absence of streptomycin, and only reveal itself when the antibiotic was added to the culture medium? The answer is that streptomycin did not induce the mutation; the mutation arose spontaneously and in the absence of streptomycin. This can be demonstrated in the following way.

A culture of *E. coli* is spread on a medium lacking streptomycin. Colonies will occur, as indicated by Fig. 28.10. If we then gently press a sterile velvet disc onto the culture plate, this disc can be removed and pressed onto another petri dish containing a medium containing streptomycin. The colonies of bacteria will be transferred by the nap of the velvet in the exact pattern as found in the original dish. Most of the colonies—those formed by nonresistant cells—will die. Occasional colonies will arise, however, and from their position on the plate they can be identified with their sister colonies on the nonstreptomycin-containing medium. These sister colonies can again be tested. They will be found to be streptomycin-resistant even though they had never previously been exposed to the antibiotic.

The resistant cells, of course, could not have anticipated that sometime they would be exposed to streptomycin. This is only one of hundreds of substances that can kill *E. coli*. In the absence of streptomycin, the resistant trait is a useless one so far as we know. Yet it appeared. Adaptation, therefore, is a makeshift event, having no purpose until an unforeseen change in the environment allowed the variation to be expressed. The frequency of such mutations is low, one in 10^8 or 10^9 cells possessing the variation. However, the fact that such random variations occur indicates that most organisms have a greater storehouse of variability than will ever find adaptive expression.

The case of adaptation just described is a simple one, easily understood in genetic terms. A somewhat more complicated instance is that known as **industrial melanism** among moths. As a phenomenon, it is not restricted to any particular species of

Fig. 28.11 These two moths, each having a different color form, inhabit an industrial area of England. The light colored form is the ancestral one; the dark form, resulting from randomly-occurring mutations, increased in numbers in the soot-covered areas. Birds, which preyed on the moths, found it more difficult to detect the dark colored form.

moth—it occurs in about 70 different species, most of which do not interbreed—and it can be described as the acquistion of protective coloration. Let us consider the case of *Biston betularia*.

As Fig. 28.11 shows, the usual moth is a light-colored specimen. When not in flight, the moth rests on lichen-encrusted trees, a position in which it blends into the background. It is not readily visible to predators. About the middle of the 19th Century, a few melanic (dark) forms made their appearance in the vicinity of the industrial city of Manchester, England. At this time, the soot from factories and homes was being deposited throughout the neighborhood, killing off the lichens and blackening the tree trunks. Against this background, the melanic form blended easily, while the normal lighter form was relatively conspicuous. The dark form spread rapidly, and was also noticed in the vicinity of other industrialized areas. It was given the variety name of *carbonaria*.

It seems clear that the melanic color is an adaptation to a changing environment. To prove this, H. B. D. Kettlewell, an English biologist, showed that the moths are eaten by birds. He released a number of light and melanic individuals in sooty and nonsooty areas, noticed their resting places, and then after a period of time counted their numbers. In the sooty areas, the light form was eaten more frequently; in nonsooty areas, the melanic form was more readily detected by birds. Ease of concealment, and thus survival, is, therefore, a function of both body coloration and the character of the physical environment.

Coloration is generally governed by a number of genes. In genetic tests that have been conducted, melanism is governed by dominant genes, and this accounts for the rapid spread of melanism. Why would the spread be less rapid if melanism were controlled by recessive genes?

Other examples of adaptation are so complicated that we stand little chance of getting at the genetic basis of them (Fig. 28.12). In Australia, for example, the male bower bird goes through an elaborate nuptial ceremony. He builds an intricate bower or nest, collects bright stones or other objects for display, sings in a special way, and performs a nuptial dance; all this to make himself acceptable to a female of the species. Is this necessary for reproduction, which is the ultimate criterion of adaptive success? Is the mating ritual of the male a means of recognition, or is it tied closely to a physiological response related to reproduction? Answers are not easy, but we can guess that if the mating ritual of the male repelled rather than attracted the female, the mating ritual would quickly disappear as harmful adaptive behavior.

In general, organisms possess a wealth of genetic variation. This arises by mutation, and without foresight or purpose. If

Fig. 28.12 This sequence of photographs shows the courtship display of the Laysan albatross on Midway Island in the Pacific. It can be assumed that behavior of this sort, which often reaches a highly complex form of ritual, is a heritable trait and a necessary prelude to successful reproduction.

the variation is in harmony with the environment, in such a way as to promote reproductive success, it will stand a chance of being perpetuated. This is the source of the diversity of life. If the variation is harmful, it will be eliminated. If it arises at the wrong time, it will probably be lost. Furthermore, an adaptive variation in one environment may be useless or even harmful in another. Evolution, therefore, appears to be governed by chance, chance that the right kind of variation appears in the right kind of environment and in the right kind of organism.

The Origin of Life

If the diversity of life now present on the Earth evolved from past forms of life, and if the unity of life, expressed through homologous structures and similar pathways of metabolism, indicates a common ancestry, then we must assume that life in all of its aspects has a developmental history. If this is so then there was a beginning some time in the past.

All of us know that the world around us can be divided into two more or less distinct systems: the living and nonliving. If there was a beginning, did these systems arise at the same time, or did one precede the other? Older theories advanced the idea that life was specially created from the nonliving. The creation myths of ancient peoples in Babylonia, Greece, Egypt, India, and the early Indian civilizations of the Americas all held that life arose spontaneously from the nonliving by supernatural or divine intervention.

Such ideas are also part of our great Judaic-Christian traditions. However, the work of Pasteur showed that life does not arise spontaneously *under present conditions,* and Darwin's theory of evolution tells us only how life evolved, not how it originated. Until recently, scientists could not deal meaningfully with the question of the origin of life. They had no way of getting beyond the point that all organic molecules—that is, all those containing carbon atoms—were the products of living cells. If this were so, they could not conceive of any way by which nonliving matter could become living matter. But in 1923, A. I. Oparin, a Russian biologist, suggested how this might be possible. His book, *The Origin of Life,* was the beginning of new approaches to this perplexing problem.

The Planet Earth

The part of the universe we can observe contains millions of galaxies. Our own galaxy, the Milky Way, contains about 100 billion stars. The Sun, around which the

Earth revolves, is a star of medium size, and is located near the edge of the Milky Way. The Solar System consists of the Sun at the center, plus the nine planets revolving around it and held in position by gravitational attraction. The Solar System is about nine billion miles in diameter, and the order of the planets, moving from the Sun outward, is Mercury, Venus, Earth, Mars, Jupiter, Saturn, Uranus, Neptune, and Pluto. There are 31 smaller satellites revolving around the planets; e.g., our Moon.

The age and origin of the universe is unknown. One idea is that it has no beginning and no end. This is a difficult idea to handle intellectually; it is also believed that about 10 billion years ago a dense core of primordial matter exploded. The effects of the explosion are still evident in the universe, for the galaxies are moving outward and away from each other at enormous speeds. The expanding material eventually thinned out, cooled, and then condensed into the present galaxies, stars, and planets. Condensation of such enormous masses of material produces heat, and our Sun and the planet Earth became molten as condensation proceeded. The Sun, with a diameter of about 850,000 miles, generated enough heat to remain in a gaseous state, with temperatures that range from 6,000°C at the surface to about 25,000,000°C at its center. The most prevalent gas is hydrogen, and at the temperature of the Sun's core hydrogen fuses and forms helium, the reaction being the same as that occurring in the explosion of a hydrogen bomb. When the hydrogen is exhausted through transformation into helium, our sun will become a dead star, no longer a source of light and heat for the Earth. Life on Earth will then cease.

The Earth is too small to generate enough heat to sustain nuclear reactions, but it achieved a molten state at first and then gradually cooled, forming a crust at its surface. As it cooled, the crust cracked, wrinkled and formed an irregular surface. This process is still going on today and accounts, in part, for the earthquakes that occur periodically. Water was formed during the cooling process, but it could not, of course, exist as a liquid until the Earth's surface cooled enough to allow it to condense and form the oceans. But once this happened, the stage was set for the beginnings of life.

In the Beginning . . .

If we compare the living with the nonliving, three things stand out: 1. life is built around the carbon atom; 2. water is a major component of all living systems; 3. life exists within a rather narrow range of temperatures.

Chapters 12 through 19 deal primarily with the chemistry of

the carbon atom in various configurations. Other chapters have dealt with cellular structures, each one of which has carbon as a central atomic ingredient. Carbon is, therefore, unique among the elements in being very much a part of living systems known to us. This uniqueness lies in the ability of the carbon atom to form stable molecular configurations. It can be bonded equally well with hydrogen, oxygen, nitrogen, and other elements. In a variety of combinations it forms molecules as small as carbon monoxide and as large as proteins and nucleic acids, which have molecular weights of many millions. The variety of combinations, considering both size and structure, into which carbon can enter is almost infinite, but of equal importance is the fact that these molecules are exceedingly versatile in function and in their ability to capture, retain, and transfer energy. If life exists on other planets in other solar systems, biologists believe that such life must be based on the carbon atom.

Of comparable significance is the water molecule. Life as we know it could not have originated in its absence, and the patterns of life we see today, and that have existed in the past, have been determined by the presence and relative abundance of water. The importance of water lies in its two major properties: it is a molecule of great stability, and it is the "universal solvent" of biological systems.

Water exists as a gas, a liquid, and a solid, but it is as a liquid that water plays a biological role. The human body is about 70 per cent water, a jellyfish is about 98 per cent water. Water is the solvent in which metabolic reactions go on; chemical substances ranging in size from ions to huge molecules exist as water solutions or suspensions. Water is the major vehicle for the transport of substances in and out of the body, from one part of the body to another, and in and out of cells. But water is not so stable that it is inert. We have already seen that it participates in numerous reactions such as the hydrolysis of carbohydrates, fats, and proteins, and in the process of photosynthesis.

Temperature is also a limiting factor for the existence of life. If the environment is too hot, molecules cannot achieve great size and complexity. Increased heat may speed up the formation of biologically important molecules, but it also increases the rate of breakdown. Decomposition of molecules cannot outrun the rate of synthesis if structures are to be built up. The environment cannot be too cool, either. Although the synthesis of molecules might go on, although slowly, breakdown would also be slow. As a result, nature would arrive at a point of stagnation. There would be a great complexity of molecules, but there would be little change. Life is ceaseless change. There is constant activity at the molecular level. The breakdown of molecules competes with synthesis; simplicity of molecular structure competes with complexity; and all the while energy is being introduced, transferred, and converted by means of molecular systems.

Above all else, life is associated with carbon atoms, water, and energy. How were these arranged, resulting in a substance said to be *living?* Since it happened in the past, why isn't it happening now? Charles Darwin considered this problem in 1871, and he wrote:

> "It is often said that all the conditions for the first production of a living organism are now present, which could ever have been present. But if (and oh! what a big if) we could conceive in some warm little pond, with all sorts of ammonia and phosphoric salts, light, heat, electricity, etc. present, that a protein compound was chemically formed ready to undergo still more complex changes, at the present day such matter would be instantly devoured or absorbed, which would not have been the case before living creatures were formed."

The point of view expressed by Darwin is not far different from that which we would offer. When the Earth was first formed conditions must have been so harsh that no living thing could have survived, had there been living things around. But the Earth was going through its own evolution, and conditions changed. What were these conditions? Darwin's "warm little pond" was probably one of them. Since oxygen and carbon dioxide are largely by-products of life, the early atmosphere probably did not contain them to any great degree. But hydrogen, ammonia (NH_3), and methane (swamp gas, CH_4) probably were plentiful. Water was also abundant, and was warmed by the heat of the gradually cooling Earth. Energy, too, was available in the form of lightning and in the form of ultraviolet light from the Sun.

If these were the conditions found at the surface of our planet during its prelife state, then a model experiment can be carried out to test the idea that organic molecules can form spontaneously in such conditions. This was done by Stanley Miller and Harold Urey at the University of Chicago in 1953. They exposed a solution of hydrogen, water, ammonia, and methane to electrical discharges and to ultraviolet light for a period of 24 hours, then analyzed the solution. A number of low molecular weight carbon compounds were formed, many of them acids, but most interestingly and surprisingly, a number of amino acids were also formed. These are the building blocks of proteins. By the addition of other materials believed to have been present on the Earth during its prelife state—particularly hydrogen cyanide

(HCN)—purines, and pyrimidines, the building blocks of nucleic acids, appeared in the solution. Only the carbohydrates and fats seemed to be lacking.

Darwin's "warm little pond" could, therefore, become a rich broth of chemical compounds, increasing in concentration as time passed. All of these compounds contained carbon in various combinations with itself and with other elements. Those that were stable persisted; those that were unstable or reactive broke down or were reconverted to other molecular forms. Natural selection took place at the chemical level, and we must assume therefore, that a long period of chemical evolution preceded biological evolution. The broth, furthermore, was sterile, since no life as yet existed.

The broth thickened with time, and then somewhere, somehow, a self-replicating system arose. The geological record indicates that this probably occurred about three billion years ago. Our knowledge at this point is fragmentary but primitive self-replicating systems with sources of energy have been produced in the laboratory. Once formed, such a system could draw from the broth the nutrients it needed for "survival," and eventually encase itself in a semipermeable membrane, thus permitting various molecules to enter and leave the system on a selective basis.

The system—we may, indeed, call it a primitive cell—was heterotrophic; its source of energy was from preformed compounds outside of the cell. However, such a cell and its descendants would soon exhaust its immediate environment of nutrient materials, and we can only assume that natural selection would lead to the evolution and survival of more complicated cellular machinery capable of making some of its own nutritional substances. Enzymes would consequently be required (but amino acids required for their formation were available). In addition, we must assume that a shift in the source of light energy took place. As a result of the use of water and carbon dioxide, molecular oxygen (O_2) was formed and released from the surface of the Earth as a gas. Forming a layer around the Earth, the oxygen would screen out much of the ultraviolet light. The evolution of compounds that could capture the energy of visible light would lead eventually to the process of photosynthesis and, consequently, to the major oxygen source for the living world to come. Autotrophy was then initiated. The fossil record supports such a succession of events. Biological evolution was now well on its way.

You might now say, and with good reason, that this story of the origin of life is highly speculative and far from proven. You would, of course, be quite correct. We have no knowledge of many of the intermediate steps; the jump from the Miller-Urey broth, or even from a primitive self-replicating system, to an integrated living cell is an enormous one. But two additional items support the argument. In the first place, there is an increasing number of experimental facts being accumulated, all of which point in this direction. And, secondly, time has been available for these processes to occur. A period of from two to three billion years passed from the time that the Earth first formed until life made its appearance. Most biologists believe that if conditions favoring the emergence of life are present, then life will inevitably arise in time as a natural consequence of changes in matter and energy. So among the many galaxies, stars, and planets of the universe, there is a high probability that life exists elsewhere, and that we are not alone in the immensity of space and time.

SUMMARY

Adaptation can be viewed as a state of being or as a process. As a state of being, adaptation is the sum total of all of the characteristics—including the element of chance—permitting an organism to live and reproduce. As a process, adaptation, or natural selection, is the manner by which evolutionary success is achieved. Since environmental changes occur, adaptation involves a continual adjustment of organisms to an environment. The bases for adaptation are the heritable variations in a population, and the fact that not all individuals in a population contribute their genes to the next generation.

Under such conditions, change is inevitable, and evolution proceeds. Therefore, all heritable adaptive changes contribute to evolution, but all evolutionary changes are not necessarily adaptive. The fossil record indicates that many species are now extinct, even though at one time these species were adaptively successful. Adaptation, consequently, is only a temporary success, and adaptation in one set of circumstances does not guarantee equal success under a different set of circumstances.

Darwin's theory of evolution through natural selection most satisfactorily explains the successive changes that have occurred as one generation of organisms succeeds another. More recent studies have demonstrated that variations result from changes in the base sequence of DNA, and from gains or losses of chromosomes or parts of chromosomes.

Life is believed to have originated from a nonliving state about three billion years ago. We are now beginning to have some understanding of how this might have happened, but we have far from a full knowledge of all the events.

FOR THOUGHT AND DISCUSSION

1 Make a list of species particularly well suited to the environment in which they exist. Try to analyze one species and show how it is adapted. What factors of the environment must be considered? Do you know of any organisms that would not be adapted if the care of man were relaxed or stopped? What features necessary for adaptation are missing in these organisms?

2 How can the rising up of a mountain chain such as the Appalachians or the Rockies affect the environment?

3 "A variable environment strongly promotes rapid evolution and may, in fact, be essential for speeding up evolutionary change." Explain this statement.

4 Evolution is essentially irreversible. Why?

5 The dolphin is a mammal, but in many of its features and behavior it is very similar to a fish. Is there any reason for this degree of parallellism?

6 The following animals all fly: butterfly, sparrow, bat, and flying fish. Are their flying organs homologous or analogous?

7 Man has been called a "lethal factor" in the environment for other organisms, and possibly for himself. What does this statement mean? What can be done about it? Could any other organism be so labeled? Explain.

8 It has been estimated that it may take a million years to form a new species of animal. What does this tell us about the causes of evolution?

9 There is much talk today about the human population explosion. What caused it in the first place? What will be the consequences if such a trend goes unchecked? Does it illustrate any facet of Darwinism?

10 We have stated that polyploidy is rare in sexually reproducing animals. Why should this be so? Why can plants tolerate polyploidy more successfully than animals?

11 Most mutations that arise are deleterious. Why should this be so?

12 What factors in the environment would keep a population constant in numbers?

13 Two similar species of birds occupy the same environment. When do they compete with each other, and when don't they compete?

14 Why do you think there are only two species of elephants and one species of man in existence, but thousands of species of *Drosophila?*

5 Why should a deletion or a duplication be likely to produce a dominant effect?
6 Why is reproductive isolation, by a geographical barrier, for instance, necessary for the formation of a new species?
7 A *rassenkreis* (race circle) is a continuous circle (often over a wide area of the range of an organism) of changing races. Assume that the races range from A to Z, with the ranges of A and Z overlapping. Assume also that A can interbreed with B, B with C, C with D, and so on, but that A cannot interbreed with Z. Can you explain what has happened?
8 In what ways can chance influence the evolutionary picture?
9 Suppose in a population breeding according to the Hardy-Weinberg principle that A for some reason now mutates to a at a slow rate, but a does not mutate back to A. What will be the consequences? What will they be if a, for some reason, breeds more rapidly than A?

SELECTED READINGS

Ehrensvard, G. *Life: Origin and Development.* University of Chicago, Phoenix Science Series, 1962.

An interesting and clear discussion of how we believe life might have originated on this Earth.

Greene, J. C. *The Death of Adam.* New York: Mentor paperback, 1959.

A history of the ideas of evolution from earliest time to Darwin.

Huxley, J. *Evolution in Action.* New York: Harper & Row, Publishers, Inc., 1953.

A series of fine essays on evolution as a process.

Simpson, G. G. *The Meaning of Evolution.* New York: Mentor paperback, 1949.

One of the finest discussions of the processes of evolution.

Stebbins, G. L. *The Processes of Organic Evolution.* Englewood Cliffs, N.J. Prentice-Hall, Inc., Concept Series paperback, 1966.

An excellent and up-to-date account of why evolution takes place.

Wallace, B., and A. M. Srb. *Adaptation.* Englewood Cliffs, N.J. Prentice-Hall, Foundations of Modern Biology Series, 1964.

A clear, well-written account of the meaning of adaptation and its relation of genetics and evolution.

29 THE ORIGINS OF MAN

Fig. 29.1 The skeleton of a man and that of a rearing horse reveal the broad similarities of bone number and arrangement. At the same time, modifications of the pelvis, tail, head, and appendages characteristic of each species can be seen.

The American Museum of Natural History

Man is an animal. If you were to dissect him, you would find the usual organs—heart, liver, lungs, stomach, and so on. They would differ very little, except in size, from similar organs in cats, horses, mice, monkeys (Fig. 29.1). At the cell level of structure it is almost impossible (except for a specialist) to distinguish among the liver cells of these animals, including man. If you were to study the physiology of these organs—their respiration and enzyme activity, for example—the same general chemical structure and behavior characterizes all of them. However, if you were to compare the nuclei of the cells of these various animals, you would find that each has a different number of chromosomes. Closer examination, by genetic techniques, would reveal that the genes in these chromosomes are also different, although many might well be similar. It is a long evolutionary road from fish to man, but these two groups share about 10 per cent of the same genes. It is the differences among the genes that make us human beings instead of some other animal.

Where does man fit into the animal kingdom? He has a vertebral column, which extends from his skull to the lower part of the back, which is made up of individual vertebrae, and through which runs the spinal cord. Man, therefore, belongs to the phylum Chordata. He is also a mammal, possessing mammary glands, and is put in the class Mammalia. It is true that he is a peculiar, almost hairless mammal, but so too is the whale, which has even less hair. Also, man walks upright on two legs, but kangaroos and some monkeys and apes are occasionally **bipedal** as well. But monkeys, apes, and men have many features in common. They are grouped in the order Primates, along with tree shrews, tarsiers, and lemurs. The common features of monkeys, apes, and man put them into a suborder Anthropoidea, and within this grouping the family of man, Hominidae, is found. The relationship of man and several of his animal relatives is shown in Fig. 29.2.

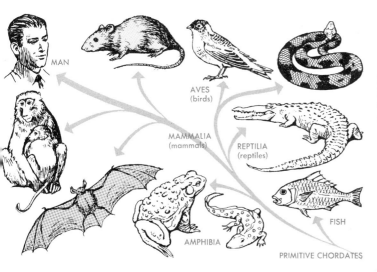

Fig. 29.2 The general evolutionary relationships among vertebrate animals, including man, are shown here.

Although man clearly is an animal, it is equally obvious that he is a unique animal. When we try to analyze this uniqueness, we can best sum it up by saying that man has developed a **culture,** something that no other living things have done. The term "culture" is, of course, a general one. Man is a maker and user of tools, a planner, inventor of symbols—spoken and written languages. Man also has a system of values. And he can do things with a purpose, not only for himself, but for future generations as well. All these add up to a cultural inheritance, a unique phenomenon which is different from the biological inheritance common to all living things.

The Beginnings of Man

Man is an adaptive creature, as much a product of evolution as any other living organism. Adam is a symbolic first man, but man's actual origin goes far back in geological history. We do not yet know all of the steps of his evolution, and the search for fossil man continues, but the main pathways are generally known and accepted.

Figure 29.3 shows some of man's primate relatives. None, however, is a direct ancestor, and it is not correct to state that present-day apes were man's ancestors. Rather both man and the modern apes had a common ancestor that was ape-like. This relation is also shown.

The first human fossil to be discovered was the skull fragments of Neanderthal man, found in 1856 near Düsseldorf, Germany (Fig. 29.4). Rudolf Virchow, the German physician, did not believe it to be a fossil man, but a skull of a modern man showing pathological deformities. Other discoveries, however, left no doubt about the antiquity of *Homo neanderthalensis,* and he is now known to have occupied parts of Europe some 70,000 to 40,000 years ago. He had a primitive culture. He

The American Museum of Natural History

Fig. 29.3 Two members of the primate group to which man belongs: the chimpanzee (above), and an Old World Rhesus monkey with her offspring.

Photo by Lynwood M. Chace

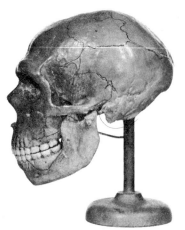

Fig. 29.4 The fossilized skull of an extinct species of man, *Homo neanderthalensis*.

Fig. 29.5 Reconstruction of the head of Neanderthal man.

buried his dead with food and weapons for an afterlife. Ultimately, he was probably overrun by other races of men who came in from eastern Asian regions. But he did not, apparently, give rise to modern man. The reconstruction of *Homo neanderthalensis* shows him as not too different from some of our more brutish appearing friends (Fig. 29.5), yet he belonged to a species different from modern man, *Homo sapiens*.

Fossil remains show that Neanderthal man was a relatively late arrival on the evolutionary stage. A long period of history preceded him. The geology of the Pleistocene Period, the Ice Age, can be used to date various fossil remains. The Ice Age was characterized by various periods when the ice advanced **(glacial)** and when it retreated **(interglacial)**. The earliest of these fossils, dated from the late Pliocene into the first interglacial period (2,000,000 to 700,000 years ago) have been called *Australopithecines* (*Australis* meaning "south"; *pithecus* meaning "ape"). Many fossils of these pre-men have been found. As a group, they were upright in stature, bipedal in their walk, about five feet tall, low browed, and had large teeth. Their cranial capacity was about 450–550 cc, just a bit larger than that of a chimpanzee (350–450 cc), but much smaller than that of modern man (1,200–1,500 cc). Were they really men, or apes, or something between? It is hard to say. The australopithecines are quite ape-like, but L. S. B. Leakey, the African anthropologist, believes that some of them, found with australopithecine fossils but differing from them, used fire and fashioned crude stone tools for hunting. These have been grouped as belonging to *Homo habilis*, meaning "man able to do things." One group of these early men has been named *Zinjanthropus*, after the area of Zinj, in Tanganyika, where fossil remains of this group have been found. Age determinations suggest that the fossils are about 1,750,000 years old, making them the oldest known "human" fossils. This may seem a very long time ago, but in geological terms it is very recent. Present records suggest that *Homo habilis* died out about 400,000 years ago (mid-Pleistocene).

Other fossilized human, or near-human, remains have been found in Java (*Pithecanthropus*, meaning "ape-man"), in China (*Sinanthropus*, meaning "Chinese man"), and in other parts of Africa and in Europe. They have been grouped as *Homo erectus*. Their relation to *H. habilis* and to modern man is still uncertain. *H. erectus* lived about 600,000 to 300,000 years ago. One thing we do know for certain is that, with the passage of time, these fossil remains show an increase in cranial capacity. *Sinanthropus* had a cranial capacity of 750 to 1,000 cc, while Cro-Magnon man, living in southern France and Spain about 50,000 to 75,000 years ago, had a brain case as large as today's adult. In addition,

The American Museum of Natural History

Fig. 29.6 Reconstruction of Cro-Magnon artists decorating the walls of a cave at Font de Gaume, in southern France. From what we can tell, cave painting of animals was performed as an act of magic intended to bring about successful hunting.

Cro-Magnon man must also have had a well-developed culture, if we are to judge from the remarkably beautiful paintings which he drew on the walls of his caves (Fig. 29.6). Certainly he must have been human in all respects.

When did modern man make his appearance? This is as hard to answer as the question of his direct ancestors, but 50,000 to 100,000 years ago is a reasonable guess. This was the time of the Wisconsin Ice Age. By then man was a skilled tool maker, had artistic ability, and was a domesticator of animals. Unfortunately, we are as much in the dark about his place of origin as we are about his time of origin. About all we can say is that he lived in many parts of Europe, Asia, Africa, and inhabited the large islands of the Pacific Ocean, and that the change from australopithecine to *H. habilis*, to *H. erectus* to *H. sapiens* was a gradual one taking place over a long period of time. Out of these have come today's races of men.

Races of Man

One useful definition of a species is that it consists of a group of populations actually or potentially capable of interbreeding, but not generally capable of cross-breeding with other species. Only one species of man, *Homo sapiens*, has survived, although there were probably other species who lived in the past. Human beings, however, fall into several races, superficially distinct from each other in skeletal and surface features but overlapping in many other characteristics, and quite capable of mixing and producing fertile offspring. The different races arose through the isolation of groups by geographic or social barriers, and because of these barriers different groups of

Swiss National Tourist Office

Ewing Galloway

Ewing Galloway

Fig. 29.7 Three types of Caucasians: a Swiss mountain man (top); an Indian (Asian) ivory carver (middle); and a Portuguese country girl.

genes were isolated. The origin and perpetuation of the races of man are explained that simply.

The number of races of man depends upon how closely one defines the term "race." It is a scientifically difficult task, and one that is made more difficult by emotions. Within each race—such as the white, or **Caucasian,** race—there is a great deal of variation. When we take all racial variations into account, it becomes clear that man is highly variable, and that the races overlap each other in many ways. This must mean that there is a tremendous store of genetic variation in mankind which is unevenly distributed among the peoples of the earth. Intermarriage between races means that distinctions are even more difficult to make.

The major, but not the only, races of man are three: **Caucasoid** or white, **Negroid** or black, and **Mongoloid** or yellow. Their distribution, despite a vast amount of migration, is basically geographical: the Caucasoids inhabit Europe and migrated to the Americas; the Negroids are African, the Mongoloids Asian. Let us consider some of the characteristics of each, and the existence of **subraces.**

The Caucasoid whites get their name from the Caucasus region of eastern Europe. They are characterized generally as follows: they have large, heavy bones, well-developed muscles, hair that is wavy or straight, but never woolly or kinky, much body hair, heavy beards in males, a straight narrow nose, blue to brown eye-color, and skin that ranges from white to light brown. Within the race there are many subraces (see Fig. 29.7): Mediterranean, Nordic, Alpine, Lapp, Irano-Afghan, Indian (Asian, not North American), and some African are among them. As Table 29.1 shows, each has certain features or groups of features that distinguish them as "types."

The Negroid race is equally variable both in stature and facial features. Very probably the Negroid race is a mixture of types. The skin varies from dark brown to black, hair is black and woolly or kinky, the nose is broad and flat, the ears are small, lips thick, upper jaw protruding, and beard and body hair are sparse. Although concentrated in central and southern Africa, the Negroid race is also found in Malaysia, the Philippines and many of the Oceanic islands. Where they encounter members of the Caucasian race, intermixture has taken place to a great extent, leading to many different blends of racial characteristics.

In stature, the Negroid race includes the extremes. The pygmy or Negrillo, of equatorial Africa, the bushman of southern Africa, and the Hottentot of western South Africa are five feet or under in height. The Negritos of Asia and the Oceanic Islands are also short. But the Negroes of the upper reaches of the Nile River—

of the Dinka tribe—average more than six feet in height, and are commonly over seven feet.

The Mongoloid race is characterized by a yellowish skin, straight black hair, flat face and nose, high cheek-bones, little development of the brow ridges and **epicanthic fold** of the upper eyelid that gives the eyes a slit-like appearance. The body tends to be short and thick and there is a sparseness of beard and body hair. The Mongoloid race includes the Chinese, Japanese,

Fig. 29.8 Two members of the Negro race: A young Zulu from Rhodesia (top); and a Maori tribal chief, with tatooed face and wearing a cloak of fine feathers, from the islands of the southwest Pacific.

TABLE 29.1 Characteristics of some Caucasoid sub-races

Source	Characteristics
Mediterranean	Main sub-race of Caucasoids. Light build, medium stature, rather long head and dark complexion. Originated near Palestine and migrated westward along the shores of the Mediterranean into Greece, Italy, southern France, Spain (north shore), and into Egypt and Arabia (south shore). The Basques differ in having a narrow face and narrow, prominent nose. The Irano-Afghan differ in having a high head, long face with high-bridged nose, and taller stature.
Alpines	Round head, thick-set build, sallow complexion, and broad nose. Represented by some of the French, Bavarians, Swiss, and northern Italians.
Nordic	Tall, blond, long head, long face, narrow nose, deep chin. Represented by inhabitants of Sweden and eastern Norway. The Anglo-Saxon type of North Germany and England differs in being heavier boned, more rugged, rounder head, broader nose, and more prominent cheek bones. Invasions of Europe by Angles, Saxons, Jutes, Danes, Goths, and Vandals spread the Nordic features widely.
Lapps	Round head, short stature, forehead steep with no brow ridges, face short flat and broad, small jaws, small teeth, projecting cheek bones, hair generally dark and straight, legs short but arms long, hands and feet small. Live in forested highlands of Sweden, northern coasts of Norway and the tundra area of northern Finland.
Indians (Asian)	Difficult to characterize because of many variants, but darker skinned than the Mediterraneans, of medium stature, small face and chin, dark somewhat wavy hair. The Ceylonese are somewhat darker, while the Sikhs are taller, more heavily built, and have strong beards. Includes most of the Indian population.

Fig. 29.9 This group of young Hawaiian women show the results of racial mixing. The group derives from Hawaiian, Japanese, Chinese, English, Scotch, and Irish ancestry.

Hawaii Visitors Bureau Photo

Koreans, Siamese, Burmese, Tartars, Tibetans, the Dyaks of Borneo, Eskimos, some Filipinos, and American Indians.

Wherever the Mongoloid has met the white or black race mixing has occurred. The American Indian, who probably crossed the Bering Strait from Asia 10,000 to 20,000 years ago, is not typically Mongoloid. Possibly he is a product of interbreeding between the Mongoloid and the Ainus of Japan, an ancient white stock that may have been the ancestor of the Mongoloid race. We must recognize that man has always been a mobile animal, moving constantly over the face of the Earth. It is not surprising, then, that races having little variation, such as the Australian aborigines, occur only where isolation has existed for long periods of time. Mixing of races has been the rule, hence the many subraces (Fig. 29.9).

Climate and Race

We have said that man is the product of evolution. He should, therefore, reflect the selective pressures exerted on him by the environment. Through natural selection, some heritable characters are favored while others selected against, particularly if the environmental conditions are severe. The white race, generally occupying the temperate zone, does not show extreme specialization. Where extremes of heat and cold are found, the races occupying these areas follow three general rules of adaptation to climate. These rules have been derived from studies of certain animal populations, but they seem to be applicable to man as well.

Gloger's Rule This states that pigmentation is greatest in warm and humid areas. Certainly the black tribes of Africa reflect this adaptation, and no dark races are found in the cold zones.

Bergman's Rule This states that animals are smaller in warmer than in cooler zones. This may seem to be contradicted by the Tungus of eastern Siberia, but if one interprets the rule in terms of body build rather than stature, agreement is found. The pygmies of Africa, for example, are a slender race with long extremities while those peoples around the Arctic Circle are thick-set, short, well padded with fat and have flat faces. It is thought that the Tungus arose during the intense cold of the last ice age, about 25,000 years ago. In terms of surface to volume, they present the least surface area, and therefore radiate very little heat. The pygmies, however, have a large surface area relative to the volume of their bodies (Fig. 29.10).

Allen's Rule This states that animals living in cold areas have shorter extremities than animals living in warm areas. This is true for man as well. The Eskimos, for example, have short toes, legs, fingers, and noses. White races of the desert, such as the Arabian Berbers, have long slender limbs and, often, prominent noses.

We should not, of course, apply these rules too closely. The body of man is relatively unspecialized. He has interbred widely, is quite mobile, and his mastery in making clothing, housing, tools and fire gives him a control over the environment possessed by no other animal.

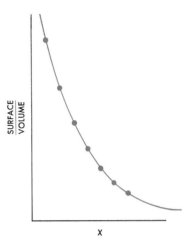

Fig. 29.10 This graph shows how the surface: volume ratio decreases as a body of constant shape (any shape, including the human body) increases in size. (X = size, and increases from left to right.)

Genetics of Man

The genetics of man are no different from that of other mammals. Human DNA is like that of other organisms, differing only in the number and kinds of genes. The rules of transmission of heritable traits apply to man as well. The methods of human genetics, however, may seem somewhat different. Although biologically possible, it is not socially possible to breed human beings selectively as we do fruit flies or mice. But studying the pedigrees of human families is no different from a study of other organisms, even though one generally has to work with few numbers of individuals, and the generation times are quite long (Fig. 29.11).

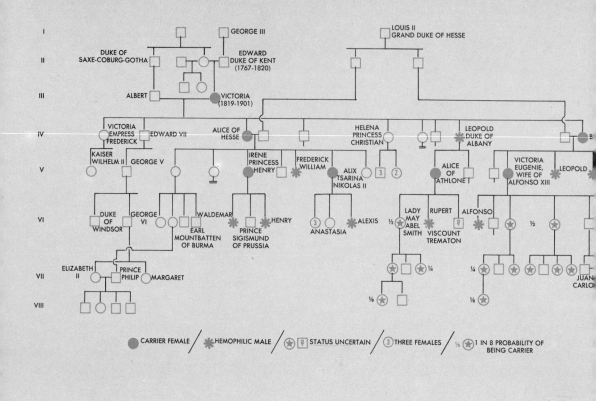

Fig. 29.11 Pedigree of European royalty and the transmission of the gene for hemophilia. Every female carrier and every hemophilic male can be traced back to Queen Victoria of England. The mutation probably originated in her. The gene responsible for the hemophilic trait is on the X chromosome. The transmission can be understood if it is remembered that the human female is XX and the male XY (the Y chromosome carries no known genes on it).

Nevertheless, much is known of man's inheritance. On a racial basis, however, details are available only for blood types. The major groupings of individuals are A, B, AB and O. These correspond to six genotypes: AA, BB, AB, AO, BO and OO. AA and AO produce the same phenotype, as does BB and BO. When plotted on a racial basis, it is found that these traits are not randomly distributed. For example, type B is not found in the American Indian, the Spanish Basque population, or among the aboriginies of Australia. These races are only remotely related, so the absence of this blood type—and the responsible gene—can be attributed to the fact that these races arose through migration by small groups lacking B. Continued isolation preserved the genetic difference. This seems a more reasonable explanation than that type B was eliminated by rigorous selection.

Other genes, as might be expected, show similar disrupted distributions. Two more might be mentioned, both similar in their effect on populations. A disease known as **thalassemia,** genetically determined, is confined to peoples of Mediterranean (generally Italian) origin. The homozygous (tt) thalassemics die early of severe anemia. One would imagine that this would tend to eliminate the responsible gene from a population, and that its frequency would be determined only by its mutation rate, $T \to t$. However, t had a selective advantage: the heterozygotes

(Tt) did not contract malaria, but the homozygous normal (TT) did. An equilibrium would tend to become established so that the rate of deaths from anemia (in the tt's) would be balanced by the rate of death from malaria (in the TT's). The heterozygote, Tt, having 50 per cent abnormal hemoglobin and 50 per cent normal hemoglobin, is a favored genetic combination. In the absence of malaria, the t gene would be selected against at a rapid rate. The situation is similar to that earlier described for hemoglobin S. This trait, however, is confined almost entirely to the Negroids, with only rare occurrences among some of the Caucasoids of India and the eastern Mediterranean countries.

The Evolution of Modern Man

Modern man had his origin in the Paleolithic, or Old Stone Age. This was a food-gathering stage. His way of life was not appreciably different from that of his primate ancestors except that man had developed tools to assist in killing game (Fig. 29.12). The population was probably made up of small groups, and the distinctiveness of their tools suggests that isolation was the rule. The flow or exchange of genes, through interbreeding, was probably slight even though man was nomadic in his habits, providing ideal conditions for rapid evolutionary change and distinct racial types. As hunters and food-gatherers, probably only a single small group occupied any given area.

The American Museum of Natural History

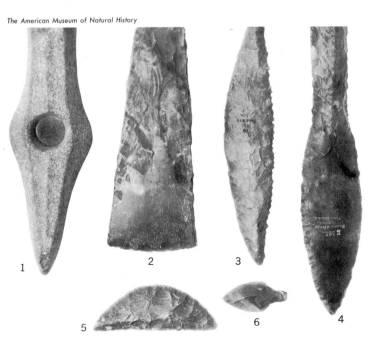

Fig. 29.12 Implements of later Stone Age man: an ax hammer (1); a flint ax with one edge polished (2); a flint saw with one edge notched (3); a flint dagger (4); a flint knife (5); an arrowhead (6).

Such groups still exist, although all are likely to lose their uniqueness. Before civilized man changed his way of life, the Eskimos, the Fuegians of South America, the Australian aborigines and the Bushman of Africa had not progressed much beyond the Paleolithic (Old Stone) stage. But ease of travel has made it difficult for isolation to persist, and these peoples will one day blend with the general population around them.

Probably by the end of the Pleistocene (10,000 to 25,000 years ago), the major races of men were established, and a gradual change from food gathering to food-producing took place. Possibly this change came about first among the inhabitants along lake shores or coastal areas, for the fisherman is a more stable inhabitant than the hunter. The time was about 7000 B.C. The principal area probably was in the Middle East, although some would argue for southeastern Asia. From these places the herdsmen and farmers spread into Europe, North Africa, India, and China, mixing with native groups as the migrations proceeded.

Food producing requires the domestication of plants as well as animals. With the exception of the cultivation of rubber and quinine trees, every domesticated plant and animal had been grown by man before recorded history. Vast improvements in quality and productiveness have been accomplished, but new domesticated species have been rare. While the dog or the pig was probably the first domesticated animal, the basis of the earliest food-producing communities was the tuber-producing perennial plants having a high starch content. Later came cultures of seed producers such as the cereals—rice, wheat, and corn. Both kinds of plants are easily grown in large quantities, can be stored easily, and are easily transported. This being so, it would appear that agriculture, as a practice, arose independently in several regions. Certainly it would be reasonable to suppose that the corn-based economy of the American Indian— first the Mayan and Incan civilizations of South America, and later that of North America—was independent of the agricultural development of the Middle East. The cereals give stability to communities, the stored grains lasting over periods of poor harvests.

The domestication of animals appears not to have been a necessary step toward the establishment of communities. The South American Indians, with their highly developed culture, had virtually no animals as either a source of food or as beasts of burden. As a food-gatherer, early man was most likely a carnivore, adding to his meat diet such roots and berries as he could find. As he became more settled he shifted to a vegetarian diet, although his taste for meat remained and was satisfied initially by game and fish.

With the establishment of communities capable of sustaining themselves through agriculture, man turned his increased leisure time to other things. Out of this has come the culture that sets man apart from his fellow animals. Central to the development of culture was the development of the art of communication, particularly that of the written language. It is this art which forms the basis of our cultural inheritance, the means whereby history and tradition are preserved, and rapid instruction becomes possible. The stored information on the magnetic tape of a computer is simply the latest step in this long process of communicative art (Fig. 29.13).

In the future the evolution of our cultural inheritance will play a greater role in the development of civilization than will the evolution of our biological inheritance. One of the most profound aspects of this cultural inheritance has been the development of science. Out of man's curiosity of the world about him, and his desire and need to control the environment, he has reached a point, where he can control his own evolution, if he so chooses. He can do this wisely or carelessly, but he cannot escape the fact that he cannot long exist in an unfavorable environment. It is for this reason that solutions are needed for the population explosion, the degree of purity of the air we breathe and the water we drink, the controlled use of our natural resources, the proper distribution of food throughout the world, and the preservation of our wilderness areas.

Fig. 29.13 Ever since man invented language he has been a storer of information. At right, a baked clay hexagonal prism covered with cuneiform (wedge) writing, describes the siege of Jerusalem by Sennacherib, king of the Assyrians, in 686 B.C. Below, many miles of magnetic tape are used to store information in a minimum of space.

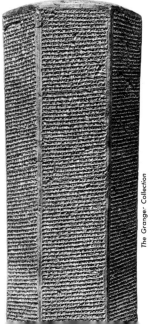

SUMMARY

Man is a vertebrate mammal belonging to the Primate groups, which include the tree shrews, lemurs, tarsiers, monkeys, and apes. He is a product of evolution like any other animal, and his evolutionary history can be traced back through fossils more than one million years. As man gradually evolved, he assumed an upright stature and a bipedal walk; he learned to use fire and to make tools. He was first a hunter, then a farmer, and later a builder of cities and nations. In the process he domesticated plants and animals for his use, and gained an increasing measure of control over his environment.

Among his unique features are his hands, which have opposable thumbs, his much enlarged brain, and his use of a complex language for communication. These features have given rise to human cultures; man today is evolving more rapidly in a cultural than in a biological way.

Man is one species consisting of many races. The races arose (probably) from small isolated groups in which certain variations were fixed by natural selection. The mobility of man, and continued interbreeding between races, tends to break down the distinction between races.

FOR THOUGHT AND DISCUSSION

1. What physical and behavioral traits does man share with his primate relatives?
2. What role does modern medicine play in affecting the genetic variability of mankind?
3. Do you consider human values such as *honesty, generosity,* and *compassion* to have an evolutionary significance? Explain. Does one find these attributes expressed in other animals?
4. Some of you may have read the novel, *The Lord of the Flies.* Can you interpret this book in biological terms?
5. Works of art are expressions of human behavior and feeling. Are they cultural or biological expressions? How do you distinguish between these aspects?
6. Our Constitution states that Americans are born "free" and "equal". Comment on this from a biological point of view. From a cultural point of view.
7. Many animals exhibit territorialism, or the tendency to defend a territory against invasion by other members of the same species. What biological significance is there in this behavior? Is man a territorial animal? If your answer is yes,

is this an acquired cultural attribute, or the retention of a biological attribute?

8 The enlargement of man's brain to its present size was a late development in human evolution. It followed the development of upright stature, grasping hand, and opposable thumb. Does this seem to be a logical sequence? Explain your answer.

9 The dolphin is an intelligent animal with a large brain. Do you think that it is capable of evolving a culture? Give your reasons.

10 Change and chance, rather than purpose and plan, are characteristic of biological evolution. What does this statement mean to you? Can you defend it if it is applied to man? Can the same be said of cultural evolution?

11 What role has science played in the development of our culture? Consider this question particularly in terms of language, education, and philosophy. Can science have any effect on our biological evolution? Explain.

12 The literary works of man are reflection of his cultural experience. They have their origin in man's ability to think abstractly, and to symbolize his thoughts in the form of words. Below are selected literary excerpts, all expressing a point of view of nature as seen through the eyes of the writer. Evaluate these in the light of your knowledge of biology in particular, and of science in general. Do you agree with the statements? Are they valid scientifically? If they are valid, what scientific thought is being expressed? How would these thoughts be expressed by a scientist? If they are not valid, to what extent do you feel that "poetic licence," is justified?

(a) *Man is incomprehensible without nature, and nature is incomprehensible apart from man. . . .* Hamilton Wright Mabie.

(b) *And a mouse is miracle enough to stagger sextillions of infidels. . . .* Walt Whitman.

(c) *One secret of success in observing nature is the capacity to take a hint; a hair may show where a lion is hid. One must put this and that together, and value bits and shreds. Much alloy exists with the truth. The gold of nature does not look like gold at first sight. It must be smelted and refined in the mind of the observer. And one must crush mountains of quartz and wash hills of sand to get to it. To know the indications is the main matter. . . .* John Burroughs.

(d) *Mountains are earth's undecaying monuments. . . .* Nathaniel Hawthorne.

(e) *Man can have but one interest in nature, namely, to see himself reflected or interpreted there. . . .* John Burroughs.

(f) *Nature, like a loving mother, is ever trying to keep land and sea, mountain and valley, each in its place, to hush the angry winds and waves, balance the extremes of heat and cold, of rain and drought, that peace, harmony and beauty may reign supreme.* . . . Elizabeth Cady Stanton.

(g) *Nature is the most thrifty thing in the world; she never wastes anything; she undergoes change, but there's no annihilation. . . . the essence remains.* . . . Thomas Binney

SELECTED READINGS

Baker, H. G. *Plants and Civilization.* Belmont, California: Wadsworth Publishing Co., Fundamentals of Botany Series, 1965.

A fine volume on the dependence and relations of plants to man.

Bates, Marston. *Man in Nature.* Englewood Cliffs, N.J.: Prentice-Hall, Inc., Foundations of Modern Biology, 1964 (2nd ed.).

Deals with man as an animal, his ancestors, the varieties of man, problems of domestication, health and disease, and the environment.

Berrill, N. J. *Man's Emerging Mind.* Greenwich, Conn.: Premier paperback, 1965.

The story of man's progress through time.

Brodrick, A. H. *Man and his Ancestry.* Greenwich, Conn.: Premier paperback, 1964.

The story of man and his primate relatives.

Dobzhansky, Th. *The Biological Basis of Human Freedom.* New York: Columbia University Press, 1956.

A philosophical discussion of the implications of modern biology in understandable nonscientific terms.

Dobzhansky, Th. *Heredity and the Nature of Man.* New York: Signet Science Library, 1964.

An excellent book dealing with the nature of beauty, the variety of human differences, the meaning of race, the genetic problems of mankind, and the future of man.

Dunn, L. C., and Th. Dobzhansky. *Heredity, Race and Society.* New York: Mentor paperback, 1946.

A sound description of the basis of human differences.

Eiseley, Loren. *The Firmament of Time.* New York: Atheneum Publishers, 1960.

The history of science as it aids in our understanding of man.

Harrison, R. J. *Man, the Peculiar Animal.* Baltimore, Md.: Pelican paperback, 1938.

Man's structures and functions are contrasted with those of other animals.

PART THREE
The Green Plant

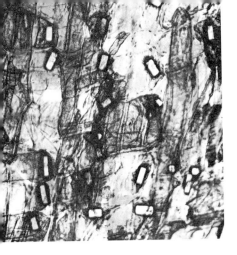

ABOUT THE PHOTOGRAPH

Photomicrograph of the dry outer skin of an onion. Under polarized light, crystals of calcium oxalate, which are embedded in the cells, stand out clearly in red and yellow-orange. They are a product of the plant's metabolism. The color difference between some crystals is not significant; it is an optical effect resulting from the position of an individual crystal in the path of the polarized light. Magnification on the cover is approximately 200×.

Photograph: Eric V. Gravé

ABOUT PART THREE

In Part 1 we took a sweeping view of the plant kingdom and saw how plants are classified—from simple one-celled algae to the giant redwoods. In Part 2 we were concerned with plants on the cellular level and on the molecular level. In this part of the book we shall take a close view of green plants.

The significance of green plants in the evolution and maintenance of living things cannot be overemphasized. Nearly all of the land-dwelling plants have descended from the green algae. With the exception of the fungi, all contain chlorophyll or are related to those that do. Through the absorption of light by chlorophyll and the conversion of carbon dioxide and water into carbohydrates by the process of photosynthesis, the life-giving energy of the Sun is trapped and made available to living things. The green plants, from the algae of the sea to the great trees of the forest, represent the first step in the conversion of energy. Then fungi, bacteria, and animals that feed on living and dead plants reuse the energy, and the energy is again made use of by animals that feed on animals. In past geological ages, much of this energy was stored in the form of fossil fuels—coal, oil, and gas—which man now uses to keep his civilization going.

Today our land masses are covered with a mantle of green. Only where there is too little water, or where man occupies the surface of the earth, are the green plants absent. They shade the ground and so moderate the temperature and reduce evaporation. Their roots retain water and prevent disastrous runoffs and erosion. Their dead remains return to the soil and, as humus, enrich our gardens, farmlands, and forests. Their wood, in all of its varieties of beauty, strength, and resistance, provide the raw materials for the building trades and the publishing of newspapers, magazines, and books. Every part of some plant is useful: the root of the carrot and beet, the stem of asparagus and potato, the leaves of cabbage and lettuce, the flowers of broccoli, hairs of cotton, bark of cinnamon, oils of spearmint and peppermint, latex of rubber trees, and the fruits of a wide variety of plants. Last, but by no means least, the green plants with their flowers and foliage grace our homes, streets, parks, and woodlands with their beauty. We would find it difficult to do without them.

30 THE GREEN PLANT'S ROLE IN NATURE

The story of life on Earth, like the story of the Earth itself, begins with the Sun. Except for man's recent use of atomic power, the Sun is the sole source of energy for almost all forms of life. All functioning machines need some source of energy to make them go: a watch uses the energy of a coiled spring; a hydroelectric plant employs the energy of falling water; an automobile runs on gasoline by releasing its chemical energy through the process of burning (oxidation). Similarly, all living cells obtain their energy from the oxidation of fuels called foods.

There are many different kinds of food molecules. We can get an idea of their nature by examining the most important one—the simple sugar, glucose ($C_6H_{12}O_6$). As the chemical formula shows, glucose is made up of 6 carbon atoms, 12 hydrogen atoms, and 6 oxygen atoms. Molecules of this simple sugar

Fig. 30.1 All green plants depend on energy of the Sun. Solar energy powers the complex process of photosynthesis.

From Rapho Guillumette pic -J. Allan Cash

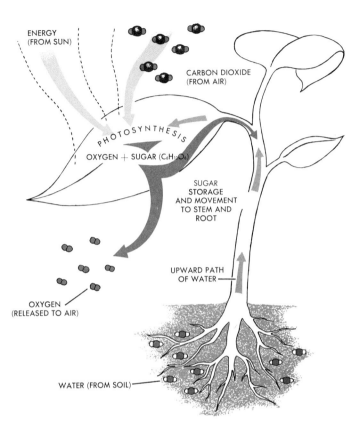

Fig. 30.2 In photosynthesis, carbon dioxide drawn from the air is combined with water drawn from the soil and produces sugar and oxygen. The sugar molecule contains in a stored form some of the energy of the light quanta which drive the reaction. The sugar formed in photosynthesis is the basic fuel of all forms of life on the Earth.

are **synthesized,** or built up, from two of the most common substances in nature—carbon dioxide and water. The cells of green plants are the "factories" in which glucose is made with the aid of energy from the Sun (Fig. 30.2).

When a green plant grows, it is, therefore, tapping solar energy. Since man eats either green plants or animals that eat green plants, he, too, is indirectly drawing on solar energy. Even the gasoline-powered automobile is using solar energy indirectly, for gasoline comes from oil which probably comes from creatures that trapped energy of the Sun and died millions of years ago. Without green plants, practically all life on our planet would cease.

THE SUN—A THERMONUCLEAR ENERGY SOURCE

Since the end of World War II, we have become aware of the tremendous amounts of energy that can be released from atomic reactions. The Sun itself is a kind of atomic reactor in which hydrogen is changed into helium. Physicists

Fig. 30.3 The energy of the Sun is derived from thermonuclear reactions in which four atoms of hydrogen are fused and form one atom of helium. In the process, energy equivalent to the mass which disappears in the reaction is liberated.

call the reaction a **thermonuclear** reaction because great heat is needed to make the nuclei of hydrogen atoms combine to form helium. Essentially this is what happens: four hydrogen atoms, each with a mass of about 1, are fused and form one helium atom with a mass of about 4. The over-all equation may be symbolized in this way:

$$4H \longrightarrow He$$

Actually, each of the four hydrogen atoms has a mass of 1.008 and the helium atom resulting from their fusion has a mass of 4.003. Since more mass is going into the reaction ($4 \times 1.008 = 4.032$) than is coming out ($1 \times 4.003 = 4.003$), the equation is unbalanced. The difference in mass (0.029 unit of mass) is converted into energy, according to the Einstein equation ($E = Mc^2$), where E is the energy produced, M is the mass of matter transformed, and c is the velocity of light. The equation shows that large quantities of energy are released by the conversion of very small amounts of mass into energy. It has been estimated that deep within the Sun some 120 million tons of matter vanish every minute, being converted into the tremendous quantities of energy radiated into space (Fig. 30.3).

About one-third of the solar energy reaching the Earth is used up in evaporating water. The rest is used in other ways. The most important of these is **photosynthesis**—the green plant's ability to convert carbon dioxide and water into sugar. Although green plants use only a tiny fraction of the solar energy available to them, photosynthesis is the most extensive chemical process on Earth. Every year, green plants convert 200 billion tons of carbon from carbon dioxide into sugar. This is about 100 times more than the total mass of all the goods that man produces in a year.

RADIANT ENERGY

As hydrogen is converted into helium in the solar thermonuclear furnace, many kinds of radiation are produced. Although the radiation differs in many ways, all of it is in the form of waves; yet each kind has a different **wavelength**. We can measure the wavelength of ocean waves in feet by measuring the distance from the crest of one wave to the crest of the wave ahead or behind. But the distances between wave crests of many kinds of radiation is so short that we need a unit of measure much smaller than feet or inches. One unit we use is the **millimicron**. One micron is one millionth of a meter, and a meter is equal to 39.37 inches; so a millimicron is a billionth of a meter. The human eye can see radiation that has wavelengths ranging from about 400 to about 700 millimicrons. The radiation within this range is called **visible light.** Each color in the visible spectrum corresponds to a particular wavelength. At one extreme, at 400 millimicrons, is the blue-violet end of the spectrum. At the other extreme, at 700 millimicrons, is the red end. In between are the other colors, ranging in order from violet, blue, green, yellow, and orange, to red. Curiously, a plant is sensitive to almost exactly the same range of radiation as is the human eye (Fig. 30.4).

The Earth's atmosphere prevents some of the Sun's radiation from reaching us at ground level. For example, a special form of oxygen, called **ozone** (O_3), absorbs much of the short-wave radiation just beyond the violet end of the spectrum. This radiation is called **ultraviolet.** It is fortunate that much of the ultraviolet is absorbed, for it can be damaging to living things. Just beyond the red end of the spectrum is a long-wave form of radiation called **infrared.** It is the invisible radiation we commonly call "heat." As this solar radiation enters the atmosphere, much of it is absorbed by water vapor (a gaseous form of water) and, to some extent, by carbon dioxide. Water vapor and carbon dioxide, then, help control the Earth's temperature. The solar energy that finally reaches us on the ground is mostly in the visible and infrared range, but it also extends somewhat into the ultraviolet. This radiation that penetrates the atmosphere furnishes the energy, directly or indirectly, for all living things. Yet the green plant alone is able to store radiant energy in the process of photosynthesis (Fig. 30.5).

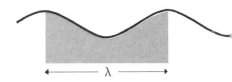

Fig. 30.4 Light may be considered as a wave motion. The length of the wave (see arrows) is different for each of the colors of light that we see. The plant is sensitive to only certain wavelengths of light in the visible region of the spectrum. Wavelength is abbreviated by the Greek symbol lambda (λ).

Fig. 30.5 Visible light has wavelengths of approximately 400 (violet) to 700 (red) millimicrons.

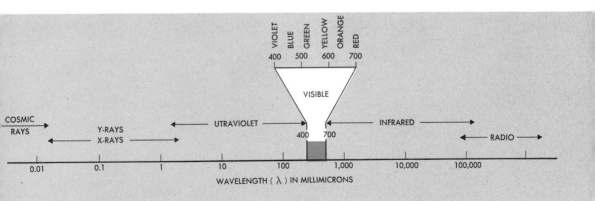

HUMAN POPULATION AND FOOD SUPPLY

We live in unusual times. The population of our planet is now about three billion people, and it is increasing at the rate of about 1.8 per cent per year. At this rate, 54 million additional consumers of food are added to the Earth every year—almost the equivalent of the population of France. The daily net increase of people (births minus deaths) is more than 100,000. Looking at it another way, for every second of time there is one extra mouth to feed. As the death rate is lowered as a result of improved and extended public health measures, we can expect the human population to just about double every 40 years. Should this continue for another 1000 years, there would be so many people that each person would have less than one-millionth of a square foot of land to stand on. Today each person has about 530,000 square feet. Obviously something must change before the "population bomb" engulfs us all.

Since the food of all animals is dependent on solar energy trapped by green plants, the number of people we can comfortably support on the Earth depends on the amount of energy that can be trapped by photosynthesis. To what extent can we increase the 200 billion tons of carbon stored each year by photosynthesis? No matter how hard we tried, we could no more

Fig. 30.6 All flesh is as grass. All human beings depend on the growth of green plants for sustenance and growth.

USDA photo *Ewing Galloway*

than double the area of land now under cultivation. But even if we did double the area of farm land, we would have to use inferior land since the best land is already in use. Where do we turn? Estimates of food production through photosynthesis show that between 50 and 80 per cent of photosynthesis takes place in the oceans and in fresh water. Does this mean that we could "farm" the seas? Or could we cultivate sea weed of various kinds for food in great highly-fertilized liquid tank cultures? Although this would be too expensive to do now, such a system might one day be necessary. In this new kind of agriculture, botanical "know-how" will play a large role.

Another way to increase food production is to improve the plant itself. Over the years we have been doing just that. Plant breeders have given us increasingly better kinds of plants to work with. Physiologists have taught us to care for the nutritional needs of the plant and to improve its growth by chemical treatment. Pathologists have shown us how to control insect and fungus pests. Soil scientists have demonstrated how to enrich and preserve the complex soil environment of weathered rock and living or decaying plants and animals. Some day, perhaps, we may understand photosynthesis well enough to control and improve its efficiency within the plant, or even copy it efficiently outside the living cell.

You may wonder what is the good, in the long run, to double the world's food production when in 40 years the gain will be completely wiped out by a doubling of the number of mouths to feed. Clearly, we must some day decide how many people can live comfortably on the Earth's surface. Then we must restrict the population to that number. Such a proposal raises a tremendous number of religious, political, and sociological problems. Even so, human populations will have to be limited in some way. The alternatives seem to be either a peaceful, planned, voluntary limit to population, or a violent and chaotic limitation imposed by starvation, disease, and war.

Fig. 30.7 The Earth's population is now more than three billion and is doubling in a span of less than 40 years. We may soon reach the practical limit of the Earth to support this increase in population. What happens in the next two generations may be crucial.

SUMMARY

The green plant, through the process of photosynthesis, captures the radiant energy of the sun and stores it in the form of foods eaten by animals and man. This energy, which is the basis for practically all life on Earth, originates in the sun through a thermonuclear reaction, in which four hydrogen atoms are fused to form helium, and excess mass is converted to energy. This energy is radiated through space at different wavelengths, some of which are absorbed by interstellar matter and by components of planetary atmospheres. Of the energy which reaches the surface of the Earth, only that part which is visible to the human eye is used by green plants.

Human beings are reproducing at an unprecedented rate, which doubles the population in less than forty years. In roughly four generations we shall probably have reached the limits of the Earth's capacity to sustain mankind. We must therefore decide to limit the human population on earth to some reasonable size. Then, a combination of increasingly scientific agriculture and limitation of reproduction will combine to stabilize the human population on Earth.

FOR THOUGHT AND DISCUSSION

1 What is the source of the energy of the Sun?
2 How does this energy reach the Earth?
3 What components of the Earth's atmosphere prevent solar energy from reaching the Earth's surface?
4 What is the role of the green plant in capturing and utilizing the energy which does reach the earth?
5 What is meant by photosynthesis?
6 How much photosynthesis occurs each year on earth?
7 Could any life on earth exist today independently of green plant photosynthesis? How might this situation change in the future?
8 What is the current human population on earth? How rapidly is it increasing? How long can this rate continue before a serious crisis confronts mankind?
9 Propose several steps which might contribute toward an amelioration or reasonable solution of the population problem on earth.
10 In what ways could research into each of the following fields lead to an increased food productivity: (a) Oceanography (b) Astronautics (c) Biochemistry (d) Nuclear physics (e) Illumination engineering?

SELECTED READINGS

Brown, H. *The Challenge of Man's Future.* New York: The Viking Press, Inc., 1956.
   A scholarly and stimulating assessment of the problems facing man in the years ahead.

Osborn, Fairfield. *The Limits of the Earth.* Boston: Little, Brown & Co. 1953.
   An interesting statement of the population-production problem.

Sax, K. *Standing Room Only.* Boston: Beacon Press, Inc., 1955.
   A vigorous presentation of the case for population planning.

31 THE GREEN PLANT CELL

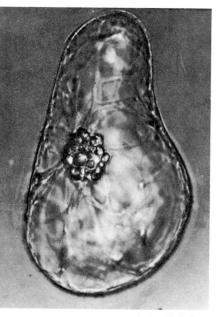

Courtesy John G. Torrey

Fig. 31.1 A cell of a pea plant grown in artificial culture medium. Note the nucleus surrounded by starch grains, the strands of cytoplasm and the large vacuole. The entire cell is encased in a cell wall.

The cell is the basic unit of life. The plants and animals you see around you are composed of billions upon billions of cells, yet there are some plants and animals that are only a single cell. The amoeba, which you have probably seen in a microscope, is one such organism. No responsible biologist claims that a knowledge of the cell automatically leads to a knowledge of the whole organism. Yet almost all biologists agree that the cell is a logical starting place for a meaningful study of an organism.

Cells vary tremendously in size. They range from bacterial cells less than a micron in diameter to cells several millimeters long. The building blocks of cells are **molecules,** which are combinations of atoms. Even a relatively small bacterial cell contains 1,000,000,000,000 molecules (abbreviated 10^{12}). The tremendous complexity of this basic biological unit should make it clear why a biologist's view of his basic unit is quite different from that of a physicist or chemist.

When we grow a plant cell in isolation, it generally takes on a spherical shape. When a cell is surrounded by other cells, however, it is squeezed this way and that and is given many sides. A cell from certain parts of a stem or root is shaped somewhat like a shoe box. Such a cell may be about 50 μ (microns) long by 20 μ wide by 10 μ deep, with a volume of about 10,000 μ^3. One hundred million such cells, if tightly packed, would fit in a volume of 1 cubic centimeter.

A plant cell has three main parts: 1. the **cell wall,** a stiff, nonliving substance produced by the rest of the cell; 2. the **protoplast,** or living substance of the cell, which is encased by surface layers called **differentially permeable membranes.** These

Courtesy W. G. Whaley and the University of Texas Electron Microscope Laboratory

Fig. 31.2 A maize root tip cell, as viewed under the electron microscope. It contains a large central nucleus with dark chromatin regions, long endoplasmic reticulum canals, smaller mitochondria and proplastids, stacks of Golgi bodies and small vacuoles.

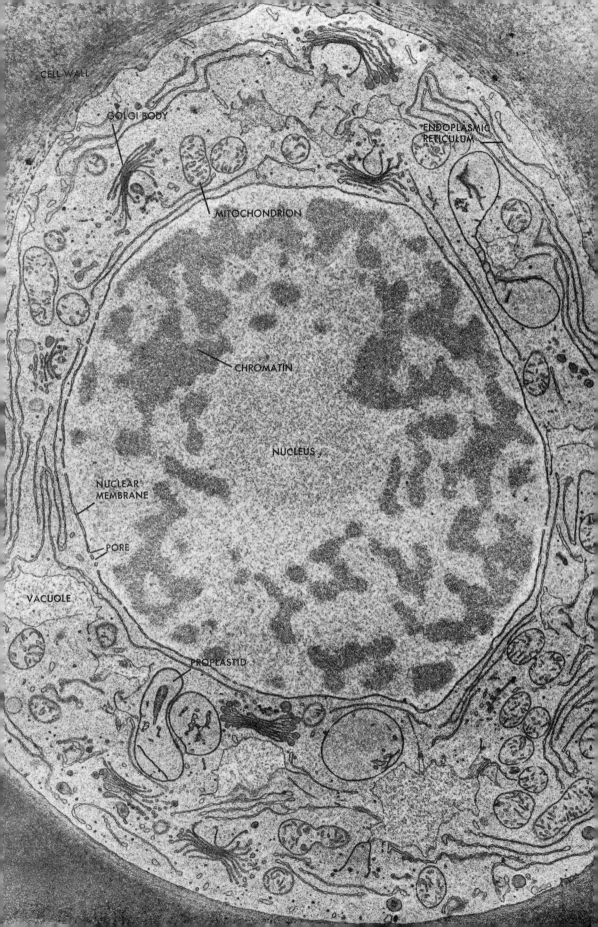

allow different types of molecules to pass through at different rates; and 3. **vacuoles,** the nonliving "storage tanks" that contain a watery solution of salts and various molecules produced by the chemical activities of the cell.

The stiff walls of the cells serve as a skeleton for the entire plant, giving the plant rigidity. The vacuoles serve as a sort of **excretory,** or waste removal, system. Material deposited within them no longer participates in the chemical activities of the cell. It is in the protoplast, then, that one finds the ceaseless activities which we call life (Fig. 31.2).

THE NUCLEUS—LIFE CENTER OF THE CELL

The microscope shows that the protoplast of a cell is composed of many different types of little bodies, called **organelles.** We are now reasonably certain about the structure, chemistry, and role of many of these bodies. The major organelle of a cell is the **nucleus.** This body, generally about 5 to 10 μ in diameter, contains the basic "blueprint" for making new cells. Each time a cell divides, its blueprint is passed on to the new generation. The blueprint occurs in the form of long strands of a complex chemical called **deoxyribonucleic acid,** or **DNA.** Molecules of DNA look like double, intertwined spirals. Each molecule is made up of four basic building blocks, called **nucleotides.** The order in which the nucleotides occur in a DNA molecule chain determines the genetic information that will be passed on to a new individual. The combination of nucleotides you inherited from the DNA of your mother and father determined that you would become a human rather than an oak

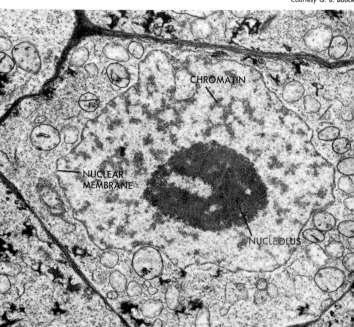

Fig. 31.3 The nucleus is surrounded by a double membrane. It contains a central dark nucleolus, chromatin strands, and a lighter nuclear sap.

Courtesy G. B. Bouck

tree, armadillo, or simply a shapeless mass of cells. If the four nucleotides are represented by A, C, G, and T, a chain in the order ACGT will carry a different blueprint, or genetic information, from AGCT or ATCG chains. The inheritance of every organism seems to be determined by repeated patterns (genes) of nucleotide units in DNA (Figs. 31.3 and 31.4).

For most of the life of a cell, the nucleus is separated from the rest of the protoplast, called **cytoplasm,** by a nuclear membrane. However, this membrane has pores and projections leading into the cytoplasm. These projections, forming long canals in the cytoplasm, are called **endoplasmic reticulum (ER).** Before a cell divides, its nucleus first undergoes a change. For one thing, the nuclear membrane appears to break down completely. This permits the nuclear material to mix with the rest of the protoplast. At this point, a dense mass called the **nucleolus** (one or several of which are found in every nucleus) disappears completely as an organized body (Figs. 31.5 and 31.6).

The nucleolus is composed mainly of a material called **ribonucleic acid,** or **RNA.** RNA differs somewhat from DNA, but like DNA it probably consists of long chains of repeated nucleotide patterns. DNA is now thought to mold RNA into some basic pattern characteristic of a given type of cell. We now believe that all biological units capable of reproduction depend on either DNA or RNA for the transfer of genetic information from one unit to another.

The main function of RNA seems to be the manufacture of protein, which is the chief constituent of protoplast. Proteins, the major body-building food of plants and animals alike, are large molecules composed of about 100 to 2,000 **amino acids.** There are about 20 different kinds of amino acids, so that

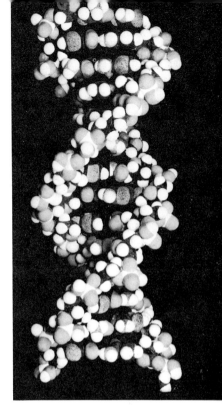

Courtesy Abbott Laboratories

Fig. 31.4 The DNA molecule consists of two long intertwined helices, each constructed of many nucleotide units.

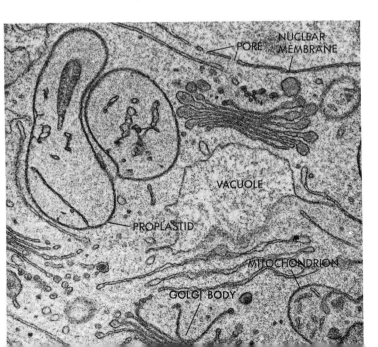

Fig. 31.5 The nuclear membrane is perforated by pores through which materials may pass. The outer layer of the nuclear double membrane is pushed out and forms the long endoplasmic reticulum canals along which ribosomes may lie. Other organelles in the picture are a proplastid, mitochondria, Golgi material and vacuole.

each one appears many times in each protein molecule. Each type of protein has its own particular pattern of amino acid units. It is this difference in amino acid sequence that determines the kind of protein; it not only distinguishes pork from beef, but also human skin from human fingernail, eyeball, and liver. Those of you who have been exposed to even a little organic chemistry will be able to appreciate the complexity of a zein molecule, a corn protein represented by the following formula:

$$C_{685}H_{1068}N_{196}O_{211}S_5$$

Many of the proteins synthesized within a cell regulate the rate of chemical reactions occurring in the cell. These proteins are called **enzymes** and their role is a very important one. In certain species of plants, for example, the difference between red and white flowers is that the petal cells of the red plant contain an enzyme that can change a colorless substance to a red pigment. The white variety lacks the enzyme, so it is unable to carry out this change (Fig. 31.7).

There are thousands of enzymes in any cell, and each enzyme controls one chemical reaction, or a group of related chemical reactions. In principle, here is how enzymes function:

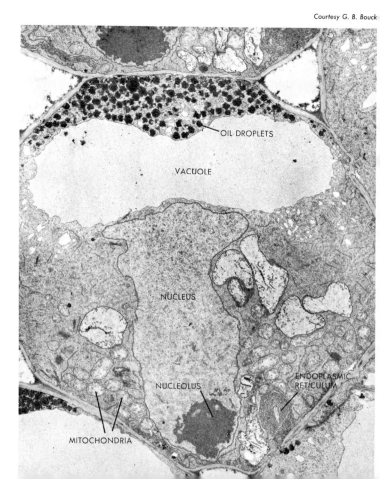

Fig. 31.6 The organelles of a cell may be stratified by exposure to a powerful centrifugal field. In this picture of a pea root cell, the heavy organelles (nucleolus, mitochondria, endoplasmic reticulum) have gone to the bottom and the lighter particles (oil droplets and vacuoles) have floated to the top. When this technique is applied to homogenized cells it results in the separation of relatively pure organelles in bulk.

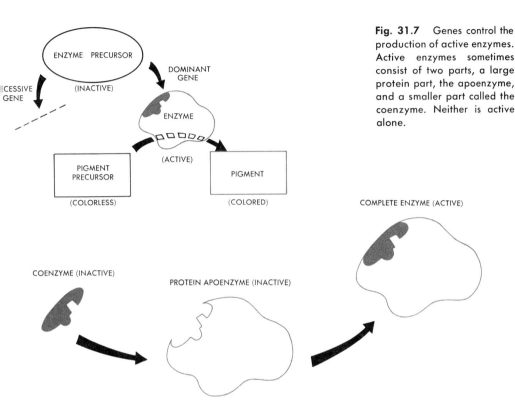

Fig. 31.7 Genes control the production of active enzymes. Active enzymes sometimes consist of two parts, a large protein part, the apoenzyme, and a smaller part called the coenzyme. Neither is active alone.

suppose that the two molecules A and B slowly unite and form a larger molecule AB:

$$A + B \longrightarrow AB$$

If enzyme E has the proper shape and chemical composition, both A and B will attach themselves to enzyme E.

$$E + A \longrightarrow EA$$
$$EA + B \longrightarrow EAB$$

Then, A and B, closely positioned on the enzyme, will unite and form molecule AB, releasing enzyme E unchanged.

$$EAB \longrightarrow E + AB$$

In short, the enzyme brings the reacting molecules together, thereby increasing the speed of the reaction. Because the enzyme is not used up in the process, it can do the same job again and again. This is why small amounts of enzyme work effectively with large amounts of reacting substances.

The enzymes of the cell, located in various cell organelles and in the other areas of the cytoplasm, are the superintendents of

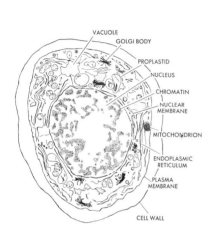

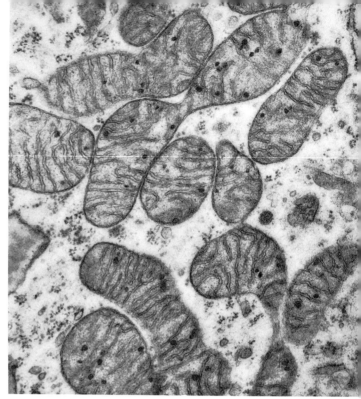

Courtesy G. E. Palade

Fig. 31.8 Mitochondria are sausage-shaped organelles surrounded by double membrane, the inner lining of which is pushed into the interior an forms *cristae*. The enzymes of the mitochondria are found along the cristae

the cell's chemical machinery. All cells are what they are because of their chemistry; their chemistry is determined by thei enzymes; the nature of the enzymes is determined by RNA in the cytoplasm; and the particular type of RNA is in turn determined by the DNA in the nucleus.

MITOCHONDRIA—THE POWERPLANTS OF CELLS

Mitochondria are sausage-shaped organelle, ranging from one to several microns in length. They are abou half a micron in width. The mitochondria form the "power plant" of a cell, for they provide the energy needed by the cel to carry on its many activities. The mitochondria do this by oxidizing certain molecules we call "food," and by releasing energy in the process. The energy is then stored in a complex chemical compound called **adenosine triphosphate,** or **ATP.** The typical cell has hundreds of mitochondria distributed throughout the cytoplasm. Cells that are especially active have more and larger mitochondria than the average cell has (Fig. 31.8).

The foods known as sugars have the general formula $[CH_2O]$, In the reaction carried out in the mitochondrion the sugar is

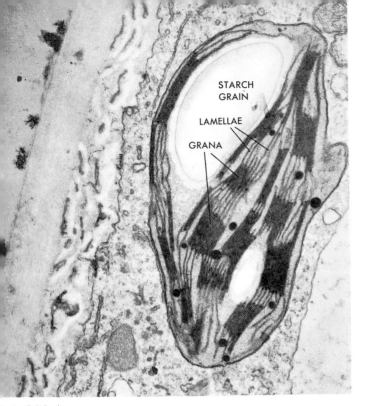
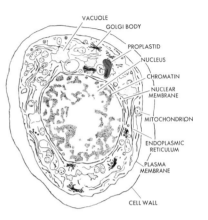

Courtesy G. B. Bouck

Fig. 31.9 The mature chloroplast contains many parallel *lamellae* occasionally thickened and forming *grana*, which contain the photosynthetic pigments. Starch grains formed from the sugar produced in photosynthesis develop inside the plastids.

"burned" (meaning that oxygen is added), producing carbon dioxide and water, and releasing energy in the process.

$$[CH_2O]_n + nO_2 \longrightarrow nCO_2 + nH_2O + \text{energy}$$

The actual process (respiration) is more complicated. Not all of the carbon of every sugar molecule is oxidized to CO_2 and water. Sometimes it is made into ethyl alcohol; sometimes it unites with ammonia (NH_3) and forms amino acids, the building blocks of proteins. While the carbons of the sugar are being converted to CO_2 and other products, the hydrogens are being transferred to oxygen to form water.

Though some ATP is used up in these various steps, the energy released in the over-all reaction is more than enough to replace the ATP used. The excess energy is stored in the form of ATP, which serves as a reservoir of power that the cell can call upon when it needs energy.

CHLOROPLASTS—ENERGY TRAPPERS IN CELLS

So far we have discussed the organelles found in plant and animal cells alike. What about those found only in the cells of *green* plants? Included in the cytoplasm are bodies called **plastids.** The most important plastids in the green plant cell are those known as **chloroplasts.** Each chloroplast is center of photosynthesis in the green cell, for the chloroplasts contain all the chlorophyll and other pigments associated with photosynthesis. In higher plant cells, the chloroplast is a disc shaped body roughly 5 to 8 μ in diameter by about 3 μ in width. There are about 50 such bodies in each cell. So far as we know, each chloroplast is formed from bodies called **proplastids.** These seem to be able to reproduce themselves by some sort of division process, thus increasing their number in the cell, but most mature chloroplasts cannot reproduce in this manner.

Photographs taken with electron microscopes show that chloroplasts have double layers of material running the length of the plastid. In some places several of these layers thicken at one point called a **granum** (plural **grana**). It is here, in the grana, that the chlorophyll is found. The nongreen areas of the plastid are called **stroma.** Enclosing the entire plastid is a double membrane that regulates exchange of materials between the plastid and the outside environment (Fig. 31.9).

In the flowering plants (**angiosperms**) chloroplasts will not develop from proplastids unless the plant is exposed to light. However, in certain cone bearing trees (**gymnosperms**) this change takes place in total darkness. In angiosperm leaves grown in the dark the proplastids do not develop to maturity. This means that the plant cells cannot make their own food— since they are without chloroplasts—so they must absorb food molecules from the outside. Only when the proplastids develop into chloroplasts by being exposed to light can the cells carry on photosynthesis and thus make their own food.

Courtesy J. L. Martens and O. J. Eige

Fig. 31.10 The development of chloroplasts may be controlled by genes in the nucleus. In the albino corn shown in this photograph, chloroplasts failed to develop. In contrast with their green siblings, such albino plants will never photosynthesize. When the reserve food of the endosperm in the corn grain has been depleted, an albino plant dies.

Some biologists regard the chloroplast as a sort of "invading" organism. By chance, it found its way into a nongreen cell, thus enabling the cell to make its own food by means of photosynthesis. (Plants that make their own food in this way are said to be **autotrophic.**) What evidence is there to support this hypothesis of invasion? If certain one-celled plants are grown at fairly high temperatures for several generations in a row, the rest of the cell reproduces faster than the chloroplast. The result is paler and paler green cells. In the end, with the loss of the last chloroplast or proplastid, the one-celled plant is totally nongreen. Such cells remain permanently nongreen and cannot regain the autotrophic habit. If they are to survive, they must absorb ready-made food molecules from the outside. (Plants that gain nourishment in this way are said to be **heterotrophic.**) Chloroplasts can also be prevented from developing by treatment with streptomycin or other chemicals. Thus, a green plant cell can be "cured" of the "invading" chloroplasts by high-temperature therapy or by chemical treatment (Fig. 31.10).

If we remove a chloroplast from its cell, for some time the chloroplast can carry out its normal photosynthetic activities. It takes up CO_2 and converts it into sugars; it also releases oxygen and produces ATP. However, the chloroplast cannot maintain itself or reproduce outside the cell. If the chloroplast is really an "invader," it has become dependent on the environment of the plant cell for many aspects of its existence.

For some reason unknown to us, chloroplasts do not usually develop in root tissues, not even in those exposed to light. In certain species, however, such as carrot and morning glory, root cells may become green when given light. Why the plastids do not develop to maturity in the cytoplasm of some cells is unknown. Recently the chloroplast has been found to contain DNA of a special kind. Since it can thus control its own heredity, at least in part, the view of it as an independent "invader" is strengthened.

In the cells of higher plants there are plastids other than chloroplasts, but these other plastids lack photosynthetic apparatus. There is the **leucoplast,** which is colorless and serves as a storage center for the cell's reserve food materials such as starch grains. There is also the **chromoplast,** which usually has bright yellow, orange, or red pigments called **carotenoids.** The function of chromoplasts is obscure. Usually they do not develop in a cell if chloroplasts are present. In the ripening process of a fruit such as a tomato, the change in fruit color from green to white to red reflects three successive stages of development: 1. the dominance of chloroplasts; 2. the decline of chloroplasts and 3. the rise of the carotenoid-filled chromoplasts. The cause of these transitions is not understood.

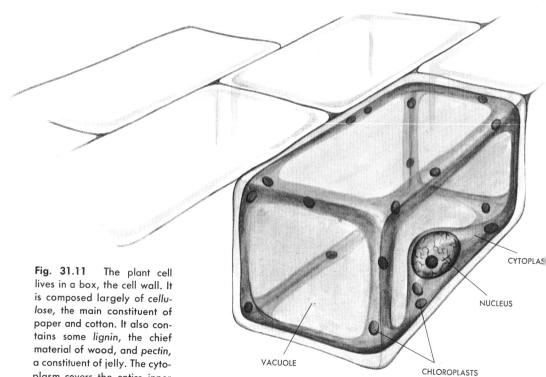

Fig. 31.11 The plant cell lives in a box, the cell wall. It is composed largely of *cellulose*, the main constituent of paper and cotton. It also contains some *lignin*, the chief material of wood, and *pectin*, a constituent of jelly. The cytoplasm covers the entire inner surface of the cell wall. For the sake of clarity, cytoplasm is omitted from the upper, lower and side faces of this cell.

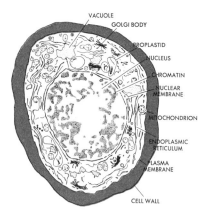

THE CELL WALL

Plant cells differ from all other cells in having a fairly rigid outer case that resembles a box. It is no great exaggeration to say that "the plant cell lives in a wooden box," for the cell wall includes those ingredients which give wood its stiffness and strength (Fig. 31.11).

The cell wall seems to be nonliving and apparently does not take part actively in any of the chemical processes going on within the cell. It is produced by the protoplast of the cell and is built up in layers as the cell develops. The first layer formed is called the **middle lamella.** In the early stages it is mostly **pectin,** a jelly-like compound. Later, **cellulose** and **lignin** are added and strengthen the wall (Fig. 31.12).

As the cell grows, its contents exert great pressure against the wall, gradually stretching it. At the same time, the cell also enlarges the wall by inserting more material into the existing wall, or by adding a new layer alongside the old. Once the cell has stopped expanding, lignin, cellulose, and other stiffening materials continue to thicken the wall, with the result that the protoplast within is squeezed into a smaller and smaller volume. In some cells the wall may become so thick that it occupies almost the entire volume of the cell. If this happens, the cell dies, leaving its wall to serve as part of the over-all skeleton, or to

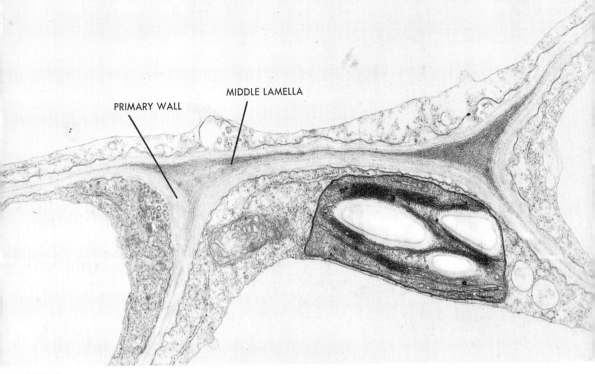

Courtesy G. B. Bouck

Fig. 31.12 A view of the cell wall separating four cells—one above, one to the right, and two below. Note the various layers. The innermost dark staining layer is the middle lamella, which forms a continuous layer throughout the plant. To each side of this is the primary wall of an individual cell. The secondary wall is then deposited inside the primary wall by each cell. Occasionally there are thin spots or holes left in the developing wall. These pits may represent areas for easy communication between one cell and another.

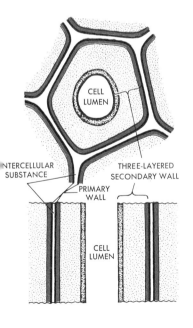

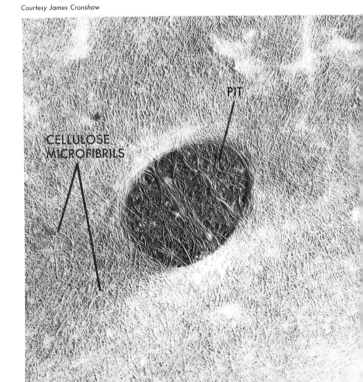

Courtesy James Cronshaw

Fig. 31.13 The shape of a leaf depends in part on the water it contains. When too much water is lost, the leaf goes limp.

The Bergman Associates

serve as part of the plant's endless fluid-carrying pipeline. In the cell walls of those cells that continue to live are holes through which strands of cytoplasm pass. These strands bind the protoplasts of neighboring cells together into one large community. They also make passage of materials from one cell to another easier by providing a continuous pathway of cytoplasm.

Because of its rigidity, the wall gives form to the plant cell and thus serves as a skeleton. Thick-walled cells that have large amounts of lignin, such as those in wood, maintain the shape of the cell without any supporting forces. Thin-walled cells such as those in a leaf, are too weak to hold their shape without the support of the contents of the cell. Lacking this support, the cell goes limp and shrinks. To the eye, such a piece of tissue seems wilted (Fig. 31.13).

THE VACUOLE AND MEMBRANES

Vacuoles appear to be "tanks" in the cytoplasm. They are, in essence, a differentially permeable membrane surrounding a watery solution of salts and waste products of the cell. It is the vacuoles that keep a cell from shrinking and going limp. When a cell is young, it has many small vacuoles. At the beginning, the vacuoles occupy only a small percentage of the total volume of the cell. As the cell grows, the vacuoles become larger and eventually join, forming one large central vacuole. This vacuole may occupy 90 per cent or more of the total volume of the cell. The differentially permeable membrane forming the surface of the vacuole is called the **tonoplast**. It permits molecules of water and certain other substances to pass through rapidly, but it slows down the passage of most other molecules, keeping some out almost entirely. This enables the vacuole to exert a force on the surrounding cytoplasm and on the cell wall. The force comes from the energy of the moving water molecules (Fig. 31.14).

To see how this works, let us examine a cell with a central vacuole containing salts, sugars, amino acids, and other materials. Assume that the cell is sitting in a tank of distilled water. As you know, the molecules of all substances are constantly in rapid motion. To begin, let us consider only the movement of the water molecules. Since the cell's vacuole contains large amounts of molecules other than water, it has fewer water molecules for each unit of volume than does the pure water outside the cell. This means that in any given area of the membrane more water molecules will enter than will leave in any given time. This rapid, unequal, two-way passage of water

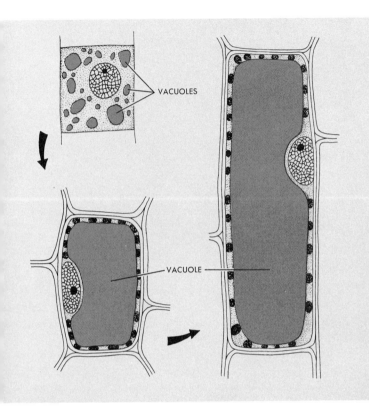

Fig. 31.14 The vacuole is a watery tank containing some dissolved materials and surrounded by a differentially permeable membrane. In a young cell the vacuoles are small and numerous. As the cell grows, it absorbs much water, which passes into the vacuoles. The vacuoles then fuse into a large central tank which ultimately occupies more than 90 per cent of the volume of the cell. The pressure of the vacuolar contents against the cell wall is responsible for the cell's expansion during growth. This pressure is referred to as *turgor pressure*.

molecules through the membrane increases the volume of the vacuole. As the vacuole swells, it presses the cytoplasm against the cell wall. Such pressure is called **turgor pressure,** and such a cell is said to be **turgid** (Fig. 31.15).

But how long can the water continue to enter the vacuole? It continues until the pressure within the vacuole becomes high enough to prevent further net entry. Much the same thing happens when you pump air into a tire at a filling station. You set the regulator on the air pump at, say, 30 pounds pressure. If the pressure in a tire is less than that, air flows into the tire from the pump. But when the pressure within the tire reaches 30 pounds, no more air can enter. Returning to our vacuole, the number of water molecules entering the cell eventually becomes equal to the number leaving it. This occurs, despite a difference in concentration of water molecules, because the pressure inside the cell effectively speeds up the bombardment of the membrane by the water molecules inside the cell. At this stage, the water content of the cell cannot increase

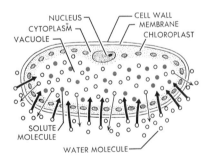

Fig. 31.15 Turgor pressure results from a greater inward than outward diffusion of water. The relative rates are determined by the concentrations of water outside and inside the cell.

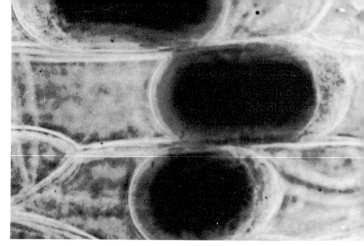

Courtesy K. Esau

Fig. 31.16 Plasmolysis of three onion cells caused by placing tissue in sugar solution. Water has left the vacuole, and the protoplast (dark color) has retreated from the cell walls. Space between the protoplast and cell wall is filled with sugar solution.

further, although water molecules still move rapidly through the membrane in both directions. The movement of water through a differentially permeable membrane from a region of higher concentration of water to a region of lower concentration of water is known as **osmosis.**

Suppose that we place the cell in a solution in which the number of water molecules per unit volume is lower than that of the vacuole. There will now be a loss of water from the vacuole since water will diffuse out of it into the surrounding medium. Eventually, the cell will lose its turgor pressure and become limp. If this process continues, the vacuolar volume will shrink so much that the cytoplasm will withdraw from the cell wall. This produces a condition known as **plasmolysis.** It is, by the way, a condition that occurs almost daily in the leaves of some plants growing where there is a shortage of water (Fig. 31.16).

How do the many separate organelles of a cell manage to exchange their material? Each is bounded by a differentially permeable membrane and separated from neighboring organelles by relatively large distances. If material is not exchanged, the cell will not function. Part of the answer is that the contents of many plant cells are in a state of fairly rapid movement, known as **cytoplasmic streaming.** The entire cytoplasm rotates around the inner surface of the cell wall, carrying the various organelles along with it. In certain instances, streaming cytoplasmic strands may pierce a vacuole. In some green cells, the disc-shaped chloroplasts can move independently and present (or withdraw) their broad surfaces to the surface of the leaf. Usually they do this in response to changing light intensity, but how such movements are made is not well known.

Now that we have surveyed the structure of cells in general, let us recall the three special features of the green plant cell:

the **chloroplast** enables the cell to convert radiant energy into chemical energy. This in turn, enables the cell to produce its own food through photosynthesis. 2. The **cell wall** serves as the skeleton of the plant body. It also encases each protoplast in a rigid framework and allows large turgor pressures to build up. 3. The central **vacuole,** which occupies 90 per cent or more of a plant cell, collects waste products produced by the cell. The central vacuole also absorbs water by osmosis and provides turgor to the cell.

TYPES OF CELLS

A plant is a community of many different types of cells with different forms and functions. The process that gives rise to differentiation is one of the major puzzles in biology. Some unknown influence, possibly from outside the nucleus, and outside the cell itself, determines what role a cell will play—whether it will become a reproductive cell, a tough supporting cell, and so on. Once a cell has differentiated and adopted a specific role, it ordinarily cannot return to the undifferentiated state or take on a new role.

From a functional point of view, a plant consists of the following types of cells (Fig. 31.17).

Meristematic Cells

These have an ability to divide repeatedly and give rise to new cells of the plant. In general, meristematic cells are cube-shaped, small, thin-walled, and have many vacuoles. Their nucleus is large in comparison with the rest of the cell.

Parenchymatous Cells

These large, usually thin-walled, many-sided cells keep their living contents. They are relatively undifferentiated, forming the bulk of softer plant parts—most parts of the fruit, soft parts of the stem and leaves, and so on. They may contain chloroplasts, or may be adapted for the storage of water and reserve foods such as starch.

Supporting and Conducting Cells

These are all relatively long, but otherwise are greatly different from one another. One group, the cells whose walls have large quantities of lignin, do not contain living mat-

MERISTEMATIC CELL

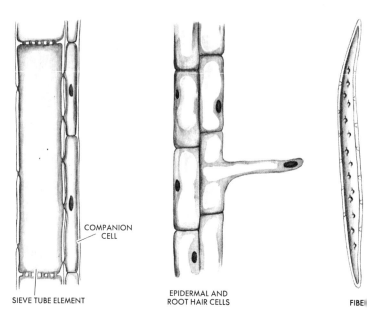

SIEVE TUBE ELEMENT — COMPANION CELL

EPIDERMAL AND ROOT HAIR CELLS

FIBER

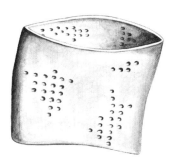

VESSEL ELEMENT

Fig. 31.17 Plants contain a variety of cell forms and types.

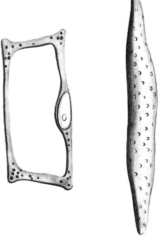

PARENCHYMA CELL   TRACHEID

ter when they mature. Another type of supporting cell, **collenchyma,** is thickened only at the corners of the cell, but otherwise resembles parenchyma. **Sieve tubes** are the main pipeline cells of the phloem. They are strings of long cells with perforated end plates. Such sieve cells do not have nuclei at maturity; their nuclei are believed to be in the small **companion cells** that generally lie alongside the sieve cells.

Protective Cells

These generally have flattened surfaces with a waxy, water proof coating which prevents too much water loss from the plant's inner tissues. There are many different kinds of specialized cells that form the outer protective "skin" of a plant. Most important among these protective cells are those forming the epidermis.

Reproductive Cells

When the temperature, length of day, and other conditions are just right, reproductive cells form in plants. These highly specialized cells ultimately give rise to sex cells called **gametes.** The gametes join in sexual reproduction and produce a **zygote,** from which a new offspring develops.

The cell, no matter how highly specialized, or how undifferentiated, is a very complex structure. Its many organelles are related to and dependent upon one another. Its various parts are rebuilt by ceaseless activity, without which the cell would die. Certain cells have the remarkable ability to make complete copies of themselves, copies that in turn grow and duplicate themselves. Others change and take on a highly specialized structure and function. Despite the tremendous recent advances in the study of the cell and its organelles, biologists are still very far from understanding in any great detail the functioning of this complex structure.

SUMMARY

All plants and animals are made up of basic units called cells, which vary greatly in size and complexity, depending on the creature. Even the simplest cell, however, probably contains more than a million million molecules. Each plant cell consists of three major regions: the **cell wall,** the living material **(protoplast),** and the **vacuole.** Within the protoplast are different **organelles,** microscopic bodies with particular functions. The largest and most important is the **nucleus,** which contains the hereditary information of the cell in the form of a chemical code. This code depends on the order of basic molecular building blocks, called **nucleotides,** in a long molecule called **deoxyribonucleic acid,** or **DNA.** This DNA is able to produce copies of itself, which are then distributed into all new cells produced by cell division.

The constitution of the cell depends on the kinds of **proteins** it contains, since many of these directly control the chemical activities on which life is based. These large molecules, made up of about 20 different kinds of amino acids, are formed as the result of the cooperative chemical activities of several different kinds of **ribonucleic acid (RNA)** molecules present outside the nucleus. The chemical code in the DNA molecules determines the nature of the RNA molecules, which in turn determine the nature of the proteins. Through this process of "information transfer" from DNA to RNA to proteins, the nucleus ultimately determines the nature of the cell.

Other important organelles are the **chloroplasts,** in which photosynthesis is carried out, and **mitochondria,** in which foods are oxidized (burned) and their energy extracted and stored in the form of molecules of **adenosine triphosphate** (ATP). In these molecules, energy is immediately available, and may be supplied for any chemical or mechanical process in the cell.

The **cell wall** is produced by the protoplasm. It is originally composed of jelly-like **pectic** substances, but later is hardened by the addition of **cellulose** and woody **lignin.** It furnishes support to the plant, acting as a kind of widely distributed skeleton. The **vacuole** is a central watery tank usually occupying more than 90 per cent of the volume of the cell. It contains a solution of raw materials and cellular waste products. Like the outer layer of the protoplast, it is surrounded by a differentially permeable membrane which governs the rate at which different molecules enter and leave the cell. Because water can pass these membrane more readily than other substances can, and because the substances dissolved in the vacuolar water can move out only with great difficulty, water tends to diffuse into the vacuole (**osmosis**) building up a great **turgor pressure** which helps give the plant its firmness and form.

The plant contains many different kinds of cells, although we do not understand how or why cells from a similar genetic background come to be different. Among the kinds of cells present in a typical plant are **meristematic** (dividing) **cells,** which produce new cells on the plant body, **protective cells, supporting and conducting cells, food and water storage cells,** and finally, the **gametes,** specialized cells for sexual reproduction, which are produced only at particular times in the plant's life cycle.

FOR THOUGHT AND DISCUSSION

1. Why is the cell called the basic unit of life?
2. Cells may be very small, yet all are exceedingly complex Discuss this statement.
3. What are the three main parts of a typical plant cell? Which are living and which nonliving?
4. What is the function of the nucleus? How does it carry out this function?
5. In the development of the cell what is the relation between protein, RNA, and DNA?
6. What is an enzyme? How do enzymes act?
7. In what organelle is food oxidized? Describe the structure of this organelle.
8. How does the cell store the energy released when food is oxidized?
9. What is the relation between the over-all processes of photosynthesis and respiration?
10. Describe the structure of a mature chloroplast. How does it develop from a proplastid?

1 Distinguish between autotrophic and heterotrophic organisms. Can any one organism have both modes of nutrition?
2 How does a young cell wall differ from a mature cell wall?
3 What function do thick cell walls serve?
4 What is a differentially permeable membrane? Where do they occur in a cell?
5 What is turgor pressure? How and where does it develop in the cell?
6 What causes a cell to become plasmolyzed? How can plasmolysis be reversed?
7 Name several types of differentiated cells occurring in higher plants and give a function for each.

SELECTED READINGS

Brachet, J. and A. E. Mirsky (eds). *The Cell.* New York: Academic Press, 1959 and later.
  A six volume (up to 1964) advanced level encyclopedia of structure and function of each of the cell's organelles.

Esau, K. *Plant Anatomy.* New York: John Wiley & Sons, Inc., 1964.
  A beautifully written and illustrated work. Clearly the best in its field.

Jensen, W. A. *The Plant Cell.* San Francisco: Wadsworth & Co., 1963.
  A college level treatment of our modern views regarding the green plant cell.

Loewy, A. G. and P. Siekevitz. *Cell Structure and Function.* New York: Holt, Rinehart, and Winston, 1962.
  A college level treatment of the architecture, biochemistry and basic physiology of cells.

Swanson, C. P. *The Cell* (2nd ed.). Englewood Cliffs, N.J.: Prentice-Hall, Inc., 1964.
  An elementary discussion of the basic facts of life, centered around cells.

32 PLANT NUTRITION

Like all organisms, green plants need three kinds of nutrients: foodstuffs, mineral elements, and water. As we saw in the previous chapter, green plants differ from most organisms in being autotrophic: they make food by converting carbon dioxide to sugar by means of the energy of light and the photosynthetic apparatus of chlorophyll contained in the chloroplast. In this chapter, we shall find out how green plants carry out photosynthesis and satisfy their various nutritional needs (Fig. 32.1).

Fig. 32.1 Plants absorb the mineral nutrients they need from the soil in which they grow. If the soil is deficient in any particular element, the plant grows poorly and develops deficiency symptoms.

The Biochemistry of Photosynthesis

When sunlight is absorbed by a chloroplast, carbon dioxide (CO_2) and water (H_2O) are converted to sugar and oxygen. The amount of oxygen released is equal in volume to the CO_2 used. Plants are important in the balance of nature because they restore to the air the oxygen (O_2) needed by most organisms. Let us use the formula [CH_2O] to stand for the basic unit of a carbohydrate molecule. (C = carbon, H = hydrogen and O = oxygen.) Six of these [CH_2O] units would yield $C_6H_{12}O_6$, or the sugar **glucose.** We can now write the equation for photosynthesis this way:

$$CO_2 + H_2O \xrightarrow[\text{Chlorophyll}]{\text{Light energy}} [CH_2O] + O_2$$

Carbon dioxide   Water   Carbohydrate   Oxygen

This equation is properly "balanced," (meaning that there are the same number of oxygen, hydrogen, and carbon atoms on the left as there are on the right), but it gives an incorrect idea of the way the reaction takes place. Looking at the equation, you would think that the O_2 on the right comes from the CO_2 on the left, but it does not. The free oxygen released in photosynthesis comes from water (H_2O), not from CO_2. Light energy is used to split the hydrogen atoms from the oxygen atom in a molecule of water. This appears to be one key to the entire process of photosynthesis, for it represents a point at which light energy is made to do chemical work. If you look at the equation again you will see that there are two atoms of free oxygen on the right, but only one atom of oxygen in the single water molecule on the left. Since the free oxygen comes from split water molecules, then we must have at least two water molecules on the left. To write a balanced equation truly representing the over-all reaction, we must add one molecule of water *to each side* of the equation.

$$CO_2 + 2H_2O \xrightarrow[\text{Chlorophyll}]{\text{Light Energy}} [CH_2O] + O_2 + H_2O$$

Since the two molecules of water on the left are completely broken down by light energy, the new water molecule formed on the right must result from a new combination of hydrogen and oxygen atoms. Perhaps this will help you visualize what happens:

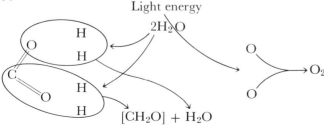

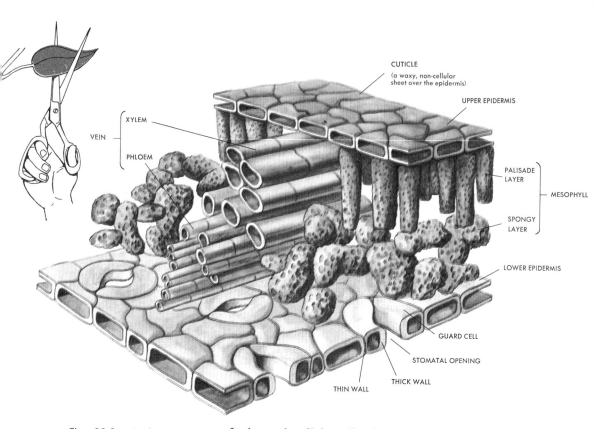

Fig. 32.2 A diagrammatic cross section of a leaf. The leaf consists of an upper and lower protective layer, the *epidermis,* between which are the actively photosynthesizing cells. Closest to the upper epidermis is the *palisade* layer, and below it is the *spongy* layer. The epidermal layers are frequently perforated by *stomatal* openings, the aperture of which is controlled by the turgor pressure of the *guard cells* surrounding them. Intercellular air spaces serve as passageways for the interchange of water, CO_2, and oxygen gases. The small veins containing xylem and phloem serve to conduct water and minerals into and synthesized sugars out of the leaf. A waxy cuticle, largely waterproof, usually covers both lower and upper epidermal cells.

It shows that light splits the two water molecules, releasing two atoms of oxygen that combine and form an oxygen molecule. The four hydrogen atoms recombine in two ways: 1. two of them join one oxygen and the single carbon atom of the CO_2 and form $[CH_2O]$; 2. the other two join the remaining oxygen atom and produce a new water molecule. This is the briefest kind of shorthand. Actually, there are many steps, some known and some unknown, that take place in each of the simple reactions shown.

The study of the photosynthesis carried out by microscopic organisms helps to demonstrate that this outline of the process in higher plants is basically correct. For example, bacteria that carry on photosynthesis use hydrogen sulfide (H_2S) instead of H_2O. If the oxygen liberated during the photosynthesis of green plants came from the CO_2, then we would expect that oxygen would also be produced by the bacteria, since they also use CO_2 in their type of photosynthesis. But no oxygen is produced; instead sulfur is.

$$CO_2 + 2\,H_2S \longrightarrow [CH_2O] + H_2O + 2\,S$$

Similarly, certain kinds of microscopic algae can be "trained" to use hydrogen (H_2) instead of water to convert CO_2 to $[CH_2O]$. In this case, no oxygen is released:

$$CO_2 + 2\,H_2 \longrightarrow [CH_2O] + H_2O$$

In this scheme, as in the others, it is clear that light energy splits (or **photolyzes**) the molecule containing hydrogen and that some of the hydrogen so released is used to convert CO_2 to $[CH_2O]$.

The Raw Materials of Photosynthesis

The CO_2 which participates in photosynthesis enters the green cells of leaves and stems through tiny openings called **stomata** (singular **stoma**) in the surface of the leaf. The stomata lead to a branched system of interconnecting air canals. Each leaf has several layers of cells that actively carry on photosynthesis. Collectively, these layers are called the **mesophyll.** Surrounding and protecting the mesophyll is the epidermis. Conducting elements (the veins) enable the leaf to carry on two-way transport; they carry raw materials to the leaf and photosynthetic and other products away from the leaf. The veins branch so profusely that no mesophyll cell is more than one or two cells away from a vein (Fig. 32.2).

The stomata open and close in response to turgor pressure of two sausage-shaped **guard cells** that surround them. The inner walls of these guard cells are much thicker than the outer

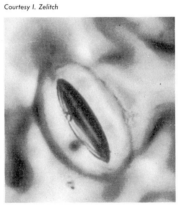

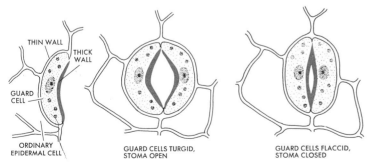

Fig. 32.3 When the guard cells are turgid, the stomata are opened (left). When the guard cells are flaccid, the stomata are closed. The pictures are photographs of silicone impressions of open and closed stomata of tobacco leaf cells. The diagrams below illustrate the mechanics of the movements.

walls. When a guard cell is under great turgor pressure, the weaker outer wall balloons out and carries the inner stronger wall along with it. This action causes the stomatal pore to open. When a guard cell goes limp (as on a hot dry day), the thick, elastic inner wall pulls the rest of the cell inward toward the pore area, closing the pore. This control of the stomatal opening helps prevent excessive water-vapor loss (Fig. 32.3).

The CO_2 molecules entering the stomatal openings flow through the interconnecting air passages to the mesophyll. When they reach the wet surface of a cell, they dissolve in water and form a reservoir of CO_2 that can be tapped later for photosynthesis. The oxygen produced during photosynthesis leaves the mesophyll, flows through the air passageways and the stomata to the outside world.

When the stomata are closed, there is a reduction of flow of CO_2 to the mesophyll, and a reduction in the amount of oxygen leaving the mesophyll. When this happens, the oxygen is probably reabsorbed in **respiration.** Just as you do, plants use oxygen in respiration. This is the reverse of photosynthesis. The reabsorbed oxygen combines with $[CH_2O]$ and produces carbon dioxide and water. The CO_2 released can then be used for further photosynthesis.

$$[CH_2O] + O_2 \longrightarrow CO_2 + H_2O$$

CARBON DIOXIDE AND THE WORLD CLIMATE

Carbon dioxide is a gas on which we all depend, although it is only one of many gases making up the atmosphere, and makes up a very small part, only three parts per 10,000 (0.03 %). This concentration varies with the place, being higher over cities where large quantities of coal, oil, and gasoline are being burned, and lower in country areas where extensive photosynthesis is proceeding. An increase in the CO_2 content of the atmosphere raises the photosynthetic rate of plants that are well supplied with light and water, but it may also injure certain sensitive leaves.

Some students of evolution believe that the CO_2 content of the atmosphere may have varied considerably in recent geological times, and may have been responsible for certain changes of vegetation and climate. For example, an increase in the CO_2 level would not only increase photosynthesis, and thus the amount of plant material, but would also cause a general warming of the Earth. This is true because the Earth, heated by the Sun, normally

Fig. 32.4 Greenhouses get warm because the visible light of the Sun enters the greenhouse and is absorbed. Soil, plants, and other objects then reradiate long-wave infrared radiation (heat), which cannot pass out through the glass. Heat is thus trapped within the greenhouse, warming it.

reradiates a portion of the absorbed energy back into space as infrared (heat) radiation. It happens that CO_2 absorbs infrared very well, thus preventing the complete escape of this heat energy and creating a sort of planet-wide "greenhouse." Warming of the Earth through such an effect could lead to partial melting of polar ice caps and glaciers, and to flooding of the low-lying land areas in which most of the world's major cities are located.

Thus, our rapid consumption of fossil fuels such as oil and coal, and the release of extra CO_2 into the atmosphere, may have profound consequences for man. This process, however, tends to limit and even reverse itself. Higher temperatures and higher CO_2 levels will result eventually in a higher rate of photosynthesis and a luxurious growth of plants, such as occurred in the Carboniferous Era, when dinosaurs abounded. This increase of absorption of CO_2 during photosynthesis should eventually lower the atmospheric CO_2 content significantly, causing a cooling of the Earth and a reversal of the cycle mentioned above.

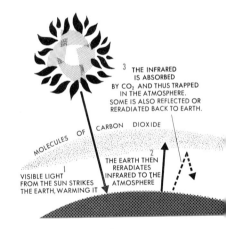

Fig. 32.5 The carbon dioxide and some other substances in the Earth's atmosphere act like the glass of the greenhouse. By preventing the escape of infrared radiation from the Earth, they cause the retention of heat within the atmosphere.

Closure of the stomata does not prevent photosynthesis or respiration, but it does largely prevent gas exchange with the outside world. Photosynthesis, which under normal conditions can occur 10 to 20 times faster than respiration, can go no faster than respiration when the stomata are closed. It is clear, then, that stomatal closure, which results from wilting, causes a serious lowering of the over-all photosynthetic activity of the plant.

The water needed for photosynthesis and other activities of the leaf is absorbed from the soil by the roots of the plant. By diffusion, the water passes into water-conducting cells of the xylem, located at the center of the root, and moves upward to the leaf veins. At the end of a vein, the water diffuses into neighboring mesophyll cells, passing from one cell to the other. A certain amount of water evaporates from the wet surfaces of the mesophyll cells and passes out of the stomata through the air passages. The amount of water required in photosynthesis is only a tiny fraction of that absorbed and evaporated by plants. For the highest rate of photosynthesis, the leaf must be turgid and the stomata open. A water shortage lowers the photosynthetic rate, since stomatal closure interferes with photosynthesis by limiting the leaf's supply of CO_2.

The sugar and other organic materials produced in photosynthesis accumulate rapidly in the mesophyll cells. Some of the sugar is made into starch (a large molecule of one to several thousand connected glucose units). Starch grains may be formed directly in the chloroplast, or in the leucoplast of non-green tissue. Most of the remainder of the sugar formed during photosynthesis is converted into **sucrose,** which consists of one molecule each of the simple hexose sugars glucose and fructose. Sucrose is the main transport sugar of the plant. It finds its way into the phloem, through which it is rapidly transported to all parts of the plant. Since, in many woody plants, the phloem is found only in the bark area, injury to the bark (by beavers, restraining wires, and so on) may interrupt the flow of sugar. This can stunt the plant's growth, or kill it.

If the main trunk of a tree is "girdled," the plant will die from starvation of the root system. If a branch is girdled, sugar transport out of it will cease. Because of sugar accumulation, any leaves or fruits on that branch will tend to develop larger than they would on an ungirdled branch. This technique is used to produce extraordinarily large fruits on single isolated branches.

Whereas water movement in the xylem occurs in the dead, hollow **tracheid** and **vessel cells,** movement of sugars through the phloem depends on the chemical activities of the phloem's sieve-tube cells. If these activities are interrupted, too much sugar and starch may accumulate in the leaf mesophyll and slow down photosynthesis.

For a high rate of photosynthesis to take place, adequate amounts of light, water, and carbon dioxide must be supplied to the leaf; and an adequate rate of transport of the products of photosynthesis must be maintained from the leaf.

We can come to understand something about photosynthesis by trying to answer the three following major questions: 1. How is light energy "captured" and then made to do chemical work? 2. Through what process is carbon dioxide transformed into sugar? 3. Through what process is oxygen released from water?

The Light Reaction

To do its photosynthetic work, light must be "absorbed" by a plant. When it is, numerous chemical reactions are set in motion. The first link in the chain of reactions is formed by the pigment molecules, of which chlorophyll is the major one. When a pigment molecule absorbs light, it is changed slightly and said to be **activated.**

In the first chapter we said that plants are sensitive to almost exactly the same range of visible radiation as the human eye is. We can now refine that statement. Suppose that we shine different wavelengths of light, one at a time, on a green leaf and measure the rate of photosynthesis. If you did this you would find that blue light and red light bring on the highest rate of photosynthesis, and green light the lowest rate (Fig. 32.6).

If we arrange the colors of the visible spectrum in such a way that those of shorter wavelengths (blue) are at the left and those of longer wavelengths (red) are at the right, and then draw a curve showing for each wavelength the rate at which photosynthesis takes place, we have an **action spectrum.** The action spectrum pictured here shows the relative effectiveness of wavelengths ranging from 400 millimicrons ($m\mu$) to 700 $m\mu$. Although red and blue are shown to be the most effective, yellow also seems to influence the rate of photosynthesis. It turns out that the yellow carotenoid pigments, present in great quantities in the chloroplast, also absorb light which is useful in photosynthesis. Since carotenoids cannot perform in photosynthesis if chlorophyll is absent, it is generally assumed that the energy which carotenoids acquire by absorbing light is passed on to chlorophyll. The chlorophyll then performs the actual photosynthetic work.

By shining each wavelength of light through a solution of

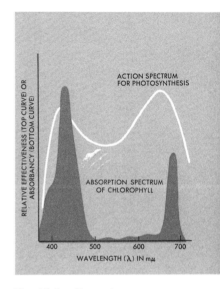

Fig. 32.6 The action spectrum for photosynthesis resembles the absorption spectrum for chlorophyll. This indicates that chlorophyll is the pigment that absorbs the light energy responsible for photosynthesis.

Fig. 32.7 In some plants, such as red algae, there are different pigments, the *phycobilins*, which absorb the light effective in photosynthesis. This leads to a different kind of action spectrum.

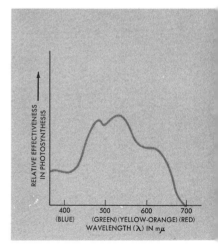

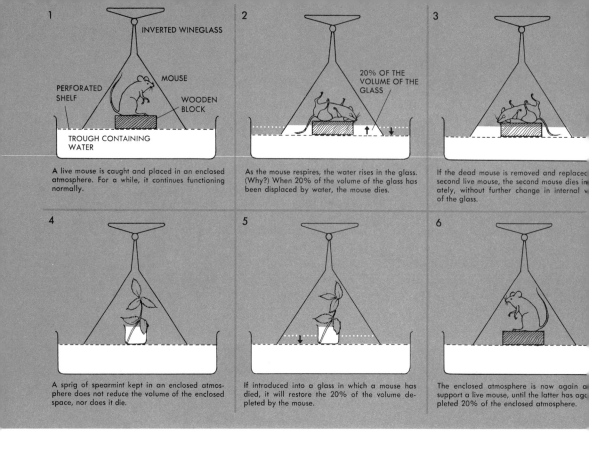

Fig. 32.8 Joseph Priestly's classical experiment showing inverse effects of plants and animals on enclosed atmospheres.

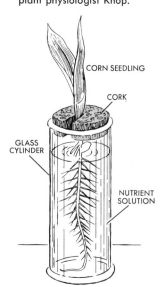

Fig. 32.9 The mineral nutrient requirements for plants are revealed by growing plants in solutions containing pure salts. This figure shows an early experiment of this type performed in the mid-nineteenth century by the plant physiologist Knop.

chlorophyll, we can also produce an **absorption spectrum,** which shows the wavelengths of light most effectively absorbed by chlorophyll. By comparing the absorption spectrum for chlorophyll and the action spectrum in Fig. 32.6, you will see how we can tell that chlorophyll is the major pigment receiving light for photosynthesis.

Several years ago botanists discovered that a combination of red and blue light produces a higher rate of photosynthesis than red light alone. This led to the theory that there are two different kinds of chlorophyll, and that both must be activated by light for photosynthesis to proceed at a high rate.

The action spectrum for photosynthesis in red, blue-green, or brown algae is very different from that for green leaves. (A diagram on page 485 shows the action spectrum for a red alga.) However, in each case there is a pigment whose absorption spectrum matches the photosynthetic action spectrum of the organism. In some of these organisms, the photosynthetic pigments are found in modified chloroplasts, or smaller bodies called **chromatophores.** In all instances, light splits water molecules, or some other material containing hydrogen, thus freeing the hydrogen atoms to convert carbon dioxide to sugar.

Energy from the Sun triggers the chemical machinery in the green plant and produces sugar and free oxygen. By the time that energy has done photosynthetic work, it has been transformed many times. We can summarize these energy transformations in the following five steps:

1. Radiant energy from the Sun activates a pigment molecule.
2. The activated pigment loses an electron. Because the electron is negatively charged, a positively charged "hole" is left in the chloroplast. The lost electron is captured by an "electron acceptor," which may then pass the electron on to another electron acceptor, and so on.
3. The positively charged "hole" in the chloroplast removes an electron from a molecule of water (actually from an OH^- unit called an **hydroxyl ion**).
4. This movement of electrons sets up a kind of "electric current." The resulting energy is used to form energy-rich phosphate bonds, such as ATP, from lower energy materials, such as ADP (see page 494).
5. The chemical energy of the ATP is used to convert CO_2 to sugar. At this stage the original radiant energy from the Sun has become potential chemical energy locked up in sugar molecules.

In this way radiant energy from the Sun is changed to the unstable physical energy of an activated pigment, then to a flow of electrons, then to the chemical energy of unstable molecules (ATP), and finally to the chemical energy of stable molecules (sugar). In a sense, the role that light plays in photosynthesis is completed when ATP is made. All subsequent reactions can take place in the dark.

From CO_2 to Sugar

In recent years we have learned much about the conversion of CO_2 to sugar during photosynthesis. One of the tools biologists have used is a form of carbon called **carbon 14**. An atom of ordinary carbon has a total of 12 protons and neutrons (six of each) in the nucleus, but carbon 14 has six protons and eight neutrons. This makes the atom unstable and, therefore, radioactive. Because carbon 14 is radioactive, we can detect it in a substance by using a radiation (Geiger) counter.

We can make radioactive carbon into CO_2 and then feed the radioactive CO_2 to cells carrying on photosynthesis. Then, by fractionation of the cells we can find out what chemical compounds the radioactive carbon has entered. It turns out that the first stable product formed from the radioactive carbon is **3-phosphoglyceric acid,** abbreviated **PGA.**

With the aid of ATP and electron acceptors formed during the light reaction, PGA is converted to sugars. These sugars may be stored in the form of starch, or they may be transformed into cellulose, organic acids, fatty acids, and amino acids. Thus, the sugar produced from CO_2 during photosynthesis is the basic organic molecule used by the cells of higher plants for energy and for making the structural building blocks required by the cell.

Production of Oxygen

The final question we have to answer about the biochemistry of photosynthesis is an unsolved one: what is the pathway along which oxygen is produced? While the intermediary steps of sugar production from CO_2 are well understood, we have yet to discover the intermediary steps involved in oxygen production.

Most investigators now feel that each water molecule is first separated into an H^+ and OH^- ion (see page 491). In photosynthesis an electron is removed from the OH^- ion, leaving behind an uncharged but unstable "free radical" of hydroxyl (OH). The OH groups (radicals) may then join and form a **peroxide,** like $HO \cdot OH$. The peroxide may then be broken down in such a way that free oxygen is released. Only through further investigation will the details of such a process become known. The process of photosynthesis is shown in the general diagrams on page 479.

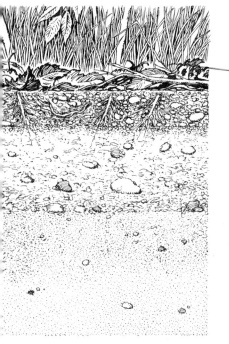

Fig. 32.10 The soil profile consists of several horizons. Near the surface there is much decayed organic matter (humus) and well pulverized rock. As one goes farther and farther down, the humus level decreases and the size of the rock particles increases. Finally one hits bed rock.

MINERAL NUTRITION

In addition to the organic materials produced during photosynthesis, the plant needs a wide variety of mineral elements. These elements are absorbed from the soil by the root system and are transported upward in both xylem and phloem. The plant uses them to build new parts of itself and to carry on the wide range of chemical activities that take place in its cells. All these elements except one, nitrogen, come from the parent rock which produces soil. Nitrogen is taken from the atmosphere, mainly through the process of "nitrogen fixation" (discussed on page 492).

An **essential element** is one without which the plant cannot complete its life cycle. To discover which elements are essential, we can transplant a seedling in a hard glass, glazed porcelain, or plastic pot containing pure quartz sand that has been

THE PERIODIC TABLE

GROUP	I	II											III	IV	V	VI	VII	VIII
1	1 H Hydrogen												METALLOIDS AND NONMETALS					2 He Helium
2	3 Li Lithium	4 Be Beryllium											5 B Boron	6 C Carbon	7 N Nitrogen	8 O Oxygen	9 F Fluorine	10 Ne Neon
3	11 Na Sodium	12 Mg Magnesium	TRANSITION METALS										13 Al Aluminum	14 Si Silicon	15 P Phosphorus	16 S Sulfur	17 Cl Chlorine	18 Ar Argon
4	19 K Potassium	20 Ca Calcium	21 Sc Scandium	22 Ti Titanium	23 V Vanadium	24 Cr Chromium	25 Mn Manganese	26 Fe Iron	27 Co Cobalt	28 Ni Nickel	29 Cu Copper	30 Zn Zinc	31 Ga Gallium	32 Ge Germanium	33 As Arsenic	34 Se Selenium	35 Br Bromine	36 Kr Krypton
5	37 Rb Rubidium	38 Sr Strontium	39 Y Yttrium	40 Zr Zirconium	41 Nb Niobium	42 Mo Molybdenum	43 Tc Technetium	44 Ru Ruthenium	45 Rh Rhodium	46 Pd Palladium	47 Ag Silver	48 Cd Cadmium	49 In Indium	50 Sn Tin	51 Sb Antimony	52 Te Tellurium	53 I Iodine	54 Xe Xenon
6	55 Cs Cesium	56 Ba Barium	57 La Lanthanum	72 Hf Hafnium	73 Ta Tantalum	74 W Tungsten	75 Re Rhenium	76 Os Osmium	77 Ir Iridium	78 Pt Platinum	79 Au Gold	80 Hg Mercury	81 Tl Thallium	82 Pb Lead	83 Bi Bismuth	84 Po Polonium	85 At Astatine	86 Rn Radon
7	87 Fr Francium	88 Ra Radium	89 Ac Actinium															
	LANTHANIDES (Rare Earth Metals)				58 Ce Cerium	59 Pr Praseodymium	60 Nd Neodymium	61 Pm Promethium	62 Sm Samarium	63 Eu Europium	64 Gd Gadolinium	65 Tb Terbium	66 Dy Dysprosium	67 Ho Holmium	68 Er Erbium	69 Tm Thulium	70 Yb Ytterbium	71 Lu Lutetium
	ACTINIDES				90 Th Thorium	91 Pa Profoactinium	92 U Uranium	93 Np Neptunium	94 Pu Plutonium	95 Am Americium	96 Cm Curium	97 Bk Berkelium	98 Cf Californium	99 Es Einsteinium	100 Fm Fermium	101 Md Mendelevium	102 No Nobelium	103 Lw Lawrencium

Fig. 32.11 Plants require approximately 16 of the 103 elements known. Most of these are the simpler elements (colored squares), which are also cosmically more abundant.

thoroughly washed. We then pour in a nutrient solution of pure salts and distilled water. Great care must be taken to exclude organic impurities, microscopic creatures, and dust, all of which may carry traces of mineral elements. Sometimes, especially with large seed plants such as the garden bean, we have to remove the part of the plant called the "cotyledon," for it may contain enough stored elements to make absorption from the outside unnecessary, at least for one generation.

When we grow a test plant in a mineral nutrient medium containing all essential elements, it develops vigorously and normally. This shows that the green plant can synthesize all the organic molecules it requires, including vitamins and hormones. Why, then, do we use organic fertilizers?

The answer lies not in the plant itself, but in the nature of soil. Soil is a complex medium for plant growth. It includes: 1. coarse sand, fine silt, and very fine clay particles, all broken down from rock; 2. organic matter, generally the decaying remains of plants and animals; 3. living organisms of various kinds, including bacteria, fungi, algae, protozoa, worms, insects, and larger animals; 4. a soil solution (containing inorganic and organic materials in water) that forms a thin film around the rock particles; and 5. a gas phase, a gaseous mixture of oxygen, nitrogen, carbon dioxide, and trace gases of the atmosphere.

The vigorous growth of a plant in soil depends on the proper physical condition of the soil. If the soil particles are too closely packed together, there will not be enough gas phase. Without oxygen, roots cannot carry out respiration. Absorption of minerals by the roots also depends on the presence of oxygen. A soil is said to be in good **tilth** when it has good **crumb structure,** meaning that the fine soil particles are cemented together to form larger crumbs which pack loosely, forming a firm yet well-aerated medium. It is here that organic fertilizers can improve soil, for the crumb-forming process is accomplished by certain soil microorganisms that produce a type of glue, or **mucilage.** These organisms live on the organic matter of the fertilizer.

Organic fertilizers, then, are required only if the physical condition of the soil is poor. Under ideal growth conditions, such as in well-aerated quartz sand watered with a mineral salt solution, organic fertilizers do not produce additional growth. Also, there is no evidence that the addition of organic fertilizers to a soil improves the quality of a plant for human or animal nutrition over that produced by growth in a well-balanced mineral solution. These facts should be remembered when one hears the spectacular claims of "organic gardening."

If we grow a plant in a mineral solution lacking adequate quantities of some essential element, the plant becomes unhealthy. It develops typical signs (symptoms) showing that the element is in short supply. A skilled botanist can learn to recognize in a particular plant the various symptoms of various element deficiencies. He can correct the situation by making suitable additions to the soil or other medium in which the plant is growing. The ideal way of keeping a plant healthy is to clip off bits of it from time to time and analyze them chemically. Any element found to be in short supply can then be added to the soil or solution, quickly improving the plant's growth and vigor.

One of the first mineral nutrient solutions botanists prepared for higher plants contained only three salts: calcium nitrate, $Ca(NO_3)_2$, potassium phosphate, KH_2PO_4, and magnesium sulfate, $MgSO_4$. These six major elements, together with the carbon, hydrogen, and oxygen involved in photosynthesis, are the ones required in largest quantities by higher plants. Most plants can use the nitrogen locked up in ammonia (NH_3) as well as, or better than, the nitrogen locked up in a nitrate, such as calcium nitrate. The phosphate may be supplied in any form that dissolves in water.

When botanists experimented with purer and purer mineral salts, it became clear that the basic three-salt solution was not really complete. Many other elements are needed by plants, but in quantities much smaller than for the six major elements. These additional elements are called **micronutrient elements.** They include iron, manganese, zinc, copper, molybdenum, boron, and chlorine. Some botanists also suspect that the plant may need very small amounts of additional elements such as cobalt, vanadium, strontium, and iodine. In addition, certain plant cells have very special needs. For example, one-celled organisms called **diatoms** need silicon to build their silica cell walls.

While plants do not *need* certain elements, their growth and vigor are improved by the presence of such elements. For example, wheat plants grown in the absence of silicon are attacked much more readily by fungi than are wheat plants grown in the presence of silicon. Beet plants grown in the presence of sodium produce larger and fleshier roots than those grown without sodium. Despite the beneficial effects of such elements, they cannot, strictly speaking, be regarded as *essential* for the plant. By our present reckoning, plants need 16 elements. Four (C, H, O, and N) are taken ultimately from the atmosphere. The remaining 12 (K, Ca, Mg, P, S, Fe, Cu, Mn, Zn, Mo, B, Cl) are taken from the soil (Fig. 32.11).

Uptake of Mineral Elements from the Soil

To enter a plant, the mineral elements in the soil must first pass through the differentially permeable membrane of a cell. Usually, the absorbing cell is in the root, but this need not be so. Fertilizers applied to leaves are readily absorbed, with prompt benefit to the growing plant.

The passage of a mineral through the differentially permeable membrane of a cell is a complex and poorly understood process. Yet, in recent years we have managed to learn about certain aspects of the problem. The membrane seems to have many specific absorption sites through which particular minerals, in the form of **ions,** enter. Potassium and lithium, for example, enter at the same site, where they interfere with each other's entry; however, sodium enters at a different site. An ion is an electrically charged atom, or group of atoms. In its "normal" state, an atom of hydrogen is electrically neutral. Its single proton forming the nucleus has a charge of plus one, and the single electron circling the nucleus has a charge of minus one. These two charges cancel each other out. If the electron is removed, the atom is left with a net charge of plus one, written H^+, and called a **cation**. An **anion** is an atom or group of atoms having a net negative charge, as a result of having more electrons than protons. An example would be chloride (Cl^-), which by gaining the electron from hydrogen acquires the extra negative charge. The molecule so formed, called hydrogen chloride (HCl), consists of H^+ and Cl^- held together by virtue of their opposite electrical charges.

Plants appear to use up energy when they absorb ions. We

think this is the case because if we slow down respiration in a plant the rate of ion absorption is also slowed down. (Respiration is the major energy liberating process in a plant.) Some elements or compounds can be accumulated in a cell against a concentration gradient or electrical charge which tends to push them out. This can be accomplished only by the expenditure of energy. If respiration is slowed down, the accumulated ions quickly pass out of the cell.

Sometimes, root cells secrete ions to the outside medium in exchange for ions absorbed. For example, H^+ ions in the cells are frequently exchanged for K^+ ions in the outside medium. This excretion of H^+ ions makes the external medium more acid. Similarly, the absorption of anions, such as NO_3^-, may be compensated for by the secretion of bicarbonate (HCO_3^-) or an organic acid anion.

Most mineral absorption occurs near the tips of the main or branch roots, where most of the root growth takes place. The older parts of the root tend to become coated with various materials which prevent water from entering.

Nitrogen Fixation

Fig. 32.12 When leguminous roots are invaded by bacteria of the genus *Rhizobium*, large nitrogen fixing nodules are formed.

Courtesy The Nitragin Company

We say that the **fixation** of nitrogen takes place when nitrogen gas is converted into a form that the plant can use. Nitrogen fixation is one of the most important reactions in all biochemistry. Atmospheric nitrogen, which makes up about 78 per cent of the air we breathe, is a very stable molecule consisting of two atoms (N_2). This molecule must somehow be made unstable and be split before it can be converted into a form usable by plants. This work is done by certain bacteria that live on the organic matter in the soil. It is also done by special bacteria living in swellings, called **nodules,** of certain plant roots. These nodule bacteria belong to the genus *Rhizobium,* and the host plant is usually a member of the family *Leguminosae.* Recently, plants other than leguminous plants have been found to fix nitrogen with the aid of bacteria. In addition, certain blue-green algae and photosynthetic bacteria can also fix atmospheric nitrogen (Fig. 32.12).

The association of two organisms for mutual benefit is called **symbiosis.** Since neither the bacteria alone nor the host plant alone can convert atmospheric nitrogen to a usable form—ammonia, NH_3—the double life of the nodule must be regarded as a symbiotic association. Recently, botanists have found that the invading bacteria enter the host plant through curiously curved root hair cells. The final result of this invasion is a massive overgrowth of the root cells producing the warty bumps called nodules.

Fig. 32.13 The element nitrogen cycles continuously in nature. The N_2 of the atmosphere is useless to living forms. It can, however, be fixed in the form of nitrates through the action of lightning and rain, or deposited in the soil in the form of ammonia, as a result of the activity of nitrogen-fixing bacteria and algae living free or symbiotically on root nodules. The nitrates and ammonia are absorbed by green plants, built into plant proteins, and finally eaten by animals whose waste and carcasses return the nitrogen to the soil. Some of the fixed nitrogen is released to the atmosphere again through the action of denitrifying bacteria.

The leguminous nodules frequently contain a red pigment called **leghemoglobin,** which is closely related to animal hemoglobin (red blood cells). It appears to be involved in nitrogen fixation. We think this is true because nodules lacking leghemoglobin cannot fix N_2, but those containing the pigment can. **Ferredoxin,** a substance containing iron, is also involved in nitrogen fixation. With its help, nitrogen fixation can now be carried out in a test tube. As far as we know, the first stable product of fixation is ammonia, but the path of its production from atmospheric nitrogen is still unknown.

As we saw earlier, most plants absorb and use nitrogen in the form of nitrate (NO^-), for example, in the form of the plant nutrient calcium nitrate. Recently, biochemists have discovered an enzyme, called **nitrate reductase,** that helps convert nitrate to ammonia. So, in addition to nitrogen-fixing bacteria supplying the plant with usable nitrogen in the form of ammonia, the enzyme nitrate reductase also is a supplier of ammonia.

ATP AND HOW IT IS MADE

A molecule of ATP

ATP stands for **adenosine triphosphate**. ATP is an energy-rich molecule consisting of a complex organic nucleus called adenosine joined to three phosphate groups in line. The last two phosphate groups are joined to the rest of the molecule by bonds of unusually high energy value. When these bonds are broken the released energy can be used to drive energy-requiring reactions, such as the joining of two amino acids (see below).

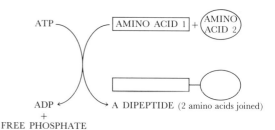

During the process **ADP** (**adenosine diphosphate**) is produced along with free inorganic phosphate. The ADP plus phosphate can now be rejoined through any reaction furnishing the required energy. In green plants this can be brought about by photosynthesis or respiration. In photosynthesis the energy of light quanta

The over-all cycle of nitrogen in nature may be summarized as follows: free nitrogen in the soil is converted to ammonia by nitrogen-fixing bacteria. Ammonia is thus supplied directly to the plant from the soil. Some of the ammonia in the soil is converted to nitrate by soil microorganisms. The nitrate is then absorbed by the plant and is converted back to ammonia. The plant then combines the ammonia with certain other substances, which leads eventually to the formation of proteins. In one way, we could say that the making of proteins, the basic "stuff" of the plant, is the end point in the nitrogen cycle. But, as with all other cycles in nature, this one, too, is endless.

Plant protein is eaten by animals and is transformed to animal protein. Urea and uric acid waste products from the animal return nitrogen to the soil. Eventually, animals and plants die and decompose in the soil. In so doing, they return still more nitrogen to the soil. These simple nitrogenous materials are thus used again and again by living organisms. Also, the nitrogen may be returned to the atmosphere as free molecular nitrogen (N_2) through the actions of certain denitrifying bacteria.

How a Plant Uses Elements

Carbon, hydrogen, and oxygen, the major elements involved in photosynthesis, plus nitrogen and phosphorus are the main building blocks of the plant body. The cell walls

excites a pigment (P) to a high-energy state (P*). This extra energy is then used to recreate ATP from ADP and phosphate.

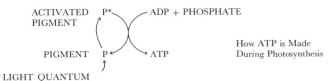

How ATP is Made During Photosynthesis

In respiration the source of energy for rebuilding ATP is the food molecule being oxidized. In oxidation electrons are removed from the molecule being oxidized and moved through intermediate chemical carriers to oxygen. The energy of this electron flow is used to resynthesize ATP.

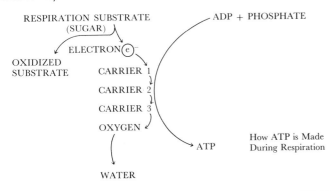

How ATP is Made During Respiration

that form the plant skeleton are almost entirely C, H, and O. The proteins that make up the organic bulk of the cytoplasm are mainly composed of C, H, O, and N. The nucleic acids making up a large part of the nucleus and some of the cytoplasm are C, H, O, N, and P. In addition, fats and carbohydrates forming a large part of the cytoplasm of some plant cells are C, H, and O.

Of the other 12 elements required by plants, four are used mainly for structural purposes:

1. **Sulfur** is contained in several amino acids, the building blocks of proteins. Even though only small quantities of sulfur are needed by plant cells, almost all of it serves an important structural function. Without the sulfur-containing amino acids, many important proteins of the cell could not be synthesized at all.

2. **Calcium** serves a variety of functions. Most of it finds its way into the walls of cells where it combines with other substances and serves to make the cell wall rigid. Some calcium is also used to neutralize and precipitate potentially harmful waste product organic acids, in the vacuoles.

3. **Magnesium** is an essential part of chlorophyll molecules. If a plant does not have enough magnesium, chlorophyll deficiency (**chlorosis**) occurs in the leaves. Magnesium is also needed for the action of several enzymes that are present in all cells.

4. **Phosphorus** is present mainly as a part of DNA and RNA, and as part of certain fatty substances, called **phospholipids.** The phospholipids are thought to play an essential role in the structure of the cell's various membranes. If a cell does not have enough phosphorus, new genetic material in the nucleus and cytoplasm cannot be formed, and new membranes around the inner surfaces of the cell and around the various organelles cannot be formed. Phosphorus is also very important for the many energy-transfer steps that take place in the cell. Compounds such as ATP, which are storage centers of energy for the cell, have three phosphate groups attached (see box). When the phosphate groups of an ATP molecule are detached, large amounts of energy are released to bring about a variety of chemical reactions; for example, joining two amino acids.

Although phosphorus, magnesium, calcium and sulfur have additional roles to play in the cell, their structural roles are the most important.

The remaining eight elements (K, Fe, Mn, Cu, Zn, Mo, B, Cl) play their major roles as essential parts of enzymes. Molybdenum, for instance, appears to be involved in the functioning of the enzyme nitrate reductase, which, as we saw earlier, converts nitrate to ammonia. Chlorine, in the form of a chloride, is known to play some role in the production of ATP during photosynthesis, but its exact role has yet to be discovered. If a plant does not have enough boron, its meristematic cells die. These cells, you will recall, have an ability to divide repeatedly and give rise to new cells of the plant. Boron may also, under certain conditions, speed the rate of sugar flow in the plant.

WATER AND TRANSPIRATION

Water makes up 90 per cent or more, by weight, of many plant tissues. Even dry and dormant seeds and spores are at least 15 per cent moisture. Plants continuously absorb water from the soil. At the same time, through a process called **transpiration,** they lose water through evaporation to the air. The amount of water passing through a plant is thus much larger than the amount contained in it at any one time. Transpiration can affect the climate of an entire geographical region.

Transpiration occurs in the leaves and other aerial parts of most plants. Essentially, the leaf is a protected layer of wet cells actively carrying on photosynthesis. It is well supplied with veins and is encased in a fairly waterproof but perforated layer, the epidermis. The wet cells of the mesophyll evaporate large quantities of water into the intercellular air passageways. Water vapor molecules move along the passageways, through the stomata, and to the outside world. When the humidity is high and the air still, water loss from a leaf may be greatly reduced. Leaves lose the most water on dry windy days when the air moving over the leaf surface quickly removes the water vapor surrounding the leaf.

In most plants, epidermal cells are protectively covered with a waxy, water-impervious layer called the **cuticle.** Water loss in those plants having a cuticle occurs almost entirely through the stomata. If the stomata close, transpiration stops completely, or is greatly reduced. We have already discussed how the stomatal opening is controlled by turgor pressure of the guard cells. When a plant lacks water, the guard cells tend to become limp, thus closing the stomatal opening and restricting water loss.

Even when a plant has enough water, the guard cells may go limp and allow the stomata to close. In recent years, botanists have discovered that the amount of CO_2 in the interconnecting air passages is an important regulator of stomatal opening in many plants. If the CO_2 concentration falls below the 0.03 per cent normally present in the air, the guard cells become turgid and the stomata open. This usually happens when the plant is well lighted and is photosynthetically active. The result is a lowering of the CO_2 content of the air chambers. In the laboratory we can cause the stomata to open by removing the CO_2 from the air passing over a leaf. This control by CO_2 helps explain why stomata are normally open during the day and closed at night.

The usual pattern of stomatal opening and of transpiration goes something like this: at dawn there is a sudden increase of stomatal opening and of the rate of transpiration, reaching a maximum near noon. Soon after noon there is a decline since the plant has dried out somewhat. As it recovers its turgor there is a small rise, then another decline as darkness approaches. Stomatal behavior and transpiration are both governed by three things operating simultaneously: water supply to the plant, light, and the CO_2 content of the atmosphere.

Plants adapted for life in dry regions (**xerophytes**) tend to have fewer stomata per unit area than do plants living in areas where rainfall is normal (**mesophytes**). Most plants have many stomata on both surfaces of the leaf. Some, however, have stomata only on the lower surfaces. The cucumber leaf has more than 400,000 stomata per square inch, but some grasses have as few as 50,000. When fully open, the stomata may occupy as much as from 1 to 3 per cent of the total area of a typical leaf. This accounts for the tremendous quantities of water lost by well-watered plants growing in bright light in warm temperatures.

All the water evaporated from leaves and other aerial parts of plants comes from the soil. The soil is a water reservoir, alternately filled and emptied. During a rain, water moves downward through the soil. After the rain, the soil is then said to be at **field capacity.** During this time roots can remove water from the ground easily. But as the soil gradually dries out, it becomes increasingly difficult for the roots to absorb water. Eventually the plant may not be able to supply itself with enough water to prevent wilting. The percentage of water in the soil at this point is called the **wilting percentage** and varies widely from soil to soil. It is low in coarse sandy soils, high in fine clay soils. Although clay soils hold more water than sandy soils do, clay soils retain water in such a way that the plant can make only poor use of it.

TRANSPORT OF NUTRIENTS

The mechanism by which water is transported to the tops of tall trees has long puzzled plant physiologists. The process is still not completely understood. Atmospheric pressure raises a column of water only about 30 feet. The tallest trees are about 300 feet high. The forces involved in moving water to this impressive height must, therefore, be equivalent to about 10 atmospheres of pressure. The theory currently most favored by plant physiologists is the **transpiration-cohesion-tension theory.** It supposes that the evaporation of water from mesophyll cells acts as a kind of "suction pump." The "pump" effect is passed on from cell to cell until the dead xylem is reached. When it reaches an open-pipe system, such as tracheids and vessels, these cells are put under tension and a cohesive water column is lifted up these pipes. The force needed for the rise of water in xylem, then, is supplied by the evaporation of water molecules from the leaf, and water rise is independent of the living activities of any cell (Fig. 32.14).

This theory is supported by at least four types of evidence: 1. living plant shoots develop lifting pressures of several atmospheres; 2. tensions in the trunk are observed during transpiration; 3. completely nonliving models can be made to work in the same way; and 4. the killing of stem cells by steam or poisons does not interfere with their ability to transport water.

Another theory is the **root pressure theory.** Botanists who favor this view explain that when stems of certain plants are cut just above the ground line, much liquid is forced out from the cut surface. A pressure-measuring device (called a **manometer**) attached to such a cut stem shows that the roots produce a pressure of several atmospheres. However, the observed pressures in trees are not high enough to transport water to the

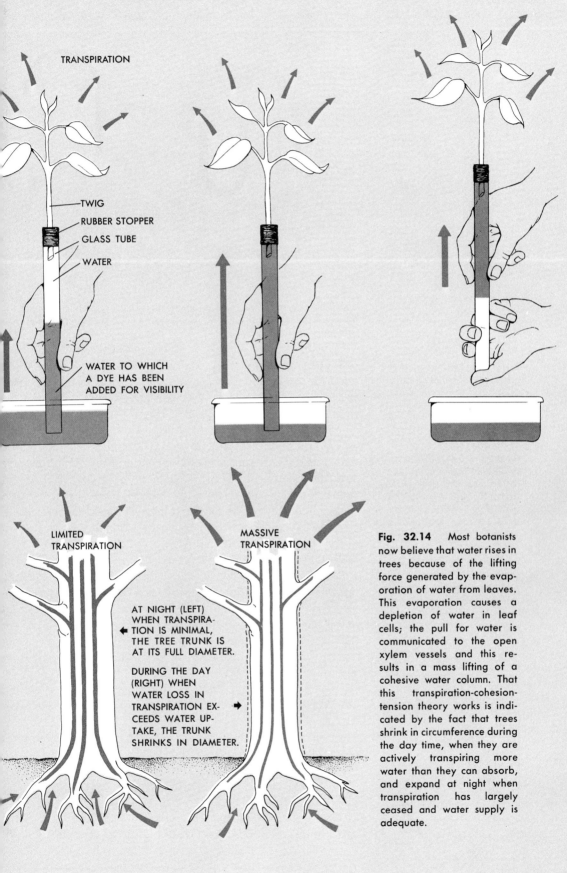

Fig. 32.14 Most botanists now believe that water rises in trees because of the lifting force generated by the evaporation of water from leaves. This evaporation causes a depletion of water in leaf cells; the pull for water is communicated to the open xylem vessels and this results in a mass lifting of a cohesive water column. That this transpiration-cohesion-tension theory works is indicated by the fact that trees shrink in circumference during the day time, when they are actively transpiring more water than they can absorb, and expand at night when transpiration has largely ceased and water supply is adequate.

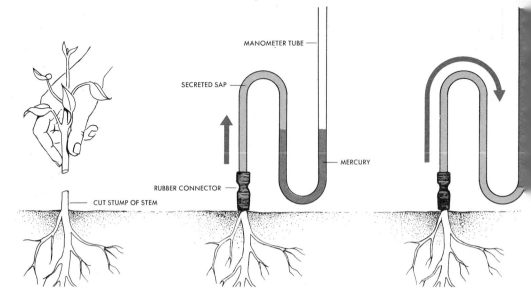

Fig. 32.15 As a result of their metabolic activities, roots can exude sap against a pressure gradient. This root pressure may be responsible for some of the rise of water in plants.

Fig. 32.16 A diagram of the pressure flow hypothesis of solute movement in plants. Sugars made in the leaf cell cause an increase in the osmotic concentration of these cells. This leads to a high turgor pressure, forcing sap through the sieve tubes down to the roots. En route, sugar and other solutes are absorbed by living cells of stem and root. The roots absorb water in response to osmotic gradients. This water is carried up to the leaves by a combination of root pressure and transpiration-generated forces. This completes the circulation of nutrients in the plant.

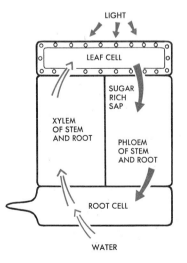

tops of tall trees. Furthermore, root pressures seem to be lowest when transpiration rates are highest, which is when trees need the most water. Most botanists feel that root pressure is not the major cause of water rise in trees (Fig. 32.15).

Salts absorbed by the roots may find their way into the xylem and be transported upward along with the water of the transpiration stream, or they may travel upward in the sieve tubes of the phloem. Many substances move in the phloem at rates of several centimeters per hour. These rates are much higher than those to be expected from diffusion alone. In addition, various substances may move up and down in the phloem at the same time, though not necessarily through the same sieve tube. The mechanism of such movement is not clearly understood. Some people believe that the photosynthetic cells of the leaf produce a high turgor pressure because their high sugar content causes a massive entry of water by osmosis. This would cause a high pressure in neighboring phloem cells. Movement downward would then occur in response to the pressure from above and result in transport of large quanties of sugar from top to bottom of the plant (Fig. 32.16).

Other people have observed that the cytoplasm of mature sieve cells streams rapidly both clockwise and counterclockwise. They feel that this streaming is of major importance in the two-way transport of materials in the phloem. Further work is needed before this problem can be resolved.

SUMMARY

Green plants require three basic classes of nutrients: organic foods, minerals and water. Green plants are completely autotrophic and can synthesize all their required organic compounds through photosynthesis and related processes. Minerals and water are normally absorbed by roots from the soil.

Photosynthesis is carried out entirely within the **chloroplast,** which can be removed from the cell and made to function in isolation. In this process, red or blue light absorbed by **chlorophyll** or other pigment in the chloroplast causes ejection of an electron from chlorophyll in the **grana** of chloroplasts. This electron is then replaced by removal of an electron from water. This causes an oxidation of water, and free oxygen is released. The ejected electrons, together with the protons (hydrogen) of water, are transferred by enzymes to carbon-containing compounds to which atmospheric CO_2 has been chemically attached. This reduces the attached CO_2 to the level of an organic acid (*phosphoglyceric acid*), later to a 3-carbon sugar, and finally to a stable 6-carbon sugar such as glucose. For this reduction process, energy in the form of **ATP** is required. This ATP is made directly in chloroplasts by the process of **photophosphorylation.**

CO_2 enters the leaf through **stomata,** openings in the leaf surrounded by sausage-shaped **guard cells.** When these guard cells are turgid, the stomata are open, when they are flaccid, the stomata are closed. The water required for photosynthesis is transported from the root to the leaf via the **xylem vessels** of the stem. The sugar and other photosynthesis products are removed from the leaf via the **sieve tubes** of the **phloem.**

The plant requires about 13 different mineral elements, most of which it absorbs from the soil via the roots. Only one element, nitrogen, does not originate from the parent rock, but comes from the atmosphere. In the process of **nitrogen fixation,** free-living bacteria of the soil, as well as certain root nodules containing associated bacteria transform free atmospheric nitrogen, which cannot be used as such by any living creature, to ammonia utilizable by the plant. In this process, an iron-containing substance, **ferredoxin,** is important.

Each essential element has a particular function in the plant. Some, like calcium and sulfur, are mainly structural. Others, like iron and copper, are parts of enzymes, where they function by becoming reversibly oxidized and reduced. Still others, like potassium, have unknown functions, though we know that they are essential. The absorption of elements from soil or solution requires energy expenditure by the plant. These elements are moved about mainly through the xylem.

Most of the water absorbed by plants is lost to the atmosphere through the stomata in the process of **transpiration.** Some plants (**xerophytes**) become adapted to minimize water loss, and may therefore exist in very dry climates. The rise of water to the tops of tall trees is probably due to evaporation from the leaves, producing a pull for water in the xylem vessels, in which a cohesive column of water is raised. Pressure from roots may also be involved. Pressures generated in leaves, due to water absorption following massive photosynthesis, may be important in causing a mass flow of organic nutrients down the phloem.

FOR THOUGHT AND DISCUSSION

1. What major classes of nutrients are required by plants?
2. In what way does the organic nutrition of plants differ from that of most other kinds of creatures?
3. What is the basic role of light in photosynthesis?
4. What is the source of the oxygen released in photosynthesis?
5. What is the path of carbon in photosynthesis? What is the first stable product formed during CO_2 fixation?
6. What evidence do we have that chlorophyll is the active pigment in most photosyntheses? What is the evidence that other pigments may also take part in the capture of light?
7. How does each of the raw materials of photosynthesis reach the chloroplast?
8. What controls the opening and closing of the stomata?
9. What is the relation between respiration and photosynthesis?
10. How can the carbon dioxide content of the atmosphere affect the climate of the earth?
11. What does the plant do with all the sugar formed in photosynthesis?
12. Why does "girdling" (removal of a ring of bark) ultimately kill a tree?
13. What are the six "macronutrient elements" required in large quantities of plants? Name a function of each.
14. What are the seven micronutrient elements required by plants? Name a function for each.
15. How would you prove that a given element is essential to a plant?
16. Of what is soil composed? In what way does organic matter in the soil benefit plant growth?
17. How can one diagnose the mineral requirements of a crop during its growth?
18. What processes are involved in the uptake of ions by roots?
19. In what way is nitrogen nutrition different from that of the other mineral elements? Explain in detail.

20 Describe the cycle of nitrogen in nature. Emphasize the roles played by microorganisms.
21 What is transpiration? What factors control its rate?
22 What is a xerophyte? How may it become adapted for life in its special environment?
23 What is the pathway for transport of (a) Water? (b) Minerals? (c) Sugars in plants? Prove your answer.
24 Propose a mechanism for (a) The transport of sugars from leaf to root. (b) The transport of water from root to leaf in a hundred-foot high tree.
25 How may cytoplasmic streaming be involved in long distance transport phenomena?

SELECTED READINGS

Arnon, D. I. "The Role of Light in Photosynthesis." *Scientific American*, **203** (1960), 104–118.

A brief, popular review of the discovery of photophosphorylation and its significance.

Bassham, J. A. & M. Calvin. *The Path of Carbon in Photosynthesis*. Englewood Cliffs, N.J.: Prentice Hall, Inc., 1957.

Hill, R., and C. P. Whittingham. *Photosynthesis*. New York: John Wiley and Sons, Inc., 1955.

A brief treatment of the biochemistry of higher-plant photosynthesis.

Hoagland, D. R. *Lectures in the Inorganic Nutrition of Plants*. Waltham, Mass.: Chronica Botanica, 1948.

A series of lucid lectures on mineral nutrition by the leader in the field several decades ago.

Kamen, M. D. *Primary Processes in Photosynthesis*. New York: Academic Press, 1963.

An advanced statement of what happens when a light quantum is absorbed by photosynthetic pigments.

Kramer, P. J. *Plant and Soil Water Relationships*. New York: McGraw-Hill, Inc., 1949.

A detailed and technical account of the physiology of water in the higher plant.

Rabinowitch, E. I. *Photosynthesis and Related Processes*. New York: Interscience, 1945 and later.

A three-volume compendium reviewing all relevant literature up to 1956.

33 HOW PLANTS GROW

The development of a seed into a mature plant is a remarkable process. It involves growth by cell division and cell extension, differentiation of new organs such as roots, stems, leaves, and flowers, and a complex series of chemical changes. The final form of the plant is a blend of the plant's genetic "blueprint" and modifying effects of the environment (Fig. 33.1).

A seed contains an **embryo plant,** surrounded and protected by a **seed coat,** and supplied with a source of stored food called the **endosperm.** The plant embryo contains a young root growing point, a shoot growing point, and embryonic leaves called

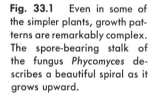

Fig. 33.1 Even in some of the simpler plants, growth patterns are remarkably complex. The spore-bearing stalk of the fungus *Phycomyces* describes a beautiful spiral as it grows upward.

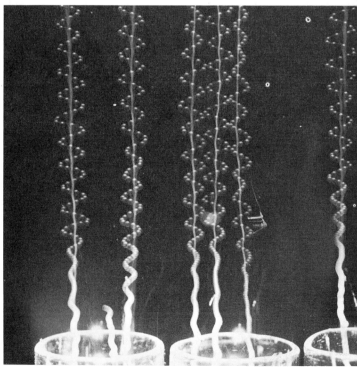

Courtesy David S. Dennison

otyledons. The cotyledons may be long, thin, and leaf-like. If they are, they serve first to digest the stored food of the endosperm tissue for use by the embryonic growing points, after which they expand into leaf-like photosynthetic organs. In other instances, the cotyledons are fleshy storage organs—above or below ground—that absorb the endosperm before the seed matures. Such cotyledons rarely become leaf-like or photosynthetic (Fig. 33.3).

When the seed begins to germinate, or sprout, it absorbs large amounts of water, and the growing points begin cell division. For reasons we do not yet understand, the root almost always begins to develop before the shoot growing point does. At both root and shoot ends of the seed, new cells are formed by the meristematic (dividing) areas of the growing points, followed by elongation and differentiation of these cells. Cell division, elongation, and differentiation in the root occur in regions that overlap. Since the root pushes its way downward through the soil, its tender growing point must be protected against abrasion. Protection is furnished by a group of cells called the **root cap.** This cap is continuously flaking off and being replaced.

One marked difference between plants and animals is that growth in plants takes place almost entirely near the meristematic areas; animals tend to have growth zones all over the body. You can observe the restricted growth in plants by marking the surface of the root or stem with lines which are equal distances apart. After several days you will see that the area just

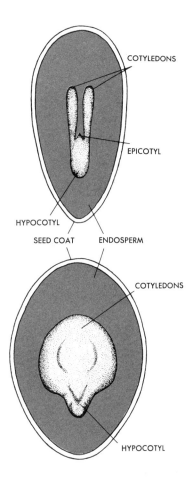

Fig. 33.3 The seed consists of a miniature plant, the embryo, encased in a protective seed coat and supplied with a source of stored food, the endosperm. The embryo itself consists of a miniature root, the hypocotyl, a miniature stem, the epicotyl and seed leaves, the cotyledons. The two views are cut perpendicular (top) and parallel (bottom) to the broad face of the seed.

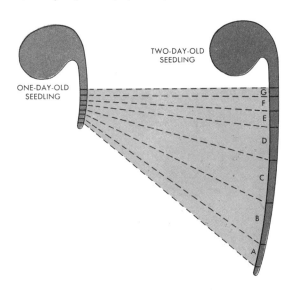

Fig. 33.2 The main region of elongation of a root is in the zone immediately behind the apex (Zones B, C, and D).

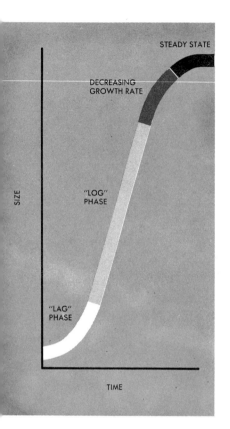

Fig. 33.4 All cells, organs, organisms, and populations describe an S-shaped or *sigmoid* growth curve.

behind the tip is the region where the most rapid growth has taken place. This is the region of cell elongation. Cell division itself does not do much to increase the size of the plant body; what it does is provide new cells, which enlarge at some later time (Figs. 33.2 and 33.7).

THE KINETICS OF GROWTH

If we measure the size of a plant at various times, after it has germinated, and then plot its size as a function of time in days, we get a very interesting curve, as shown on this page. This S-shaped, or **sigmoid,** curve is typical of the growth of all organs, plants, populations of plants or animals, and even of civilizations of men. It has at least four distinct sections: 1. an initial lag period during which internal changes are preparing the organism for growth; 2. a phase during which there is a constant rate of growth; 3. a phase during which growth rate gradually diminishes; and 4. a point at which the organism reaches maturity and growth stops. If the curve is prolonged further, a time will arrive when aging and death of the organisms occur, giving rise to one or two other sections of the growth curve (Fig. 33.4).

Aging and death, which afflict all animals as a part of their developmental cycle, seem not to be a necessary part of the developmental cycle of plants. We know, for example, that some pine trees and Sequoia trees of the western United States reach ages of well over 3,000 years. The fact that they one day die is due probably to infection, or to the weakening of the mechanical base from which they spring. If there were some way of preventing such conditions, the trees might continue to grow forever.

Plant tissues nourished on artifical solutions demonstrate the potential immortality of plant cells. In 1937 a research worker in France removed portions of carrot roots and placed them in chemical nutrients. The result was a rapidly growing undifferentiated mass of callus tissue. Next, the tissue was subdivided and transferred to new flasks at frequent intervals. Growth continued at a constant rate and showed no sign of diminishing, even after 30 years. The plant from which the original carrot tissue was taken certainly must have died many years previously. The normal halting of growth in a plant, then, must be due to some inhibitory effect. If we knew what it is, and if we could remove or prevent it from acting, we might be able to produce a potentially immortal plant.

Growth curves have given us evidence of physiological growth controls of various kinds. By examining the length of lag periods, for instance, we are given clues about changes that must occur before growth begins. The lag period is only a few hours long in many seeds, yet in others it may be days, weeks, or even months. Seeds with long lag periods probably have inhibitory substances that delay growth until the substances have been destroyed or removed. The rate of growth during the rapid growth phase is often determined by hormonal substances, which we shall discuss later in the chapter.

The slope of the sigmoid curve can also give us clues about the genetic background of the plant's growth potential, as well as the environment in which the plant is growing. The total height of the plant and the time of the onset of the steady state phase are also frequently genetically controlled, but are also susceptible to control by the environment. Finally, the aging and death of the organism are not determined entirely by the genetics of the organism, but are under the control of the experimenter.

MERISTEMS AND TISSUE ORGANIZATION

In the highly organized plant root and shoot, each cell goes through an orderly series of developmental phases. The cubical cell produced in the meristematic region, or **meristem,** of a growing plant has many vacuoles. Its increase in size, especially in length, is due mainly to the uptake of water into these vacuoles. The vacuoles grow in size, then ultimately fuse into one large central vacuolar tank. The rest of the cell keeps pace with the increase in size by making additional cell wall material, cytoplasmic material, and the various types of cell organelles.

Usually, together with elongation, but sometimes after it, differentiation occurs. The cells on the exterior of the root, for example, adopt one of two final forms. They become either flattened epidermal cells, or they become root hair cells. The latter consist of epidermal cells with tremendously long extensions which readily absorb water and minerals. During the rapid growth phase of these cells, the nucleus is almost always at the tip of the growing root hair and appears to be the center of great metabolic activity. Root hair cells are short-lived, but are rapidly produced in large numbers as the root tip continues to push its way through the soil. The root hairs greatly increase the surface area of root that is brought into contact with the soil (Fig. 33.5).

The central tissues of the root differentiate into the vascular elements, since roots characteristically do not have pith. The same pattern occurs even in tissue culture. Deep within the

The Bergman Associates

Fig. 33.5 Root hairs develop in profusion immediately behind the growing point.

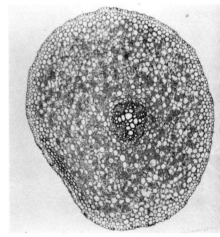

Photo by M. S. Fuller

Fig. 33.6 A cross section of a young root showing the central cylinder surrounded by the endodermis, the cortex, and outer epidermal layer.

rapidly growing masses of an undifferentiated tissue culture we find little groups of tracheids. We therefore suppose that there is something about the interior of a mass of tissue that favor the xylem or tracheid type of differentiation. This "something" could be either the absence of oxygen deep within a mass of cells or a lack of contact with the soil or other outside medium.

Surrounding the root xylem cells are three structures: 1. bundles of phloem; 2. the meristematic **pericycle** that gives rise to branch roots; and 3. an **endodermis** that surrounds the entire central vascular cylinder. The endodermis has a curious thick structure, the **Casparian strip,** which represents a band-like thickening of the walls of the cells of the endodermis. Some botanists think the water-impervious Casparian strip acts as a sort of dam that prevents diffusion of water along the wall and forces movement of all materials through the differentially permeable membrane of the endodermal cells. This theory is still somewhat in doubt.

Between the internal vascular cylinder and the epidermis lies a group of loosely packed undifferentiated cells called the **cortex.** They are large, thin-walled, have nuclei and large central vacuoles. Their function probably is to store reserve materials in the root (Fig. 33.6).

As a dividing **cambium** develops between the xylem and the phloem of the central cylinder, and as the roots thicken because of its radial cell divisions, the cortex becomes smaller and smaller, cracking and flaking off at the outside of the root. Finally, in an older root, the epidermis and cortex are completely lost. The new outside layer is built up from corky cells known as **periderm tissue.** These corky cells develop from a secondary meristem called the **cork cambium.**

This pattern of growth and development is particularly apparent in stems that thicken as they age. A stem tip, like a root tip, has a meristematic zone of cells that divide rapidly, and behind this zone is a region of cells that elongate rapidly. A stem tip is more complicated than a root tip since in addition to its stem tissue, a stem tip also must form leaves and buds.

The buds are first visible as minute projections of tissue, which develops into either vegetative buds or flower buds. (Fig. 33.8). In many plants, the nature of the bud is controlled by environmental conditions such as temperature and light (which will be discussed in Chapter 34).

Behind the region of cell elongation is the region of differentiation. Here, too, one can see quite clearly the development of epidermal tissues, of a central vascular cylinder, and of cortical cells between the two. Probably the main anatomical difference between stems and roots is that stems generally have a central pith. The xylem is found around the pith, and the phloem around the xylem. Stems grown in the light generally do not have an endodermis; stems grown in the dark usually do. In stems, as in roots, the cambial layer develops between the xylem and the phloem. By rapid divisions inward and outward, the cambium gives rise to cells that differentiate into xylem on the inward side, and into phloem on the outward side.

Eventually, great pressures are set up by this growth from within, causing external layers of the stem to crack and flake off. As this occurs, the plant produces new protective cells under the areas that flake off. Here, again, it is a cork cambium that appears, and the cells produced by this cambium are the heavy-walled waterproof cells typical of corky tissues comprising the bark of a tree or shrub.

The **annual rings** of the stems of trees result from different climatic conditions in different periods of the year. In the spring, when water is plentiful and other conditions are favorable, the cambium produces cells that are thin-walled and contain a large central cavity. In summer and fall, when conditions tend to be less favorable, the tracheids formed have thicker walls with smaller cavities. This regular alternation of spring and summer wood produces an annual ring. The transition from spring to summer wood is usually gradual, but the abrupt halt at the end of the growing season is sharply distinguished from the spring wood of the following year (Fig. 33.9).

The regularity of annual rings enables us to date trees and, therefore, civilizations in which remains of trees have been found. For instance, we know that certain climatic cycles have occurred in various regions. If a year is especially favorable to growth, a very thick annual ring is produced. In drought years very small annual rings appear. The sequence of large and

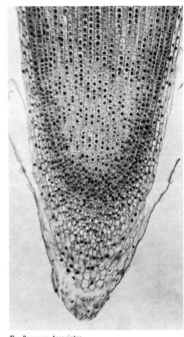

The Bergman Associates

Fig. 33.7 A longitudinal section of a root, showing root cap and meristematic zone.

Fig. 33.8 Left: A section of cells of the stem apex of wheat. The other three photos are surface views of the stem apex as it produces leaf primordia (left), elongates (center), and matures flower primordia (right).

Courtesy Dr. C. Barnard

Courtesy Dr. C. Barnard

Courtesy Dr. C. Barnard

Courtesy Dr. C. Barnard

Eric V. Gravé

Fig. 33.9 The annual rings in the wood of a tree are caused by the changing size of tracheids formed by the cambium. In the spring large tracheids are formed. As the year progresses, smaller and smaller tracheids are formed. Thus the ring itself is the line between the summer tracheids marking the end of one year and the spring tracheids marking the beginning of the next year.

small rings, forming a pattern reflecting climatic changes of past years, tends to be constant in trees inhabiting a given region. Thus, a piece of wood forming the supporting timber of a house in an extinct civilization could be compared with other materials whose age is known and the civilization dated by this technique. Although extremely useful, the method is not always reliable, because trees sometimes produce several growth rings in one year, and the annual rings of successive years are not always sharply separated.

How rapidly or slowly a plant grows depends both on its genetic constitution (**genotype**) and the environment. Suppose that we have a small group of plants that have been self-pollinated, or crossed only within their limited group, for several generations. The offspring of these plants are said to be **inbred**. If, on the other hand, the plants are pollinated by plants of different groups ("stocks"), the offspring are **hybrids**. For many years geneticists have known that successive inbreeding diminishes the vigor of a stock, but cross-breeding between weakened inbred lines often produces a very vigorous hybrid. This condition is called **hybrid vigor** or **heterosis**. The causes of heterosis are not yet clearly understood. However, we do know that such a plant uses elements of its environment more effectively for growth than inbred plants do.

Elements of the environment exert tremendous control over a plant. For example, too little water slows down growth. Too little nitrogen, potassium, phosphorus, or of any essential element retards growth or kills the plant. The intensity of light falling on a photosynthesizing plant also determines its growth rate, and whether or not it will survive. Without enough light and carbon dioxide, a photosynthetic plant cannot store enough energy to meet its needs for growth and development.

The temperature of the environment may also be extremely important in determining the nature and rate of plant growth. With most chemical processes, the rate of reaction increases steadily with an increase in temperature. In fact over most ranges, the rate of a chemical reaction is doubled by a 10°C increase in temperature. For reasons that we do not understand, different plants have vastly different temperatures (*temperature optima*) at which they grow best. This indicates that some fundamental biochemical process in their make-up is adversely affected by high temperature. Thus, if we steadily raise the temperature we eventually reach a temperature at which the positive and negative effects combine to make it best for growth. If we continue to raise the temperature above this point, the growth rate declines, sometimes very dramatically, because the deleterious effects outweigh the beneficial effects. For most plants, the optimum temperature is in the range of 28–32°C.

We do not know why plants are injured by temperatures of about 35°C, but they are. As far as we can tell, enzymes obtained from plants are not damaged by this temperature. One guess is that certain essential "growth" chemicals produced by the plant may either be destroyed or prevented from forming in adequate quantity at elevated temperatures. For example, the red bread mold, *Neurospora,* has "temperature sensitive" genes. The gene responsible for the production of vitamin B_2 in one strain of *Neurospora* works quite well when the organism is grown at low temperatures. It does not function well when the organism is grown at higher temperatures. At 35°C the organism requires an outside source of B_2, but at 25°C it produces this material for itself. This same general situation probably holds for higher plants. If we knew why high temperature slows down growth rate, we might greatly improve growth at high temperatures by supplying the plant with the material it needs.

GROWTH HORMONES

In addition to the water, light, carbon dioxide, and various minerals, a plant requires other chemicals for growth. These substances, called **hormones,** are formed in one

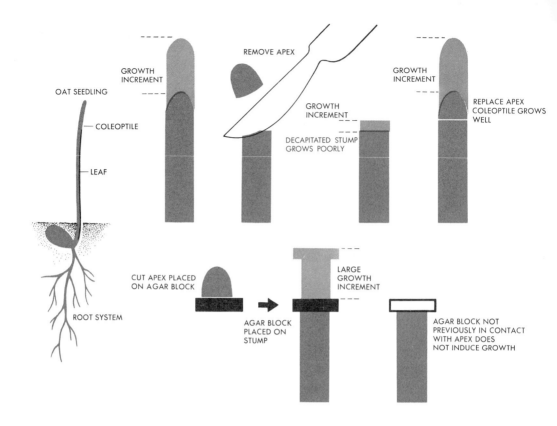

Fig. 33.10 The elongation of grass coleoptiles is controlled by growth hormone produced at the tip of the coleoptile. If the apex is removed, growth slows down due to depletion of this hormone. If the apex is replaced, growth resumes. The growth-promoting properties of the apex can be duplicated by hormone permitted to diffuse out of the apex into an agar block.

part of an organism and transported to another part where they produce some special effect. Generally, they are needed only in extremely small quantities. In most instances, they are produced in adequate amounts by the plant itself. Three major classes of growth-regulatory hormones exist in most, if not all, higher plants—the **auxins, gibberellins,** and **cytokinins.**

The auxins are produced by the growing tips of stems and roots. They migrate from the tip (also called the **apex**) to the zone of elongation, where they are needed for the elongation process. If the tip of a rapidly growing stem is removed, growth in the region below the cut will slow down very quickly. Within several hours or days, depending on the type of plant, growth of the affected part will come to a complete halt. If the removed tip is replaced, growth of the stem continues almost normally, showing that some influence coming from the tip is conducted across the wound to the growing cells (Fig. 33.10).

If the tip is placed on a block of gelatin or agar for several hours, and the block without the tip is then transferred to the cut stump of a decapitated stem, the block will partially substitute for the tip in promoting growth in the decapitated stem. From this experiment, we deduce that a substance, called auxin, moves from the tip to the block and from the block down to the stem. We now know of the existence of many substances with auxin activity. Several of them have been isolated in pure form from plant tissues, and are thus native plant growth hormones. The most common one is a material known as **indole-3-acetic acid.**

Laboratory experiments show that indoleacetic acid produces some rather interesting effects on the growth of stems and roots. If the growing portions of a stem, such as those of a pea plant, are cut off and put in petri dishes of sucrose and a mineral salt solution, growth is very slow. If, however, we add small amounts of indoleacetic acid (**IAA**), growth is speeded up greatly. Within certain limits, the more IAA we add, the faster the growth. When we reach a particular concentration of IAA, however, growth is once again somewhat slower, and if we add still stronger concentrations, growth may stop (Fig. 33.11).

If we apply auxin to the stem of an intact plant, extra growth usually does not occur. From this we conclude that a stem is normally saturated with auxin produced by its own tip. But if we apply auxin to the roots of a normally growing plant, something interesting happens. Growth usually slows down; however, if we cut off the root tip, thus cutting off the root's own supply of auxin, growth then sometimes speeds up. Since the amounts of auxin in stem and root are about the same, we conclude that the root is more sensitive to auxin than the stem is.

We can tell how much auxin a plant tissue has by using a solvent, such as ether, to extract the auxin. Then, by applying the extract to some other plant tissue, we can note the growth effects produced. Usually, the leaf sheath (**coleoptile**) of dark-

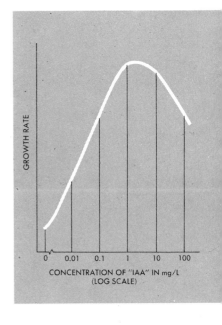

Fig. 33.11 The amount of elongation produced by the auxin indole-3-acetic acid depends on the concentration applied. The optimum concentration is frequently near one milligram per liter.

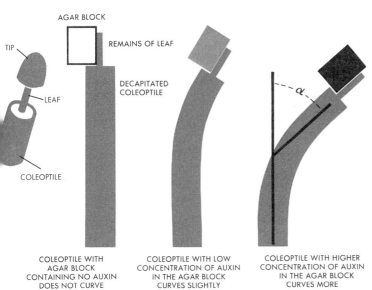

Fig. 33.12 The amount of auxin in any extract may be determined by measuring the angle of curvature produced in grass coleoptiles to which the extract has been unilaterally applied in an agar block.

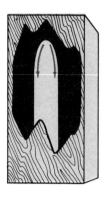  

A COLEOPTILE GROWING IN THE DARK OR IN SYMMETRICAL LIGHT HAS SYMMETRICAL AUXIN DISTRIBUTION AND SYMMETRICAL GROWTH

A COLEOPTILE EXPOSED TO UNILATERAL LIGHT DEVELOPS AN ASYMMETRY IN AUXIN DISTRIBUTION

THIS LEADS TO AN ASYMMETRY IN GROWTH, WHICH RESULTS IN CURVATURE

Fig. 33.13 Phototropic curvature is caused by asymmetry in distribution of the growth hormone produced at the tip of the coleoptile.

grown oat plants is used. When the extract has been prepared, it is added to an agar block that is then placed on the decapitated stump of the coleoptile, so that only half of the stump is in contact with the block. The auxin then stimulates the growth of tissue only directly below the block. This unequal growth on the two sides of the coleoptile causes a curvature, as shown in the diagram. The more auxin contained in the agar block, the greater the curvature. To find out how much auxin a plant tissue contains, all we do is compare the amount of curvature with that produced by known quantities of auxin (Fig. 33.12).

Fig. 33.14 The stronger the light intensity, the more a coleoptile curves toward it. Coleoptiles at the left are close to the light source, those at the right are far from it. Notice the decrease in curvature from left to right.

Richard F. Trump

Various plant organs also curve in response to light and gravity. We now know that such curvature (called a **tropism**) is due to an unbalanced distribution of auxin. For example, if we shine light on one side of an oat coleoptile, the organ will curve toward the light (**phototropism**). What happens is that growth on the lighted side is depressed by the light, while growth on the shaded side is accelerated. If we expose a coleoptile tip to light in this way, then extract auxin from the two halves (light and dark) and assess them by the curvature test, we find that the shaded side of the coleoptile has about twice as much auxin as the lighted side. Light, then, seems to produce curvature by affecting the distribution of auxin in a plant organ; this auxin concentration, then, controls growth (Figs. 33.13 and 33.14).

Similarly, after some time a plant stem arranged horizontally along the ground accumulates more auxin on the lower surface than on the upper surface. This results in accelerated growth on the lower side, hence a curvature upward (**geotropism**). In a root that is growing horizontally the auxin also accumulates in the lower tissue. Why, then, dosen't the root also curve up instead of down? As already mentioned, in a normal root the auxin concentration is already at or beyond the concentration best for growth. This greater concentration of auxin on the under side of the root slows down growth, thus producing a downward curvature of the root. The forces that cause auxin to accumulate as it does are not understood.

Another unsolved problem in tropism is how a plant detects light and gravitational stimuli. The action spectrum (see page 485) for phototropism indicates that only blue light and some ultraviolet produce curvature. Some yellow pigment, possibly related either to carotene or to riboflavin (Vitamin B_2), could be involved as the receiver of light. The gravitational sensing mechanism appears to involve **statoliths.** These are small particles in the cell that fall from one wall to another as the cell is tipped.

Auxin produced in the apex flows away from the tip at a rate of about one centimeter per hour at normal temperatures. For reasons we do not understand, this flow is mainly in one direction—from apex to base. If we place a block of agar containing auxin on top of a decapitated stem, auxin from the block flows down into the tissue. If, however, the base of a decapitated stem is placed on top of a block af agar, the auxin does not move up through the base toward the apex. Inverting the entire assembly does not affect this *polarity*. Nothing in the stem structure seems to account for this one-way-only flow; the polarity seems to be caused by chemical factors (Fig. 33.15).

So far we have been describing auxin action pertaining only to cell elongation. Auxin may also influence cell division. For

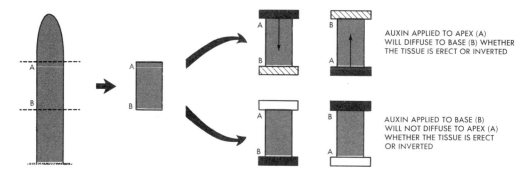

Fig. 33.15 Auxin is transported only from the morphological apex to the morphological base. This is an aspect of *polarity*. It is not affected by inverting the tissue.

example, in the spring, the beginning of cambial cell division in trees seems to be controlled by auxin diffusing downward from developing buds. In addition, the formation of branch roots (which involves initially cell division in the pericycle) can be started by the application of auxins (Fig. 33.16).

Auxin also determines whether the apical bud or a lateral bud on a stem will develop. In many plants, only the apical bud grows if the plant is undisturbed; however, if the apex is removed, one or several of the buds lower down begin to develop. If the tip of the branch is removed and the cut surface is covered with an auxin paste, the buds lower down will not develop. We conclude from this type of experiment that auxins produced by the apical bud inhibit lateral bud growth.

Even though we cannot say exactly how auxins act, we can make them work for us in agriculture. The potato tuber is a modified stem, and the "eyes" of the potato are buds. A freshly dug potato is dormant, but after some months of storage its buds begin to sprout. If we add synthetic auxins to such potatoes, the buds are prevented from developing for very long periods. This

Fig. 33.16 Auxins are used for many purposes, including formation of roots on cuttings. The geraniums on the right have been dipped in auxin. Those on the left are untreated.

Courtesy Boyce Thompson Institute

has permitted the storage of potatoes for up to three years instead of one year.

Auxins are also important in regulating the fall of leaves and fruits from plants. A leaf blade is attached to a stem by a **petiole.** The petiole persists during the growing season but falls off at some time later in the year. As long as the leaf blade produces adequate quantities of auxin, the petiole remains firmly attached to the stem. When the leaf blade produces too little auxin, however, the petiole weakens and the leaf falls from the plant.

This knowledge has been put to good use in agriculture. Say that we want an apple or orange tree to keep its leaves and fruit a bit longer than the tree normally would. All we have to do is spray the tree with a dilute solution of one or another of certain auxins. This simple process has saved millions of dollars for orchardists whose fruits normally fall off the tree before they are ready for harvest. This process can also be reversed by spraying the tree with chemicals that interfere with the action of auxin. This will cause leaves or fruits to fall prematurely from a plant. Cotton picking by machine has been greatly aided by this process. The mechanical cotton picker does a good job when the plants do not have too many leaves. If there are too many leaves, "antiauxins" can be sprayed over the cotton field several days before harvest. This is followed, usually within 48 hours, by a massive shedding of leaves. The machine then can run through the field easily, harvesting the bolls without encountering too much interference.

Pineapple plants can be made to flower by the application of a synthetic auxin (**α-naphthaleneacetic acid**). The value of this knowledge in pineapple agriculture is obvious. The plants can all be grown to a uniform size and naphthaleneacetic acid can be applied at any desired time. The fruits will then develop uniformly, making mass methods of harvest possible. The flowering of the litchi nut and several other fruits can also be controlled by the use of auxin sprays.

Auxin can also be applied to the pistil of a flower to produce artificial, or **parthenocarpic,** fruits. Normally, most fruits are formed as a result of pollination and fertilization of the ovary of the flower. The ovary then develops into the fruit. If, however, pollen does not reach the pistil, the development of the ovary into a fruit can be stimulated by the application of fairly large quantities of auxin-type materials. Naphthaleneacetic acid, for example, can be sprayed or smeared onto a tomato ovary to produce a fairly typical fruit that is large, red, and tasty, but lacking in viable seeds. So far, no one has discovered a way of bringing about development of the seeds without normal fertilization.

Earlier we said that high concentrations of auxin slow down

518 The Green Plant

or stop the growth of certain kinds of cells. It is also true that some of the synthetic auxins (such as **2,4-dichlorophenoxyacetic acid,** abbreviated **2,4-D**) kill some plants but not others. Two, 4-D kills **dicotyledonous,** or broad-leaved, plants, but does not kill **monocotyledonous,** or narrow-leaved, plants. In a lawn infested with dandelions, therefore, 2,4-D kills the dandelions while leaving the grass intact. Similarly, when 2,4-D is sprayed on a cornfield infected with bindweed, bindweed is killed but the corn is left intact.

The Gibberellins

These form the second group of important plant growth hormones. In the 1890's Japanese farmers noticed that extraordinary elongated seedlings were appearing in their rice paddies. They watched these seedlings closely. Any alert farmer knows that a large-sized plant might be valuable as breeding stock. These tall seedlings, however, never lived to maturity, and only rarely did they flower. The "disease" was aptly named *bakanae* or "foolish seedling." In 1926, a botanist discovered

Fig. 33.17 The plant growth hormone *gibberellin* was discovered as a result of the chance observation of excessive elongation in rice plants infected with the fungus *Gibberella*.

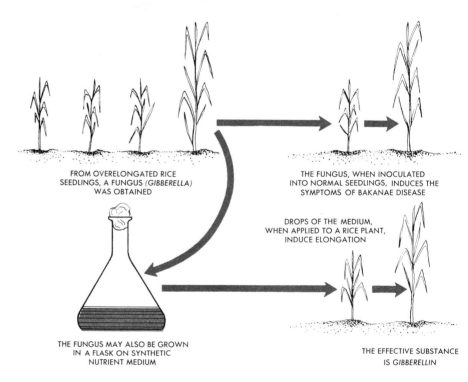

that these seedlings were all infected with a fungus called *Gibberella fujikuroi*. If the fungus was transferred from an infected seedling to a healthy plant, the healthy plant became diseased. The Japanese also found that if the fungus were grown on an artificial medium in a flask, the nutrient medium accumulated a certain substance. When this substance was transferred to a healthy plant, the plant developed overgrowth symptoms typical of the "foolish seedling" disease. The substance was named gibberellin, after the fungus that produces it (Fig. 33.17).

In the 1930's, Japanese physiologists and chemists working together isolated several other substances from the *Gibberella* growth medium. Some of them inhibited plant growth, others improved it. World War II interrupted further work, and the exciting story of gibberellin remained unknown in the Western world until about 1950. By 1955 British scientists had confirmed the original Japanese observation and had, in addition, isolated a substance they named **gibberellic acid,** which differed slightly from the substance isolated by the Japanese. Gibberellic acid produces tremendous elongation effects on stems. In some instances it also decreases the leaf area. Its most dramatic effect is its prompt stimulation of flowering in "long-day" plants (see also page 537). Any chemical that causes a vegetative plant to start flowering is a potentially important one in agriculture. At the moment, the most important agricultural use of the gibberellins is in the grape industry. When gibberellins are applied to seedless grape clusters, the grapes become larger and more numerous (Figs. 33.18 and 33.19).

Gibberellin is also useful in celery growing. It produces large and succulent plants in a relatively short time. It can also be used to stimulate seed germination and growth of early seedling grass, including the barley used as malt in the brewing industry. Its potential in agriculture is very great, and is only now beginning to be explored.

Not all plants respond to gibberellin, however. If we apply gibberellin to dwarf peas and tall peas, or to dwarf corn and tall corn, the dwarf varieties become tall, but the normally tall varieties show little or no change. An attractive hypothesis is that dwarfness is caused by the plant's inability to produce enough gibberellin to permit it to grow tall by itself. We know that dwarfness, which may be due to a single gene, is passed on from parent to offspring. We also know that dwarf plants, although they can be given a tall appearance by gibberellin treatment, nevertheless produce dwarf offspring. No matter how much gibberellin is added, and no matter how many generations of a dwarf plant we treat, we do not change the gene that produces dwarfness (Fig. 33.20).

It would seem logical that tall plants should be richer in

Courtesy S. H. Wittwer

Fig. 33.18 When gibberellin is applied to certain rosette plants it induces bolting and flowering of these plants. This is shown for spinach (above) and cabbage (below). Notice the "control" plant beside each of the treated plants.

gibberellin than dwarf plants. But they do not seem to be. At this stage, about all we know is that gibberellin-like materials exist in plants. So far, 11 different gibberellins have been isolated, both from fungi and from the young seeds of various types of plants. Exactly how gibberellin works biochemically is still a puzzle. Possibly, it stimulates the production of certain enzymes. We do know that gibberellic acid greatly stimulates the appearance of the starch-digesting enzyme **amylase** in the endosperm of barley grains. It also starts the production of auxin in stem apexes.

Cytokinins and Other Growth Regulators

In addition to the auxins and gibberellins, there are other promoters of the growth of plant cells. Coconut milk, the liquid endosperm of the coconut fruit, stimulates the growth of many plants. Among the materials coconut milk contains are auxins, amino acids, and **inositol** (a sugar alcohol).

Another growth-promoting material is the synthetic substance **kinetin**. When it is applied to certain plant cells, such as some calluses or roots grown in tissue culture, it acts together with auxin and produces a tremendous increase in cell division activity. Auxin alone produces only a swelling and an enlargement of the existing cells, while kinetin stimulates cell division (mitosis) and bud formation (Fig. 33.21).

Kinetin produces several other interesting effects. Leaves that have been cut off a plant tend to yellow and die quickly, but when they are supplied with kinetin they survive much longer. If kinetin is applied to a small area of a freshly cut leaf, only that area will remain green. If nutrient materials are supplied to another area of the leaf, they are transported to and accumulated at the site of the kinetin application. Recently,

Fig. 33.19 When gibberellin is applied to certain varieties of seedless grapes, the yield of fruit is greatly increased. From left to right, the concentrations of applied gibberellin increases.

Courtesy R. J. Weaver and S. B. McCune   Courtesy R. J. Weaver and S. B. McCune

chemical analogs of kinetin have been found to occur naturally in certain plant tissues. These substances, referred to as **cytokinins,** probably constitute a new group of plant hormones. They will be discussed further in the next chapter.

GROWTH INHIBITORS

In addition to substances that speed up growth, plant tissues may contain substances that slow down or prevent growth. For example, many seeds will not germinate if they are placed on moist filter paper in a petri dish at room temperature. But if the seeds are placed in running tap water for several hours, they will germinate. The running water removes substances that prevent germination. This can be important to seeds of plants living in arid zones. On the desert, if a light rain were to fall, such seeds would not germinate. This is fortunate, because germination would produce a tender seedling that would die in the absence of additional moisture. During a heavy rainfall, however, the inhibitor is washed out of the seed, germination promptly occurs, and the plant can survive because of the abundance of water in the soil around the root of the tender germinating seedling.

Inhibitors controlling seed and bud dormancy can also be removed by temperature treatment. It now appears that inhibitory substances are made during the growing season and accumulate in the bud, causing it to become dormant. As cold weather sets in, low temperatures gradually begin to destroy the inhibitor. By spring, after a sufficiently long cold period, the buds are free of the inhibitor and can grow. In a sense, then, it is not the coming of spring that stimulates bud growth, but the completion of winter.

The survival value of this mechanism is obvious. If a few warm days come very early in the spring, an unprotected bud might start to grow and then be killed by a return of cold weather. If, however, the bud is kept in check by the inhibitor, it will not grow until a "safe" date later in the season, because a sufficient number of cold days would have insured the complete destruction of the inhibitor. Such buds and seeds contain a "chemical clock," the inhibitor, that runs to a stop only when the inhibitor has been destroyed.

Research on chemical regulators of plant growth and development frequently turns up many surprises. For example, British workers were recently able to isolate from the buds of dormant beech trees a substance which they believed was responsible for the onset of dormancy. They named this substance **dormin.** At the same time, researchers in California were investigating a substance produced in cotton plants which causes

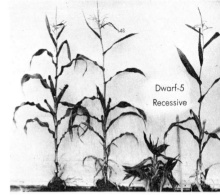

Courtesy B. O. Phinney and Charles A. West

Courtesy Anton Lang

Fig. 33.20 Gibberellin, when applied to certain dwarfs, as in corn (above) and peas (below), causes an approximation of the tall habit. In the figure above the four plants represent normal corn, normal plus gibberellin, dwarf corn, and dwarf plus gibberellin. In the bottom figure the plant at the left is an untreated dwarf pea. The pea plant at the right was treated with gibberellin.

Fig. 33.21 The growth of plants may be stunted by substances that antagonize the action of one of the hormones. From left to right, the plants are untreated controls, treated with a high concentration of CCC, and treated with a low concentration.

Dept. Biochemistry, Mich. State University

Fig. 33.22 Kinetin causes the development of buds on undifferentiated callus tissues. Note the buds formed on this geranium callus.

Courtesy H. R. Chen

leaves to fall off (absciss) late in the season. They, too, were successful in isolating pure crystalline material, which they named **abscisin**. These discoveries were made independently and practically simultaneously. Imagine the surprise of the two groups of experimenters when they discovered that they had both discovered the same substance! Abscisin and dormin, derived from different plants in the study of different processes, have exactly the same chemical constitution!

When the structural formulae for these two identical substances were published, biochemists noted that they seemed to be composed of isoprenoid units (5-carbon branched chain units containing one double bond) and that they therefore bore some structural relation to the gibberellins, which are similarly constructed. Since the gibberellins normally promote growth while dormin-abscisin inhibits growth, the question naturally arose "Is growth controlled by the ratio of gibberellin to dormin-abscisin?" The answer seems to be "yes," for the application of abscisin seems to inhibit the activation of amylase in barley endosperm by gibberellin (see p. 518). Also, while gibberellin causes the flowering of long-day plants, even those kept on short days, (see p. 537) abscisin prevents the flowering of such plants, even when they are properly stimulated by the correct photoperiod. Thus, a single type of substance, built of isoprenoid units, seems to control such widely different processes as bud dormancy, leaf fall, enzyme activation during germination, and flowering.

We have already discussed the synthesis of artificial inhibitors of plant growth in the laboratory, and the use of these inhibitors in the control of weeds and other plants. Certain auxins are a case in point. Other interesting inhibitors (such as **Amo-1618** and **CCC**) have recently been used as "antigibberellins." When applied to chrysanthemums, these substances cause inhibition of stem elongation without much effect on leaf size, on flowering, or on the size of the flower. Thus, it is possible to grow large-flowered chrysanthemum plants of very small total size. Another substance (**monuron**) is a photosynthetic poison. It can be used to clear an area, such as a railway embankment, of all green vegetation, without damaging any animal inhabitants and without danger to humans (Fig. 33.21).

The widespread use of chemicals for the control of weeds, fungi, and insects in agriculture poses difficult problems for man. Some of these chemicals are quite poisonous. Thus, if they accumulate in the soil, or in plants or in animals (such as man) which eat the plants, many people can be made ill. It is, therefore, important that we test all new chemicals carefully before spreading them over the countryside by the tons. The ideal chemical would be toxic only to one undesirable organism, and would be broken down rapidly by soil bacteria, so that dangerous levels did not continue to rise as a result of repeated use.

Even a well-designed insecticide, fungicide, or herbicide can have unfortunate side effects which make its use inadvisable. For example, if we exterminate insects too effectively, we will starve many birds which bring us pleasure and economic benefits. In our attempt to kill mosquitoes, we may also decimate the population of bees, which pollinate our commercial fruit crops. If we exterminate certain weeds too ruthlessly, we may cause a deterioration of the bacterial population of the soil, on which we depend for the maintenance of good crumb structure and soil fertility.

The rapid growth of the world's population and our increased dependence on intensive, high-yield agriculture make it necessary for us to continue to use chemical agents to combat plant and animal pests. We must thus expect an increase in the use of such substances, many of which will find their way into our bodies. At the same time we must safeguard our health by seeing to it, through appropriate government and industrial cooperation, that only well-tested and safe chemicals are used.

SUMMARY

All seeds contain an embryonic plant, together with a supply of stored food (sometimes called the endosperm) and a protective coat. During germination, water is absorbed, the stored food is digested and transported to the growing points, and growth starts. Cell divisions occur in **meristems** at the apex of stems and roots and in the **cambium**. Most increase in size occurs by cell elongation some distance behind the apex. Over-all growth starts slowly, becomes more rapid, then tapers off and stops. The cessation of growth is due to internal factors, for if bits of growing tissue are cut off the plant and transferred aseptically to artificial nutrient solutions, they can keep growing indefinitely. Differentiation patterns become visible after the newly-formed cells age somewhat. In roots, the central cells differentiate into **xylem,** around which are found **phloem** bundles, **pericycle** and **endodermis.** Around this central cylinder are found a storage **cortex** and **epidermis.** In some roots, a cambium develops between xylem and phloem. Its radial divisions, producing secondary xylem and phloem, cause an increase in diameter, cracking off of outer tissues, and their replacement by corky **periderm** tissue. Branch roots arise by divisions in the pericycle, and push their way out through the cortex and epidermis. Stems resemble roots, except that they have a central pith, and produce branches from exterior, rather than interior layers. The annual rings in woody stems result from differences in cell wall thickness between tracheid cells produced in the later summer (thick walls) and early spring (thin walls).

The rate of growth is governed by genetic constitution, various environmental factors and by internal mobile substances called hormones. Among the classes of growth regulatory chemicals in plants are **auxins, gibberellins, cytokinins** and growth inhibitors. **Auxins,** produced at stem and root apices, migrate polarly away from the apex and influence cell elongation. One auxin has been identified as **indole-3-acetic-acid** (IAA). The symmetry of auxin distribution in stems and roots may be disturbed by light or gravity acting from one side; in such stems or roots, growth curvatures occur which tend to restore the plant to a normal erect orientation. Auxins also inhibit the growth of lateral buds, promote root initiation and cambial division and in some plants, affect differentiation of flower buds. In agriculture, synthetic auxins are used to kill weeds, prevent leaf and fruit drop, prevent buds from sprouting prematurely to form roots on stem cuttings, and to produce artificial (parthenocarpic) fruits. **Gibberellins** are complicated substances found in many plants, which greatly affect over-all growth and flowering. They especially improve the growth of certain genetic dwarfs and induce long-day type plants (see Chapter 34) to flower. They are widely used in agriculture to speed up growth (malted barley), produce larger plants (celery) or fruit yield (grape). **Cytokinins,** related to a nucleic acid component, promote cell division, prevent senescence of cut leaves and promote the growth of inhibited buds. Inhibitors of various types exist in plants. They may control such phenomena as dormancy of seeds and buds, rate of growth and flowering. Synthetic analogs of all of these regulators are now widely used in agriculture. Some of them may be harmful, but it is probably possible to devise safe ones. Since modern intensive agriculture depends on such chemicals, we must continue to try to improve them.

FOR THOUGHT AND DISCUSSION

1. What are the major structural components of any seed?
2. What processes are involved in the onset of seed germination?
3. How does growth in plants differ from growth in most animals?
4. What is meant by the sigmoid curve of growth? What significance does it have? How may it be experimentally altered?
5. What does tissue culture tell us about the aging process?
6. What is a meristem? Where do meristems occur in plants?

7 What contributes most to elongation of stems and roots?
8 Contrast growth and differentiation. Present one plausible explanation of the origin of differences in cells of identical genetic constitution.
9 Describe briefly the internal structure of a typical root.
10 How does internal stem structure differ from that of a root?
11 Describe what happens when stems and roots increase in diameter.
12 What produces the appearance of annual rings?
13 Present one explanation for the existence of fairly low optimum temperatures for the growth of plants.
14 What is a hormone? Describe the classes of plant hormones.
15 What is the basic effect of an auxin? How can this effect help explain the role of auxin in tropistic curvatures?
16 Name several uses of synthetic auxins in agriculture and describe the basic physiological effect on which each use is based.
17 How can one tell how much auxin is present in or produced by a piece of plant tissue?
18 How does a plant sense light and gravitational stimuli?
19 What is the relation between certain kinds of dwarfness and gibberellin?
20 Name several effects produced by cytokinins in plants.
21 What role can inhibitors have in controlling plant growth?
22 Discuss the problems surrounding the rise of synthetic plant growth regulators in agriculture.

SELECTED READINGS

Audus, L. J. *Plant Growth Substances*. London: Leonard Hill, 1959.
  A well-written, scholarly, detailed treatment of the massive literature on this subject.

Leopold, A. C. *Auxins and Plant Growth*. Berkeley: University of California Press, 1955.
  A useful supplement to the fourth reference. Brings the field up to date as of the middle 1950's.

Leopold, A. C. *Plant Growth and Development*. New York: McGraw-Hill & Co., Inc., 1965.
  A readable, selected review of the research literature in fields related to Chapters 4 and 5.

Went, F. W. and K. V. Thimann. *Phytohormones*. New York: The Macmillan Co., 1937.
  A highly readable synthesis of the early literature on plant hormones by the discoverer of auxins and a distinguished colleague.

34 DEVELOPMENT OF THE PLANT BODY

If all of the cells of an organism arise by mitotic division of the fertilized egg, then the cells should all receive the same genetic material. In turn, their development and final form should be the same. Yet we know that the various cells of a plant or an animal are different in appearance, in function, and in their basic chemistry. How this comes about is one of the major unsolved problems in biology today.

Some researchers working on the differentiation puzzle think that in all cells only part of the genetic material influences development. Some genes seem to be turned "on" while others are turned "off." The problem then becomes: what turns genes "on" and "off?" Although we can control certain aspects of differentiation by chemical and physical means, in doing so we are like someone who puts a key into a door and opens it without knowing anything about the intricate mechanism of the lock (Fig. 34.1).

The word **morphogenesis** means the origin of form or structure. It also includes the problem of differentiation. We can choose many biological systems to illustrate the basic problem in morphogenesis—for example, the apparently simple problem of the origin of two different "poles" in the fertilized eggs of *Fucus*. Commonly called the "rockweed," *Fucus* is a brown alga that grows on rocks along the ocean front. The plant body consists of an attachment disc at the base called the **holdfast,** and a flattened branched body called a **thallus.** At certain times of the year, spherical structures called **conceptacles** are produced at the end of the thallus. These contain the male and female sex organs, called, respectively, **antheridia** and **oogonia.** The single-celled eggs and sperms found in these sex organs are released into the ocean water, where fertilization occurs. The

Richard F. Trump

Fig. 34.1 The daisy is not a single flower. It is rather an *inflorescence* or bouquet, composed of small central *disc* flowers and large peripheral *ray* flowers. The origin of this complex structure from the growing point of the stem is an example of the complexity of morphogenetic processes in plants

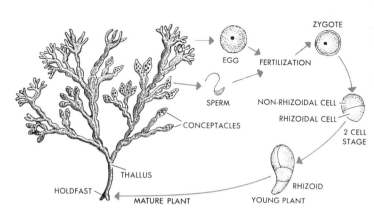

Fig. 34.2 The photograph shows the brown alga *Fucus*; the diagram shows the early stages of its reproductive cycle. Differentiation into rhizoidal and nonrhizoidal poles occurs by the first division.

fertilized egg (zygote) is a spherical cell floating free in the water. In its early stage it has no characteristics that would indicate any morphological differences between regions.

Soon after fertilization, however, the zygote no longer appears symmetrical. With the first cell division, definite and permanent structural and functional roles become apparent. At one pole **rhizoidal** (root-like) tissue forms; at the other pole nonrhizoidal tissue forms. The cells produced by the rhizoidal pole respond to gravity by growing downward to produce a stalk and holdfast. The nonrhizoidal pole gives rise to the horizontal flattened thallus (Fig. 34.2).

Why does the zygote develop in this two-pole way? Some years ago botanists found that the rhizoidal pole developed: 1. on the side away from visible light, when zygotes are exposed to light on one side; 2. on the warm side when one side is kept warmer than another; 3. on the side toward neighboring zygotes, when several are present in a cluster; 4. toward a more acid medium, when one side is exposed to a more acid medium than another; and 5. at the heavy end when a zygote is centrifuged. Botanists have also found that some kinds of light induce more than one rhizoidal pole. Although we can observe these morphological changes in the zygote, we do not understand the biochemical or genetic mechanisms that govern differentiation.

Let us take another example. You will recall that if a block of agar containing auxin is placed at the upper end of a cut *Avena* (oat) coleoptile, the auxin is rapidly transported downward and influences the growth of cells below the cut. If, however, a cylinder of a coleoptile is removed and the agar block containing the auxin is placed at the base of this cylinder, there is no movement of auxin up into the tissue (see page 516). This phenomenon of polarity is not restricted to *Avena*. It is found in many plant tissues. It obviously means there is some kind of difference between the apex and base of each individual cell in the *Avena* coleoptile. But we cannot see any difference, nor can we find any chemical difference between cell apex and base. To date, the problem of polarity, which is part of the larger problem of differentiation, remains unsolved.

In the remainder of this chapter, we shall consider some of the more obvious problems of differentiation and morphogenesis in higher green plants, and some of the techniques being used to attack them.

ORGAN AND TISSUE CULTURE IN THE STUDY OF MORPHOGENESIS

If you wanted to find out what makes an automobile go, you could not learn very much by walking around it and just looking. You would have to examine the engine, the drive shaft, and other components. And to understand how a certain part of the engine works, you would have to break it down into its components. In a similar way, a botanist interested in those factors that influence differentiation in a plant cannot learn very much by just looking at the plant as it grows in its natural environment. What he usually does is remove, or excise, from the plant a group of cells constituting either an entire organ or a particular kind of tissue. He can then transfer these cells to a nutrient medium free of bacteria and then study them under a wide variety of controlled conditions. These excised parts grown in a nutrient medium are called **tissue cultures** or **organ cultures.**

Culture of Excised Roots

If we study a pea root while it is still part of the plant, we can discover only that its nutritional requirements are the same as those of all plant cells. If, however, we cut a one-centimeter piece off the tip of a sterilized seedling root and place the tip in a nutrient solution containing only sucrose and mineral salts, the root will cease to grow. In the late 1930's, when this experiment was first done, botanists added yeast extract to the basic salt and sugar nutrient medium. The yeast contained substances which made possible the continued and potentially limitless growth of these roots. The extra nutritional requirements found in yeast turned out to be the vitamins thiamin (B_1) and nicotinic acid. When these vitamins are supplied to the nutrient medium at the proper concentration, the growth of the excised root continues indefinitely (Fig. 34.3).

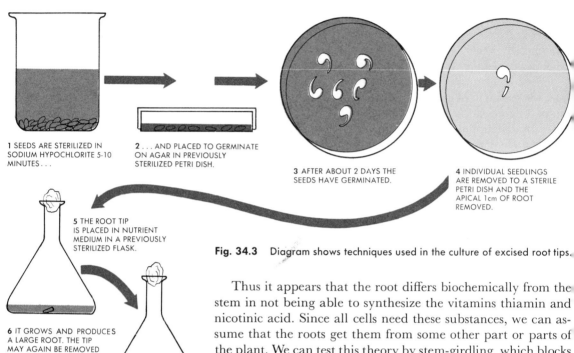

Fig. 34.3 Diagram shows techniques used in the culture of excised root tips.

Thus it appears that the root differs biochemically from the stem in not being able to synthesize the vitamins thiamin and nicotinic acid. Since all cells need these substances, we can assume that the roots get them from some other part or parts of the plant. We can test this theory by stem-girdling, which blocks transport in the phloem of the stem. The transport of materials in the xylem is not interrupted by such treatment. In such a girdled stem, the vitamins thiamin and nicotinic acid accumulate above the girdle but are reduced below the girdle. We may conclude, therefore, that these vitamins normally are synthesized in the aerial parts of plants (leaves and stems) and are transported down to the roots where they are used in chemical reactions essential for growth.

When grown in a proper medium, the excised tip of a root continues indefinitely to produce new organized root tissue. With some roots, branch roots may appear. These secondary roots may grow as vigorously as the primary root, although in some instances they fail to grow if excised and placed in the same medium that is adequate for the primary root. The application of auxin to root cultures tend to increase the number of branch roots. Branch-root initiation and root growth generally proceed best in darkness and are inhibited by very small amounts of visible light.

Some roots, such as those of morning-glory (*Convolvulus*), develop buds under certain conditions—an absence of auxin, the presence of light, and, in some instances, the addition of substances similar to certain parts of nucleic acid. Kinetin, discussed in the previous chapter, is quite effective in this process. Differentiation into root tissue, therefore, does not, at least in the case of *Convolvulus*, mean a complete loss of the potential to form stem and leaf tissue. It is generally true, however, that stem tissues give rise to roots much more readily than root tissues give rise to stems.

Culture of Excised Stems

The tip of a stem, like the tip of a root, can be cut from a plant and placed on a synthetic nutrient medium, where it can be made to grow. Stem tissues are remarkably autotrophic. They require only sucrose and the usual mineral nutrients discussed earlier. When cultured in the light, stems usually become photosynthetic. When they do, they no longer need an outside source of sucrose, since they manufacture this source of energy themselves. Almost all stem tissues, without any special urging, form roots, especially if they are cultured in the dark. In effect, a stem tip regenerates an entire plant. This obviously means that the stem tissue has the potential to produce roots, just as some root tissue has the capacity to form stems. In fact, we know of only two types of stems that can be grown in culture indefinitely without forming root tissue. These are the stem tips of asparagus and of dodder (Fig. 34.4).

Asparagus stem tips can be grown completely autotrophically in the light, since they are capable of photosynthesis. In bright light, they rarely give rise to anything but stem tissue plus flattened stem parts resembling leaves. But if such stems are transferred to darkness and supplied with sucrose, they grow very rapidly, and occasionally, roots begin to form. If the culture is kept in the dark, and if auxin is added, root growth may be increased quite a bit.

Dodder stem cultures, which do not develop roots, give rise to normal flowers. This is one of the very unusual instances in which an excised and cultured stem tip is able to produce floral organs. Another case involves the pith of the upper part of tobacco stems, which also may give rise to a flower.

In general, stems can be grown as isolated systems only in unusual instances because their normal tendency is to produce roots. This is probably due to the accumulation of auxin at the base of the stem. The tendency to form roots is somehow inhibited by light.

Culture of Leaves

Some young leaves can be cut from the parent plant and grown well in culture to a mature stage. The culture medium must contain an energy source plus the usual mineral nutrients and generally certain organic materials. The addition of kinetin can greatly increase growth. It helps the

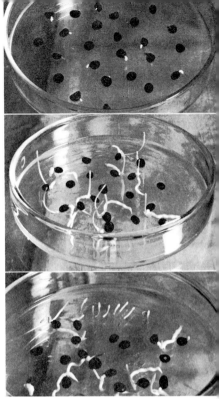

From A. W. Galston

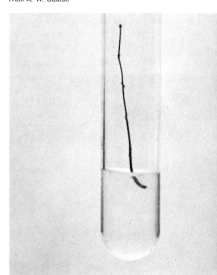

A. W. Galston

Fig. 34.4 Techniques used in the cultivation of excised asparagus tips. From top to bottom, the photos show sterilized seeds germinating on agar, erect stems excised and laid out on the agar surface, and a rapidly-growing stem implanted in nutrient agar in a test tube.

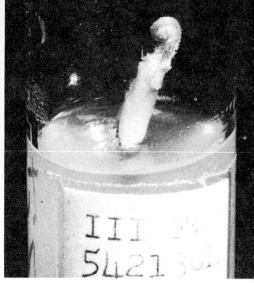

Courtesy T. A. Steeves and I. M. Sussex   Courtesy T. A. Steeves and I. M. Sussex

Fig. 34.5 Fern leaves of small size (right) may be excised and planted in nutrient agar where they will grow and resemble small normal fronds.

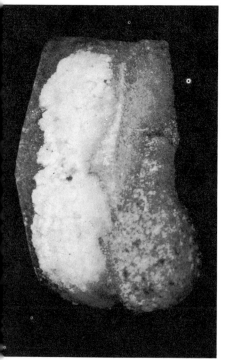

Fig. 34.6 A piece of carrot root excised and placed in a proper nutrient medium develops large *calluses* as a result of renewed meristematic activity.

A. W. Galston

leaf along by helping it accumulate nitrogen and by retarding protein loss.

In some plants, such as the African violet, new organs can be regenerated from the base of the petiole of the excised leaf. The first step in such a process is usually the formation of roots, followed by a growth of buds to form stem material. Thus, an entire plant can be regenerated from a single excised leaf. Once again, this shows that plant organs, even though they are specialized, retain an ability to develop other kinds of structures.

Culture of Callus Tissue

One of the most interesting culture experiments we can perform involves callus tissues—masses of thin-walled, undifferentiated cells. We can obtain calluses by culturing a bit of the fleshy tissues of plants such as potato tubers and carrot roots (Fig. 34.6). The calluses may then be grown for a potentially indefinite time on a very simple medium of sugar, mineral salts, auxin, kinetin, and, sometimes other organic additives. The cells of such callus cultures are large, have giant vacuoles, and are capable of many remarkable changes. For example, deep within the mass of such tissue, we often find isolated areas of lignified cells resembling the tracheids of a normal plant. Although these cells seem to have normal structure, they clearly have nothing to do with transport. The fact that they always arise deep within the fleshy parts of the tissue implies that there is something about an interior location that favors the differentiation of cells into lignified, tracheid-like forms. But what that "something" is eludes us.

Auxin seems to favor the development of lignified cells. If we graft a normal lilac bud onto an undifferentiated callus of lilac tissue, the bud causes the callus tissue to develop an organized strand of vascular tissue that links itself with the vascular tissue of the bud. This occurs because the lilac bud produces some substance that moves into the callus. If we remove the bud, xylem tissue ceases to develop within the callus. If we then substitute for the bud a high concentration of auxin in agar, xylem tissue again develops, but it is not as well oriented as when normal tissue of the bud is present. A certain concentration of sugar also appears to be very important for the formation of xylem tissue in the callus. We may thus conclude that buds influence callus tissue to form lignified cells by supplying auxin, together with sugar, to cells in the interior of the tissue mass.

The fact that callus tissue may not have differentiated cells does not mean that the tissue is without the ability to develop

Fig. 34.7 Sometimes huge tumors (calluses) develop on trees as a result of the upset of normal regulatory patterns.

normal tissue patterns. The development of xylem cells is one example of this ability. Another is that in the presence of a high concentration of auxin, callus tissue frequently develops organized roots. In still other instances, such as in tobacco pith callus, kinetin or similar materials stimulate the development of buds. The type of specialized tissue developed seems to be determined by the relative quantities of the two substances (auxin type and kinetin type) applied to the callus. If there is more of the auxin type, roots tend to develop; if there is more of the kinetin type, buds will develop. If balanced quantities of the two are applied, growth tends to continue as a callus. Although this seems to be true for tobacco pith cells, it does not appear to apply to other callus tissues that have been cultured.

Sometimes callus-like growths develop normally on plants. If we excise and culture them (Fig. 34.9), they continue their unusual development. A familiar example is the leguminous root nodule (see page 492) that forms when the bacterium *Rhizobium* invades a root hair. The long "infection thread" which the bacterium produces somehow alters the metabolism and growth potential of the cells in the region of the infection. Normal morphogenetic patterns are upset and knotty masses of nodules are formed instead. This new tissue is quite different from either the bacterium or host.

If we inject an actively multiplying culture of the bacterium *Agrobacterium tumefaciens* into certain cells of certain plants such as a geranium, a tumor-like mass develops (Fig. 34.8). Much of the growth-promoting materials normally used by the plant are diverted to the tumor. As a result, the plant's growth may be slowed down. After the tumor has been growing for some time, we can recover from it the bacteria that caused it originally, although the bacteria themselves have changed in form.

Courtesy G. Morel and R. J. Gautheret

Fig. 34.8 Tumors may be produced on plants through the action of various microorganisms, such as Agrobacterium. Such tumors are malignant and may spread, producing secondary tumors all over the plant.

As this **crown gall** disease develops, secondary tumors form at some distance from the original site of inoculation. If we excise and transfer these to a nutrient medium for culture, they continue to grow, as the primary tumor does if it is excised and cultured. All attempts to recover the bacterium from such secondary tumors have been unsuccessful. This indicates that the bacterium is not the direct cause of the secondary tumors. It seems to be the "package" that carries some "tumor-inducing agent," which can be transported about the plant independently of the bacterium.

Crown gall cells differ from normal callus cells in several ways. First, they do not need an outside supply of auxins. Normal callus requires a source of auxin for continued growth. But if we add auxin to crown gall cells, their growth is inhibited. They seem to produce enough auxin on their own to meet growth needs. This may explain why they are able to grow competitively with other cells, which are dependent on the normal auxin supply from the apex of the plant. But the difference between normal cells and crown gall tumor cells does not lie solely in their auxin metabolism. If we apply auxin to a normal cell, the cell does not become a crown gall cell. But if we graft crown gall cells to normal cells, additional tumors are produced. We have yet to discover what the "tumor-inducing agent" is and how crown gall cells transmit it to normal cells.

The crown gall disease in plants is, in many respects, similar to animal cancer. Many investigations of cancerous growths are being performed with crown gall tissue. If we can discover the chemical nature of the tumor-inducing agent, the information may shed some light on the causes of animal cancer.

The Culture of Single Cells

As botanists came to know more about morphogenesis, they asked a fascinating question: If one cell of a plant is isolated and grown in a nutrient medium, will that cell develop into a complete plant? To put it another way: Does each individual cell of a mature plant have the morphogenetic potential to restore the intact plant? The answer seems to be yes.

Isolating and growing an individual plant cell is difficult because plant cells tend to grow in masses. One way of separating a single cell from a mass is to apply an enzyme that dissolves the "cement" binding the cells together. Individual, free cells can then be obtained. The next problem is to keep the single cell alive. A single cell shows great reluctance to grow when it is placed in a medium that normally is sufficient for the growth

Fig. 34.9 Single cells can be grown in artificial media with some difficulty. Survival is improved if the cell is near some mature tissue, which seems to "condition" the medium by making it more appropriate for growth of the single cell.

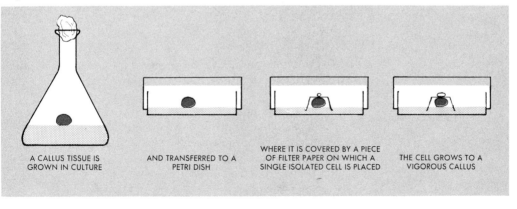

A CALLUS TISSUE IS GROWN IN CULTURE

AND TRANSFERRED TO A PETRI DISH

WHERE IT IS COVERED BY A PIECE OF FILTER PAPER ON WHICH A SINGLE ISOLATED CELL IS PLACED

THE CELL GROWS TO A VIGOROUS CALLUS

of a mass of cells. Apparently the single cell loses to the medium materials that it needs for growth. The easiest way to overcome this difficulty is to provide the cell with "nurse tissue"—masses of cells of its own type that are kept nearby but are still physically isolated from the single cell. We can do this by placing a single cell on a filter paper that sits on top of the "nurse tissue." When properly arranged, the single cell grows and divides. Single cells also have been made to grow and divide in "conditioned" media—media that have previously been in contact with growing tissue masses. These tissue masses either add unknown essential nutrients to a medium, or they remove or alter some harmful substance (Fig. 34.9).

When single cells are thus grown in isolation, they may form many-celled masses. The masses then give rise to different cell types and ultimately to all the differentiated organs of the intact plant. This cycle, from single cell to intact plant, has been carried out most successfully with the carrot, but it may be generally true for all or most plants. If it is, it may mean that during the process of differentiation there is never a complete loss of morphogenetic potential by any of the cells of a plant.

THE DIFFERENTIATION OF REPRODUCTIVE ORGANS

One of the most remarkable of all changes that take place in a plant is the change from the vegetative to the reproductive stage. Many flowering plants produce roots, stems, and leaves for a long period of time. Then, at one point in their life some chemical messenger causes the plant to stop producing and enlarging these vegetative parts and to begin a series of changes that lead to the production of the reproductive organ—the flower.

Within the last several decades, we have learned much about the environmental factors that control the time at which reproduction begins. We also have learned something about the chemical changes resulting from environmental stimulation. Some of these trigger reproduction. There is good evidence that in certain algae and fungi the production and maturation of sex organs, the release of gametes, and the attraction of the sperm to the egg are all regulated by particular substances produced by the organisms at different stages of their development. This reinforces the view that development consists of a series of stages, each of which may be triggered by a specific chemical event. We now know that the flowering of many plants is controlled by two major factors of the environment, **photoperiod** and **temperature.**

Photoperiod and Plant Growth

In 1920, two U.S. Department of Agriculture research workers reported that plants respond in important ways to the length of their daily light period. They were trying to grow a new type of large-leaf tobacco, called Maryland Mammoth. One such plant had arisen by chance in a field of other tobacco plants. As the season progressed, the other plants flowered, but the Maryland Mammoth did not. Wishing to obtain seeds of this valuable new type, and fearful that the plant might not flower before the autumn frost, they removed the plant from the field and transferred it to a greenhouse. Despite every urging, however, the plant failed to flower until about mid-December, many months after the other plants had completed seed production. When the seeds of the self-pollinated Maryland Mammoth were planted in the field the next year, the same thing happened. The Maryland Mammoths grew vigorously in the field but failed to flower when the other plants did. Again they were moved to the greenhouse, where they flowered around Christmas time.

After eliminating several other possibilities, Garner and Allard concluded that the plant would flower only during the very short days characteristic of the northern hemisphere at Christmas time. They found that they could induce the plant

Fig. 34.10 The reproductive and vegetative tobacco plants used by Garner and Allard. Reproduction was induced by short-day treatments.

Courtesy Ronald Press Co.

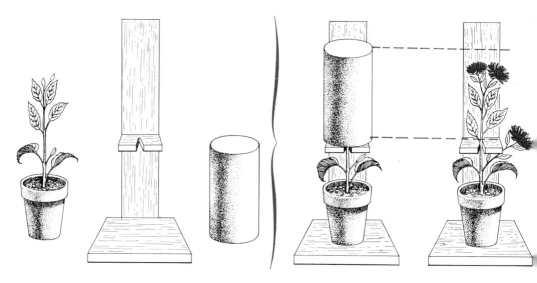

Fig. 34.11 An experiment showing movement of a flower-producing stimulus in a short-day plant. A darkened paper carton shield is placed over the upper part of the plant which is then induced to form flowers by the shortened days. But the lower part of the plant, which was never exposed to short days, also flowers. Something must have moved from top to bottom.

to flower early by transferring it to special chambers in which the length of day could be artificially shortened. They named this response of plants to length of day **photoperiodism.** Maryland Mammoth tobacco, which flowers only if the day length is reduced below a certain critical value, was called a **short-day** plant. Other plants of this type are soybeans and chrysanthemums (Fig. 34.10).

Another class of plant is the **long-day** type, such as spinach and certain cereals. In this type, flowering occurs only if the day length exceeds a certain critical value. There are also **day-neutral** plants. The photoperiod does not have a major effect on the time at which these plants flower. An example of this type is the tomato plant, in which floral organs begin to develop when the plant reaches a certain size. This situation cannot be controlled by photoperiod, although certain synthetic chemicals related to auxins can cause the tomato plant to flower prematurely.

Since 1920, botanists have learned much about photoperiodism. We now know that the leaf is the organ that receives the photoperiodic stimulus. If we enclose a single leaf of a Maryland Mammoth tobacco plant in a black bag and then expose it to a typical short photoperiod, the terminal bud some distance away from the leaf begins to develop floral organs. This means that there must be some kind of link between leaf and the distant bud (Fig. 34.11).

In another interesting experiment we can graft one plant, which has been exposed to a short-day period, to a plant which has been kept in the vegetative state by exposure to a long-day period. The leaves of the short-day-period plant then transmit some hormone (tentatively named **florigen**) to the vegetative plant and cause the vegetative plant to flower. We can also stop the flow of florigen into the vegetative plant by stem-girdling, or by other means of interrupting phloem transport from the plant exposed to a short day.

Many long-day plants may, if supplied with gibberellin, develop floral organs even if the plants are not exposed to the long-day period which they normally need in order to produce flower. In this group of plants there appears to be some relation between gibberellin and florigen. Gibberellin, however, does not promote flowering in short-day plants. Since there is very good evidence from grafting experiments that the florigen of long-day and of short-day plants works in the same way (though it may not be the same chemically), the exact nature of the relation between gibberellin and florigen is not clear. In still other plants, such as the pineapple and the litchi, the application of certain synthetic auxins causes floral organs to develop.

In the series of events leading to the production of reproductive organs, then, *various* substances may well become the controlling factor in different plants. This substance may be gibberellin in some plants, auxin in other plants, and perhaps additional, unknown substances in still other plants.

A so-called short-day plant is really a "long-night" plant that requires an uninterrupted dark period of a certain minimal length to bring about the development of floral organs. In the same way, a long-day plant is in reality a "short-night" plant—a plant that will flower only if the night period is not longer than a certain critical maximum (Fig. 34.12).

We can show that this is so by shortening the night period (even a few minutes will do), or by briefly flashing a light during the middle of the dark period. For example, in the short-day cocklebur plant, flowering occurs when the day-night period is regulated to 15 hours of light and nine hours of dark. If the nine-hour dark period is shortened to about $8\frac{1}{2}$ hours, flowering will not occur. But one single cycle of exposure to 15 hours of light and nine hours of dark is enough to cause the plant to develop floral organs—even if the plant is immediately put back on a light-dark schedule unfavorable for flowering. If the single long dark period of nine hours is interrupted at its midpoint by a flash of light, the plant will not flower.

Certain chemical processes that are very sensitive to small amounts of light must be taking place in darkness within the leaf. If light interrupts these reactions, the entire sequence of

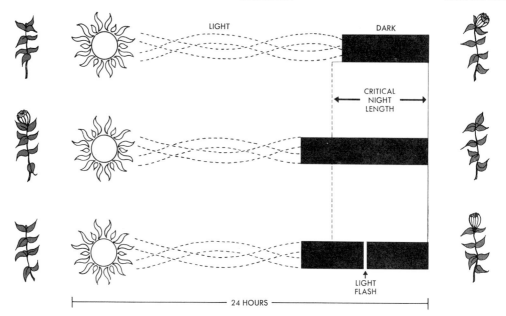

Fig. 34.12 Flowering in both short-day and long-day plants is controlled by night length. If the uninterrupted night length exceeds the critical length, short-day plants flower and long-day plants do not.

events is thrown off and the plant must start over again. With the long-day plants, the situation is just the reverse. The interruption of an unfavorably long dark period by a brief flash of light causes floral organs to develop. Long-day plants and short-day plants seem to have the same kind of photoperiodic mechanism, but they somehow work in reverse.

We should mention that the same kind of light that inhibits the flowering of short-day plants promotes the flowering of long-day plants. Recent experiments have revealed that many types of plants respond best to red light in the region near 660 mμ. It has been surprising to find that the effect of red light may be instantly and completely wiped out by shining "far-red" light, in the region of 730 mμ on the plant. These experiments are best interpreted by assuming that there is in plants a pigment (named **phytochrome**) existing in two forms, a red-absorbing form and a far-red-absorbing form. This pigment has recently been isolated from plant tissue. It consists of a protein bound to a blue pigment similar chemically to the phycocyanin found in blue-green algae. Light of 660 mμ wavelength transforms the red-absorbing form to the far-red absorbing form. Light of 730 mμ wavelength does the reverse (Fig. 34.13).

The discovery of the existence of this reversible light reaction governing flowering has cleared up several problems in plant physiology. For example, we know that the germination of many seeds is affected greatly by light. Seeds of the Grand Rapids variety of lettuce do not germinate when placed in darkness on moist filter paper at room temperature, but they germinate promptly when given small amounts of red light. But if the red-light-treated seeds are promptly exposed to far-red

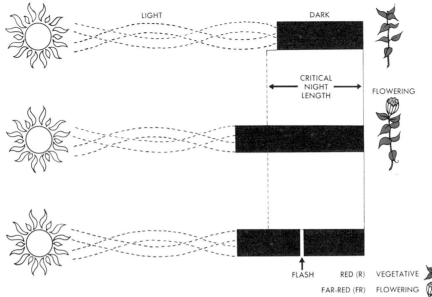

Fig. 34.13 Interruption of the critical night length by a flash of red light interrupts flower initiation in short-day plants. The effect of red can be annulled by far-red applied immediately after the red. This red, far-red reversibility can be extended through several cycles. The plant "remembers" only the last quality of light to which it has been exposed.

light, the effect of the red is completely canceled out and the seeds remain dormant. Here again, the growth of the plant is apparently controlled by a two-way switch, which in turn is controlled by the mysterious pigment phytochrome.

The same red and far-red light reaction operates in the stems and leaves of plants that have been grown in darkness. A seed germinated in total darkness gives rise to a seedling with a very long, slender, colorless stem, and to leaves that never expand greatly. We find that red light is the most effective among all the colors in transforming the **etiolated** (dark-grown) plant into a normal plant. But the effectiveness of the red light can be immediately canceled by far-red light. The exact nature of the response, however, depends on the tissue. If the red light is given to stem tissue, the stem's growth is greatly inhibited. But a leaf exposed to the same red light grows more vigorously. Yet the effects of red light on both stem and leaf are immediately stopped when far-red light is applied after the red (Fig. 34.13).

The differentiation processes that give rise to various types of cells determine how a specific tissue will respond to visible light. The kind of response each tissue makes is important in the life of the plant, for it affects such varied processes as the germination of seeds, the growth of roots, stems, and leaves, and the formation of floral organs. Increased knowledge of the nature of phytochrome, and of how it works, is greatly needed and must be an important aim of research in plant physiology.

The Importance of Temperature

Temperature may also control the flowering of plants. As any gardener knows, annual plants begin to grow in the spring, flower in the summer, and produce ripe fruit and

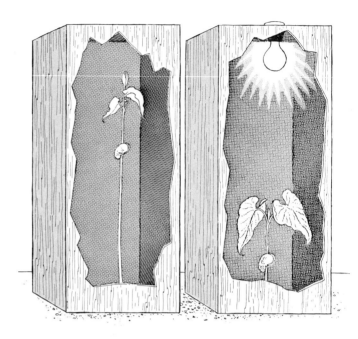

Fig. 34.14 In the darkness bean plants develop a long, unpigmented stem and smaller leaves. In the light, the growth of the stem is inhibited while the growth of the leaves is promoted.

seeds in the fall of the same year. Biennial plants, on the other hand, produce only vegetative organs during their first growing season. They do not flower until the following year, after they have been exposed to prolonged periods of low temperature. Only after exposure to such a cold period are they able to respond to a proper photoperiodic stimulus.

This need for low temperature can be satisfied at any point in the development of the plant after germination. For example, all you have to do is soak a biennial seed in water to start germination and then expose it for about six weeks to low temperatures of from $2°$ to $5°C$. It will then behave as if it had gone through the cold winter after a year of normal growth, and it will flower in the first season if exposed to the proper photoperiod. Such cold treatment of a plant to hasten its reproduction is called **vernalization.** Since the difference between the annual and biennial races of most plants seems to be due to a single gene, low temperature treatment of a biennial seems to be a substitute for some biochemical event that is produced under genetic control in an annual.

Here, again, gibberellin seems to enter the picture. There are several biennials that can be made to flower in the first season by the application of gibberellin alone. Whether cold treatment has anything to do with the synthesis of gibberellin in the plant is not yet clear. At any rate, we know that the vernalization stimulus (probably a hormone) like the photoperiodic stimulus, can be transported from a plant which has received low temperature to a plant which has not.

ENDOGENOUS RHYTHMS

Any event that occurs with a regular periodicity, such as day and night, the swing of a pendulum, the beat of a heart, is said to be rhythmic. When the rhythm in an organism continues independently of outside influences, the rhythm is said to be *endogenous*. The sleep movements of the first leaves of the common bean are an example. During the day the blades are more or less horizontal, and the petioles are spread wide from the stem. At night, however, the leaf blades droop to a vertical ("sleep") position and the petioles close toward the stem. We can record these movements by connecting the leaf tip to a pen by a fine thread, and by permitting the movements to be recorded on a rotating drum. If such a plant is placed in a darkened room where the temperature and humidity are constant, the leaf rhythm continues for several days, until starvation sets in (Fig. 34.15). The rhythmic period continues to be about 24 hours. Such rhythms of approximately one day's duration have been named **circadian rhythms,** and the time-keeping machinery has been called a **biological clock.**

The cause or causes of such a rhythm remain a mystery, although we do know that the time-keeping mechanism is almost completely independent of temperature. We also know that the rhythm is sensitive to narcotics and to various agents that interfere with membrane integrity and with certain aspects of normal cell metabolism. But instead of helping to clear away the mystery surrounding circadian rhythms, those bits of knowledge seem at present to cloud the picture.

Some people believe that endogenous rhythms are in some way related to the photoperiodic system. Those varieties of soybeans that have large scale endogenous leaf movements have specific short-day periods. But those without such rhythms tend to be day-neutral.

The one-celled marine algae *Gonyaulax polyedra* has at least three different rhythms: 1. photosynthetic; 2. bioluminescent (light emission); and 3. cell division. Although these rhythms may reach their peak at different times of day, they have equal periods, and are probably governed by the same clock. Rhythms and clocks are found in virtually all types of organisms—algae, fiddler crabs, man, cockroaches, and higher plants. One day

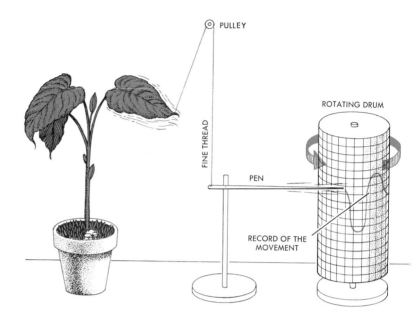

Fig. 34.15 The rhythmic movements of a bean leaf may be recorded on a rotating drum.

they may well provide vital new information on the nature of biological systems. Since certain microorganisms and animals are also sensitive to photoperiodic stimuli, comparative studies in this area will probably be valuable. In the study of endogenous rhythms, as with all biological problems, experimentation with any biological system may be expected to yield information which will further our understanding of the basic unit of all living things—the cell.

SUMMARY

Differentiation is the origin of differences in cells having the same genetic information. It is part of the general problem of *morphogenesis*, the origin of form. Both problems are still poorly understood. One approach is to separate the plant into individual organs, grow them in artificial nutrient solutions and observe their behavior. Excised root tips continue to grow indefinitely as roots in solutions supplied with sugar, salts, and two of the B vitamins. The latter, which are normally transported down the phloem by the leaves, may be considered as plant hormones. Cultured roots only rarely give rise to stem tissue, but cultured stems almost always give rise to roots. In various tissue culture systems, auxin leads to root formation, while cytokinins lead to bud formation. Excised leaves require cytokinins for survival, some will form roots at the base of the petiole when supplied with auxin. Certain fleshy tissues, when removed to an artificial medium containing sugar, salts, auxin and cytokinin, will grow indefinitely as undifferentiated **callus** tissue. Raising the auxin concentration often leads to root formation, while raising the cytokinin concentration leads to bud formation. Thus, a piece of undifferentiated tissue, properly stimulated by chemicals, can give rise to an entire plant. Even a single cultured cell can do this, but only if grown in the vicinity of other cells. Mature, differentiated tissues can influence differentiation of neighboring cells. Thus, buds grafted onto a callus may cause formation of a string of xylem cells.

Bacteria can influence differentiation of plant cells. Invasion of root hairs by **Rhizobium** causes formation of a nodule; invasion of many plants by **Agrobacterium** causes crown gall disease, or "plant cancer." Crown gall cells are completely independent of external auxins and cytokinins, which may explain why they "grow wild" on the plant. They can induce other cells to become tumorous like themselves, possibly because they contain a virus introduced by the entering bacterium.

The differentiation of flowers on vegetative plants is controlled primarily by **photoperiod.** Long-day, short-day and day-neutral types of plants exist. The leaf is the organ which perceives the stimulus, which is really the dark period. Properly stimulated, the leaf starts to produce a flower-induced hormone, still chemically unknown, but called **florigen.** This stimulus can be transmitted through grafts from plant to plant. In long-day plants only, gibberellin appears to substitute for florigen. Photoperiodism is connected with a pigment called **phytochrome,** which occurs in the plant in two interconvertible forms, one absorbing red light, the other absorbing "far-red" light. Thus, giving red light to a short-day plant in the middle of a dark period will inhibit flowering, but the effect of red is reversed immediately by far red. Phytochrome also controls seed germination, stem and leaf growth, and other processes. Temperature may also control floral differentiation. All biennial plants require cold treatment (winter) before they are able to flower. In the process of **vernalization** the cold may be administered artificially to slightly germinated seeds, which then produce plants flowering in the first year. This effect, too, can be duplicated by gibberellin.

Most organisms have built-in timekeeping devices, which probably have some biochemical basis. These "biological clocks" cause repeated recurrence of events, such as sleep movements of leaves, even without external stimuli. Frequently these occur roughly in 24-hour periods and are called **circadian** rhythms. They probably indicate a complexity in cellular architecture and processes which we cannot yet define.

FOR THOUGHT AND DISCUSSION

1. Define (a) differentiation; (b) morphogenesis. Which is the broader term?
2. What is meant by polarity? Give an example in plants.
3. Describe how you would produce a culture of a root.
4. Mention several biochemical differences between roots and stems.
5. What chemical treatment would you use to cause initiation of roots on stems? Stems on roots?
6. What is a plant callus? How could you produce one?
7. What is the role of bacteria in producing (a) nodules? (b) crown galls?
8. Describe the main characteristics of crown gall cells.
9. Why do single cells have great difficulty in commencing growth, while large groups of cells do quite well?
10. What environmental factors control the initiation of flowers?
11. Distinguish between long-day, short-day, and day-neutral plants.
12. Describe an experiment to prove that the leaf is the organ which receives the photoperiodic stimulus.
13. Describe an experiment to prove that the dark period, not the light period, is responsible for the photoperiodic effect.
14. What is meant by the term *florigen?* What is its relation to gibberellin?
15. What is *phytochrome?* Describe some of its properties and activities within the plant.
16. Why must some seeds be exposed to light before germination can occur?
17. What is meant by etiolation? How can it be corrected?
18. What is the difference between annual and biennial plants? What is the physiological basis for this difference?
19. What role can gibberellin play in flower induction in biennials?
20. What is meant by vernalization?
21. What is meant by an endogenous rhythm? Give one example in a plant.
22. What is meant by a biological clock?
23. What do you think is the significance of endogenous rhythms in plants?

SELECTED READINGS

Borthwick, H. A. and S. B. Hendricks. "Photoperiodism in Plants." *Science,* **132** (1960), 1223–1228.

A brief, semipopular account of the development of the phytochrome story.

Bünning, E. *The Physiological Clock.* Berlin: Springer, 1964.

A monograph by the man who has done the most to bring biological rhythms to our attention.

Hillman, W. S. *The Physiology of Flowering.* New York: Holt, Rinehart & Winston, 1962.

A reasoned and scholarly account of the external and internal events governing the transition from vegetation to reproduction.

Salisbury, F. B. *The Flowering Process.* Oxford: Pergamon Press, 1963.

A lively, personalized account of floral initiation, with emphasis on the cocklebur plant.

Sinnott, E. W. *Plant Morphogenesis.* New York: McGraw-Hill, Inc., 1960.

A competent, modern treatment of some fascinating problems in the development of the higher plant.

Sweeney, B. M. "Biological Clocks in Plants." *Ann. Rev. of Pl. Physiol.,* **14** (1963), 411–440.

An authoritative brief statement.

Thompson, d'A. *On Growth and Form.* Cambridge: Cambridge University Press, 1952.

A beautifully written and elegant statement of the problems and fascination surrounding organic form.

35 PLANTS AND MAN

Man's complete dependence on the green plant for photosynthetically-produced food and other products makes it essential that he learn to live permanently with stable populations of certain green plants. Ironically, many such stable populations of green plants have been artificially developed by man and depend on man for their continued existence. Left to themselves, they would disappear, the victims of weeds, insects, fungi, and other predators; or perhaps they would become less violent victims of impoverished soils.

This situation dictates that man must understand the needs of green plants. Also, he must furnish these needs; but in doing so he must not so alter the environment that the healthy and vigorous development of natural plant communities is prevented. In the past, man has not shown much wisdom in these matters, and many of his practices have made life impossible in large areas of the world. If he is to continue to flourish on this planet, man must come to understand and to correct his errors of the past. Partly, at least, this will be brought about through an in-

Fig. 35.1 Each of the stable communities shown here (grass and trees) dominates a particular habitat and is in a state of *dynamic equilibrium*. Any change in the environment or in the conditions governing competition may lead to a change in the community.

U.S. Forest Service   U.S. Forest Service

creased study and understanding of the science of **ecology,** which deals with the interactions of organisms with each other and with their environment.

PLANT COMMUNITIES

Plants, like animals, are organized into **communities** in the wild. A community may be defined as any group of organisms having mutual relations with each other and with the environment. Familiar examples would be a stand of pine trees, a grassy plain, or a group of seashore seaweeds. Each of these communities has arisen in its particular location as a result of a long historical development, and as a result of the interaction of successive groups of plants and animals (including man) with changing environmental forces. Some of these communities are stable, some are in transition; all are very sensitive to man's manipulations (Fig. 35.1).

In the wild state, communities of plants compete with one another for space, nutrients, water, and light. This is most easily seen in such environments as deserts and forests. In the desert, where moisture limits plant growth, the only plants that survive are those with an efficient water economy. Among the devices employed by desert plants for efficient competition in this extreme environment is an extensive root system adapted for deep penetration of the substratum, as well as for extensive gathering of water penetrating just below the soil surface (Fig. 35.2).

Because desert plants must tap large volumes of soil to get the water they need for maintenance and growth, desert populations tend to be spaced quite openly. Frequently, mature plants are found only at a distance of several meters from their nearest neighbor. Such wide spacing is sometimes guaranteed by a kind of chemical warfare among the plants. Inhibitory substances from one species may inhibit growth of another species, or older plants may inhibit younger plants of the same species. Efficient desert competitors also have adaptations for minimizing transpirational water loss, such as absence of expanded leaves, thick cuticles, sunken stomata, and thickened water storage tissues.

By contrast, a moist forest will consist of a crowded community of plants struggling for light (Fig. 35.3). The most rapidly growing trees form the top canopy and, therefore, receive the most incident radiation. Frequently, these species must have full sunlight for vigorous growth. Below them are other species of trees, well adapted for growth in partial shade. Still farther down, on the forest floor, are ferns, mosses, lichens, herbs, and small woody plants. All are very efficient users of even small

Trans World Airlines, Inc.

Fig. 35.2 A giant saguaro cactus of the southwestern American desert. This plant has no expanded leaves. It photosynthesizes with its fleshy stem. The stem is adapted for water storage and can swell or shrink conveniently through its accordion-like structure. The saguaro has a very extensive root system permitting it to tap soil water over a large area. Neighboring saguaros are found no closer than 25 or 30 feet away, probably due to some mutually inhibitory action. Other small shrubs may, however, grow in the region of the saguaro.

Fig. 35.3 In this complex forest community tall hardwoods dominate the upper story. Evergreens and herbs compete for space on the forest floor. Available light for photosynthesis is the main competitive factor of the environment here.

Fig. 35.4 An eroded gullied hillside in the Tehachapi region of California. This condition was caused by overgrazing, leading to a depletion of the surface cover of plants whose roots held the hillside soil in place.

USDA Photo

amounts of radiant energy. Thus, a complex community of interacting and competing individuals and species may exist in a stable form as long as the environment is reasonably constant.

What might be the result of building a paved road through such a forest? Excavation would disturb water tables and seriously alter the competitive advantages held by the various dominant species. Interruption of gently rolling slopes, stabilized by interlaced roots of many species, could lead to erosion of the soil and successive denudation of large areas. Opening up of the stabilized canopy layers would give to the previously disadvantaged members of the lower canopies a chance to outstrip their higher rivals. The intrusion of automobile traffic into a forest can kill off, or at least seriously disadvantage, species which are sensitive to the uncombusted hydrocarbons in exhaust fumes, or to the lead of the tetraethyl lead added to gasoline to cut down engine knocking. It is for these reasons that many conservationists, foresters, and ecologists decry the increasing tendency to bulldoze major roads into existing virgin forest stands. Their point is one frequently overlooked by the public—you cannot make wild areas easily available without destroying their very nature.

Animals in general, and man in particular, have had a major effect on the distribution of plants in many areas of the world. Excessive grazing by domesticated goats has denuded previously fertile hillsides in many parts of the Mediterranean countryside, leaving eroded, unproductive hillsides (Fig. 35.4). Excessive grazing by deer and rabbits, and excessive cutting or girdling of trees by beavers and porcupines, have also produced major upheavals in natural forests. But of all animals, man has been and continues to be the prime destroyer of the natural environment.

Pouring the waste products of his technology into rivers, streams, and the atmosphere, he has killed off many species of plants. This not only changes the plant population, but may also seriously alter the population of wild animals, including economically valuable fish, birds, insects, and mammals. The massive use of agricultural insecticides and herbicides has also had many deleterious side effects. While depletion of the wild insect population is beneficial to man in some respects, it is also harmful if it leads to decreased pollination and fruitset, and to decreased bird populations. Similarly, killing dicotyledonous weeds in a grain field with 2, 4-D is beneficial, but if the drift of the herbicidal mist to a neighboring forest kills off the trees, that becomes a highly damaging by-product of the operation.

In the United States, as in most other industrialized regions of the world, we have belatedly come to realize that our soil, our water, and our air are not inexhaustible resources. The dust storms of the 1930's in the Great Plains, which carried off great quantities of fertile topsoil in the form of choking black clouds, were mute testimony to man's lack of wisdom. Dust storms have now largely disappeared thanks to more sensible agricultural practices (Fig. 35.5). But erosion by water is still an enemy, and the continued existence of a "muddy Missouri" means that somebody's topsoil is drifting down the river to be added, in part, to the Mississippi Delta. Improved cultivational practices, including contour plowing, strip cropping, crop rotation, and permitting fields to lie fallow can help this situation. Artificial control of water runoff by dams, culverts, and reforestation must also be built into any rational system of planned agriculture.

Lake Tahoe, nestled between California and Nevada, used to be one of the major beauty spots of America. Unhappily its overdevelopment as a resort area has led to a great diminution in its natural charm. Still worse, careless waste disposal practices have added considerable amounts of nitrogen, phosphorus,

Fig. 35.5 Photo at bottom right shows natural reseeding of lodgepole pine following a burn in the Malheur National Forest of Oregon. This reseeding process was aided by the planting of grass on the forest floor, resulting in retention of the soil. Strip cropping in Wisconsin is shown in the middle photo. This field contains alternate strips of corn and hay. The practice, especially if the strips follow the contour of the land, diminishes run-off and consequent erosion. Photo at left shows contour planting of early tomatoes under hot caps in the hilly frost-free land near San Diego, California. The tractor-dug trenches adjacent to the planted seedlings retard the run-off of water.

USDA Photo

USDA Photo

U.S. Forest Service

Fig. 35.6 Los Angeles smog, known the world over, is complex, acrid, smoky air pollution resulting from the interaction of organic wastes of gasoline motor exhausts and industrial plants with the high intensity ultraviolet radiation in the air. The resulting compounds are harmful to the growth of plants and to the health of animals and humans. Smog accumulates in the Los Angeles area because of a peculiar basin-like effect created by the San Gabriel mountains seen in the background, giving rise to a temperature inversion of the atmosphere which prevents turbulence and escape of the pollution.

Fred Lyon

and other elements to the water, thus causing choking growths of assorted algae and other organisms in the once clear waters. Fortunately, the danger in this situation has been recognized, and cooperative action of the various communities and states ringing the lake can still reverse the present trend. In the same way, the once beautiful Connecticut River has suffered great despoliation through indiscriminate dumping of industrial waste products and sewage. Cooperative action by several New England states, stimulated by a conservation-minded Department of the Interior, may lead to the restoration of the river's former beauty. Certain troubles are even being anticipated. Thus, proposals for building an atomic power plant on the river are being carefully scrutinized in advance, since the use of the river water to cool the reactors might raise the water temperature by one or more degrees, and the entire ecology of the river could be upset. Algal populations could change and lead to changes in populations of animals, including the famous Connecticut River shad, a fish deservedly celebrated for its delectable eating qualities.

The word **smog** has recently been introduced into our language to describe a noxious condition of the atmosphere. It results from overloading the atmosphere with gaseous and smoky waste products. Some of this comes from industrial plants—oil refineries, ore smelters, synthetic chemical and cement-making plants. Some also comes from the activities of individual men—exhaust fumes of cars, heating systems in homes, the burning of leaves. In regions where there is little air turbulence, or where geographical conditions cause a temperature inversion layer of air, the normal disappearance of these waste products into the great ocean of the atmosphere is prevented. Thus, the waste products accumulate and are sometimes altered by sunlight and oxidation. Frequently they cause discomfort, and sometimes death, to humans, domestic animals, and plants. In fact, damage to certain plants is one of the most sensitive indicators of smog build-up. Like other kinds of pollution, smog can be controlled through cooperative action involving control of factory and automobile exhausts, prohibition of open-air burning, and the planning of the size of communities. The terrible smog which periodically blankets Los Angeles (Fig. 35.6) will most likely always be a problem, at least for as long as man is around. The fact is that the region cannot comfortably contain the many millions of people who have flocked there because of the desirable climate. Yet, people continue to pour into this overtaxed area.

In general, it seems clear that the activities of men will have to be increasingly regulated if our planet is to continue to furnish a suitable abode for the most destructive of all species, *Homo sapiens*.

PLANTS USEFUL TO MAN

Early man was a nomad, hunting animals and gathering edible wild plants for food in the course of his random wanderings. For his fires he used wood, a product of plant cell wall growth. To protect himself against the elements he used leaves, straw, branches and other products of vegetation. Through chance and random observations, he came to know that certain wild plants had medicinal values and that others produced odoriferous or spicy components which appealed to him. Still others produced curious products like rubber, resin, and fibers which could be extracted and then put to great use. We do not know exactly how or when man began to cultivate plants for his needs. The first man to realize that seeds could be collected, stored, and planted *en masse* to guarantee a supply of food to a stable, non-nomadic population brought about a revolution no less important than that caused by the control of fire, the industrialization of manufacturing, and the introduction of atomic energy.

For the remainder of this chapter, we shall consider the history and botanical nature of one plant which has shaped man's history and development over a large part of the world, especially in the New World. This plant is the familiar maize, or Indian corn.

Maize is known scientifically as *Zea mays* L. (the L. indicates that it was named by Linnaeus, the great Swedish botanist who founded the binomial system of nomenclature). It is a member of the grass family (Gramineae). Superficially, it resembles other grasses in several respects. The modern corn plant is a tall, annual grass, bearing numerous strap-shaped leaves on a hard, thick stem. The leaves are borne at nodes and emerge alternately from opposite sides of the stem. The root system is fibrous and much-branched. "Prop roots" arise from the lower nodes and broaden the base of support for the plant. The plant produces grains rather similar in structure to those of wheat, rye, barley, oats, and rice; yet it differs from the others in that its grains are gathered in a complex organ called the **ear** (Figs. 35.7 and 35.8).

The ancestors of our present corn plant are not known with certainty. What does seem clear is that the modern corn plant could not survive without man's constant intervention; it is entirely a creature of cultivated agriculture. In searching for its ancestor, we have to imagine a more typically grassy plant, one from which corn could have been developed by selection and breeding. We assume that such an ancestor would have been native to the Indians of North and South America long before waves of Europeans settled here after 1492. There are still some people who believe that corn could have arisen in the Malaysian

Fig. 35.7 The first known picture of corn, appearing in a mediaeval herbal. Note the *tassel* (male inflorescence) above, the *ears* (female inflorescences) in the axils of the leaves, *prop roots* and general grass like appearance.

Fig. 35.8 An ear of corn on the plant with husks enclosing the ear and the remains of the silks, each one leading to a grain.

Fig. 35.9 An early drawing showing the importance of corn in the life of the pre-Columbian indians of the New World.

Peninsula and been brought across the Pacific by early seafaring Asiatics making the treacherous journey on rafts. Though this theory cannot be ruled out, it is not favored by modern biologists. The most popular candidate for the role of corn's ancestor is a Mexican grass, which is commonly called teosinte (*Euchlaena*); another possibility is a related genus, *Tripsacum*. By the use of the radiocarbon dating technique, corn grains dating to 1000 B.C. have been identified in South America, and corn grains dating to 2000 B.C. have been found in southern North America. Corn pollen has been found in deep cores of earth removed from under the present Mexico City (Fig. 35.9).

When the Europeans first came to the new world, Indians were cultivating maize as their major food plant. Typically, they planted several seeds in a hill into which several holes had been poked with a stick, and into which a fish or part of a fish had also been placed. We now know that the decaying fish was an excellent source of nitrogen and other mineral elements, but it would be interesting to know the reason given by the Indians for including a fish in a hill meant for corn grains. Cultivation of the soil was generally performed with a hoe-like object made from bone, generally the shoulder blade of the then abundant bison. The Indians knew about the desirability of limiting weed growth and used their crude hoes to chop away at the weeds.

Once harvested, the fresh maize was eaten as roasted ears or as succotash. More generally the grains were removed from the ear and stored in communal centers, being withdrawn as needed. Once removed from the ear, the grain was pounded to yield coarse hominy, or it was ground in a metate to yield a finer corn flour. The corn flour was then made into mush, tortillas, tamales, and other dishes still in favor today. Corn flour was also used for baking bread, but since the gluten content of such flour is low, the bread did not hang together well. Thus, crumbly corn bread never became very popular with Europeans.

The European settlers soon realized the superior quality of maize for stock-feeding purposes and rapidly developed certain strains especially suited for such use. Other corn, with a higher sugar content ("sweet corn") was adapted for direct use by man, usually in the immature or "milk" stage of the grain. Other types of modern corn handed down to us by the Indians included popcorn, whose hard, inedible kernels burst upon heating; flour corns, whose kernels have soft starch and are, therefore, especially useful for hand grinding; dent corns, whose grains have a mixture of hard and soft starch, and which are the highest yielding strains today; and, finally, flint corns which have exclusively hard starch (Fig. 35.10). All these were developed by selection from "pod-corn," in which each grain is surrounded by

Fig. 35.10 Various types of corn, all developed from a common primitive ancestor. From left to right they are popcorn, sweet corn, flour corn, flint corn, dent corn, and pod corn.

its own husk. Ears of such pod corn have been found in some of the earliest archeological sites of the New World. How it developed from the ancestors of *Zea mays* can only be conjectured.

Today, about one-fourth of the cropland of the United States is given over to corn. The yield in the United States represents well over half the world's annual total of about $\frac{1}{4}$ billion tons. Thanks to sophisticated breeding techniques, strains of corn have been developed which, when adequately fertilized, yield more than 200 bushels per acre. The average yield is much less, of course, perhaps 30 bushels per acre. Corn must be grown in almost totally flat soil since it tends to deplete soil drastically and cause erosion of the topsoil. Therefore, it should be grown

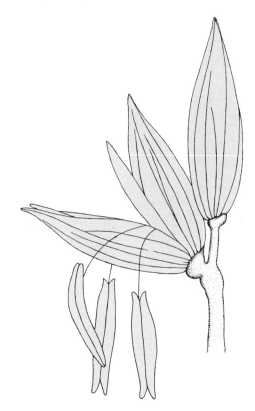

Fig. 35.11 Male flowers from the tassel of a corn plant. Note the dangling *stamens* which contain the *pollen*.

only with great care in hilly areas. From the corn grain harvested in the various parts of the United States many products are obtained, including corn oil, corn starch, dextrins, corn syrup, corn sugar (glucose) and fermented liquors. The vegetative part of the plant is fed to animals in the form of silage or stover. Corn stalks are also processed into wallboard and paper; corn husks are used as filling material in a variety of operations; corn cobs are used as fuel, as raw material in the manufacture of charcoal, and for the production of corncob pipes. All this has developed in man's hands from a wild grassy plant with a few small clustered grains!

Corn is a **monoecious** plant, that is, its flowers are unisexual, but both sexes are found on the same plant. The male, or staminate, flowers are clustered in a **tassel** at the apex of the stem. In the tassel the male **florets** occur in pairs, forming a cluster called a **spikelet.** Each male flower contains three stamens, and each anther contains several thousand pollen grains. When the tassel is ripe, the pollen grains, each about 0.1 mm in diameter, are carried by the wind to the surface of the stigma. Several million pollen grains may be produced by a single tassel (Fig. 35.11).

The female flowers occur in pairs in longitudinal rows in the axils of several leaves in median positions on the stem. This collection of female flowers is called a **spike,** and the thickened axis of this spike forms the cob of the ear. Because of the paired nature of the female florets, ears of corn always have an even number of rows of grain, from about eight to more than 30.

The female **spike** is enclosed by modified leaves which become the shucks of the mature ear. Each female flower contains one **pistil,** consisting of an **ovary** attached to a long **style** and feathery **stigma.** The style and stigma are the familiar "corn silks." The pollen grains lodge in the protrusions of the stigma, germinate, and send their pollen tubes down the style to the ovary, where fertilization occurs. Only one pollen grain per silk does the job of fertilization, though many grains may start germinating on each silk. The total number of seeds per ear cannot exceed the number of silks (Fig. 35.12).

After fertilization, the internal economy of the plant changes in such a way that there is a massive flow of nutrients to the developing ears. As the ears mature and fill out, the remainder of the plant goes into senescence and ultimately dies. Senescence is at least partially a consequence of fruit maturation, and can be delayed by preventing the ear from maturing.

Modern corn is **hybrid,** that is, it contains genes from various inbred lines which have been crossed. Geneticists have long known that successive inbreeding diminishes the vigor and productivity of most agricultural species, both plant and animal. However, if certain highly inbred lines are crossed, a phenomenon known as **hybrid vigor** appears. The offspring are much larger, more vigorous, and more productive than their parents, presumably because of the mixing of particular genes within a single cell. In modern hybrid corn, four distinct strains are involved: these are hybridized two by two, and the final step

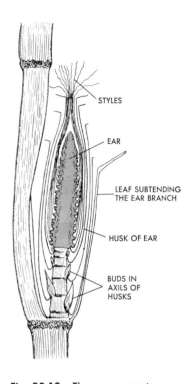

Fig. 35.12 The arrangement of female flowers in the ear of corn. Each *style* is attached to an ovary located in a bud found in the axils of the husks. The other end of the style ("silk") is a feathery stigma on which the pollen grains fall and germinate.

Fig. 35.13 An example of hybrid vigor in corn. Each of the small plants on the left and right are the inbred parents of the hybrid corn shown in the middle. In the production of modern hybrid corn, four strains are used, each of them the result of a long process of inbreeding. They are first hybridized two by two to produce two vigorous hybrids and then the hybrids are themselves interbred to produce "double cross" corn which has extraordinary vigor and high productivity.

Redrawn from Eric Ashby

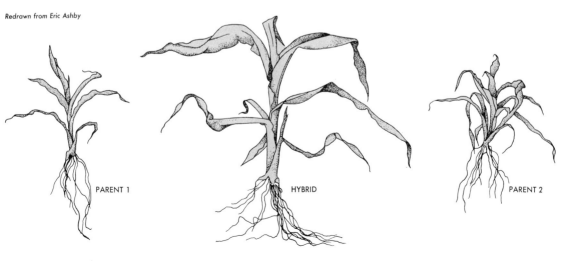

involves a hybridization of the two hybrids, producing the final "double cross" corn. Clearly, the seed produced by self-pollination of such stock will not "breed true," but will be highly variable. Thus, new seed, produced on specialized seed farms must be purchased each year by the farmer. Despite the added expense, the farmer has learned to follow this practice since it pays off well in increased yield.

Thus, the modern corn plant, which we use here as but one example of a general situation, is in many ways a creation of man. Its genetics, its growth, and its continued existence depend on man. Like most plants, it has responded well to man's effort to improve it. As dramatic as this improvement has been, its future may be even more spectacular. All will depend on man's continued devoted study of his most remarkable servant and ally, the green plant.

SUMMARY

Man is dependent for food and other products on plants which he has artificially developed, and which could not exist in the wild when left to their own devices. To insure an optimal existence for these artificially developed types, man must understand the basic laws of **ecology,** the branch of biology which deals with the interactions between organisms and of organisms with their environment.

Plants are organized into **communities,** some stable and some in transition. Stable communities exist in equilibrium with their environment, and any change in the environment may result in a change in the plant population. This is true because plant species compete with each other for water, nutrients, light and space, and the margin of superiority which one species enjoys over another may be very slight. Man's activities frequently do disturb delicate natural equilibria, and result in greatly changed plant populations or in the creation of completely denuded areas such as deserts.

Man, by virtue of his great numbers and prodigious chemical activities has produced massive changes in the nature of his environment, especially in the agricultural topsoil, the waters of lakes, rivers, and streams, and in the atmosphere. Pollution or denudation of these regions of the biosphere could profoundly alter man's life, or even make it ultimately impossible.

Corn, one of the most important food plants of the world, was probably developed in the new world from a wild Mexican grassy ancestor long before the European emigration. The modern plant is the result of a long program of selection, breedng and the application of knowledge of plant pathology and physiology. Although this plant has been completely altered by man's intervention, its future alteration may be as profound as that which has gone before. It all depends on man's continued acquisition and application of knowledge about plants.

FOR THOUGHT AND DISCUSSION

1 Compare the survival ability of wild and cultivated plants.
2 What is ecology? What is its importance to modern, industrialized society?
3 What is meant by a plant community? Name several different types of communities.
4 Describe several structural features which suit desert plants to their habitat.
5 What necessary environmental factor is in short supply in a typical temperate forest?
6 Describe several ways in which civilization may affect plant communities.
7 Mention some undesirable consequences of the use of chemical herbicides on a large scale.
8 Discuss several ways in which damage to natural resources can be reversed.
9 Is it ever practicable to anticipate and thus prevent ecological damage?
10 What is smog? What are some problems associated with combatting it?
11 Describe the appearance of *Zea mays* L.
12 Botanically, what are the tassel, the silks, the ear and the cob?
13 Why does an ear have an even number of rows of kernels?
14 Discuss the possible origin and age of corn.
15 Describe some of the practical uses to which corn was put by the Indians, and some of its main uses today.
16 What is the difference between field corn, sweet corn, popcorn, dent corn, flint corn and pod corn?
17 What are some problems associated with growing corn?
18 What is meant by a monoecious plant?
19 In corn, define a spike and a spikelet.
20 What is meant by hybrid vigor? How is it employed in corn breeding?

SELECTED READINGS

Baker, H. G. *Plants and Civilization.* Belmont, Calif.: Wadsworth Publishing Co., 1965.

    An attractively written paperback, summarizing the impact of plants on man.

Billings, W. D. *Plants and the Ecosystem.* Belmont, Calif.: Wadsworth Publishing Co., 1965.

    A brief discussion, in paperback form, of the principles of plant ecology.

Carson, Rachel L. *The Silent Spring.* Boston, Mass.: Houghton Mifflin Co., 1962.

    A passionate plea for the preservation of nature.

Hayward, H. E. *The Structure of Economic Plants.* New York: The Macmillan Co., 1938.

    The classic treatment of the anatomy of useful plants.

Mangelsdorf, P. C. *Reconstructing the Ancestor of Corn.* Washington, D.C. Annual Report of the Smithsonian Institution, 1959.

    An exciting botanical detective story.

Schery, R. W. *Plants for Man.* Englewood Cliffs, N. J.: Prentice-Hall Inc., 1952.

    An authoritative discussion of the history, botany, agronomy and economics of many commercially important plants.

PART FOUR
Human Physiology

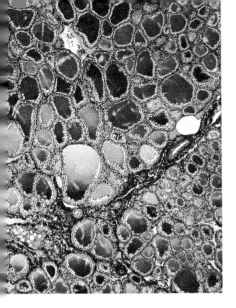

ABOUT THE PHOTOGRAPH

Microscopic section through the thyroid gland. This section shows thyroid follicles, which are hollow, spheroid structures with cellular walls. The interior of the follicle is a storage site for the thyroid hormone, which is attached to a protein called thyroglobulin. The interiors of the more active follicles are stained blue, the least active ones red. The stain is Mallory-Azan.

Section: E. Evans   Photograph: A. Blaker

ABOUT PART FOUR

In Parts 1, 2, and 3 of this volume we have dealt with the biology of man in several ways: his position of dominance in the animal world as a member of the order Primates; his origin through the process of evolution; and his relation to, and dependence on, the plant kingdom. Here we deal with man exclusively, and enlarge our understanding of him through a study of human physiology.

The most certain knowledge we have of man is that he is an animal. With equal certainty, we know that man is also a unique animal, whose rise to dominance and whose control over his environment has resulted from a combination of talents and abilities. Among these are his development of languages which enables him to store, transmit, and make use of facts and ideas; his powers of imagination and learning; and his ability to make and use the tools needed to exercise control over his environment. His similarities and dissimilarities with other animals reveal themselves in his physiology and his structures. By knowing man as an animal we do not lessen his humanity. Rather we believe this knowledge to be vitally important, for only by making use of this information can we hope to understand our physical, mental, and social well-being as individuals and as members of a complex human culture.

36 SURVIVAL AND THE INTERNAL ENVIRONMENT

The face value of the chemical elements of your body are worth only a few dollars. Yet this few dollars worth of chemicals can lift weights, convert food and air into bones, muscles, and nerves, work mathematical problems and study the very chemicals which make it up. What remarkable transformations have the chemicals undergone to enable them to do these complex and diverse tasks? You, as one of the most recent links in the evolutionary chain going back millions and millions of years, reap the benefits of eons of unceasing biological development.

When you stop to think about the variety of things your body is capable of doing—simple everyday things that we all take for granted—you begin to see what an astounding thing your body really is. For example, what are the processes involved in reach-

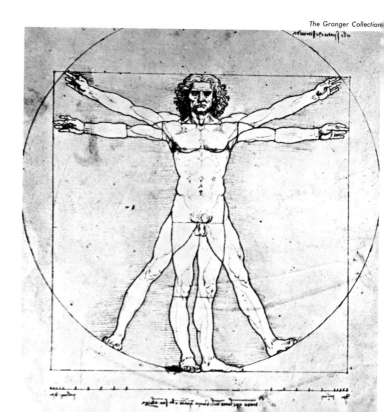

Fig. 36.1 Leonardo da Vinci's drawing "The Proportions of the Human Figure after Vitruvius."

ing out for an object and picking it up? How can you ride a bicycle without concentrating on keeping your balance? When you lift a very heavy weight, why does your heart seem to beat faster? What keeps your heart beating day after day, year after year? When you cut your finger, what chemical processes stop the bleeding?

These are exactly the kinds of questions a **physiologist** is constantly asking, and most of the questions are only partially answered. He wants to know: How does the heart beat? How does a muscle contract? How does a nerve carry messages? And where does the energy come from to perform all of these, and other, tasks? In short, physiologists want to know how living organisms work, yet they find it difficult even to define "living." However, the physiologist knows that the processes he is interested in take place in anything that grows and reproduces itself by undergoing chemical changes. Although human physiology—the subject of this book—is concerned with man's body, many studies of other animals cast some light on the human organism.

The Internal Environment

You have probably always thought that air is the external environment of your body. In a way, of course, it is, yet if you examine a piece of living tissue from your body under a microscope, you will find that its living cells are surrounded by a watery fluid. All of the living parts of your body are inhabitants of water, not air. Those parts that are not moist—the outermost skin, fingernails and hair—are not composed of living cells; they are essentially dead tissue. The fluid medium surrounding cells is called the **extracellular fluid** or, sometimes, the **internal environment**.

Extracellular fluid is about the same in all parts of your body. The fluid surrounding the cells that make up your toes has virtually the same composition as that surrounding the cells of your brain. These fluids have a composition very similar to that of blood plasma, the fluid that circulates in the blood stream. Further, a similar fluid bathes the cells of a variety of different species of animals. Although men, dogs, frogs, lobsters, and crabs are in many ways quite different from one another, the internal environment of each is strikingly similar. The physiologist wants to know why.

Life seems to be possible in these animals only under highly restricted conditions. For one, their cells can stay alive only if they are bathed in an internal environment of specific chemical composition. If the calcium content in a man's blood (normally about 10 milligrams per hundred milliliters of blood) drops to one-half its normal level, he begins to twitch and eventually goes into convulsion, which can be fatal. On the other hand, if the calcium content is increased by one-half, he shows severe depression and is likely to go into a coma, which can also be fatal. There are many other similar examples. Human life, then, is possible only when calcium (and many other substances) are present within certain rigidly fixed proportions. When the proportions are upset, the result can be sickness or death.

Steady State

Knowing that the composition of the internal environment must remain fairly constant, we might now ask how it is kept that way. In the simplest case, we can imagine a container which allows an exchange of materials and energy with the external environment. Sugar, salt, water, and heat enter the container by certain routes and leave it by certain other routes (Fig. 36.2). If sugar enters at exactly the same rate as it leaves the amount of sugar in the container remains *constant*, despite the fact that it is *continually being renewed*. We would say that sugar is in a **steady state,** meaning that its concentration in the container does not vary, even though it is continually being exchanged with the external environment.

This steady state example corresponds to the internal environment in living systems. For example, certain elements in the food we eat are absorbed into the internal environment through the walls of the intestinal tract. They are removed by processes taking place in the kidney, lungs, and skin. Intake and removal are balanced so that the concentrations of many substances (such as salts, sugars, and water) in the internal environment do not change very much.

Although a steady state normally tends to be maintained, the interchange between the internal and external environments varies from time to time. If we move from a warm room to a cold room, the heat exchange between the air and our internal environment alters quite a bit, yet our body temperature remains close to 98.6°F. Again, we may eat an excessive amount of sweets on one day and none at all on the next. The amount of sugar transferred between the external and the internal environments is quite different on the two days, but the concentration of sugar within the blood remains remarkably constant.

Homeostasis and the Steady State

Since the rate of exchange between external and internal environment *does* change from time to time, the

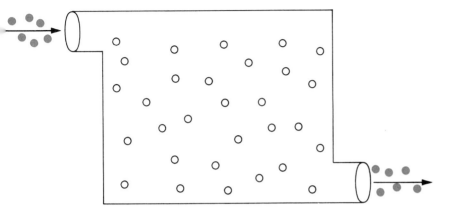

Fig. 36.2 The idea of a steady state is shown in this diagram. During the time six molecules are entering the container, six others are leaving it. The total number of molecules in the container does not change.

body must have some mechanisms for keeping the composition of the internal environment constant. Otherwise, animals could not survive. This tendency to regulate the internal environment, keeping it in a steady state, is called **homeostasis**. The higher we go on the evolutionary tree (moving from simpler to more complex animals), the more vital this process becomes. When homeostasis fails, the central nervous system is often affected first; without homeostasis, the highly developed nervous system and the intelligence of man probably could not have evolved. The many mechanisms involved in regulation—not all of which have yet been discovered—influence the behavior of all the living things we will be studying in this book.

Storage and Feedback as Regulators

Most of us are involved in problems of regulation of one sort or another throughout our lives. For example, say that you want to raise tadpoles in order to observe their stages of development. You need a barrel of rain water, so you put a barrel out in rainy weather, and when it is partly filled you put in some fertilized frog eggs. Rain is pouring into the barrel at the top and draining out through a hole equipped with a valve at the bottom, as shown in Fig. 36.3. To keep the barrel from overflowing and spilling the eggs out, you have to open the drain, yet you want a nearly full barrel at all times. In other words, you want to allow fresh rain to enter, yet want to be able to control the level of the water. The simplest solution is to set the drain valve so that the overflow at the bottom just balances the inflow of rain. But this will work only as long

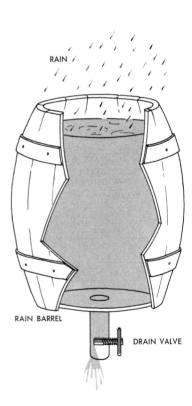

Fig. 36.3 A steady state is maintained because the amount of water running out of the barrel is equal to the amount of rain entering.

Fig. 36.4 If the amount of rain entering the barrel varies, the automatic drain valve compensates for the change, maintaining a steady state. This is an example of *negative feedback* (Fig. 36.5).

as the rate of rainfall remains constant. Each time there is a change in the rain, you have to readjust the drain valve. The same problem arises in most regulatory systems; you have to devise ways to compensate for *changes* that are not under your control.

How could you concoct a simple device to keep the water level in the barrel constant? As Fig. 36.4 shows, you could float a cork on the water, then attach it to a metal rod with a cone-shaped piece of wood at the end. The length of the rod is exactly equal to the depth of water you want to keep in the rain barrel. When the water is this deep, the piece of wood closes off the drain. If it begins to rain again, the water rises and lifts the cork, which, in turn, pulls the wooden plug out of the drain. When the water level and plug sink downward again, the drain opening becomes smaller, and less water flows out. In rainy weather you always have a full barrel of water, but no overflow.

We can represent the response of your system to an increase in rainfall by the diagram in Fig. 36.5.

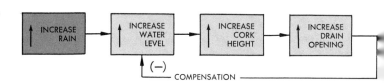

Fig. 36.5

In Figs. 36.5 and 36.7 we use a vertical arrow (↑) to mean *increase*, a descending arrow (↓) for *decrease*, and a horizontal arrow (→) for *leads to*. Figure 36.5 can then be read as:

> An increase in rain leads to an increased water level in the barrel, which leads to an increase in cork height, which leads to an increase in drainage, which compensates for (negates) the increased water level.

The essential feature of the system is that it can be represented as a circle or **closed loop**. Closed-loop systems are called **feedback** systems; information about the response of the system (in this case the water level) is "fed back" to change the response. The feedback in our system is called **negative** feedback because it compensates for, or negates, the increase (change) in water level of the barrel, hence the (−) minus symbol at the end of the feedback loop. Negative feedback tends to *stabilize* a system.

Suppose, as shown in Fig. 36.6, you make a mistake and turn the cone-shaped drain plug the wrong way. When there is a downpour, the water level rises, but this time it closes the drain and the barrel overflows. You have created a vicious circle, which can be represented as shown in Fig. 36.7.

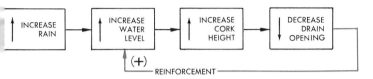

Fig. 36.7

Again we have a feedback loop, but now instead of negating the increase in water level, the feedback reinforces it. We call this a **positive** feedback system, and hence the (+) plus symbol at the end of the feedback loop. In contrast to negative feedback, positive feedback leads to *instability*.

Let us return to the original negative feedback system, the one that kept the water level constant. Although it works fine in rainy weather, you have made no provision for a dry spell, during which the water will evaporate. You need a storage system to supply water if the level is to remain constant.

Figure 36.8 shows one system which fills the need. For dry periods, its response is illustrated in Fig. 36.9. Notice that the

Fig. 36.6 The steady state is abolished if the amount of rain increases, thus raising the cork float and closing the valve. It is also abolished if the amount of rain decreases, thus lowering the float and opening the valve. Both cases are examples of positive feedback (Fig. 36.7).

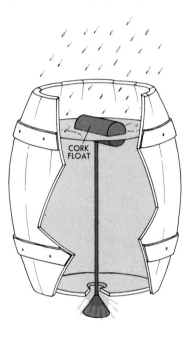

Fig. 36.8 If a storage tank is added to the system, a steady state can be maintained during periods of drought as well as during periods of rain. Rain causes Drain B to open, keeping the water level constant. During drought, water evaporation from the barrel causes Drain A to open, allowing water from the storage tank to enter. At the same time the valve at right closes. (See Figs. 36.9 and 36.10.)

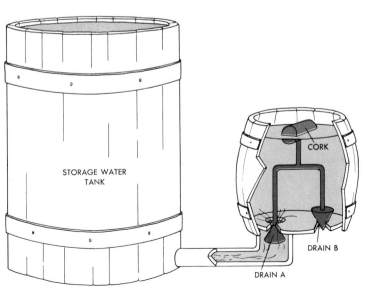

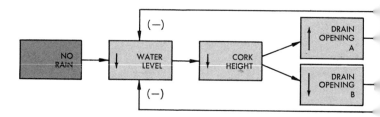

Fig. 36.9

diagram is similar to Fig. 36.5, except that now there is an additional feedback loop for Drain A. A decrease in rainfall decreases the water level. As a result, the float falls, lowering Plug A into the open position and permitting an inflow of water from the storage tank. At the same time, Plug B is dropped into the closed position, preventing water from draining out. The pair of feedback loops (Fig. 36.9) shows how the water level tends to be stabilized during drought.

Figure 36.10 shows what happens in case of heavy rainfall. The water level rises and the float drops. This opens Drain B, allowing water to leak out, and closes Drain A, cutting off the inflow of water from the storage tank. Again, both loops tend to stabilize the water level.

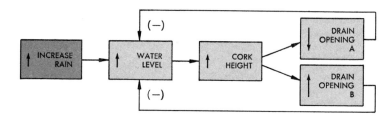

Fig. 36.10

Biological Feedback Systems

Our bodies have countless feedback systems that regulate our internal environment. They frequently make use of storage depots and often use more than one feedback loop. The regulation of plasma calcium is a good example of negative feedback.

Calcium is a common element in many foods. When it enters the intestinal tract it is absorbed into the blood plasma. It leaves the plasma by way of the kidney, which excretes the cal-

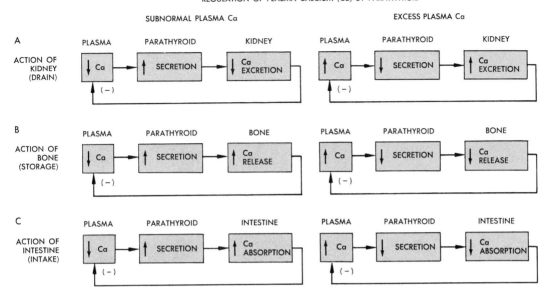

Fig. 36.11 The body tends to maintain itself in a steady state. Here we show how the level of calcium is maintained in a steady state by the parathyroid hormone. The system shown here includes the regulatory actions of drain, storage, and intake, similar to the system shown in Fig. 36.8.

cium in the urine. A steady state of calcium may be obtained by balancing the dietary intake with urinary excretion. However, these two are not the only factors in maintaining the proper concentration of calcium.

Bones form large *storage* depots for calcium, and the plasma can tap these storage supplies in times of need. Regulation of the dietary intake, urinary excretion, and calcium storage is provided by tiny glands located in the neck—the **parathyroid** glands. These glands secrete a substance (a **hormone**) into the blood stream and the blood stream carries the hormone to all organs in the body. Low levels of plasma calcium increase secretion of this hormone. The action of the hormone on the kidney, intestinal tract, and bone provides the three feedback loops illustrated in Fig. 36.11.

Figure 36.11A shows the negative feedback resulting from the action of the parathyroid hormone in regulating the urinary excretion by the kidney. The kidney in this instance is similar to the drain on our water tank. The second loop, shown in Fig. 36.11B, results from the action of the hormone which regulates the transfer of calcium from bone to plasma. In this instance, bone provides an internal storage system, similar to our reservoir tank, which can be tapped when there is too little or no calcium in our diet. The third loop, shown in Fig. 36.11C, results from the action of the hormone which increases calcium absorption from the intestinal tract. The dietary intake of calcium corresponds to rain entering the barrel in our example.

Each of the three loops in the diagram tends to stabilize the plasma calcium level. Notice that the response to low calcium is directly opposite to the response to high calcium. All we have to do to make one diagram in a pair like the other is reverse the direction of the vertical arrows.

To understand other regulatory systems in the body, we ask the same kinds of questions we ask about calcium regulation. How does the substance enter the internal environment, and how does it leave? Is there an internal storage system? How is the level of material sensed? Then, how is this information relayed to something corresponding to the drain valve, or kidney, or storage depot? In other words, how does each step in the feedback loop work?

However, a description of feedback loops is not enough. Ideally, we want a complete description of the entire machinery. In our example of calcium regulation, we would like to know what process the parathyroid gland goes through when it synthesizes and secretes its hormone, and how this activity is governed by calcium. Further, we would like to know how it is possible for the parathyroid hormone to stimulate the kidney to retain calcium, or to stimulate bone to release calcium. Finally, we would like to know why calcium is indispensible. What is its function? How does it act?

The physiologist studies not only the parts of the body in isolation, but also the way various parts work together and keep the animal alive by keeping its internal environment constant. Most of our questions are not fully answered. In this book we will study some of the partial answers discovered by forming and testing theories. These partial answers give us a basis for asking further questions, for forming new hypotheses, and for planning experiments to test new theories.

SUMMARY

Living cells are fragile and very sensitive to changes in their immediate environment. This is especially true of animal cells, which are surrounded by a fluid called the **internal environment.** The sensitive and fragile cells cannot be expected to survive unless the internal environment remains fairly constant.

The gastrointestinal tract, the lungs, the kidneys, and skin are continually involved in the exchange of materials and energy between the internal and the external environments. Nevertheless, the composition of the internal environment is not changed drastically; it is subjected to **regulatory mechanisms** that balance the inflow and outflow of the internal environment, resulting in a **steady state.** The tendency to regulate the internal environment so that it is maintained in a steady state is called **homeostasis.**

Homeostatic regulatory systems can be represented by **feedback loops.** The feedback is *negative* when it compensates for, or negates, any change. Negative feedback tends to stabilize a system. When feedback reinforces a change, we call it *positive* feedback. Positive feedback creates a vicious circle and leads to instability.

FOR THOUGHT AND DISCUSSION

1. What is meant by a **steady state?** What are the **inflows** and **outflows** that could be balanced to maintain the following systems in a steady state: (a) The volume of fluid in a river? (b) The population of a city? (c) The weight of your body? (d) The number of cars on a super highway?
2. Recent experiments suggest that our description of calcium regulation illustrated in Fig. 36.11 is not complete. These discoveries involve a new hormone called **thyrocalcitonin.** Thyrocalcitonin is secreted by the thyroid gland and is believed to decrease the level of circulating blood calcium. If this hormone is important in maintaining a constant level of blood calcium, what do you suppose the effect of calcium might be on the secretion of the thyrocalcitonin by the thyroid gland? Draw a feedback loop.

SELECTED READINGS

Baldwin, E. *An Introduction to Comparative Biochemistry.* New York: Cambridge University Press.

Cannon, W. B. *The Wisdom of the Body.* New York: W. W. Norton and Co., 1939 (reprinted 1963).

Langley, L. L. *Homeostasis.* New York: Reinhold Publishing Co., 1965.

Rasmussen, H. "The Parathyroid Hormone," *Scientific American* (April 1961) (*Scientific American* reprint #86).

Tustin, A. "Feedback," *Scientific American,* **187** (September 1952), 48.

37 TRANSPORT

Fig. 37.1 A cell membrane is the major barrier to an exchange of materials between cells and their environment. The cell membrane in the above photo is believed to be limited by the two dark stained lines which give it a railroad track appearance.

Courtesy J. David Robertson, M.D.

The exchange of material between the external and the internal environments will be one of the focal points of our study. In addition, we shall be concerned with the exchange of materials between cells and the fluids making up the internal environment (extracellular fluid). All of the substances taking part in these continual exchanges must be transported back and forth in some way. Certain physical forces cause this movement of materials. The most important of these forces arise from differences in **pressure, concentration,** and **electrical charge.**

A ball rolling downhill illustrates some important aspects of transport. We would expect that in the diagram below the ball at left rolling down to B would not need to be pushed; but to go

from Point B to Point A, it would have to be pushed all the way. At right the *barrier* would prevent the ball from rolling to Point B, unless it were given a push. Nevertheless, A is still higher than B, so once the barrier is removed the ball will move to B.

At Point A in both situations, the ball has **potential** energy. The force acting on the ball (gravity) pulls the ball from a point where its potential energy is high toward a point where its potential energy is low (B). To describe the transport of a ball from A to B, we must specify two things: 1. the relative heights of A and B; and 2. the nature of the path. A similar description holds for other types of transport. Although the difference between two given points in a biological transport system may be easy to obtain, the nature of the path may be very difficult to discover.

Pressure Forces and Transport

To show how a difference in pressure causes transport, we could build a device like that shown in Fig. 37.4. We begin by filling a tube with a solution of small sugar molecules and large protein molecules dissolved in water. Then we place one piston at Point A, and another at Point B. When we push piston A harder than B, we would expect the fluid to move along the tube to the right. In other words, if the push, or **pressure**, at A is greater than the pressure at B, the fluid will move in the direction A to B. The fluid will always move from a region of high pressure to a region of low pressure. Since the entire solution (water, sugar, and protein) flows, we call the movement **bulk flow**.

However, we can stop the flow by changing the nature of the path (Fig. 37.3). If we place a solid wall at B, then no matter how hard we push at A, there will be no flow. But if we poke holes in the barrier, we will again have a flow from left to right. The extent of the flow will depend on how many holes we poke and how big they are. Like the solid barrier, the porous barrier poses resistance, but only partial resistance—the smaller the holes, the greater the resistance will be, and the more slowly the fluid will move under given pressure. A similar resistance arises in the blood stream when the blood is forced through narrow vessels.

Now, imagine that the barrier at B has *very* small holes in it—holes just a little larger than the sugar molecules. The protein molecules are too large to get through. We describe this barrier by saying it is **permeable** to sugar and water, but **impermeable** to protein; that is, sugar and water can penetrate, but protein cannot. When we apply pressure at A, only the sugar and water move; the protein stays behind. This process of separating small particles from large particles by pushing them through a sieve-like structure is called **filtration**. We shall see later that filtration is the first step in the formation of urine by the kidney.

We can measure pressure by using a device like the one shown in Fig. 37.4. We place some mercury (Hg) in a U-shaped

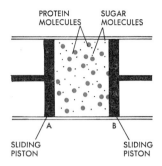

Fig. 37.2 If the pressure on Piston A is greater than the pressure supporting Piston B, there will be a movement of sugar and protein molecules toward the right. This is an example of bulk flow.

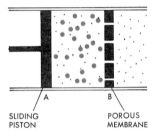

Fig. 37.3 If the stationary barrier at B were a sieve, only the sugar molecules would be permitted to flow through. No matter how great the pressure at A, the protein molecules could not get through. This is an example of filtration.

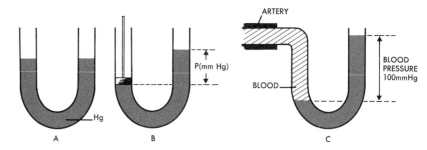

Fig. 37.4 The height of the mercury columns in Tube A is the same because the pressure is the same on each column. In Tube B the heights of the columns are different because of the pressure exerted by the piston. This pressure can be measured by measuring the height indicated by P. In Tube C, the pressure is exerted by the blood and the height of the mercury column gives us a measure of the blood pressure.

tube and let it settle so that the levels in the two limbs of the tube are equal (Fig. 37.4A). Then we place a piston in the left-hand limb of the tube and push down so that the mercury begins to rise in the right-hand limb (Fig. 37.4B). As the mercury rises, its additional weight in the right-hand column exerts a downward pressure. The higher the mercury rises in the tube, the greater the downward pressure becomes. Eventually, the downward pressure exerted by the mercury just balances the force exerted by the piston so that the movement of the mercury stops. The pressure exerted by the piston can be represented by the difference in height between the right-hand column and the left-hand column, as indicated by P. This difference is usually measured in millimeters (mm) and, accordingly, we speak of pressure in terms of millimeters of mercury (mm Hg).

The pressure exerted by the heart on the blood in the arteries in a normal human averages about 100 mm Hg. If an artery of an animal were cut and one end of the U-tube were inserted into the artery (Fig. 37.4C), the blood would then push directly on the mercury, just as the piston did in Fig. 37.4B. The difference between the heights of the mercury in the two columns would then give us a direct measure of the animal's blood pressure. When a doctor takes your blood pressure, he does not, of course, cut an artery. However, the indirect (and less precise) method he uses does involve balancing the pressure of your blood against the force exerted by a column of mercury.

In our study of motion caused by differences in pressure, there are two essential factors: 1. the forces which arise from differences in pressure; and 2. the nature of the pathway (for example, the presence of barriers). Similar factors determine the motion resulting from differences in the concentration of liquids separated by a porous barrier.

Concentration Differences and Transport

Whenever there is a difference in concentration of molecules in two regions, there will be a movement of molecules from the region of high concentration to regions of low concentration. This movement is called **diffusion.** It results simply from the fact that molecules are in continuous random motion.

For example, suppose that we fill a tank with water and are able to dissolve 18 molecules of sugar into the water on the left-hand side of the tank and 10 molecules on the right-hand side. The broken line in Fig. 37.5 represents a mark on the glass of the tank and visually divides the tank into two halves, Sides A and B. The sugar is free to move about in the whole tank. Eventually, we count the molecules in each side of the tank again. Now we find that there are 14 sugar molecules in each half. How do we account for this change?

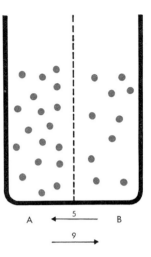

Fig. 37.5 The molecules in this tank (divided by an imaginary line) are more concentrated in Side A than in Side B. In time the concentration will become uniform. Molecules move from regions of higher concentration to regions of lower concentration. This is an example of *diffusion.*

We know that the molecules in the tank are bouncing around completely at random. We repeat the experiment many times, keeping track of all the molecules and averaging the results of our experiments. On the average, half the molecules in Side A have been moving to the right, and half have been moving to the left. At the beginning, an average of nine sugar molecules (half of 18) from Side A were moving to the right. Of the 10 sugar molecules in Side B, five were moving to the left. In other words, while nine sugar molecules were moving across to the right from Side A, only five were moving across to the left from Side B. Side A of the tank had a net *loss* of four molecules, Side B a net *gain* of four. Thus, when we take our count at the end of the waiting period, on the average we find the number of molecules equal in both sides of the tank.

Our description of diffusion is based only on the *average* motion of the molecules. This means that our rule (net movement from higher to lower concentrations) may occasionally fail—that is, we must be prepared to find motion that is occasionally different from the average motion. However, 18 molecules was a very small sample. In any tank that we might use, there would be billions upon billions of molecules. Whenever we deal with such huge numbers, significant deviations from average behavior are extremely rare so that our rule about diffusion turns out to be very precise.

Molecules diffuse from a region of high concentration to a region of lower concentration. Eventually the concentrations will be the same in both regions. The time required for the concentrations to equalize depends on how far the molecules have to travel; that is, on the size of the container. Molecules in solution take a long time to diffuse unless the distances are very short.

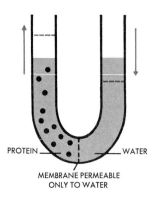

Fig. 37.6 Water is leaving the region where the concentration of dissolved molecules is low (right-hand column), in favor of the region where the concentration is high (left-hand column). The colored broken lines represent the change in water levels after *osmosis* has taken place. We can stop the motion of the rising water column by applying a back pressure (Fig. 37.7). The magnitude of the back pressure is a measure of the *osmotic pressure.*

Over a distance of 1 centimeter (cm), diffusion takes about 13 hours before it is nearly complete. Over a distance of 10 cm, the time is about 53 days. Over a distance of only 0.001 cm (10 microns), the diameter of some cells, diffusion is nearly complete in about 0.05 second. If oxygen had to diffuse from the lungs to our feet, it would take years. But oxygen does not travel from the lungs to the body tissues by diffusion. Instead, it is transported by the blood stream (bulk flow). Once the oxygen reaches the tissues, diffusion takes over as the mechanism of transport.

In contrast to bulk flow, which carries everything dissolved in a fluid regardless of concentration, diffusion may transport only one kind of dissolved molecule. In our bodies, for example, salt may be equally distributed throughout a cell, but at a given moment sugar may be more concentrated in one half of the cell than in the other half. In this case, there is a **concentration difference** for sugar, but none for salt. In the absence of a barrier, the sugar diffuses and its concentration becomes equal in all parts of the cell.

If there is a barrier present, such as a cell membrane, the diffusion process is greatly affected by the number and size of the holes in the barrier. The barrier can slow down the diffusion of some molecules and completely prevent other molecules from passing through.

Osmosis

Figure 37.6 shows an example of *water* movement resulting from differences in concentration of dissolved substances (**solutes**). Suppose that we have separated the limbs of a U-tube by a membrane that allows the passage of water but not proteins. The membrane is water permeable, but it is protein impermeable. The left-hand limb has proteins dissolved in it, but the right-hand limb does not. We soon notice that the water in the left-hand column begins to rise as the water in the right-hand column falls. Water is leaving the region where the concentration of *dissolved molecules* is low, in favor of the region where the concentration is high. This process is called **osmosis.** Although we know the conditions under which osmosis takes place, we do not understand its precise mechanisms.

The forces that arise during osmosis can be measured very simply. We insert a piston in the left-hand compartment of a U-tube (Fig. 37.7). Then we apply pressure to the piston by pouring mercury on top of it. The pressure on the piston begins to counteract the pressure of the rising water. If we pour just enough mercury on the piston to exactly balance the force of

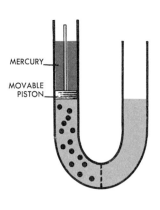

Fig. 37.7

osmosis, so that the water does not move, we then have a measure of the **osmotic pressure** of the solution on the left. The more molecules dissolved on the left-hand side, the greater will be the osmotic pressure. In a general way, the osmotic pressure may be regarded as the tendency of a solution to draw water to itself. Water will tend to flow from regions of low osmotic pressure to regions of high osmotic pressure.

You can see the effect of osmotic pressure by doing a very simple experiment. Place some animal cells in distilled water and observe the results under a microscope. The cells will begin to stretch and expand until their membranes can no longer contain the contents of the cells. The cause is a difference between the osmotic pressure inside the cells and that outside the cells. There are many molecules dissolved in the fluid inside the cells, so the osmotic pressure there is high and tends to draw water in. But the fluid outside the cell is pure water. Since it is without dissolved substances, its osmotic pressure is zero. Water continues to flow into the cells, stretching their membranes, until they become porous enough to release the contents of the cells.

The results of this simple experiment have obvious applications to experimental work with animal cells. When cells are placed in a solution, we try to make sure that the osmotic pressure of the solution matches the osmotic pressure within the cells. A solution that allows cells to retain their original size is called an **isotonic** solution. A common isotonic solution is isotonic saline. For humans and other mammals isotonic saline is a solution of sodium chloride containing 9 g of sodium chloride per 1000 milliliters (ml) of water.

To study the processes that we have been describing—bulk flow, diffusion, and osmosis—in any specific example, we must know what forces are present and what barriers these forces meet. We must also know whether one force counteracts another. For instance, the pressure on the U-tube shown in Fig. 37.7 counteracted the osmotic pressure of the protein solution, preventing any fluid movement.

Electrical Forces and Transport

Some atoms or molecules carry either a positive or negative electrical charge. Any charged atom or molecule is called an **ion.** Two positively charged ions repel each other, as do two negatively charged ions; however, a positive ion and a negative ion attract each other. The attraction or repulsion of ions in a solution sets up forces in addition to those we have been considering. Salt solutions, in particular, show electrical forces. Sodium (Na^+) and potassium (K^+) ions, for

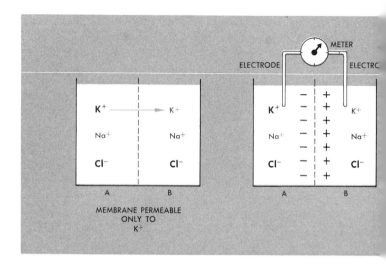

Fig. 37.8 The movement of potassium ions through the membrane upsets the electrical balance. Side A is more negative than Side B.

example, repel each other in solution because both are positively charged. However, each attracts chloride (Cl⁻) ions, which are negatively charged. In a solution, ions are under the influence of pressure, concentration, *and* electrical forces.

The solutions that we ordinarily deal with are electrically neutral. Nevertheless, in some instances, the fact that the solution is made up of charged particles (ions) leads to the development of electrical forces which we cannot ignore. This is particularly true when a solution containing ions must pass through a membrane. The origin of an electrical force across a membrane is shown in Fig. 37.8.

We have a tank partitioned by a membrane. Say that we fill Side *A* with an electrically neutral solution consisting mainly of potassium and chloride ions, plus a small amount of sodium ions. (By electrical "neutrality" we mean that the number of positive charges equals the number of negative charges.) We fill Side *B* with another electrically neutral solution, one consisting mainly of sodium and chloride ions with a small amount of potassium ions. Assume that the membrane allows potassium to pass through, but not sodium or chloride. If we could see the individual ions moving, the first thing we notice is that potassium ions are moving from Side *A* through the membrane to Side *B*. The reason for this movement is that potassium ions are more concentrated on Side *A* than on Side *B*, and they tend to move from an area of high concentration to an area of low concentration.

The movement of potassium ions across the membrane upsets

the electrical balance on both sides of the membrane. Side A is now left with more negative than positive charges, and Side B is gaining positive charges. As more potassium ions move across the membrane, the excess positive charge accumulating on Side B acts as a repelling force. This opposing electrical force becomes larger with each new potassium ion that moves from Side A to Side B. Finally, the electrical force built up along the membrane is so strong that it counteracts the force of diffusion—no further net transfer of potassium ions can occur.

On Side A negative charges have been building up because the electrical effects of chloride ions which have been left behind are no longer compensated for by potassium ions. There is an excess of negative ions on Side A and an excess of positive ions on Side B. Since negative and positive charges attract each other, these excess ions line up along the membrane. Whenever a membrane separates layers of opposite charges, we say that the membrane is **polarized.**

If we try to measure the amount of K^+ that has crossed the membrane, we find it *so small* that ordinary chemical tests cannot detect it! Yet, even a small number of excess ions produces a significant electrical force. We can measure this electrical force simply by bridging the two sides of the tank with suitable wires (Fig. 37.8). Some electrons (negative charges) making up the metal of the wire are free to move under the influence of electrical forces. When the two ends of the wires, called **electrodes,** are dipped into the solution, one on each side of the membrane, the electrons flow from Side A to Side B. This flow of electrons (or **current**) can easily be detected with sensitive meters.

Suppose now that we use a membrane permeable to more than one ion, say, to both sodium and potassium, but which is much more permeable to potassium than to sodium. At first, we see the ions lining up just as they did in Fig. 37.8. But gradually a change takes place. Although the potassium ions move very rapidly from Side A to Side B, occasionally we notice a sodium ion moving from Side B to Side A (Fig. 37.9). It, too, is moving toward a region of lower concentration (since there are many more sodium ions in Side B than in Side A) but at a much slower rate. The excess positive charge along the membrane in Side B never builds up to the point where it completely stops the diffusion of potassium ions because sodium, also carrying a positive charge, is slowly diffusing through to Side A. Each time a sodium ion passes from Side B to Side A, another potassium ion can pass from Side A to Side B without affecting the charge distribution across the membrane.

This exchange of Na^+ for K^+ will continue until the concentration differences tend to disappear, and at the end of the

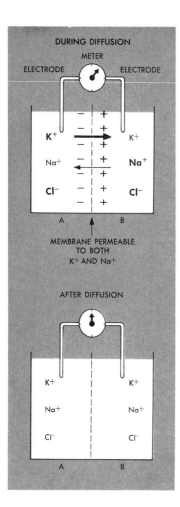

Fig. 37.9 In this case there is an exchange of sodium and potassium ions across the membrane because the membrane is permeable to both. However, the membrane is more permeable to K^+ than to Na^+ and this accounts for the membrane polarization during diffusion. After diffusion is complete (bottom diagram), there is no longer a difference in concentration of K^+ and Na^+. The membrane is no longer polarized.

experiment, we have an equal concentration of both Na^+ and K^+ ions on each side of the membrane. At this time the membrane is no longer polarized. Diffusion caused by the concentration differences has ceased, and the conditions required for polarization of the membrane no longer exist (Fig. 37.9).

Almost every living cell we will be studying has a polarized membrane. Usually the inside of the cell is negative.

Cell Membranes

A living cell membrane is the site of continuous activity; molecules are constantly being transported to and fro across it. This transport is controlled by special properties of the membrane barrier. For one thing, if we surround a cell with both small and large molecules, we notice that only the small ones enter the cell. The large ones seem to be excluded. The cell membrane may have a porous structure with pores so small that they permit only small molecules to squeeze through.

Molecules easily dissolved in oil also penetrate the cell membrane very rapidly. On the other hand, large molecules which do not dissolve in oil are excluded. The cell membrane itself contains an oil-like material called **lipid**. Evidently, molecules which are easily dissolved in oil are also easily dissolved in the lipid structure of the membrane and, therefore, can penetrate the cell. It has been proposed that the lipid is spread over the surface of the cell as a film-like structure, and that the film is supported by a protein network superimposed on both its outer and inner surfaces.

When we begin a more detailed study of cell membranes, we find some things that are puzzling. Some molecules seem to go "uphill," that is, they move through the cell membrane from a region of low concentration to one of high concentration. For instance, the intestinal tract may contain very little glucose (sugar) in its digested foodstuffs, while the bloodstream may have a much higher concentration of glucose. Nevertheless, the cells lining the walls of the intestine can transport the little bit of glucose from the intestine toward the blood.

If we study an individual cell, we find other *apparent* inconsistencies. The inside of the cell, for example, is rich in potassium, and the fluid bathing the outer cell membrane is rich in sodium (Fig. 37.10). The cell membrane is permeable to potassium, but sodium ions permeate much more slowly. We would expect that eventually the potassium and sodium ions would exchange to the point where each is equally concentrated inside and outside the cell. But this does not happen under normal conditions.

However, if we cool the cell (or interfere with the cell's activity in some other way), potassium ions begin to leak out of the cell, and sodium ions begin to leak in. If we wait long enough, the concentrations of both potassium and sodium on both sides of the cell membrane reach equilibrium, just as we would have expected originally.

If we warm the cell to its normal temperature, sodium ions begin moving out of the cell and potassium ions begin moving in. Both are now moving toward the region of high concentration, in other words, "uphill." Eventually, the cell returns to its original state with a high concentration of potassium inside and a high concentration of sodium outside. This cannot be explained by the influence of electrical forces, because both potassium and sodium ions are positively charged, yet they are moving in opposite directions. If one were attracted to a certain region (or repelled from that region) by electrical forces, the other should be also. That is, electrical forces should make them both move in the same direction.

Evidently, cell membranes have the capacity to "pump" solute particles in an "uphill" direction. The exact mechanism of this pumping action is unknown, but energy is required to pump the molecules uphill. The energy is obtained from chemical reactions taking place in the cell (metabolism). Transport which moves in an "uphill" direction is called **active transport.** When we first study a cell, we would expect sodium ions to leak in and potassium ions to leak out. But, apparently as fast as some sodium ions leak into the cell, an equal number of them are extruded by a "pump;" and as fast as potassium ions leak out, an equal number are pumped back in. When we cool the cell, or curb its metabolism by poison, we interfere with the energy supply to the pumping mechanism shutting it down. As a result, the sodium ions which leak in, and the potassium ions which leak out are no longer compensated for by active transport. When we rewarm the cell (or remove the poison), the pumping mechanism begins to operate again, and finally a new steady state is reached. The rate of pumping is just equal to the rate of leakage.

All living cells are capable of active transport in one way or another. Some cells actively transport glucose; some cells actively transport amino acids. Almost all cells transport sodium and potassium ions in the way we have described. Although biologists have built experimental models that transport solutes uphill, and in some ways mimic the actions of cells, the detailed mechanism used by the cells to accomplish uphill transport still remains a mystery.

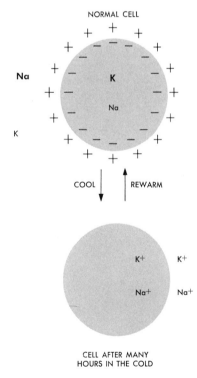

Fig. 37.10 In a normal cell sodium is pumped out as fast as it leaks in, and potassium is pumped in as fast as it leaks out (top). Cooling the cell depresses the "pump" so that it no longer compensates for the leakage. The cell loses potassium and gains sodium (bottom).

The active transport of sodium and potassium ions requires a substantial portion of the cell's total energy and plays an important part in its total economy. For instance, a rich supply of potassium ions within a cell facilitates some chemical reactions involved in protein synthesis, so its accumulation inside the cell may be vital to the life of the cell. Maintaining osmotic equilibrium with the cell's outside environment is also vital. Each cell contains many large solute molecules that cannot be transported out; and because they add to the total osmotic pressure, they tend to draw water into the cell. To counteract this pull, the cell can pump sodium ions out to help equalize the osmotic pressure. Finally, sodium and potassium ions must be present in different concentrations across the membrane of nerve and muscle cells if these cells are to function properly.

SUMMARY

Some of the physical forces involved in movements of molecules arise from differences in **pressure, concentration,** and **electrical potential. Bulk flow** takes place from regions of high pressure toward regions of lower pressure; molecules **diffuse** from regions where they are highly concentrated toward more dilute regions; water flows toward regions where the dissolved molecules are more concentrated **(osmosis).** Positive ions tend to move away from positively charged regions (high electrical potential) toward negatively charged regions (low electrical potential); negative ions tend to flow in the opposite direction. Transport is affected by the presence of barriers. Cell membranes form effective barriers for transport, permitting some molecules to pass through while excluding others.

In many cases, more than one type of force is involved. Osmotic water flow can be prevented, or even reversed, by an oppositely-directed pressure force. The diffusion of ions results in electrical forces that affect the diffusion process. Living cell membranes are able to transport some dissolved substances "uphill," from regions of low concentration to regions of high concentration. The energy required for this **active** ("uphill") **transport** is obtained from chemical reactions taking place within the cell (metabolism). All cells can actively transport potassium ions into the cell and sodium ions out of the cell.

FOR THOUGHT AND DISCUSSION

A membrane separates two solutions of solute S. The solute on the right-hand side is more concentrated than that on the left.

1 If the membrane is permeable to S, then immediately after the experiment begins, in what direction will S move?
2 If the membrane is permeable to water, but impermeable to S, then immediately after the experiment begins, in which direction will the water move?
3 If you wanted to prevent water movement, on which side of the membrane would you apply a pressure (for instance, by pushing on a piston)?
4 Assume that S is the salt, sodium chloride (Na^+Cl^-). Further assume that the membrane is permeable to Na^+ but not to Cl^-. Would you expect electrical differences between the two sides of the membrane? Which side would be positive?

SELECTED READINGS

Carlson, A. J., V. Johnson, and H. M. Cavert. *The Machinery of the Body*, 5th Ed. Chicago: University of Chicago Press, 1961.

Hardin, G. *Biology, Its Principles and Implications*. San Francisco: W. H. Freeman and Co., 1961.

Holter, H. "How Things Get into Cells," *Scientific American* (September 1961) (*Scientific American* reprint #96).

Kimball, J. W. *Biology*. Palo Alto, California: Addison-Wesley Publishing Co., Inc., 1966.

Ponder, E. "The Red Blood Cell," *Scientific American* (January 1957).

Weisz, P. B. *The Science of Biology*. New York: McGraw-Hill Book Company, 1963.

38 MOTION: MUSCULAR CONTRACTION

If you touch a live animal it usually responds by moving. Motion is one of the most common signs of animal life. Many cells are capable of some form of movement, but there are certain cells specialized to move various parts of the body, or the whole animal. Breathing movements, heart beat movements, and many, many other forms of motion necessary for homeostasis and survival are caused by the contraction of muscle cells.

Types of Muscle

There are three types of muscle: 1 **skeletal** muscle; 2. **cardiac** muscle; and 3. **smooth** muscle. **Skeletal** muscle consists of bundles of muscle cells—long, fibrous cylinders with diameters ranging from 0.01 to 0.1 mm, and a length up to 40 mm. These muscles are under **voluntary** control; that is, we can

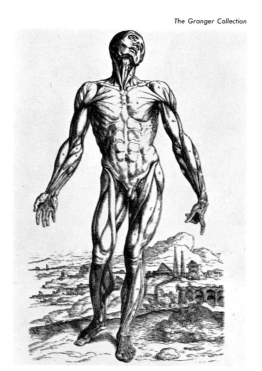

The Granger Collection

Fig. 38.1 This plate illustrates "the anterior view of the body from which I have cut away the skin, together with the fat and all the sinews, veins, and arteries existing on the surface." It displays "a total view of the scheme of muscles such as only painters and sculptors are wont to consider." Andreas Vesalius (1514–1564).

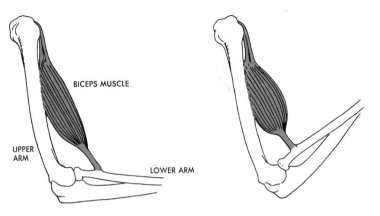

Fig. 38.2 Contraction of skeletal muscle moves the limbs and other parts of the skeleton. These muscles are under *voluntary* control.

command them to work for us whenever we wish them to. Most skeletal muscles are attached at each end to bones; when the muscles contract, the bones move on their joints. Figure 38.2 shows how contraction of muscles in the upper arm moves the lower arm. In this case, the elbow serves as a pivot. Compared to the other two types of muscle, skeletal muscle is capable of rapid and short bursts of activity. **Cardiac** muscle is the major component of the heart. Its alternating contraction and relaxation is responsible for pumping blood. **Smooth** muscle is found imbedded in the walls of hollow internal organs, for example, in the intestinal tract, in the bladder, and in blood vessels. In contrast to the other two types, smooth muscle is most often involved in slow, sustained contractions. Cardiac and smooth muscle are not under voluntary control.

Muscle Contraction

There are two classifications of muscle contraction—**isotonic** contraction, and **isometric** contraction. An isotonic contraction involves actual shortening of the whole muscle. When you pick up a weight, your limbs move. It is the shortening of muscle which is responsible for the motion of the limbs. On the other hand, you may try to pick up a weight which is too heavy, say, something that weighs 300 pounds. In this case there is no motion. Nevertheless, your muscles become tense; they pull on the weight but not enough to move it. When muscles exert a pull or tension but do not move, we say that they have undergone **isometric contraction**.

How can we imagine a muscle exerting tension, but not shortening? One possibility is shown in Fig. 38.4. Gross muscle is composed of many different parts, but not all of the parts shorten or change dimension when a muscle contracts. Those

Fig. 38.3 A microscopic view shows that skeletal muscle is made of long, fibrous cells which contain many nuclei. Note the stripes running perpendicular to the surface.

William Windle, M.D.

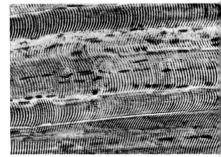

Fig. 38.4 This diagram shows the relationship between the *series elasticity* and *contractile element* when a skeletal muscle is relaxed, and in isometric contraction.

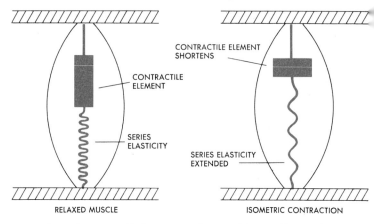

parts that do are called **contractile machinery**. The contractile machinery must be attached to other parts of the muscle, called the **series elasticity**. The series elasticity has the properties of a stiff rubber band. When we stretch the series elasticity, it exerts some tension, but it in itself does not *cause* the fundamental change in dimension. In the second part of Fig. 38.4, the contractile machinery shortens and stretches the series elasticity, thus creating tension. In an isometric contraction some shortening may actually occur in the contractile machinery, but all the shortening goes into stretching the series elastic component. As a result, there is no gross movement, only a certain amount of tension develops.

The most interesting part of muscle is the contractile machinery. The physiologist wants to know the structure of this machinery, and its immediate source of energy. We now think that the contractile machinery is composed largely of two proteins, **actin** and **myosin**. The immediate source of energy for contraction seems to be a very highly reactive compound called **ATP (adenosine triphosphate)**. This compound contains 3 phosphate groups, as illustrated in Fig. 38.5. It is a product of the cell's

Fig. 38.5 When phosphate is split off from ATP, energy is released. This energy can be used by the cell to do work.

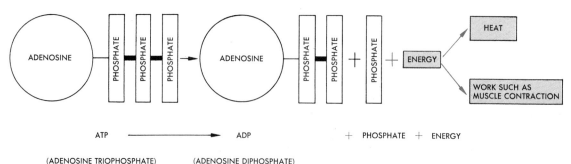

metabolism and is formed, for example, when glucose (blood sugar) is broken down to carbon dioxide and water. A sizable fraction of the energy released by the breakdown of glucose is stored in small packets in the form of ATP molecules. This energy, in turn, can be released when phosphate groups are split off from the ATP, as shown in Fig. 38.5. Among the many chemical reactions which liberate energy, the splitting of ATP plays a unique role—energy liberated by the splitting off of phosphate groups can be used by the cellular machinery to do work.

What evidence allows us to make these statements? First of all, muscle does contain large amounts of actin and myosin. Secondly, by acting as a **catalyst,** myosin speeds up the breakdown of ATP to ADP. More evidence comes from a study of muscle fibers which have been soaked in a solution of glycerol for several days. This treatment leaches out most of the soluble constituents of muscle, including salts, sugar, and ATP. The framework which is left behind consists primarily of actin and myosin. This framework is called a **glycerinated** muscle fiber. If now we wash off a glycerinated muscle fiber and add ATP to it, it will contract and lift the same weights that an intact muscle will. Furthermore, we can show that during the process of contraction ATP is split to ADP. These glycerinated fibers may be kept in a deep freeze for several months without losing their capacity to respond to ATP.

An intact muscle not only contracts, but it also relaxes. When ATP is added to a glycerinated muscle, the muscle fiber contracts, but does not relax. This poses a new problem—how does relaxation take place? Recent experiments have shown that addition of juices from a minced, fresh muscle to a contracted glycerinated muscle causes the contracted muscle to relax. Apparently, the intact muscle has some **relaxation factor** which keeps the muscle in a relaxed state until it is excited. Further experiments have shown that the relaxation factor contains essential components, tiny fragments that are parts of a muscle cell structure called **sarcoplasmic reticulum,** which bind calcium ions. Now free calcium ions are needed for the reaction of glycerinated muscle with ATP to take place. The effect of the relaxing factor, then, may be simply the result of its binding with calcium. If this is so, then during contractions somehow or other the calcium must be released from these fragments; just how this takes place is not known.

Structure of the Contractile Machinery

In the past 10 years, we have learned a great deal about the structure of intact skeletal muscle, how actin and

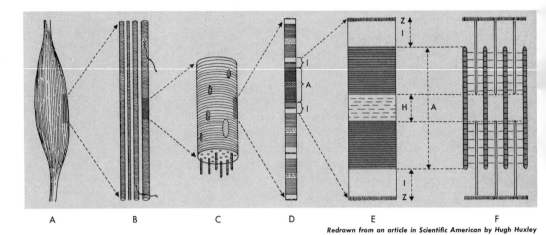

Fig. 38.6 Schematic drawing of skeletal muscle as it is dissected and seen under higher and higher powers of magnification. A represents a whole muscle (similar to Fig. 38.2). B represents muscle fibers (similar to Fig. 38.3). C shows the fibrils that run through a single fiber, while D and E show details of the striped fibril (similar to Fig. 38.7). The overlapping relation between the actin and myosin filaments is shown at F.

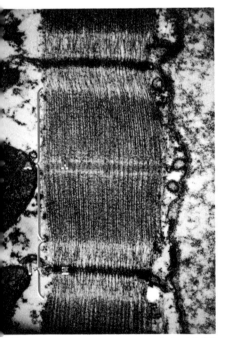

Fig. 38.7 This electron micrograph (enlarged 24,000 times) shows some of the structure of muscle fibrils. The A bands, the I bands and the Z lines are very prominent. Compare this figure with Figs. 38.6D and E.

myosin are arranged, and how they move during contraction. The series of illustrations in Fig. 38.6 shows the details of muscle structure as they are brought out by observing muscle under greater and greater magnification. Part A of the diagram represents muscle as we see it with the unaided eye. Whole muscle is composed of a great number of fibers or cells. Each fiber is about 100 microns (0.1 mm) in diameter. If we look at the single cells shown in Part B, we see that each cell has stripes. Further magnification of a single muscle cell (Part C) shows that the cell also has very fine fiber-like structures running lengthwise throughout the muscle. These very fine structures are called **fibrils.** The fibrils are about one micron thick. They are striped, and their stripes are aligned in such a way that they give the cells a striped appearance. Details of the stripes are shown in Parts D and E, representing a single fibril. The wide, dark striped portion is called the **A band,** and the light portions between the A bands are called the **I bands.** Frequently, a light region is seen in the middle of the A band; it is called the **H zone.**

The fibrils seem to make up the contractile machinery. In between the fibrils are many of the usual chemical constituents found in most types of cells. In addition, structures that seem to be part or all of the relaxing factor structure can be found in these spaces.

When we examine a fibril under an electron microscope, we see even smaller fiber structures. They are called **filaments.** These filaments are 50 or 100 angstroms (0.005 to 0.01 micron) thick. Their arrangement is shown in Part F of Fig. 38.6. The A bands n the fibril seem to be made of relatively thick filaments which overlap with thinner filaments. Bridges can be seen reaching out from the thick filaments toward the thin filaments. Toward the middle region of the thick filaments there is no overlap with the thin filaments, and it is in this region that the H zone appears.

By placing muscle in special salt solutions, we can dissolve the myosin without destroying the actin. Whenever this is done, the thick filaments disappear. In other words, we identify the thick filaments with myosin. However, it is also possible to dissolve out the actin. Whenever this is done, the thin filaments disappear; for this reason we identify the thin filaments with actin.

Our current ideas are that when muscle contracts, or when muscle stretches, the myosin components and the actin components do not crumple or stretch out, but remain fixed in their lengths. What does happen is that the filaments slide over each other, with the result that the ends of the actin filaments come closer together (contraction). Parts A, B, and C of Fig. 38.8 show muscle in the stretched, relaxed, and contracted states according to this idea. If the idea is correct, then we would expect that in an intact muscle the length of the A bands does not change whenever the muscle is contracted or stretched. This is, in fact, found to be the case. Further, we would expect that the length of the I bands would decrease when the muscle contracts, and increase when it stretches. This also is found to be so. Finally, the H zone begins to disappear as the muscle contracts, and increases in length during stretching. This is just what we would predict on the basis of this sliding filament hypothesis.

Since the small bridges seen in the electron micrographs provide the only contact between the thick and thin filaments, we now think that the actual work of contraction and the use of energy take place in these elements. Somehow or other, these small bridges are supposed to make contact with the thin filaments and propel them toward one another during contraction. Although we have traced the mystery of contraction to these small bridges, we have yet to learn the details of their actions.

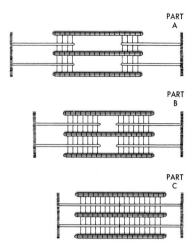

Fig. 38.8 This diagram shows the relationship between the actin and myosin filaments of skeletal muscle in the stretched (A), relaxed (B), and contracted (C) states.

SUMMARY

Most body movements are caused by the contraction of muscle. There are three types of muscle: **Skeletal** muscle (responsible for moving the bones of the skeleton) is frequently involved in rapid, short bursts of activity. **Cardiac** muscle (responsible for the pumping of blood by the heart) contracts and relaxes with each beat of the heart. **Smooth** muscle (found in the walls of hollow internal organs and blood vessels) is often involved in slow, sustained contractions. Skeletal muscle is under **voluntary** control. Smooth and cardiac muscle are not.

Muscle contraction can be classified into two different types. In an **isotonic contraction** the muscle shortens. In an **isometric contraction** the muscle does not shorten but becomes tense.

When the contractile machinery is examined under the electron microscope, it appears to be made up of two types of filaments. The thinner filaments are made of protein called **actin**. The thicker filaments are made of the protein **myosin**. The actin and myosin filaments overlap one another. When the muscle contracts, these filaments slide upon each other and the extent of overlapping increases. When the muscle is stretched, the extent of overlapping decreases. The interaction between the actin and myosin filaments is believed to occur through regularly spaced bridges that can be seen extending from the thick filaments toward the thin ones. The energy for contraction is made available when a phosphate is split off from **ATP**. Intact muscle appears to contain a "relaxation factor," which keeps the muscle in a relaxed state until it is excited. The relaxation factor is probably a part of the muscle structure able to bind calcium ions.

FOR THOUGHT AND DISCUSSION

1 When you pick up a weight, your muscles undergo both isometric and isotonic contractions. Which comes first? Explain
2 The two ends of a muscle are tied to a rigid support so that the muscle cannot change its length. The muscle is then stimulated. How could you tell the difference between this muscle and another similar muscle that was not being stimulated? (What tests would you perform?)
3 Suppose that you examine two muscles of the same size and weight and find that one muscle can lift heavier weights than the other. Can you suggest any possible differences (in the muscles) that could account for this difference in their ability to do work?
4 The amount of force that can be exerted by a muscle during contraction and the amount of energy that the muscle uses depend (among other things) on the length of the muscle. For example, if the muscle is stretched enough and then stimulated, it does not exert any force. From your knowledge of

the fine structure of muscle, and of what happens during stretching, can you suggest a reason why stretching a muscle beyond a certain length prevents it from exerting any additional force when it is stimulated?

SELECTED READINGS

Asimov, I. *The Human Body, its Structure and Operation.* Boston: Houghton Mifflin Co., 1963.

Hardin, G. *Biology, its Principles and Implications,* 2nd Ed. San Francisco: W. H. Freeman and Co., 1966.

Hayashi, T. "How Cells Move," *Scientific American* (September 1961) (*Scientific American* reprint #97).

Huxley, H. E. "The Mechanism of Muscular Contraction," *Scientific American* (December 1965) (*Scientific American* reprint #1026).

Huxley, H. E. "The Contraction of Muscle," *Scientific American* (November 1958) (*Scientific American* reprint #19).

Satir, P. "Cilia," *Scientific American* (February 1961) (*Scientific American* reprint #79).

Stumpf, P. K. "ATP," *Scientific American.* (April 1953) (*Scientific American* reprint #41).

Winton, F. R. and L. E. Bayliss. *Human Physiology,* 5th Ed. Boston: Little, Brown, and Co., 1962.

39 INFORMATION TRANSFER: NERVES

The cork float in our water tank in Chapter 36 was useful because it responded to the change in the water level; however, this response alone was not enough. Somehow or other, the float had to be in communication with the valve regulating the outflow. The communication linkage was provided by a rod. In our other example of feedback and control in regulating systems, the parathyroid gland responded to the level of calcium in the blood. It communicated with the gastro-intestinal tract, the kidney, and bone by secreting a chemical into the blood stream. Once the chemical arrived at these organs, the action we described took place. In this chapter we are going to take a close look at ways in which information is transferred from one part of the body to the other. In general, there are two ways: 1. through the secretion of **hormones;** and 2. through "messages" carried by **nerves.**

The Role of Hormones

Any chemical that is released by one organ and affects another organ can be regarded as a **hormone**. It acts as a chemical messenger that coordinates the activities of different organs. Usually a hormone is carried from one organ to another by the blood stream.

The method of hormonal communication is illustrated in Fig. 39.1. The gland releases hormone into the blood stream, and the blood stream carries the hormone to all organs in the body. Even though all organs receive the hormone, they are not all influenced by it. The parathyroid hormone, for example, affects the kidney, the intestinal tract, and bone, but has little effect on other organs, such as the heart and lungs.

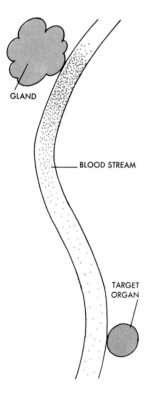

Fig. 39.1 In general, there are two ways in which information is transferred from one part of the body to another. One way is through the action of hormones, which are "chemical messengers."

THE NERVOUS SYSTEM

Figure 39.2 shows the nervous system of man. The three major components are the **brain, spinal cord,** and

1 BRAIN
2 SPINAL CORD
3 SYMPATHETIC GANGLIA
4 VAGUS NERVES (PARASYMPATHETIC)
5 SPINAL NERVES
6 VERTEBRA
7 SPINAL CORD
8 SENSORY NERVES
   (CARRY IMPULSES TO THE CNS)
9 MOTOR NERVES
   (CARRY IMPULSES FROM THE CNS)
10 SPINAL NERVES
   (MIXED, SENSORY AND MOTOR NERVES)

THE NERVOUS SYSTEM

Fig. 39.2 Another way information is transferred from one part of the body to another is through the action of nerve impulses. The large illustration shows three key parts of the nervous system: the brain, spinal cord, and the nerves that link these structures to the rest of the body. The chains of sympathetic ganglia are shown in solid red, parasympathetic nerves in black, and spinal nerves gray.

The inset shows a cross section of the spinal cord as it lies incased in the vertebral column. The central gray area is heavily populated with nerve cell bodies. The lighter area surrounding the central gray is made up of nerve axons running to and from the brain.

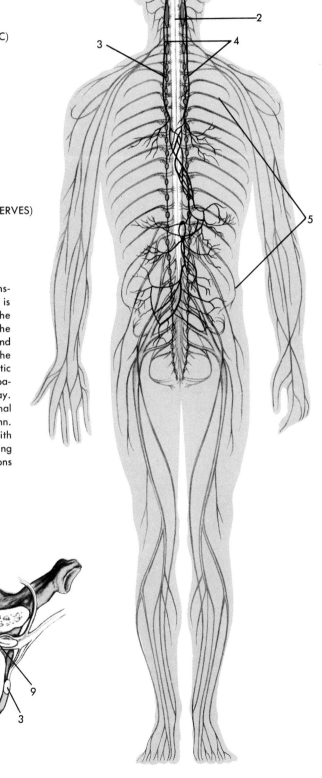

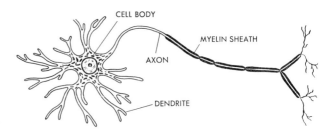

Fig. 39.3 A nerve cell consists of a cell body, dendrites, and a single long axon. Axons may be up to a meter long.

nerves, which link these two structures to other parts of the body. Nerves linking the brain with the body are called **cranial** nerves. Those linking the spinal cord with the body are called **spinal** nerves. If we sliced any one of these nerves we would find that it is made up of cylinders of very fine structure. These cylindrical structures are called **axons.** They are actually portions of nerve cells (Fig. 39.3). Nerve cells are called **neurons** and, like any cell, they have a nucleus and a cell membrane. In addition, they may have several very short structures called **dendrites,** which extend from the cell in virtually every direction. The single long fiber, the axon, is specialized to carry messages. Frequently, it is covered with a thick fatty sheath called **myelin,** which gives the characteristic white color to nerve fibers. The nerves in your body are cable-like structures, each containing a great many axons.

The combination of brain and spinal cord is called the **central nervous system** (abbreviated **CNS**). The CNS coordinates many of the body's activities. Some nerve axons carry messages from the surface and interior of the body to the CNS. These axons are called **sensory fibers.** Other axons carry messages from the CNS to muscles and glands. These axons are called **motor fibers.** Messages carried by motor fibers can be considered as commands that regulate the activity of muscles and glands.

Nature of a Message

Suppose that you were able to remove a nerve and the muscle it is attached to. You place the nerve and muscle in a special solution, called **Ringer's Solution,** whose salt composition resembles that of the internal environment. If you now give the nerve an electric shock (or heat it, hit it, or pinch it), the muscle will shorten. When you stimulate, or excite, the nerve, information of some sort, which we will call a nerve **impulse,** travels along the axon to the muscle, resulting in a muscle response. What, exactly, is this "impulse" that travels along the nerve fiber?

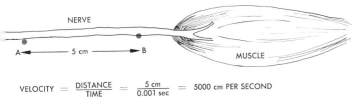

$$\text{VELOCITY} = \frac{\text{DISTANCE}}{\text{TIME}} = \frac{5 \text{ cm}}{0.001 \text{ sec}} = 5000 \text{ cm PER SECOND}$$

Fig. 39.4 This diagram shows one method of measuring the velocity of nerve impulse. (See text for details.)

We can study the properties of a nerve impulse by setting up in a laboratory the nerve muscle preparation just described. For simplicity, assume that we have a single axon connected to the muscle. We can now measure how fast the nerve impulse travels by the simple arrangement shown in Fig. 39.4. Suppose that we stimulate the axon at Point A and measure the time it takes for the muscle to respond. Next, we stimulate the axon at Point B, five centimeters closer to the muscle. We find that it takes less time for the muscle to respond when Point B is stimulated. Evidently, the difference in the time required for the muscle to respond in these two experiments must be due to the time required for the impulse to pass along the axon from the Point A to B.

In a typical experiment, if A and B are five centimeters apart, the muscle responds $1/1000$ second sooner when we stimulate B than when we stimulate A. In other words, it takes the nerve impulse $1/1000$ second to travel five centimeters. In one second the impulse would have traveled 5000 cm, so the velocity of the nerve impulse is 5000 cm per second (50 meters per second, or about 110 miles per hour). When we study other axons, we find that the velocity may vary, but it will almost always fall somewhere between one and 100 meters per second. In general, axons with large diameters conduct impulses faster than axons with small diameters.

In this type of experiment, we stimulate the nerve axon and use the response of the muscle simply to indicate whether or not the impulse has traveled along the axon. Occasionally we find it difficult to tell whether we are really studying the axon or the muscle. It would be very useful if we could find in the axon itself some change taking place during activity that serves not only as an indicator of the nerve impulse, but that also gives us information about the nature of the impulse. We do have some clues.

When a nerve-muscle preparation is stimulated, many changes occur in the axon just before the muscle contracts. For example, a small amount of heat is produced, there is a very

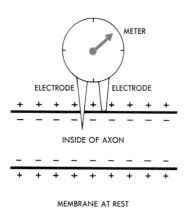

Fig. 39.5 When at rest, the membrane of a nerve axon is electrically polarized, with the inside of the axon negative in relation to the outside.

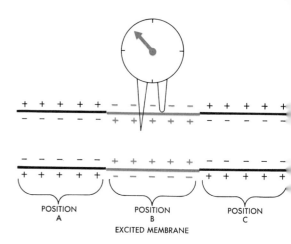

Fig. 39.6 When a nerve impulse moves along the axon, it is accompanied by a wave of reversed polarity (colored area). The inside of the axon at the excited site of the membrane is positive in relation to the outside.

small volume change, and there are electrical changes. The electrical changes are most striking and are easily measured.

Figures 39.5 and 39.6 show how electrical changes may be measured along the axon. We embed an electrode in the axon and connect it through some measuring meter to an electrode resting along the outside of the membrane. The first measurement shows that the axon membrane is polarized when at rest; the inside is negative with respect to the outside. Now we stimulate the axon. Since we know the speed of the nerve impulse, we can estimate where it will be at each moment following stimulation. When the impulse passes the recording electrodes at Position B, the polarity is reversed; in this part of the axon the inside is positive with respect to the outside (Fig. 39.6). If we move the electrodes to Points A or C, the result is the same when the impulse passes these points.

As the impulse moves along the axon, the polarity is reversed. In other words, the passage of a nerve impulse down an axon is accompanied by a wave of reversed polarity. This wave of reversed polarity traveling down the nerve fiber is called the **action potential**. Whenever we stimulate the axon of a nerve-muscle preparation and find the muscle responding, we also find the action potential traveling along the axon. Further, the action potential travels at a speed exactly equal to the speed of the nerve impulse mentioned earlier. All of the properties of the nerve impulse that we can measure by using the nerve-muscle preparation are shared by the action potential. The action potential seems to be a clear sign of the passage of the nerve impulse, and frequently we tend to think of the two as a single event.

What happens to the action potential when we change the strength of the stimulus? We find that weak stimuli do not produce any response (action potential) in the axon. As we increase the strength of the stimulus, there is still no response until we reach a certain level called the **threshold.** Stimuli whose strength lie above the threshold all produce the same action potential. In a way, the same thing happens with a stick of dynamite. If you gradually raise the temperature of a stick of dynamite by heating it, the dynamite eventually explodes, but increasing the temperature beyond this point (the threshold) does not increase the intensity of the explosion. You either get the explosion or you don't. Similarly, in a nerve fiber we get an action potential or we do not. For this reason, we say that the nerve fiber behaves in an **all-or-none** way. The size of the action potential is the same, no matter how large the stimulus might become.

After being stimulated, a nerve fiber can recover and carry an impulse over and over again. If two stimuli are given to a nerve fiber a few seconds apart, each stimulus produces a nerve impulse. If, however, the second stimulus is given within about $1/1000$ of a second after the first stimulus, only one impulse occurs. Recovery from the first stimulus requires a very short, but measurable time. This short recovery time is called the **refractory period.** The existence of a refractory period of about $1/1000$ of a second means that there is a maximum of about 1000 impulses per second that can be carried along a nerve fiber.

A *threshold* stimulus intensity, an *all-or-none response,* and a *refractory period* are three of the most striking properties of nerve axons. In recent times, physiologists have been able to describe these and other properties in terms of ion transport across the axon membrane.

Polarization of Nerve Membranes

The inside of a nerve cell, like all cells, is rich in potassium and poor in sodium. As you saw in Chapter 37, the active transport process located in the membrane pumps sodium ions out and potassium ions into the cell. The fluids outside the cell, however, are rich in sodium and poor in potassium ions. Considering the concentration forces across the cell membrane, we see that potassium ions are in position to diffuse out of the cell, while sodium ions are in position to move inward. What happens depends on the permeability of the membrane to potassium and sodium ions.

When resting, a nerve membrane is much more permeable to potassium. As a result, small amounts of potassium ions tend to diffuse outward, leaving the cell and building up positive charge on the outside. The resting membrane is polarized—the outside is positive and the inside is negative.

Fig. 39.7 The polarity is reversed at the excited site of an axon because the excited part of the membrane is more permeable to sodium than to potassium.

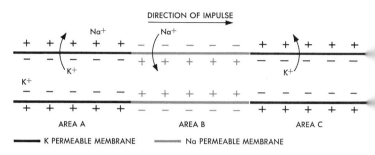

During the passage of an impulse, however, the membrane polarity reverses. The inside is now more positive than the outside. This happens because the membrane suddenly becomes more permeable to sodium than to potassium, and sodium enters the cell faster than potassium leaves. In Fig. 39.7, Areas *A* and *C* of the membrane are inactive, while Area *B* is active. In this figure, the impulse (Area *B*) is traveling from left to right. Fractions of a second ago it was at Area *A*, but *A* has recovered and is a resting cell membrane—negative on the inside, positive on the outside. Compare Area *A* with Area *B* (where the impulse is at this moment). Then compare it with Area *C* (where the impulse will be a fraction of a second from now).

The size of the action potential is limited by the concentrations of sodium and potassium ions on the two sides of the cell membrane. Normally these concentrations do not vary to any appreciable extent. As a result, every time the nerve is excited, the action potential is of the same size. In other words, it behaves in an **all-or-none** way.

Whenever a nerve is stimulated, the effect of the stimulus is to make the cell membrane more permeable to sodium than to potassium. It is this property of a nerve membrane that distinguishes it from ordinary membranes. Although we know that it happens, we do not yet know *how* a nerve cell membrane changes its permeability in response to a stimulus. However, we do know that for a stimulus to be effective, it must in one way or another begin to *weaken (decrease) the polarity of the membrane*. Then the membrane responds by increasing its sodium permeability. Once the sodium permeability has increased, sodium ions rush in and the membrane polarity will be reversed. The reversed polarity of an excited membrane lasts for less than $\frac{1}{1000}$ of a second. Apparently, the highly sodium-permeable state cannot persist. Potassium ions once again leak out much faster than sodium ions enter, and the membrane soon returns to its normal condition. During rest the tiny amount of sodium ions that have leaked into the cell are pumped out by the active transport mechanism, and the tiny amount of potassium ions that have leaked out are pumped back in.

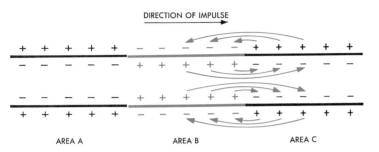

Fig. 39.8 The excited part of an axon membrane (Area B) affects adjacent Area C by decreasing its polarity. The decrease in polarity stimulates Area C. The impulse "moves" to Area C and thus travels along the axon.

A question we now must ask is how does an impulse *travel* along an axon? As shown in Fig. 39.8, Area B is the active portion of the nerve cell membrane, being positive on the inside and negative on the outside. As the impulse moves along to the right, the polarity of the membrane is reversed. Now exactly what happens to the membrane's inside and outside charge at Area C as a result of the reversed polarity that now exists at B? On the outside of the membrane, the positive charge at C is attracted to and moves toward B. On the inside, the positive charge at B is attracted to and moves toward C. The net result at C is that the positive charge is taken away from the outside of the membrane; at the same time, positive charge is added to the inside of the membrane. In other words, the polarity at C has been weakened, *just as though we had stimulated it!* The membrane at C responds to this slight loss of polarity by increasing its sodium permeability. As a result, positively charged sodium ions enter the cell and the polarity at C becomes reversed, that is, Area C is excited.

Notice that the conditions for change in membrane polarity at C also occur at Area A. In other words, the active area at B can excite adjacent areas in *both* directions. If we excite a nerve in the middle of the fiber, we indeed find that the impulse spreads in both directions. In the body, however, axons are never stimulated in their middle regions; they are always excited at one end, so that the impulse travels in only one direction.

We see from the discussion above that "messages" are carried from one part of the body to another in a form called nerve impulses. Since each axon responds in an all-or-none way, all of its impulses are identical. If each impulse is no different from any other impulse, then by itself an impulse cannot convey much information. We must, then, look for *patterns* of impulses in order to detect different types of information.

This situation is like the Morse code. For the most part, single dots and dashes do not mean much by themselves; however, certain patterns of dots and dashes indicate letters that can form words. The code used by the nervous system depends on both the *number of nerve impulses* arriving at a given place during each unit of time, and on the *specific axons* that carry the impulses. Somehow, the brain is able to interpret these patterns and to convey its commands to muscles and glands by a similar code.

Sensory Receptors

Impulses sent toward the CNS normally originate in structures called **sensory receptors,** which are distributed throughout the body. These receptors are especially sensitive to particular types of stimuli. For example, the eye is sensitive to light, the ear to mechanical vibrations, or sounds, and the tongue to chemical stimuli, or taste. Other examples of receptors are those which are sensitive to touch, or to heat, or to cold, or to pressure, or pain.

These receptors are always associated with a sensory nerve axon. Sometimes the receptor simply consists of free nerve endings, as in the case of axons that enable us to feel pain. Other times the axons are embedded in an elaborate structure which makes the receptor sensitive to a specific kind of stimulus. An example of this would be the ear. In all cases, however, stimulating the receptor results in the same type of response: action potentials are set up on the sensory nerve axons and carried to the CNS. But remember, as far as we can tell, the impulses are the *same* in all sensory axons, whether they come from the ear, or the eye, or from pain receptors in the toe.

Although the stimulus for the eye is light, the messages, or impulses, which are perceived by the brain reach the brain in total darkness. Messages of cold are carried along nerves which are just as warm as any body structure. Even though all the messages are simple action potentials, you still know whether you have been stimulated by light or by cold.

Apparently, the messages that reach your brain can be interpreted primarily on the basis of where the nerve axon carrying the impulse is located. Thus, you may shine a flashlight in your eye and "see" light, or you may get hit in the eye and see flashes of light. When you are hit, the blow might be strong enough to stimulate the deeply embedded sensory nerve axons leading from the eye to the brain. All the brain "knows" is that impulses were on sensory nerve axons coming from the eye, so it interprets the impulses as light, and you "see stars." If a person's arm has been amputated, he may complain of pain in the arm that is no longer there. The stump of the arm still contains the severed nerves which originally led from the amputated arm to the brain. These severed nerve ends may be highly irritable.

When impulses are sent along these nerves the brain responds as it always has and interprets the messages as signaling pain from the arm that is no longer there.

The brain not only has to interpret *where* the message comes from, so that it can assign the proper sensation to it, but it must also be able to tell how *intense* the stimulus is. It does this in two ways: 1. As the stimulus intensity grows larger and larger, the frequency at which the sensory axons send out impulses becomes greater. 2. The intensity of the stimulus will determine how many axons are made active. The more intense the stimulus, the more axons will be sending in impulses. Thus, the type of information that is fed into the brain is simply what axons are "firing" and how fast they are firing. It is on the basis of this simple type of coding system that the brain must act.

Motor Nerves

A similar code conveys "command" information from the CNS to muscles along motor nerves. If we stimulated a single nerve axon attached to a muscle, we would find a single action potential followed by a very brief contraction lasting less than $\frac{1}{10}$ of a second. This very brief contraction and subsequent relaxation which follows a single stimulus is called a **twitch**. With a possible exception of blinking of an eyelid, we normally do not see twitches in a healthy person. Most muscle contractions last much longer and result in smooth, sustained motions. This is because the motion of a muscle in a healthy person is generally not the result of a single nerve impulse going toward it. Instead it is the result of a whole train of impulses, one following the other in rapid succession.

The first impulse in such a train reaches the muscle and causes the muscle to contract, but before the muscle has time to relax, a second impulse arrives; and before the muscle can relax from the second impulse, a third impulse arrives. In this way a rapid succession of impulses keeps a muscle contracted. Smooth and sustained muscle contractions result from the super-position of many simple twitches and are called **tetanizing** contractions. The CNS can increase the size of contraction in two ways: 1. by increasing the number of axons carrying impulses, so that more muscle fibers are activated; and 2. by increasing the frequency of impulses, so that each contracting muscle fiber is more effective.

Nerve axons are not the only tissues that carry action potentials and that can be excited by electric shock. If you apply a threshold electrical shock directly to the surface of a muscle

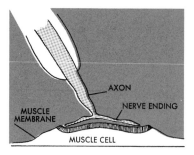

Fig. 39.9 At the neuromuscular junction there is a tiny but distinct gap. When a nerve impulse reaches the end of an axon, the axon releases a chemical transmitter that diffuses across the gap and excites the muscle.

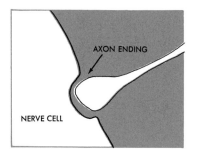

Fig. 39.10 There is also a tiny but distinct gap at the synaptic junction. When a nerve impulse reaches the end of an axon, the axon releases a chemical transmitter that crosses the gap and stimulates or inhibits the neighboring nerve cell.

cell, it will result in an action potential that travels over the entire surface. This is invariably followed by a contraction. The steps leading from an action potential on the muscle surface to contraction of the muscle are not known. However, the properties of the action potential on the surface of a muscle cell or nerve axon do not seem to differ very much.

Passage of Action Potentials Between Cells

When a muscle is excited by nerve impulses, action potentials travel along the nerve fiber to the junction between nerve and muscle (*neuromuscular* junction). Arrival of the impulse excites the muscle fiber so that an action potential spreads over the muscle surface, resulting in contraction of the fiber. The transmission of an impulse from nerve to muscle is called **neuromuscular transmission.**

Impulses are also transmitted from neuron to neuron. For example, the axon of a sensory neuron enters the central nervous system and connects to the dendrites or cell bodies of other neurons. They in turn connect to still others. The junction between the ending of a neuron and a cell body (or dendrite) is called the **synapse,** and transmission of the impulse at this point is called **synaptic transmission.** A nerve cell of the CNS receives many synaptic endings on its dendrites and cell body—sometimes thousands of them. This means that each neuron receives information from many other neurons. The endings look similar, like little buttons pressed on the surface membrane. Actually, a small space separates the axon ending from the surface membrane of the next neuron.

The junction between cells plays an important role in the transmission of excitation. One special property of these junctions is that *conduction takes place only in one direction.* When a muscle is excited directly by an electrical shock, an action potential spreads over the surface of the muscle but does not go backward up the nerve fiber. On the other hand, the action potential does spread down the nerve fiber to the junction and from the junction to the muscle. In synaptic transmission, the action potential always goes from the axon to the cell body of the next neuron, never in the reverse direction.

Another special property of these junctions is that the junctional region itself seems subject to fatigue or subject to the action of certain drugs. For example, a drug called **curare** (which was used by the Indians to poison their arrows) prevents the passage of an impulse from a nerve to the muscle. When the junction is poisoned with this drug, stimulation of the nerve does not make the muscle contract. Yet, in spite of the drug, the nerve carries action potentials; and if the muscle itself is stimulated directly, it contracts. This means that the nerve is intact and that the muscle is intact. Both are capable of carrying an impulse, but somehow or other the impulse does not pass across the junction.

In Figs. 39.9 and 39.10 you can see that at both the neuromuscular and synaptic junctions, there is a tiny, but distinct gap between the ending of the axon and the muscle or nerve cell. Many years ago it was thought that nerve endings release a chemical and that this chemical some how bridged the gap between the nerve ending and the adjacent cell. At first the idea seemed far-fetched, but in 1921, Otto Loewi showed that it was not at all far-fetched when he performed an ingenious experiment on heart muscle.

Loewi knew that stimulation of the **vagus** nerve slowed the heart's beat. His problem was to show that a chemical was released when the vagus nerve was stimulated, but he was not certain of which chemical he was looking for. He used two frog hearts, one with an intact vagus nerve, the other without. The first heart with the intact vagus nerve was arranged so that fluid bathing it could be collected. When he stimulated the vagus nerve of the first heart, it began to beat at a slower rate. This was expected. However, if he now placed the fluid from the first heart into the second heart, the beat of the second heart also slowed—even though he did not stimulate any nerve leading to the second heart. The only contact between the two hearts was the fluid. Some chemical released by stimulating the vagus nerve of the first heart accumulated in its bathing fluid. When this fluid was applied to the second heart, the second heart behaved exactly as if its own vagus nerve had been stimulated (Fig. 39.11).

Sometime after Loewi had performed the experiment, he identified the chemical as **acetylcholine.** Afterwards, an enzyme which destroys the acetylcholine was found. This was called **cholinesterase.**

Each impulse traveling along the vagus nerve releases a small amount of acetylcholine. The acetylcholine diffuses from the nerve ending to the adjacent heart muscle and slows the heart. This action does not last very long because cholinesterase found in the junctional regions soon destroys the acetylcholine. The effect of a single impulse is terminated very rapidly and the junctional tissue is returned to its normal state, ready for the arrival of more impulses.

Experiments have also shown that acetylcholine is an **excitatory** transmitter for nerve impulses arriving at skeletal muscle junctions. Eash time an impulse arrives at the junction, a small

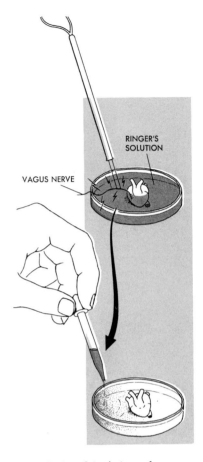

Fig. 39.11 Stimulation of the vagus nerve releases a chemical transmitter (acetylcholine), which slows the heart. We can recover some of the transmitter and transfer it to a second heart. The second heart slows just as if its own vagus nerve had been stimulated.

amount of acetylcholine is released and diffuses across the gap between the nerve and the muscle. Once across the gap, the acetylcholine increases the permeability of the muscle membrane to ions. This results in a depolarization capable of stimulating those parts of the muscle membrane near the neuromuscular junction. Again, this action does not last very long. Cholinesterase located in the junctional region quickly destroys the acetylcholine. The rapid release and subsequent rapid removal of acetylcholine is required if the characteristics of an impulse are to be maintained in crossing the junction.

Inhibition

In Loewi's experiments, we saw an example where stimulation of a nerve decreases, or **inhibits,** a response. Stimulation of the vagus nerve caused the heart to slow down. Stimulation of some nerves leading to smooth muscle causes the muscle to relax rather than to contract. This is another example of inhibitory action of nerves.

We find a very important example of inhibition by nerve axons in synaptic transmission. Figure 39.12 shows how we might study it. Suppose that Nerve Axon 1 and Nerve Axon 2 both lead to the cell body of Nerve Axon 3. Nerve Axon 1 is an ordinary excitatory axon. As Part A shows, when we stimulate Nerve Axon 1 impulses are transmitted out along Axon 3. If we stimulate Axon 2 (Part B), nothing happens. With no other information, we might conclude that Axon 2 has no effect. However, if we stimulate Axon 1 and Axon 2 at the same time (Part

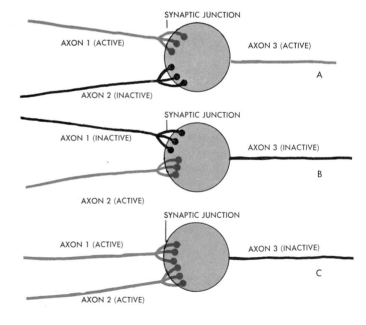

Fig. 39.12 This diagram shows how the inhibition of synaptic transmission might be demonstrated. Axon 1 is an excitatory fiber. Axon 2 is an inhibitory fiber. If both fibers are stimulated at the same time (Part C), Axon 2 inhibits the excitatory effect of Axon 1.

C), we still get no response. The action of Axon 2 somehow prevented, or inhibited, the excitatory action of Axon 1. In this case, we say that Axon 2 is an **inhibitory** fiber.

We now think that inhibition is caused by some inhibitory chemical that is released at the nerve ending. In contrast to excitatory chemicals that depolarize the structure which they excite, inhibitory chemicals do the reverse. They *increase* the polarization of the structure. This increased polarization raises the threshold. Although we know the chemical substance involved in neuromuscular transmission, we have yet to identify the chemicals involved in most cases of synaptic transmission.

Reflex Actions

The simplest coordinated movements that you make—blinking, sneezing, or suddenly withdrawing your hand from a hot stove—seem to be automatic reactions that are brought about almost immediately when there are certain changes in the environment. You withdraw your hand from a hot stove even before you know that you have been burned. These involuntary reactions involve nerve impulses and are called **reflexes.**

In all reflexes, the transfer of information follows a pathway known as the **reflex arc.** As Fig. 39.13 shows, a reflex arc has five components: 1. a **receptor,** which is excited by a stimulus and starts a nerve impulse on its way; 2. a **sensory nerve,** which carries the impulse to the CNS; 3. the **CNS,** where the sensory nerve may branch and make many synaptic connections with other neurons; 4. a **motor nerve,** which carries the impulse to the muscle or gland; 5. the **effector,** which is the muscle or gland activated by the impulse.

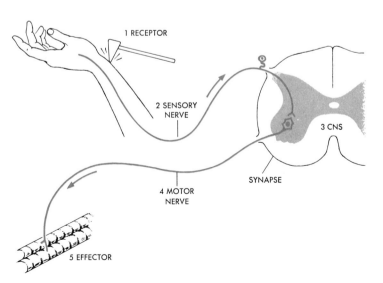

Fig. 39.13 This diagram shows the components of a simple reflex arc that has a synapse in the spinal cord.

Autonomic Nervous System

Many reflexes that help control the internal environment involve a special part of the nervous system called the **autonomic nervous system (ANS)**. This system consists of nerves making up the motor pathway to the heart, stomach, intestine, and other internal organs. These nerves are not under voluntary control; that is, you cannot make your heart beat faster on command. The nerve axons of the autonomic nervous system leave the spinal cord and brain and are mixed with other nerve axons in the cranial and spinal nerves. They differ from other nerves that go directly to skeletal muscle. Autonomic nerves, unlike motor nerves which go directly to the skeletal muscles they control, first make synaptic connections with other neurons which relay impulses to the organs (Fig. 39.14).

The two major divisions of the autonomic nervous system are called the **sympathetic** and **parasympathetic** nervous systems. Sympathetic nerves leave the middle regions of the spinal cord. Parasympathetic nerves leave the central nervous system from the upper regions where they travel in cranial nerves, and from the lowermost regions of the spinal cord where they travel in spinal nerves. Most internal organs of your body are supplied with both parasympathetic and sympathetic nerve fibers.

In general, these two types of nerves work in opposite ways. For example, impulses traveling along the sympathetic nerve fibers toward the heart increase your heart beat rate, whereas impulses traveling along the parasympathetic nerves leading to the heart decrease its rate. When impulses traveling along the autonomic nerve fibers reach the end of the line—the contact point with some organ—they exert their action by releasing a chemical transmitter. Parasympathetic nerve fibers release acetylcholine. Sympathetic nerve fibers release a transmitter called **noradrenalin.**

Some of the nerve fibers of the sympathetic nervous system do not follow the general rule of ending at a relay station. Instead, they go directly to the **adrenal** gland. There they release acetylcholine which stimulates the adrenal gland, causing it to secrete a mixture of adrenalin and noradrenalin. This mixture is then carried by the blood stream to all parts of the body and acts on various organs. Thus, adrenalin and noradrenalin are hormones. Adrenalin and noradrenalin have very similar actions. Noradrenalin, remember, is the transmitter of the sympathetic nervous system. So the secretion of the adrenal glands produces the same physiological effects as massive stimulation of the sympathetic nervous system. The heart beat becomes stronger when the sympathetic nerves are stimulated because noradrenalin

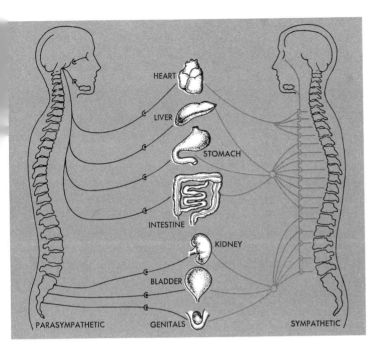

Fig. 39.14 The *autonomic nervous system* conveys impulses to the internal organs and regulates their activities.

is released at the sympathetic nerve endings. The same thing happens when the adrenal gland secretes hormones, at which time noradrenalin and adrenalin are carried to the heart by the blood stream. In both cases, the heart is responding to the same or almost the same substances.

When we examine the effects of sympathetic stimulation of various organs, a pattern begins to emerge. We find that sympathetic stimulation causes a number of events: 1. widening of air passages leading to the lungs; 2. constriction or narrowing of blood vessels in the skin and in the intestinal tract, resulting in blood being shifted from the digestive organs and skin to the muscles; 3. a general slowing down of movements in the intestinal tract; 4. an increase in the force and rate of heart beat; 5. release of the blood sugar from the liver. All of these activities prepare the animal for emergencies: such as running or fighting. The widening of the air passages makes it easier for the animal to breathe faster and get more oxygen. Blood is shifted from regions where it will not be needed during the emergency (the intestinal tract, for example) to skeletal and heart muscle which will need oxygen and blood sugar. The heart beats faster and stronger so that blood circulates through the muscles at a greater rate. Blood sugar is released from storage in the liver into the blood stream where it will be available to supply muscles with energy.

In the chapters that follow we shall see many examples of the influence of the autonomic nervous system in the control of the internal environment of an animal.

Fig. 39.15 The *sympathetic nervous system* readies the animal for action.

Courtesy Black Star

SUMMARY

Coordination of the activities of the various organs within the body is brought about principally by the secretion of **hormones** and by the transmission of **nerve impulses**. A hormone is a chemical that is liberated by one organ and has a physiological effect on another organ. Nerve impulses travel along **axons**. **Sensory** nerve axons carry impulses from different parts of the body to the **central nervous system** (brain and spinal cord). **Motor** axons carry impulses from the central nervous system to muscles and glands. In a simple **reflex,** an impulse is initiated in a **receptor** and carried along a sensory nerve to the CNS. At this point, **synaptic** connections with other neurons are made and impulses are sent along motor nerves to a muscle or gland.

The membrane of a resting nerve axon is electrically **polarized**. The passage of an impulse along the axon is accompanied by a wave of reversed polarity called the **action potential**. The electrical properties of nerve fibers have been described in terms of the movements of sodium and potassium across the axon membrane.

Nerve impulses behave in an **all-or-none** way and are practically identical. A single impulse cannot convey much information. The brain and body rely on *patterns* of impulses (the number of impulses arriving during each second at a given place) to provide meaningful information.

Action potentials pass between different cells at the **neuromuscular** and **synaptic junctions**. When the action potential arrives at the junction, a **chemical transmitter** is released. It diffuses across the junction and acts on the adjacent muscle or nerve cell. Some transmitters have an **excitatory** effect; others **inhibit**.

The motor pathway to the internal organs is known as the **autonomic nervous system**. This pathway is not under voluntary control. It is divided into two systems: the *sympathetic* and *parasympathetic* nervous systems. Most organs are supplied by both sympathetic and parasympathetic nerves. The actions of the sympathetic and parasympathetic nerves are usually antagonistic. In general, synpathetic nerve impulses help prepare the animal for emergencies.

FOR THOUGHT AND DISCUSSION

1 Acetylcholine and noradrenaline are sometimes called *neuro hormones*. Why is this name appropriate?

2   If parasympathetic nerves leading to the intestine are stimulated, the contractile activity of the intestinal muscle is increased. If the vagus nerve to an isolated heart is stimulated and fluid bathing the heart is removed and placed on a strip of intestinal muscle, the contractile activity of the intestinal strip is also increased. Why?

3   If calcium is removed from the fluid bathing a nerve or muscle, the nerve (or muscle) becomes very irritable and may become excited, even in the absence of any stimulus. If the parathyroid glands are removed from an animal, it very often begins to twitch. Can you relate these two findings? (Recall the discussion of the parathyroid gland in Chapter 36.)

4   If sodium is removed from the fluid bathing a nerve fiber, what would you expect to happen to the action potential?

5   Compare the excitation and conduction of a nerve impulse with the ignition and transmission of a spark along a fuse. In what ways are they similar? In what ways do they differ?

6   When a nerve impulse travels down a nerve axon, that portion of the axon lying immediately behind the action potential is in its refractory period. How does this help to ensure one-way conduction of the nerve impulse?

SELECTED READINGS

Carlson, A. J., V. Johnson, and H. M. Cavert. *The Machinery of the Body*, 5th Ed. Chicago: University of Chicago Press, 1961.

Eccles, J. C. "The Synapse," *Scientific American* (January 1965) (*Scientific American* reprint #1001).

Katz, B. "The Nerve Impulse," *Scientific American* (November 1952) (*Scientific American* reprint #20).

———. "How Cells Communicate," *Scientific American* (September 1961) (*Scientific American* reprint #98).

Keynes, R. "The Nerve Impulse and the Squid," *Scientific American* (December 1958) (*Scientific American* reprint #58).

Loewenstein, W. R. "Biological Transducers," *Scientific American* (August 1960) (*Scientific American* reprint #70).

Loewi, O. "On the Humoral Transmission of Heart Nerves," in *Great Experiments in Biology*, ed. by M. L. Gabriel and S. Fogel. Englewood Cliffs, N.J.: Prentice-Hall, Inc., 1955.

Winton, F. R. and L. E. Bayliss. *Human Physiology*, 5th Ed. Boston: Little, Brown, and Co., 1962.

40 CIRCULATION

Unlike simpler animals, complex animals have certain groupings of cells that become specialized in their ability to perform certain tasks. These "communities" of cells are called **organs.** The lungs, for example, are organs that exchange oxygen and carbon dioxide with the external environment; and the intestinal tract processes food so that it can be used by the cells of the body. In order for a system of organs to function in harmony, the material products of each organ must be available to all cells of the body. The brain, for example, is completely dependent on the lungs for its oxygen supply. If the brain is deprived of oxygen for more than a few minutes, the brain cells begin to

Fig. 40.1 Transport of various materials between different parts of the body is essential to the organism. Transport is performed by circulation of the blood, here visible in the blood vessels in the human mesentery.

William Windle, M.D.

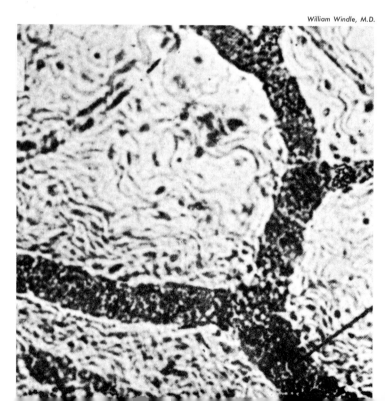

degenerate. Transport of various materials between different parts of the body, then, is essential. The task of transport from organ to organ is performed by circulation of the blood.

The circulatory system consists of a complex network of tubes, or blood vessels. Each of the trillions of cells in the body is close to at least one of these vessels. Blood circulates continuously, picking up products from one organ and delivering them to others. In a way, we can think of the circulatory system as "stirring" the internal environment so that its composition does not differ greatly from one place to another in the body.

The circulatory system has several features enabling it to perform these tasks effectively. One is a pumping system to keep the blood moving. This job is performed by the heart. Second, there must be some means for materials carried by the blood to reach the cells, and for materials in the cells to reach the blood. This exchange takes place in minute blood vessels called **capillaries.** Third, the circulatory system is able to adjust itself to the changing needs of the body. During exercise, muscles have a greater need for oxygen and blood sugar; this need is met by increasing the blood supply to the muscles.

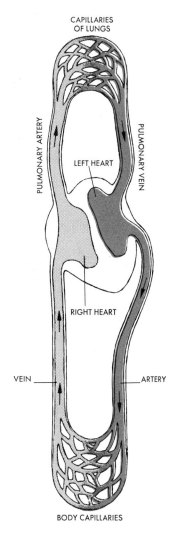

Fig. 40.2   The motion of blood through the body is continuous. It travels in a circular path and is driven by rhythmic contraction and relaxation of the heart.

Heart Muscle

The continuous flow of blood is brought about by the rhythmic contraction and relaxation of heart muscles. The heart is a hollow cavity with muscular walls. A thick partition divides the heart cavity into a right and left half. When the muscular walls relax, both the right and left half fill with blood. When the heart beats, the muscles contract; blood is squeezed into vessels called **arteries,** which direct the blood away from the heart toward the organs of the body (Fig. 40.2).

Like any muscle, the heart can be stimulated, and it will conduct action potentials. In many ways, it behaves like a skeletal muscle, but there are some exceptions. Skeletal muscles contract only if they receive some external stimulus. Ordinarily, the stimulus is a nerve impulse leading to the muscle. This is not true of heart muscle, which seems to be capable of exciting itself. Even if we cut all of the nerves leading to the heart, it will continue to beat. This capacity of self-excitation is common to all heart tissue.

If we remove the heart of a cold blooded animal (a frog, say), place it in a dish and cover it with Ringer's Solution, the heart continues to beat—even when it is completely disconnected from the body. If we now cut the heart into pieces, even the pieces continue to beat. However, some pieces beat faster than others. Those from the upper parts of the heart (the **atrium**) beat faster than those from further down (the **ventricle**).

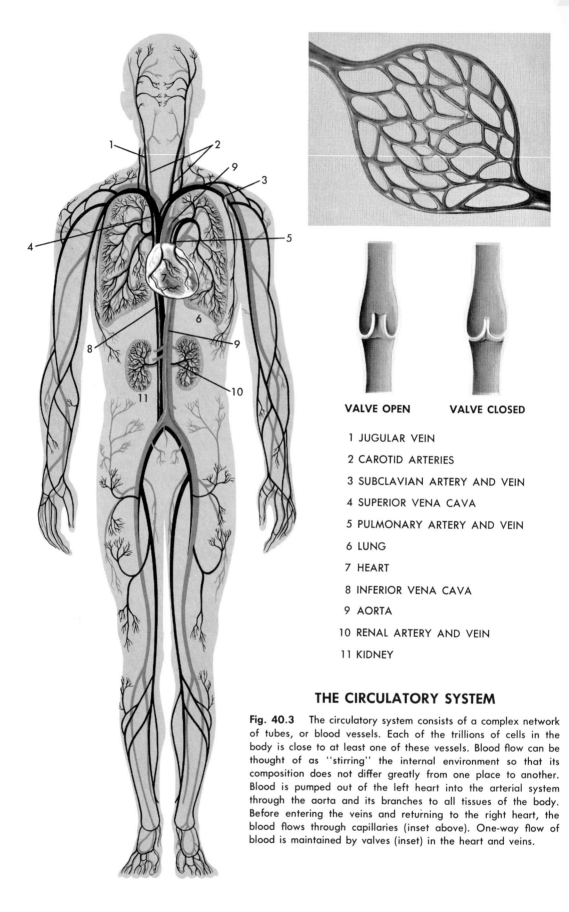

VALVE OPEN   **VALVE CLOSED**

1. JUGULAR VEIN
2. CAROTID ARTERIES
3. SUBCLAVIAN ARTERY AND VEIN
4. SUPERIOR VENA CAVA
5. PULMONARY ARTERY AND VEIN
6. LUNG
7. HEART
8. INFERIOR VENA CAVA
9. AORTA
10. RENAL ARTERY AND VEIN
11. KIDNEY

THE CIRCULATORY SYSTEM

Fig. 40.3 The circulatory system consists of a complex network of tubes, or blood vessels. Each of the trillions of cells in the body is close to at least one of these vessels. Blood flow can be thought of as "stirring" the internal environment so that its composition does not differ greatly from one place to another. Blood is pumped out of the left heart into the arterial system through the aorta and its branches to all tissues of the body. Before entering the veins and returning to the right heart, the blood flows through capillaries (inset above). One-way flow of blood is maintained by valves (inset) in the heart and veins.

We do not know what causes this built-in rhythm of the heart. In a normal heart, the various parts do not beat at different times and with independent rhythms. This is because there is an excellent conduction system in the heart. The first piece of tissue that becomes excited generates an action potential. The action potential is then quickly transmitted to all parts of the heart, exciting the entire tissue. As a result, the entire heart beat is coordinated, pumping with maximum force and sending the blood surging into the arteries.

The Pathway of Blood Flow

Blood leaving the left half of the heart enters a single artery, the **aorta** (Fig. 40.3). The aorta branches like a tree into smaller and smaller arteries. Each of the smaller arteries carries blood to the different tissues of the body. Within the tissues, the smallest arteries empty into a network of even smaller tubes—the **capillaries.**

Capillaries can be seen only with the aid of a microscope. They have very thin porous walls which are easily penetrated by small molecules like sugars, oxygen, carbon dioxide, salts, and amino acids. These molecules simply diffuse through the capillary walls and enter the fluids surrounding the cells. Oxygen, for example, is consumed so rapidly by the cells that the concentration of oxygen in the cells is low compared with the concentration of oxygen in arterial blood. As a result, oxygen diffuses from the capillary vessels to the cells.

After passing through the capillaries, blood enters the **veins.** Small veins merge into larger veins which carry the blood back to the right side of the heart (Figs. 40.2 and 40.3). From here the blood is pumped by the right side of the heart into the pulmonary artery which carries the blood to the lungs. In the lung capillaries the blood recharges with oxygen and rids itself of excess carbon dioxide. It then collects in the pulmonary vein and is carried back to the left side of the heart.

Figure 40.3 shows the heart in more detail. In addition to being divided into a right and left side, each side is subdivided into two chambers—the atrium and the ventricle. At rest the atrium serves as a storage depot for blood returning from the veins toward the heart. When the heart begins its beat the atruim contracts first. Although it may help fill the ventricles with blood, it plays a very minor role in the pumping of blood. A moment later the ventricles contract sending the blood into the arteries. The ventricles contribute most of the pumping action of the heart. The right ventricle is responsible for pumping blood through the lungs; the left ventricle is responsible for pumping blood through the rest of the body.

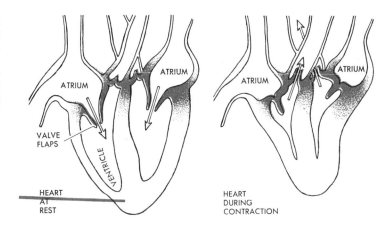

Fig. 40.4 Blood flows into the heart from the veins when the heart is at rest. When the heart contracts, blood is forced into the arteries. Valves in the heart prevent blood from flowing in the reverse direction.

Drawn after Carlson, Johnson, and Cavert

One-Way Flow

When the heart muscles contract, why isn't blood squirted backward into the veins as well as forward into the arteries? And when the heart relaxes, why doesn't blood flow into it from the veins and arteries?

Imagine that the heart is transparent and that we can watch the action of blood flowing in and out of it (Fig. 40.4). First we see the heart at rest and notice **valve flaps** between the atrium and ventricle (the **A-V** valves) on each side of the heart. Blood is pushing down on them from above. Below the flap there is very little pressure, because the heart is relaxed. This means that the pressure of the blood from above pushes the flaps open and fills the ventricle.

Now the muscles in the walls of the heart start pumping. They begin contracting and squeezing the blood in the ventricles. This is the time when we might expect blood to flow back into the veins through which it entered, but as we watch, we notice something happening to the valve flaps. The pressure below the flaps is now much greater than that above. This forces the flaps of the valve toward one another until they close up tight. Blood cannot push its way back into the atria. Instead, it is forced into the arteries. The opening into the arteries is guarded by two other sets of valves, located between the ventricles and their arteries. When the heart was at rest, these valves were closed tight. The pressure in the arteries was greater than the pressure in the ventricles; this kept the valves shut and prevented blood backing up from arteries into the ventricles. (Notice that the flaps of these valves do not hang down into the ventricle like the A-V valves. Instead, they point upward into the arteries.) When the heart makes its pumping stroke, the high pressure of the blood in the ventricle pushes on the flaps of the valves

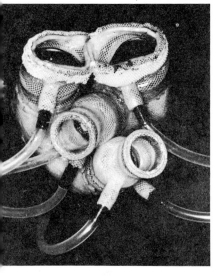

Fig. 40.5 Total artificial hearts, like the one shown here, are still in the early experimental stage.

Photograph by Harold Friedman, work by Dr. Adrian Kantrowitz

guarding the arteries and forces them open. Blood now flows through the open valves because the pressure of blood in the ventricle is now greater than the pressure in the artery. Each time the valves open and close they produce a sound. These are the sounds you hear when you listen to your heart beat.

Cardiac Output

The amount of blood pumped by the heart is staggering. When you are at complete rest, your heart pumps enough blood to fill four automobile gasoline tanks each hour. Let's break this down into more precise figures. During rest, the heart beats about 70 times per minute. During each beat, each side of the heart pumps roughly 70 ml of blood. The amount of blood pumped during each minute would then equal 70 ml per beat × 70 beats per minute, or 4900 ml per minute (almost five liters, or $5\frac{1}{4}$ quarts, per minute).

The amount of blood pumped by each side of the heart during each minute is called the **cardiac output.** During activity, the cardiac output changes. When you exercise strenuously, your cardiac output may rise to as much as 25 liters per minute. When a trained athlete exercises, his output may go as high as 40 liters per minute.

The cardiac output is controlled in part by nerves of the autonomic nervous system. Impulses carried by sympathetic nerves to the heart tend to increase cardiac output by increasing both the rate of the heart beat and the strength of each beat. Impulses carried by the parasympathetic nerves to the heart tend to decrease cardiac output by slowing the rate of heart beat.

BLOOD PRESSURE AND BLOOD FLOW

Blood flows from the arteries through the capillaries to the veins because the pressure of blood in the arteries is higher than the pressure of blood in the veins. By the time the blood enters the veins, its pressure has been reduced to only a few millimeters of mercury (mm Hg). This creates a problem because the blood still has a long distance to travel to return to the heart. Furthermore, the veins expand, so blood has a tendency to pool in them.

These difficulties are overcome by a system of valves in the veins themselves, which work in the following way. Any movement you make, particularly a muscular contraction, compresses portions of your veins (Fig. 40.6). The compression squeezes

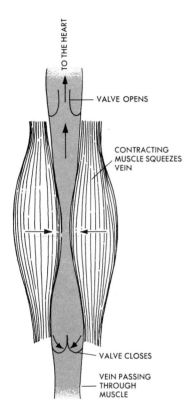

Fig. 40.6 Contraction of muscle squeezes the veins, forcing blood toward the heart. A closed valve prevents the blood from flowing backward.

blood out of those portions of the vein and into other portions. The valves in the veins are arranged like the heart valves and allow the blood to travel in one direction only. (In Chapter 42 we shall see how respiration also aids venous return. Each time you breathe in, blood is drawn into the vessels in your chest from more distant parts of your body.)

All main arteries seem to have about the same average pressure throughout their length. If we think of them as a group, we might say that they form a blood reservoir maintained at high pressure by activity of the heart (Fig. 40.7). The extent to which body tissues draw blood from this reservoir is governed by the resistance offered by the terminal arteries which lead into the tissues. These terminal arteries are called **arterioles.**

From moment to moment the pressure in the arterial reservoir pulsates. The pulsation is caused by the heart beat. Each time the heart contracts it thrusts blood into the arteries. This expands the arterial reservoir by stretching the elastic walls of the arteries. An increase in pressure results. When the heart is relaxed, no blood enters the arteries, but some leaves them through the arterioles. As a result, the amount of blood in the arterial reservoir is reduced and the pressure tends to fall. Thus, with each contraction of the heart the pressure tends to rise, and during relaxation of the heart the pressure falls. In a healthy, resting person the blood pressure rises to about 120 mm Hg during each contraction of the heart. It falls to about 80 mm Hg each time the heart relaxes. The actual pressure oscillates between these two figures—80 and 120 mm Hg. (Figure 37.4 shows how arterial pressure could be measured.)

This pulsation can be felt at several points on the surface of the body—the inner wrist, ankle, and temple—where arteries are close to the surface. The average pressure is usually about 100 mm Hg, and it is this average pressure that we shall be concerned with. The amount of blood in the arteries determines to a large extent the average pressure. If the amount is large, this means that the arterial walls are stretched and the pressure is high. On the other hand, when the amount is low the pressure is also low.

Fig. 40.7 Average blood pressure in the arteries depends on the inflow and outflow of the arterial reservoir.

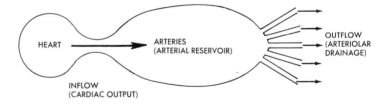

The average pressure depends on the inflow and outflow of blood to and from the arterial reservoir. As Fig. 40.7 shows in a schematic way, blood comes in from the heart and leaves through the arterioles. The arterioles are tiny narrow vessels, consequently they offer a good deal of resistance to the passage of blood. Further, their walls are surrounded with smooth muscle. When the muscle contracts, the arterioles constrict and offer even more resistance. Thus, two things govern the amount of blood in the arterial reservoir, and hence the average pressure of the arterial reservoir: 1. The size of the arteriolar passageway can change and thereby alter the resistance to blood flowing out of the reservoir. 2. The cardiac output (inflow to the reservoir) can also change. As a general rule, the amount of blood in the reservoir, and hence the average arterial pressure, increase with an increase in cardiac output (increased inflow), and with an increase in arteriolar resistance (decreased outflow). On the other hand, the pressure decreases if the cardiac output or the arteriolar resistance decrease. This can be summarized by a simple expression:

average arterial pressure = cardiac output × arteriolar resistance

We have already seen that muscles of the heart are controlled by the autonomic nervous system. What controls the smooth muscle in the arterioles? Normally they are partly contracted. When in this state, they can either relax, or contract more. The smooth muscles which control the size of the arterioles are controlled by two things: 1. *Sympathetic nerve impulses*—the smooth muscles are partially contracted because they are normally bombarded by excitatory sympathetic nerve impulses. Increasing the frequency of these impulses causes further contraction, and the arterioles constrict. Decreasing the frequency allows the muscles to relax, and the arterioles dilate. 2. *Metabolic products* such as acids and carbon dioxide also cause a dilation of the blood vessels.

Regulation of Local Blood Flow by Metabolic Products

The regulation of local blood flow by metabolic products provides us with a striking example of a **homeostatic** mechanism. First, let us show that metabolic products affect blood flow by performing a simple demonstration. Suppose a tourniquet is tied around your arm so tightly that it closes off all the blood vessels lying under it. The blood flow to your arm below the level of the tourniquet is stopped completely. When the tourniquet is released, the blood flow to the lower part of your arm does not simply go back to normal; it exceeds the

Fig. 40.8 This negative feedback loop shows how accumulation of metabolites in a tissue helps regulate the tissue's blood supply.

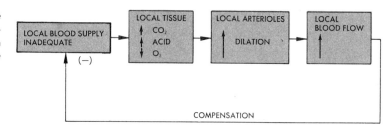

normal level. Although the average arterial pressure does not change significantly, the blood vessels in the lower part of your arm have enlarged a bit, thus allowing more than the normal amount of blood to flow into the stricken region when the tourniquet is released. This response is advantageous to the surrounding tissue in the following way. During the period of cutoff, metabolism continued; waste products were building up since there was no blood flow to carry them away. Oxygen was badly needed since none was being delivered. The increase in blood flow after the tourniquet was removed enabled the cells to meet their metabolic needs. The question now is "how do the blood vessels dilate?"

If we cut the nerves leading to the deprived tissues, we could repeat the demonstration and get exactly the same results. This tells us that the nerves are not involved in the blood vessel dilation of this experiment. If we take blood from that part of the arm where the circulation has been cut off, and then inject it into some other part of the body, we get a startling result. *The vessels in the area of the injection dilate;* therefore, some substance produced by the tissue deprived of fresh blood seems to be responsible for the dilation. It turns out that many products of normal metabolism bring about the action. Carbon dioxide and acids are among them. Thus, blood vessel dilation stimulated by metabolic products during times of intense activity is an automatic safeguard assuring that active tissues receive an adequate supply of blood. The local environment of these tissues is restored and normal metabolism proceeds. The feedback loop describing this regulation of local blood supply is illustrated in Fig. 40.8.

Regulation of Blood Pressure

The action of **metabolites** (metabolic products) on the arteriolar walls provides us with an automatic mechanism for increasing the blood flow through an active tissue. But this works only if the blood pressure in the arterial reservoir is high enough to drive the blood. The mechanism fails if the blood pressure drops too low. On the other hand, an extremely high blood pressure also has disadvantages. If the blood pressure is too high, the heart has to pump harder to expel its contents each time it beats. Also, a high pressure may be transmitted to the capillaries where it will force fluid out into the tissue spaces. As a result, the tissues swell. Thus, if the blood pressure is either too high or too low, the constancy of the internal environment and survival is threatened. The regulation of blood pressure, so that it remains reasonably constant despite the changing demands of the tissues, provides us with another illustration of the principle of homeostasis.

Actually, we know of more than one mechanism that regulates blood pressure. Of these, one of the most important involves a reflex which has its receptors located in the walls of the aorta and in the walls of branches of the aorta leading to the brain. These branches are called the **carotid arteries** (Fig. 40.9). The receptors are sensitive to stretch. If, for example, the blood pressure is increased, the walls of the aorta and the carotid arteries are stretched, and so are the receptors. Stretching the receptors increases the frequency of impulses on sensory nerves leading from these receptors to a reflex center in the lower part of the brain (the **medulla**). This results in an inhibition of the supply of sympathetic nerve impulses to the heart and the arterioles, together with an excitation of the parasympathetic supply to the heart. In turn, this leads to decreased cardiac output, and dilation of the arterioles, which reduces the resistance to blood flow out of the arteries. Both the decreased cardiac output and the decreased resistance reduce the elevated blood pressure, bringing it back down toward normal. This feedback loop for regulating blood pressure is illustrated in Fig. 40.10.

If the blood pressure is reduced for some reason, the opposite response occurs. The frequency of impulses on sensory nerves leading from the aorta and carotid arteries is reduced. The sympathetic nerve supply to the heart and arterioles is excited.

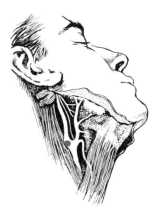

Fig. 40.9 The red dot shows the location of stretch receptors in the wall of the carotid artery. These stretch receptors respond to changes in blood pressure and take part in the feedback loop in Fig. 40.10.

Fig. 40.10 The negative feedback system below shows how the heart and arterioles compensate for an increase in blood pressure. If the blood pressure were *decreased*, the heart and arterioles would again compensate for the change. Can you reconstruct the diagram to show what happens?

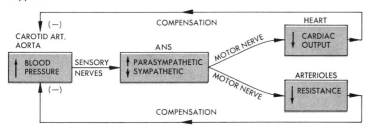

The parasympathetic supply to the heart is inhibited. The cardiac output and arteriolar resistance increase. The blood pressure is raised back toward normal.

The importance of this homeostatic reflex becomes apparent, for example, when a person or animal is injured and loses blood. As blood is lost, the total volume of fluid in the circulatory tree is reduced and the pressure falls. The fall in pressure is minimized by the reflex increase in cardiac activity and blood vessel constriction, which are initiated by receptors in the carotid arteries and aorta. Other compensating adjustments that occur in response to blood loss are revealed when we study the properties of blood itself. These will be discussed in Chapter 41.

SUMMARY

The organs of the body are specialized to perform certain tasks. Each organ produces materials that are carried by the circulating blood to all cells of the body. The blood flow to various organs is adjusted by **homeostatic** mechanisms, with the result that active tissue is provided with an adequate supply of blood.

The circulatory system consists of the heart and a complex network of blood vessels. The direction of blood flow from arteries through the capillaries to the veins is determined by a difference in pressure. The pressure is relatively high in the arteries and low in the veins. The term **blood pressure** refers to the pressure in the main arteries. This pressure pulsates, rising with each beat of the heart and falling during relaxation of the heart. The average pressure in the major arteries depends on two things: **cardiac output** and **arteriolar resistance.**

The cardiac output is controlled by the nerves of the autonomic nervous system. The resistance offered by arterioles is determined by smooth muscles, which control the diameter of the arterioles. These smooth muscles are in turn controlled by *sympathetic nerve impulses* and *metabolic products.* The effect of metabolic products helps to *regulate the local blood supply* so that blood flow through active tissues is increased.

Many mechanisms are involved in the regulation of blood pressure. One of the most important involves a reflex with receptors located in the walls of the **aorta** and in the **carotid** arteries. When the pressure is increased in these areas, the receptors are stimulated so that they send more impulses to the brain. This results in an inhibition of sympathetic nerves leading to the heart and arterioles; it also results in an excitation of parasympathetic nerves leading to the heart. Cardiac output is then decreased and the arterioles are dilated. Both of these factors reduce the elevated blood pressure back toward normal. When the blood pressure drops below normal, the opposite responses occur.

FOR THOUGHT AND DISCUSSION

1. How will the pressure in the arteries compare to the pressure in the veins if the heart stops beating for a long time?
2. If the sympathetic nerves that carry impulses to the hand are cut, what effect will this have on blood flow to the hand? Does it increase or decrease? Explain.
3. If the sensory nerves that carry impulses from the aortic and carotid artery stretch receptors are cut, how will blood pressure be affected? Explain.
4. Heart failure is sometimes caused by damaged heart valves. Sometimes the aortic valve (between the left ventricle and aorta) is damaged and does not close properly. Why would this be a disadvantage? What physiological adjustments might partially overcome the disadvantages?
5. When noradrenalin is placed on an isolated heart, it beats faster. When noradrenalin is injected into an intact animal, the heart rate may increase slightly at first, but very soon after it decreases! Can you offer an explanation for these results? (HINT: consider the inter-relations of noradrenalin, blood pressure, and heart rate.)

SELECTED READINGS

Carlson, A. J., V. Johnson, and H. M. Cavert. *The Machinery of the Body*, 5th Ed. Chicago: University of Chicago Press, 1961.

Harvey, W. *Anatomical Studies on the Motion of the Heart and Blood.* Springfield, Illinois: Charles C. Thomas Publisher, 1931.

Kilgour, F. G. and W. Harvey. *Scientific American* (June 1952).

Lillehei, C. W. and L. Engel. "Open-heart Surgery," *Scientific American* (February 1960).

Scher, A. M. "The Electrocardiogram," *Scientific American* (November 1961).

Scholander, P. F. "The Master Switch of Life," *Scientific American* (December 1963) (*Scientific American* reprint #172).

Wiggers, C. J. "The Heart," *Scientific American* (May 1957) (*Scientific American* reprint #62).

Zweifach, B. W. "The Microcirculation of the Blood," *Scientific American* (January 1959) (*Scientific American* reprint #64).

41 BLOOD

In the last chapter we saw some of the ways in which circulation of the blood is controlled. Let us now take a close look at the circulating fluid itself—blood. An average sized adult has about five liters of blood. As blood travels through the circulatory tree, it carries food to the tissues and carries away waste products. Some food stuffs and waste products are simply dissolved in the blood, much as sugar is dissolved in a cup of tea. However, blood is not a simple watery solution. It consists of cells suspended in a clear fluid called **plasma.** These cells, along with some of the proteins dissolved in the plasma, give the blood a number of remarkable properties. Some of these properties are revealed whenever a severe **hemorrhage** (bleeding) occurs.

Soon after bleeding starts, the leak in the cut blood vessel is sealed off with blood that seems to harden, or **clot,** near the

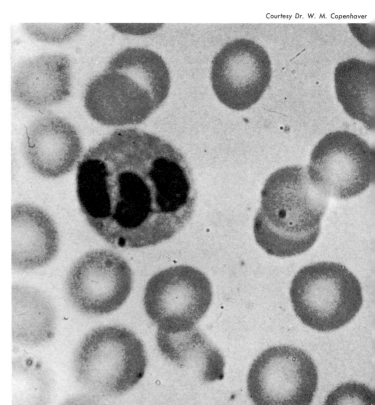

Fig. 41.1 Blood is not a simple watery solution, but is made up of cells suspended in a clear plasma fluid. Both the cells and plasma are specialized to perform a variety of tasks.

Courtesy Dr. W. M. Copenhaver

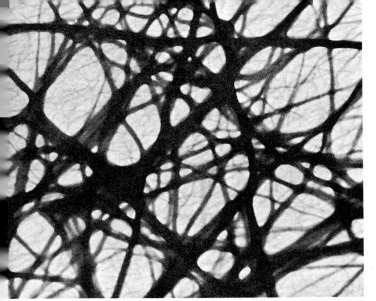

Courtesy Keith Porter

Fig. 41.2 Blood clots are formed by networks of long fibers of fibrin. The network shown in the electronmicrograph above has been magnified 46,000 times.

wound. During the next few hours or so, quite a bit of fluid bathing the tissue cells of the body (extracellular fluid) enters the blood stream. This replaces some of the fluid that was lost during the hemorrhage. Within a few days the lost plasma proteins are replaced. Finally, after a few weeks the number of circulating blood cells returns to normal. In other words, new blood cells have been produced to replace those lost through bleeding. All of these responses help to preserve or restore the internal environment; they are homeostatic responses.

Preventing Loss of Blood

If you cut your hand, it will bleed rapidly at first. Then the bleeding slows down. If the cut is not too deep, the bleeding will probably stop in a few minutes. What causes the blood to stop flowing? The bleeding stops because blood clotted near the wound and sealed the leak. You can demonstrate blood clotting simply by collecting fresh blood in a test tube. If you tip the tube upside down immediately, the blood will flow out. However, if you wait a few minutes before tipping the tube, the blood will not flow out. It is as though you poured some hot gelatin into a test tube and allowed it to cool and set.

We know that as the blood circulates through the body it is not clotted. If the blood does clot, it clogs up important blood vessels, say to the brain, and the results are disastrous. What prevents blood from clotting as it circulates through the body, and yet allows it to clot at the site of a wound?

If we examine a blood clot under a microscope, we find that

Fig. 41.3 Three major steps in blood clot formation are shown here. The substances in color are normally present in plasma (or tissues). Thromboplastin and Ca^{++} are necessary for Step 2 to proceed. Thrombin is required for Step 3.

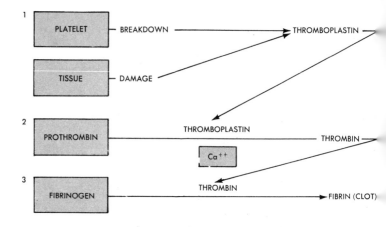

it is composed of a number of threadlike structures entangled in a network (Fig. 41.2). This fibrous network is made up of an insoluble protein called **fibrin**. Blood cells become trapped in the meshes of this network. If we separate the cells from the plasma we still get clot formation. Clotting is a property of plasma and is independent of the cells. Where does fibrin, which forms the clot, come from?

Fibrin is formed from a protein called **fibrinogen** which is dissolved in the plasma. However the formation of fibrin takes place only under special circumstances. A blood clot that forms at a wound, or in a glass test tube, is triggered by the liberation of substances called **thromboplastin**. Thromboplastin is liberated either by injured cells or by the rupture of tiny fragile cell fragments called **platelets**, which are normally present in the blood. The presence of thromboplastin starts a series of reactions producing fibrin from fibrinogen. A simplified outline of this series of reactions is given in Fig. 41.3. You will see from the figure that an important intermediary step in the series of reactions is the conversion of a substance called **prothrombin** into **thrombin**. This step requires free calcium (Ca^{++}), which is normally contained in the blood.

Although most of the substances needed for clotting are normally present in the circulating blood, they are not activated. That is why circulating blood does not clot. It is only when thromboplastin is liberated, and thrombin is formed, that we get clotting. So long as the platelets remain intact they do not rupture and liberate the thromboplastin, which begins the process of clotting. Clot formation occurs in a test tube because the fragile platelets rupture upon contacting glass surfaces. It is possible to prevent clotting by interrupting any one of the steps outlined in Fig. 41.3. Blood banks commonly remove Ca^{++} from the blood, or they block the action of thrombin, to prevent clotting of stored blood. Successful clotting depends on many other substances besides the major ones mentioned here. Some of the substances have only recently been discovered, and quite likely still others remain to be discovered.

Capillary Fluid Exchange

Substances involved in blood clotting make up only a small fraction of the total protein dissolved in the plasma. The entire pool of plasma protein plays an important role in the transfer of fluid across the capillary walls. Fluid can move through the capillary walls in two directions. It can pass from the blood stream into the fluid that bathes body tissues, or it can leave the tissues and enter the blood stream. The pressure of the blood in the capillaries, and the concentration of protein in the blood, govern which way the fluid goes.

Two forces normally balance each other and prevent any net movement of fluid into or out of the capillaries:

1. **BLOOD PRESSURE**: We know that the pumping action of the heart sends blood into the arteries under strong pressure. Although some of this pressure is lost in the arterioles, it is still about 35 mm Hg when it enters the capillaries. By the time the blood leaves the capillaries the pressure has dropped to about 15 mm Hg. Thus, the average pressure in the capillaries is about 25 mm Hg. This pressure is strong enough to force fluid out of the capillaries—which would mean that all of the fluid would drain into the tissue spaces—but a second force prevents such drainage. This second force is osmotic pressure.

2. **OSMOTIC PRESSURE:** As you saw in Chapter 37, when one region has a higher osmotic pressure than another, it tends to draw fluid toward it. If we examine the chemical composition of the extracellular tissue fluids and of the plasma, we find that they are almost identical, except for large protein molecules found in the plasma. The capillary walls are easily permeable to all of the substances dissolved in the blood except for these huge proteins. The proteins keep the osmotic pressure of the blood plasma in the capillaries higher than the osmotic pressure in the tissues. This means that fluid is drawn into the capillaries by osmotic pressure. The magnitude of the osmotic pressure exerted by the plasma proteins is about 25 mm Hg.

The two forces—blood pressure, which tends to push fluid out of the capillaries, and osmotic pressure, which tends to draw fluid into the capillaries—are thus kept in balance. So long as these two forces are equal, there is no net fluid movement. This means that the tissues do not swell or shrink. Actually, at the

Fig. 41.4 This diagram shows why there is no net exchange of fluid (protein-free plasma) between capillaries and surrounding tissue spaces. Because blood pressure at the arterial end of the capillary (colored arrow) is greater than the osmotic pressure (shorter black arrow), fluid leaks out of the capillary into the tissue space. At the venous end of the capillary the osmotic pressure is greater than the blood pressure in the capillary, hence there is movement of fluid into the capillary. The amount of fluid lost and regained by the capillary and tissue space is equal, hence a steady state is maintained.

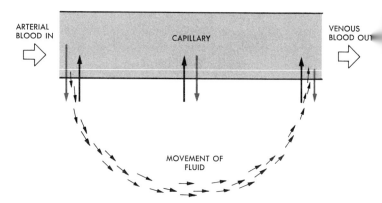

arteriole end of the capillaries the blood pressure is a little higher than the average, so some fluid does seep out into the tissue spaces. However, at the venous end of the capillaries, the blood pressure is lower than the osmotic pressure, and an equivalent amount of fluid tends to seep back into the capillaries. This means that there is no net movement of fluid; the fluids in the tissue space are in a **steady state.** (See Fig. 41.4.)

During hemorrhage, when a large volume of blood is lost, this balance is upset. The volume of blood goes down, which means that the pressure also drops. In turn, this means that the force pushing the blood out of the capillaries is weakened. However, the osmotic pressure—the pressure drawing fluid from the tissues into the capillaries—does not drop. This is because immediately after the hemorrhage the *concentration* of plasma proteins in the circulating blood remains the same. Although there is less blood circulating, it is just as concentrated as before the hemorrhage.

The force drawing fluid from the tissues into the capillaries is now greater than the force pushing fluid out of the capillaries into the tissues. Thus, fluid begins to move from the tissue spaces into the capillaries. This fluid helps to make up for the volume of blood lost during hemorrhage. When the volume has built up toward normal, the plasma proteins are then less concentrated. This means that the osmotic pressure drops so that a balance is again brought about between the blood pressure and the osmotic pressure. Net fluid transfer across the capillaries again comes to a virtual standstill.

During inflammation, and in some allergic reactions, the walls of the capillaries become leaky and permit plasma proteins to pass through. The proteins leak out into the tissue spaces and carry fluid with them. In other words, the osmotic pressure difference across the capillary wall drops and, since the blood pressure

emains the same, it forces fluid into the tissue spaces. The result
is a swelling of the tissues.

The Lymphatic System

Normally, some protein does leak out very slowly from the capillaries to the tissue spaces. Although the leak is slow, if continued indefinitely the difference in protein osmotic pressure across the capillary membrane would vanish and the blood pressure would force fluid from the circulatory system into the tissue spaces. This does not happen because the small amount of protein that leaks into tissue spaces is continually drained off by a system of thin vessels called **lymphatics** (Fig. 41.5).

The lymphatics originate in nearly all tissue spaces. They are very tiny tube structures which join into larger and larger vessels. They finally lead to the neck region where they drain into the venous system (Fig. 41.5). The lymphatic system contains one-way valves similar to those in the veins. The transport of fluid through the lymphatic system is in many ways similar to the transport of fluid in the venous system. But flow through the lymph system is very slow compared with flow through the circulatory system. Nevertheless, it is sufficient to keep the fluid in the tissue spaces in a steady state, draining off the small amount of protein that may leak out of the capillaries, and draining off other foreign bodies that enter the tissue spaces. Fig. 41.6 shows an example of tissue swelling resulting from deficient lymphatic drainage.

All along the lymph system are a number of enlargements called **nodes.** These important structures are responsible for the filtering out of foreign particles. Specialized cells, called **phagocytes,** are able to engulf and remove foreign particles from the lymph fluid, thus preventing them from entering the general

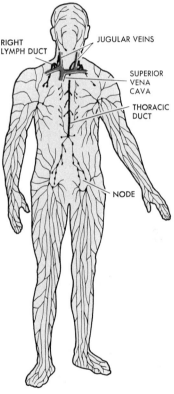

Fig. 41.5 The lymphatic system is constantly but slowly draining fluids (and protein) from the tissue spaces and carrying them to the circulatory system. Enlargements called *nodes* filter out and destroy foreign particles in the fluids. The lymphatic fluid enters the veins at the two sites in the neck region.

Armed Forces Institute of Pathology

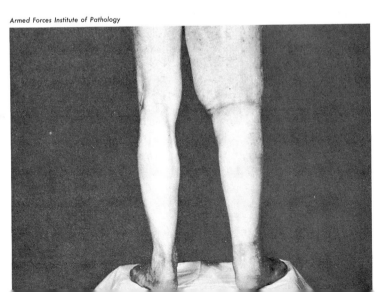

Fig. 41.6 When the lymphatic system fails, the body tissues swell. Note the swelling which has resulted from impaired lymphatic drainage in the right leg.

circulatory system. In addition, the lymph tissue (as well as white blood cells, spleen, and bone marrow) produces substances called **antibodies.** Antibodies also play a role in the control of foreign particles.

Antibodies

Whenever bacteria or foreign proteins find their way into the body they stimulate the production of antibodies. Particular kinds of antibodies react with particular kinds of invading particles, neutralizing or destroying the invading particles.

Some of the plasma proteins (**globulin**) are antibodies. They react with the invading particles (called **antigens**) in such a way that the antigen no longer has any harmful action. This inactivation may take any one of several forms. It may make the particle insoluble, or it may make it more susceptible to engulfment (**phagocytosis**) and subsequent digestion by other cells. Or if the protein (antigen) with which the antibody reacts is attached to a cell, it may cause these cells to clump together or to disrupt. These antibody reactions are responsible for immunity to many diseases.

These same reactions are also one of the chief difficulties that stand in the way of transplanting tissue—a skin graft, for example—from one person to another. Apparently, when tissue is transplanted the donor cells are recognized by the host as being "foreign" to the host's body. Antibodies specific for the donor cells are then produced by the host and react with the donor cells and destroy them.

The problem of antibody reactions must be dealt with whenever blood transfusions are made. If a person is given the wrong type of blood, antibodies in the plasma of the recipient react with antigens of the donor's blood cells. When they do, the cells of the donor's blood clump together. This clumping, called **agglutination,** can lead to severe illness or death of the person receiving the blood transfusion.

Differing Blood Types

There are several types of antigens that may be present on the donor's red blood cells. Such antigens form the basis for classification of blood types. Two of the antigens are called simply **A** and **B**. Red blood cells may have either A or B, both, or neither antigen. If your red cells contain both antigens, then you have type **AB** blood. If they contain only the A antigen, then you have type **A** blood; if only the B antigen, type **B**. If

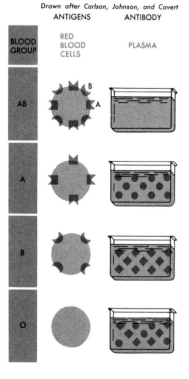

Fig. 41.7 Schematic diagram of blood cell antigens and plasma antibodies that correspond to different blood group types.

Drawn after Carlson, Johnson, and Cavert

they do not contain either of the antigens, then you have type O blood.

A person whose blood type is AB does not have any antibodies for either the A or the B antigen in his plasma. If he did, these antibodies would react with the antigens and cause the cells to agglutinate, and he would not survive. Similarly, a person whose blood type is A cannot have the A antibody in his plasma, but he may, and usually does, have the B antibody. A person whose blood type is B has only the A antibody. And, finally, the person whose blood type is O usually has both the A and the B antibodies in his plasma. No reaction occurs here because there is no antigen on his red blood cells.

Suppose we want to transfuse blood into a person whose type is B. We must assume that he has A antibodies in his plasma. This means that he cannot receive blood from anyone who has the A antigen on his red cells. In other words, he cannot receive blood cells from a donor of type A or of type AB. However, he can receive blood from a person with the same blood type (type B) or from a donor whose blood type is O. Using the same arguments, we see that a person with type A blood can receive cells from another person whose blood type is either A or O. A person whose blood type is AB has no antibodies in his plasma, so he can receive blood from any of the types A, B, AB, or O. Finally, a person whose blood type is O has both antibodies in his plasma. Consequently, he can receive blood only from a type O donor. The A and B antigens are not the only ones carried by red blood cells. Sometimes, although less frequently, reactions that do not involve A and B antigens occur. Figure 41.8 shows the clumping reactions that occur when red cells are placed in different plasma types.

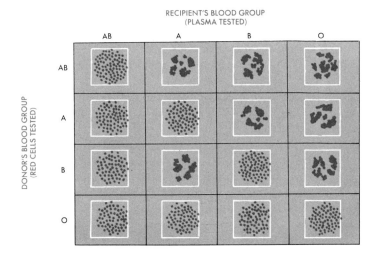

Fig. 41.8 This figure shows reactions that take place when different types of blood are mixed.

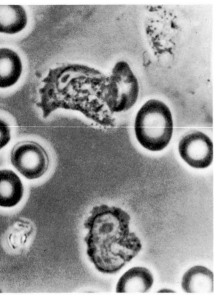

Lester V. Bergman and Associates, Inc.

Fig. 41.9 Some white blood cells are capable of engulfing foreign particles. The process is called *phago-cystosis*.

BLOOD CELLS

There are several kinds of cells in the circulating blood. Those that have little or no color are collectively called white blood cells. The white blood cells play an important role in the removal of foreign particles. They accomplish this in several ways. For example, some white blood cells are capable of engulfing foreign particles or bacteria (Fig. 41.9). In an infected wound, such cells tend to accumulate near the infected region and take up bacteria and dead tissue. Once inside the white cell, the bacteria and dead tissue are broken down. These white cells are limited in the number of particles they can engulf, and eventually they die. Pus that accumulates near a wound consists largely of dead white cells. Other white cells are involved in the production of specific antibodies which react with foreign particles or bacteria and inactivate them.

The red cells in the circulating blood are almost 1000 times more numerous than white blood cells and are responsible for the red color of blood. The red color is due to an iron-containing protein called **hemoglobin,** which makes up about 30 per cent of the weight of the red blood cell. Hemoglobin is very important in the carriage of oxygen, carbon dioxide, and acid by the blood. These matters will be discussed in the next chapter.

SUMMARY

Blood consists of red and white blood cells suspended in a liquid called **plasma.** White blood cells play an important role by removing foreign particles from the blood. Red blood cells contain **hemoglobin,** a protein that is very important in the transport of oxygen, carbon dioxide, and acid by the blood. Proteins that are dissolved in the plasma play an important role in blood clotting, in the exchange of fluid across capillary walls, and in antigen-antibody reactions.

Blood clots that form near a wound are made from a fibrous network of an insoluble protein called **fibrin.** The blood clot forms as a result of a series of reactions initiated by the liberation of **thromboplastin** from injured tissue or **platelets.**

The fluid exchanged between the capillaries and the tissue spaces is governed by two forces: capillary blood pressure and osmotic pressure of the plasma proteins. The blood pressure tends to force fluid out of the capillaries and into the tissue spaces; osmotic pressure exerted by the plasma proteins tends to move fluid in the opposite direction. Normally, these two forces balance one another.

When **antigens** gain entry into the body, they stimulate the production of **antibodies.** The antibodies react with the antigens and destroy or inactivate the antigens. These antigen-antibody reactions are responsible for immunity to many diseases. Incompatible blood transfusions result from antigen-antibody reactions if the antibodies in the plasma of the recipient react with the antigens of the donor's blood cells and cause them to **agglutinate.**

FOR THOUGHT AND DISCUSSION

1 During an accident, a large blood vessel is cut and bleeding is severe. What are the threats to survival? Outline the physiological responses that will help the person survive.

2 In some forms of heart failure the left side of the heart is the weakest and fails to perform properly while the right side continues to pump blood into the lungs with near normal vigor. Under these conditions fluid flows from the lung capillaries into the spaces of the lungs, resulting in a condition called *pulmonary edema.* Explain. (HINT: What do you think happens to the blood pressure in the lung capillaries under these conditions?)

3 During starvation, there may be inadequate formation of plasma proteins. Would you expect tissue spaces to swell or shrink?

4 A clever student, who knows he has blood type B, is given eight red blood cell suspensions and the *serum* (blood plasma in which fibrinogen has been removed) from each sample. He manages to identify each of the four blood types present. To test these samples, the only materials he has at his disposal are his own blood cells and his own serum. How did he make the identifications? Why was serum used instead of plasma?

SELECTED READINGS

Burnet, Sir McFarlane. "The Mechanism of Immunity," *Scientific American* (January 1961) (*Scientific American* reprint #78).

Carlson, A. J., V. Johnson, and H. M. Cavert. *The Machinery of the Body,* 5th Ed. Chicago: University of Chicago Press, 1961.

Guyton, A. C. *Textbook of Medical Physiology,* 3rd Ed. Philadelphia: W. B. Saunders Co., 1966.

Mayerson, H. S. "The Lymphatic System," *Scientific American* (June 1963) (*Scientific American* reprint #158).

Zucker, M. B. "Blood Platelets," *Scientific American* (February 1961).

42 OXYGEN

We live at the bottom of a great sea of air composed mostly of the gases nitrogen and oxygen. As we move through it, we constantly exchange materials with the air. Because this great reservoir of air has been present on the Earth all during the evolution of man, the human body was able to evolve without developing special storage depots for oxygen. Storage depots are necessary when an essential substance, sugar, say, is not always available.

The action of breathing is the major way in which we constantly exchange materials with the air around us, our external environment. Breathing movements keep the air in our lungs thoroughly mixed with the atmosphere. In turn, the air in our lungs is in close contact with blood contained in capillaries of the pulmonary circulation.

The act of breathing is only one link in the chain of events enabling us to use oxygen. We must also consider the problem of how the blood carries enough oxygen to satisfy the demands of our numerous body tissues. If oxygen were simply dissolved in plasma, as sugar dissolves in water, the amount of oxygen carried by the blood would be inadequate. We shall see that red blood cells increase the capacity of blood to carry oxygen 70 times. We must also ask why oxygen is essential, and what are some of the homeostatic mechanisms that adjust the supply of oxygen to the changing demands of body tissues.

How We Breathe

Although air enters our bodies through the nose and mouth, and travels down the windpipe (**trachea**), it is not until it reaches the lungs that the real exchange of gases between our bodies and the external environment takes place. The trachea is divided in two, and each division subdivides several more times. The subdivisions end in tiny dead end sacs called

THE RESPIRATORY SYSTEM

Fig. 42.1 Breathing is the major way in which we constantly exchange materials with the air around us. During inspiration, air moves into the body through the nose and mouth, passes through the trachea and its branches (bronchi and bronchioles, shown at lower right), into the alveoli, (lower left), which are tiny sac-like structures in close proximity to capillaries. It is here that oxygen diffuses into the blood and carbon dioxide diffuses out. The cycle of breathing is completed during expiration, when carbon dioxide and other gases are expelled from the alveoli, travel through the respiratory passages, and are emptied into the atmosphere.

1 TONGUE
2 PHARYNX
3 EPIGLOTTIS
4 LARYNX
5 ESOPHAGUS
6 TRACHEA
7 BRONCHUS
8 LUNGS
9 BRONCHIOLE
10 ALVEOLUS
11 PULMONARY ARTERIOLE

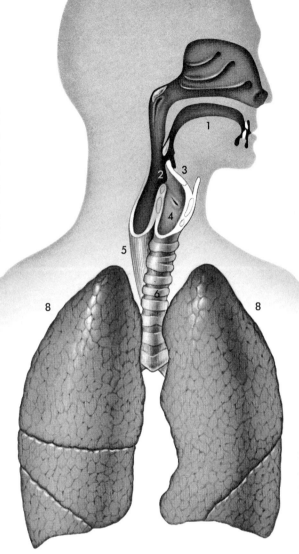
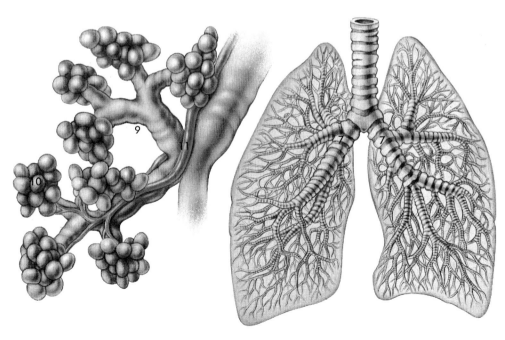

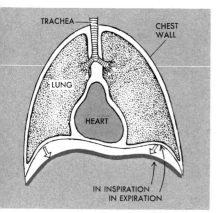

Redrawn from Carlson, Johnson, and Cavert

Fig. 42.2 Breathing involves a rhythmic change in the volume of the chest cage. The diaphragm acts like a piston that moves up and down. During the downward movement (contraction), air is drawn into the lungs; during upward movement (relaxation), air is forced out.

alveoli (Fig. 42.1). Alveoli have very thin walls and a rich supply of blood capillaries. Thus, air in the alveoli is brought in very close contact with capillary blood, enabling a rapid exchange between alveolar air and blood to take place.

The act of breathing maintains a supply of fresh air in the alveoli of the lungs. Each time we inhale, the lungs expand and suck in air. Each time we exhale, the lungs contract and push out air. What is responsible for this rhythmic expansion and contraction of the lungs? We know that the lungs do not contain any muscle tissue. If we open the chest of an animal to look at the lungs, we find that they are collapsed. With the chest open, the lungs remain collapsed despite the fact that the animal continues its breathing movements. The animal's chest cage expands and contracts rhythmically just as though it were breathing normally. Nevertheless the lungs do not move. Apparently the breathing movements are due to muscles which are not directly connected to the lungs.

If we look closely, we find that they are due to movements of muscles connected to the rib cage, and due to a large sheet of muscle called the **diaphragm** which separates the chest cavity from the abdominal cavity.

Figure 42.2 is a diagram of the chest with the right and left lungs inside the chest wall and the diaphragm below. We can think of the chest cavity as a closed box. The lungs are like two balloons, connected at the neck. The neck of the balloon would be the trachea, which allows air to pass into or out of the chest cavity. The diaphragm acts as a piston which can move up or down. When the diaphragm is relaxed, it assumes a curved dome position and rests up against the bottom of the lungs. When it contracts, its curvature decreases and it moves downward, just as a piston would (Fig. 42.2). This creates a partial vacuum in the closed box (and in the lungs). Since the pressure in the lungs is now reduced, air moves from the outside (high pressure) into the lungs (low pressure). This is the intake of breath (**inspiration**). The diaphragm then relaxes. In doing so, it pushes up against the lungs, causing them to expel air to the outside.

Muscles connected to the rib cage also expand and contract the volume of the chest cage. The ribs are attached to the spinal column by hinge-like joints which allow them to move up and down. During relaxation of the rib muscles, the ribs are slanted downward (Fig. 42.3). When the muscles contract the rib cage is pulled upward in a more horizontal position. You can see from the figure that this motion will increase the volume of the chest cage.

Thus, when we breathe in, the rib cage expands because the muscles attached to the ribs pull the ribs upward, and the diaphragm contracts, lowering the floor of the chest cage. As a result, air moves into the lungs. When we exhale (**expiration**) the rib and diaphragm muscles relax, the diaphragm goes back to its original dome shape, pushing up against the lungs, and the ribs drop back into place. As a result, the size of the chest cage decreases and air is pushed out of the lungs. During rest, expiration results from the simple passive relaxation of the diaphragm and rib muscles. However, during strenuous breathing, air can be expelled more forcibly from the lungs by contraction of another set of rib muscles, which pull the rib cage down (to a smaller volume).

What is responsible for the rhythmic activity of the rib and diaphragm muscles? Apparently, the contraction of muscles during respiration is completely controlled by the nervous system. If we cut the motor nerves leading to the diaphragm and ribs of an experimental animal, respiration ceases immediately. Contrast this with the heart, which continues to beat in the absence of any nerve supply.

Nerve impulses that stimulate breathing movements come from regions in the lower parts of the brain (the **medulla**). These regions are called **respiratory centers.** If these centers are destroyed, breathing ceases immediately. If we stimulate certain areas within these centers, prolonged inspiration results. If other areas in the centers are stimulated, prolonged expiration takes place. These respiratory centers send out patterns of nerve impulses to the diaphragm and rib muscles. The result is a regular pattern of inspiration followed by expiration.

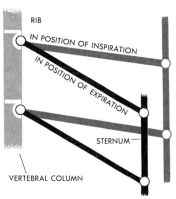

Redrawn from Carlson, Johnson, and Cavert

Fig. 42.3 While the diaphragm expands the rib cage in a vertical direction, the upward movement of ribs expands the cage in a horizontal direction.

Oxygen Transport

The concentration of oxygen in blood entering the lungs is low because the blood has just come from other parts of the body which have extracted substantial amounts of oxygen from it. On the other hand, the concentration of oxygen in the alveolar air is relatively high because it is flushed with fresh air with each breath. As a result, oxygen diffuses from alveolar air (high concentration) to capillary blood (low concentration). In the process, oxygen is dissolved in the blood plasma.

The amount of oxygen that can be dissolved in plasma is severely limited and cannot possibly supply the needs of the body. However, once oxygen enters the plasma, it passes readily into the red blood cells where it becomes loosely attached to **hemoglobin** molecules. Hemoglobin performs a vital function in getting oxygen to the tissues of the body. We can think of each hemoglobin molecule as a large mass containing four sites, or positions, where oxygen can be attached. If the concentration of oxygen in contact with hemoglobin is high, nearly all of the sites will be filled with oxygen. In this case, any particular oxygen molecule can shake loose from the hemoglobin. But, since the oxygen concentration in the surrounding fluid is also high, there is always an oxygen molecule waiting to take the place of one that becomes dislodged. On the other hand, if the oxygen concentration is low, there are very few oxygen molecules available to take the place of any that come loose. In this case, then, many of the hemoglobin sites are unoccupied by oxygen.

Blood leaving the lungs, where it has been in contact with a high concentration of oxygen, has about 99 per cent of its hemoglobin sites filled with oxygen. From the lungs, the oxygen-rich blood enters the left heart and is then pumped to the various body organs. All organs consume oxygen. As a result, the oxygen concentration in the tissues is low. Oxygen diffuses from the plasma directly to the tissue cells. This lowers the plasma oxygen concentration, so additional oxygen is now released by hemoglobin. By far most oxygen consumed by tissue is transported by hemoglobin, not by blood plasma. In order to complete our story, we must at this point ask a basic question: *Why* do the cells need oxygen?

The Role of Oxygen

A cell's need for oxygen is directly linked to the capacity of the cell to use the chemical energy stored in foods. This energy enables muscle contraction, active transport, the building of cellular structure, and other processes to take place. Energy is released from cellular foods (such as glucose) when the foods are broken down into smaller molecules. The most complete and efficient breakdown requires oxygen. The net result can be represented as follows:

$$C_6H_{12}O_6 + 6O_2 \rightarrow 6CO_2 + 6H_2O + energy \begin{cases} ATP \\ Heat \end{cases}$$

$$\text{1 glucose} + \text{6 oxygen} \quad \text{6 carbon} + \text{6 water}$$
$$\text{dioxide}$$

The actual chemical reactions that take place in cells are more complicated than our simple representation indicates. For instance, the glucose is split apart in several steps. In the process, hydrogen atoms are stripped off the glucose molecule, and eventually they are transferred to oxygen, resulting in the formation of water. The role of oxygen is simply to remove the hydrogen. During this process a sizable portion of released energy is again stored—but this time through the formation of ATP (see Chapter 38).

In the absence of oxygen, glucose can still be broken down,

but only partially because there is no longer any substantial means for removal of hydrogen. Instead of forming CO_2 as an end product, a partial breakdown product, **lactic acid** ($C_3H_6O_3$), is formed. The release of energy in this case is less complete and far less efficient. In the absence of oxygen the partial breakdown of glucose yields only about 6 per cent of the ATP that is normally produced during complete breakdown in the presence of oxygen.

During the complete breakdown of glucose in the presence of oxygen, for each molecule of oxygen that is used, one molecule of CO_2 and one molecule of H_2O are formed. In addition to glucose, cells can use other foods (such as proteins or fat) for energy. However, the same pattern emerges: oxygen is consumed, ATP, CO_2, and H_2O are produced. The amount of water formed by this process is no problem because it is very small compared with the large amounts normally contained within the cells. But the formation of CO_2 *does* present a problem. The CO_2 must be removed. The problems of oxygen usage and CO_2 removal, then, are very closely linked.

Transport of Carbon Dioxide

The concentration of CO_2 is higher in the cells than in the capillary blood stream, simply because CO_2 is constantly being produced in the cells. When CO_2 diffuses from the cells into the blood, only a small amount of it (about 9 per cent) reaching the blood is held in simple solution. Another fraction (about 27 per cent) attaches directly to the hemoglobin. The remaining (major) portion (64 per cent) combines with water, forming bicarbonate ions and hydrogen ions.

Hydrogen ions are very reactive. They are responsible for the very strong chemical activity of acids. Normally the amount of hydrogen ions in blood is tiny—about 0.00000004 gram per liter. Any substantial increase or decrease in this concentration is fatal. Each time blood passes through the tissues, it picks up large quantities of carbon dioxide. This then reacts with water, forming **bicarbonate** (HCO_3^-) and **hydrogen** (H^+) **ions.** If the story ended here, each time the blood passed through tissues its free hydrogen ion content would increase intolerably. The free hydrogen ions must be removed in some way.

There are many substances in the blood capable of binding the excess free hydrogen ions. Hemoglobin is one of the most important of these substances. Hydrogen ions, oxygen, and hemoglobin interact in a remarkable way so that just the "right thing" seems to happen at just the "right time." When hydrogen ions combine with hemoglobin, the hemoglobin releases some

Fig. 42.4 This diagram shows chemical interactions that take place in the blood as oxygen and carbon dioxide are exchanged between capillaries and cells. Reactions 2 and 3 take place inside red blood cells. The products HCO_3^- and HHb are carried to the lungs by the venous blood.

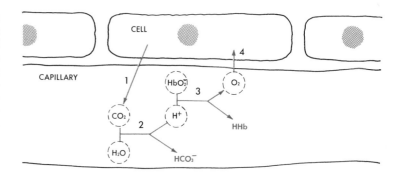

of the oxygen attached to it. Conversely, when oxygen combines with hemoglobin, the hemoglobin releases some of the hydrogen attached to it.

With the aid of Fig. 42.4, let us follow the details as blood passes through a tissue:

1. Carbon dioxide is produced in the cells of the tissue and diffuses into the plasma and into the blood cells.

2. Blood cells contain an enzyme (**carbonic anhydrase**) which accelerates the reaction of CO_2 and H_2O. As a result, the CO_2 combines with H_2O, forming bicarbonate (HCO_3^-) and hydrogen (H^+) ions.

3. Most of the released H^+ is taken up by the combined form of oxygen and hemoglobin which is called **oxy-hemoglobin** (HbO_2^-). The oxy-hemoglobin is present in blood arriving at the tissue. The binding of H^+ by HbO_2^- aids in the release of O_2 by HbO_2^-.

4. Oxygen diffuses into the tissue cells where it is consumed. Thus, we see that free H^+ does not accumulate because it is bound to the hemoglobin.

The blood leaving the tissues contains large quantities of HCO_3^- and large quantities of hemoglobin which is free of oxygen and is called **reduced hemoglobin** (HHb). No further changes take place until the blood reaches the capillaries of the lungs. Details of the changes taking place in the lungs are illustrated in Fig. 42.5 as follows:

Fig. 42.5 Chemical interactions that take place in the blood as oxygen and carbon dioxide are exchanged between capillaries and the lungs are shown here. Reactions 2 and 3 take place inside the red blood cells. The product HbO_2^- is carried to the tissues by arterial blood.

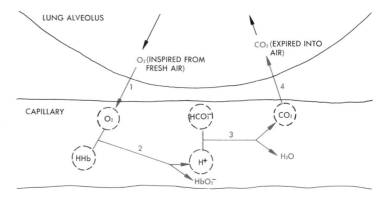

1. The concentration of oxygen in the lung alveoli is high because of breathing movements. The O_2 diffuses from the alveoli into the plasma and blood cells.
2. The O_2 combines with HHb to form HbO_2^-, and this releases H^+.
3. The H^+ combines with HCO_3^- and forms H_2O and CO_2.
4. The CO_2 diffuses into the lung alveoli where it is expelled in the process of normal breathing.

Again, H^+ does not accumulate, because as soon as it is released from HHb it combines with HCO_3^- which releases CO_2. Hemoglobin is essential because it serves as a carrier within the circulation for O_2, CO_2, and H^+.

Control of Oxygen Delivery

The demands of the tissue for oxygen are not always the same. During exercise, for example, you need more energy, and your tissues consume more oxygen than when you are at rest. In addition to the changing demands of the tissue, there are times when the supply of oxygen from the external environment changes. If you climb a mountain the air becomes "thinner" and the concentration of oxygen reaching the lung alveoli is reduced. Nevertheless, the cells of the body require just as much oxygen at the top of a mountain as they do at its base. How does the body adjust to variations in supply and demand of oxygen?

The amount of oxygen delivered to the cells can be altered by the body in two ways: 1. *The amount of blood reaching the tissue may be changed.* If each ml of blood has the same amount of oxygen dissolved in it, and if the blood flow to the tissue is doubled, then the amount of oxygen delivered to the tissue will also double. We saw examples of changes in local blood flow in Chapter 40. 2. A second method of altering the amount of oxygen delivered to the tissue is *to alter the amount of oxygen contained in each ml. of blood.* If the amount of blood reaching a tissue does not change, but if the concentration of oxygen in the blood doubles, then the total amount of oxygen delivered to the tissue will also double.

In this chapter we are concerned primarily with *regulation* of the oxygen delivery through regulation of the oxygen concentration in the blood. Regulation of the oxygen concentration can be brought about in two ways.

1. The concentration of oxygen in the blood can be altered by *changing the concentration of oxygen in the air* that is in direct contact with the blood in the alveoli of the lungs. This can be accomplished by changes in the rate and depth of breathing.

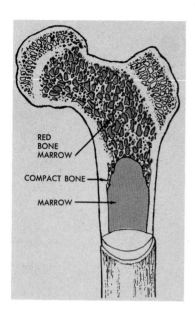

Fig. 42.6 Since the average life-span of a red blood cell is about 120 days, these cells must constantly be replaced. Red bone marrow is the site of formation of the red cells.

When you increase your rate and depth of breathing, you bring about a more efficient mixing of the alveolar air with the air of the external environment. As a result, the oxygen content of alveolar air increases. An increase in breathing is an important response when body tissues become active and demand more and more oxygen from the blood. The oxygen-poor venous blood returning to the lungs, in turn, requires more and more oxygen from the alveoli. Unless you increase your breathing rate, the alveolar oxygen concentration will fall and be unable to supply your body's needs.

2. The capacity of blood to take up oxygen can be altered by a *change in the amount of hemoglobin* in each ml of blood. This can be brought about by a *change in the number of red blood cells* in each ml of blood. Let us consider this latter possibility in some detail.

Regulation of Hemoglobin—The Life of the Red Cell

The average red blood cell stays in the circulation for only about 120 days before it is destroyed in the liver, spleen, or bone. This means that $\frac{1}{120}$ or almost 1 per cent of all the body's red blood cells are destroyed each day! Yet, under normal circumstances, the number of these cells in the circulation does not change; that is, they are in a steady state. It follows that new cells must be produced and added to the circulation just as fast as old cells are destroyed.

Where do the new cells come from, and how is their production rate geared to the body's need for oxygen? Since red blood cells do not have a nucleus, they certainly cannot divide and form new cells. Rather, they are formed from nucleated cells found in the interior of bones—in the **bone marrow** (Fig. 42.6). In the adult this takes place mainly in the vertebrae, and in bones of the chest, skull, and pelvis.

The rate of production of red cells (and, consequently, of hemoglobin) is not rigidly fixed. It depends on the oxygen con-

Fig. 42.7 Red blood cell production is stimulated by a lack of oxygen. The feedback loop shows how this response helps maintain the oxygen content in the blood.

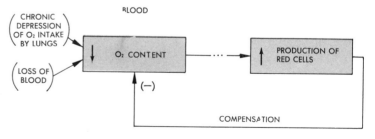

tent of the blood. Whenever the oxygen content of blood is low, the production rate of red cells increases. This provides the body with an effective homeostatic mechanism for regulation of blood oxygen. The feedback loop for this process is illustrated in Fig. 42.7.

When you climb a mountain, for example, as you go higher the air becomes less dense. This means that less and less oxygen is reaching your lungs, so the oxygen content in the blood falls. You find yourself short of breath. However, after a few weeks of living at high altitude, a process of acclimatization sets in. You are more comfortable. Why? Each ml of your blood contains a greater number of red cells and, therefore, a greater concentration of hemoglobin which can carry oxygen. The oxygen content of the blood has risen back toward normal. We account for this by the scheme illustrated in Fig. 42.7. The decreased blood oxygen stimulated an increase in the rate of red cell production.

Another example is found in the body's response to severe bleeding. Red cells are lost, which means that the capacity of blood to carry oxygen is decreased. The resulting decrease in oxygen content stimulates red cell production, which compensates for the original loss.

It is important to emphasize that this response (increased red cell production) to a decreased oxygen content is a slow one. It takes many days before the increased cell production fully compensates for the low oxygen state. Although this long-term adjustment is essential, the body tissues cannot wait for days before the oxygen delivery returns to normal. A quicker response is necessary. This is accomplished by an increase in the amount of blood flowing to the tissues (see Chapter 40) and an increase in breathing.

Control of Breathing

Whenever your cells become unusually active, you breathe faster. This provides the extra oxygen needed by the more active tissues and expels the excess CO_2 that has been produced. But what is responsible for the increase in breathing when the tissues become more active? In other words, why do the muscles involved in breathing (diaphragm and rib muscles) become more active when other parts of the body begin consuming more oxygen?

Since the blood O_2 and CO_2 are controlled by breathing, it is reasonable to ask whether or not blood O_2 and CO_2, in turn, have any influence on breathing. In other words, is there any simple control loop involving breathing and O_2 and CO_2? The answer is yes in both cases. A *low* concentration of O_2 and

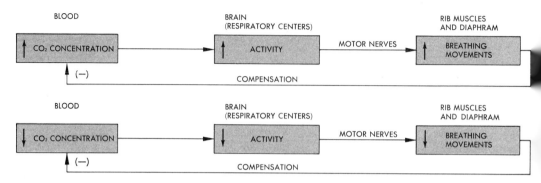

Fig. 42.8 The level of carbon dioxide in the blood is of primary importance in regulating breathing. A high concentration of carbon dioxide stimulates breathing. This response reduces carbon dioxide concentration back toward normal (top diagram). A low concentration of carbon dioxide decreases breathing. This response increases the carbon dioxide concentration back toward normal (bottom diagram).

a *high* concentration of CO_2 both stimulate breathing. The increase in concentration of CO_2 in blood carried to the brain acts directly on the medulla. The respiratory centers respond by discharging a greater number of motor nerve impulses to the rib muscles and diaphragm. This produces an increase in breathing movements, more CO_2 is expelled, and more O_2 may be picked up by the blood. This homeostatic response, which guards against large changes in CO_2 concentrations in the body fluids is summarized in Fig. 42.8. For example, this response prevents you from holding your breath for an indefinite period. While you hold your breath, CO_2 is not removed from your body. Its concentration quickly builds up and it begins to stimulate respiration so strongly that you are no longer able to hold your breath—you gasp for air.

The breathing response to a low oxygen concentration differs in detail from the response to CO_2. Unlike CO_2, O_2 does not act *directly* on the brain. Instead, it acts on special nerve receptors called **chemoreceptors**, which are located near the aortic arch and carotid sinus (close to the pressure receptors responsible for regulating blood pressure). The chemoreceptors are stimulated by low O_2 concentrations. Nerve impulses are sent to

Fig. 42.9 A low concentration of oxygen in the blood stimulates breathing. The feedback loop shows how this response guards against serious lowering of the concentration of oxygen.

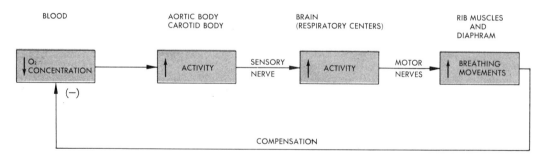

the respiratory centers, which in turn send more motor nerve impulses to the muscles controlling breathing. As a result, breathing movements are increased and we have a homeostatic reflex which specifically guards against low O_2 concentrations. The feedback loop is shown in Fig. 42.9. This reflex operates when you climb a mountain. The higher you go, the more you breathe. The chemoreceptor reflex is responsible for the rapid breathing.

Although the concentrations of O_2 and CO_2 both influence breathing, under normal conditions the response to CO_2 is much stronger. It is only when the level of O_2 is greatly lowered that O_2 begins to play a role.

In addition to O_2 and CO_2 levels, there are other factors that influence your breathing. For example, you can breathe faster any time you choose, so the respiratory centers must be influenced by higher centers of the brain. We have found it difficult to locate and sort out the relative importance of all of the factors that control breathing. One outstanding puzzle is "What causes the pronounced increase in breathing during exercise?" As a first guess you might say something like this: "During exercise muscles use more O_2 and produce more CO_2. This results in a low blood O_2 concentration, and high CO_2 concentration; and, as we know, both of these factors increase breathing." Logical as this seems, it is not the whole answer. During severe exercise, the extent of the increase in breathing is so great that the level of CO_2 and O_2 in the blood may not show *any* substantial change! This large increase in breathing could not be produced by the simple responses to low O_2 and high CO_2 that we have outlined. Something else *must* be involved, but what? It remains to be discovered.

During the act of breathing, water is lost from the body. It

Wide World Photos

Fig. 42.10 This runner is breathing hard. Why? The physiological causes are not yet fully understood.

simply evaporates from the moist surfaces of the lungs and is expelled with each breath. The amount of water lost in this way has no relation to the needs of the body for water, but occurs incidentally with the body's exchange of O_2 and CO_2 with the external environment. Water is also lost through the skin (for example, by sweating), the intestinal tract, and the kidney. Of these various routes for water loss, only the kidney responds to the body's needs for water and acts to regulate it. In the next chapter we shall turn our attention to the role of the kidney in the regulation of salt and water.

SUMMARY

Oxygen enters the body through the lungs, is picked up by **hemoglobin** in the red blood cells, and is carried to the body tissues where it is used. The carbon dioxide produced by the tissues is carried (as bicarbonate, or in combination with hemoglobin) by the blood back to the lungs, where it is eliminated. Hemoglobin also helps stabilize the concentration of hydrogen ions in the blood.

The amount of hemoglobin contained in the blood is regulated by the production of red blood cells. The rate of red cell production depends on the oxygen content of the blood. Whenever the oxygen content is low, the rate of production of red cells increases. This slowly increases the hemoglobin concentration and compensates for the original reduction in oxygen content.

Faster responses are brought about by the regulation of breathing. Breathing is increased whenever the level of oxygen in the blood is lowered, or whenever the carbon dioxide level is raised. **Chemoreceptors,** which are sensitive to the reduced level of oxygen, respond by sending impulses to the respiratory centers, which in turn stimulate breathing. This compensates for the original oxygen deficit. Excess carbon dioxide acts directly on the brain; the respiratory centers again respond and breathing is increased. This compensates for the original carbon dioxide excess. Normally, the response to an excess of carbon dioxide is much stronger than the response to a deficit in oxygen.

FOR THOUGHT AND DISCUSSION

1 The term *hypoxia* usually refers to a condition in which the availability or utilization of oxygen is depressed. The data listed below illustrate four different types of hypoxia compared with the state of a "normal" person breathing fresh room air. (The weight, sex, and age of all subjects are the same.)

Subject	Hemoglobin (grams Hb per 100 ml blood)	O_2 content of arterial blood (ml O_2 per 100 cc blood)	O_2 content of venous blood (ml O_2 per 100 ml blood)	Cardiac output (liters/min)
A NORMAL	15	19	15	5.0
B HYPOXIA	15	15	12	6.6
C HYPOXIA	8	9.5	6.5	7.0
D HYPOXIA	16	20	13	3.0
E HYPOXIA	15	19	18	no information

(a) Which subject is suffering from a dietary iron deficiency?
(b) Which subject is suffering from heart failure and poor blood circulation?
(c) Which subject has recently climbed a mountain, where the air is "thin" and atmospheric oxygen low?
(d) Which subject is suffering from a poison (for example, cyanide) which prevents his cells from using oxygen?
(e) Subject B has increased respiration. Briefly describe the physiological mechanism that is responsible.
(f) In subject A how much blood is flowing through the lungs each minute? Using this figure, and data from the table, calculate how many cc of oxygen are carried *to* the lungs each minute. How many cc of oxygen are carried *away* from the lungs each minute. Using these last two figures, calculate the oxygen consumed each minute.
(g) A standard method of measuring cardiac output is to measure oxygen consumption during each minute along with the oxygen content of both the arterial and venous blood. Cardiac output is calculated from these figures. Explain. [HINT: study your answer to question 1(f).]

2 A large blood vessel is cut and the bleeding is severe. Outline the physiological responses (discussed in this and the preceding two chapters) that will help the person to survive.

SELECTED READINGS

Carlson, A. J., V. Johnson, and H. M. Cavert. *The Machinery of the Body*, 5th Ed. Chicago: University of Chicago Press, 1961.

Chapman, C. B. and J. H. Mitchell. "The Physiology of Exercise," *Scientific American* (May 1965) (*Scientific American* reprint #1011).

Comroe, J. H. "The Lung," *Scientific American* (February 1966).

———. *Physiology of Respiration*. Chicago: Yearbook Medical Publishers, Inc., 1965.

McDermott, W. "Air Pollution and Public Health," *Scientific American* (October 1961) (*Scientific American* reprint #612).

43 SALT AND WATER: THE KIDNEY

At that time in the long course of evolution when animals first left fresh water and began to spread over the land, they were faced with a new danger—the threat of drying up. In order to survive on the land, they had to develop a way of conserving water. Man still faces this threat. The living cells making up your body must be bathed in a watery fluid to survive. The regulation of the salt and water content of this fluid is largely the responsibility of the kidney. If we drink too little water, the kidney excretes less water in the urine. If we drink too much, the kidney excretes more water in the urine.

The work of the kidney, however, is not limited to the regulation of salt and water alone. It also plays a very important role in the regulation of the acidity of the blood, and in the excretion of waste products. The waste product **urea,** for example, which is formed when proteins are broken down, is eliminated from the body primarily by the kidney.

Fig. 43.1 The living cells of your body must be bathed in a watery solution if they are to survive. Although this man is surrounded by sea water, he must find other water to drink.

Wide World Photos

Structure of the Kidney

In Fig. 43.2 we see the two kidneys as they rest in the abdominal cavity. Three important tubes are attached: 1. The **renal artery,** which conducts blood from the aorta to the kidney. 2. The **renal vein,** which conducts blood away from the kidney. 3. A **ureter** from each kidney conducting urine from the kidneys to the bladder. Somehow, the kidneys are able to extract materials from the blood that passes through them, and then pass these materials into the ureter, where we identify them as components of urine.

If we use a microscope to take a closer look, we find that each kidney contains about a million tiny tubes, called **nephrons.** These tubes are the actual site of urine formation. A nephron is a long structure with walls composed of a single layer of cells. One end of the nephron is enlarged into a funnel-like structure called the **capsule.** The other end of this tube empties into a larger tube called the **collecting duct** (see Fig. 43.3).

Blood comes in contact with the nephrons by means of two capillary beds arranged in series. The first capillary bed, called the **glomerulus,** is contained within the funnel-like structure of the capsule, and is shown in detail in Fig. 43.4. Blood leaving the glomerulus does not return directly to the larger veins, as we would normally expect. Rather, it flows out through the **efferent arteriole** and enters another capillary bed, which supplies the cells lining the tubules (Fig. 43.3).

The Formation of Urine

The formation of urine from blood takes place in two major steps. First, the blood cells and plasma proteins are separated from a portion of blood. Second, the composition of this portion is changed until it becomes urine. Let us trace the details of these two steps with the aid of Fig. 43.4. Blood enters the kidney from the renal artery and finds its way to the glomerular capillaries. Here, one-fifth of the plasma *filters* through the capillary walls and is collected in the funnel-like structure. The other four-fifths continues on toward the tubule capillaries. The pores in the glomerular walls are too small to permit large protein molecules or cells to pass through them. The fluid that filters through is like plasma, except that it does not contain protein or cells.

The filtered fluid flows from the capsule into the tubule and on toward the collecting ducts and finally into the ureter. Fluid entering the tubule is very much like plasma. Fluid leaving the tubule resembles urine. It follows that the composition of the

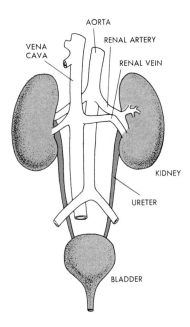

Fig. 43.2 The kidneys form urine by extracting material from the blood passing through them.

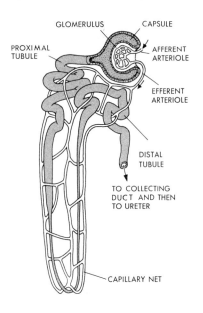

Fig. 43.3 The nephron shown in color is the functional unit of the kidney.

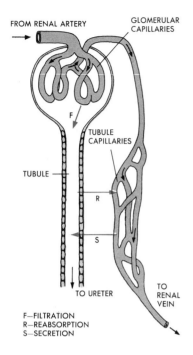

Fig. 43.4 Urine formation takes place in two stages. First, water and small molecules filter through the *glomerulus* into the tubules; second, the composition of filtered fluid is changed by tubular reabsorption and secretion.

fluid changes as it flows down the tubule. Many substances are transported from the tubular fluid through the cells of the tubular walls and redeposited into the blood capillaries that supply these cells. This process is called **reabsorption**. A few substances are transported in the reverse direction—from the blood to the tubular fluid. This is called **tubular secretion.**

The process of urine formation, then, takes place in two steps. 1. *Filtration* in the glomerulus involves a large volume of fluid which is filtered from the blood and sent into the tubules. Proteins and blood cells are held back and, consequently, never appear in the urine. 2. *Reabsorption and tubular secretion* in the tubules is the second step. Cells lining the tubular walls transport materials either from the tubules to the blood (reabsorption) or from the blood to the tubules (tubular secretion). The altered fluid reaching the end of the collecting duct enters the ureter as urine.

Evidence for filtration: We have uncovered several facts that make us believe that filtration is the first step in urine formation. 1. Large molecules are never found in normal urine. Evidently, they cannot pass through the filter. 2. Pressure must be relatively high to force fluids through a filter. If blood pressure dropped sufficiently, we would expect the glomerular filter to stop working. This seems to happen. When the arterial blood pressure falls below 70 mm Hg, urine formation stops. 3. Samples of fluid have been taken from the capsule and chemically analyzed. The chemical composition of the capsule fluid is identical to that of the plasma, with the exception that the fluid does not contain proteins. This is just what we would expect in a filtration process.

Mechanisms of reabsorption: Reabsorption is a key process in urine formation. As an example, consider the reabsorption of water. During each minute, a normal person has roughly 600 ml of plasma flowing through his kidneys. Of this 600 ml, one-fifth, or 120 ml, is filtered. On the other hand, under average conditions only about 1 ml of urine is formed per minute. In other words, during each minute 120 ml of fluid enter the tubules through the filtration process but only 1 ml of fluid leaves. This means that more than 99 per cent of the fluid (which is mostly water) is reabsorbed. These figures are surprisingly large. There are only about 3000 ml of plasma in the entire body. In about 25 minutes a volume of fluid equivalent to the entire plasma volume filters through the glomerulus. In other words, the kidney samples the entire plasma volume about every 25 minutes! Imagine what would happen if the reabsorption mechanism failed. We would simply excrete our body fluids into the urine and the whole living system would collapse.

What is responsible for the reabsorption of such a large volume of fluid? Experiments indicate that the prime mover in reabsorption of fluid is the active transport of sodium from the tubular fluid to the blood. Sodium is one of the most abundant solutes in the plasma. As fluid progresses down the tubules, sodium is pumped back into the blood by the cells lining the walls of the tubule. As we saw earlier, sodium contains a net-positive charge, so it attracts negatively charged particles. Each time a sodium ion is pumped out of the tubule, it upsets the charge balance. As a result a negatively charged particle will follow it. Usually, this is chloride, since it is the most abundant negatively charged ion in the tubular fluid. In other words, salt (sodium chloride) moves from the tubular fluid back into the plasma.

The water movements that take place during this process can be explained by the osmotic pressure of the tubular fluid, which decreases each time a solute molecule is reabsorbed. As a result, water is drawn out to balance the osmotic pressure. It follows the movement of sodium chloride. Thus, simply by pumping sodium out of the tubule, both chloride and water are reabsorbed into the blood plasma. In fact, the major portion of water that is reabsorbed is due to the initial pumping of sodium. The reabsorption of water influences other solutes in the tubules. This is because whenever water leaves the tubular fluids, any solutes that remain behind become more concentrated. If these solutes can pass through the tubular walls, they may simply diffuse back into the blood.

Sodium is not the only substance that is actively transported in the reabsorption process. Glucose and amino acids, for example, are pumped from the tubular fluid back into the blood. Ordinarily, all of the glucose is transported in this way; none appears in the urine. However, in some rare cases the blood may contain so much glucose that when it is filtered into the tubules the walls become overloaded. The transport system cannot handle all of the glucose that comes to it. Some of the glucose remains behind in the tubular fluids and eventually finds its way into the urine.

In summary, the reabsorption of huge amounts of water during each minute occurs by simple osmosis. The osmotic forces responsible for this are a direct result of the reabsorption of sodium chloride. This withdrawal (by reabsorption) of water from the tubules concentrates any solutes left behind. Any of these concentrated solutes which cannot penetrate the tubular walls easily will be found in high concentrations in the urine. This is true of some waste products. Any solute that can penetrate the tubular walls will be reabsorbed to a certain extent

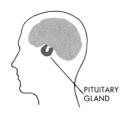

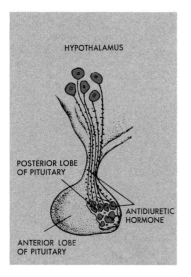

Fig. 43.5 ADH is made in the *hypothalamus* and migrates to the *pituitary* gland, where it may be stored, or released into the blood.

because it will simply diffuse from tubular fluid (where the solute is concentrated) to blood. Finally, we emphasize that many important solutes like sodium, glucose, and amino acids are reabsorbed by an active process requiring energy supplied by the metabolism of the tubular cells.

Regulating Osmotic Pressure in Body Fluids

The body gains and loses water and salts in a variety of ways. As you found in the last chapter, some body water is lost with each breath, but the rate of breathing (which depends on blood carbon dioxide and oxygen concentrations) is not governed by the body's needs for water. We also lose water through sweating, but this is also not regulated by the body's water requirements. Nevertheless, in spite of large variations in water loss by these two routes, the changes in osmotic pressure of the body fluids are very small. This is because the kidney *is* responsive to the body's water requirements. By altering the amount of water excreted in the urine, the kidney regulates the osmotic pressure.

A simple example will demonstrate how accurately the kidney works. If you drink a quart of water quickly you find that you produce much more urine than normal over the next three or four hours. If you collect the excess urine during this period, you find that there is just about a quart of it. But after three or four hours, urine production returns to normal. Somehow or other, the kidney "knew" just how much extra water you drank. It excreted almost precisely that amount so that the body's water content would remain constant. How does the kidney "know" when to excrete extra water and when to conserve it? What are the details of this homeostatic response?

A hormone called **antidiuretic hormone** (**ADH**) plays a central role in the regulation of body water. It acts directly on the kidney tubules causing an increase in water reabsorption. When the body's water supply is low, a large amount of ADH is present in the blood. This causes the kidney to conserve water because most of the tubular water is reabsorbed and cannot pass into the urine. The output of urine may be as little as $\frac{1}{2}$ ml per minute at this time.

When the body contains excessive water, very little ADH is present in the blood, so the capacity of the kidney to reabsorb water is reduced. Some of the tubular water that normally would have been reabsorbed now passes into the urine. Urine output may rise to 16 ml per minute. How does ADH increase in the blood when we need to conserve water? Where does the ADH come from?

ADH is manufactured by specialized nerve cells lying in the

hypothalamus—a small region at the base of the brain. These cells send fibers a very short distance to the posterior **pituitary** gland (Fig. 43.5). ADH produced by these cells migrates along the fibers to the pituitary gland where it may be stored or released into the blood stream. The release of this hormone is controlled to a large extent by specialized nerve cells called **osmoreceptors** (located in the hypothalamus), which are sensitive to the osmotic pressure of the fluids that surround them. When the body has too little water, the osmotic pressure of the fluids goes up. The osmoreceptors respond and the rate of release of ADH is increased. As a result, more water is reabsorbed by the kidney. When too much fluid is taken in, the osmotic pressure of body fluids goes down. Less ADH is released by the gland, and more water passes into the urine. This homeostatic feedback system is shown in Fig. 43.6.

Many experiments support the scheme that we have outlined above. In one of the most dramatic tests, an animal's pituitary glands may be removed, which means that there is no longer a source of ADH. Urine flow increases dangerously because the tubules can no longer reabsorb a normal amount of water. If we now prepare an extract of the gland and inject it into the animal, the urine flow returns to normal. Apparently the extract of the gland contains enough ADH to restore the ability of the tubules to reabsorb water. This same dramatic increase in urine flow occurs in humans when the back (posterior) part of the pituitary gland is damaged.

What does ADH do to the tubules to allow them to reabsorb more water? Before we can pursue this question, we will need more information about the structure of the nephron (Fig. 43.7).

Fig. 43.6 ADH, released when the osmotic pressure of the blood plasma increases, acts on the kidney by promoting an increase in water reabsorption. The feedback loop shows how this response helps stabilize the osmotic pressure of the body fluids.

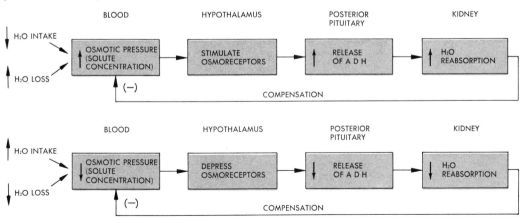

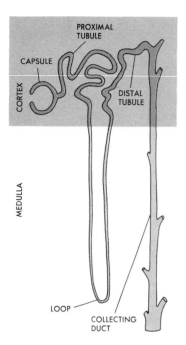

Fig. 43.7 The nephron consists of three main parts—the *proximal tubule*, the *loop*, and the *distal tubule*.

Structure of the Nephron

For convenience, we divide the tubule into three parts. 1. That part of the tubule that is attached to the funnel-like capsule is called the **proximal tubule.** It lies almost entirely in the surface layers, or **cortex** (cortex means bark), of the kidney. 2. The second portion of the tubule is called the **loop.** It takes a straight course and dips deeply into the core (or **medulla**) of the kidney. Then it bends back out toward the cortex. 3. When the tubule re-enters the cortex it becomes the **distal tubule.** The distal tubule enters into the collecting duct, which again proceeds downward into the core of the kidney.

Under most circumstances the proximal tubule does the same job. It reabsorbs about 85 per cent of the water that has been filtered in the glomerulus. In addition, it reabsorbs all of the glucose, and a large number of other substances, such as amino acids and vitamins, which are essential to the internal environment. The remaining 15 per cent of the fluid passes down through the loop and on toward the distal tubule and the collecting ducts. By varying the reabsorption of the last 15 per cent of the water, the kidney regulates the fluid volume of the internal environment.

When we study the role of the kidney in water regulation, we must not lose sight of the fact that the kidney always excretes some solutes. For example, it excretes waste products like urea whether it is conserving or excreting water. When excess water is excreted it is as though the excess water were added to the load of waste products normally excreted, thus making the urine very dilute. On the other hand, when water is conserved, the urine is highly concentrated.

Mechanism of Action of ADH

How does the kidney excrete a concentrated urine? This is a problem that has puzzled scientists for many years. They reasoned that water normally follows the movements of solutes by simple osmosis. If this were the case, it would be difficult to see how the solute concentration of urine can change so drastically under different circumstances. Within the last 10 years, the answer to this riddle has been clarified.

The most important clue was the startling observation that solutes contained in the tissue spaces surrounding the nephron in the medulla of the kidney are four times more concentrated than anywhere else in the body. Notice (Fig. 43.7) that in passing from the capsule to the ureter, fluid must pass through the medulla twice, first in the loop and second in the collecting ducts. The first time through, sodium is pumped out of the as-

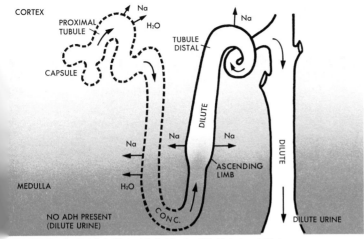

Best and Taylor, The Physiological Basis of Medical Practice, Holt, Rinehart, and Winston.

Fig. 43.8 If ADH secretion stops or is low, the urine will be dilute because the walls of the distal tubule and collecting duct are impermeable to water. (Broken lines represent water permeable walls of the nephron; solid lines represent water impermeable walls.)

cending limb of the loop, but water cannot follow it because the tubular walls of this segment of the loop are impermeable to water. Sodium collects in the extracellular spaces of the medulla, accounting for the fact that it is four times more concentrated than other body fluids.

The loop of the nephron has loaded the spaces with solute in preparation for the second passage through the medulla. Since sodium has been removed from the loop, the fluid presented to the distal tubule will be dilute. The fate of the tubular fluid from this point onward depends on whether or not the hormone ADH is present. If there is little or no ADH, the remaining parts of the nephron—from distal tubule through the collecting duct—are impermeable to water. Although some solute may be reabsorbed from the fluid in these parts, water is not. This makes the fluid more dilute (by losing solute) and the result is very dilute urine (Fig. 43.8).

However, if much ADH is present, it acts in some unknown

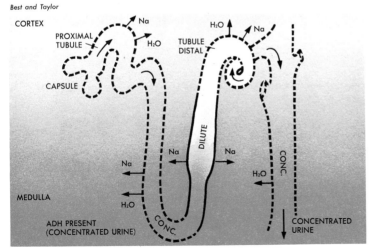

Best and Taylor

Fig. 43.9 If there is an abundant supply of ADH, the urine will be concentrated because ADH makes the walls of the distal tubules and collecting ducts permeable to water. (Broken lines represent water permeable walls of the nephron; solid lines represent water impermeable walls.)

manner on the distal tubule and collecting duct, making them permeable to water. As a result, water leaves the distal tubule and collecting duct by osmosis (Fig. 43.9). But we have just seen that the collecting ducts must pass through the medulla. Because solutes are highly concentrated in the medulla, water will leave the ducts until fluid within the ducts is just as concentrated as the outside fluid. In short, when a large quantity of ADH is present, the urine will be concentrated.

The transfer of water in the kidney is a passive process. Water flows from regions of low osmotic pressure to regions of higher osmotic pressure. It is completely controlled by the solute distribution. But the kidney can pump sodium, and water will follow the sodium. So, by pumping sodium the kidney can regulate the water content of urine. This also regulates the water content of the body. When sodium is pumped into the medulla, the fluid in the extracellular spaces become concentrated. The fluid in the collecting duct and, therefore, the urine never gets more concentrated than the fluid in the medullary extracellular spaces.

This limitation on the concentration of solutes in the urine explains why it is not wise to drink sea water when you are thirsty. The concentration of salt in sea water is usually larger than the highest concentration of the salt that the kidney can excrete. The effect of drinking sea water, then, is to increase the solute content of the body fluids, and it is this increased solute content which created the original need for water. This will increase your thirst. Instead of alleviating the problem you only aggravate it by drinking sea water. You have set the *positive feedback* loop illustrated in Fig. 43.10 into operation. If you continue to drink sea water, the increase in body salt content will become a threat to survival.

Fig. 43.10 The ability of the kidney to excrete a high concentration of salt in the urine is limited. Drinking salt water that is even more concentrated can only increase the concentration of solutes in the body fluids. This initiates sensations of thirst which may set the *positive* feedback loop shown here into operation.

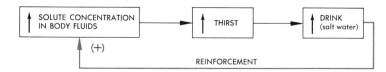

Regulation of Body Fluid Volume

We have seen that the pituitary gland and kidney interact and regulate the *concentration* of solutes. The kidney also interacts with another gland, the adrenal cortex (Fig. 45.6) to control the *amount* or *volume* of the body fluids. The adrenal gland secretes many hormones. One of them called **aldosterone** may be involved in regulation of body fluid volume. It acts on

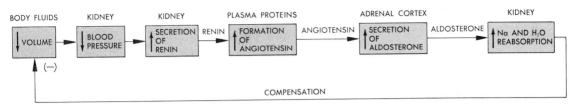

kidney tubules by increasing the reabsorption of sodium. The more the tubules reabsorb sodium, the more water they will reabsorb to maintain an osmotic balance. In other words, the secretion of aldosterone acts to conserve body fluids. It is secreted when the volume of body fluids drops. This probably occurs in the following way (see Fig. 43.11): a reduction in body fluid volume is usually accompanied by a reduction in blood volume. A reduction in blood volume, in turn, may be accompanied by a reduction in blood pressure in the kidneys. The kidneys respond to this lowered blood pressure by releasing a substance called **renin** into the blood. Renin acts on one of the plasma proteins and produces a new substance called **angiotensin.** Finally, angiotensin stimulates the release of aldosterone by the adrenal gland. The details of this complicated feedback loop are illustrated in Fig. 43.11.

Fig. 43.11 When the body fluid volume is decreased, the secretion of aldosterone is increased. Aldosterone stimulates an increase in sodium (and water) reabsorption by the kidney. This response helps compensate for the decrease in body fluid volume. Recent experiments suggest that the feedback loop shown above is involved in stabilizing the volume of body fluids.

Other Functions of the Kidney

Besides regulating salt and water, the kidney helps regulate the acidity of the blood. This process occurs in the distal tubules, where the cells are able to actively transport H^+ from the blood to the tubular fluid and thus dispose of excess acid.

The role of the kidney in excreting waste products is an extremely important one. Some wastes are simply filtered through the glomerulus and are not reabsorbed as rapidly as water. They are thus more concentrated in the urine than in the blood. As a result, there is a net loss of these materials from the body. Urea, which is produced in the liver from waste products of protein breakdown, is a good example of this type of excretion. Other substances are more actively removed. In addition to being filtered through the glomerulus, they are transported from the blood to the tubular fluid (the exact opposite of reabsorption) by the tubular cells.

In this chapter we have been concerned with homeostatic mechanisms that operate on the outflow of materials from the internal environment. In the next chapter we turn our attention to the gastro-intestinal tract, a major avenue for the inflow of materials into the internal environment.

SUMMARY

The kidney helps regulate the internal environment by forming and excreting urine. Urine contains waste products and variable amounts of salt, water, and acid. The formation of urine in the kidney takes place in two stages—**filtration,** then selective **reabsorption** (or the opposite, **secretion**).

The kidney is responsible for regulating the osmotic pressure of the body fluids. When the osmotic pressure of the blood reaching the hypothalamus is increased, special cells (osmoreceptors) respond and the hormone ADH is released in greater quantities by the **posterior pituitary gland.** ADH acts on the kidney and causes an increase in the reabsorption of water; the body fluids are then diluted back toward normal.

The action of ADH: the fluids surrounding the nephrons in the medulla are very highly concentrated because of the active transport of sodium, which takes place in the loop of the nephron. In the presence of ADH, the walls of the collecting duct and distal tubules become permeable to water, fluid is reabsorbed (by osmosis) in large quantities from the collecting duct into the medulla; the urine becomes concentrated. In the absence of ADH, the walls of the collecting ducts and distal tubules are relatively impermeable to water. Fluid reabsorption is diminished and large quantities of water are excreted in the urine.

The reabsorption of sodium by the kidneys is controlled in part by a hormone called *aldosterone*. Aldosterone acts on the kidney tubules and increases the reabsorption of sodium. The more the tubules reabsorb sodium, the more water they will reabsorb, thus maintaining an osmotic balance. In this way, aldosterone is involved in the control of the volume of body fluids.

FOR THOUGHT AND DISCUSSION

1 Trace the possible routes taken by a glucose molecule as it enters the kidney through the renal artery. How do these routes differ from the route taken by a urea molecule that finds its way into the urine?

2 In some abnormal cases, the blood may contain so much glucose that when it is filtered into the nephron the tubules are not able to reabsorb all of it, and some glucose appears in the urine. In these cases, the volume of water excreted in the urine may be abnormally high. Explain.

3 In some cases of kidney disease, the glomerular membrane becomes very much more permeable than it is normally. The

body tissues of patients (who have the disease) usually swell and become distended with fluid (*edema*). Can you suggest an explanation? (HINT: what substance normally excluded from the urine by the filtration process may now find its way through the highly permeable glomerular membrane?)

4 A man is lost in the desert without any water supply. What physiological responses described in this chapter will tend to help him survive?

SELECTED READINGS

Best, C. H. and N. B. Taylor. Human Body: Its Anatomy and Physiology. New York: Holt, Rinehart and Winston, 1963.

Carlson, A. J., V. Johnson, and H. M. Cavert. *The Machinery of the Body*, 5th Ed. Chicago: University of Chicago Press, 1961.

Merrill, J. P. "The Artificial Kidney," *Scientific American* (July 1961).

Pitts, R. F., *Physiology of the Kidney and Body Fluids*. Chicago: Yearbook Medical Publishers, 1963.

Schmidt-Nielsen, K. and B. Schmidt-Nielsen. "The Desert Rat," *Scientific American* (July 1953).

Smith, H. W. *From Fish to Philosopher*. Boston: Little, Brown, and Co., 1953.

Solomon, A. K. "Pumps in the Living Cell," *Scientific American* (August 1962).

Wolf, A. V. "Thirst," *Scientific American* (January 1956).

44 FOOD

When cellular life began on the Earth, every cell was in direct contact with the sea. From this liquid environment the cell was able to obtain all of its nutritional requirements. As animal life became more complex, tissues and organs developed. Because the cells of certain tissues and organs were neither on the surface of the body nor in direct contact with the sea, the cells had to be supplied with food from the external environment in a new way. The evolution of the **gastro-intestinal tract** helped satisfy this requirement.

The gastro-intestinal tract (GI tract) is a long, hollow tube lying deep within the body. Both ends of this tube open onto the surface of the body (Fig. 44.2). The digestive system is made up of the various organs shown in Fig. 44.3: the **mouth** (including teeth and salivary glands), the **esophagus,** which conveys food to the **stomach** where it is stored and mixed, the **small intestine** into which the liver and the pancreas secrete materials, and the **large intestine** with its terminal opening, the **anus.** The

Fig. 44.1 Breugel's painting "The Peasant Wedding."

Courtesy Vienna Kunst Museum

GI tract, which is about 15 feet long in the living body, is contained within a body (head and trunk) only about four feet long. As a result, the GI tract is folded over on itself many times. You can see from both Figs. 44.2 and 44.3 that swallowed food within the GI tract is still *outside* of the internal environment. To enter the internal environment, it must be transported across the cellular walls of the GI tract and into the bloodstream. The process by which this transport occurs is called **absorption.**

Only certain types of molecules are absorbed by the cellular walls of the GI tract. This is an advantage because our diet may contain many things that would be harmful if they entered the blood. Most of the foods that we eat are complex molecules which must be broken down into simpler building blocks before they can be absorbed. This breakdown process is called **digestion.** In this chapter, we investigate the nature of the foods we eat. We also ask how they are broken down by the processes of digestion, how the simpler products of digestion are absorbed, and how these various processes are controlled.

FOOD AND DIGESTION

The foods we normally eat fall into one of the following categories: carbohydrates, fats, proteins, minerals, vitamins, or water. Vitamins, minerals, and water do not have to be digested in order to be absorbed. Other compounds, such as cellulose (a plant carbohydrate) pass right through the human GI tract without being absorbed, since there are no enzymes for digesting these compounds into simpler building blocks. Carbohydrates, fats, and proteins are complicated molecules and must be broken down into simpler molecules if they are to be absorbed.

The breakdown of food into simple molecules occurs because the inner walls of the GI tract, and certain large glands associated with the tract, are able to secrete enzymes. Enzymes act as **catalysts**—they increase the rate of chemical reactions but they do not influence the direction of those reactions. They seem to work by providing a surface upon which the reaction may take place easily. Enzymes are *specific* in their action; that is, a particular enzyme is required for a particular reaction. The various digestive enzymes can be classed according to the basic foods upon which they work.

Classes of Food

Carbohydrates: The class of food called carbohydrates includes the sugars and the starches. These compounds

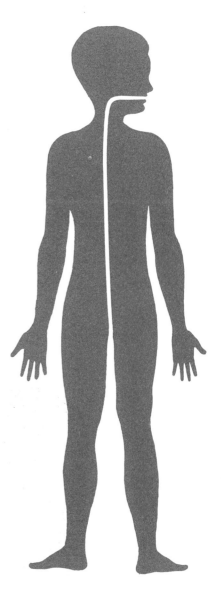

Fig. 44.2 The gastro-intestinal tract is a hollow tube, both ends of which open to the external environment. Actually the GI tract is part of the body's surface. Even though food is within the GI tract, it is outside of the internal environment.

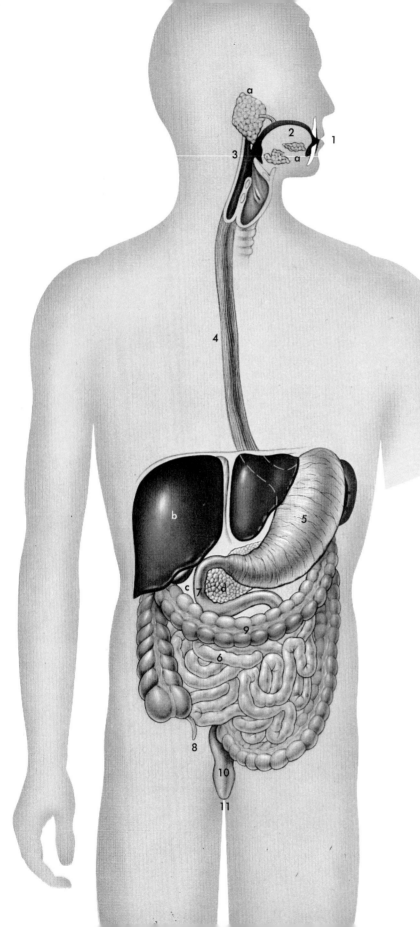

GASTRO-INTESTINAL TRACT

1 MOUTH
2 TONGUE
3 PHARYNX
4 ESOPHAGUS
5 STOMACH
6 SMALL INTESTINE
7 DUODENUM
8 APPENDIX
9 LARGE INTESTINE
10 RECTUM
11 ANUS

ASSOCIATED STRUCTURES

a SALIVARY GLANDS
b LIVER
c GALLBLADDER
d PANCREAS

THE DIGESTIVE SYSTEM

Fig. 44.3 The gastro-intestinal tract is a specialized surface of the body. When food enters the GI tract it is chewed, then mixed, and lubricated by salivary secretions. Swallowing carries it to the stomach, where digestive secretions begin to attack the proteins, but the food is not yet ready for absorption by the blood stream. When moved into the small intestine the food is subjected to the action of new secretions (from the liver, pancreas, and small intestine walls) that complete the job of digestion. It is in the small intestine that absorption into the blood stream takes place. Food that has not been absorbed is moved into the large intestine. Waste is finally eliminated through the anus.

TABLE 44.1

Food	Products: Building Blocks Which Can Be Absorbed
Fat (Lipid)	→ Glycerol, Fatty Acids,
Carbohydrates	→ Simple Sugars,
Proteins	→ Amino Acids

all contain carbon, hydrogen, and oxygen. Usually there are two hydrogen atoms for every oxygen atom. The building blocks of the common carbohydrates in our food are the simple, single sugars which contain six carbon atoms $C_6H_{12}O_6$. Two simple sugars can be joined by a chemical bond, forming a double sugar. In the process two atoms of hydrogen and one atom of oxygen are removed, so that the formula of a double sugar is $C_{12}H_{22}O_{11}$. Common table sugar (**sucrose**) is a double sugar. The sugar of the blood (**glucose**) is a simple six-carbon sugar.

Proteins: These are extremely large compounds made up of anywhere from about 100 to many thousands of smaller building blocks called **amino acids.** There are only about 20 different kinds of amino acids commonly found in proteins; but like the letters of the alphabet, they can be joined together in an endless variety of combinations and sequences, thus forming different protein molecules.

Amino acids are the constituents of protein that are absorbed by the digestive system. The proteins we eat must be broken down by a class of enzymes called **proteases** into amino acids. Of the 20 different amino acids, eight are termed **essential** because they cannot be formed by the body and, therefore, must be included in the proteins we eat. The remaining 12 amino acids can be formed from other substances in the body.

Once the amino acids are absorbed, they can be assembled into many new kinds of proteins by the cells of the body. However, the sequence in which the amino acids are joined to form some of these proteins is different for each cell type. Some of the proteins found in muscle cells, for example, differ from those found in kidney cells. When you eat beef, you swallow cattle-protein, but the beef protein never enters your blood. Instead, it is first broken down (digested) into its component amino acids; the amino acids are then absorbed into the blood, and finally new protein is constructed from these amino acids. This new protein is used in many different ways—for example, in the formation of enzymes, and in the formation of various cellular structures. Some of the proteins that your body builds are different from those of any other person! This is one of the things that make you different from every other person in the world.

Fats: These substances are made up of any three of the many different fatty acids chemically bonded to one glycerol molecule. Fats must also be broken down into simpler units in order to be absorbed. The lipases form the general class of enzymes responsible for this breakdown.

TABLE 44.2   Chemical Factors in Digestion

Gland	Secretes	Digestive Juice Acts Upon	Digestion Products
Salivary	Salivary amylase (ptyalin)	Cooked Starch	Double sugar
Stomach (gastric glands in stomach walls)	HCl* and pepsin (protease)	Proteins	Intermediate stages between proteins and amino acids
Liver	Bile salts**	Large fat droplets	Emulsified fat (small droplets)
Pancreas	Pancreatic amylase	Intact or partially digested starches	Double sugar
	Trypsin*** (protease)	Intact or partially digested proteins	Small amino acid groups (peptides)
	Steapsin (Lipase)	Fats	**Fatty acids†** **glycerol†**
Small intestine (glands in intestinal wall)	Erepsin (peptidase)	Split products of gastric and pancreatic digestion of proteins	**Amino acids†**
	(Enterokinase)	(Inactive trypsinogen)	(Active trypsin)
	Several carbohydrate-splitting enzymes	Double sugars	**Simple sugar†**

** Not enzymes
*** Secreted as inactive trypsinogen, which is converted to active trypsin by enterokinase
† In a state which can be absorbed

Adapted from *The Machinery of The Body*, by Carlson, Johnson, and Cavert

Enzymes: Table 44.2 shows a list of enzymes which are released into the GI tract and act upon the various foods we eat. You might wonder why the digestive enzymes do not attack and digest the walls of the stomach and the intestines themselves. The proteins that make up the structural components of the GI tract walls are protected by a coating of mucous which also lubricates the GI tract. But this is not the whole answer. Somewhere the digestive enzymes have to be formed. If they are formed in cells, then why don't they digest each cell that forms them? One possible explanation is that some enzymes are not present in an *active* form inside the cell. They are converted into an active form only after they have been secreted into the GI tract. Thus, the stomach enzyme **pepsin** is secreted in an inactive form known as **pepsinogen.** Pepsinogen is converted to an active form by the hydrochloric acid which is secreted in other parts of the stomach. Similarly, the enzyme **trypsin,** which comes from the pancreas, is secreted in an inactive form called **trypsinogen.** Trypsinogen is converted to the active trypsin by another enzyme, one which is liberated by the intestinal walls.

The enzymes listed in Table 44.2 are not secreted continuously into the gastro-intestinal tract. They are secreted in a more economical way only when food is present. Some are secreted only when specific types of food are present. Even though secretion is not continuous, on the average a person secretes four to nine liters of fluid into the GI tract each day!

Control of Digestive Secretions

Digestive enzymes and fluids are usually secreted in response to foods in the GI tract. The foods act by 1. stimulating sensory nerve endings which are involved in reflex control of secretion; 2. causing the release of hormones which in turn act on the secreting cells of the GI tract; and 3. simple mechanical stimulation of the walls of the tract. In the mouth, secretion of saliva is primarily under the control of nerve reflexes, but further down the tract, in the stomach and intestines, secretion is controlled by all three mechanisms—nerve reflex, hormonal, and mechanical stimulation.

A study of digestion in the stomach offers a good example of how the various types of control mechanisms have been discovered. The cells of the stomach secrete pepsin (which digests proteins), hydrochloric acid (which activates pepsin), and mucous.

The different types of control are most dramatically seen in animals which have had their gastro-intestinal tract surgically altered. Fig. 44.4 shows an experimental animal whose esophagus has been severed and the top portion sewn to the outside of its

Fig. 44.4 When food is placed in the mouth, gastric juices are secreted in the stomach. If the vagus nerves are cut, this response is lost, demonstrating *neural control* of gastric secretion.

body. When the animal eats, the food never enters the stomach, it just drops back into the dish. The lower cut end of the esophagus is also sewn to the outside, enabling the animal to be fed artificially. Finally, part of the stomach is opened to the outside so that its contents can be examined.

This animal can be used to demonstrate that the stomach is under nervous control before food enters it. For example, when the animal eats, gastric juices (enzymes, HCl) drip into the funnel from the stomach even though food never reaches the stomach. This same response, a nervous response, results from the sight, smell, or taste of food. The response produces one-half of the gastric juice that is secreted during a meal. When the two vagus nerves (the nerves of the stomach and most of the digestive system) are cut, this response is lost.

Even when all of the nerves of the stomach are cut, gastric juice is still released when food enters the stomach. This response, which is independent of the nervous system, seems to be controlled by at least two things: 1. when food enters the stomach and is churned and mixed there, the cells that secrete gastric juice are stimulated by the churning action. 2. There is also evidence that a hormone called **gastrin** is liberated when the stomach contains partially digested proteins. Gastrin, being a hormone, is not liberated into the stomach, but into the blood stream and is carried by the blood to the cells which produce gastric juice. We can show this hormonal effect by transplanting a piece of stomach to the skin (Fig. 44.5). When proteins enter the intact stomach the small piece of transplanted stomach begins to produce a gastric juice. Since the piece of transplanted stomach now has no nerve supply, it must have been stimulated by a hormone.

We now have some idea of what causes the stomach to secrete. We want to ask how stomach movements and secretions are coordinated with processes taking place in other parts of the GI tract. When we place fat in the small intestine of an animal with a transplanted piece of stomach, the transplanted tissue stops producing gastric juice. Since the transplanted piece of stomach is not attached to any intact nervous structure, it is reasonable to suspect that its secretion is controlled by a hormone. Further experiments indicate that the presence of fat in the small intestine stimulates the secretion of a hormone called **enterogastrone** into the blood. Upon reaching the stomach, enterogastrone inhibits the secretion of gastric juices and also inhibits movements of materials in the stomach. Slowing the stomach activity whenever food (for example, fat) arrives in the intestine is an advantage; digestion in the intestine will now have time to take place without being displaced by any further emptying of the stomach.

The Liver and Pancreas

The liver and the pancreas both deliver secretions to the digestive system. The secretions of the pancreas are particularly important. They include enzymes of the three basic types: **proteases, amylases,** and **lipases.** In addition to the enzyme portion of its secretions, the pancreas also adds a large quantity of the basic bicarbonate ion which is important in the neutralization of the hydrochloric acid secreted by the stomach. The secretions of the pancreas are controlled by hormones and nerve reflexes. They are carried to the gastro-intestinal tract by a duct (Fig. 45.2) which leads to the upper part of the small intestine (**duodenum**). In addition to these digestive secretions, the pancreas also produces endocrine hormones which it delivers without a duct into the blood stream (see the next chapter).

The common bile duct of the liver enters the duodenum, along with the pancreatic duct. Through this duct the liver secretes a material called **bile.** Bile helps break up large collections of fat molecules into smaller ones. This is called **emulsification**

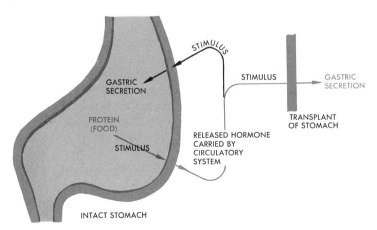

Fig. 44.5 When food is placed in the stomach, the stomach transplant secretes gastric juices, even though the nerve supply to the transplant no longer exists. This demonstrates *hormonal control* of gastric secretion.

It is the same general process that allows oil and water to mix when you wash your hands with soap (since the soap is an emulsifier). When occurring in the intestine, emulsification allows more lipase molecules to come into close association with the lipid, thus increasing the breakdown of lipid into fatty acids and glycerol.

The Small Intestine

A series of enzymes secreted by the cells of the small intestine complete the digestion of food. From Table 44.2, you will see that these include 1. erepsin which acts on small clusters of amino acids (called **peptides**), splitting them into the individual amino acids; and 2. other enzymes that split double sugars into simple sugars.

ABSORPTION AND MOTILITY

The secretion of digestive juices which split food into elementary building blocks is only one of the processes taking place in the GI tract. Next, the building blocks must be absorbed into the blood, and material within the tract must be moved along the length of the tube. A closer look at the structure of the tubular walls of the GI tract will help us understand these processes.

In Fig. 44.6 (top), we see a section made by cutting across the intestine (cross-section). The intestine is composed of three basic layers: 1. the innermost layer, or **mucosal layer,** which is richly supplied with blood vessels and contains cells responsible for the secretion of enzymes, the production of mucous, and the absorption of digested food. 2. The middle layer is a **muscular layer.** Some smooth muscles wrap around the cylindrical tube in a circular manner, while other smooth muscle cells run longitudinally up and down the length of the intestine. These muscles, which are influenced by autonomic nerve fibers, are responsible for the movement of material through the gastro-intestinal tract. 3. The outer layer of the intestines is composed of **fibrous connective tissue** which is covered for the most part by thin, flattened cells lining the inside of the abdominal cavity and covering many of the internal organs.

Absorption: In the bottom part of Fig. 44.6 we see that in order to reach the blood stream, materials must travel across the mucosal cells. This important process occurs almost exclusively in the intestines, but the mechanisms of absorption are poorly

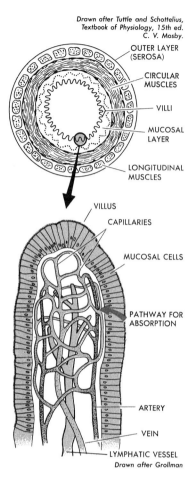

Fig. 44.6 *Muscosal cells form the inner lining of the small intestine (cross-section at top). These cells are specialized to absorb digested foods and provide the pathway for food to enter the blood circulation.*

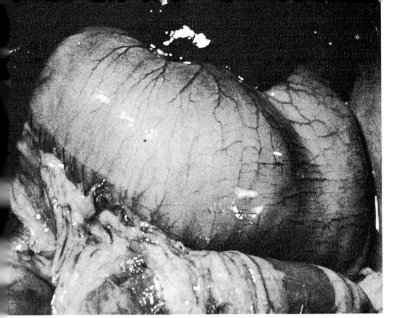

Encyclopedia Britannica Films

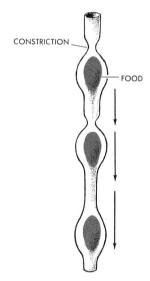

Fig. 44.7 *Peristalsis (in the esophagus, stomach, and intestines) moves food along the GI tract by waves of constriction that squeeze the food forward. The photograph shows one such wave in the stomach.*

understood. Some substances simply diffuse into the blood stream whenever their concentrations in the intestine is higher than in the blood. Other substances, such as simple sugars and amino acids, are actively transported so that they are absorbed even in low concentrations. In addition to nutritional components, the large amounts of fluid that are secreted into the tract must be reabsorbed into the body.

Whatever the mechanism, the speed of absorption is increased many times because of the peculiar structure of the mucosal surface. You can see that the intestinal surface is not smooth because of the many finger-like projections called **villi**. The villi increase the amount of mucosal surface available to the digested material. After the nutrients have been absorbed into the blood they are further processed by other cells of the body.

Motility: Material not absorbed by the GI tract must be removed to make room for fresh food. On the other hand, movement of material through the tract must not be so fast that the complex foods we eat are removed before they can be digested and absorbed. In other words, food must be moved through the system slowly enough so that each process has time enough to take place. On the average, food remains in the stomach from about two to four hours. After it is passed to the intestines, it takes many more hours before the undigested material is eliminated.

Food is moved along the GI tract by a process called **peristalsis.** This occurs in the esophagus, stomach, and intestines. It is caused by constriction resulting from contraction of the circularly arranged smooth muscles. As Fig. 44.7 shows, this constriction is a local one. It progresses down the length of the gastro-intestinal tract and, in a sense, squeezes the food along the length of the tract much in the same way as toothpaste is squeezed from a tube. The net effect of peristalsis is that the contents are pushed down the length of the GI tract toward the anus.

Other types of movements besides peristalsis also occur. One of the most important of these motions is the churning and the mixing that occurs in the stomach. The speed and the strength of gastric and intestinal contractions is controlled by the autonomic nervous system, by a slow spread of nerve impulses originating in a network of nerve fibers located entirely within the GI tract, and by the hormonal environment. These controlling mechanisms coordinate movements of the different parts of the tract. For example, the digestion of a large meal may be followed by an urge to defecate. This is partly due to a reflex initiated by mechanical distention of the stomach and duodenum, resulting in massive peristaltic movements in the lower parts of the large intestine. In general, excitation of parasympathetic nerves speeds the propulsion of material through the tract, while excitation of sympathetic nerves does the opposite.

Absorbed nutrients are transported by the blood to various organs of the body. Some of these nutrients are used by the cells for their energy content; others are used for growth and repair of cell structures; still others are stored for future use. These many processes do not occur in a haphazard way. They are balanced and delicately controlled. We shall see some examples of this in the next chapter.

SUMMARY

To enter the internal environment, food must be transported across the cellular walls of the GI tract into the blood stream—it must be **absorbed.** There is only a small variety of molecules absorbed by the GI tract. **Carbohydrates** must be broken into **simple sugars; proteins** into **amino acids;** and **fats** into **glycerol** and **fatty acids.** This digestion of food is accelerated by specific **enzymes** secreted by the walls of the GI tract and certain glands associated with the tract.

Digestive enzymes are not secreted continuously. They are controlled by nerve reflexes, hormones, and mechanical stimulation of the walls of the GI tract. Absorption occurs in the intestines. While some substances are absorbed by diffusion, others, such as simple sugars and amino acids, are actively transported into the blood stream, so they are absorbed even if they are present at low concentrations.

Food is moved along the GI tract by **peristalsis,** which occurs in the esophagus, stomach, and intestines. It is caused by a progressive constriction, which results from the contraction of the circularly arranged smooth muscles. This constriction progresses down the length of the tract and squeezes the food along. The movements of the GI tract are controlled by nerve impulses and by hormones. **Parasympathetic** nerve impulses speed movements through the tract, while **sympathetic** nerve impulses do the opposite.

FOR THOUGHT AND DISCUSSION

1 Suppose that you eat a hamburger. The bun is rich in carbohydrate; the meat is rich in protein, fat, and some indigestable matter. Trace the fate of the hamburger from the time it enters your mouth until it leaves the GI tract. What enzymes are involved, where are they secreted, how are they controlled?

2 A peptic ulcer is an open sore in the wall of the stomach or duodenum. It occurs more frequently in persons subjected to continual stress and anxiety. Peptic ulcers are believed to be caused by the digestive action of excessive gastric juice secretions. It is sometimes treated by cutting the vagus nerve supply to the stomach (vagotomy). Discuss the basis and physiological implications of this treatment.

SELECTED READINGS

Bayliss, W. P. and E. H. Starling. "The Mechanism of Pancreatic Secretion," in *Great Experiments in Biology*. Englewood Cliffs, New Jersey: Prentice-Hall, Inc., 1955, p. 60.

Carlson, A. J., V. Johnson, and H. M. Cavert. *The Machinery of the Body*, 5th Ed. Chicago: University of Chicago Press, 1961.

Davenport, H. W. *The Physiology of the Digestive Tract*, 2nd Ed. Chicago: Year Book Medical Publishers Inc., 1961.

Guyton, A. C. *Textbook of Medical Physiology*, 3rd Ed. Philadelphia: W. B. Saunders Co., 1966.

45 METABOLISM AND HORMONES

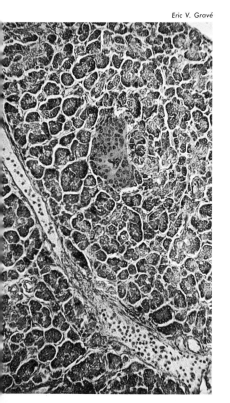

Fig. 45.1 The islet cells of the pancreas (color) secrete a small protein called *insulin*. Without this vital protein the body starves in the midst of plenty, for it cannot convert glucose into useful energy. People who cannot produce their own insulin have the disease *diabetes mellitus*.

Eric V. Gravé

In 1889, the two German physiologists Joseph von Mering and Oscar Minkowski began a study that led to the solution of an age-old mystery. They wanted to know what, if any, disturbances in digestion would result if they removed the pancreas glands from dogs. After they operated on several dogs the animals began to show many deficiencies, and after a short time they died. Why? The operation had been a simple one. The pancreas glands secrete digestive juices, but they are not the only source of digestive juices. Apparently the glands had some special function. Although the dogs had hearty appetites after the operation, they lost weight and lacked energy and muscular strength. Further, they began to use stored fat and cellular protein for the energy required just to keep alive. These and other symptoms were very similar to those of starvation.

von Mering and Minkowski soon found that the problems facing these animals were far more profound than simple starvation. One curious thing they noticed was that ants gathered in the kennels where the sick dogs were kept, and the ants were attracted in greatest numbers to those places where the dogs had urinated. Why? There were no gatherings of ants in the kennels of the healthy dogs. von Mering and Minkowski found that the blood sugar (glucose) level in the sick animals was abnormally high—so high that the animals were excreting large amounts of it into their urine. Although there was plenty of blood sugar available, these animals were not able to use it; they were "starving" in the midst of plenty. This is precisely what happens to people who have the disease called **diabetes mellitus** if they are not treated. The experiments of von Mering and Minkowski were to prove that the pancreas was associated with this disease.

Further research has shown that most of the "starvation" symptoms which occur either in diabetes or following removal of the pancreas are not due to disturbances in digestion; instead they result from an inability to use glucose. This can be demonstrated simply by tying-off the pancreatic ducts leading from the

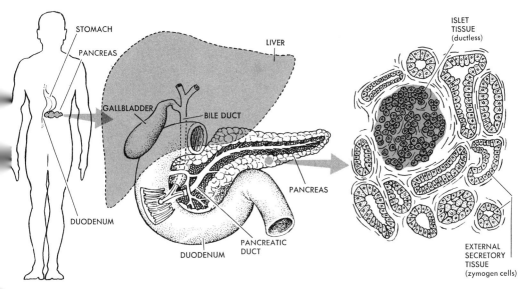

Fig. 45.2 The pancreas is a dual organ. Zymogen cells secrete digestive enzymes into the pancreatic duct, which conveys the enzymes to the duodenum. Islet cells secrete insulin into the circulating blood.

pancreas to the duodenum. This prevents the pancreatic secretions from reaching the GI tract. Although the animal will have digestive disturbances, he can still use glucose, he does not develop diabetic symptoms, and he will not die. However, if the pancreas is removed, the animal loses the ability to use glucose, diabetic symptoms appear, and the animal dies.

Apparently, in addition to secreting digestive enzymes, the pancreas is involved with the animal's capacity to use glucose. This dual role of the pancreas is better understood if we examine its microscopic structure. Figure 45.2 shows that the pancreas is made up of two types of cells. One type, called **zymogen** cells, makes up a tubular network ending in dilated sacs called **alveoli**. The ducts of the many alveoli contained within the pancreas unite and give rise to the ducts of the pancreas which carry the pancreatic digestive enzymes to the small intestine. The other type of cells is found among the alveoli in small clumps, or "islands." They are called **islet** cells. When the ducts are tied off, the zymogen cells degenerate, but the islet cells remain healthy. The fact that animals in which the ducts have been tied off do not develop diabetes suggests that the islet cells are required for the normal use of glucose by other cells of the body.

How do the islet cells exert an influence on other cells? One possibility is that they secrete into the circulating blood a hormone which is needed for glucose utilization. If this is true, then we might expect to find high concentrations of the hormone within the islet cells, where it is produced. This can be demonstrated by first tying off the pancreatic ducts in an animal so that the zymogen cells degenerate, but leaving healthy islet cells. We then remove the pancreas, mince it, and soak it in a salt solution. The resulting solution, containing many materials which have escaped from the islet cells, is called an **extract.** If we inject some of the extract into a diabetic animal, all symptoms of diabetes disappear! One of the substances in the extract is the hormone we are looking for; it is a small protein called **insulin.**

We have just described a common pattern for the study of **endocrine** (hormone producing) glands: 1. the gland is removed and the resulting symptoms are recorded. 2. Various extracts of the gland are injected back into the animal until we find an active one that relieves the symptoms. 3. We then examine the various components of the extract in hopes of finding which chemical is responsible for the activity of the extract—that is, which chemical is the hormone. But this is not the end of our search. We also want to know how the hormone influences cellular activities, how a lack of hormones leads to the gross symptoms that we observe, and how the secretion of hormones is controlled.

Function of Insulin

Many experiments suggest that one function of insulin is to make cells more permeable to glucose. This would explain why the diabetic cannot use glucose. If the sugar cannot get into these cells, the cells cannot use it and have to turn to other sources for energy—fat and protein. Despite the fact that little or no sugar can enter many of the body cells in the diabetic, glucose continues to be absorbed into the blood from the GI tract. As a result, the concentration of blood sugar rises to such high values that it cannot be reabsorbed by the kidney tubules, and large amounts of glucose escape into the urine. The excessive urine sugar is accompanied by an equivalent excess of water. (Recall from Chapter 44 how osmotic forces determine water excretion by the kidney.) This extra excretion of water explains the large flow of urine invariably found in the diabetic. In fact, the loss of body fluid (through excessive urine formation) can become so great in the untreated diabetic that it can lead to a collapse of the circulation, and death.

Control of Insulin Secretion

In the normal animal, insulin secretion is controlled by the circulating blood glucose as shown by the feedback loop in Fig. 45.3. Blood glucose stimulates insulin secretion by the pancreas so that when blood glucose levels are high, more insulin is secreted. The extra insulin increases glucose utilization (probably because it increases cell membrane permeability to glucose) and the blood glucose level falls back toward normal.

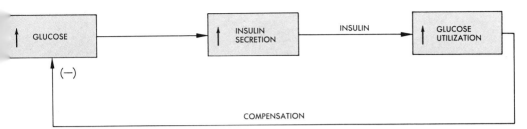

Fig. 45.3 A high concentration of glucose in the blood stimulates the secretion of insulin. The feedback loop shows how this response helps to stabilize the blood glucose concentration.

However, if the blood glucose level falls, less insulin is secreted, resulting in a smaller amount of glucose used and a consequent rise in the blood glucose level back toward normal. Here we see a simple feedback system which, in this instance, helps stabilize the level of blood glucose. However, blood glucose as well as tissue fat and protein are strongly influenced by a number of hormones and other factors, in addition to insulin. In order to understand them, we must first examine some of the inter-relations between carbohydrates, fats, and proteins.

METABOLISM

Figures 45.4 A, B, C are simplified diagrams of some of the chemical transformations that take place within body cells. These reactions plus all the other chemical reactions taking place in the body are collectively called **metabolism**.

There are several reactions within the cell that liberate useful energy. Figure 45.4A shows that each of the three primary building blocks—fatty acids (and glycerol), glucose, and amino acids—can be broken down within the cell. If oxygen is present, carbon dioxide and water are formed. During the process, energy is liberated, part of it is lost as heat but part is trapped and stored through the formation of ATP. These steps are indicated by processes 1, 2, and 3 in the figure. The ATP can be used by the "machinery" of the cell (step 4) for a variety of purposes: for example, to supply energy for muscle contraction, secretion, and synthesis of complicated molecules.

In contrast to energy-yielding reactions that break molecules into smaller pieces, cells are capable of many synthetic reactions that require energy. In these reactions the primary building blocks are combined into larger molecules. Step 7 in Fig. 45.4B shows that many molecules of glucose can be combined and stored in the form of **glycogen** (animal starch). This occurs primarily in liver and in muscle. In times of need, the liver glycogen breaks down (step 8) and liberates glucose to the blood.

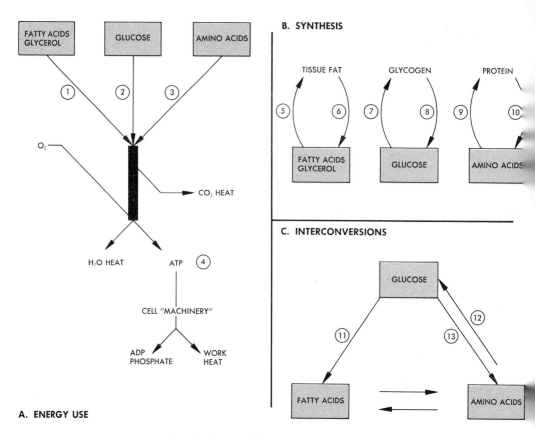

Fig. 45.4 In addition to breaking down molecules, cells can build up, or *synthesize*, large molecules. This diagram shows that the three principal foodstuffs—carbohydrates, fats, and proteins—are broken down, built up, and reconverted, depending on the cell's energy needs.

Similarly, fatty acids are combined with glycerol to form fat, which is stored in fat tissues distributed throughout the body (step 5). Finally (step 9), the amino acids can be strung together into huge protein molecules which form structural parts of cells, enzymes, plasma proteins, and other body constituents.

Figure 45.4C also shows that the three principal foodstuffs—carbohydrates, fats, and amino acids—can be interconverted. Through a sequence of steps (11 and 5), glucose can be used to form tissue fat. This often occurs when you eat too much carbohydrate. The amount of carbohydrate that can be stored as glycogen is not very large, but this is not true of tissue fat. Accordingly, excessive amounts of glucose are converted into tissue fat. Finally, steps 13 and 12 show that glucose can also be converted to amino acids, and amino acids into glucose.

From Fig. 45.4 we can again see some of the problems faced by the diabetic, who is not able to make use of glucose. Step 2 is blocked, yet the work of the body must continue. Since glucose cannot be tapped as a source of energy, fatty acids (step 1) and amino acids (step 3) must be broken down, and eventually so must tissue fat and tissue protein.

Hormones and Energy Release

The metabolic scheme that we have outlined in Fig. 45.4 is influenced by several hormones. We can get some idea of the complexities involved by examining how they control the blood sugar, or glucose, level. In a normal person the blood sugar remains relatively constant despite the fact that his dietary intake of carbohydrate is very irregular. The liver (Fig. 45.5) which serves as a storage depot for carbohydrates plays an important role in this regulation. Immediately after a meal, while food is being absorbed, blood sugar is carried by the blood from the gastrointestinal tract to the liver, where a large part of it is stored as glycogen (step 7 in Fig. 45.4). At this time the blood entering the liver contains more glucose than does the blood leaving the liver. Later, when absorption is complete, the reverse is true. Blood leaving the liver is richer in glucose than is blood entering it. We conclude that the liver is now breaking down some of its stored glycogen (step 8 in Fig. 45.4), releasing it as glucose, thus helping to replenish blood glucose as it is used by other tissues.

The release of glucose from the liver is accelerated by the action of adrenalin (**epinepherine**), a hormone secreted by the adrenal medulla. You will recall from previous chapters that adrenalin is secreted in times of stress. During such times it aids the animal by providing him with glucose from the liver (step 8 in Fig. 45.4).

The liver contains enough glycogen to supply body tissues with glucose for several hours. However, when the supply of glycogen becomes low, protein and fat are broken down. This use of protein and fat is thought to be controlled by secretions from the adrenal cortex (Fig. 45.6) and the anterior pituitary (Fig. 45.7). **Cortisol,** one of many hormones secreted by the adrenal cortex, promotes the use of protein. For example, during fasting, cortisol induces the passage of amino acids from tissue protein into the liver where they are converted to glucose (steps 10 and 12 in Fig. 45.4). Cortisol is also necessary for the effective breakdown of fat. If your adrenal glands were removed, you would develop a low blood sugar within 12 to 24 hours after you stopped eating. However, the presence of adrenal glands helps you to maintain a normal level of blood glucose, even during many days of fasting.

One of the hormones secreted by the anterior pituitary gland, called **growth hormone,** plays an equally important role in the regulation of blood glucose. This hormone promotes the mobilization of fat from tissue deposits, making the fat available as an energy source (steps 6 and 1 in Fig. 45.4). It also inhibits the use of glucose. The action of growth hormone seems to switch the source of cellular fuel from carbohydrate to fat. Thus, its action

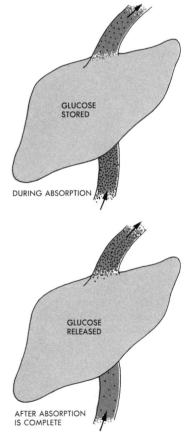

Fig. 45.5 The liver plays an important role in regulating the blood sugar (glucose) level. In a normal person, glucose is stored or released as it is needed.

GLUCOSE STORED

DURING ABSORPTION

GLUCOSE RELEASED

AFTER ABSORPTION IS COMPLETE

DOTS REPRESENT GLUCOSE

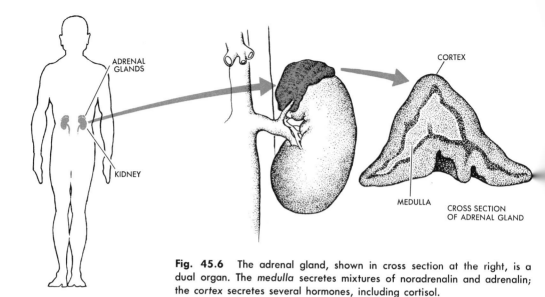

Fig. 45.6 The adrenal gland, shown in cross section at the right, is a dual organ. The *medulla* secretes mixtures of noradrenalin and adrenalin; the *cortex* secretes several hormones, including cortisol.

seems to be antagonistic to insulin. We can demonstrate this in a dramatic way: if we remove the pancreas from an animal, the animal develops diabetes. However, if we next remove the anterior pituitary gland, so that the source of growth hormone is cut off, the diabetic symptoms are greatly diminished! This experiment does not mean that removal of the pituitary gland is a cure for diabetes, but simply that under such conditions the diabetic symptoms disappear. An animal that has had both glands removed becomes very unhealthy.

Despite the antagonistic actions of insulin and growth hormone, the secretion of both hormones is *controlled* so that they work together and help stabilize the blood glucose. Recent experiments suggest that a low blood glucose stimulates the secretion of growth hormone. This promotes cellular use of fat rather than glucose and the metabolic machinery is no longer as dependent on the low level of glucose. One result is that the blood glucose level may increase back toward normal. If these ideas are correct, the interaction between blood glucose and growth hormone provides us with the homeostatic feedback

Fig. 45.7 Secretions of the anterior pituitary gland stimulate the activity of several other glands; it also secretes growth hormone.

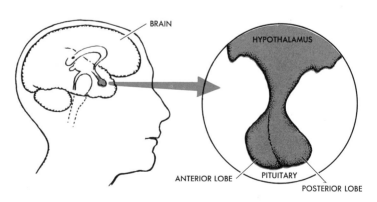

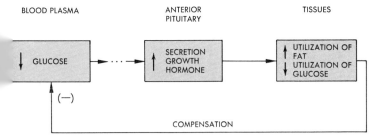

Fig. 45.8 Growth hormone is secreted in response to a low concentration of blood glucose. The feedback loop shows how this response may help stabilize the blood glucose concentration.

loop illustrated in Fig. 45.8. Compare this with the control of insulin illustrated in Fig. 45.3.

One other hormone affecting the processes outlined in Fig. 45.4 is **thyroxin**. This hormone is secreted by the thyroid gland (Fig. 45.9) and seems to play a role in the regulation of the overall metabolic rate. With excessive thyroid activity, oxygen consumption of most cells is increased above normal. With subnormal levels of thyroid hormone oxygen consumption is reduced.

Hormones and Growth

We have already mentioned two hormones which are very important in normal growth, one of them being growth hormone. Growth hormone plays a very important role by promoting the entry of amino acids into cells and by further promoting the incorporation of these amino acids into proteins (step 9 in Fig. 45.4). In short, it strongly influences the growth of cellular structures. Its effects are dramatically revealed if growth hormone secretion during childhood is abnormal. With subnormal amounts of growth hormone a child will never reach adult size; he will be a dwarf. However, excessive secretion during

Fig. 45.9 Thyroxin helps regulate the over-all metabolic rate. It is secreted by cells of the thyroid gland and stored in the follicles (see cover photograph).

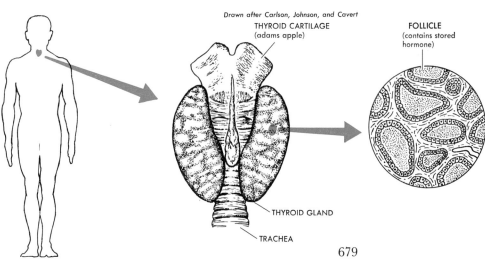

childhood and adolescence results in abnormally fast growth and produces a giant. Figure 45.10 shows a dwarf compared with a giant. Both suffered from abnormal growth hormone secretions during childhood.

Thyroxin also plays a role in incorporating amino acids into protein and so influences growth in the child and adolescent. In children with very poor thyroid function, growth will be stunted. There will also be a retardation in mental, physical, and sexual development. If the condition remains untreated, the person will be a sterile dwarf with an intelligence that may never exceed that of a five-year-old.

Hormone Control: the "Master Gland" and Hypothalamus

The anterior pituitary is often referred to as the "master gland" because it secretes several hormones required by other glands. These include the thyroid, the adrenal cortex, and the sex glands (discussed in Chapter 47). However, recent experiments have shown that the anterior pituitary, in turn, seems to be controlled by secretions originating in the central nervous system, specifically from the **hypothalamus,** which is located at the base of the brain, very close to the pituitary (see Fig. 45.7).

Let us consider interactions between the pituitary gland and thyroid as an illustration of this type of control. The anterior pituitary secretes a hormone called **thyroid-stimulating-hormone,** abbreviated **TSH.** As its name implies, TSH stimulates the thyroid glands, promoting the glands' growth and secretory activity. The resulting secretion of thyroxin in turn acts on the pituitary gland, depressing further TSH secretion. Thus, we have a control loop which stabilizes the level of circulating thyroxin.

If, for some reason, the thyroxin level gets too high, it will depress the secretion of TSH, which in turn will depress the production and secretion of thyroxin so that its concentration in the blood will return to normal. However, if the thyroxin level is reduced, then its inhibiting effect on the secretion of TSH will also be reduced; the production of TSH will then be increased, resulting in a greater secretion of thyroxin. This feedback regulation is shown in Fig. 45.11.

The description we have given explains how a constant level of thyroxin can be maintained in the circulating blood. However, it does not explain how the secretion of the hormone changes and meets changing demands of the body. A good guess would be that the changing needs of the body are "sensed" by the hypothalamus, which, in turn, influences the release of TSH. For example, in a cold environment the hypothalamus secretes a substance which stimulates the secretion of TSH. This, in turn,

Fig. 45.10 During childhood, an abnormally high secretion of growth hormone causes excessive growth (giantism). An abnormally low secretion inhibits growth (dwarfism).

Ewing Galloway

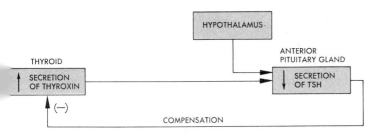

Fig. 45.11 Thyroxin inhibits the secretion of TSH by the anterior pituitary. As the feedback loop shows, this response tends to stabilize the thyroxin level in the blood. However, the level of thyroxin can be changed by hypothalamus secretions, thus stimulating release of TSH.

stimulates the secretion of thyroxin, and, finally, thyroxin stimulates general body metabolism. The net result is greater heat production within the body, thus enabling the animal to withstand the cold environment and maintain homeostasis.

A similar process controls the secretion of cortisol by the adrenal cortex. The anterior pituitary secretes another hormone called **ACTH**, which promotes the growth and secretion of the adrenal cortex. The secretion of ACTH, in turn, is inhibited by cortisol, and stimulated by secretions from the hypothalamus.

Hormonal Mechanisms

In recent years physiologists have done a great deal of research to find out how hormones exert such a strong influence on metabolism. Some hormones, such as insulin, are known to change the permeability of cell membranes and, therefore, regulate the amount of materials made available to the metabolic machinery. However, this explanation does not apply to all hormones, nor even to all of the actions of insulin. This means that we must look for other possibilities.

A promising suggestion is that hormones act by promoting the activities of cellular enzymes. Most of the chemical reactions occurring within a cell take place on the surface of large protein enzyme molecules. Without these enzymes the reactions would be too slow to be of any importance. The presence of hormones may be necessary for certain of these enzymes to work, or the hormones may influence the actual synthesis of specific protein enzymes. Evidence supporting this latter possibility has come out of experiments in which physiologists have used specific antibiotics known to interfere with protein synthesis. These same antibiotics have inhibited the actions of a number of hormones.

HEAT AND BODY TEMPERATURE

Roughly one-half of the energy liberated by decomposition of carbohydrate, protein, or fat is captured in the form of ATP. The rest of the energy is lost as heat. The ATP formed eventually is used up as the body does work—lifting running, and so on. Since no process is 100 per cent efficient in converting all of the available energy into work, we can expect to find some heat produced during most physical activity.

Despite the continuous heat production within your body your temperature remains constant. This means that heat is lost from the body just as fast as it is produced. This ability to regulate body temperature is advantageous, because changes in temperature cause changes in the rates of chemical reactions. Cooling or heating could upset a number of chemical reaction sequences which must be delicately balanced if the body is to function well. Animals like frogs, which cannot regulate their body temperature, are at the mercy of their environment. When the weather cools, their body temperature falls, their metabolism slows, and they become very sluggish.

What, actually, is body temperature? If you hold a thermometer against your skin you get a reading quite different from the reading when you hold the thermometer under your tongue. The temperature that we measure under the tongue corresponds fairly well with the temperature that we would measure in any of the vital organs—heart, liver, or the intestines. It is this inner, or **core temperature** (temperature of the vital organs) that we mean by "body temperature." The core temperature is well regulated by the body, rarely falling below 97 or rising above 104 Fahrenheit degrees.

The temperature measured on the skin is called the **surface temperature.** It fluctuates much more than the core temperature, rising and falling with the temperature of the environment. The correspondence between the core temperature and the surface temperature depends on the blood flow to the skin. If the blood flow to the skin is large, then the surface temperature will begin to approach the core temperature. When the blood flow to the skin diminishes, the surface temperature tends to approximate the environmental temperature.

If the temperature of the body is to remain constant the heat produced must just balance the heat lost. The processes whereby heat can be produced and heat can be lost are illustrated in Fig. 45.12.

Heat Production

Your body produces heat even when you are resting. The heat that is produced by the metabolism required just to keep you alive and awake is called the **basal metabolic heat.** Over and above the basal metabolic heat, muscular activity such as exercising or shivering also produces heat. In addition

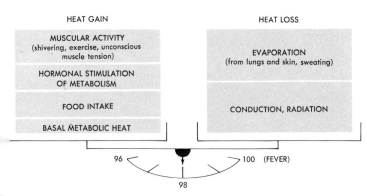

Fig. 45.12 Because heat gain is balanced by heat loss, our body temperature tends to remain constant.

to muscular activity, hormones can influence the production of heat through their influence on metabolism. This is particularly true of thyroxin and adrenalin. It is by variations in muscular activity and the quantity of circulating hormones that heat production in your body frequently changes.

Heat Loss

You may lose heat by radiation and conduction from the surface of your body. The amount of heat you lose this way depends on the difference in temperature between the surface of your body and the environment. The greater the difference, the greater the heat loss will be. In order for heat to be lost, of course, the surface of your body must be at a higher temperature than that of the environment. If the environmental temperature rises above your body temperature, you will not lose heat by radiation or conduction; instead, your body will gain heat from the environment.

A second very important method of losing heat is through evaporation. When you come out of the water after swimming, water evaporates from your skin and you feel cool. In order for water to evaporate, the water molecules must have a certain minimum amount of energy. The faster moving molecules can overcome the forces holding them in the liquid state and bound off into the air as gas (water vapor) molecules. The slower and, therefore, cooler molecules are left behind. Heat then flows from the warmer surface of your skin to the cooler water molecules. This flow of heat transfers energy to the water, speeding the water molecules up so that more of them escape. This cooling of your skin surface also cools any blood which tends to flow through that part of your body. Sweating is an obvious way to lose heat by evaporation. However, in addition to sweating, some water is continually lost by processes that we are normally not aware of. Water continuously evaporates from the skin. There is also a small loss of water from the surface of the lungs when you breathe. The amount of water that evaporates—when you breathe or sweat—depends on the humidity of the air. When the humidity is high, water evaporates much more slowly and, therefore contributes less to the cooling process.

Temperature Regulation

Control of body temperature depends on the **hypothalamus.** It acts as a thermostat and responds to the temperature of the blood bathing it, and to nerve impulses that come from temperature receptors in the skin. It is the front part, or anterior hypothalamus, that responds to heat. We can either stimulate this part electrically or heat it up and get several responses which increase heat loss. These include 1. sweat secretion, which increases heat loss by evaporation; and 2. dilation of blood vessels in the skin, which causes the surface temperature of the body to approximate the core temperature, with the result that more heat is lost by conduction and radiation. Signals for these responses travel over sympathetic nerves. If we destroy the anterior hypothalamus there will be no response to heat.

The rear portion, or posterior hypothalamus, responds to cold. If we chill the blood surrounding this part of the hypothalamus, any sweating that may be occurring at the time stops. Also, the blood vessels in the skin constrict, minimizing the heat exchange between the core and the cool surface of the body. In addition to preventing excessive heat loss, the hypothalamus is involved in regulating heat production. Chilling, for example, causes shivering and a hormonal discharge which increases body metabolism. These two responses increase heat production. The feedback loop which is involved in controlling temperature is illustrated in Fig. 45.13.

Fever and the Body's Thermostat

The hypothalamus acts as a thermostat which is set to maintain a temperature of 98.6°F. Sometimes this thermostat setting will be altered. Although the precise mechanism is obscure, we know that this occurs during infection and when products of tissue destruction are introduced into the blood stream. In both cases a fever results. Although the body's temperature regulation machinery is operating, the thermostat seems to be set at an abnormal level.

At the onset of fever, even though your body temperature is normal, the body responds as if it were too cold. You begin to shiver and the blood vessels in the skin constrict causing pale, dry skin, and decreased heat loss. Finally your metabolism may step up, with a resulting increase in heat production. All of

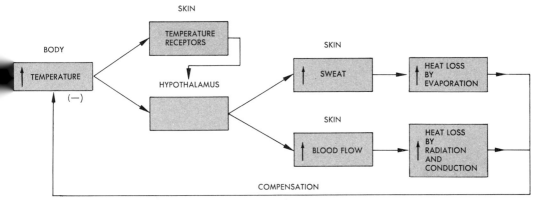

Fig. 45.13 An increase in temperature of the body surface or blood acts on the hypothalamus, which controls the sweat glands and blood vessels in the skin. The feedback loop shows how the hypothalamus responds and stabilizes body temperature. Can you draw the corresponding feedback loop to show the response to cold?

these factors tend to raise your temperature. This continues until the temperature settles at the new high-level setting. When fever is on the decline the opposite response occurs. Your skin becomes flushed, indicating that the blood vessels of the skin have expanded and you begin to sweat. Both of these responses lower your body temperature by increasing its heat loss.

Response to Intense Heat

Suppose that you go into a room that has an air temperature of about 110°F. Radiation and conduction do not work in your favor. Instead of losing heat from the surface of your body to the surroundings, you gain heat. You can survive, but now sweating is the only mechanism you have for losing heat.

The normal response to intense heat strains the circulatory system. This follows because the hypothalamus responds to the increased heat by causing the blood vessels in your skin to expand. This leads to a decreased resistance to blood flow and your blood pressure tends to fall. Reflexes which prevent large changes in blood pressure (see Chapter 40) then begin to operate, and the decreased resistance to blood flow is compensated for by the heart working harder. The expanded blood vessels make it possible for large amounts of blood to pool in the vessels of your skin at the expense of other organs. If, as a result, the blood supply to your brain becomes sufficiently low, you will faint.

Sweating may also create a circulatory problem because of the salt and water loss. Excessive fluid loss causes a decreased plasma volume. This may slow down the output of blood from the heart, which could lead to decreased blood flow to the skin which, in turn, could reduce sweating. If this happened, your main avenue for heat loss would be closed. In that event heat production would continue and your body temperature would rise until your whole system collapsed.

The body's ability to control heat loss is limited. *When heat cannot be lost rapidly enough to prevent a rise in body temperature,* a vicious

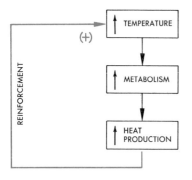

Fig. 45.14 The body's ability to control heat loss is limited. If heat gain increases beyond the body's ability to compensate, positive feedback throws the system out of control, as shown by this feedback loop.

circle may occur. This follows because the rate of metabolism (like most chemical reactions) increases with a rise in temperature. As the body temperature rises, the rate of metabolism and, consequently, the rate of heat production increase. This accelerates the rise in body temperature. When heat regulation fails, the positive feedback loop illustrated in Fig. 45.14 goes into operation; if unchecked it ends in heat stroke and death.

The best defense against this intense heat is simple: leave the room! You use your brain and decide to manipulate or change the external environment until it no longer poses a threat. Interactions between brain and environment are taken up in the next chapter.

SUMMARY

Many of the chemical transformations taking place within the body are controlled by hormones. Hormone secretions, in turn, are subjected to complex feedback regulation.

Insulin promotes the use of glucose by making some cells more permeable to glucose. A high blood glucose stimulates secretion of insulin and, as a result of its action, insulin helps maintain a stable concentration of glucose in the blood.

Growth hormone promotes the use of fat, inhibits the use of glucose, and helps regulate the level of blood glucose. Further, it promotes the incorporation of amino acids into proteins and is essential for growth.

Thyroxin stimulates oxygen consumption and plays an important role in regulating the metabolic rate. It is also important in growth. Children with deficient thyroid secretions will be retarded in their mental and physical development. The growth and secretion of the thyroid gland is controlled by TSH (secreted by the anterior pituitary). The secretion of TSH is, in turn, controlled by thyroxin and secretions from the hypothalamus.

Cortisol promotes the formation of glucose from amino acids in the liver. Secretion of cortisol is controlled by ACTH (secreted by the anterior pituitary). The secretion of ACTH is, in turn, controlled by cortisol and secretions from the hypothalamus.

The work and metabolism of the body is always accompanied by the liberation of heat. Nevertheless, the body temperature (core temperature) remains constant; heat production balances heat loss. The hypothalamus is a center for temperature regulation. It responds to the temperature of the blood that comes to it and to nerve impulses that come from temperature receptors in the skin. Heating the hypothalamus results in sweat secretion and dilation of blood vessels in the skin. Chilling the hypohalamus results in decreased heat loss (inhibition of sweat secretion and constriction of blood vessels in the skin) and increased heat production (shivering and an increase in body metabolism).

FOR THOUGHT AND DISCUSSION

1 If the liver is removed from an animal, how will its blood glucose be affected? Explain.
2 Around 1900, researchers tried to make hormonal extracts of the entire pancreas gland, which contained both islet cells and zymogen cells, but they all failed. Can you suggest a reason for their failure? (HINT: in addition to secreting insulin what are the other functions of the pancreas?)
3 Experimental research procedures used in the study of hormones include: (a) removal of glands (elimination of hormones); (b) replacement of hormones; and (c) overdoses of hormones. Keeping these procedures in mind, design experiments to demonstrate the dependence of the thyroid on the pituitary gland.
4 A man is lost in the hot desert without water. What physiological responses have you studied in this and in preceding chapters that will tend to help him survive?
5 After exposure to cold, your lips may turn blue. The color is due to the high content of reduced hemoglobin in the blood supplying the skin. Why does cold bring about this response?

SELECTED READINGS

Baldwin, E. B. *The Nature of Biochemistry.* New York: Cambridge University Press, 1962.

Benziger, T. H. "The Human Thermostat," *Scientific American* (January 1961).

Davidson, E. H. "Hormones and Genes," *Scientific American* (June 1965).

Irving, L. "Adaptations to Cold," *Scientific American* (January 1966).

Levine, R. and M. S. Goldstein. "The Action of Insulin," *Scientific American.* (May 1955).

Scholander, P. F. "The Wonderful Net," *Scientific American* (April 1957).

Tepperman, J. *Metabolic and Endocrine Physiology.* Chicago: Yearbook Medical Publishers, Inc. 1963.

Wilkins, L. "The Thyroid Gland," *Scientific American* (March 1960).

46 INTERPRETING THE ENVIRONMENT: THE BRAIN

Bettman Archive

Fig. 46.1 Self-portrait of Leonardo da Vinci: man as a creative animal stands alone in the animal kingdom. No other organism is capable of contemplating itself in an analytical way.

Up to now we have considered how man survives by regulating his internal environment. But a significant part of man's success in surviving depends on his ability to change his external environment. This involves a large variety of behavior patterns ranging from the reflex withdrawal of a hand from a hot stove to the invention of agriculture, the use of shelter and clothing, the storage of food, and interaction with other men. All of these activities depend on the central nervous system, the brain, and spinal cord.

Suppose that you see a bunch of grapes on a table, walk over and pick one up. You see the grapes with your eyes, then use muscles in your legs and arms to get to the grapes. Nerve impulses that originate in your eye touch off a chain of events leading to the activation of many different muscles. Still other muscles would have to be activated if you were going to catch a grape tossed to you. We could imagine direct pathways (nerves) leading from the eye (and every other receptor) to every muscle in the body. If this were the case, then all motions would be possible, but the number of nerves and connections would have to be enormous; and the actions that took place would probably be chaotic. Instead, the eye and other receptors send their impulses to a central station, the brain and spinal cord. From here connections are made with nerves that lead to different muscles. This is like a telephone switchboard. Each telephone has a wire leading into the central switchboard, and if appropriate connections are made there, any telephone user can call any other user. There is no need for each telephone to have a direct line to every other telephone in the country.

The nervous system is much more than a simple switchboard. The messages (nerve impulses) sent from the central nervous system to the effectors are determined not only by the incoming messages from all parts of the body, but also by the past history of incoming messages. In other words, the central nervous system has a "memory;" in addition, it can learn and is capable of comining in unique ways the information it gathers. The result of such gathering and storage of information may give rise to a work of art, a scientific theory, or a "nervous breakdown." Today we have very little understanding of these higher mental activities, and we are only beginning to appreciate some of the simpler forms of behavior.

DEVELOPMENT AND STRUCTURE OF THE CNS

The central nervous system of an adult human being is an extremely complex structure, yet it begins as a simple line of cells on the back of the developing embryo. The cells that form the line begin to divide, then form a groove which later develops into a hollow tube, the **neural tube** (Fig. 46.2). The brain and spinal cord develop out of this primitive neural tube. At the head end of the embryo, the brain develops from three bulging structures, the **forebrain,** the **midbrain,** and the **hindbrain** (Fig. 46.2). The remaining portion of the neural tube becomes the spinal cord. Portions of the forebrain and hindbrain greatly increase in size and give rise to two very prominent structures of the adult brain—the **cerebrum** and the **cerebellum** (Fig. 46.2). Both of these structures are covered by an outer layer of cells which gives them a gray color. This outer gray layer is called the **cortex.**

The brain develops from a hollow tubular structure. In the adult the hollow interior persists in the form of an intricate system of chambers (or **ventricles**) which contain the **cerebrospinal fluid.**

In the human, the cerebrum is the most massive part of the brain and is divided into two hemispheres. The cerebral cortex grows so large that it folds in on itself in wrinkles and is thus contained in the available space within the skull. When you look down on an open skull the exposed part of the brain consists primarily of the wrinkled cortex, looking something like a large, soft walnut.

Most of the mental abilities that set man apart from other animals are thought to be due to his more highly developed cortex. If a man loses his cortex he becomes an absolute idiot. On the other hand, however, a frog is not nearly so dependent on its cerebral cortex. Even if the animal loses its cortex it may be difficult to distinguish the frog from a normal frog. Apparently the primitive behavior characteristic of a normal frog can be carried out to a large extent by structures lying beneath the cortex.

Nerve fibers carrying impulses to and from the cells of the

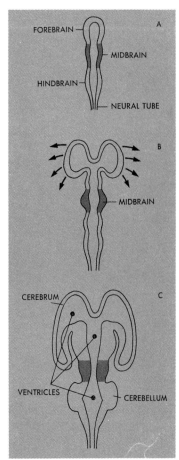

Drawn after E. D. Gardner "Fundamentals of Neurology," Scientific American.

Fig. 46.2 The nervous system develops from a hollow tube (A). At the head-end of the tube three bulging structures become prominent and develop into the brain (B and C).

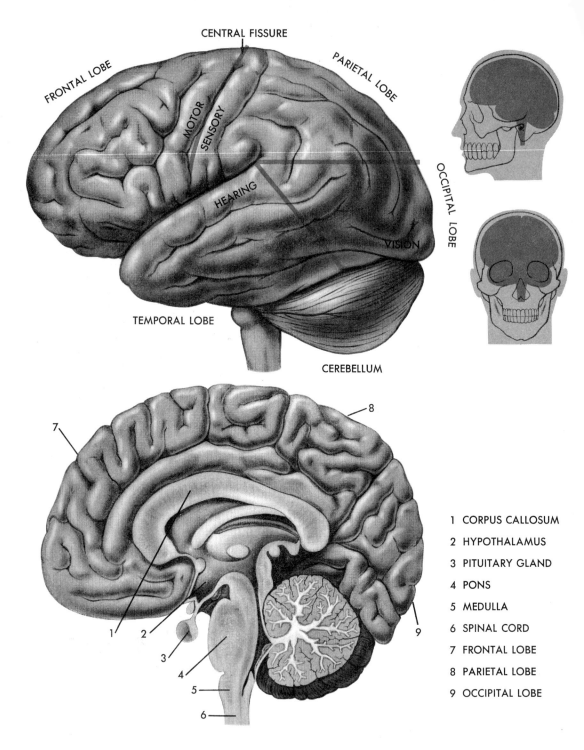

THE HUMAN BRAIN

Fig. 46.3 Man's brain is the single organ that most sets him apart from all other organisms. The highly developed cerebrum has a surface consisting of many folds, which increase its area. On the scale of evolution, the cerebral cortex, which forms a thin layer over the cerebrum, is a relatively new refinement. We are just now beginning to learn something about its role in determining our behavior and how it interacts with other parts of the brain. The bottom illustration shows a longitudinal cut through the brain, exposing the inner surface of the right hemisphere.

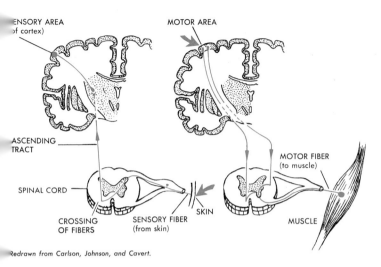

Fig. 46.4 At left, skin is stimulated on the left side of the body. An impulse travels to the right side of the brain, where it arrives in the sensory area of the cortex. At right, a motor impulse leaves the motor area of the cortex on the right-hand side of the brain and travels to a muscle on the left side of the body.

gray cortex make connections only with other cells of the brain and spinal cord. These cells, in turn, may be connected to the receptors or muscles and glands in the body. The more developed the cortex, the more it seems to overshadow and dominate these deeper brain and spinal cord structures.

Mapping the Cortex

We can begin studying the brain by stimulating different parts of the cortex. If we choose a certain area and stimulate it, a specific movement is produced. For example, stimulating one small spot makes a finger move; if an adjacent area is stimulated the whole arm may move. These areas, called **motor areas,** have been discovered during experiments with animals and during brain operations on humans. The motor areas are clustered in the region (indicated in Fig. 46.3) just in front of a prominent groove known as the **central fissure.** Just about any part of the body can be made to move if we stimulate some spot within this area (Fig. 46.5). Stimulating the motor areas on the left cerebral hemisphere always causes movements on the right side of the body; similarly, stimulating the right motor cortex produces movements on the left side. This also seems to be true for many other brain activities. The left side of the brain controls the right side of the body; the right side of the brain controls the left side of the body (Fig. 46.4).

The size of the brain controlling a particular part of the body is not related to the size of the body part; instead, it seems to be proportional to the skill or complexity of movement of which that body part is capable. The cortical areas devoted to finger

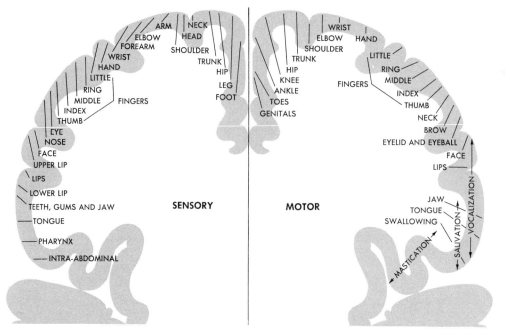

Fig. 46.5 These diagrams show certain brain areas that are associated with various activities. The sensory cortex shown here is located just behind the central fissure; the motor cortex is just in front.

Redrawn from Scientific American.

movements, for example, are much larger than areas devoted to movements of muscles in the back.

If we move our stimulating electrodes backwards from the motor area to the area labeled "sensory cortex" in Fig. 46.3, we find that movements are no longer produced; instead, tingling sensations are felt. These sensations are not felt on the brain but in specific parts of the body such as the fingertips or face. Apparently, sensations coming from different parts of the body are sent to specific regions on the sensory cortex. We can verify this by another type of experiment. If we stimulate different parts of the body with bristles of a stiff brush we find that there is an increase in electrical activity in the sensory cortex. When we stimulate an area on the right side of the body the sensory cortex on the left cerebral hemisphere shows increased activity and *vice versa*. There is some sensory information, however, that is not directed to this part of the cortex. Vision, hearing, smell, and taste all have special cortex areas that receive incoming impulses.

Association Areas

So far, we have described those areas of the brain where sensory information arrives in the cortex, and other areas where motor commands leave. These areas occupy only a small fraction of the total cortex. The remaining areas making up the greater mass of the cortex are loosely referred to as **association areas.** The space taken up by association areas in man is relatively larger than it is in lower animals. Both men and lower animals seem to receive similar types of information from their environments. They both perform similar movements and have comparable sensory and motor areas of their brains. Nevertheless, man is the superior animal because of his ability to store vast amounts of information and use it in an endless variety of combinations whenever he wishes. His ability to think reflectively seems to involve the association areas. Without these areas it would be impossible for your brain to be wondering how it works.

Let's return to our bunch of grapes on the table to illustrate the role of association areas. First you see the grapes. The image is registered on your retina, which is the light-sensitive portion of your eye, and is transmitted by the optic nerves to the brain and finally to the visual area of your cerebral cortex. However, you have yet to *recognize* what you see as grapes, and here is where the association areas play a role. The concept "grape" implies many things—for example, certain colors, texture, size, shape, taste, and smell. You gain information and form your own idea of what a grape is through many different sensations. Impulses that convey these different sensations arrive at different areas of the cortex. Yet all of this information must be integrated into a single concept, "grape." Somehow or other, association areas seem to be involved in this integration process. We get some idea of their importance by considering what happens when they are injured.

If some of the association area near the visual region of the cortex were destroyed, you would see the grapes, but you would not be able to identify them as grapes. The visual clues you would be receiving would no longer be meaningful. However, if you handled the grapes, or tasted one, or used any other sense whose association areas were still intact, you would be able to identify the objects as grapes. The association areas in some way relate sensory clues to stored (memory) information. Patients whose visual association area has been damaged have been shown keys, for example. They shook their heads, unable to recognize the objects as anything in particular. When the keys were rattled, however, the patients were able to identify them immediately.

The perception-recognition process is an extremely complex one. It not only involves what is sensed at the moment, it also involves memory of past associations and our response to them. We can imagine a mentally ill person who, upon being shown a key, reacts violently because of some unpleasant event the key causes the patient to recall. We can also imagine the response of a person who has never seen a key before. Essentially the same process of sensing and recognizing is operating in each person, but the visual association areas of the cortex along with other parts of the brain cause the two people to respond to the key stimulus in entirely different ways.

Fig. 46.6 The cerebellum enables us to operate with precision and with smooth coordination.

The Cerebellum

Now suppose that you pick up the grapes with a smooth, swift, motion. The accuracy of your arm motion is aided by the sight of your hand during its journey, also by sensations arising from your muscles, tendons, and joints. The cerebellum coordinates this information and plays an important role in producing smooth, accurate motion. If your cerebellum were damaged and you sat quietly, your handicap would not be noticed. But the moment you moved, your condition would be obvious. Your motion would be jerky and uncoordinated. When you reached out for the grapes your hand might overshoot them. If you attempted to correct this and bring your hand back, you might over correct and withdraw your hand too far. The result would be a series of jerky motions which eventually would "zero in" on the target.

The cerebellum seems to be involved in assessing the body's position in space and in sending out signals that enable you to make rapid corrections of faulty motions. The cerebellum receives information from the cerebral cortex, from sensory organs, and from deep sensations coming from receptors in muscles, tendons, and joints. This deep sense, called **proprioception,** provides information about the position and performance of the limbs and body. Whenever the motor cortex commands muscles to move, it sends information at the same time to the cerebellum. The cerebellum, in turn, sends impulses back to the cerebral cortex (and to other motor centers in the brain) correcting any errors with the outcome that the resulting motion is smooth and well-timed. The cerebellum is able to do this on the basis of information it is always receiving from muscles and joints; that is, it compares the "commands" of the motor cortex with the "performance" of the muscles. The cerebellum does not initiate movements. It acts only to make them precise in time and space.

Our description of the cerebellum illustrates how the cerebral cortex interacts with, and is influenced by, other parts of the brain. Even the simplest behavior patterns are not wrapped up in neat little anatomical packages in the brain, where we can stimulate them or cut them out. The dependence of the cerebral cortex on deeper brain structures is further emphasized in the process of sleep and attention.

Sleep and Attention: the Reticular Formation

The possibility of awakening from sleep and remaining conscious seems to depend on a diffuse network of nerve cells called the **reticular formation** (Fig. 46.7). Stimulation of one part of the reticular formation awakens a sleeping animal, more or less as though it had been gently patted on the head. If the reticular formation is destroyed, the animal immediately lapses into a deep sleep, or coma, never again to awaken. The reticular formation is necessary if the awakened conscious state is to persist. We can interpret this by assuming that when the reticular formation receives sufficient sensory impulses from the environment it showers the cortex with activating impulses which are necessary to alert it.

Recent experiments suggest that lower brain centers can filter information that is passed along to the higher centers. This filtering process may be important in directing our attention to specific subjects. When you are reading, for example, and are unaware of noises around you, it *seems* as though impulses that would ordinarily convey noise sensations do not reach your cortex. Experiments on cats suggest that not only does it *seem* as though these impulses do not reach the cortex, but they *actually* may not.

In an experiment, each cat had an electrode implanted in the medulla at the point where the nerve carrying impulses from the ear first makes synaptic connections with nerve cells in the brain. From here, impulses are sent on toward the cerebral cortex. The electrode was used to record the cat's responses to noise. Whenever a clicking noise was made near the cat's ear a wave of electrical activity was received by the electrode. Each time there was a click, there was a wave of activity. Next, the cat's attention was diverted to something of special interest, such as some live mice. As soon as the cat became interested in the mice, the electrical activity produced by the continuing series of clicks decreased. The cat no longer "heard" the clicks because nerve impulses from the ear were not reaching the cortex. Apparently the lower parts of the brain, and in some cases the receptors, are capable of filtering information before it is passed on to the cortex.

Fig. 46.7 The *reticular formation* receives sensory impulses from the environment and showers the cortex with activating impulses, thus alerting it.

Drawn after Scientific American

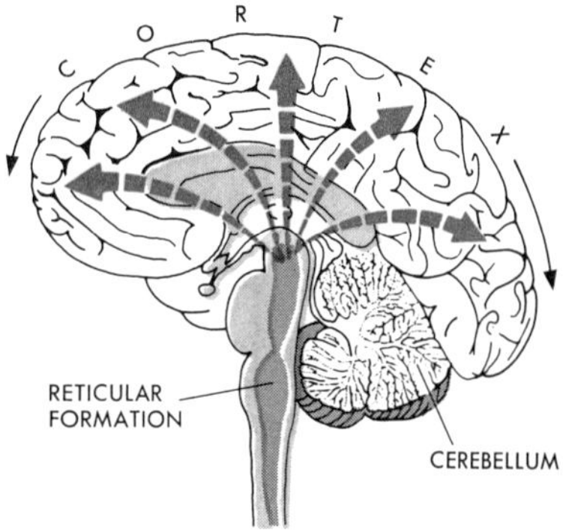

Memory

The human brain has the remarkable capacity of storing impressions of the past and calling them into consciousness at appropriate times. Are there specific memory areas in the brain? And what physical changes take place as a result of this storage of information? Here again our information is meager; the best we can do is offer some of the observations and experiments which may provide clues.

Some of our memories are fleeting, yet others endure. After recovering from a blow on the head a person may suffer from loss of memory, but as time passes he regains his memory. However, he may never remember what happened just a few moments before the blow was struck. It seems that a certain length of time is needed for an event to be transferred to our permanent memory. This can be shown very clearly in laboratory animals. Say that we train an animal to perform a certain task and within five minutes after each training session we give the animal a severe electric shock. The animal then shows no signs of remembering what it was taught during the session. However, if the animal is given a shock several hours after each training session, the shock has no effect on its training; that is, it learns to perform the task and remembers what it has learned just as well as animals which are not given a shock.

What do these experiments suggest about the processes involved in the storage of information? Possibly there are two stages: first, as soon as information—a new telephone number say—reaches the brain it is stored, but only temporarily. Some time later, and this would be the second stage, the information is more permanently stored. If an electrical shock, or a blow on the head is delivered during the first stage, the information is lost so that when the brain recovers, the information cannot be processed and delivered to a permanent memory storage. Some scientists now think that the permanent memory mechanism may ultimately be traced to chemical changes in some of the structures of the nerve cells, and that the process of information storage in the nervous system is related to protein synthesis. It should be emphasized, however, that we still have very little evidence to construct a theory of memory, so these last remarks can be considered only a guess about what may be revealed in the future.

Perhaps the most startling discovery made about memory is that the brain can be stimulated to recall events long past. Some of the first comprehensive reports of this phenomenon came from Wilder Penfield, a neurosurgeon in Montreal. Penfield was interested in the surgical removal of diseased parts of the brain, and he wanted to be sure that he did not remove any brain part that was vital to normal activities, such as speech. His technique was to expose the surface of the brain under local anesthesia. Even though the patient was fully conscious, the exposed brain caused no discomfort because there are no pain receptors on the surface. Penfield would then stimulate different parts of the brain with electrodes. Next he would ask the patient what he felt and would observe any motions. In this way he could associate the patient's response with certain brain areas stimulated.

...nd then carefully select the diseased areas which could be removed safely. By recording all of his observations Penfield made significant contributions to our knowledge of the localization of functions in the human brain.

During the course of one of these operations an electrode was applied to the grey matter on the face of one of the temporal lobes (see Fig. 46.3). Here is how Penfield described what happened:

". . . the patient observed: 'I hear some music.' Fifteen minutes later, the electrode was applied to the same spot again without her knowledge. 'I hear music again,' she said. 'It is like radio.' Again and again, then, the electrode tip was applied to this point. Each time, she heard an orchestra playing the same piece of music. . . . Seeing the electrical stimulator box, from where she lay under the surgical coverings, she thought it was a gramophone that someone was turning on from time to time. She was asked to describe the music. When the electrode was applied again, she began to hum a tune, and all in the operating room listened in astonished silence. She was obviously humming along with the orchestra at about the tempo that would be expected. . . .

"After the patient returned home, she wrote to me on April 16, 1950. The letter was, in part, as follows: . . . 'I heard the song right from the beginning and you know I could remember much more of it right in the operating room. . . . There were instruments . . . . It was as though it were being played by an orchestra. Definitely it was *not* as though I were imagining the tune to myself. I actually heard it. It is not one of my favorite songs, so I don't know why I heard that song.' "

[After similar observations Penfield concludes] ". . . that there is, hidden away in the brain, a record of the stream of consciousness. It seems to hold the detail of that stream as laid down during each man's waking conscious hours. Contained in this record are all those things of which the individual was once aware—such detail as a man might hope to remember for a few seconds or minutes afterwards, but which are largely lost to voluntary recall after that time. The things that he ignored are absent from the record." [From: *Proceedings of the National Academy of Science,* Volume 44, 1958, page 57.]

Conditioned Reflex

The process of learning is a good example of how stored information is put to use. One of the most primitive types of learning, called **conditioned reflex,** was studied by the Russian physiologist Ivan Pavlov. Pavlov knew that whenever he placed meat in the mouth of a dog the animal would reflexly salivate; in other words, its mouth would "water" automatically. He experimented with this reflex in dogs by ringing a bell each time he gave meat to the dog. After many trials, Pavlov found that if he withheld the meat and simply rang the bell the dog would still salivate. The dog had been **conditioned** to associate the sound of the bell with the presence of meat in his mouth.

The flow of saliva that usually follows the sight of meat is in itself a conditioned reflex. Pavlov proved this by raising puppies without ever giving them meat. When the puppies were shown meat, they did not salivate. However, after the meat had been placed in their mouth a few times, they began to salivate whenever they saw meat.

It is possible to combine different types of stimuli into conditioned reflexes in novel ways. Normally you cannot voluntarily control the size of the pupil in your eye, but if a bright light shines in your eye, the pupil will automatically (reflexly) constrict. If someone rings a bell each time the light is turned on, you may become conditioned to the sound of the bell. Now you have some control. Each time you ring the bell, your pupil will constrict, even though the light is not on.

Conditioned reflexes are a useful tool to study animal behavior. Dogs, for example, can be conditioned to distinguish between different shades of grey, but not between different colors. From this we infer that dogs are color blind. Interesting behavior in animals results when they are forced to make difficult discriminations. A dog can be conditioned to salivate whenever he sees a circle, but not to salivate when an ellipse appears. Now, the ellipse is made more and more circular until the dog is not able to tell the difference between the ellipse and the circle. At this point the dog begins to show signs of anxiety, squealing and wiggling about. He may whine and howl and have violent temper tantrums—he becomes "neurotic."

The conditioned reflexes we have described thus far may not be of any particular advantage to the animal. The response is fully automatic and not under voluntary control. More complicated and useful forms of learning occur when an animal is punished or rewarded for responses that are under voluntary control. An animal may be placed in a situation where he must choose between two alternatives, say, turning to the right or turning to the left. By turning to the right he is rewarded; by turning to the left he is punished. Here, the voluntary action of turning to the right and the reward are associated. In short, the animal is taught to perform some task, or to operate within the environment, in order to obtain a reward or avoid punishment.

The physical or neuronal basis for even primitive learning such as conditioning is not known. A search for specific areas on the cortex which might be involved in learning has not been fruitful. Rather than finding specific "learning" areas, the whole brain seems to be involved. This problem has been studied extensively in white rats. Different parts of the cortex have been removed to find out what, if any, effect the surgery would have on the amount of practice required to learn to run a maze. No matter what part of the cortex was removed, the learning process was slowed. The size of the piece of cortex removed seemed to be more important than the area from which it was taken.

The only information that reaches our brain is sent there by our sensory receptors. Let us now turn to two of man's most important special senses—sight and hearing.

HEARING AND THE EAR

Think of the variety of sounds you hear during just about any five-minute period of a typical day: loud noise and soft noise, high-pitched tones and low-pitched tones. Among the many aspects of hearing that physiologists study, two have been particularly interesting: 1. how the ear translates sound into action potentials along the sensory nerve which leads to the brain; and 2. how high-pitched tones are discriminated from low-pitched tones.

The sound a book makes when you drop it onto a table is a complex one because it consists of many different pitches. Simpler sounds, pure tones, can be produced by hitting a tuning fork so that the tines vibrate. As each tine moves outward, it compresses the air in front of it. The region of compressed air in front of the tine tends to compress the air just adjacent to it, and this new region of compressed air also tends to compress adjacent air (Fig. 46.8). In this way the area of compression travels outward in all directions from the tine of the fork, at 1129 feet per second (770 miles per hour). Since the fork is vibrating, this means that it is moving forward many times per second and producing many waves of compression, which constitute **sound waves.**

Now suppose that we put a membrane in the path of these sound waves. As each wave of compressed air hits the membrane it pushes against the membrane, and stretches it. Each time a wave of compressed air hits the membrane, the same thing happens. The membrane vibrates. The **tympanum,** or ear drum (Fig. 46.9), is just such a membrane that initiates a series of events leading to our perception of sound. When the strength of the vibration is increased, we hear a louder sound. When the frequency of vibrations (the number of them per second) is increased, the pitch becomes higher. The human ear can hear between about 20 and 20,000 vibrations (cycles) per second. Middle C corresponds to 256 c.p.s.

The tympanum separates the external ear from the middle ear (Fig. 46.9). The external and middle ear are filled with air; the inner ear is filled with fluid. The coiled portion of the inner

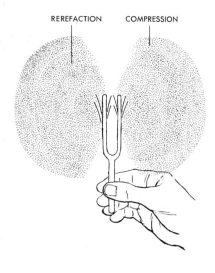

Fig. 46.8 Areas of compressed air travel outward as waves from the tines of a tuning fork. The ear and brain perceive the waves as sound.

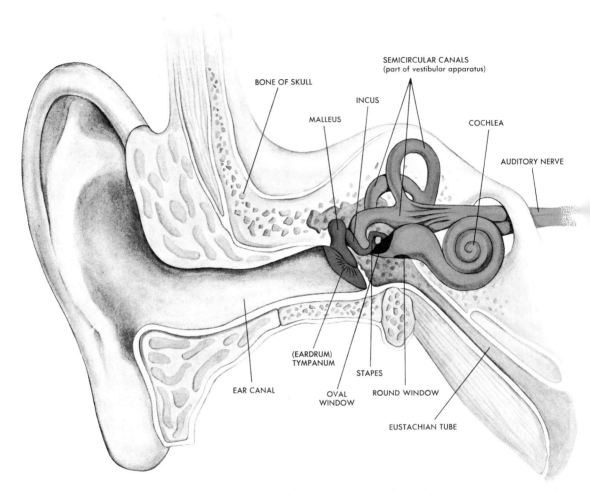

Fig. 46.9 This diagram shows the various parts of the outer, middle, and inner ear.

ear, called the **cochlea,** is very important for hearing. The fluid-filled cochlea is divided into three major compartments by two flexible membranes. One of them—the basilar membrane—connects with nerve endings that transmit impulses from the ear to the brain.

The middle and inner ear are separated by rigid walls except at two places, the oval window and the round window, as shown in Fig. 46.10. Here the separating walls consist of membranous material which can stretch. The bones of the middle ear form a bridge between the ear drum and the oval window. If we were able to take hold of the ear drum and move it back and forth the bones would also move back and forth, and since they are connected to the oval window it, too, would move back and forth. Because the inner ear is filled with fluid, every time the oval window is pushed in, the round window bulges outward. In other words, the fluid in the inner ear pulsates with the ear drum.

The membranes separating the compartments of the inner ear are pliable. When the fluid vibrates, these membranes also vibrate and move up and down. These movements of the basilar membrane are responsible for stimulation of the nerve endings.

If we follow a compression wave from the moment the oval window is pushed in, we find an important clue to how the ear discriminates pitch. The wave of compression does not necessarily go all the way down to the end of the inner ear, bounce off the wall and then back on the other side to the round window (broken line in Fig. 46.10). Rather it tends to take a shorter path and goes through the basilar membrane (solid line in Fig. 46.10). This means that some parts of the basilar membrane will move and be deformed more than others. With low frequencies the wave tends to travel farther on down before it turns and goes through the basilar membrane. With high frequencies (more rapid vibrations) a shorter path is taken and the maximum deformation occurs closer and closer to the middle ear. This is the basis for pitch discrimination. For each frequency there is a particular part of the basilar membrane that vibrates most strongly. When the nerve impulses from that region dominate we get the characteristic sensation of that particular pitch. Thus when most of the impulses come from a specific region of the basilar membrane, they give you the sensation of hearing, say, middle C. A pitch higher than middle C would come from nerve fibers located closer to the middle ear, and likewise a pitch lower than middle C would come from nerve fibers located farther away.

Our evidence for this theory of pitch discrimination comes from several sources. Some people are deaf only to specific tones. Autopsy shows that in such people specific regions of the basilar membrane had been injured. These sites correspond to the sites which we would predict on the basis of the theory. Second, we can produce lesions in laboratory animals by exposing them to loud noises. By selecting a particular pitch, and subjecting the animal to an intense sound of that pitch, we can cause a particular part of the basilar membrane to vibrate so violently that it becomes damaged. These animals will then be deaf to that pitch. (How would you test this in an animal?)

As Fig. 46.9 or 46.10 shows, the air in the middle ear is not sealed off from the air surrounding it. A tube called the **eustachian tube** leads from the middle ear to the mouth. This opening is very important in preventing the development of any pressure differences across the ear drum. If a large pressure difference did

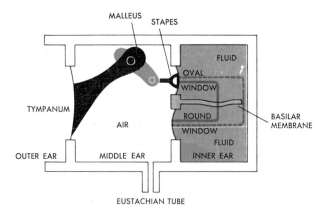

Fig. 46.10 A wave of compression transmitted by the ear drum does not travel all the way to the end of the inner ear (dotted line). Instead, it tends to take a shorter path, passing through the basilar membrane (solid line).

develop, the ear drum could shatter. For example, when you climb a mountain the air around you becomes less and less dense. If there were no eustachian tube, the pressure in the middle ear would remain at its sea-level value. Pressure in the middle ear, then, would be greater than in the external ear. This would cause the ear drum to bulge outward. If the pressure difference became great enough, the ear drum would break. The sensation you notice when you take off or land in an airplane is due to mild pressure differences before your ears are adjusted by the eustachian tube. Sometimes you can help their adjustment by yawning, which helps clear the opening of the eustachian tube.

In Fig. 46.9 you may have noticed structures in the inner ear which we have not considered. These make up the **vestibular apparatus.** They are not concerned at all with hearing, but enable you to detect the motion of your head and thus orient yourself in space. The inner ear, then, consists of two organs, the cochlear part, which is concerned with hearing, and the vestibular apparatus which is concerned with motion and orientation.

SEEING AND THE EYE

A number of problems concerned with vision are resolved in the eye—focusing, seeing in bright or dim light, and so on. The main problem is how light energy is transformed into action potentials.

As Fig. 46.11 shows, the eye is covered on the outside with a tough white tissue called the **sclera.** Toward the front of the eye, where it bulges, the sclera becomes transparent and is known as the **cornea.** An inner layer, called the **retina,** contains cells sensitive to light. These cells are connected to nerve fibers that transmit impulses to the brain. If the eye were a camera, the retina would correspond to the film.

As you can see from Fig. 46.11, the eye lens divides the eye into two chambers, both of which are filled with fluid. Like a lens on a camera, the lens of the eye must be brought into focus if the eye is to get a sharp image of what we are looking at. In

Interpreting the Environment: The Brain 703

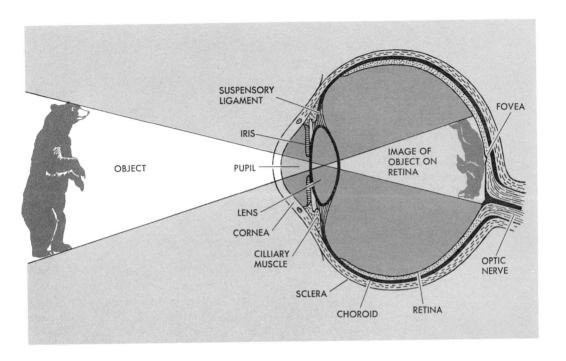

Fig. 46.11 The human eye and a camera are similar in many respects.

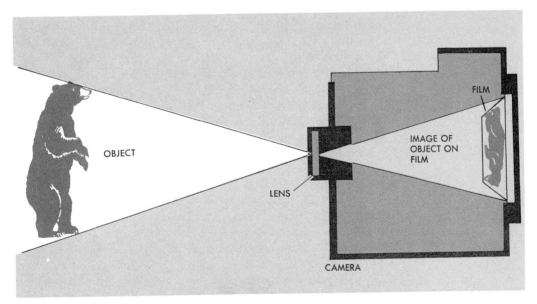

Fig. 46.12 The lens of the eye is elastic. When the *ciliary* muscles contract, they move toward the point of attachment (see diagram at right), relaxing the tension on the lens and permitting it to take a more rounded shape. Diagram at left shows the lens in a stretched position (solid line), when the eye is focused on a distant object. The red line shows how the lens *accommodates* and thus keeps an object in focus as the object moves closer.

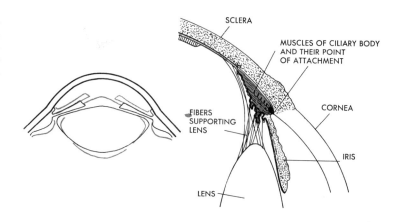

Drawn after Ganong, Lange Medical Publications; Carlson, Johnson, and Cavert.

order to do this, we can change the strength of the eye's lens because it is an elastic body. Normally the eye lens is stretched in such a way that it is flattened and elongated. It is held in this shape by its attachments to the **ciliary body.** A person with normal eyes can see an object 20 feet away in clear focus when the ciliary muscles are relaxed. But if the object is brought closer, and the ciliary muscles remain relaxed, the object becomes blurred. For the object to be kept in focus, your ciliary muscles must contract, in the process pulling the lens attachments closer together and thereby releasing tension on the lens. This action permits the lens to get rounder and more powerful (Fig. 46.12). In this way you are able to get sharp images of objects as they are moved as close as $3\frac{1}{2}$ inches away from your eye. Try looking at your finger and notice that as you bring it closer and closer your finger eventually reaches a point where it becomes blurred. This is because your eye lens has become as round as it can get.

This process of adjusting the eyes so that they can focus on an object as it is brought closer is called **accommodation.** In addition to changing the shape of the lens, accommodation also causes a constriction of the pupil (the opening of the eye which permits light to enter and reach the retina) and a change in the position of the eye so that it begins to point toward the nose. The power of accommodation usually diminishes with age. This is because the lenses become more rigid, hence it is difficult for them to take on a round shape when the ciliary muscles contract and reduce tension on the lens. Many defects in the lens system of the eye can be corrected with glasses. This simply involves adding another lens to the system, just as you might add an extra lens to a camera to increase its performance.

A good photographer is not only concerned with the lens system in his camera, but also is careful in selecting his film. In bright sunlight he uses a relatively insensitive film, which he replaces with a sensitive film for dimly lit subjects. The eye employs the same strategy by using different cells of the retina. The retina has two types of light-sensitive cells. One type consists of **rods,** the other of **cones.** We can illustrate the functions of the rods and cones by some simple examples. If you want to see fine detail in the daytime it is best to look directly at the object. By doing so, you are causing light reflected from the object to fall on the central part of your retina, the **fovea.** At night the situation is quite different. On entering a dark room it is best not to look *directly* at an object. If you want to see it more clearly, look at it out of the side of your eye, so that light reflected from the object falls toward the edges of the retina (Fig. 46.13). When we compare the distribution of the rod cells and the cone cells in the retina, we find that the cones are most concentrated toward the center and the rods toward the edges. This leads us to believe that the cones are primarily concerned with day vision and the rods are mostly concerned with night vision. This is confirmed when we study animals that have poor vision in the dark and contrast them with animals, like bats or owls, which see very well at night. Animals with little or no night vision have retinas containing mostly cones. Bats, owls, and other animals with good night vision have more rods in their retinas.

What happens in the cones and the rods when light falls on them? Our information is not complete in either case, but we do know more about rods than cones. The rods contain a pigment called **visual purple.** When light falls on visual purple the visual purple is bleached. During the bleaching process action potentials are set up on the nerve fibers connected to the rod cells. It is at this point that light energy is in some way translated into action potentials. When you are in a bright room, there is little visual purple in the rods, so your rods are ineffective. This is because the visual purple has been bleached by the bright light. Now if you go into a dark room, your rods are of little use to you immediately; but if you wait a few minutes, visual purple will be regenerated and will enable your rod cells to function. Gradually you will see objects in the room more and more clearly. Now suppose that you have become completely adapted to the dark and suddenly go into the light. At this point the light "hurts" your eyes and you have difficulty seeing. Now your rod cells, with a full supply of visual purple, are too effective. They are so effective that *everything* looks very bright; there is no contrast. You have to remain in the light a few minutes until the visual purple has been bleached so that the highly sensitive rods no longer respond. Information originating in the cones is then no longer "swamped" by the great activity of the rods.

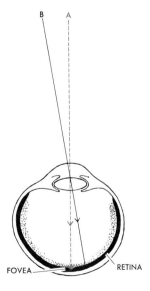

Fig. 46.13 In a dark room it is easier to see objects if you don't look directly at them. The object at B is easier to see if you look directly at Point A. Light from the object at B will then fall on a region of the retina rich in rods.

SUMMARY

The **cerebrum** is the most massive part of the brain. Its gray surface, or **cortex,** is particularly well developed in man and it is essential for man's unique mental abilities. Although parts of the brain involved in various physiological functions have been located, attempts to localize complex behavior patterns in areas have failed. The cerebral cortex interacts with itself and with other parts of the brain in a complex way. Our present efforts to describe these interactions are at an early stage of development.

Sensory areas, where impulses first arrive on the cortex, have been located. **Motor areas,** where "command" impulses leave the cortex, have also been located. Other areas of the cortex are thought to be involved in *associations* of sensory information. Stimulating some parts of the cortex produces a detailed recall of past events.

The **cerebellum** receives impulses from sensory receptors and interacts with the motor cortex, insuring smooth muscular movements. Damage to the cerebellum results in a loss of coordination. The **reticular formation** receives sensory impulses from the environment and sends impulses to the cortex. If this part of the brain is destroyed, the animal will sleep and never awaken.

The ear has three parts—the **external ear,** the **middle ear,** and the **inner ear.** Vibrations caused by sound waves striking the tympanum are transmitted (by the bones of the middle ear) to the oval window of the cochlea, then through the fluid of the cochlea to the basilar membrane. Deformations in the basilar membrane stimulate sensory nerve endings, which transmit impulses to the brain. The pattern of vibrations of the basilar membrane depends on the frequency of sound waves. This provides a basis for pitch discrimination.

The light-sensitive portion of the eye is called the **retina.** Light reflected from objects is focused on the retina by the lens. The retina is composed of **rods** (most numerous toward the periphery of the retina) and **cones** (most numerous at the center of the retina). Cones are stimulated by bright light (day vision); the rods are stimulated by dim light (night vision). Rods contain a pigment, **visual purple,** which is bleached by light. During the bleaching process, action potentials are set up on nerve fibers which carry impulses to the brain.

FOR THOUGHT AND DISCUSSION

1 When you take an examination, you read a question, pick up a pencil, and begin writing. Your performance depends on

sensory receptors, nerves, the central nervous system, and muscles. Outline the interactions that take place between your eyes, brain, and muscles in your arm.

2 When you first enter a movie theater, it is difficult to find your way to an empty seat. However, after a short time your vision improves (explain) and you can see the person in the next seat in detail; but can you distinguish the color of his clothing? What does this suggest about the ability of rods and cones to discriminate colors?

3 What handicaps would result from: (a) Damage to the cerebellum? (b) Damage to the peripheral parts of the retina? (c) A clogged eustachian tube? (d) Damage to a small portion of the basilar membrane lying close to the oval window? (e) Damage to the motor areas on the cortex of the right cerebral hemisphere? (f) Damage to sensory areas on the cortex of the left cerebral hemisphere? Would either or both of the last two injuries [cases (e) or (f)] interfere with the withdrawal of a hand from a hot stove? Explain.

SELECTED READINGS

French, J. D. "The Reticular Formation," *Scientific American* (May 1957) (reprint #66).

Grey, W. W. "The Electrical Activity of the Brain," *Scientific American* (June 1954) (reprint #73).

Hubel, D. H. "The Visual Cortex of the Brain," *Scientific American* (November 1963) (reprint #168).

McGaugh, J. L., Weinberger, N. M., and Whalen, R. E. *Psychobiology—Readings from Scientific American.* San Francisco: W. H. Freeman and Co., 1967.

Olds, J. "Pleasure Centers in the Brain," *Scientific American* (October 1956) (reprint #30).

Rushton, W. A. H. "Visual Pigments in Man," *Scientific American* (November 1962) (reprint #139).

Snider, R. S. "The Cerebellum," *Scientific American* (August 1958) (reprint #38).

Sperry, R. W. "The Great Cerebral Commissure," *Scientific American* (January 1964) (reprint #174).

von Bekesy, G. "The Ear," *Scientific American* (August, 1957) (reprint #44).

Wald, G. "Eye and Camera," *Scientific American* (August 1950) (reprint #46).

Woolderidge, D. E. *Machinery of the Brain.* New York: McGraw-Hill, 1963.

47 REPRODUCTION

Throughout this book we have emphasized problems related to the physical survival of an individual man. We have seen many examples of delicately balanced regulatory devices that protect the environment of living cells, enabling them to flourish. Nevertheless, eventually each man dies, and it is our ability to reproduce that sustains the life process of our species.

In man, as well as in other animals and plants, reproduction involves the division of cells. Hereditary material (contained in chromosomes) which directs cellular activities is replicated and distributed in such a way that each cell receives an identical complement. In some primitive organisms new individuals sometimes arise by a simple process of duplication—a single cell divides into two individuals. However, in most species of plants and animals, including man, new individuals arise only out of the union of two separate individuals; that is, through sexual reproduction.

Fig. 47.1 Of the millions of sperm cells that surround an ovum only one penetrates and fertilizes the egg cell.

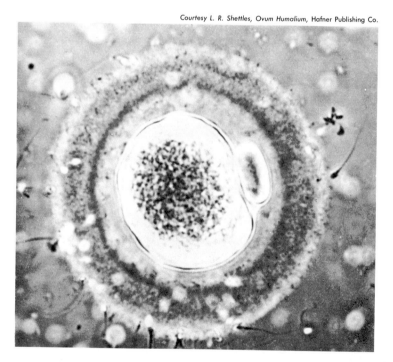

Courtesy L. R. Shettles, Ovum Humalium, Hafner Publishing Co.

The Biological Significance of Sexual Reproduction

The development of each living organism follows a "blueprint" (determined by genes) inherited from its parents. If reproduction required only one parent then the offspring would inherit one set of genes and would resemble his parent in every detail. In sexual reproduction, however, each new individual inherits two sets of genes, one from each parent. He is different from both parents, having inherited some traits from one and certain traits from the other. This raises the possibility that some of the offspring may be stronger than either parent, and better able to survive. It does not necessarily follow that the offspring *will* be stronger than either of his parents, but at least the possibility is open.

One advantage of sexual reproduction is that it produces variety. Within any given species each individual differs from all others. Some may be small and fast, others large and sluggish, and still others may be more intelligent. Without knowing what environmental changes are apt to occur, it is impossible to predict which are the best possible traits that would enable a species to flourish a thousand years hence. Nevertheless, for any environmental change that is not drastic, there is a good chance that at least some members of the species would have just the right traits to enable them, and the species, to survive. The chance of this occurring would be much smaller if each offspring stemmed from a single parent and was an exact replica of it.

Sexual reproduction always involves specialized cells called **gametes.** There are two types of gametes; 1. **sperm** (produced by the male) and 2. **ova,** or eggs (produced by the female). A new individual will develop only after a sperm cell has united with an ovum (one egg cell). The fusion of a sperm cell with an ovum is called **fertilization.** A newly fertilized egg contains two sets of genes, a set from each parent. Before discussing some of the problems faced by this egg, we should first find out how the formation of sperm and ova is controlled in the male and female.

PRODUCTION OF SPERM

The organs involved in the production and liberation of human sperm are illustrated in Fig. 47.2. Sperm are produced in the **testes,** which are contained in the **scrotum,** located outside of the abdominal cavity. The testes consist of thousands of small tubules which lead through a series of tubes to the urethra and to the outside. These tubules, called **seminiferous tubules,** produce the sperm. The location of the testes outside of the body in the scrotum is very important, because the develop-

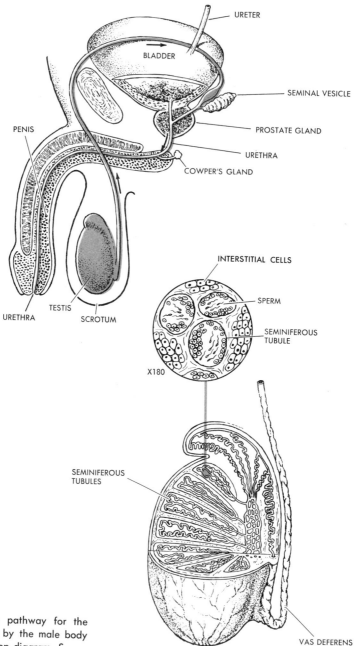

Fig. 47.2 The pathway for the release of sperm by the male body is shown in the top diagram. Sperm cells are produced in the *seminiferous tubules* (bottom diagram) of the testes. The production of sperm depends in part on the presence of the male hormone *testosterone*, which is supplied by the interstitial cells.

Redrawn from *Illustrated Physiology* by McNaught and Callander.
E. & S. Livingstone, Edinburgh

ment of healthy sperm takes place only at temperatures slightly lower than body temperature. In some abnormal cases, the testes do not descend into the scrotum but remain in the abdominal cavity. Sperm cannot be produced in such cases, and the men are sterile.

In going from the testes to the opening of the penis, the sperm cells pass through the **vas deferens** and pass by the **seminal vesicle, prostate,** and **Cowper's** glands. These glands secrete material which mixes with the sperm and is thought to provide the sperm with nourishment, to offer chemical protection, and to support swimming motions. These movements result when the tail of the sperm cell waves back and forth, propelling it along. The fluid, consisting of sperm cells mixed with the glandular secretions, is called **semen.**

Interspersed between the seminiferous tubules of the testes are a number of cells called **interstitial cells.** The interstitial cells secrete the male hormone **testosterone.** Testosterone is essential to the development of mature sex organs and for the development of sperm. It also promotes skeletal and muscular growth during adolescence and is necessary for the development of secondary sexual characteristics—for example, the deepening of the voice during the early teens and the growth of pubic hair, both of which are characteristic of masculinity. The secretion of testosterone and the normal development of sperm are in turn controlled by anterior pituitary secretions called **gonadotropic hormones.**

There is little or no testosterone secreted until some time between the ages of 12 and 16 years (the age of puberty). This is probably due to a lack of secretion of gonadotropic hormones by the pituitary gland. There is evidence that the gonadotropic hormones are stored in the pituitary up until this age but are not released. We do not know why they are suddenly released at this age, but we believe that they are controlled by the central nervous system.

Cycle of Ova Production

The female sex organs are illustrated in Fig. 47.3. They consist of the **ovaries** together with the **fallopian tubes** (oviducts), **uterus** (womb), and **vagina** (birth canal). The ovum begins to develop within the ovary. As it matures, it migrates toward the surface and becomes surrounded by a fluid-filled cavity called a **follicle.** Every 28 days, on the average, one of the most highly developed follicles ruptures and releases a single egg. This process is called **ovulation.** The liberated ovum then enters the fallopian tubes and after about three to four days enters the uterus.

Following ovulation, the ruptured follicle is transformed into a new structure, called the **corpus luteum.** The corpus luteum has two possible fates. If the egg has been fertilized, the corpus luteum persists for several months. Most often, however, the egg is not fertilized, in which event the corpus luteum then degenerates in about 14 days after ovulation. The cyclic changes taking place in the ovaries are illustrated in Fig. 47.4.

Cycle of Changes in the Uterus

Cyclic changes also take place in the uterus about every 28 days. The lining of the uterus, called the **endometrium,** thickens and begins to soften (Fig. 47.5). In addition, many blood vessels and small glands which store and secrete nutrient materials develop within the endometrium. This development of the uterus prepares it to house and nourish the embryo if pregnancy occurs (if the egg is fertilized). If pregnancy does not occur, endometrium development stops when the corpus luteum begins to degenerate. At this time there is a local spasm of blood vessels, which starves the cells of the thickened portions of the endometrium for oxygen and nutrients. These cells do not survive and, as a result, the thickened portions of the endometrium become detached and are discharged along with a small amount of blood (perhaps 50–250 ml) through the vagina.

Fig. 47.3 The ovum which has developed within the ovarian follicle, is released from the surface of the ovary and migrates through the fallopian tube toward the uterus. The ruptured follicle is transformed into the corpus luteum.

Drawn after Scientific American

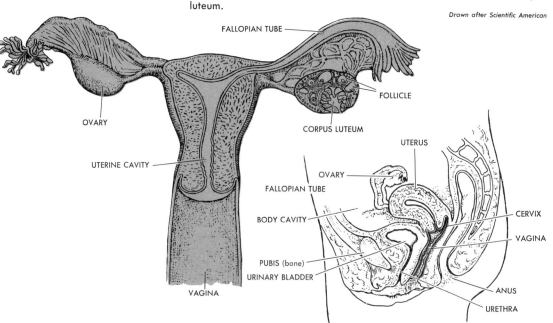

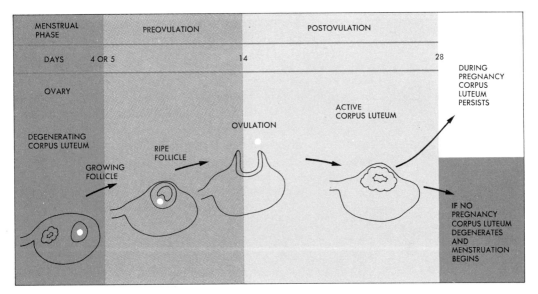

Fig. 47.4 The ovaries undergo cyclic changes each month. Ovulation occurs around the 14th day of the cycle. The ruptured follicle is then transformed into the corpus luteum, which degenerates around the 28th day. If pregnancy occurs, the corpus luteum persists and the cycles are interrupted.

Fig. 47.5 Cyclic changes also take place in the uterus, shown here in cross section. The cycle begins with a poorly developed endometrium (top photo) which thickens (bottom photo), softens, and becomes enriched with blood vessels and glands.

This process, the discharge of detached endometrium and blood, is called **menstruation.** On the average, it takes three to five days for completion, but there is a great deal of individual variation. The "cramps" sometimes felt during menstruation are caused by contractions of the muscular walls of the uterus. After menstruation the uterus is ready to begin a new cycle, in preparation for the next ovulation.

How the Ovarian and Uterine Cycles Work Together

Changes taking place in the uterus are coordinated with changes taking place in the ovaries. This must be so in order for pregnancy to proceed. The fertilized egg must find a well developed endometrium when it enters the uterus. We have already indicated that the time of menstruation corresponds with the time when the corpus luteum degenerates, and that ovulation takes place about 14 days earlier. These relationships are shown in Fig. 47.4. Ovulation occurs when the endometrium is just about midway in its development. Since it takes a few days for the fertilized egg to reach the uterus, the egg enters the uterus when the endometrium is just approaching its peak of development.

This synchronization of ovarian and uterine cycles is not just a matter of chance. If a piece of the uterus is transplanted to another part of the body, it continues normal development and menstruation, despite the fact that all of the nerves to the transplanted tissue have been severed. This suggests that the menstrual cycle is under hormonal control. Further experiments show that hormones controlling the menstrual cycle come from the ovaries!

Courtesy Dr. Hisaw, Howard University

Courtesy Dr. Hisaw, Howard University

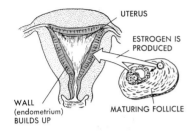

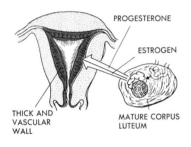

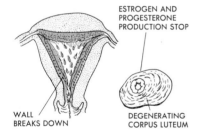

Fig. 47.6 The uterine cycles depend on the cyclic secretion of estrogen and progesterone by the ovaries.

When the ovaries are removed, the endometrium does not develop, and menstruation never occurs. But if extracts of the follicle and corpus luteum are injected into the blood, the endometrium begins to thicken and develop as it normally does. Finally, the endometrium remains in the thick developed state as long as the extracts are administered, but when this treatment is stopped, menstruation occurs and the endometrium remains dormant until it receives more extract. Recall that normal menstruation begins when the corpus luteum begins to degenerate.

The ovaries secrete two types of hormone that control the uterine cycle. Cells of the follicle and the corpus luteum both secrete hormones called **estrogens.** These hormones stimulate growth in the endometrium. The estrogens are a general female sex hormone. They are necessary for the development of the female sex organs and breasts, for the distribution of fat and hair on the body, and for general body growth during adolescence. The other important ovarian hormone, secreted primarily by the corpus luteum is called **progesterone.** This hormone is sometimes called the hormone of pregnancy. Along with estrogen it induces a thickening of the endometrium, it stimulates the development of glands in the endometrium and in the breasts, and it *inhibits* contraction of the muscular walls of the uterus. Here, then (Fig. 47.6), is the sequence of steps in the uterine cycles. 1. Estrogen is secreted by the developing follicle and initiates growth of the endometrium. By the time ovulation occurs the endometrium is well on its way toward full development. 2. The corpus luteum now secretes a combination of estrogen and progesterone. This completes endometrial development. 3. When the corpus luteum degenerates, the source of progesterone, and a good deal of estrogens, is withdrawn. The thick endometrium can no longer be maintained and menstruation begins. Although this explains the coordination of ovarian and uterine cycles, it poses further questions. For instance, what controls the development of the follicles and corpus luteum?

Gonadotropic Hormones

The activities of the ovary (follicle and corpus luteum) are controlled by gonadotropic hormones secreted by the anterior pituitary gland. One of them—called **follicle stimulating hormone (FSH)**—stimulates the maturation process in the follicles. However, FSH by itself does not induce ovulation. A second hormone of the pituitary—called **luteinising hormone (LH)**—is required. This hormone, in combination with FSH, stimulates secretion of estrogens by the follicle and brings the follicle to the stage at which ovulation occurs. In addition LH is required for the development of the corpus luteum and the secretion of progesterone. A third pituitary hormone—called **luteotropic hormone** (**LTH**)—may also be necessary in some species for the secretion of progesterone. We see that the gonadotropic hormones have two roles: 1. they maintain the production of gametes (sperm and ova), and 2. they stimulate the secretion of sex hormones.

Secretion of the gonadotropic hormones are, in turn, influenced by the central nervous system and by sex hormones. The inter-relations between the central nervous system, the gonadotropins, and the sex hormones are complex. Although there has been much speculation, the precise details of these interactions, and the basis for the rhythmic activity, have not been completely worked out. One of the difficulties in filling in the details arises from our lack of a good method for measuring gonadotropic hormones in the blood. Another difficulty arises when we try to work out the role of the central nervous system. Also, because the nature of the ovulation cycle varies from species to species, experiments performed on one animal do not necessarily apply to another. Rabbits, for example, ovulate each time they mate.

WHEN THE UTERINE CYCLES STOP

Menopause: The menstrual cycle can be interrupted in either of two ways—by menopause or by pregnancy. When a woman reaches the age of 45 to 55 years the uterine cycles begin to change. At first they become irregular, then after a few months to a few years the cycles cease. During this time FSH is still produced but the ovaries do not respond; estrogen is no longer produced and ovulation ceases. During this period the woman must adjust both physiologically and psychologically to the withdrawal of estrogens.

Pregnancy: The uterine cycles are also stopped by pregnancy, which begins with fertilization of the ovum. At the climax of the sex act male sperm are deposited in the upper region of the vagina. From here some of the sperm may be transported through the uterus and into the oviducts, or fallopian tubes (see Fig. 47.6). The sperm are carried into the oviducts by their own swimming action and perhaps by movements of the uterus and oviducts. The sperm can probably remain in a healthy, fertile state for as long as 24 hours. The ovum probably remains fertilizable for a shorter time, perhaps for

Fig. 47.7 Fertilization seems to take place in the fallopian tube (oviduct). The fertilized egg undergoes several cell divisions before it enters the uterus and implants in the endometrium.

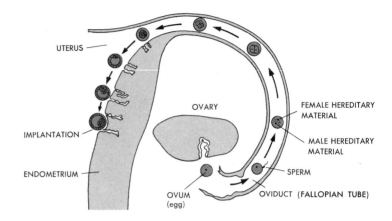

only a few hours. These two periods must overlap if fertilization is to take place.

Although millions of sperm are liberated by each ejaculation (release of male sperm), only one sperm out of the millions enters and fertilizes the ovum. Once the sperm enters the ovum (Fig. 47.7), its head expands and the hereditary material which it contains combines with the hereditary material of the ovum. Next, the egg begins to divide into a two-cell stage, then into a four-cell stage, eight, and so on. The process continues and forms a small ball of cells. By this time the developing mass of cells has moved through the oviduct and entered the uterus.

Cellular Specialization—Differentiation

The fertilized cell contains all of the information required to form a human being. When it divides it results in two cells which are exact replicas of one another. If these cells should happen to separate, not one but two human beings would result. Identical twins would be born. Apparently both cells in the two-cell stage are just alike. Nevertheless, we know that in the fully developed human being the cells are very different. Muscle cells differ from nerve cells which differ from liver cells. It follows that at some stage the cells in the developing embryo must begin to specialize. This process of specialization is called **differentiation.** The mechanism of how differentiation takes place is one of the great unsolved problems in biology. In attacking this problem two opposing ideas have been proposed. One idea is that at some stage of development, all cells do not receive the same genes. Thus the genes present in a cell which develops into a muscle cell are different from those in cells which become kidney cells. The alternative idea suggests that all cells in the body have the same genes, but that all of the genes are not active. A muscle cell, for example, would have the same set of genes as a kidney cell, but a different set of genes has been activated in the muscle cell. Today most scientists consider the second alternative as the better guess.

Implantation and the Placenta

Once the developing mass of cells has reached the uterus, it generally takes four or five days more before it firmly implants or attaches itself to the lining of the uterus. It begins this process by digesting away some of the endometrium. At this stage the embryo gets its nutrition from the endometrium. When implantation is complete the cells of the embryo and the endometrium form an important structure called the **placenta.** This structure enables the mother and embryo to exchange materials throughout the rest of the pregnancy period. It is through the placenta that the embryo sustains its parasitic life in the uterus, getting nutrients from the mother's body and depositing waste products to be expelled by the mother's kidney and lungs. The structure of the placenta is illustrated in Fig. 47.8. It consists of loops of blood capillaries supplied by and connected to the embryo through the **umbilical cord.** These capillary loops are bathed in tiny pools of blood supplied by the mother. The blood of the developing embryo and the mother are separated by thin membranes and do not mix. The exchange of nutrients and waste products takes place primarily by diffusion through these membranes.

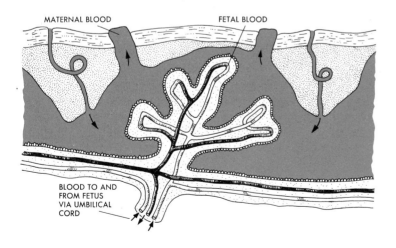

Fig. 47.8 The placenta is an organ of exchange between mother and fetus. An exchange of materials takes place through the loops of blood capillaries, which carry fetal blood and lie in close proximity to small "pools" of maternal blood. The maternal and fetal blood do not mix.

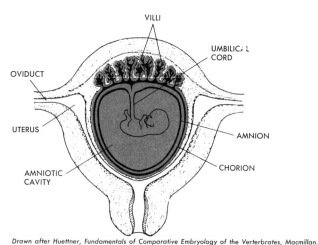

Drawn after Huettner, Fundamentals of Comparative Embryology of the Verterbrates, Macmillan.

Fig. 47.9 The fetus develops within a pool of fluid contained within the amnion.

The developing embryo forms two enveloping coats, the outer is called the **chorion,** and the inner, the **amnion** (Fig. 47.9). The amnion encloses the embryo within a pool of fluid. The chorion invades the endometrium with finger-like projections called **villi,** forming the embryo's contribution to the placenta.

Placenta Hormones

The placenta, in addition to performing the important function of an organ of exchange between mother and fetus, is also an important source of hormones. It secretes estrogens and progesterone; and in the early months of pregnancy it secretes a hormone called **chorionic gonadotropin.** This hormone is responsible for maintaining the corpus luteum before the placenta is capable of producing estrogens and progesterone. Thus, the corpus luteum persists during the early months of pregnancy, secreting estrogens and progesterone which maintain the endometrium. Progesterone also inhibits muscular contractions in the uterine wall. As a result, the menstruation that would ordinarily occur is prevented.

Chorionic gonadotropin which is secreted by the placenta in the early months of pregnancy is also excreted into the urine of the pregnant woman. This is the basis of pregnancy tests; the urine is tested for the presence of gonadotropins.

Birth and the Challenge of the New Environment

About 280 days after fertilization birth takes place. For some unknown reason, just before birth the proges-

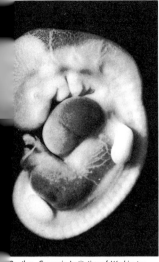 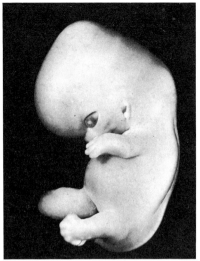 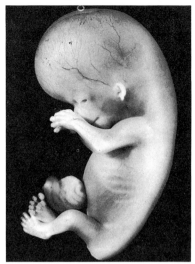

Reather, Carnegie Institution of Washington Chester F. Reather, Carnegie Institution of Washington Chester F. Reather, Carnegie Institution of Washington

Fig. 47.10 The human embryo in successive stages of development. (A) shows the embryo at about the fourth week. The heart is very prominent in this photo; it is located just adjacent to the primitive arms. (B) shows the embryo in the sixth week. The limbs and the eyes and ears can be easily recognized. (C) shows the embryo after eight weeks of development, while (D) shows the infant lying within fetal membranes just prior to birth.

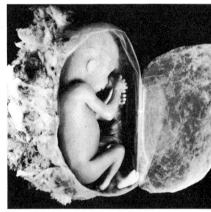

Richard Grill, Carnegie Institution of Washington

Fig. 47.11 These figures show the infant propelled through the birth canal at birth.

"Birth Atlas" by Maternity Center Association, N.Y.

terone and estrogen hormone levels suddenly drop and the birth process begins with uterine contractions which become more and more regularly spaced. The uterine contractions give rise to the pains commonly called labor pains. In the first stage of birth, the fetus is pushed toward the opening of the uterus (the cervix), which begins to dilate. The membranes surrounding the fetus burst and release fluid. In the next stage the fetus is expelled, but it is still connected to the mother by the umbilical cord. The cord must then be cut. Finally, the placenta (the after-birth) is expelled. Figures 47.10 and 47.11 show some of the stages in the development and birth of a human being.

The new baby is no longer protected by its mother's body. It must breathe its own air, digest and absorb its own food, regulate its own temperature, salt and water content, and must ward off foreign invaders if it is to survive. Having lost access to the mother's internal environment, the infant must now begin to rely on its own homeostatic devices.

SUMMARY

Survival of the species depends on the ability of individuals to reproduce. Sexual reproduction takes place when a female gamete (ovum) is fertilized by a male gamete (sperm cell). The production of gametes and sex hormones by the ovaries (female) and testes (male) is controlled by **gonadotropic hormones** of the anterior pituitary. These hormones in turn are regulated by the nervous system and by sex hormones.

The normal production of sperm takes place in the seminiferous tubules of the testes and depends on the presence of the male hormone **testosterone** and on gonadotropic secretions from the anterior pituitary. During and following puberty, testosterone is secreted in significant amounts by interstitial cells of the testes. In addition to its effect on sperm production, testosterone promotes skeletal and muscular growth and the development of secondary sex characteristics. The supply of testosterone is controlled by anterior pituitary gonadotropic secretions.

The ovaries, which produce ova and sex hormones, and the uterus go through cyclic variations each month. About 14 days (on the average) following the onset of **menstruation,** an ovarian follicle ruptures and liberates a mature ovum. The ovum passes through the fallopian tubes to the uterus, where it finds a rapidly developing endometrium. The ruptured follicle becomes the secretory corpus luteum. If fertilization has not occurred, the corpus luteum degenerates and the thickened portions of the endometrium detach (about 14 days following ovulation), resulting in menstruation.

The development of the endometrium depends on the ovarian hormones, **estrogens** and **progesterone,** which provide a link between ovarian and uterine activities. Estrogens are also a general female sex hormone and are necessary for the development of feminine characteristics. The secretion of both estrogen and progesterone are controlled by gonadotropic hormones of the anterior pituitary. Gonadotropic hormones, in turn, are controlled by the nervous system and are subject to some feedback regulation by sex hormones.

Menstrual cycles stop during menopause and pregnancy. Fertilization probably takes place in the oviducts (fallopian tubes) when one of the millions of sperm, deposited in the vagina at the climax of the sex act, penetrates the ovum. The fertilized ovum divides several times during its migration to the uterus, where it becomes implanted in the endometrium and the **placenta** begins to develop. The placenta, which is connected to the embryo through the umbilical cord, performs two functions: 1. it is an *organ of exchange* between mother and embryo; 2. it first secretes *chorionic gonadotropin,* which maintains a secretory corpus luteum, and later it secretes estrogens and progesterone, which maintain the endometrium. Birth takes place about 280 days following fertilization.

FOR THOUGHT AND DISCUSSION

1 During the later months of pregnancy (beyond the first three months) the ovaries may be removed without ending pregnancy. However, during the early months the ovaries must remain intact if pregnancy is to continue. Can you suggest an explanation?

2 The hormone *oxytocin* is secreted by the posterior pituitary gland. Oxytocin stimulates contraction of the pregnant uterus. Mechanical stimulation of the cervix (opening of the uterus) and of the birth canal are believed to cause a reflex secretion of oxytocin. Do these facts suggest that a positive feedback loop may operate during childbirth? (Draw the loop.) How would this help the process of childbirth?

3 On very rare occasions an ovum is fertilized in the abdominal cavity and never enters the fallopian tube or uterus. To what extent do you think such a fertilized ovum would develop?

4 Some animals, such as human beings, ovulate periodically. Other animals, such as rabbits, cats, and raccoons, ovulate only after they have mated. Discuss the relative advantages and disadvantages of these two types of ovulation for the survival of the species.

SELECTED READINGS

Allen, R. D. "The Moment of Fertilization," *Scientific American* (July 1959).

Carlson, A. J., V. Johnson, and H. M. Cavert. *The Machinery of the Body,* 5th Ed. Chicago: University of Chicago Press, 1961.

Csapo, A. "Progesterone," *Scientific American* (April 1958) (reprint #163).

Guyton, A. C. *Textbook of Medical Physiology,* 3rd Ed. Philadelphia: W. B. Saunders Co., 1966.

Moog, F. "Up from the Embryo," *Scientific American* (February 1950).

Sussman, M. *Growth and Development,* 2nd Ed. Englewood Cliffs, New Jersey: Prentice-Hall, Inc., 1964.

Tepperman, J. *Metabolic and Endocrine Physiology.* Chicago: Yearbook Medical Publishers, Inc., 1963.

Winton, F. R. and L. E. Bayliss. *Human Physiology,* 5th Ed. Boston: Little, Brown, and Co., 1962.

EPILOGUE

Earlier, we cited a quotation from the French philosopher-scientist Blaise Pascal. He recognized both the immense size and complexity of the universe, and his relative insignificance within it. Pascal was clearly awed by all around him, yet the universe of his time was much smaller than yours is today. Pascal remained undaunted, for in his mind's eye he could embrace all he saw. If this book has had a purpose beyond teaching you some biology, it has been to enlarge your "embrace" so that a larger part of the world of living things is now included in your realm of knowledge. You are a part of this living world, and the place you occupy is unique. As your consciousness of that place grows, and as you contemplate it, you are both the observed and the observer.

Who Is Man?

As an object of study, man is observed for the simple reason that he wishes to understand himself as a human being, and in relation to other organisms. Man is an animal; of all his knowledge about himself, this is the most certain. But as we have tried to point out, man is also a very special animal. It was the Greek poet Sophocles who said, *Wonders are many, but none is more wonderful than man.* He is like all animals in that he grows, reproduces, and dies. But he is also a dreamer and a planner; a creator of myths, gods, and demons; a developer of values that include such abstractions as beauty, justice, and love; a writer of poetry, textbooks, and comic strips; a builder of mud hovels, schools, and cathedrals; a dancer of the ballet and the frug; an explorer of outer space, of his own inner space, and of the silent depths of the sea. He is alternately cruel, compassionate, and indifferent to himself, to his fellow men, and to his environment. Man is all of these things and many more. Stemming from his animal inheritance, his attributes have been refined, added to, and often forgotten as he has evolved toward human status. In order to understand what he is, man must observe himself as part of the biological scene. He feels compelled to know how he is similar to and different from other living things in structure, organization, and function, from conception to death. And he is a constant experimenter, seeking knowledge of his dependence on, relation to, and control of the environments in which he lives.

Above all else, man is curious. He has an unquenchable desire to understand. He begins with things known to him and then challenges the unknown. Through trial and error he has discovered the value of observation and experimentation; of collecting, ordering, and using data; of formulating, applying, and continually testing hypotheses, theories, and laws. He has found that the details of nature are significant, and that through these details the variety and unity of nature are revealed. He has relied on his imagination to carry him beyond the isolated fact and beyond his immediate surroundings and sensations. He has also relied on his remarkable toolmaking ability to extend his vision across space and his understanding of matter. His greatest invention of all—language—has enabled him to record, store, and communicate his discoveries and thoughts. In doing so, he has acquired the "habit of truth," even though he remains skeptical of absolute truths. Neither science nor society could exist or progress if this were not so.

Time and Change

Like all things, our attitudes toward biology have changed and will continue to change with time. That we can predict with certainty. The information contained within these pages—important though we think it is, and as accurate as we could make it at the moment—is but an introduction to the life sciences, which grow and change with every passing day. Each of the topics covered in this book is a door opened only a crack, an invitation to you to explore more deeply and more broadly until you see beyond that door to the edge of the unknown. It is there, in the realm of the unknown, where such giants as Galileo, Darwin, Pasteur, Fleming, and many others trod. And it is there where the challenge and excitement of original discovery awaits you.

In addition to being a learner, you may be a contributor of knowledge as well. But this role demands an intelligent use of time, energy, and talents. If your lot in life is to enter biology, or one of the related fields of medicine, agriculture, psychology, or public health, you will find it a world of stimulation and fun. We who have written this book know this to be so.

Not all of you will become biologists, doctors, nurses, or farmers. Other careers are equally challenging and meaningful. But it is our hope that this book has expanded your horizons as well, that your eye has become a more perceptive and discriminating one, that the details of nature now appear in greater richness, variety, and beauty than before.

It is also our hope that you are now more aware of yourself as an individual, unlike any other individual who has ever lived, but at the same time sharing with all other living things many of the same structures and functions. You are linked through evolution to the whole organic world. As part of that world, you have a history that extends far back into time and a future that extends far ahead. In some way, you as an individual will help mold that future. We believe that the more you know about yourself and your environment, the more responsive and alert you will be to the world about you. Your "embrace" has no limits except those you set for yourself.

SUPPLEMENTARY GLOSSARY

Most of the technical terms in this volume have been defined within the text. To locate a particular definition, the reader can look up the word in the index and find the sub-entry marked "defined." This appears in **boldface**.

The words listed here are basic terms that, for the most part, have not been previously explained. Thus, this glossary is a supplement to the information contained in the text.

Absorption Passage of a substance, or substances, into a cell or the blood system.

Acid A substance having a sour taste, and with hydrogens which can be replaced by a metal to form a salt.

Anatomical Pertaining to the structure of an organism, or any of its parts.

Anemia (sickle cell and **pernicious)** A condition caused by a deficiency of red blood cells and/or hemoglobin.

Angstroms A measure of distance equal to one ten-millionth of a millimeter.

Antibiotic A chemical produced by microorganisms and which inhibits the growth of, or kills, bacteria or other microorganisms.

Arithmetic progression A sequence formed by the addition of a constant term to the preceding term, e.g., 2, 4, 6, 8, etc.

-ase A suffix denoting an enzyme, e.g., maltase, an enzyme which breaks down the sugar maltose.

Asexual reproduction Formation of new individuals without sexual processes being involved, e.g., division of a bacterium or the rooting of a cutting.

Autotrophic The ability of an organism to form its own nutritive substances out of inorganic materials in the environment, as by photosynthesis.

Catalysis Causing a chemical change by the addition of a **catalyst,** which itself is not permanently affected by the reaction.

Cellular differentiation Process whereby cells acquire specialized structures and junctions.

Chiasma A point in a bivalent where nonsister chromatids have crossed-over.

Cold blooded Referring to animals whose body temperature changes with the temperature of the environment.

Differentiation The process of changing structure and/or function, usually from generalized to specialized, in cells, tissues, or organs.

Diffusion (simple laws of) Gradual penetration and uniform distribution of a substance through a fluid, caused by the thermal energy of the molecules.

Digestion Process by which food is broken down into units which can be absorbed into the blood stream.

Diploid An organism having double the basic (haploid) number of chromosomes.

Dissociation (as in water) The breaking down in solution of a compound into simpler units.

Distal Away from the point of attachment; the region of a chromosome arm farthest from the centromere.

Dormancy (as in bud dormancy) State of rest, or suspended animation; not growing.

Ecology Science dealing with the relation of organisms to their environment.

Electron micrograph Photograph taken through an electron microscope.

Electron microscope A microscope in which electrons instead of visible light are used as a source of "illumination."

Element (chemical) The simplest form of a substance which cannot be separated into smaller units by chemical means.

Embryo An organism in the very early stages of development.

Endogenous Originating or taking place within.

Enzyme A biological catalyst, made of protein, which promotes specific chemical reactions, and which is not itself permanently altered in doing so.

Ester A compound resulting from the interaction of an acid and an alcohol, with the release of a molecule of water.

Etiolated Whitened, as a plant growing in darkness.

Fertilization Union of egg and sperm in sexual reproduction.

Fission The splitting of a cell or an organism into two parts.

Fissure A narrow opening.

Fixation (as in CO_2 fixation) The incorporation and transformation of a gas into a solid form.

Fluorescent Having the property of emitting light when exposed to external radiation.

Fossilize To replace an organic substance with minerals; to preserve the remains of an organism or its parts, usually in rock formations.

Free-living Existing by itself, or outside of another cell or organism.

Fusion Union of cells or nuclei in the act of fertilization.

Genus A category of classification higher than a species and lower than a family.

Geometric progression Sequence of terms in which the relation of any given term to its predecessor is determined by a constant factor of multiplication or division; e.g., 2, 4, 8, 16, 32, etc., 100, 20, 4, 4/5, 4/25, etc.

Germination The beginning of growth and development of a seed or spore.

Gland Organ which produces a secretion.

Grafting Transplanting, or insertion, of tissue from one part of the body to another, or from one organism to another.

Haploid Containing one basic set of chromosomes.

Hemoglobin Colored protein found in red blood cells and which conveys oxygen to, and carbon dioxide away from, cells and tissues.

Heterosis Increase in growth, size, or yield characterizing certain hybrid organisms.

Heterotrophic The inability of an organism to form its own nutritive substances out of inorganic materials in the environment.

Heterozygous Containing two genes of unlike or distinguishable characteristics.

Homologous Similar in structural origin, as the arm of a man and the foreleg of a horse; similar, as the members of a pair of chromosomes in a diploid organism.

Hydrolysis Dissociation of a compound by taking up the elements of water.

Hypothesis Tentative explanation or guideline to serve as a working basis for argument or for further investigation.

Influorescence Collection or cluster of flowers on an axis.

Infrared That region of the electromagnetic spectrum just beyond visible red and longer in wavelength than any part of the visible spectrum.

Ionization Separation of a compound into positively and negatively charged particles.

Life cycle Course of development and change from fertilization of the egg to the production of new germ cells.

-lysis (suffix) Destruction or decomposition of cells (cytolysis) or compounds (e.g., hydrolysis).

Meiosis A kind of cell division that leads to the production of gametes or asexual spores, and during which the chromosome number is reduced to a haploid state, and the genes are generally segregated into new combinations.

Membrane A thin layer of tissue that covers a surface or divides a space or organ.

Menopause Period of irregular menstrual cycles prior to the final cessation of menstruation.

Mesentery Tissue which attaches a portion of the intestine to the posterior (back) wall of the abdomen.

Metabolism Sum total of the chemical reactions in a cell or an organism, by which food substances are used for building structures or yielding energy.

Metric system Decimal system of weights and measures, with the meter (39.37 inches) as the basic unit of length, and the gram (.035 ounce) the basic unit of weight.

Microorganisms Organisms too small to be seen except in a microscope (includes bacteria, protozoa, viruses and many algae).

Mitosis Division of the nucleus of a cell, during which the chromatin resolves itself into thread-like chromosomes (from *mitos*, Greek for thread).

Morphological Referring to form and structure of a cell or an organism, without regard to function.

Mutualism Cooperative existence of two or more kinds of organisms, with benefit to each.

Nutrients An organic or inorganic substance that supports the function and growth of organisms.

Optimum The conditon (temperature, pH, etc.) favorable for maximum activity or function.

Osmotic equilibrium Osmotic equilibrium is reached when the movement of the solvent across a membrane is equal in both directions.

Parasite A plant or animal living in, on, or with some other living organism at whose expense it obtains food, shelter, etc.

Permeable Capable of being passed through, such as passage of fluids through membranes.

Photoperiod The optimum length of day or period of daily illumination required for normal growth and maturity of a plant.

Photosynthesis Synthesis of chemical compounds energized with the aid of light and in the presence of chlorophyll, as in plant synthesis of carbohydrates.

Phylum One of the primary divisions of the animal or plant kingdom.

-plast (suffix) A form denoting an organized particle or granule, as in chloroplast.

Polarity An inherent condition in a body, a condition exhibiting opposite properties, parts, or direction, i.e., the presence of poles.

Pollinated The addition of microspores (pollen) in seed plants to the stigma part of the flower.

Primate (as in primate ancestors) One of an order of mammals consisting of man and the apes, monkeys, marmosets and lemurs.

Primordial Originally or earliest formed in the growth of an individual or organ.

Progeny Offspring.

Proto- (prefix) A combining form, meaning first. For example, first in time, as in protocol.

Proximal nearest, opposite to distant.

Pulmonary Pertaining to or resembling the lungs.

Quantum An elementary unit of energy according to the quantum theory.

Radiation The process by which energy is emitted from molecules and atoms—to emit rays.

Radical Botanical: Proceeding from the root or from a root-like stem. Chemical: A fundamental constituent of a compound.

Reflex arc A movement due to a nervous impulse transmitted inward from a receptor to a nerve center and outward to an effector.

Refraction The deflection from a straight path suffered by a ray of light, heat, etc. in passing obliquely from one medium into another.

Secondary sex character Characteristics associated with the factors that determine sex (male or female).

Sedimentation The settling to the bottom of substances that are heavier than the surrounding material.

Senescence growing old; aging.

Sexual reproduction Reproduction by sexual activity, in contrast to asexual reproduction.

Soluble Susceptible to being dissolved, capable of passing into solution.

Solution The condition of being dissolved, a homogeneous mixture.

Spectrum The series of images formed when a beam of radiant energy is resolved into its component waves in the order of wavelengths.

Spore Any of various primitive reproductive bodies, typically unicellular, produced by plants, and by some protozoans.

Steroid hormones Any of a class of compounds containing the carbon ring system of the steroids; sex hormones, for example.

Stimulus Any agent or environmental change capable of influencing the activity of living protoplasm.

Sugars Any of a class of soluble compounds comprising the simpler carbohydrates.

Symptoms Any perceptible change in the body or its functions indicating disease, or the kind or phases of disease.

Syndrome A group of signs and symptoms that occur together and characterize a disease.

Synthesis The art or process of making or "building up" a compound by the union of simpler substances.

Taxonomy Classification of animals and plants according to their natural relationships.

Tissue An aggregate of cells and intercellular substances forming one of the structural materials of a plant or animal.

Toxic Poisonous.

Tumor An abnormal mass of tissue arising without obvious cause from cells of pre-existent tissue.

Ultraviolet Radiation beyond the violet end of the visible spectrum, of shorter wavelength than visible light.

Valence The degree of combining power of an element.

Vascular Of or pertaining to a vessel or vessels for the conveyance of liquids.

Warm blooded Having a relative high and constant body temperature, as birds and mamals.

Wavelength The distance measured from one wave crest to the next, or from one wave trough to the next.

INDEX

A

A bands, 590–591
Abscisin, discovery of, 522
Abscission, 521–522
Absorption, 661, 668–669
Absorption spectrum, 266
  defined, 486
Acclimatization, 643
Accommodation, 704
  defined, 704
Acetabularia, 158, 180
Acetaldehyde, 198, 245, 248
Acetic acid, 199, 202, 250, 251
Acetoacetic acid, 248
Acetone, 198, 248
Acetylcholine, 605–606
Achillea lanulosa, 366
Acidity of blood, 657
Acids, 200–201, 248, 253–254; see also names of acids
  oxidation to an aldehyde, 238
Acrosome, 320
Actin, 246, 588–591
Action potential, 598–599
  defined, 598
  of the heart, 613
  of nerves, 598–599
  passing between cells, 604–606
Active transport, 226–227, 298–299, 583–584
  of sodium, 651, 654
Actomyosin, 246
Adaptation, 18–35, 108
  defined, 18
  evolution, 20–31
  special examples of, 32–34
    camouflage, 32
    convergence, 32
    mimicry, 32
Adenine, 275
Adenosine diphosphate (ADP), 239, 240, 243, 244, 487, 494, 495
  production of, 246
  structure of, 239
Adenosine monophosphate, 239

Adenosine triphosphate (ATP), 246, 251–254, 296, 298–299, 464–465, 467, 475, 487, 495, 588–589, 638, 639, 681–682
  defined, 494
  formation of, 239–240, 243–244, 675
  in muscle contraction, 247
  synthesized under aerobic conditions, 253
Adenylic acid, 239
Adrenal cortex, 656
Adrenal gland, 608, 677, 678
Adrenalin, 677, 678
  in heat production, 683
Adrenocorticotropic hormone (ACTH), 681
Aerobic metabolism, 252
Aerobic organisms, 215–216
Aerobic respiration, 253
Afferent arterioles, 649
After-birth, 720
Agar-agar, 79
Agglutination, 630
Agrobacterium tumefaciens, 534
Alaska brown bear, 230
Albatross, 421
Albino plants, 466
Alcohol(s), 198, 253
  ethyl, 197, 198, 245, 248
  fermentation, 240–245
    key reactions, 245
  formation of, 245
  isopropyl, 248
Alcohol dehydrogenase, 245, 265
Aldehyde, 198
  oxidation to an acid, 238
Aldehyde phosphates, 242
Aldosterone, 656–657
Algae, 16, 51, 74–82, 85, 91, 102, 104, 109, 142, 145, 152, 162, 166, 480
  blue-green, 15, 75, 103
  brown, 75, 77, 79, 91
  green, 75, 76, 79–81, 104
  lichens, 29, 86–87
  red, 77, 79
  single-celled, 180
Algal fungi, 84, 86

Algin, 79
Allard, 537, 538
Alleles, 307, 412
  defined, 304
  multiple, 412
Allen's rule, **defined, 439**
Allium cepa, 54, 170
All-or-none response, 599, 600
Alveoli, 635, 636, 673
Ameba, 138, 220, 222, 227
  shape of, 158
Amino acid(s), 147, 209–210, 384, 427, 461–462, 465, 470, 494, 583, 651
  code, 284
  defined, 461
  essential, 663
  metabolism, 254
Ammonia, 83, 427, 465, 493
Ammonium salts, 255
Amnion, 60, 61, 718
  defined, 60
Amo-1618 (inhibitor), 522
Amoebic dysentery, 38
Amphibians, 58–60, 69, 102
  embryo, 157
  testes, 313–315
Amphioxus, 56–57
Amylases, 520, 663, 667
  pancreatic, 664
  salivary, 664
Anabolism, **defined, 253**
Anaerobic organisms, 215–216, 240
Anaerobic respiration, 253
Anaphase, 171, 173, 174, 175
  of meiosis, 316, 317
Anatomy, **defined, 124**
Anemia:
  pernicious, 166
  sickle-cell, 166, 286, 393
Angiosperms, 37, 97, 98–103, 104, 466
Angiotensin, 657
Angstroms (A), 139
Animal cells, 145–149
  compared to plant cells, 149
  division in, 173–175

Animal kingdom, 36–70
  classification of, 37
  major types, 38–66
    amphibians, 58–60
    annelids, 46–47
    arachnids, 51–52
    birds, 62–63
    centipedes, 52
    cnidarians, 41–42
    crustaceans, 49–51
    echinoderms, 53–55
    fishes, 57–58
    flatworms, 42–44
    insects, 48–49
    mammals, 64–66
    millipedes, 52–53
    mollusks, 45–46
    nematodes, 44
    protochordates, 55–57
    protozoans, 38–40
    reptiles, 60–62
    sponges, 40–41
Animalia, 8, 11
  difference between, 14–16
  mobility, 16
  phagotrophic, 16
  mammalia, 4, 10, 11
  reproduction in, 318–321
Anion, **defined, 491**
Annelida, 11, 46–47
Annelids, 46–47
Annual plants, 541–542
Annual rings, 509–510
Ant, fossilized, 397
Antennae, 49
Antennules, 49
Anther, **defined, 323**
Antheridia, 526
Anthropoidea, 432
Antiauxins, 517
Antibiotics, 397, 418
  discovery of, 119–120, 130
  in hormone inhibition, 681
Antibodies, 209, 630
Antidiuretic hormone (ADH), 652–653
  manufacture of, 652
  mechanism of action, 654–656
Antigen, 630
Anus, 660

Anxiety, 698
Aorta, 614, 615, 621, 649
Aortic arches, 47
Apoenzyme, 463
Applied sciences, 119–120
Arachnids, 51–52
Archaeopteryx, 402
Archeozoic era, 399
Arginine, 287
Aristotle, 7, 36, 72
Aromatic ring compounds, 198
Arterial reservoir, 618
Arterial system, 614
Arteries, 613
  carotid, 621
  pressure in, 616
  pulmonary, 615
  renal, 649
Arterioles, 618–619
  afferent, 649
  defined, 618
  efferent, 649
  pulmonary, 635
  resistance of, 619
  smooth muscle in, 619
Arthropods, 37
  arachnida, 51–52
  chilopoda, 52, 53
  crustacea, 49–51
  diplopoda, 52–53
  insecta, 48–49
Aschelminthes, 44
Ascomycete, **defined, 303**
Ascospores, 304, 307
Ascus, 85, 304, 306, 308
  defined, 303
Asexual reproduction, 40, 42, 43, 108, 376
  in *neurospora,* 303, 304
Asparagus stem tips, 531
Aspartic acid, 254
Aspergillus niger, 145
Association areas of the brain, 692–693
Astral rays, 172, 174
Atmosphere, 453
  carbon dioxide content, 482–483
  photosynthesis and, 260–263
Atomic mass unit, 188
Atomic number of elements, 189
Atoms, 26, 134
  combining, 190–192
  diagrammatic sketch, 187
  and molecules, 186–193
  reactive, 190–191
  structure of, 187–190
Atria, 60, 613–615
Attached-*X* inheritance, 356–357
Auditory nerve, 701
Australopithecines, 435
  defined, 434
Autonomic nervous system (ANS), 608–609

Autonomic nervous system (ANS) (*cont.*)
  cardiac output and, 617
  major divisions of, 608
Autoradiography, 278
Autosomes, 350, 358–359
Autotrophic nutrition, 83
Autotrophic organisms, 213–214
  chemosynthetic, 213–214
  defined, 16, 467
  photosynthetic, 214
Auxins, 268, 512–518, 524, 532–533, 544
A-V valves, 616
Avery, Oswald, 182
Avirulent bacteria, 180–181
Axis, *see* Rachis
Axons, 156, 596–599
  defined, 596
  electrical changes measured along, 598
  polarity reversed, 600

B

B vitamin complex, 212
Bacilli bacteria, 82, 83
Bacillus subtilis, 150
Backcross, 329–330
Bacteria, 15, 82–83, 85, 103–104, 142, 152, 181–182
  bacilli, 82, 83
  cocci, 82–83
  killed by penicillin, 120
  protoplasts of, 226–227
  resistance of, 418, 420–421
  Rhizobium, 534, 545
  size, 154
  spirilla, 82, 83
  tumor-inducing agent and, 534–535
Bacterial cells, 150–151
  time of division, 175
Bacterial transformation, 181–182
Bacteriology, **defined, 124**
Bakanae, 518–519
Basal metabolic heat, 682
Basic dyes, 144
Basic sciences, 119–120
Basidia, 85
Basilar membrane, 700–702
Bateson, William, 169
Beadle, G. W., 286
Beagle (ship), 129
Bean roottip, cell division, 170–173
Benzene ring, 198
Bergman's rule, **defined, 439**
Beta-galactosidase, 384
Bicarbonate, 639
Bicarbonate ions, 639, 667
Biennials, 542–543, 545

Bilateral symmetry, 53
  defined, 43
Bile, 667
Bile duct, 667
Bile salts, 664
Binomial nomenclature, 36
Biochemistry, 123, 128, 187
Biogenesis, 170
Biological compounds, chemistry of, 194–203
  acids and bases, 200–201
  carbon, 196–197
  chemical groups, 198–199
  chemical reactions, 199–200
  oxidation-reduction, 200
  pH, 201–202
  water, 194–196
Biology, 118–133
  defined, 123
  molecular, **defined, 128**
  stages of, 126–128
  subdivisions of, 123, 124–126
Biology study, 107
Bioluminescence, 267–268
Bioluminescent rhythms, 543
Biotin, 212
Bipeds, 432
Birds, 62–63, 102
Birth, 718–720
Bison, 124
Biston betularia, 422
Bivalents, 305, 315, 340
  defined, 314
Bivalve, 45
Bladder, 649, 710
Blastula, 377, 378
Blood, 624–633; *see also* Circulation
  antibodies, 630
  calcium in, 626–627
  capillary fluid exchange, 627–629
  clotting, 624–626
  loss of, 621–622
  preventing, 625–627
  the lymphatic system, 629–630
  regulation of acidity of, 657
  sugar in, 663, 672
  liver in regulating, 677
  transfusions, 630, 631
  types, 440, 630–631
  volume of, 624
Blood banks, 626–627
Blood cells, 163, 165, 624, 625, 632
  red, 142, 147, 165–166, 226, 630, 632, 642
  nucleus loss, 178
  shape of, 159, 160
  white, 134, 148, 165 630 632

Blood cells, white (*cont.*)
  size of, 154
Blood flow, 615–622
  intense heat response, 685
  metabolic products regulation, 619–620
  one-way, 614, 616–617
  resistance to, 619
Blood pressure, 576, 617–622
  average, 618
  in capillary fluid exchange, 627
  intense heat response, 685
  internal environment and, 620–621
  regulation of, 620–622
  urine formation and, 650
Blood vessels, 613
Blue-green algae, 15, 75, 103
Body fluids:
  osmotic pressure, 652–653
  regulation of volume of, 656–657
Body temperature, 681–686
  core, 682
  fever, 684–685
  loss of heat in, 683–684
  production of heat in, 682–683
  regulation of, 684
  response to intense heat, 685–686
  surface, 686
Bombyx mori, 359
Bond, chemical, 190–192
Bond energy, 236, 237
Bone marrow, 642
Boron, 490, 496
Botany, **defined, 124**
Botany study, 72
Brain, the, 594–596, 688–707
  association areas, 692–693
  the cerebellum, 689, 690, 694
  coordination and, 694
  conditioned reflex, 697–699
  cortex, the, 689, 691–692
    central fissure, 691
    motor areas, 691
  in hearing, 699–702
  importance of oxygen to, 612–613
  intense heat response, 685
  medulla of, 621, 637, 690
  memory, 695–697
  reticular formation, the, 694–695
  in seeing, 702–705
  ventricles of, 689
Branch-root initiation, 530

Breathing, 634–637
  control of, 643–646
  strenuous, 637
  water loss during, 645–646, 683–684
Bridges, C. B., 355–357
Bromouridylic deoxyriboside (BUDR), 415
Bronchiole, 635
Bronchus, 635
Brown algae, 75, 77, 79, 91
Brownian motion, 221–222, 224
Bryophytes, 87–91
  liverworts, 87–89
  mosses, 89–91
Buchner, Eduard, 241
Buds, 508
  apical, 516
  callus tissue and, 533–534
  dormancy, 521–522
  kinetin and, 522
  lateral, 516, 524
  potato, 516–517
  sprouting, 516, 524
Buffers, 202
Bulk flow, 575
Butterfly, 48–49
  metamorphosis of, 49
Butyric acid, 248, 253–254
  metabolism of, 254

C

Cactus plant, 194
Calcium, 570–572
  in blood, 626–627
  functions of, 495
Calcium nitrate, 490
Callus tissue, 506–522, 535, 545
  culture of, 532–535
Cambium, 508, 510, 523
Cambrian Period, 69, 70
Camouflage, 32, 33
  defined, 33
Cancer, 144
  cell, 140
  treatment of, 120
Canidae, 9, 10, 11
Canis, 10, 12
Capillaries, 159, 613, 614, 615
  average pressure in, 627
  fluid exchange, 627–629
  permeability of, 627
Capsule of the nephron, 649
Carbohydrates, 495, 661–663
  cell's energy needs and, 676
  energy from, 205
  role of, 204–207
  storage depot for, 677
Carbon 14, 487
  atom, 188
  compounds, 196–197

Carbon (cont.)
  cycle, 215
  and life, 425–426
  radioactive, 487–488
Carbon dioxide, 197, 427, 428, 451, 612
  in the atmosphere, 482–483
  climate and, 482–483
  converted to sugar, 487–488
  diffusion, 639, 640
  fixation of, 250
  photosynthesis and, 263, 452
  plant growth and, 511
  stomatal opening and, 497
  transport, 639–641
Carbonaria, 422
Carbonic anhydrase, 640
Carboniferous Period, 69, 402, 483
Carbonyl group, 198
Carboxyl group, 198–199
Cardiac muscle, 586–587, 613–615
Cardiac output, 617, 618–619, 621
  defined, 617
Carnivora, 9, 10, 11
Carnivorous, defined, 53
Carotenoids, 467, 485
Carotid arteries, 621
Carotine, 212, 515
Carrot roots, 532
Casparian strip, 508
Cat family, 362–363
Catabolism, defined, 253
Catalysts, 661
  defined, 208
Cation, defined, 491
Caucasoid race, 436–438
  sub-races, 436–437
Cave paintings, 124, 435
CCC (inhibitor), 522
Cell division, 39, 75, 169
Cell membranes, 141, 142, 208, 221–228, 574 582–584
  active transport, 226–227
  electronmicrograph of, 146
  osmosis, 223–226
  permeability, 223
  phagocytosis, 227–228
  pinocytosis, 227–228
Cell plate, 173
Cell theory, 38, 136
  exceptions, 144–145
Cell wall, 143, 147, 156
Cells, 26, 458–476
  action potentials passing between, 604–606
  analysis, beginnings of, 128
  as a basic unit, 134

Cells (cont.)
  blood, 163, 165, 624, 625, 632
    red, 142, 147, 159, 160, 165–166, 178, 226 630, 632, 642
    white, 134, 148, 154, 165, 630, 632
  conducting, 476
  defined, 473–474
  death of, 163–166
  development as an inherited pattern, 374–389
  differentiation of, 532
  discovery, 137
  division, 168–177, 543
  egg, 708, 711–712
  energy needs of, 675–676
  function and shape related, 159–160
  growth of, 376–379
  idealized, 141
  immortality of, 506
  interstitial, 710, 711
  islet, 672–674
  influence of, 673
  light-sensitive, 705
  major compounds, 204–219
  metabolic properties of, 220–236
  muscosal, 668
  nerve, 154–156, 159–160, 163, 596
  number, 161–163
  oxygen in, 155, 156
  parts of, 458–460
  replacement, 164–166
  shape, 157–161, 458
  single, culture of, 535–536
  size, 138, 153–157
  specialization of, 716–717
  sperm, 708–711
    production of, 709–711
  structure, 138–143, 145–149, 225
  synthetic reactions, 675
  turgid, 471
  types of, 473–475
  wall of, 468–472, 476
    defined, 458
    function of, 473
    silica, 491
  zymogen, 673
Cellular metabolism, 107
  biological membranes, 297–299
  control systems, 297–299
  enzyme activity, 294–295
  enzyme complexes, 295–297
  enzyme repression, 292–294
  feedback inhibition of enzymes, 292–294

Cellular metabolism (cont.)
  effect of nutrients on enzyme synthesis, 291–292
  protein structure, 294–295
Cellulose, 143, 206–207, 468, 476, 661
  microfibrils, 469
Cenozoic Era, 398, 402
Centipedes, 52, 53
Central nervous system (CNS), 596–607
  action potentials, 604–606
  defined, 596
  development and stucture of, 689–699
    association areas, 692–693
    the cerebellum, 694
    conditioned reflex, 697–699
    the cortex, 691–692
    memory, 695–697
    the reticular formation, 694–695
  inhibition by nerve fibers, 606–607
  motor fibers, 603–604
  polarization of nerve membranes, 599–602
  reflex actions, 607
  sensory receptors, 602–603
Centriole, 174, 321
Centromere, defined, 173
Centrosome, 174
Cephalin, 208
Cerebellum, 689, 690, 694
  coordination and, 694
Cerebrospinal fluid, 689
Cerebrum, 689
Chameleon, 160
Chargaff, Irving, 275
Charge balance, 651
Chemical bond, 190–192
Chemical energy, 240
  transformations, 237
Chemical evolution, 403
Chemical groups, 198–199
Chemical reactions, 199–200
Chemistry, 119
  of biological compounds, 194–203
  of chromosomes, 182–184
Chemoreceptors, 644–645
Chemosynthetic autotrophs, 213–214
Chest cage, 636
Chiasmata, 315, 340, 342
Chilomonas, 159
Chimpanzee, 433
Chipmunk, 230
Chloride ions, 191, 580
Chlorine, 192, 490, 496

Chlorophyll, 72, 75, 76, 77, 91, 143, 149, 258, 262, 466, 478, 485, 501
  absorption spectrum for, 485
  chlorosis, 495
  defined, 16
  kinds of, 486
  production of, 263–264
Chlorophyta, 76
Chloroplasts, 143–144, 259, 468, 478
  defined, 76, 466, 475
  function of, 473
  the grana of, 501
  as invaders, 467
  movement of, 472
  as multienzyme complex, 296–297
  structure of, 297
Chlorosis, 495
Cholinesterase, 605–606
Chordates, 11, 37, 53, 55
  agnatha, 57
  amphibia, 58–60
  aves, 62–63
  chondrichthyes, 57
  defined, 8
  mammalia, 64–66
  osteichthyes, 58
  protochordata, 56
  reptilia, 60–62
  urochordata, 55–57
  vertebrata, 56
Chorion, 718
Chorionic gonadotropin, 718
Choroid, 703
Chromatids, 171
  cross-over, 339
Chromatin, 141, 142, 341, 458, 459, 460
Chromatophores, **defined, 486**
Chromomeres, 315
  defined, 314
Chromoplasts, 467
Chromosomes, 26, 171–174, 708
  behavior in inheritance test, 305–306
  chemistry of, 182–184
  as controlling element in nucleus, 180–182
  defined, 12
  described, 179–180
  in a dividing cell, 178
  division of, 305–306
  genetic maps, 346
  haploid, 305
  homologous, 305, 313, 339, 340, 350
  of a human male, 350
  knobbed, 341
  in maize, 341
  maps, 342–346
  number of, 312–313

Chromosomes (cont.)
  segregation, 383
  sex determination, 358–359
  structural changes, 415
  triploid, 323
  X-, 349–360
  Y-, 350–360
Cilia, **defined, 39**
Ciliary muscle, 703, 704
Cinquefoil, 364–365
Circadian rhythms, 543, 545
Circulation, 127, 612–623; see also Blood; Transport
  blood flow, 615–622
    intense heat response, 685
    metabolic products regulation, 619–620
    one-way, 614, 616–617
    resistance to, 619
  blood pressure in, 617–622
    regulation of, 620–622
  cardiac output, 617
  heart muscle in, 613–615
Citric acid, synthesis of, 251
Citric acid cycle, 251–254
Citruline, 288
Clams, 45–46
  luminous, 268
Class, 7
Classification:
  of organisms, 7–18
    artificial, 7
    natural, 7–8, 11–12
Clay particles in soil, 489
Cleavage, 377, 378
Climate:
  carbon dioxide and, 482–483
  plant growth and, 521
  race and, 438–439
  tree dating and, 509–510
Closed circulatory system, 48
Closed-loop systems, see Feedback systems
Club fungi, 85
Cnidarians, 41–42, 53
Cobaea, 391
Cobalt, 491
Cocci bacteria, 82, 83
Cochlea, 700, 701
Coconut milk, 520
Code words, 283
Codosiga, 159
Coelenterata, see Cnidarians
Coezyme, 228, 463
Cofactors, 228
Coiling, **defined, 171–172**
Colchicine, 172
Coleoptiles, 512–514
  unequal growth, 514
Collagen, 209

Collecting duct, 649, 650
Collenchyma cell, 474
Colloidal systems, and protoplasm, 221–223
Color-blindness, 357–358
Coma, 695
Communities, 548–552, 558
  competition in, 549, 558
  defined, 549
  desert, 549
  equilibrium in, 548, 558
  forest, 549–551
Companion cells, 474
Comparative anatomy, 392
Compounds:
  biological, chemistry of, 194–203
    acids and bases, 200–201
    carbon, 196–197
    chemical groups, 198–199
    chemical reactions, 199–200
    oxidation-reduction, 200
    pH, 201–202
    water, 194–196
  major, 204–219
    carbohydrates, 204–207
    growth requirements, 211–216
    lipids, 207–208
    proteins, 208–211
    ring, 198
Concentration difference, transport and, 574, 577–578
Concentration gradient, 298
Conceptacles, 526
Conditioned reflex, 697–699
  in study of animal behavior, 698
Conducting cells, 476
  defined, 473–474
Conduction, heat loss by, 683
Cones, 265, 705
Conifers, 97–98, 102, 104
Conservation, 109–110
Constitutive enzyme, 291
Contractile machinery, 588–591
Contraction (muscle), 587–589
  tetanizing, 603
Control systems, 297–299
Convergence, 32, 33–34
  defined, 33
Convolvulus, 530–531
Copernicus, Nicolaus, 129, 395
Copper, 490

Coral, 42
Core temperature, 682
Cork cambium, 508
Corn, 99, 553–558, 558–559
  ancestry of, 553–555
  ear of, 553, 556, 557
  female flowers, 556–557
  hybrid, 557–558
  male flowers, 556
  pollen tubes, 556–557
Cornea, 702, 703
Cornified cells, 164–165
Corpus callosum, 690
Corpus luteum, 712
Cortex, 351, 352
  of the brain, 689, 691–692
    central fissure, 691
    motor areas, 691
    cells, 508
  of the kidney, 654
Cortisol, 677, 678
  control of secretion, 681
  defined, 677
Cotyledons, 489, 504–505
  defined, 505
Covalent bond, 191–192
Cowper's glands, 711
Crab Nebula, 118
Cranial nerves, 596
Crayfish, 49–50
Creatine phosphate, 246
Creighton, Harriett, 341
Crick, Francis, 130, 276
Cristae, 147, 148
Cro-Magnon man, 434–435
Cross-breeding, 435
Cross-fertilization, 40
Crossing-over, 336–347
  frequency of, 342–345
  proofs of, 340–342
Cross-over chromatids, 339
Cross-pollination, 99
Crown gall disease, 534–535
Crumb structure (soil), **defined, 490**
Crustaceans, 49–51
Crystals, 120
Culture of man, 433
Cultures, 529–536
  callus, 532–535
  leaf, 531
  root, 529
  single cells, 535–536
  stem, 531
Cup fungi, 85
Curare, 604–605
Current, 581
Curvature, 514–515
  auxin and, 514–515
Cuticle, 47, 480
Cuvier, Georges L., 392
Cyanophyta, 75
Cycads, 97, 98, 102, 104

Cytochromes, 249, 252
Cytokinins, 507, 521–523, 544–545
Cytology:
  defined, 124, 136
  golden age of, 128
Cytoplasm, 134, 141–144, 150, 221, 468
  cell division, 169
  defined, 75, 141, 461
Cytoplasmic streaming, 472
Cytosine, 275,

D

Daisies, 526–528
Darwin, Charles Robert, 22–26, 27, 127, 129, 130, 409–411, 427
  theory of natural selection, 23–25
Day-neutral plants, 538
Death of cells, 163–166
Decarboxylases, 251
Decarboxylation, 245
Deciduous trees, 103
  defined, 97
Dehydration, 245
Dehydrogenases, 251
Dehydrogenation, 237
Dendrites, 156, 596
Dent corn, 554, 555
Deoxyadenylic acid, 275
Deoxycytidylic acid, 275
Deoxyguanilic acid, 275
Deoxyribonucleic acid (DNA), 14, 26–28, 30, 34, 75, 79, 107, 181–184, 302, 304, 412–415, 439, 460–461, 464
  defined, 27, 460, 475
  enzymes, 286–288
  genes, 286–288
  molecule, 27–29
    of inheritance, 130
    of life, 272–289
  nucleotide composition of, 392–393
  structure of, 274–286
    discovery, 130
    DNA-RNA protein chain of relationships, 281–286
    molecular model, 276–277
    replication, 277–281, 383
Deoxyribose, 205–206, 274–275
Deoxythymidilic acid, 275
Deplasmolysis, 226
Descent of Man, The (Darwin), 25
Desert plants, competition among, 549

Development as an inherited pattern, 374–389
  differentiation, 376, 380–385
  growth, 376, 377–379
  integration, 376, 386
Developmental biology, 123
Devonian Period, 69, 402
Diabetes mellitus, 672, 676, 678
  large flow of urine in, 674
Diakinesis, 317
Diaphragm, 636
Diatomic molecules, 191–192
Diatoms, 122, 491
Dicotyledonous weeds, 518
  killing of, 551
Differentiation, 532, 544, 716–717
  defined, 716
  problems of, 526–529
  of reproductive organs, 536
  root tissue, 530–531
Diffusion, 577–578
  carbon dioxide, 639, 640
  defined, 577
  of oxygen, 637, 638, 640
  through capillary walls, 615
Digestion, 661–668
  chemical factors in, 664
  defined, 661
  hormones in, 665–667
  liver in, 664
  pancreas in, 664
  salivary gland in, 664
  small intestine in, 664, 668
  stomach in, 664–667
Digestive secretions, 665–668
  reflex control of, 665
Digestive system, 662
Dihybrid crosses, 330, 331, 337
Dinosaurs, 418
Dioxyacetone phosphate, 242
Dipeptide, 210
Diphosphoglyceric acid, 243
Diphosphopyridine nucleotide (DPN), 244, 247, 252–253
  oxidation of, 249
Diploid organism, 305, 306
  defined, 87
  inheritance in, 326–335
    chance, 331–333
    probabliity, 331–333
    ratios, 326–330
Diplonema, 317
Disaccharides, 205–206
Disc flowers, 526

Distal tubule of the kidney, 654
Disulfide, 211
Diversity:
  defined, 3
  order in, 107–109
Division of cells, 168–177, 543
  animal, 173–175
  roottip, 170–173
  time sequence, 175–176
Dizygotic (DZ) twins, 368–371
Doctrine of Special Creation, 395
Dogs:
  breeds, 396
  classification, 8–12
Dominance, 330
Dormancy, 230–231
Dormin, 521, 522
Dorsal cells, 377, 378
Dorsal surface, 8
Dotted (Dt), 415
Double bond, 197
Down's syndrome, 179
Drosophila (Stern), 341
Drosophila melanogaster, 336–346, 353–360, 384, 385
  genetic maps of chromosomes, 346
  mutations in, 415
DuBois, Raphial, 268
Ducts:
  bile, 667
  collecting, 649, 650
  lymph, 629
  pancreatic, 672–673
  thoracic, 629
Duodenum, 667
Dust storms, 551
Dwarfism, 161, 326–330, 679, 680

E

Ear, 699–702
  external, 699–702
  inner, 701, 702
  middle, 699–702
  bones of, 700
Ear canal, 701
Ear drum, *see* Tympanum
Earth, 118, 424–425, 427–428
  origin of, 122
Earthworm, 47
Echinoderms, 11, 53–55, 69, 102
Ecology, 4–5, 18, 108
  defined, 3, 124, 549
Effectors, 607
Efferent arterioles, 649
Eggs, 152, 169, 180, 313, 318, 370

Eggs (*cont.*)
  activation of, 380
  fertilization of, 321, 526–529
  fish, 319
  fucus, 381
  hen's, 154
  human, 154, 162–163, 318–319, 321, 374
  ostrich, 154
  sizes of, 138
Egg cells, 708, 711–712
Einstein equation, 452
Electrical energy, 240
Electrical forces, transport and, 574, 579–582
Electrodes, 581
Electromagnetic attraction, 190
Electromagnetic lens, 140
Electron acceptor, 487
Electron mass, 188
Electron microscope, 128, 139–142, 145–149
Electron transport, 249, 253
Electrons, 139–140
  defined, 187–188
  number of, 190–192
  orbitals, 189–190
  valence, 190
Elements, 187, 488
  in the atmosphere, 491, 493
  atomic number, 189
  deficiencies, 490
  essential, 488–491
  micronutrient, 490
  mineral, 478, 491–494
  plant growth and, 511
  for structural purposes, 495–496
Embryo, 152
  amphibian, 159
  development, 351, 375
  frog, 379
  of whitefish, 168, 173
Embryo sac, 321–322
Embryonic plants, 504–506, 523
Emulsification, **defined, 667–668**
Endocrine glands, 674
Endodermis:
  defined, 508
  light and, 509
Endogenous rhythms, 543–544
  defined, 543
Endometrium, 712
Endoplasmic reticulum (ER), 141, 147, 148, 221, 458, 459, 461
Endoskeleton, 56
Endosperm, 504, 505
  food of, 466
  nucleus, 101
Energy:
  ATP used to supply, 675

Energy (cont.)
  availability on Earth, 427
  basic units, 134
  bond, 236
  from carbohydrates, 205
  cell's needs of, 675–676
  chemical, 237, 240
  chloroplasts, 465–468
  electrical, 240
  free, **defined, 236**
  hormones and, 677–679
  ion absorption and, 491–492
  levels, 189
  light, 240
  mechanical, 240
  and metabolism, 236–257
  order in, 119
  osmotic, 240
  and oxidation, 237
  photosynthesis as a source of, 264–265
  potential, 574–575
  defined, 236
  sun and, 264–265, 450–456, 487
  transfer, 245
  transformations in, 487
  use of, 676
Energy of activation, 229
Enterogastrone, 667
Enterokinase, 664
Entomologist, 119
Entomology, **defined, 124**
Environment:
  biological changes, 409
  city-street, 4
  country, 4
  in evolution, 29–31
  heredity and, 362–373
    temperature variation, 367
    twin studies, 368–371
  and organisms, 406–409
  and phenotype, 363–368
  seasonal aspects, 408
  and shape variation, 160
Enzyme systems, 107
Enzyme-catalyzed reactions:
  effect of substrate concentration on, 232
  effect of temperature on, 230
Enzyme-substrate complex, 231
Enzymes, 143, 208–209, 228–234, 462–464, 605, 640, 663–664, 665, 667
  active, 231, 463
  activity, 294–295
  amylase, 520
  as catalysts, 661
  complexes, 295–297
  constitutive, 291
  defined, 208, 462

Enzymes (cont.)
  effect on chemical reaction, 229
  essential parts of, 496
  feedback inhibition of, 292–294
  formation of, 384
  and genes, 286–288
  gibberellin and, 520
  hormones and, 681
  inducible, 291
  Kornberg, 277
  naming of, 229
  nature of, 464
  repression, 292–294
  synthesis, effect of nutrients on, 291–292
Eocene Period, 398–399
Epicanthic fold, 437
Epicotyl, 505
Epidermal cell, 474
  cuticle layer, 497
Epidermal tissues, development of, 508–509, 523
Epiglottis, 635
Epinepherine, 677
  defined, 677
Epithelium, 320
Erepsin, 664
Escherichia coli, 384, 418, 420–421
Esophagus, 635, 660
  peristalsis in, 669
Essay On the Principle of Population, An (Malthus), 23, 25
Estrogens, 714, 718, 720
Ethane, 197, 198
Ethyl alcohol, 197, 198, 245, 248, 465
Ethylene, 197
Etiolated plants, 541
Euchlaena, 554
Eumycophata, 84–87
Eustachian tube, 701–702
Evaporation, 484
  heat loss by, 683
  humidity in, 683–684
Evolution, 13, 20–31, 53, 55, 102, 108
  of animals, 66–69
  biological basis of, 26–31
  causes, 406–417
  cells and, 169
  chemical, 403
  course of, 399–403
  defined, 21
  environmental effects on, 29–31
  evidence of, 396–399
  failures of, 420
  of inherited patterns, 390–405
    continuity, 394
    diversity, 391, 392–394
    unity, 390–392

Evolution (cont.)
  of man, 66, 433–435, 441–444
  natural selection, theory of, 409–412
  origin of life, 424–429
  of plants, 103–104
  results of, 406–417
  science of, 128
  theory of, 127, 129, 130, 394–399
  variation, 412–417
    adaptations, 419–424
    fate of, 417–419
    source of, 412–417
  of vertebrates, 433
Excitatory transmitters, 605–606
Excretory system, 460
Exercise:
  cardiac output and, 617
  heat produced by, 682
Exoskeleton, 48, 161
Expiration, 636, 637
External ear, 699–702
Extracellular fluid, *see* Internal environment
Extract, 674
Eyes, 265–267, 702–705
  color, 353–357
  compared to a camera, 703
  resolving power, 139, 140

F

Fallopian tubes, 321, 711
Family, 8, 36
Fats, 495, 663, 674, 675
  breakdown of, 677
  cell's energy needs and, 676
  emulsified, 664
Fatty acids, 147, 207, 663, 664, 668
  oxidation, 253
Feedback inhibition of enzymes, 292–294
Feedback systems, 131, 567–570, 643, 644, 645, 674, 675
  biological, 570–572
  defined, 568–569
  homeostatic, 653, 678–679
  negative, 568, 569, 571, 620, 621
  positive, 568, 569, 656
  temperature control, 684–686
Felidae, 9
Fermentation, alcoholic, 240–245
Fermentative metabolism, products of, 248

Ferns, 91–94, 102, 104
Ferredoxin, 501
  defined, 493
Fertilization, 138, 370, 526–529, 708, 715–716
  defined, 709
  genes and, 306
  in humans, 321
  of lilies, 323
  in *Neurospora,* 305
Fertilizers, organic, 490
Fetus, 717–718
Feulgen stain, 171, 172
Fever, 684–685
Fibers, 471
  inhibitory, 607
  motor, 596, 603–604, 691
  nerve, 606–607
  sensory, 596, 691
Fibrils, 590
Fibrin, 625, 626
Fibrinogen, 626
Fibrous connective tissue, 668
Fiddleheads, 92
Filaments, 75, 590–591
Filtration:
  defined, 575
  evidence for, 650
Fish, 57, 58, 59, 69, 102
  egg, 319
  fossilized, 398
Fixatives, 140
Flagella, 81
  defined, 39
Flatworms, 42–44, 69, 376, 394
  flukes, 42
  free-living forms, 42, 43
  tapeworms, 42
Flavin, 249, 252
Fleming, Alexander, 119–120, 130
Flint corn, 554, 555
Florets in tassels of corn, 556
Florigen, 538, 539
Flounder, 160
Flour corn, 554, 555
Flowers, 526–528, 536–541
  accessory parts, 99
  essential parts, 99
  leaves of, 98
Fluids, in steady state, 628; *see also* Blood; Body fluids
Flukes, 42, 44
Fly, fruit, 336–347, 353–360, 384, 385
Focus, 704
Folic acid, 212
Follicle, 319, 711
Follicle stimulating hormone (FSH), 714, 715
Food, 660–671
  classes of, 665

Food (cont.)
  digestion of, 661–668
    chemical factors in, 664
    defined, 661
    hormones in, 665–667
    liver in, 664
    pancreas in, 664
    salivary gland in, 664
    small intestine in, 664, 668
    stomach in, 664–667
  supply, 454–456
Food web, 78
Forebrain, 689
Forests, 549–551
Formic acid, 195, 199
Fossil fuels, 483
Fossils, 19, 21, 397–399, 420
Fovea, 703
Fraternal twins, 368–371
Free energy, **defined, 236**
Frog:
  cleavage stage, 378
  embryo, 379
Fronds, 92
Frontal lobe, 690
Fructose, 205
Fructose diphosphate, 242, 243
Fructose phosphate, 242, 243
Fruit, ripening process of, 467
Fruit fly, 336–347, 353–360, 384, 385
Fruit growth, 101
Fucus, 526–528
  defined, 526
Fucus egg, 381
Fumaric acid, 250, 252
Fungi, 15, 82–86, 102, 104, 109, 120, 145, 166
  bacteria, 82–83
  lichens, 86–87
  luminous, 267
  slime molds, 83
Fungicides, 523
Furrowing, 174

G

Galactose, 384
Galapagos Islands, 23, 25
Galen, 125
Gametes, 81, 88, 108, 331, 474, 476, 709
  defined, 40, 318
Gametophyte:
  defined, 88
  of ferns, 93
  of liverworts, 88
  of mosses, 89–91
  of seed plants, 95
Garner, 537–538
Gastric mill, 50

Gastric mucosa, 383
Gastrin, 666
Gastro-intestinal (GI) tract, 660, 662, 668, 670
  fluid secretions of, 665
  length of, 661
  lubrication of, 665
  as outside the internal environment, 661
Gastrula, 379
Gastrulation, 377–379
Gay-Lussac, Joseph Louis, 240
Geiger counters, 487
Gel-like substances, 222–223
Generic name usage, 13
Genes, 26, 27, 107, 461, 463, 466, 716–717
  crossing-over, 337–340
    interference, 345
    proofs of, 340–342
  defined, 12, 26
  differential action, 383
  and enzymes, 286–288
  law of independent assortment, 307–308, 330
    exceptions to, 336
  law of segregation, 306
  linkage, 337–340
  and meiosis, 306
  random assortment, 307–308
  rate of mutation, 412–413
  recombinations, 338
  regulator, 292
  structural, 292
  temperature sensitive, 511
Genetic alphabet, 412
Genetic distance, 343
Genetic kinship, 17
Genetic maps, 342–346
Genetic ratios, 336
Genetics, 128
  defined, 25, 124
  of man, 439–441
Genotype, 330, 351, 364, 367
  defined, 510
Genus, 7
Geologists, 396
Geology, 21, 23
Geotropism, 515
Geranium stem, 376
Germinal epithelium, 319
Germination, 505, 521, 530, 540, 545
Giantism, 161, 680
Gibberella fujikuroi, 518–519
Gibberellic acid, 519, 520
Gibberellin, 518–520, 524
  biennials and, 542–543
  discovery of, 518
  long-day plants and, 539, 545
Gill clefts, 55–56
Glass lens, 140

Globin, 392
Globulin, 630
Glochidia, 46
Gloger's rule, **defined, 439**
Glomerulus, 649
  filtration in, 650
Glucose, 205, 253, 450–451, 479, 583, 638, 639, 663, 672
  in liver, 675
  permeability of, 674
  phosphorylation of, 242
  in urine, 651, 674
Glucose phosphate, 242
Glutamic acid, 209
Glyceraldehyde phosphate, 242
  formation of ATP from, 243
Glyceraldehyde-3-phosphate, 243
Glyceraldehyde-3-phosphate dehydrogenase, 244
Glycerol, 207, 208, 663, 664, 668
Glycogen, 246, 253, 674, 677
  defined, 206
  in liver, 674, 677
Glycolysis, 246–248
Golgi body, 458, 459, 461
Golgi material, 141, 142, 147–149
Gonadal hormones, 352
Gonadotropic hormones, 711, 714–715
Gonads, 351, 352
Gonyaulax polyedra, 543
Graafian follicle, 319
Grafting, 539
Grana, 149, 466, 501
Grand Canyon, 398
Grass family (Gramineae), 553
Grasses, 99, 102
Grasshopper, 349
Green algae, 75, 76, 79–81, 104
Griffiths, Frederick, 181
Growth, 504–524, 526–545
  of cells, 376–379
    inhibitors, 507, 521–523, 524
  kinetics of, 506–507
    tissue organization and, 507–511
  meristems, 507–511
  plant body, 526–545
    cultures, 529–536
    endogenous rhythms, 543–544
  photoperiod, 537–541
  temperature and, 536, 541–543
Growth hormones, 511–521, 677–680
  antagonistic actions of, 678

Growth hormones (cont.)
  important role of, 679
Growth requirements of organisms, 211–216
Guanine, 275
Guard cells, 480, 481
  turgor pressure, 482
Gymnosperms, 97–98, 104, 466
  conifers, 97–98
  cycads, 98

H

H zones, 590–591
Haematoxylin, 142, 171, 172
Hale telescope, Mount Wilson, California, 121, 138
Hämmerling, Max, 180
Haploid, **defined, 87**
Haploid cells, 318
Haploid chromosomes, 305
Haploid nuclei, 306
Hapsburg family, 286
Harden, Sir Arthur, 242
Hardy-Weinberg principle, 417, 418
Harvey, William, 127
Hearing, 692–693, 699–702; *see also* Ear
Heart, 127
  action potential of, 613
  artificial, 616
  chambers of, 615
  excitation of, 613
  intense heat response, 685
  muscle, 586–587, 613–615
  pressure within, 616
  as a pump, 615
  sounds, 617
  valves in, 614, 616
  ventricles of, 615
Heart cells, 381
Heart muscle, 128, 613–615
Heat, 681–686
  basal metabolic, 682
  loss of, 683–684
  production of, 682–683
  response to intense, 685–686
Helium, 187, 188, 456
Helmont, Jean Baptiste van, 258
Heme groups, 393
Hemoglobin, 147, 286, 392, 632, 637, 639, 642
  oxy-, 640
  reduced, 640
  regulation of, 642–643
Hemoglobin A, 285, 286, 393
Hemoglobin S, 285, 286
Hemophilia, 440
Hemorrhage, 624, 628, 643
Herbicides, 523
Herbivorous, **defined, 53**

Heredity, 708; *see also* Inheritance
chemical basis, 182
development pattern, 374–389
  differentiation, 376, 380–385
  growth, 376, 377–379
  integration, 376, 386
  and environment, 362–373
  temperature variation, 367
  twin studies, 368–371
molecular basis of, 180–182
Heterosis, 510
Heterotrophic nutrition, 82, 103, 104
Heterotrophic organisms, 214–215
Heterotrophic plants, **defined, 467**
Heterozygous state, 330
Hindbrain, 689
Histone, 175, 182, 183
Holdfast, 526–528
Homarus vulgaris, 390
Homeostasis, 621
  defined, 567
  steady state and, 566–567
Homeostatic mechanism, 619
Hominidae, 432
Homo erectus, 434, 435
Homo habilis, 435
  defined, 434
Homo neanderthalensis, 433–434
Homo sapiens, 409, 434, 435
Homologous chromosomes, 305, 313, 339, 340, 350
Homologous structures, 392
Homologues, 314–315
Homology study, 67
Homozygous dominant, 330
Homozygous recessive, 330
Hooke, Robert, 137
Hookworms, 44
Hormones, 511–521, 523, 524, 538, 539, 543, 544, 571, 667, 672, 687
  adrenocorticotropic (ACTH), 681
  antidiuretic (ADH), 652–653
    manufacture of, 652
    mechanism of action, 654–656
  control, 680–681
  defined, 511, 594
  in digestion, 665–667
  energy release and, 677–679

Hormones (*cont.*)
  enzymes and, 681
  follicle stimulating (FSH), 714, 715
  gonadal, 352
  gonadotropic, 711, 714–715
  growth, 511–521, 677–680
    antagonistic actions of, 678
    important role of, 679
    in heat production, 683
    in information transfer, 594
  inhibition of, 681
  luteinising (LH), 714
  luteotrophic (LTH), 715
  mechanisms of, 681
  parathyroid, 571
  placenta, 718
  in regulation of body water, 652–653
  in reproduction, 710, 711, 714–715, 718–720
  thyroid-stimulating (TSH), 680
Horse, skeleton, 432
Humidity, 496–497
  in water evaporation, 683–684
Humus, 488
Husks (corn), 553, 557
Hybrid vigor, 510, 557
Hybrids, 510
Hydra, 41, 43
Hydrocarbons, 197
Hydrochloric acid, 383, 665, 667
Hydrogen, 192, 427, 480
  atoms, 187, 188, 194, 479, 486, 491, 638
  nuclei of, 452
Hydrogen bond, 195
Hydrogen cyanide (HCN), 427–428
Hydrogen ion concentration (pH), 201–202
Hydrogen ions, 639
Hydrogen sulfide, 196
  photosynthesis and, 480
Hydrogen transport, 249
Hydrolyzed, **defined, 209**
Hydroxyl, free radical of, 488
Hydroxyl group, 198
Hydroxyl ion, **defined, 487**
Hypertonic solution, 226
Hyphae, 303, 304
Hypocotyl stem, 505
Hypothalamus, 652–653, 680–681, 690
  anterior, 684
  in control of body temperature, 684–685
  posterior, 684
Hypotonic solution, 226

I
I bands, 590
Ice Age, 434
Identical twins, 368–371
Impermeability, 575
Implantation, 717–720
Inbred plants, 510
Incus, 701
Independent assortment, law of, 307–308, 330
  exceptions to, 336
Indole, 384
Indole-3-acetic acid (IAA), 513, 524
Inducible enzyme, 291
Influenza virus, 144
Information transfer, 594–611
  hormones, 594
  nervous system, 594–609
    action potentials, 604–606
    autonomic, 608–609
    central, 596–607
    code used by, 601–602
    inhibition by nerve fibers, 606–607
    main components of, 594–595
    motor fibers, 603–604
    polarization of nerve membranes, 599–602
    reflex actions, 607
    sensory receptors, 602–603
Infrared (heat) radiation, 453, 483
Inheritance, *see also* Heredity
  attached-X, 356–357
  in a diploid organism, 326–335
  chance, 331–333
  probability, 331–333
  ratios, 326–330
  Mendel's law of, 306, 326–333
  of a trait, 302–311
    inheritance test, 303–306
    Neurospora studies, 303–308
    random assortment of genes, 307–308
  sex, 348–361
Inheritance tests, 303–306, 336, 362
  chromosome behavior, 305–306
Inherited patterns, evolution of, 390–405
  continuity, 394
  diversity, 391, 392–394
  unity, 390–392
Inhibition, 606–607

Inhibitors, competitive, 233
Inhibitory fibers, 607
Inner ear, 701, 702
Inorganic salts, 216
Inositol, 520
Insecticides, 523
Insectivores, 66
Insects, 48–49, 102
Inspiration, 636, 637
Insulin, 672–674
  antagonistic actions of, 678
  function of, 674
  permeability of cell membranes and, 681
  secretion control, 674–675
Integration, 376, 386
Intelligence tests, 371
Interbreeding, 13, 435
Internal environment, 564–573
  blood pressure and, 620–621
  circulatory system and, 613
  defined, 565
  fluid regulation, 654
Interphase, 171, 174
  stages of, 175
  time of, 175
Intersexuality, 352, 359
Interstitial cells, 710, 711
Intestinal tract, water loss through, 646
Intestinal wall, 156
Intestines, *see* Large intestine; Small intestine
Invagination, 227, 379
Invertebrates, 399
Iodine, 491
Ionic bonding, 191
Ions, 579–581
  absorption, 491–492
  bicarbonate, 639, 667
  chloride, 580
  defined, 191, 579
  hydrogen, 639
  potassium, 579–582, 599–600
    active transport of, 584
  secretion of, 492
  sodium, 579–580, 582, 599–600
    active transport of, 584
Ipomopsis, 391
Iris, 703
Iron, 490
Islet cells, 672–674
  influence of, 673
Isocitric acid, 252
Isomers, **defined, 205**
Isometric contraction, 587–588
Iso-osmotic solution, 225
Isoprenoid units, 522
Isopropyl alcohol, 248

Index   **739**

sotonic contraction, 587
sotonic solution, **defined, 579**
sozymes, 294–295

J

Janus Green B, 142
Jello, 222
Jellyfish, 41
Jurassic Period, 402

K

Kelps, 75, 77
Keratin, 209
Ketoglutaric acid, 252
Ketones, 198
Kettlewell, H.B.D., 29–30, 422
Kidney cells, 140, 148, 381
Kidneys, 648–659
   ADH action, 654–656
   body fluid volume regulation, 656–657
   cortex of, 654
   distal tubule of, 654
   in excreting waste products, 657
   loop of, 654
   medulla of, 654
   nephron structure, 654
   osmotic pressure and, 652–653
   structure of, 649
   urine formation, 649–652
   water loss through, 646
Kinetic energy, 190
Kinetics of growth, 506–511
Kinetin, 520–521, 532, 534
   in bud development, 522
   culture of leaves and, 531–532
Kingdoms, 7
   animalia, 8
   monera, 15
   protista, 15
Kornberg, Arthur, 277
Kornberg enzyme, 277

L

Labor pains, 720
Lactic acid, 253, 639
   formation of, 248
   production of, 246
Lactic acid dehydrogenase (LDH), 247, 294–295
Lactose, 291
Lamarck, Jean Baptiste, 21–22, 29
   theory of inheritance of acquired characteristics, 21
Lange, Johannes, 371

Large intestine, 660
Larvae, 46
Larynx, 635
Leaf cell, 143
Leaflets, 92
Leakey, L.S.B., 434
Leaves:
   abscission, 522
   culture of, 531–532
   the fall of, 517
   photoperiodic stimulus, 538, 545
   structure of, 480
   tobacco, 537–538
   water loss from, 497
Lecithin, 208
Leech, 47
Leghemoglobin, 493
Leguminosae, 492
Lens, 703, 704
Leptonema, 317
Leucocyte, 134
   size of, 154
Leucoplast, **defined, 467**
Lichens, 29, 86–87
Life:
   carbon as necessity, 423–424
   and light, 258–271
      bioluminescence, 151–152
      photosynthesis, 258–265
   origin of, 424–429
   temperature range, 425, 426–427
   vision, 265–267
   water as major component, 425, 426, 427–428
Life cycle:
   of ferns, 93–94
   of flowering plants, 101
   of liverworts, 88
   of mosses, 90
   of seed plants, 95–96
Light, 451
   branch-root initiation and, 530
   competition for, 550, 551
   effect on biological processes, 266
   endodermis and, 509
   and life, 258–271
      bioluminescence, 267–268
      photosynthesis, 258–265
      vision, 265–267
   photosynthesis and, 485–487
   plant growth and, 511
   role of, 260
   visible, **defined, 453**
Light energy, 240
Light microscope, 138–139, 141–143

Light-sensitive cells, 705
Lignified cells, 532–533
Lignin, 468, 470, 476
Lilies, 314
   female parts of, 321
   fertilization of, 323
   male parts of, 323
   reproduction in, 321–323
Lilium regale, 314
Linanthus, 391
Linkage, 336–347
   sex, 352–358
      color blindness, 357–358
      eye color, 353–357
      secondary character, 352
Linnaeus, Carolus, 7–8, 12, 13, 36, 72, 73, 107, 127, 553
Linné, Karl von, *see* Linnaeus, Carolus
Lipases, 663, 664, 667
Lipids, 146–148, 182, 207–208, 582, 663
Lipoic acid, 250
Lithium, 491
Liver, 660, 667–668
   in digestion, 664
   glucose in, 675
   glycogen in, 675, 677
   in regulating blood sugar, 677
Liver cells, 140, 142, 381
Liverworts, 87–89, 91, 93, 104
   life cycle of, 88
Lobe-finned fishes, 58
Lobster claw, 390
Locus, 305
Loewi, Otto, 605
Long-day plants, 538–539
   behavior, 540–541
   critical night length, 540–541
   flowering of, 522
Loop of the kidney, 654
Luciferase, 268
Luciferin, 268
Luminous clam, 268
Luminous fungi, 267
Lungs, 156, 634–637
   water loss from surface of, 683–684
Luteinising hormone (LH), 714
Luteotrophic hormone (LYH), 715
Lycopods, 402
Lyell, Charles, 21
Lymph duct, 629
Lymph nodes, 629
Lymph valves, 629
Lymphatic system, 629–630
Lymphatics, 629
Lysosomes, 141, 149, 221

M

McClintock, Barbara, 341
Macromolecules, **defined, 146, 281**
Macronutrients, 216
Magnesium, 495
Magnesium sulfate, 490
Magnifying lens, 139
Maize, 553–558
   chromosomes in, 341
   meiosis of, 315
   mutations in, 415
Malacology, **defined, 124**
Malaria, 38, 39
Malic acid, 250, 252
Mallards, 348
Malleus, 701, 702
Malonic acid, 233
Malthus, Thomas R., 23, 129
Maltose, 206, 231
Mammalia, 11, 64–66, 432
   defined, 4, 10
Mammals, 64–66, 69
Mammals, Age of, 398
Man:
   culture, 433
   evolution of, 433–435, 441–444
   genetics of, 439–441
   origins of, 432–446
   races of, 435–438
      climate and, 438–439
   sexual reproduction in, 318–321
   skeleton, 432
   as a vertebrate, 444
Manganese, 490
Manganese deficiency, 212
Manometers, 498
Map unit of genetic distance, **defined, 343**
Marsupials, 65
Maryland Mammoth (tobacco), 537–538
Matter:
   basic units, 134
   defined, 119
   in motion, 119
   order in, 119
Mayer, Robert, 259
Mechanical energy, 240
Medulla, 351, 352
   of the brain, 621, 637, 690
   of the kidney, 654
Megaspore, 95
Megasporocyte, 321–322
Meiosis, 80, 88, 108, 305, 306, 312–325, 329–330, 349
   defined, 81, 313
   genes and, 306
   of maize, 315
   relation to sexual reproduction, 312–325

Meiosis (cont.)
  stages of, 313–318
    anaphase, 316, 317
    metaphase, 316, 317
    prophase, 313–316
    telophase, 316, 317
Meiotic cell, 340, 349
Meiotic metaphase, 316
Melanin, 160, 367
Melanism, industrial, 421–422
Melanocyte, 160
Membranes, 297–299
  basilar, 700–702
  bimolecular lipid-protein structure for, 298
  cell, 574, 582–584
  differentially permeable, 458
  metabolic properties, 223–228
  nerve:
    polarization of, 599–602
    resting, 599
  nuclear, 141, 142
  plasma, 221
  polarized, 581–582
  site of entry in, 491
Memory, 695–697
  long term, 696
  short term, 696
Mendel, Gregor, 25–26, 306, 308, 326–333
Mendel's laws of inheritance, 306–308, 326–333
Menopause, 715
Menstruation, 713
Meristematic cells, 474, 476
  defined, 473
Meristematic zone, 508, 509
Meristems:
  defined, 507
  tissue organization and, 507–511, 523
Meselson, Matthew, 279–281
Mesophyll, 480, 481
  air passages to, 482
Mesophytes, defined, 497
Mesozoic Era, 69, 402
Messenger RNA (m-RNA), 282
Metabolic products, 619–620
Metabolic properties of cells, 220–236
  enzymes, 228–234
    effect on a chemical reaction, 229
    effect of temperature on, 229–231
  the membrane, 223–228
  protoplasm, nature of, 220–223
Metabolic pumps, 299
Metabolism, 156–157, 672–687

Metabolism (cont.)
  aerobic, 252
  amino acid, 254
  anaerobic and aerobic phases compared, 253
  of butyric acid, 254
  cellular, control of:
    biological membranes, 297–299
    control systems, 297–299
    enzyme activity, 294–295
    enzyme complexes, 295–297
    enzyme repression, 292–294
    feedback inhibition of enzymes, 292–294
    nutrients' effect on enzyme synthesis, 291–292
    protein structure, 294–295
  defined, 155, 187, 675
  and energy, 236–257
  fermentative, products of, 248
  intense heat and, 686
  muscle, glycolysis, 246–248
  oxidative, 250
Metabolites, action of, 620–621
Metamorphosis, of butterfly, 48, 49
Metaphase, 171–175
Metazoa, 69
  defined, 38
Methane, 197, 427
Methyl group, 198
Micrographia (Hooke), 137
Microns, 139
Micronutrients, 216
Microscopes, 121, 122
  electron, 128
    cells under, 145–149
    described, 139–140
  lens, 139–140
  light:
    cells under, 141–143
    described, 139
    parts of, 138
  primitive, 137
  resolving power, 139, 140
Microspore, 95, 98, 101
Microsporidian, 3
Microsporocytes, 323
Midbrain, 689
Middle ear, 699–702
  bones of, 700
Middle lamella, 468–469
  defined, 468
Mildews, 82, 85
Milky Way, the, 395, 424–425
Miller, Stanley, 427

Miller-Urey broth, 427–429
Millimicrons, 139
  defined, 452–453
Millipedes, 52–53
Mimicry, 32–33
  defined, 32
Minerals, 478, 491–494
  absorption, 492
  as food, 661
  nutrients, 216, 488–496
Minkowski, Oscar, 672
Minute mutants, 415
Mites, 51–52
Mitochondria, 141, 142, 147–149, 221, 251–252, 458–461, 464–465
  defined, 464, 475
  as multienzyme complex, 296–297
Mitosis, 80
  defined, 81
Molds, 82, 85
Molecular basis of heredity, 180–182
Molecular biology, 123
  defined, 128
Molecule of inheritance (DNA), 130
Molecules, 134
  and atoms, 186–193
  defined, 26, 458
  diatomic, 191–192
  diffusion of, 577–578
Mollusca, 37, 45–46
Mollusks, 45–46, 69
  bivalve, 45
  univalve, 45
Molybdenum, 490, 496
Monera kingdom, 15
Mongoloid race, 436–438
Monocotyledons, 518
Monoecious plants, 556
Monohybrid cross, 330
Monosaccharides, 205–206
Monozygotic (MZ) twins, 368–371
Monuron, 522
Morgan, L. V., 357
Morgan, T. H., 353
Morphogenesis, 526–529, 544
  culture and, 529–536
  defined, 526
Mosses, 89–91, 92, 93, 102, 104
Moths, 421–422
Motile cells, 79, 81
Motility, 669–670
Motion, 586–593
Motor fibers, 596, 603–604, 691
Motor nerves, 607
Mouth, 660, 662
Mucilage, 490
Mucous cells, 383

Mucus, 383, 665
Müllerian duct, 351
Multicellular organisms, 152–153, 158–159
  division of, 168
Multienzyme complex, 296
Multiple alleles, 412
Muscle(s), 586–593; see also names of muscles
  cardiac, 586–587, 613–615
  ciliary, 703, 704
  contractile machinery, 588–591
  contraction of, 587–589, 603
  series elasticity, 588
  skeletal, 586–587
    in a relaxed state, 590
  smooth, 586–587
    arterioles in, 619
  types of, 586–587
  voluntary control of, 586–587
Muscle cells, 136, 154, 163, 381–383
  shape of, 159–160
  skeletal, 247
Muscle metabolism, glycolysis, 246–248
Muscosal cells, 668
Mushrooms, 82, 85, 86
Mustela erminea, 236
Mutation(s), 27–30, 32, 34, 418
  chemically induced, 415
  defined, 27–28, 285
  hemophilia, 440
  in maize, 415
  rate of, 412–413
  spontaneous, 413–414
Mutualism, defined, 39
Mycelia, 84, 86, 303, 304
Mycology, defined, 124
Myelin, defined, 596
Myoblasts, 382
Myosin, 246, 588, 589–591
Myxomycophyta, 83–86

N

Naphthaleneacetic acid, 517
Natural classifications, 11–12
  defined, 7–8
Natural resources, 109
Natural selection theory, 23–25, 29–31, 53, 108, 129, 409–412
Nature:
  defined, 118–119
  man's place in, 109–111
Neanderthal man, 433–434
Negroid race, 436–438
Nematodes, 44
  roundworms, 44

Nephrons, 649
  structure of, 654
Nerve cells, 163, 596
  shape of, 159–160
  size of, 154–156
Nerve fibers, inhibition by, 606–607
Nerve membranes:
  polarization of, 599–602
  resting, 599
Nerves, 594–611
  action potential of, 598–599
  auditory, 701
  cranial, 596
  impulses of, 596–602
  motor, 607
  optic, 703
  parasympathetic, 617, 621
    excitation of, 670
  polarized membranes, 599–602
  sensory, 607
  spinal, 596
  sympathetic, 617, 621, 684
    excitation of, 670
  vagus, 605, 666
Nervous system, 594–609
  action potentials, 604–606
  autonomic, 608–609
    cardiac output and, 617
    major divisions of, 608
  central, see Central nervous system
  code used by, 601–602
  inhibition by nerve fibers, 606–607
  main components of, 594–595
  motor fibers, 603–604
  parasympathetic, 608–609
  polarization of nerve membranes, 599–602
  reflex actions, 607
  sensory receptors, 602–603
  sympathetic, 608–609
Neural tube, 689
Neuromuscular junction, 604–605
Neurons, **defined, 596**
Neurospora, 287, 316, 337, 341–342, 360
  asexual reproduction in, 303, 304
  fertilization in, 305
  life cycle of, 303–304
  sexual reproduction in, 303–304, 305
Neurula, 378, 379
Neutrons, **defined, 187–188**

Nicotinamide adenine dinucleotide (NAD), 244
Nicotinic acid (niacin), 212, 244, 529
Nirenberg, Marshall, 284
Nitrate reductase, **defined, 493**
Nitrates, 83, 493
Nitrogen, 488, 493
  over-all cycle, 494
Nitrogen fixation, 488, 492–494, 501
  defined, 492
Nitrosocystis oceanus, 150
Nitrosomonas, 213–214
Noctiluca, 159
Nodes, 629
Nodules:
  defined, 492
  leguminous, 492–493
Nomenclature, binomial system of, 553
Nondisjunction, 355–356
Nonvascular plants, 92
Noradrenalin, 608–609, 678
Notch effect, 415
Notochord, 55–56
  defined, 8
Nuclear membrane, 141, 142
Nucleic acids, 14, 142, 146, 428, 495
Nucleolar organizers, 314
Nucleolus, 141–142
  defined, 461
Nucleoside, 239
Nucleotides, 274–275, 460–461
  defined, 460, 475
  sequence, 343
Nucleus, 15, 134, 142, 149, 221, 458, 459, 460–464, 466, 468, 491
  cell division, 169
  chromosomes as controlling element, 180–182
  as control center, 178–185
  defined, 141, 187–188, 475
  haploid, 306
  polar, 321–322
Nutrients, 186–187, 478–502, 531
  classes of, 478–502
    foods, 478–488, 500
    mineral, 216, 488–496, 500
    water, 496–502
  effect on enzyme synthesis, 291–292
  oxygen as, 215–216
  transport of, 498–501
Nutrition, **defined, 211**
Nymph stage, 48

O

Oat coleoptile, 528
Objective lens, 139
Occipital lobe, 690
Oceans, food supply and, 455
Octopus, 45
Ocular lens, 139
Omnivorous, **defined, 49**
On The Origin of Species By Means of Natural Selection (Darwin), 23, 34, 409–410
Onion roottip, cell division, 170–173
Oocytes, 319
Oogonia, 526
Oogonial cells, 319
Oparin, A. I., 424
Open circulatory system, 48
Opsin, 265
Optic nerve, 703
Orbitals of electrons, 189–190
Order, 7, 119
Ordovicean Period, 399
Organ cultures, **defined, 529**
Organelles, 462, 463
  defined, 460, 475
Organic chemistry, 197
Organic gardening, 490
Organic matter in soil, 489
Organisms, 6
  adaptation of, 20, 34
  base ratios, 276, 393
  chemical energy transformations in, 237
  classification of, 7–18
  continuity among, 390, 394
  defined, 3
  diversity among, 390–394
  and environment, 406–409
  growth, and growth of knowledge, compared, 123
  growth requirements, 211–216
    aerobic, 215–216
    anaerobic, 215–216
    autotrophic, 213–214
    heterotrophic, 214–215
  interbreeding, 13
  life spans, 164
  multicellular, 152–153, 158–159
    division of, 168
    in soil, 489
  unicellular, 152, 157–158, 163
    division of, 168
  unity among, 390–392

Organs, **defined, 612**; *see also* names of organs
Origin of Life, The (Oparin), 424
Ornithine, 288
Osmoreceptors, 653
Osmosis, 223–226, 476, 578–579, 651
  defined, 223, 472, 578
  schematic representation of, 224
Osmotic energy, 240
Osmotic pressure, 224, 579
  in body fluids, 652–653
  in capillary fluid exchange, 627–629
Ova cells, *see* Egg cells
Ovaries, 42, 98, 711
  corn, 557
  cycles, 713–714
  human, 318–319
  of lily, 321–322
Oviparity, **defined, 65**
Ovoviparity, **defined, 65**
Ovulation, 711, 713
Ovule, 95, 98, 101, 321–322
Ovum, 319
Oxaloacetic acid, 250–252, 254
Oxalosuccinic acid, 252
Oxidation, 237–239, 245
  of an aldehyde to an acid, 238
  defined, 237
  of DPN, 249
  fatty acid, 253
  of glyceraldehyde phosphate, formation of ATP, 243
Oxidation-reduction, 200
Oxidative decarboxylation, of pyruvic acid, 250
Oxidative metabolism, 250
Oxidative phosphorylation, 251–252
Oxidizing agent, 200
Oxygen, 197, 427, 428, 451, 465, 634–647
  atom, 194, 479
  brain dependency on, 612–613
  in cells, 155, 156
  control of delivery, 641–642
  diffusion of, 637, 638, 640
  as a nutrient, 215–216
  during photosynthesis, 482
  production of, 488
  regulation of hemoglobin, 642–643
  role of, 638–639
  transport, 637–638
Oxygen debt, 253
Oxy-hemoglobin, 640
Ozone, 453

P

P-aminobenzoic acid (PABA), 233
Pachynema, 317
Paleolithic Age, 441–442
Paleozoic Era, 399, 400
Palisade layer (leaf), 480
Pancreas, 660, 667, 672–674
    in digestion, 664
    dual role of, 673
    removal of, 672
Pancreatic amylase, 664
Pancreatic ducts, tying-off, 672–673
Pandorina charkowiensis, 162
Pandorina morum, 162
Paramecium, 145, 158
Parasites, 38, 39, 44, 109
Parasympathetic nervous system, 608–609, 617, 621
    excitation of, 670
Parathyroid glands, 571
Parathyroid hormones, 571
Parenchyma, cell, 474
    defined, 473
Parietal cells, 383
Parietal lobe, 690
Parthenocarpic fruits, 517
Pascal, Blaise, 131
Pasteur, Louis, 153, 170, 240, 424
Pathogenic bacteria, 83
Pathogens, 153
    defined, 397
Pathology, **defined, 124**
Pavlov, Ivan, 697–698
Peas, 336–337
    gibberellin and, 519, 521
    Mendel's studies of, 326–327
    roots, 529
Pectin, 468, 476
Penfield, Wilder, 696–697
Penicillin, 85, 119–120
Penicillium notatum, 120
Penis, 710
Pentose, 205
Pepsin, 383, 664, 665
Peptidase, 664
Peptides, 210, 664, 668
Perennial plants, **defined, 92**
Pericycle, 508, 523
Periderm tissue, 508, 523
Perinuclear space, 149
Peristalsis, 669–670
    reflex initiation of, 670
Permeability, 223, 575
    of capillary walls, 627
    of glucose, 674
    insulin and, 681
    of water, 655
Permian Period, 402
Pernicious anemia, 166
Peroxide, **defined, 488**
Petals, 98–99
Petiole, **defined, 517**
Phacus, 159
Phaeophyta, 77
Phagocytosis, 227–228, 629, 630, 632
Phagotrophs, 16
Pharynx, 635
Phenocopy, **defined, 367**
Phenol, 198
Phenotype, 330, 331
    environment and, 363–368
    sex as, 351
Phloem, 91–92, 480, 484, 488, 501, 523
    movement in, 500, 530
Pholas dactylus, 268
Phosphoglyceric acid, 243
3-Phosphoglyceric acid, (PGA), 488
Phospholipids, **defined, 496**
Phosphoric acid, 274–275
Phosphorus, 496
Phosphorylation, 245
    of glucose, 242
    oxidative, 251–252
Phosphotides, 208
Photobacterium fischeri, 267
Photolysis, 481
Photons, 260
Photoperiodism, 268, 538
    temperature and, 536
Photophosphorylation, 262, 501
Photosynthesis, 78, 204, 258–265, 451, 453, 478–485, 490, 496, 501
    action spectrum, 485
    and the atmosphere, 260–263
    biochemistry of, 479–481, 488
    carbon dioxide in, 263, 487–488
    defined, 72, 452
    as an energy source, 264–265
    food supply and, 454–456
    hydrogen sulfide in, 480
    light reaction, 485–487
    oxygen in, 482
    rate of, 263
    raw materials of, 481–485
Photosynthetic autotrophs, 214
Phototropism, 268, 514, 515
Phycobilins, 485
Phylum, 8, 36
    chordata, 8, 11
Physics, 119
Physiologists, 565
Physiology, 126
    defined, 124
Phytochrome, 269, 540, 545
Pigment molecules, 485–486, 487
    activated, 485
Pigments, accessory, 264
Pinocytosis, 227–228
    defined, 138
Pistils, 98–99, 321–322, 557
Pitch, 699–700
    discrimination, 700
Pith, the, 509
Pithecanthropus, **defined, 434**
Pituitary gland, 652, 653, 690
    anterior, 677, 678, 680
    removal of, 653, 678
Placenta, 717–720
    hormones of, 718
Planarians, 42, 43, 44, 376, 394
Plant cells, 143–144, 156
    and animal cells, contrasted, 149
Plants, 72–104
    annual, 541–542
    autotrophic, 16, 467
    classification of, 73
    differences between, 14–16
    divisions in, 74–103
        bryophyta, 87–91
        pteridophyta, 91–94
        spermatophyta, 94–103
        thallophyta, 74–87
    heterotrophic, 16
    reproduction in, 321–323
Plasma, 165, 624, 625, 628
    membrane, 142, 221
    proteins, 628, 629, 630
Plasmodium, 83
Plasmolysis, 226
    defined, 472
Plastids, 143, 149, 466, 467
Platelets, 626
    rupture of, 626
Platyhelminthes, 42–44
Platypus, 402
Pleistocene Period, 434, 442
Pleuro-pneumonia-like organisms (PPLO), 153, 154
Pneumococcus, 180
Pod corn, 554, 555
Polar body, 319, 320
Polar molecule, 195
Polar nuclei, 101, 321–322
Polarized membranes, 581–582
    nerve, 599–602
Polemonium, 391
Pollen grain, *see* Microspore
Pollen tube, 101, 323
Pollination, 99–101
    cross-, 99
    self-, 100, 101
Pollution, 111
Polymer, **defined, 274**
Polymerase, 277
Polymorphism, **defined, 419**
Polypeptide, 210
Polypeptide chains, 392–393
Polyploids, **defined, 417**
Polyribosomes, 148
Polysaccharides, 146, 180–181, 206–207
Polysomes, 148
Polyuridylic acid (polyU), 284
Pons, 690
Popcorn, 554, 555
Population, 111
    food supply and, 454–456
Potassium, 491
    ions, 579–582, 599–600
        active transport of, 584
    phosphate, 490
    and sodium pump, 299
Potential energy, 574–575
    defined, 236
Potentilla glandulosa, 364–365
Pregnancy, 715–716
    tests for, 718
Pressure:
    in arteries, 616
    blood, 576, 617–622
        average, 518
        in capillary fluid exchange, 627
        intense heat response, 685
        internal environment and, 620–621
        regulation of, 620–622
        urine formation and, 650
    measuring, 575–576
    osmotic, 579
        in body fluids, 652–653
        in capillary fluid exchange, 627–629
    transport and, 574–576
    turgor, 471–472, 476, 480, 482
    within heart, 616
Pressure flow hypothesis, 500
Priestley, Joseph, 259, 486
Primates, 65–66, 102
Progesterone, 714, 718–720
Propane, 197
Prophase, 171–172, 173
    first stage of meiosis, 313–316
    time of, 175
Proplastids, 458, 459, 461, 466
Proprioception, 694
Propylene, 197
Prostate gland, 710, 711
Proteases, 663, 664, 667

Protective cells, 476
defined, 474
Protein(s), 14, 107, 142, 146–148, 208–211, 254, 427, 461–462, 495, 663–664, 674, 675
cell's energy needs and, 676
chemical formula, 462
composition of, 461
defined, 475
DNA-RNA chain, 281–286
enzymes, 462–464, 520
residual, 182
structure of, 209–211, 294–295
Protentor, 350
Proteozoic Era, 399
Prothrombin, 626
Protista kingdom, 15
Protists, **defined, 399**
Protochlorophyll, 263
Protochordates, 55–57
Protons, **defined, 187–188**
Protoperithecium, 303
Protoplasm, 157–158
and colloidal systems, 221–223
nature of, 220–223
Protoplast, **defined, 226–227, 458**
Protozoa, 15, 38–40, 69, 145, 152, 158–159
amoeba, 39
cilia, 39
flagella, 39, 69
sporozoa, 39
Proximal tubule of the kidney, 654
Pseudopodia, 39, 227
Ptarmigan, 408
Pteridophytes, 91–94
ferns, 91
Ptolemaeus, Claudius, 129
Ptolemaic theory, 129, 395
Ptyalin, 664
Puffballs, 85
Pulmonary artery, 615, 635
Pulmonary vein, 615
Pumps:
ATPase as, 298–299
metabolic, 299
Punnett square, 330–332
Pupa, 49
Pupil, 703
Purines, 275, 428
Pyrenoids, **defined, 76**
Pyridoxine, 212
Pyrimidines, 275, 428
Pyrophosphate, 239
Pyruvic acid, 248, 252, 253
alcohol formed from, 245
formation of, 244
oxidative decarboxylation of, 250

R

Rabbits, 367
heart muscle of, 128
Races of man, 435–438
Rachis, 92
Radial symmetry, 53
defined, 43
Radiation:
of absorbed energy, 483
heat loss by, 683
kinds of, 452–453
ultraviolet, 414, 453
Radio waves, 121–122
Random assortment of genes, 307–308
Ray flowers, 526
Reabsorption, 650
mechanisms of, 650–652
of water, 650
Reactive atoms, 190–191
Receptors, 607
sensory, 602–603
stretching, 621
Recessiveness, 330
Recombinant spores, 308
Recombinations of genes, 338
Red algae, 77, 79
Red blood cells, 142, 147, 165–166, 226, 630, 632, 642
nucleus loss, 178
shape of, 159, 160
Reduced hemoglobin, 640
Reducing agent, 200
Reduction division of chromosomes, 305
Reflex actions, 607
conditioned, 697–699
in study of animal behavior, 698
Reflex arc, 607
Refractory period, 599
Regeneration, 54, 55
defined, 41
Regulator gene, 292
Regulators, 567–570
Renal artery, 649
Renal veins, 649
Renin, 657
Replication, 394
defined, 274
of DNA, 277–281, 383
Reproduction, 708–722
in animals, 318–321
asexual, in *Neurospora*, 303, 304
biological significance of, 709
birth, 718–720
cellular specialization, 716–717
differentiation, 716–717
hormones in, 710, 711, 714–715, 718–720
implantation, 717–720
menopause and, 715

Reproduction (*cont.*)
ova production, 711–712
the placenta, 717–720
in plants, 321–323
pregnancy, 715–716
sexual, 40, 41, 42, 43, 108, 376–377
in man, 318–321
and meiosis, 312–325
in *Neurospora*, 303–304, 305
sperm production, 709–711
uterus cycle changes, 712–713
Reproductive cells, **defined, 474**
Reproductive organs, 44, 536
Reptiles, 60–62, 69
Reptiles, Age of, 402
Residual protein, 182
Resolving power, 139, 140
Respiration, 253, 482, 492
defined, 465
roots and, 490
Respiratory system, 296, 634–637
Reticular formation, 694–695
defined, 695
Retina, 693, 702, 703
Retinene, 265–266
Rhesus monkey, 433
Rhizobium, 492
root nodule and, 534, 545
Rhizoid, 158, 528
Rhodophyta, 77
Rhodopsin, 265–266
Rhyncosciara angelae, 385
Rib cage, 636
Riboflavin, 228, 249, 515
Ribonuclease, 282
Ribonucleic acid (RNA), 14, 182–183, 385, 412
defined, 461
DNA-RNA protein chain of relationships, 281–286
function of, 461–462
Ribose, 205–206, 239
Ribosomal RNA (r-RNA), 282
Ribosomes, 141, 148, 153, 221
Ring compounds, 198
Ringers solution, 596
Rings, annual, 509–510
RNA, *see* Ribonucleic acid
Robin, 171, 312
Rods, 265, 705
Root cap, 505, 509
Root hair cells, 474, 492, 507
Root nodules, 534, 545
Root pressure theory, 498–500

Roots:
anatomy of, 508–509
carrot, 532
central cylinder, 507
cortex, 507, 508, 523
culture of excised, 529–531
desert plants, 549
endodermis, 507, 508, 523
formation of, 516, 531, 544–545
leguminous, 492
morning-glory, 530–531
outer epidermal layer, 507
pea, 529
pressure, 500
region of elongation, 505
respiration and, 490
Roottip cells, in division, 170–173
time of, 175
Rosette plants, gibberellin and, 520
Rothschild, Lord, 36–37
Roundworms, 44
Rusts, 85

S

Sac fungi, 85
Saguaro cactus, 549
Salamander, 179, 290, 315
Saliva, secretion of, 665
Salivary amylase, 664
Salivary gland in digestion, 664
Salt(s), 490, 648–659
bile, 664
concentration difference for, 578
loss from sweating, 685
transported, 500
Salt water, drinking, 656
Sand in soil, 489
Schizomycophyta, 82–83
Schleiden, M. J., 136–137
Schwann, Theodor, 136–137
Science(s):
applied, 119–120
as an attitude, 122
basic, 119–120
as body of knowledge, 121
growth of, 119–121
progressive, 122–123
purpose of, 121–122
Scientific method, **defined, 128**
Sclera, 702, 703
Scorpions, 51
Scrotum, 709–711
Sea squirts, 56–57
Seaweeds, *see* Kelps
Seed ferns, 95
Seed plants, 94–103

Index

Seeds:
   coat of, 504, 505
   germination of, 505, 521, 530, 540, 545
   growth patterns, 504–506
Seeing, 692–693, 702–705; see also Eye, the
   day vision, 705
   night vision, 705
Segmented worms, see Annelids
Segregation, law of, 306
Selective breeding, 396
Self-pollination, 100, 101
Semen, 711
Seminal vesicles, 710, 711
Seminiferous tubules, 709, 710
Sensory fibers, 596, 691
Sensory nerves, 607
Sensory receptors, 602–603
Sensory structures, 48, 49
Sepals, 98–99
Sequoia trees, 163–164
Series elasticity, 588
Serine, 384
Sex:
   cells, 169
   differentiation, 351–352
   as inherited trait, 348–361
   linkage, 352–358
      color-blindness, 357–358
      eye color, 353–357
   secondary character, 352
Sexual reproduction, 40, 41, 42, 43, 108, 376–377
   in man, 318–321
   and meiosis, 312–325
   in *Neurospora*, 303–304, 305
   of plants, 79–82
      algae, 79–81
      fungi, 85–86
Sharks, 57
Shelf fungi, 85
Shivering, heat produced by, 682
Short-day plants, 538–539
   behavior of, 540–541
   critical night length, 540–541
Siamese twins, 368–369
Sickle-cell anemia, 166, 286, 393
Sieve tubes, 474, 485, 500, 501
Sigmoid curve, 506
Silicon, 491
Silks (corn), 553, 557
Silkworm, 359
Silt in soil, 489
Silurian Period, 69, 402
Sinanthropus, **defined, 434**
Siphons, 45

Skeletal muscle, 247, 586–587
   in related state, 590
Skeleton, of man and horse compared, 432
Skin (human), 222
   flushed, 685
   intense heat response, 685
   water loss through, 646, 683–684
Sleep, 694–695
Sliding filament hypothesis, 591
Slime molds, 15, 83–86
Small intestine, 660
   in digestion, 664, 668
   layers of, 668
   peristalsis in, 669
Smelling, 692–693
Smog, plant communities and, 552
Smooth muscle, 586–587
   in arterioles, 619
Smuts, 85
Snow crystal, 122
Sodium:
   active transport of, 651, 654
   and potassium pump, 299
Sodium atom, 188
Sodium hydroxide, 202
Sodium hypochlorite, 530
Sodium ions, 191, 579–580, 582, 599–600
   active transport of, 584
Soil, 478, 497–498, 555–556
   contents of, 489
   erosion, 551
   at field capacity, 498
   physical conditions, 489
   profile of, 488
   tilth, 490
Solar energy, 264–265
Solar system, 118, 129, 395, 425
   origin of, 122
Soluble RNA (s-RNA), 282
Solutes, 224
   defined, 578
Solvent, 224
Sorus, 93
Sound waves, 699
   frequency of vibrations, 699
Space walk, 407
Special Creation, Doctrine of, 395
Species, 7, 8–13
   defined, 6, 435
Spectrometer, 121–122
Spectrum, 260
   absorption, 266
Sperm, 138, 169, 180, 313, 318, 350, 370, 526
   cells, 708–711
      in lilies, 323
      parts of, 320

Sperm, cells (*cont.*)
      production of, 709–711
   human, 320–321
   size, 374
Sperm nuclei, 101
Spermatids, 350
Spermatocytes, 320
Spermatophytes, 94–103
   seed plants, 94–97
Spiders, 51
Spikelets (corn), 556–557
Spinal cord, 594–596, 689, 690
Spinal nerves, 596
Spindle, 172–173
Spinifex, 4
Spirilla bacteria, 82, 83
Sponges, 16, 40–41, 53
Spongin, 40
Spongy layer (leaf), 480
Sporangium:
   of ferns, 93
   of mosses, 90
Spores, 85, 88, 108
   defined, 83
   recombinant, 308
Sporophyte:
   defined, 88
   of ferns, 93
   of liverworts, 88
   of mosses, 89–91
   of seed plants, 95
Sporozoa, 39
Spring wood, 509
Squid, 45
Srb, Adrian, 411
Stahl, Frank, 279–281
Stamens, 98–99
Stapes, 701, 702
Staphylococcus aureus, 397, 420
Starches, 484, 661–663
   defined, 206
Starfish, 54
Statoliths, **defined, 515**
Steady state, 566–568, 628
   defined, 566
   homeostasis and, 556–567
Steapsin, 664
Stems:
   anatomy, 508–509
   apex, 512, 520, 523
   culture of excised, 531
   of trees, 509–510
Stern, Curt, 341
Stigma, 101, 321–322, 557
Stomach, 660
   in digestion, 664–667
   peristalsis in, 669
Stomata, 480, 481–484, 496, 497, 501, 549
   defined, 481
Stone Age, 441–442
Storage, 567–570
Strawberry plant, 376
Streptomycin, 397, 421
Stroma, 466

Strontium, 491
Structural gene, 292
Structural stage of biology, 126
Sturtevant, A. H., 343
Style, 321–322
Style (corn), 557
Subclass, 8
Subcutaneous connective tissue, 208
Subphylum, 8
   vertebrata, 10
Subspecies, 13
Substrate, 231
Succession of life, 398
Succinic acid, 233, 250, 252
Succinic dehydrogenase, 233
Sucrose, 206, 531, 663
   defined, 484
Sugar, 450–451, 464–465, 661–663
   in blood, 663, 672
   liver in regulating, 677
   concentration difference for, 578
   double, 664
   photosynthesis and, 484
   simple, 664
   in steady state, 566
Sulfanilamide, 233
Sulfhdryl, 238
Sulfur, 480, 495
Sulfur-35, 186
Summer wood, 509
Sun, the:
   energy and, 450–456, 487
   thermonuclear reactions and, 456
Supporting cells, 476
   defined, 473–474
Surface temperature, 682
Sweat glands, 685
Sweating, 683–685
   salt and water loss from, 685
Sweet corn, 554, 555
Symbiosis, 123, 207
   defined, 492
Sympathetic nerves, 617, 621, 684
   excitation of, 670
Sympathetic nervous system, 608–609
Synapse, 604
Synaptic transmission, 604–605
Synthesis, 127
   of aspartic acid, 254
   of enzymes, 291–292
Systema Naturae (Linnaeus), 125

T

Tadpoles, 59
Tall versus dwarf plants, 326–330

Tapeworms, 42, 43, 44, 46
Tassels (corn), 553, 556
Tasting, 692–693
Tatum, E.L., 286
Taxonomic biology, 126
Taxonomy, 8–9, 127–128
  defined, 3, 124
  and species problem, 8–13
Taylor, J.H., 278–279
Telescopes, 121–122
  Hale, Mount Wilson, California, 121, 138
Telophase, 171, 173, 174
  of meiosis, 316
  time of, 175
Temperature(s):
  effect on enzyme reaction, 229–231
  and life, 425, 426–427
  photoperiod and, 536
  in plant growth, 511, 521, 536, 541–543
  sensitivity to, 367
Temperature, body, 681–686
  care, 682
  fever, 684–685
  loss of heat in, 683–684
  production of heat in, 682–683
  regulation of, 684
  response to intense heat, 685–686
  surface, 682
Teosinte, 554
Terrestrial vertebrates, 60
Testcross, 329–330
Testes, 42, 318, 350, 709, 710
  amphibian, 313–315
  of grasshopper, 349
  mammalian, 320
  of salamander, 179
Testosterone, 710, 711
Tetanizing contractions, 603
Thalassemia, 440
Thallophytes, 74–87
  algae, 74–82
  fungi, 82–87
Thallus, 526–528
Theophrastus, 7, 72
Thermonuclear reaction:
  defined, 452
  sun and, 456
Thiamin B_1, 529
Thiamine, 228, 250
Thirst, 656
Thoracic duct, 629
Threshold, 599
Thrombin, 626
Thromboplastin, 626
Thymine, 275, 415
Thymidine, radioactive, 278–280
Thyroid gland, 679

Thyroid-stimulating-hormone (TSH), 680
Thyroxin, 679–680, 681
  control of secretion, 680
  in heat production, 683
Ticks, 51–52
Tissue analysis, beginnings of, 128
Tissue cultures, **defined, 529**
Tissue organization, 507–511, 523
Tobacco mosaic virus (TMV), 144, 182–183
Tobacco plants, 537–538
  pith, 534
Tongue, 635
Tonoplast, 149, 470
Toynbee, Arnold, 121
Trachea, 634, 635
Tracheids, 474, 508, 509, 510, 532
  cells, 484, 524
  spring, 510
  summer, 510
Tracheophyta, 91–103
  angiospermae, 98–103
  gymnospermae, 97–98
Transfer RNA (t-RNA), 282
Transmission:
  neuromuscular, 604
  synoptic, 604–605
Transpiration, 496–498, 501–502
  defined, 496
Transpiration-cohesion-tension theory, 498–499
Transplanting tissue, 630
Transport, 574–585; see also Circulation
  absorption, 661, 668–669
  active, 226–227, 298–299, 583–584
  of sodium, 651, 654
  carbon dioxide, 639–641
  cell membranes, 574, 582–584
  concentration differences and, 574
  electrical forces and, 574, 579–582
  osmosis, 578–579
  oxygen, 637–638
  pressure and, 574–576
Trees:
  animal girding of, 550
  calluses, 533
  dating, 509–510
  ferns, 91
  girdled trunk, 484
  stems of, 509–510
  transpiration-cohesion-tension theory and, 499
Triglyceride, 207

Trillium erectum, 171, 179, 312
Triose phosphate compounds, 243
Triose phosphate dehydrogenase, 244
Tripeptide, 210
Triplet code, 282, 283
Triploid endosperm nucleus, 323
Tropism, **defined, 515**
True fungi, 84–87
  algal, 84
  club, 85
  lichens, 86–87
  sac, 85
Trypsinogen, 664, 665
Tryptophan synthetase, 384
Tube feet, 54
Tube nucleus, 101
Tubular secretion, 650
Tubules, 649, 650
  of the kidney, 654
  reabsorption in, 650
  secretion in, 650
  seminiferous, 709, 710
Tumors, 533–535
  inducing agent, 534–535
  secondary, 534–535
Tung seedling, 212
Turgor pressure, 471–472, 476, 480
  defined, 471
  guard cells under, 482
  21-keto steroid, 364
Twins, 302, 368–371
  identical, 716
  Siamese, 368–369
Twitches, 603
2, 4-Dichlorophenoxyacetic acid (2, 4-D), 518, 551
Tympanum, 699, 701, 702
  defined, 699

U

Ultraviolet radiation, 414, 453
Umbilical cord, 717, 718, 720
Ungulates, 65
Unicellular organisms, 152, 157–158, 163
  division of, 168
U.S. Department of Agriculture, 537
Unity among organisms, 390–392
Univalve, 45
Universe, 129
  origin of, 122, 403
Uracil, 282, 284
Urea, 648, 657
Ureter, 649, 710
Urethra, 709, 710
Urey, Harold, 427

Urine:
  concentrated, 654
  of diabetics, 674
  dilute, 654, 655
  formation of, 649, 652
  glucose in, 651, 674
Ursidae, 9
Ussher, Archbishop Henry, 396
Uterus, 370, 711
  cycles, 712–714

V

Vacuoles, 141, 143, 149, 458, 459, 468, 476
  defined, 460
  function of, 473
  growth, 507
  membranes, 470–473
  pressure within, 471
Vagina, 711
Vagus nerves, 605, 666
Valence electrons, 190
Valonia, 143
Valve flaps, 616
Valves:
  A-V, 616
  in the heart, 614, 616
  lymph, 629
  in veins, 614, 617–618
Vanadium, 491
Vas deferens, 710, 711
Vascular plants, 91–92
  defined, 91
Vascular tissues:
  defined, 91
  phloem, 91–92
  xylem, 91
Veins, 480, 484, 615–616
  pulmonary, 615
  renal, 649
  valves in, 614, 617–618
Ventral cells, 377, 378
Ventricles:
  of the brain, 689
  of the heart, 615
Vernalization, 542–543, 545
Vertebrata, 11
  defined, 10
Vertebrates, 56
  development of, 392
  evolutionary relationship, 433
  humans as, 444
Vesicular stomatitis, 119
Vessel cells, 474, 484
Vestibular apparatus, 701, 702
Vicia faba, 170
Victoria, Queen, 440
Villi, 156, 157, 669, 718
Virchow, Rudolf, 137, 169, 394, 433
Virology, **defined, 124**
Viruses, 14, 107, 144–145, 153, 274
  contents, 182–183

Visible light, **defined, 453**
Vision, 265–267
Visual purple, 705
Vitamins, 212–213, 228, 233, 265–266, 515, 529, 544
  fat-soluble, 212
  as food, 661
Viviparity, **defined, 65**
Von Mering, Joseph, 672

W

Wallace, Alfred, 129–130
Wallace, Bruce, 411
Warburg, Otto, 249
Water, 451, 478, 480, 648–659
  chemistry of, 192–194
  evaporation of, 484
  as food, 661
  and life, 425–428
  loss during breathing, 645–646, 683–684
  loss from sweating, 685

Water (*cont.*)
  molecules, 194–196, 470–472, 479, 488
  geometry of, 193
  permeability of, 655
  photosynthesis and, 452, 484
  reabsorption of, 650
  salt, drinking of, 656
  soil erosion and, 551
  storage cells, 476
  transpiration and, 496–498, 501–502
    cohesion-tension theory, 498–499
  turgor pressure, 471
  vapor, 453, 482
  wilting percentage, 498
Water pollution, 550–552
Watson, J. D., 130, 276
Watson-Crick model, 277
Wavelengths, 456
  defined, 452
Waves, sound, 699
  frequency of vibrations, 699

Weasel, 236
Weeds, control of, 522–523
Weiss, Paul, 123
White, E. H., II, 407
White blood cells, 134, 148, 165, 630, 632
  size of, 154
  cells, 173–174
  embryo, 168, 173
Wilkins, M. H. F., 130, 276
Windpipe, 634
Wolffian duct, 351
Womb, *see* Uterus

X

X-chromosome, 349–360, 440
Xerophytes, 501
  defined, 497
X-rays, 121–122, 414
Xylem, 91–92, 480, 488, 523
  cells, 508, 534
  location of, 509

Xylem (*cont.*)
  salts in, 500
  tissue, 533
  vessels, 499, 501, 502

Y

Yarrow, 364
Y-chromosome, 350–360, 440
Yeasts, 85, 241
Yolk, 319, 378
Young, Sydney, 242

Z

Zea mays L., 553
Zein molecule, 462
Zinc, 490
Zinjanthropus, 432
Zoology, **defined, 124**
Zygonema, 317
Zygote, 42, 81, 86, 108, 321, 474, 528
  defined, 40
Zymogen cells, 673